IMPORTANT:

HERE IS YOUR REGISTRATION CODE TO ACCESS
YOUR PREMIUM McGRAW-HILL ONLINE RESOURCES.

MCGRAW-HILL
ONLINE RESOURCES

For key premium online resources you need THIS CODE to gain access. Once the code is entered, you will be able to use the Web resources for the length of your course.

If your course is using **WebCT** or **Blackboard**, you'll be able to use this code to access the McGraw-Hill content within your instructor's online course.

Access is provided if you have purchased a new book. If the registration code is missing from this book, the registration screen on our Website, and within your WebCT or Blackboard course, will tell you how to obtain your new code.

Registering for McGraw-Hill Online Resources

TO gain access to your McGraw-Hill web resources simply follow the steps below:

(1) USE YOUR WEB BROWSER TO GO TO: **www.mhhe.com/griffith**

(2) CLICK ON **FIRST TIME USER**.

(3) ENTER THE REGISTRATION CODE* PRINTED ON THE TEAR-OFF BOOKMARK ON THE RIGHT.

(4) AFTER YOU HAVE ENTERED YOUR REGISTRATION CODE, CLICK **REGISTER**.

(5) FOLLOW THE INSTRUCTIONS TO SET-UP YOUR PERSONAL UserID AND PASSWORD.

(6) WRITE YOUR UserID AND PASSWORD DOWN FOR FUTURE REFERENCE. KEEP IT IN A SAFE PLACE.

TO GAIN ACCESS to the McGraw-Hill content in your instructor's **WebCT** or **Blackboard** course simply log in to the course with the UserID and Password provided by your instructor. Enter the registration code exactly as it appears in the box to the right when prompted by the system. You will only need to use the code the first time you click on McGraw-Hill content.

saturation-17132873

REGISTRATION CODE

Thank you, and welcome to your McGraw-Hill online Resources!

* YOUR REGISTRATION CODE CAN BE USED ONLY ONCE TO ESTABLISH ACCESS. IT IS NOT TRANSFERABLE.

0-07-291883-7 T/A GRIFFITH: THE PHYSICS OF EVERYDAY PHENOMENA, 4E

the physics
of everyday phenomena

W. Thomas Griffith

Pacific University

the physics

of everyday phenomena

fourth edition

A Conceptual

Introduction to

Physics

 **Higher Education**

Boston   Burr Ridge, IL   Dubuque, IA   Madison, WI   New York   San Francisco   St. Louis
Bangkok   Bogotá   Caracas   Kuala Lumpur   Lisbon   London   Madrid   Mexico City
Milan   Montreal   New Delhi   Santiago   Seoul   Singapore   Sydney   Taipei   Toronto

Higher Education

THE PHYSICS OF EVERYDAY PHENOMENA: A CONCEPTUAL INTRODUCTION
TO PHYSICS, FOURTH EDITION

Published by McGraw-Hill, a business unit of The McGraw-Hill Companies, Inc., 1221 Avenue of the Americas, New York, NY 10020. Copyright © 2004, 2001, 1998, 1992 by The McGraw-Hill Companies, Inc. All rights reserved. No part of this publication may be reproduced or distributed in any form or by any means, or stored in a database or retrieval system, without the prior written consent of The McGraw-Hill Companies, Inc., including, but not limited to, in any network or other electronic storage or transmission, or broadcast for distance learning.

Some ancillaries, including electronic and print components, may not be available to customers outside the United States.

This book is printed on acid-free paper.

International 1 2 3 4 5 6 7 8 9 0 DOW/DOW 0 9 8 7 6 5 4 3
Domestic   1 2 3 4 5 6 7 8 9 0 DOW/DOW 0 9 8 7 6 5 4 3

ISBN 0–07–250977–5
ISBN 0–07–121465–8 (ISE)

Publisher: *Kent A. Peterson*
Sponsoring editor: *Daryl Bruflodt*
Developmental editor: *Brian S. Loehr*
Marketing manager: *Debra B. Hash*
Lead project manager: *Peggy J. Selle*
Production supervisor: *Kara Kudronowicz*
Lead media project manager: *Judi David*
Senior media technology producer: *Jeffry Schmitt*
Designer: *David W. Hash*
Cover/interior designer: *Kaye Farmer*
Cover image: *©Getty Images/Reza Estakhrian*
Lead photo research coordinator: *Carrie K. Burger*
Photo research: *Chris Hammond/PhotoFind, LLC*
Compositor: *Lachina Publishing Services*
Typeface: *10/12 Times*
Printer: *R. R. Donnelley Willard, OH*

The credits section for this book begins on page 491 and is considered an extension of the copyright page.

Library of Congress Cataloging-in-Publication Data

Griffith, W. Thomas.
  The physics of everyday phenomena : a conceptual introduction
to physics / W. Thomas Griffith.—4th ed.
    p. cm.
Includes bibliographical references and index.
ISBN 0–07–250977–5 (hard copy : alk. paper)
1. Physics. I. Title.

  QC23.2.G75   2004
  530—dc21

                          2003005449
                          CIP

INTERNATIONAL EDITION ISBN 0–07–121465–8
Copyright © 2004. Exclusive rights by The McGraw-Hill Companies, Inc., for manufacture and export. This book cannot be re-exported from the country to which it is sold by McGraw-Hill. The International Edition is not available in North America.

www.mhhe.com

brief contents

contents

Unit Two   Fluids and Heat

preface

The satisfaction of understanding how rainbows are formed, how ice skaters spin, or why ocean tides roll in and out—phenomena that we have all seen or experienced—is one of the best motivators available for building scientific literacy. This book attempts to make that sense of satisfaction accessible to non-science majors. Intended for use in a one-semester or two-quarter course in conceptual physics, this book is written in a narrative style, frequently using questions designed to draw the reader into a dialogue about the ideas of physics. This inclusive style allows the book to be used by anyone interested in exploring the nature of physics and explanations of everyday physical phenomena.

> *"Griffith has done a very respectable job in presenting his conceptual physics course in a clear, useable fashion. It is a fine work that is evidently quickly evolving into a top-notch textbook."*
> —*Michael Bretz, University of Michigan*

How This Book Is Organized

With the exception of the reorganization of chapters 15, 16, and 17 dictated by the addition of new material, we have retained the same order of topics as in the previous three editions. It is traditional with some minor variations. The chapter on energy (chapter 6) appears prior to that on momentum (chapter 7) so that energy ideas can be used in the discussion of collisions. Wave motion is found in chapter 15, following electricity and magnetism and prior to chapters 16 and 17 on optics. The chapter on fluids (chapter 9) follows mechanics and leads into the chapters on thermodynamics. The first 17 chapters are designed to introduce students to the major ideas of classical physics and can be covered in a one-semester course with some judicious paring.

The complete 21 chapters could easily support a two-quarter course, and even a two-semester course in which the ideas are treated thoroughly and carefully. Chapters

18 and 19 on atomic and nuclear phenomena, are considered essential by many instructors, even in a one-semester course. If included in such a course, we recommend curtailing coverage in other areas to avoid student overload. Sample syllabi for these different types of courses can be found on the Instructor Center of the Online Learning Center.

Some instructors would prefer to put chapter 20 on relativity at the end of the mechanics section or just prior to the modern physics material. Relativity has little to do with everyday phenomena, of course, but is included because of the high interest that it generally holds for students. The final chapter (21) introduces a variety of topics in modern physics—including particle physics, cosmology, semiconductors, computers, and superconductivity—that could be used to stimulate interest at various points in a course.

One plea to instructors, as well as to students using this book: Don't try to cram too much material into too short a time! We have worked diligently to keep this book to a reasonable length while still covering the core concepts usually found in an introduction to physics. These ideas are most enjoyable when enough time is spent in lively discussion and in consideration of questions so that a real understanding develops. Trying to cover material too quickly defeats the conceptual learning and leaves students in a dense haze of words and definitions. Less can be more if a good understanding results.

Mathematics in a Conceptual Physics Course

The use of mathematics in a physics course is a formidable block for many students, particularly non-science majors. Although there have been attempts to teach conceptual physics without any mathematics, these attempts miss an opportunity to help students gain confidence in using and manipulating simple quantitative relationships.

Clearly mathematics is a powerful tool for expressing the quantitative relationships of physics. The use of mathematics can be carefully limited, however, and subordinated to the physical concepts being addressed. Many users of the first edition of this text felt that mathematical expressions appeared too frequently for the comfort of some students. In response, we substantially reduced the use of mathematics in the body of the text in the second edition. Most users have indicated that the current level is about right, so we have not changed the mathematics level in this edition.

Logical coherence is a strong feature of this book. Formulas are introduced carefully after conceptual arguments are provided, and statements in words of these relationships generally accompany their introduction. We have continued to fine tune the Try This Boxes that present sample exercises and questions. Most of these provide simple numerical illustrations of the ideas discussed. No mathematics prerequisite beyond high school algebra should be necessary. A discussion of the basic ideas of very simple algebra is found in appendix A, together with some practice exercises, for students who need help with these ideas.

try this box 3.4

Sample Exercise: Projectile Motion

A ball rolls off a tabletop with an initial velocity of 3 m/s. If the tabletop is 1.25 m above the floor,

a. How long does it take for the ball to hit the floor?
b. How far does the ball travel horizontally?

a. In figure 3.7, we saw that a ball will fall a distance of 1.25 m in approximately half a second. This could be found directly from

$$d_{\text{vertical}} = 1.25 \text{ m} \qquad d_{\text{vertical}} = \tfrac{1}{2} at^2$$
$$a = g = 10 \text{ m/s}^2 \qquad \text{Solving for } t^2:$$
$$t = ? \qquad\qquad t^2 = \frac{d}{\tfrac{1}{2}a}$$
$$= \frac{1.25 \text{ m}}{5 \text{ m/s}^2}$$
$$= 0.25 \text{ s}^2$$

New to This Edition

We have made some significant additions and changes to the fourth edition. As this book has evolved, however, we have tried to remain faithful to the principles that have guided the writing of the book from the outset. One of these has been to keep the book to a manageable length, both in the number of chapters and in the overall content. Many books become bloated as users and reviewers request more and more pet topics. We have strived to maintain a carefully organized framework for building an understanding of basic physics. The changes include:

■ **A new chapter on Light Waves and Color** has been added. It discusses the nature of electromagnetic waves and physical optics topics including color perception, interference, diffraction, and polarization. Although some of this material had been covered in other places, it has been pulled together and substantially expanded. The sections on color perception and polarization are completely new.

■ **A section on the Physics of Music** has been added to chapter 15, which covers wave phenomena. This section builds upon concepts of standing waves and sound waves introduced in sections 15.3 and 15.4. The new section discusses harmonic analysis, harmony and dissonance (including beats), and the structure of musical scales. It should be of particular interest to students with some background in music, but is intended to be accessible to all students.

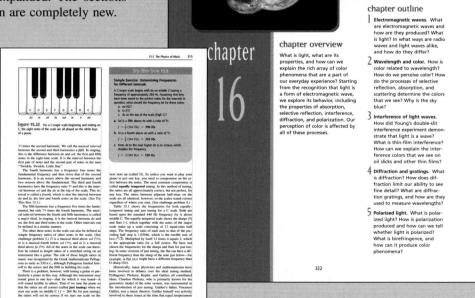

■ **A new section called Focusing Light with Curved Mirrors** has been added to chapter 17 (formerly chapter 16), which covers topics from geometric optics. This had also been requested by many users.

> *"I have taught mirrors in this course while using Griffith and I think this will make the students happier to have a few more examples of the image formation. This is a nice addition."*
> —Matt Evans, University of Wisconsin, Eau Claire

■ **New Everyday Phenomenon Boxes.** Each chapter includes one or more Everyday Phenomenon Boxes. These boxes have been a popular feature of the book and we have had repeated requests to add more. In response, we have produced nine new Everyday Phenomenon Boxes. See list on page xvi.

■ **We have added more open-ended questions.** Most of the questions at the end of each chapter call for brief, well-defined answers and short explanations. In response to requests from some reviewers, we have added a few questions at the end of each chapter that are more open-ended. They are intended to be useful for generating discussion and are marked with an asterisk.

■ **We have added more exercises and challenge problems.** As in previous revisions, we have modified most of the exercises and challenge problems and have added some new exercises.

■ **New design.** We have designed this edition to be more inviting by modernizing it with the use of soft and inviting colors, and breaking up lengthy chunks of text with new designs of Everyday Phenomenon Boxes, Try This Boxes, and Running Summary Paragraphs. This new design also offers students an improved art program to learn with by isolating the physics concept in several art pieces.

> *"One of the strengths of the book is its professional look and earnest attempt to explain the physics of everyday phenomena."*
> —Dan Bruton, Stephen F. Austin State University

Chapter Tools

The overriding theme of this book is to introduce physical concepts by appealing to everyday phenomena whenever possible. Each chapter begins with an illustration from everyday experience and then proceeds to use it as a theme for introducing relevant physical concepts. Physics can seem abstract to many students, but using everyday phenomena and concrete examples reduces that abstractness.

The chapter outlines, questions, and summaries provide a clear framework for the ideas discussed in each chapter. One of the difficulties that students have in learning physics (or any subject) is that they fail to construct the big picture of how things fit together. A consistent chapter framework can be a powerful tool in helping students see how ideas mesh. Running Summary Paragraphs are found at the end of each chapter section to supplement the more general summary at the end of the chapter. The subsection headings are often cast in the form of questions to motivate the reader and pique curiosity. A few key concepts form the basis for understanding physics and these textual features reinforce this structure so that the reader will not be lost in a flurry of definitions and formulas.

> *"Very good chapter overview and chapter outline for each chapter and for each unit. Very clear introduction and illustration of physics phenomena, concepts, and principles, and excellent exercises, problems, and home experiments/observations at the end of each chapter."*
> —Hai-Sheng Wu, Minnesota State University, Mankato

Everyday Phenomenon box 17.2

Laser Refractive Surgery

The Situation. Megan Evans has been nearsighted since her early teens. She has worn contact lenses for several years after first using spectacle lenses. Now in her twenties, she has heard friends talk about a new procedure called laser refractive surgery that can allow people to see well without corrective lenses. She is intrigued and wants to know more about it.

How can bombarding her eye with a laser beam improve her vision? She knows that lasers can be dangerous in other situations. Is this procedure safe? How does it work, and can it help her situation?

The Analysis. In our culture, myopia or nearsightedness is the most common visual problem. It may develop from doing a lot of near work such as reading during childhood, although there are also hereditary factors. As is described in figure 17.26, the lens system of the myopic eye is too strong, which causes light from distant objects to focus in front of the retina rather than on the retina.

Most of the optical power of the eye is produced by the front surface of the cornea. Optical power is measured in diopters, which is the reciprocal of the focal length measured in meters ($P = 1/f$) when the lens is surrounded by air. The shorter the focal length, the stronger the optical power because a short focal length implies that the light rays are being strongly bent by the lens. The overall power of the lens system of the eye is about 60 diopters, but the front surface of the cornea produces 40 to 50 diopters by itself.

The optical power of the cornea (or of any lens) is determined by two things—how strongly the surface is curved and the difference in index of refraction on either side of the surface. For a nearsighted person, the surface of the cornea is too strongly curved for the length of the eyeball. It is not unusual for a person like Megan to have an optical power of the cornea that is too strong by 4 to 5 diopters. She then requires a corrective lens of −4 to −5 diopters to allow her to see distant objects clearly.

The purpose of laser refractive surgery is to reshape the cornea by vaporizing different portions of the cornea by different amounts. The most commonly used procedure is called LASIK, which is an acronym for *laser* assisted in situ *keratomileusis*. In this procedure, the surgeon cuts a circular flap of the outer layer of the cornea with a surgical scalpel and pulls this flap to the side as shown in the drawing. She then uses a pulsed *excimer laser* to vaporize small amounts of corneal tissue to produce a predetermined new shape for the central portion of the cornea. When finished, the flap of the outer layer is replaced.

The excimer laser used has a wavelength of 192 nm, which lies in the ultraviolet portion of the spectrum. This wavelength

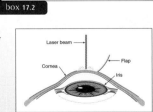

In the LASIK procedure a circular flap of the outer layer of the cornea is pulled aside. Controlled pulses from the laser reshape the central region of the cornea.

is strongly absorbed by corneal tissue, so it vaporizes or *ablates* this tissue without heating the surrounding tissue. The laser operates in a pulsed mode, with each pulse delivering a definite amount of energy. The surgeon can then control how much tissue is ablated by the number of pulses that are delivered to each section of the cornea. This is all controlled by a computer program to achieve the desired new shape.

The LASIK procedure is done on an outpatient basis, and the cornea heals in just a few days. When successful, the reshaped cornea generally allows a person to discard their glasses or contact lenses. Sometimes a weak correction is still needed because the cornea does not heal to quite the desired power. Older people who have lost the ability to accommodate will generally still need reading glasses unless one eye is shaped to have a stronger power than the other. The LASIK procedure is most commonly used to cure myopia where the goal is to flatten the shape of the cornea. It can also be used, though, for farsightedness (hyperopia) or astigmatism. In the case of astigmatism, the cornea is not spherical and this can also be addressed by reshaping with the laser.

Is the procedure safe? The jury is still out on possible long-term effects, but most patients experience only minor problems, if any. There is always a small risk of infection or poor healing, as with any surgical procedure. People sometimes experience problems with night vision after undergoing LASIK. This is because only the central portion of the cornea is reshaped so there is then a circular boundary between the reshaped and untreated portions of the cornea. At night when light levels are low, the pupil of the eye opens more widely and some light may get through this boundary region producing blurring of the image.

Here is a closer look at the Chapter Tools:

■ **Chapter Overview**

■ **Chapter Outline and Chapter Summary Correlation**

■ **Study Hints**

■ **Try This Boxes**

■ **Everyday Phenomenon Boxes**

The Case of the Malfunctioning Coffee Pot
   (chapter 1)
NEW! Transitions in Traffic Flow (chapter 2)
The 100-m Dash (chapter 2)
Shooting a Basketball (chapter 3)
NEW! The Tablecloth Trick (chapter 4)
Riding an Elevator (chapter 4)
NEW! Seat Belts, Air Bags, and Accident Dynamics
   (chapter 5)
Explaining the Tides (chapter 5)
Energy and the Pole Vault (chapter 6)
An Automobile Collision (chapter 7)
Achieving the State of Yo (chapter 8)
NEW! Bicycle Gears (chapter 8)
Throwing a Curveball (chapter 9)
Solar Collectors and the Greenhouse Effect
   (chapter 10)
NEW! Hybrid Automobile Engines (chapter 11)
A Productive Pond (chapter 11)
Lightning (chapter 12)
The Hidden Switch in Your Toaster (chapter 13)
Direct-Current Motors (chapter 14)
NEW! Vehicle Sensors at Traffic Lights
   (chapter 14)
A Moving Car Horn and the Doppler Effect
   (chapter 15)
NEW! Why Is the Sky Blue? (chapter 16)
NEW! Antireflection Coatings on Eyeglasses
   (chapter 16)
Rainbows (chapter 17)
NEW! Laser Refractive Surgery (chapter 17)
Electrons and Television (chapter 18)
Radiation Exposure (chapter 19)
What Happened at Chernobyl? (chapter 19)
The Twin Paradox (chapter 20)
Holograms (chapter 21)

■ **Running Summary Paragraphs**

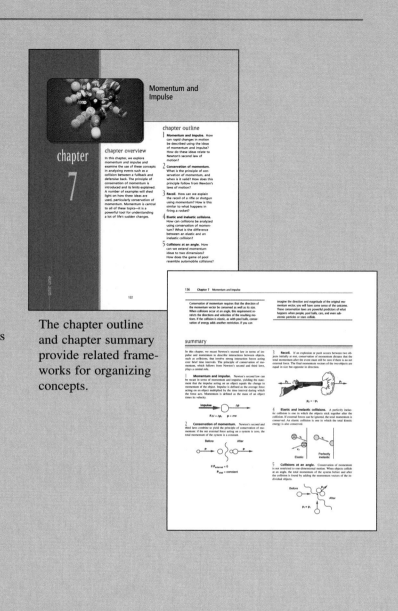

The chapter outline and chapter summary provide related frameworks for organizing concepts.

Rotational displacement, rotational velocity, and rotational acceleration are the quantities that we need to fully describe the motion of a rotating object. They describe how far the object has rotated (rotational displacement), how fast it is rotating (rotational velocity), and the rate at which the rotation may be changing (rotational acceleration). These definitions are analogous to similar quantities used to describe linear motion. They tell us how the object is rotating, but not why. Causes of rotation are considered next.

■ End of Chapter Tools

Key Terms
Questions
Exercises
Challenge Problems
Home Experiments and
Observations

Key Terms

Challenge Problems

Exercises

Questions

Home Experiments and Observations

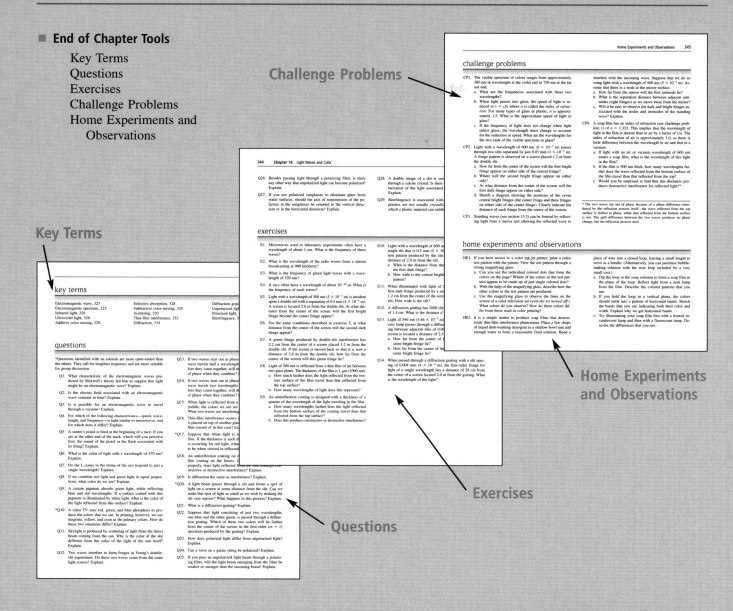

Since many courses for non-science majors do not have a laboratory component, we have continued to develop the home experiments and observations found at the end of each chapter. The spirit of these home experiments is to enable students to explore the behavior of physical phenomena using easily available rulers, string, paper clips, balls,

toy cars, flashlight batteries, and so on. Many instructors have found them useful for putting students into the exploratory and observational frame of mind that is important to scientific thinking. This is certainly one of our objectives in developing scientific literacy.

Supplements

■ Instructor's Testing and Resource CD

Available on CD-ROM, in both Mac and Windows platforms, this test bank of questions utilizes Brownstone Diploma software to quickly create customized exams. This user-friendly program allows instructors to search for questions by topic, format, or difficulty level; edit existing questions or add new ones; and scramble questions and answer keys for multiple versions of the same test. For your convenience, the Instructor's Manual and set of test bank questions are available in Word and pdf formats.

■ Digital Content Manager

This multimedia collection of visual resources allows instructors to utilize artwork from the text in multiple formats to create customized classroom presentations, visually based tests and quizzes, dynamic course website content, or attractive printed support materials. The digital assets on this cross-platform CD-ROM are grouped by chapter within easy-to-use folders.

■ Online Learning Center (OLC)

The OLC is a text-specific website designed to provide students with useful study tools that take advantage of the power of computers to improve their understanding of the material presented in the text and class. For the instructor, the OLC is designed to help ease the time burdens of the course by providing valuable presentation and preparation tools.

For Students

Student Study Guide Integration
- Study Objectives
- Study Hints
- Working with Equations: Practice Problems
- Solutions to Practice Problems
- Mastery Quiz

Animations
Links Library
New York Times News Links
Glossary
Career Opportunities
Ask the Author

Chapter Level Resources
Chapter Summary
Crossword Puzzles
Animations
Chapter Objectives
Key Terms

For Instructors

All Student Content
PowerPoint images
Instructor's Manual
PageOut

Acknowledgments

A large number of people have contributed to this fourth edition, either directly or indirectly. I extend particular thanks to those who participated in reviews of the third edition and of the manuscript for the fourth edition. Their thoughtful suggestions have had direct impact upon the clarity and accuracy of this edition, even when it was not possible to fully incorporate all of their ideas due to space limitations or other constraints. I also thank the contributors to the website, Joseph Schaeffer, and the Instructor's Manual and Test Bank, Art Braundmeier, Virgil Stubblefield, and Mikolaj Sawicki. The reviewers include:

List of Reviewers

Reviewers of Previous Editions

Murty A. Akundi, *Xavier University of Louisiana*
Charles Ardary, *Edmonds Community College*
James W. Arrison, *Villanova University*
Richard A. Atneosen, *Western Washington University*
Dr. Jean-Claude Ba, *Columbus State Community College*
Maria Bautista, *University of Hawaii*
Richard L. Bobst, *LaSierra University*
Ferdinando Borsa, *Iowa State University*
Richard A. Cannon, *Southeast Missouri State University*
Edward H. Carlson, *Michigan State University*
Cary Caruso, *Fayetteville State University*
Rory Coker, *University of Texas, Austin*
Doug Davis, *Eastern Illinois University*
Renee D. Diehl, *Penn State University*
David Donnelly, *Sam Houston State University*
Abbas M. Faridi, *Orange Coast College*
Clarence W. Fette, *Penn State University*
Lyle Ford, *University of Wisconsin, Eau Claire*
Bernard Gilpin, *Golden West College*
Kenneth D. Hahn, *Truman State University*
John M. Hauptman, *Iowa State University*
H. James Harmon, *Oklahoma State University*
Lionel D. Hewett, *Texas A & M University, Kingsville*
Robert C. Hudson, *Roanoke College*
Stanley T. Jones, *University of Alabama*
Sanford Kern, *Colorado State University*
James Kernohan, *Milton Academy*
John B. Laird, *Bowling Green State University*
Paul L. Lee, *California State University, Northridge*
Joel M. Levine, *Orange Coast College*
John Lowenstein, *New York University*
Robert R. Marchini, *University of Memphis*
Paul Middents, *Olympic College*
Joseph Mottillo, *Henry Ford Community College*
William J. Mullin, *University of Massachusetts, Amherst*
Dr. Arnold Pagnamenta, *University of Illinois at Chicago*

Russell L. Palma, *Sam Houston State University*
Ervin Poduska, *Kirkwood Community College*
Joseph A. Schaefer, *Loras College*
Rahim Setoodeh, *Milwaukee Area Technical College*
Elwood Shapanasky, *Santa Barbara City College*
Lawrence C. Shepley, *University of Texas, Austin*
Bradley M. Sherrill, *Michigan State University*
Cecil G. Shugart, *University of Memphis*
John W. Snyder, *Southern Connecticut State University*
Thor F. Stromberg, *New Mexico State University*
Charles R. Taylor, *Western Oregon State College*
Fred Thomas, *Sinclair Community College*
Jeffrey S. Thompson, *University of Nevada, Reno*
Paul Varlashkin, *East Carolina University*
Douglas Wendel, *Snow College*
John Yelton, *University of Florida*
Mike Young, *Santa Barbara City College*

Reviewers of the Fourth Edition

Art Braundmeier, *Southern Illinois State University, Edwardsville*
Michael Bretz, *University of Michigan*
Dan Bruton, *Stephen F. Austin State University*
Matt Evans, *University of Wisconsin, Eau Claire*
Robert Grubel, *Stephen F. Austin State University*
James Gundlach, *John A. Logan College*
Scott Johnson, *Idaho State University*
Ian M. Littlewood, *California State University, Stanislaus*
Jeffrey S. Olafsen, *University of Kansas*
Michael Read, *College of the Siskiyous*
Mohammad Samiullah, *Truman State University*
Michael Thorensen, *University of Northern Iowa*
Hai-Sheng Wu, *Minnesota State University, Mankato*
Jens Zorn, *University of Michigan*

I also wish to acknowledge the contributions of the editorial staff and book team members at McGraw-Hill Higher Education. Their commitment of time and enthusiasm for this work have helped enormously in pushing this project forward. I also owe a huge debt of thanks to my colleagues at Pacific University for helpful suggestions as well as for their forbearance when this project limited my time for other activities. Many other users of the third edition have also provided constructive criticisms or suggestions, such as Jerry Clifford, Seton Hall University, Mikolaj Sawicki, John A. Logan College and Mike Crivello, San Diego Mesa College.

Finally, I owe a debt of gratitude to members of my family, who have suffered without complaint the time that this project has stolen from other activities. My wife Adelia and my boys have often served as guinea pigs as I have tested demonstrations or ideas. My son Mark assisted in providing preliminary digital photographs for some of the new material in chapter 16. The support of my entire family has been constant and essential to this work.

Physics, the Fundamental Science

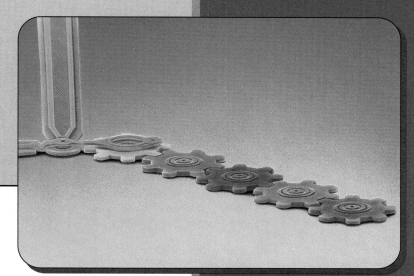

chapter overview

The main objective of this chapter is to help you understand what physics is and where it fits in the broader scheme of the sciences. A secondary purpose is to acquaint you with some of the features of this book and give some tips on how to use them most effectively.

chapter 1

Imagine that you are riding your bike on a country road on an Indian-summer afternoon. The sun has come out after a brief shower, and as the rain clouds move on, a rainbow appears in the east (fig. 1.1). A leaf flutters to the ground, and an acorn, shaken loose by a squirrel, misses your head by only a few inches. The sun is warm on your back, and you are at peace with the world around you.

No knowledge of physics is needed to savor the moment, but your curiosity may bring some questions to mind. Why does the rainbow appear in the east rather than in the west, where it may also be raining? What causes the colors to appear? Why does the acorn fall more rapidly than the leaf? Why is it easier to keep your bicycle upright while you are moving than when you are standing still?

Your curiosity about questions like these is similar to what motivates scientists. Learning to devise and apply theories or models that can be used to understand, explain, and predict such phenomena can be a rewarding intellectual game. Crafting an explanation and testing it with simple experiments or observations is fun. That enjoyment is often missed when the focus of a science course is on accumulating facts.

This book can enhance your ability to enjoy the phenomena that are part of everyday experience. Learning to produce your own explanations and to perform simple experimental tests can be gratifying. The questions posed here lie in the realm of physics, but the spirit of inquiry and explanation is found throughout science and in many other areas of human activity. The greatest rewards of scientific study are the fun and excitement that come from understanding something that has not been understood before. This is true whether we are talking about a physicist making a major scientific breakthrough or about a bike rider understanding how rainbows are formed.

figure **1.1**   A rainbow appears to the east in the Columbia River Gorge on a summer afternoon. How can this phenomenon be explained? (See pages 354–355.)

study hint

If you have a clear idea of what you want to accomplish before you begin to read a chapter, your reading will be more effective. The questions in the chapter outline—as well as those in the subheadings of each section—can serve as a checklist for measuring your progress as you read. A clear picture of what questions are going to be addressed and where the answers will be found forms a mental road map to guide you through the chapter. Take a few minutes to study the outline and fix this road map in your mind. It will be time well spent.

1.1 The Scientific Enterprise

How do scientists go about explaining something like the rainbow described in the introduction to this chapter? How do scientific explanations differ from other types of explanations? Can we count on the scientific method to explain almost anything? It is important to understand what science can and cannot do.

Philosophers have devoted countless hours and pages to questions about the nature of knowledge, and of scientific knowledge in particular. Many issues are still being refined and debated. Science grew rapidly during the twentieth century and has had a tremendous impact on our lives. What is it about science that explains its impressive advances and steady expansion?

How are scientific explanations developed?

Let's consider a specific example of how a scientific explanation comes to be. Where would you turn for an explanation of how rainbows are formed? If you returned from your bike ride with that question on your mind, you might turn to an encyclopedia or a textbook on physics, look up *rainbow* in the index, and read the explanation found there. Are you behaving like a scientist?

The answer is both yes and no. Many scientists would do the same if they were unfamiliar with the explanation. When we do this, we appeal to the authority of the textbook author and to those who preceded the author in inventing the explanation. Appeal to authority is one way of gaining knowledge, but you are at the mercy of your source for the validity of your explanation. You are also hoping that someone has already raised the same question and done the work to create and test an explanation.

Suppose you go back three hundred years or more and try the same approach. One book might tell you that a rainbow is a painting of the angels. Another might speculate on the nature of light and its interactions with raindrops but be quite tentative in its conclusions. All of these books might have seemed authoritative in their day. Where, then, do you turn? Which explanation will you accept?

If you are behaving like a scientist, you might begin by reading the ideas of other scientists about light and then test these ideas against your own observations of rainbows. You would carefully note the conditions when rainbows appear, the position of the sun relative to you and the rainbow, and the position of the rain shower. What is the order of the colors in the rainbow? Have you observed that order in other phenomena?

You would then invent an explanation or **hypothesis** using current ideas on light and your own guess about what happens as light passes through a raindrop. You could devise experiments with water drops or glass beads to test your hypothesis. (See chapter 17 for a modern view of how rainbows are formed.)

If your explanation is consistent with your observations and experiments, you could report it by giving a paper or talk to scientific colleagues. They may criticize your explanation, suggest modifications, and perform their own experiments to confirm or refute your claims. If others confirm your results, your explanation will gain support and eventually become part of a broader **theory** about phenomena involving light. The experiments that you and others do may also lead to the discovery of new phenomena, which will call for refined explanations and theories.

What is critical to the process just described? First is the importance of careful observation. Another aspect is the idea of testability. An acceptable scientific explanation should suggest some means to test its predictions by observations or experiment. Saying that rainbows are the paintings of angels may be poetic, but it certainly is not testable by mere humans. It is not a scientific explanation.

Another important part of the process is a social one, the communication of your theory and experiments to colleagues (fig. 1.2). Submitting your ideas to the criticism (at times blunt) of your peers is crucial to the advancement of science. Communication is also important in assuring your own care in performing the experiments and interpreting the results. A scathing attack by someone who has found an important error or omission in your work is a strong incentive for being more careful in the future. One person working alone cannot hope to think of all of the possible ramifications, alternative explanations, or potential mistakes in an argument or theory. The explosive growth of science has depended heavily on cooperation and communication.

figure **1.2**   A scientific meeting. Communication and debate are important to the development of scientific explanations. The speaker is Albert Einstein.

What is the scientific method?

Is there something we could call **scientific method** within this description, and if so, what is it? The process just described is a sketch of how the scientific method works. Although there are variations on the theme, this method is often described as shown in table 1.1.

The steps in table 1.1 are all involved in our description of how to develop an explanation of rainbows. Careful observation may lead to **empirical laws** for when and where rainbows appear. An empirical law is a generalization derived from experiments or observations. An example of an empirical law is the statement that we see rainbows with the sun at our backs as we look at the rainbow. This is an important clue for developing our hypothesis, which must be consistent with this rule. The hypothesis, in turn, suggests ways of producing rainbows artificially that could lead to experimental tests and, eventually, to a broader theory.

table **1.1**
Steps in the Scientific Method
1. Careful observation of natural phenomena.
2. Formulation of rules or empirical laws based on generalizations from these observations and experiences.
3. Development of hypotheses to explain the observations and empirical laws, and the refinement of hypotheses into theories.
4. Testing of the hypotheses or theories by further experiment or observation.

This description of the scientific method is not bad, although it ignores the critical process of communication. Few scientists are engaged in the full cycle that these steps suggest. Theoretical physicists, for example, may spend all of their time with step 3. Although they have some interest in experimental results, they may never do any experimental work themselves. Today, little science is done simply by observing, as implied by step 1. Most experiments and observations take place to test a hypothesis or existing theory. Although the scientific method is presented here as a step-wise process, in reality these steps often happen simultaneously with much cycling back and forth between steps (fig. 1.3).

The scientific method is a way of testing and refining ideas. Note that the method only applies when experimental tests or other consistent observations of phenomena are feasible. Testing is crucial for weeding out unproductive hypotheses; without tests, rival theories may compete endlessly for acceptance.

How should science be presented?

Traditional science courses focus on presenting the results of the scientific process rather than the story of how scientists arrived at these results. This is why the general public often sees science as a collection of facts and established theories. To some extent, that charge could be made against this book since it describes theories that have resulted from the work of others without giving the full picture of their development. Building on the work of others, without needing to repeat their mistakes and unproductive approaches, is a necessary condition for human and scientific progress.

This book attempts to engage you in the process of making your own observations and developing and testing your own explanations of everyday phenomena. By doing home experiments or observations, constructing explanations of the results, and debating your interpretations with your friends, you will appreciate the give-and-take that is the essence of science.

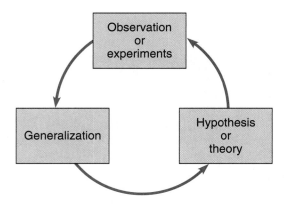

figure **1.3** The scientific method cycles back to observations or experiments as we seek to test our hypotheses or theories.

Whether or not we are aware of it, we all use the scientific method in our everyday activities. The case of the malfunctioning coffee pot described in Everyday Phenomenon Box 1.1 provides an example of scientific reasoning applied to ordinary troubleshooting.

The process of science begins with, and returns to, observations of or experiments on natural phenomena. Observations may suggest empirical laws, and these generalizations may be incorporated into a more comprehensive hypothesis. The hypothesis is then tested against more observations or by controlled experiments to form a theory. Working scientists are engaged in one or more of these activities, and we all use the scientific method on everyday problems.

1.2 The Scope of Physics

Where does physics fit within the sciences? Since this book is about physics, rather than biology, chemistry, geology, or some other science, it is reasonable to ask where we draw the lines between the disciplines. It is not possible, however, to make sharp distinctions among the disciplines or to provide a definition of physics that will satisfy everyone. The easiest way to give a sense of what physics is and does is by example, that is, by listing some of its subfields and exploring their content. First, let's consider a definition, however incomplete.

How is physics defined?

Physics can be defined as the *study of the basic nature of matter and the interactions that govern its behavior.* It is the most fundamental of the sciences. The principles and theories of physics can be used to explain the fundamental interactions involved in chemistry, biology, and other sciences at the atomic or molecular level. Modern chemistry, for example, uses the physical theory of *quantum mechanics* to explain how atoms combine to form molecules. Quantum mechanics was developed primarily by physicists in the early part of this century, but chemists and chemical knowledge also played important roles. Ideas about energy that arose initially in physics are now used extensively in chemistry, biology, and other sciences.

The general realm of science is often divided into the life sciences and the physical sciences. The life sciences include the various subfields of biology and the health-related disciplines that deal with living organisms. The physical sciences deal with the behavior of matter in both living and nonliving systems. In addition to physics, the physical sciences include chemistry, geology, astronomy, oceanography, and meteorology (the study of weather). Physics underlies all of them.

Everyday Phenomenon

box 1.1

The Case of the Malfunctioning Coffee Pot

The Situation. It is Monday morning, and you are, as usual, only half-awake and feeling at odds with the world. You are looking forward to reviving yourself with a freshly brewed cup of coffee when you discover that your coffeemaker refuses to function. Which of these alternatives is most likely to work?

1. Pound on the appliance with the heel of your hand.
2. Search desperately for the instruction manual that you probably threw away two years ago.
3. Call a friend who knows about these things.
4. Apply the scientific method to troubleshoot the problem.

Fixing a malfunctioning coffee pot—alternative 1.

The Analysis. All of these alternatives have some chance of success. The sometimes positive response of electrical or mechanical appliances to physical abuse is well documented. The second two alternatives are both forms of appeal to authority that could produce results. The fourth alternative, however, may be the most productive and quickest, barring success with alternative 1.

How would we apply scientific method as outlined in table 1.1 to this problem? Step 1 involves calmly observing the symptoms of the malfunction. Suppose that the coffeemaker

simply refuses to heat up. When the switch is turned on, no sounds of warming water are heard. You notice that no matter how many times you turn the switch on or off, no heat results. This is the kind of simple generalization called for in step 2.

We can now generate some hypotheses about the cause of the malfunction, as suggested in step 3. Here are some candidates:

a. The coffee pot is not plugged in.
b. The external circuit breaker or fuse has tripped.
c. The power is off in the entire house or neighborhood.
d. An internal fuse in the coffee pot has blown.
e. A wire has come loose or burned through inside the coffeemaker.
f. The internal thermostat of the coffeemaker is broken.

No detailed knowledge of electrical circuits is needed to check these possibilities, although the last three call for more sophistication (and are more trouble to check) than the first three. The first three possibilities are the easiest to check and should be tested first (step 4 in our method). A simple remedy such as plugging in the pot or flipping on a circuit breaker may put you back in business. If the power is off in the building, other appliances (lights, clocks, and so on) will not work either, which provides an easy test. There may be little that you can do in this case, but at least you have identified the problem. Abusing the coffee pot will not help.

The pot may or may not have an internal fuse. If it is blown, a trip to the hardware store may be necessary. A problem like a loose wire or a burnt-out connection often becomes obvious by looking inside after you remove the bottom of the pot or the panel where the power cord comes in. (You must unplug the pot before making such an inspection!) If one of these alternatives is the case, you have identified the problem, but the repair is likely to take more time or expertise. The same is true of the last alternative.

Regardless of what you find, this systematic (and calm) approach to the problem is likely to be more productive and satisfying than the other approaches. Troubleshooting, if done this way, is an example of applying the scientific method on a small scale to an ordinary problem. We are all scientists if we approach problems in this manner.

Physics is also generally regarded as the most quantitative of the sciences. It makes heavy use of mathematics and numerical measurements to develop and test its theories. This aspect of physics has often made it seem less accessible to students, even though the models and ideas of physics can be described more simply and cleanly than those of other sciences. As we will discuss in section 1.3, mathematics serves as a compact language, allowing briefer and more precise statements than would be possible without its use.

What are the major subfields of physics?

The primary subfields of physics are listed and identified in table 1.2. Mechanics, which deals with the motion (or lack of motion) of objects under the influence of forces, was the first subfield to be explained with a comprehensive theory. Newton's theory of mechanics, which he developed in the last half of the seventeenth century, was the first full-fledged physical theory that made extensive use of mathematics. It became a prototype for subsequent theories in physics.

The first four subfields listed in table 1.2 were well developed by the beginning of the twentieth century, although all have continued to advance since then. These subfields—mechanics, thermodynamics, electricity and magnetism, and optics—are sometimes grouped as **classical physics.** The last four subfields—atomic physics, nuclear physics, particle physics, and condensed-matter physics—are often under the heading of **modern physics,** even though all of the subfields are part of the modern practice of physics. The distinction is made because the last four subfields all emerged during the twentieth century and only existed in rudimentary forms before the turn of that century.

The photographs in this section (figs. 1.4–1.7) illustrate characteristic activities or applications of the subfields. The invention of the laser has been an extremely important factor in the rapid advances now taking place in optics (fig. 1.4). The development of the infrared camera has provided a tool for the study of heat flow from buildings, which involves thermodynamics (fig. 1.5). The rapid growth in consumer electronics, as seen in the availability of home computers, pocket calculators, and many other gadgets, has been made possible by developments in condensed-matter physics (fig. 1.6). Particle physicists use particle accelerators (fig. 1.7) to study the interactions of subatomic particles in high-energy collisions.

figure **1.4**   An optics experiment using a laser.

figure **1.5**   An infrared photograph showing patterns of heat loss from a house is an application of thermodynamics.

table **1.2**
The Major Subfields of Physics
Mechanics. The study of forces and motion.
Thermodynamics. The study of temperature, heat, and energy.
Electricity and Magnetism. The study of electric and magnetic forces and electric current.
Optics. The study of light.
Atomic Physics. The study of the structure and behavior of atoms.
Nuclear Physics. The study of the nucleus of the atom.
Particle Physics. The study of subatomic particles (quarks, etc.).
Condensed-Matter Physics. The study of the properties of matter in the solid and liquid states.

Science and technology depend on each other for progress. Physics plays an important role in the education and work of engineers, whether they specialize in electrical, mechanical, nuclear, or other engineering fields. In fact, people with physics degrees often work as engineers when they are employed in industry. The lines between physics and engineering, or research and development, often blur. Physicists are generally concerned with developing a fundamental understanding of phenomena, and engineers with applying that understanding to practical tasks or products, but these functions often overlap.

One final point: physics is fun. Understanding how a bicycle works or how a rainbow is formed has an appeal that

figure **1.6** An integrated circuit employing semiconductor devices developed from knowledge of condensed-matter physics. Magnification: ×50.

figure **1.7** A Super–Proton–Synchrotron (SPS) particle accelerator used to study interactions of subatomic particles at high energies. It is located at CERN, the European particle-physics laboratory in Switzerland.

anyone can appreciate. The thrill of gaining insight into the workings of the universe can be experienced at any level. In this sense, we can all be physicists.

Physics is the study of the basic characteristics of matter and its interactions. It is the most fundamental of the sciences; many other sciences build on ideas from physics. The major subfields of physics are mechanics, electricity and magnetism, optics, thermodynamics, atomic and nuclear physics, particle physics, and condensed-matter physics. Physics plays an important role in engineering and technology, but the real fun of physics comes from understanding how the universe works.

1.3 The Role of Measurement and Mathematics in Physics

If you go into your college library, find a volume of *Physical Review* or some other major physics journal, and open it at random, you are likely to find a page with many mathematical symbols and formulas. It would probably be incomprehensible to you. In fact, even many physicists who are not specialists in the particular subfield covered by the article might have difficulty making sense of that page, because they would not be familiar with the particular symbols and definitions.

Why do physicists make such extensive use of mathematics in their work? Is knowledge of mathematics essential to understanding the ideas being discussed? Mathematics is a compact language for representing the ideas of physics that makes it easier to precisely state and manipulate the relationships between the quantities that we measure in physics. Once you are familiar with the language, its mystery disappears and its usefulness becomes more obvious. Still, this book uses mathematics in a very limited manner, because most ideas of physics can be discussed without extensive use of mathematics.

Why are measurements so important?

How do we test theories in physics? Without careful measurements, vague predictions and explanations may seem reasonable, and making choices between competing explanations may not be possible. A quantitative prediction, on the other hand, can be tested against reality, and an explanation or theory can be accepted or rejected based on the results of measurements. If, for example, one hypothesis predicts that a cannonball will land 100 meters from us and another predicts a distance of 200 meters under the same conditions, firing the cannon and measuring the actual distance provides persuasive evidence for one hypothesis or the other (fig. 1.8). The rapid growth and successes of physics began when the idea of making precise measurements as a test was accepted.

Everyday life is full of situations in which measurements, as well as the ability to express relationships between measurements, are important. Suppose, for example, that you normally prepare pancakes on Sunday morning for three people, but on a particular Sunday there is an extra mouth to feed. What will you do—double the recipe and feed the rest to the dog? Or will you figure out just how much the quantities in the recipe should be increased to come out right?

Let's say that the normal recipe calls for 1 cup of milk. How much milk will you use if you are increasing the recipe to feed four people instead of three? Perhaps you can solve this problem in your head, but some might find that process dangerous. (Let's see, 1 cup is enough for three people, so $\frac{1}{3}$ cup is needed for each person, and 4 times $\frac{1}{3}$

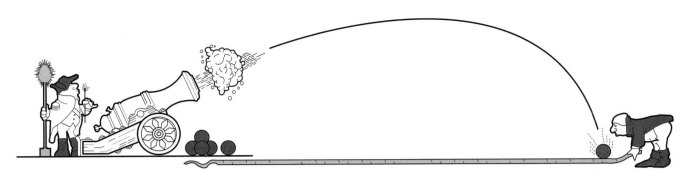

figure **1.8**   Cannonballs and a measuring tape: the proof lies in the measurement.

equals ⁴⁄₃ or 1⅓ cups. See figure 1.9.) If you had to describe this operation to someone else, for the milk and all the other ingredients, you might find yourself using a lot of words. If you looked closely at the person you were talking to, you might also notice his eyes glazing over and confusion setting in.

How can mathematics help?

You can reduce the confusion by creating a statement that works for all of the ingredients in the recipe, thus avoiding the need to repeat yourself. You could say, "The quantity of each ingredient needed for four people is related to the quantity needed for three people as 4 is to 3." That still takes quite a few words and might not be clear unless the person you were talking to was familiar with this way of stating a **proportion.** If a piece of paper was handy, you might communicate this statement in writing as:

Quantity for four : Quantity for three = 4 : 3.

To make the statement even briefer, you could use the symbol Q_4 to represent the quantity of any given ingredient needed to feed four people, and the symbol Q_3 to represent the quantity needed for three people. Then the statement can be expressed as a mathematical equation,

$$\frac{Q_4}{Q_3} = \frac{4}{3} \, .$$

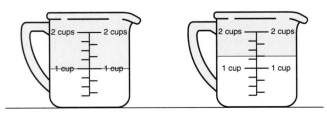

figure **1.9**   Two measuring cups, one containing enough milk to make pancakes to feed three people and the other enough for four people.

Using symbols is simply a compact way of saying the same thing that we expressed in words earlier. This compact statement also has the advantage of making manipulations of the relationship easier. For example, if you multiply both sides of this equation by Q_3, it takes the form

$$Q_4 = \frac{4}{3} \, Q_3 \, ,$$

which in words says that the quantity needed for four people is ⁴⁄₃ times what is needed for three people. If you are comfortable with fractions, you could use this relationship to find the proper amount for any ingredient quickly.

There are two points to this example. The first is that making measurements is both a routine and important part of everyday experience. The second is that using symbols to represent quantities in a mathematical statement is a shorter way of expressing an idea involving numbers than the same statement in words would be. Using mathematics also makes it easier to manipulate relationships to construct concise arguments. These are the reasons that physicists (and many other people) find mathematical statements useful.

Despite the brevity and apparent clarity of mathematical statements, many people are still more comfortable with words. This is a matter of personal choice and experience, although some fear of mathematics may also be involved. For this reason, word statements are provided in this book with most of the simple mathematical expressions that we will use. Together with the mathematical statement and the drawings, these word statements will help you to understand the concepts we will be discussing.

Why are metric units used?

Units of measurement are an essential part of any measurement. We do not communicate clearly if we just state a number. If you just talked about adding 1⅓ of milk, for example, your statement would be incomplete. You need to indicate whether you are talking about cups, pints, or milliliters.

The liter and milliliter are *metric* units of volume. Cups, pints, quarts, and gallons are holdovers from the older

English system of units. Most countries have now adopted the **metric system,** which has several advantages over the English system still used in the United States. The main advantage of the metric system is its use of standard prefixes to represent multiples of 10, making unit conversion within the system quite easy. The fact that a kilometer (km) is 1000 meters and a centimeter (cm) is $1/100$ of a meter, and that the prefixes *kilo* and *centi* always mean 1000 and $1/100$, makes these conversions easy to remember (see table 1.3). To convert 30 centimeters to meters, all we have to do is move the decimal point two places to the left to get 0.30 meter. Moving the decimal point two places to the left is equivalent to dividing by 100.

Table 1.3 is a list of the common prefixes used in the metric system. (See appendix B for a discussion of the **powers of 10** or **scientific notation** used for describing very large and very small numbers.) The basic unit of volume in the metric system is the liter (L), which is slightly larger than a quart (1 liter = 1.057 quarts). A milliliter (mL) is $1/1000$ of a liter, a convenient size for quantities in recipes. One milliliter is also equal to 1 cm^3, or 1 cubic centimeter, so there is a simple relationship between the length and volume measurements in the metric system. Such simple relationships are hard to find in the English system, where 1 cup is $1/4$ of quart, and a quart is 67.2 cubic inches.

The metric system predominates in this book. English units will be used occasionally because they are familiar and can help in learning new concepts. Most of us still relate more readily to distances in miles than in kilometers, for example. That there are 5280 feet in a mile is a nuisance, however, compared to the tidy 1000 meters in 1 kilometer. Becoming familiar with the metric system is a worthy objective. Your ability to participate in international trade (for business or pleasure) will be enhanced if you are familiar with the system of units used in most of the world.

> Stating a result or prediction in numbers lends precision to otherwise vague claims. Measurement is an essential part of science and of everyday life. Using mathematical symbols and statements is an efficient way of stating the results of measurements and eases manipulating the relationships between quantities. Units of measurement are an essential part of any measurement, and the metric system of units used in most of the world has a number of advantages over the older English system.

1.4 Physics and Everyday Phenomena

Studying physics can and will lead us to ideas as earth-shaking as the fundamental nature of matter and the structure of the universe. With ideas like these available, why spend time on more mundane matters like explaining how a bicycle stays upright or how a flashlight works? Why not just plunge into far-reaching discussions of the fundamental nature of reality?

Why study everyday phenomena?

Our understanding of the fundamental nature of the universe is based on concepts such as mass, energy, and electric charge that are abstract and not directly accessible to our senses. It is possible to learn some of the words associated with these concepts and to read and discuss ideas involving them without ever acquiring a good understanding of their meaning. This is one risk of playing with the grand ideas without laying the proper foundation.

Using everyday experience to raise questions, introduce concepts, and practice devising physical explanations has the advantage of dealing with examples that are familiar and concrete. These examples also appeal to your natural curiosity about how things work, which, in turn, can motivate you to understand the underlying concepts. If you can clearly describe and explain common events, you gain confidence in dealing with more abstract concepts. With familiar examples, the concepts are set on firmer ground, and their meaning becomes more real.

For example, why a bicycle (or a top) stays upright while moving but falls over when at rest involves the concept of angular momentum, which is discussed in chapter 8. Angular momentum also plays a role in our understanding of atoms and the atomic nucleus—both in the realm of the very small—and the structure of galaxies at the opposite end of the scale (fig. 1.10). You are more likely to understand angular momentum, though, by discussing it first in the context of bicycle wheels or tops.

The principles explaining falling bodies, such as the acorn mentioned in the chapter introduction, involve the concepts of velocity, acceleration, force, and mass, which are discussed in chapters 2, 3, and 4. Like angular momentum, these concepts are also important to our understanding

table **1.3**				
Commonly Used Metric Prefixes				
		Meaning		
Prefix	in figures		in scientific notation	in words
giga	1 000 000 000		$= 10^9$	= 1 billion
mega	1 000 000		$= 10^6$	= 1 million
kilo	1000		$= 10^3$	= 1 thousand
centi	$1/100$	$= 0.01$	$= 10^{-2}$	= 1 hundredth
milli	$1/1000$	$= 0.001$	$= 10^{-3}$	= 1 thousandth
micro	$1/1\,000\,000$	$= 1/10^6$	$= 10^{-6}$	= 1 millionth
nano	$1/1\,000\,000\,000$	$= 1/10^9$	$= 10^{-9}$	= 1 billionth

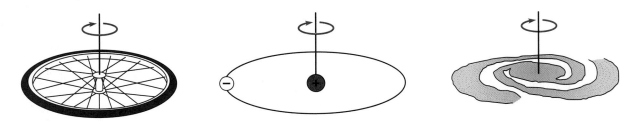

figure **1.10**   A bicycle wheel, an atom, and a galaxy all involve the concept of angular momentum.

of atoms and the universe. Understanding how rainbows are formed involves the behavior of light, discussed in chapter 17. The behavior of light also plays a major role in how we think about atoms and the universe.

Our "common sense" sometimes misleads us in our understanding of everyday phenomena. Adjusting common sense to incorporate well-established physical principles is one of the challenges we face in dealing with everyday experience. By performing simple experiments, either at home (as is often suggested in this book) or in laboratories and demonstrations associated with your course in physics, you can take an active part in building your own scientific worldview.

Although it may seem like an oxymoron, everyday experience is extraordinary. A bright rainbow is an incredible sight. Understanding how rainbows originate does not detract from the experience. It adds excitement to explain such a beautiful display with just a few elegant concepts. In fact, people who understand these ideas see more rainbows because they know where to look. This excitement, and the added appreciation of nature that is a part of it, is accessible to all of us.

Studying everyday phenomena can make abstract ideas more accessible. These ideas are needed to understand the fundamental nature of matter and the universe, but they are best encountered first in familiar examples. Being able to explain common phenomena builds confidence in using the ideas and enhances our appreciation of what happens around us.

1.5 How to Use the Features of This Book

This book has a number of features designed to make it easier for you to organize and grasp the concepts that we will explore. These features include the chapter overview and outline at the beginning of each chapter and the summary at the end of each chapter, as well as the structure of individual sections of the chapters. The questions, exercises, and challenge problems at the end of each chapter

also play an important role. How can these features be used to best advantage?

Chapter outlines and summaries

Knowing where you are heading before you set out on a journey can be the key to the success of your mission. Students get a better grasp of concepts if they have some structure or framework to help them to organize the ideas. Both the chapter overview and outline at the beginning of each chapter and the summary at the end are designed to provide such a framework. Having a clear idea of what you are trying to accomplish before you invest time in reading a chapter will make your reading more effective and enjoyable.

The list of topics and questions in the chapter outline can be used as a checklist for measuring your progress as you read. Each numbered topic in the outline, with its associated questions, pertains to a section of the chapter. The outline is designed to stimulate your curiosity by providing some blanks (unanswered questions) to be filled in by your reading. Without the blanks, your mind has no organizational structure to store the information. Without structure, recall is more difficult. You can use the questions in the outline to check the effectiveness of your reading. Can you answer all of the questions when you are done? Each section of a chapter also begins with questions, and the section subheadings are likewise often cast as questions. At the end of each section there is also an indented summary paragraph designed to help you tie the ideas in that section together.

The end-of-chapter summary gives a short description of the key ideas in each section, often cast in the form of answers to the questions raised in the outline (fig. 1.11). Summaries provide a quick review, but they are no substitute for a careful reading of the main text. By following the same organizational structure as the outline, the summary reminds you where to find a more complete discussion of these ideas. The purpose of both the outlines and the summaries is to make your reading more organized and effective.

Studying any new discipline requires forming new patterns of thought that can take time to gel. The summaries at the end of each section, as well as at the end of the

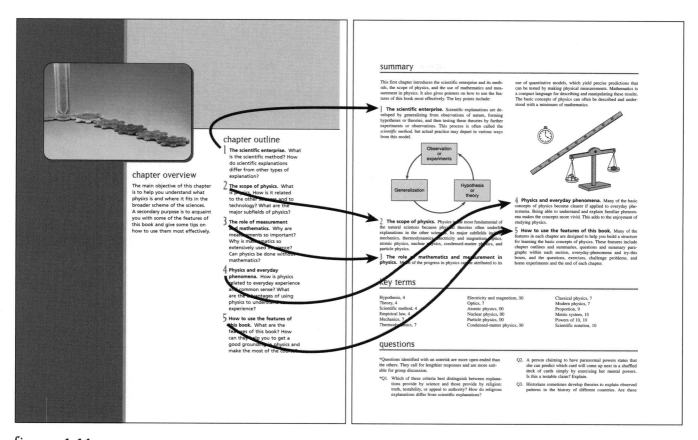

figure **1.11** The chapter outline and chapter summary provide related frameworks for organizing concepts.

chapter, can help this gelling to take place. A structure is often built layer by layer, and the later layers will be shaky if the base is unstable.

How should the questions and exercises be used?

At the end of each chapter you will find a group of questions, followed by a group of exercises, and, finally, by a small number of challenge problems. Your grasp of the chapter will improve if you write out answers to the questions and exercises, either as assigned by your instructor or in independent study. The ideas contained in each chapter cannot be thoroughly mastered without this kind of practice.

The questions are crucial to helping you fix the important concepts and distinctions in your mind. Most of the questions call for a short answer as well as an explanation. A few of the questions, marked with asterisks, are more open-ended and call for lengthier responses. It is a good idea to write out the explanations in clear sentences when you answer these questions, because it is only through reinforcement that ideas become a part of you. Also, if you can explain something clearly to someone else, you understand it. A sample question and answer appears in Try This Box 1.1.

The exercises are designed to give you practice in using the ideas and the related formulas to do simple computations. The exercises also help to solidify your understanding of concepts by giving you a sense of the units and the sizes of the quantities involved. Even though many of the exercises are straightforward enough to work in your head without writing much down, we recommend writing out the information given, the information sought, and the solution in the manner shown in Try This Box 1.2. This develops careful work habits that will help you avoid careless

---

try this box **1.1**

Sample Question: How Reliable Is Astrology?

Question: Astrologers claim that many events in our lives are determined by the positions of the planets relative to the stars. Is this a testable hypothesis?

Answer: Yes, it could be tested if astrologers were willing to make explicit predictions about future events that could be verified by independent observers. In fact, astrologers usually carefully avoid doing this, preferring to cast their predictions as vague statements subject to broad interpretation. This prevents clean tests.

try this box 1.2

Sample Exercise: Conversions

Exercise: If you are told that there are 2.54 cm in 1 inch,
  a. How many centimeters are there in 1 foot (12 inches)?
  b. How many meters does 1 foot represent?

a. Given:

1 inch = 2.54 cm   1 ft = 12 in. $\times$ 2.54 cm/in.
1 foot = 12 inches   1 ft = **30.5 cm**
1 foot = ? (in cm)

b. 1 ft = 30.5 cm
    1 m = 100 cm $1 \text{ ft} = \dfrac{30.5 \text{ cm}}{100 \text{ cm/m}}$
    1 ft = ? (in m)   1 ft = **0.305 m**

mistakes. Most students find the exercises easier than the questions. The sample exercises scattered through each chapter can help you get started. They are marked "try this" as an invitation—and a bit of a dare.

The challenge problems are more wide-ranging than the questions or exercises. They often involve features of both. Although not necessarily harder than the questions or exercises, they do take more time and are sometimes used to extend ideas beyond what was discussed in the chapter. Doing one or two of these in each chapter should build your confidence. They are particularly recommended for those students who have worked the exercises and want to explore the topic in more depth.

Answers to the odd-numbered exercises and challenge problems are found in the back of the book in appendix D. Looking up the answer before attempting the problem is self-defeating. It deprives you of practice in thinking things through on your own. Checking answers *after* you have

worked an exercise can be a confidence builder. Answers should be used only to confirm or improve your own thinking.

Home experiments and everyday phenomenon boxes

Reading or talking about physical ideas is useful, but there is no substitute for hands-on experience with the phenomena. You already have a wealth of experience with many of these phenomena, but you probably have not related it to the physical concepts you will be learning. Seeing things in new ways will make you a more astute observer.

In addition to the home experiments at the end of each chapter, we often suggest some simple experiments in the main text or in the study hints. We strongly recommend making these observations and doing the experiments. Lecture demonstrations can help, but doing something yourself imprints it vividly on your mind. There is excitement in discovering things yourself and seeing them in a new light.

The boxes that discuss everyday phenomena also give you practice in applying physical concepts. Most of the phenomena discussed in these boxes are familiar. The boxes allow us to explore these examples more thoroughly. Participating in these investigations of everyday phenomena can help bring the ideas home.

Having a framework on which to hang new ideas is critical to your success in grasping the ideas and putting them in context. This book has features explicitly designed to help you build that structure, including chapter overviews, outlines, and summaries, as well as the structure and summaries in each chapter section. How much you engage yourself in the questions, exercises, challenge problems, and home experiments will determine your enjoyment and your success. Learning science, particularly physics, is not a passive process. It needs lively participation.

summary

This first chapter introduces the scientific enterprise and its methods, the scope of physics, and the use of mathematics and measurement in physics. It also gives pointers on how to use the features of this book most effectively. The key points include:

1   The scientific enterprise.   Scientific explanations are developed by generalizing from observations of nature, forming hypotheses or theories, and then testing these theories by further experiments or observations. This process is often called the *scientific method,* but actual practice may depart in various ways from this model.

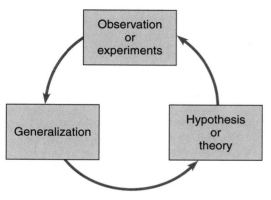

2   The scope of physics.   Physics is the most fundamental of the natural sciences because physical theories often underlie explanations in the other sciences. Its major subfields include mechanics, thermodynamics, electricity and magnetism, optics, atomic physics, nuclear physics, condensed-matter physics, and particle physics.

3   The role of measurement and mathematics in physics.   Much of the progress in physics can be attributed to its use of quantitative models, which yield precise predictions that

can be tested by making physical measurements. Mathematics is a compact language for describing and manipulating these results. The basic concepts of physics can often be described and understood with a minimum of mathematics.

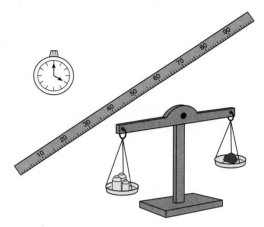

4   Physics and everyday phenomena.   Many of the basic concepts of physics become clearer if applied to everyday phenomena. Being able to understand and explain familiar phenomena makes the concepts more vivid. This adds to the enjoyment of studying physics.

5   How to use the features of this book.   Many of the features in each chapter are designed to help you build a structure for learning the basic concepts of physics. These features include chapter outlines and summaries, questions and summary paragraphs within each section, everyday phenomena and try this boxes, and the questions, exercises, challenge problems, and home experiments at the end of each chapter.

key terms

Hypothesis, 3
Theory, 3
Scientific method, 3
Empirical law, 3
Classical physics, 6
Modern physics, 6

Mechanics, 6
Thermodynamics, 6
Electricity and magnetism, 6
Optics, 6
Atomic physics, 6
Nuclear physics, 6

Particle physics, 6
Condensed-matter physics, 6
Proportion, 8
Metric system, 9
Powers of 10, 9
Scientific notation, 9

questions

*Q1. Which of these criteria best distinguish between explanations provided by science and those provided by religion: truth, testability, or appeal to authority? How do religious explanations differ from scientific explanations?

Q2. A person claiming to have paranormal powers states that she can predict which card will come up next in a shuffled deck of cards simply by exercising her mental powers. Is this a testable claim? Explain.

Q3. Historians sometimes develop theories to explain observed patterns in the history of different countries. Are these theories testable in the same sense as a theory in physics? Explain.

*Q4. Over the years, there have been several credible claims by experienced observers of sightings of Unidentified Flying Objects (UFOs). Despite this, scientists have shied away from taking up serious study of UFOs, although there are ongoing searches for signals from extraterrestrial intelligent beings. Can you think of reasons why scientists have not taken UFOs seriously? What problems can you see in trying to study UFOs?

Q5. Suppose that your car will not start and you form the hypothesis that the battery is dead. How would you test this hypothesis? Explain.

Q6. Suppose that your phone has not rung in several days, but a friend tells you he has tried to call. Develop two hypotheses that could explain why the phone has not rung and state how you would test these hypotheses.

*Q7. Suppose that a friend states the hypothesis that the color of socks that he wears on a given day, brown or black, will determine whether the stock market will go up or down. He can cite several instances in which this hypothesis has been apparently verified. How would you go about evaluating this hypothesis?

Q8. Which of the three science fields: biology, chemistry, or physics, would you say is the most fundamental? Explain by describing in what sense one of these fields may be more fundamental than the others.

Q9. Based upon the brief descriptions provided in table 1.2, which subfield of physics would you say is involved in the explanation of rainbows? Which subfield is involved in describing how an acorn falls? Explain.

Q10. Based upon the descriptions provided in table 1.2, which subfields of physics are involved in explaining why an ice cube melts? Which subfields are involved in explaining how an airplane flies? Explain.

Q11. Suppose that you are told that speed is defined by the relationship $s = d/t$, where s represents speed, d represents distance, and t represents time. State this relationship in words, using no mathematical symbols.

Q12. Impulse is defined as the average force acting on an object multiplied by the time the force acts. If we let I represent impulse, F the average force, and t the time, is $I = F/t$ a correct way of expressing this definition? Explain.

Q13. The distance that an object travels when it starts from rest and undergoes constant acceleration is one-half the acceleration multiplied by the square of the time. Invent your own symbols and express this statement in symbolic form.

Q14. What are the primary advantages of the metric system of units over the older English system of units? Explain.

Q15. What are the advantages, if any, of continuing to use the English system of units in industry and commerce rather than converting to the metric system? Explain.

Q16. Which system of units, the metric system or English system, is used more widely throughout the world? Explain.

Q17. The width of a man's hand was used as a common unit of length several hundred years ago. What are the advantages and disadvantages of using such a unit? Explain.

exercises

E1. Suppose that a pancake recipe designed to feed three people calls for 600 mL of flour. How many milliliters of flour would you use if you wanted to extend the recipe to feed five people?

E2. Suppose that a cupcake recipe designed to produce twelve cupcakes calls for 1000 mL of flour. How many milliliters of flour would you use if you wanted to make only eight cupcakes?

E3. It is estimated that six large pizzas are about right to serve a physics club meeting of 30 students. How many pizzas would be required if the group grows to 50 students?

E4. A man uses his hand to measure the width of a tabletop. If his hand has a width of 12 cm at its widest point, and he finds the tabletop to be 10.5 hands wide, what is the width of the tabletop in cm? In meters?

E5. A woman's foot is 9 inches long. If she steps off the length of a room by placing one foot directly in front of the other, and finds the room to be 15 foot-lengths long, what is the length of the room in inches? In feet?

E6. A book is 265 mm in length. What is this length in centimeters? In meters?

E7. A crate has a mass of 8.60 kg (kilograms). What is this mass in grams? In milligrams?

E8. A tank holds 1.24 kL (kiloliters) of water. How many liters is this? How many milliliters?

E9. A mile is 5280 ft long. The sample exercise in Try This Box 1.2 shows that 1 foot is approximately 0.305 m. How many meters are there in a mile? How many kilometers (km) are there in a mile?

E10. If a mile is 5280 ft long and a yard contains 3 ft, how many yards are there in a mile?

E11. Area is found by multiplying the length of a surface times the width. If a floor measures 6.25 m^2, how many square centimeters does this represent? How many square centimeters are there in 1 m^2?

challenge problems

CP1. Astrologers claim that they can predict important events in your life by the configuration of the planets and the astrological sign under which you were born. Astrological predictions, called horoscopes, can be found in most daily newspapers. Find these predictions in a newspaper and address the questions:

a. Are the astrological predictions testable?
b. Choosing the prediction for your own sign, how would you go about testing its accuracy over the next month or so?
c. Why do newspapers print these readings? What is their appeal?

home experiments and observations

HE1. Look around your house, car, or dormitory room to see what measuring tools (rulers, measuring cups, speedometers, etc.) you have handy. Which of these tools, if any, provides both English and metric units? For those that do, determine the conversion factor needed to convert the English units to metric units.

unit One

The Newtonian Revolution

In 1687, Isaac Newton published his *Philosophiae Naturalis Principia Mathematica* or *Mathematical Principles of Natural Philosophy.* This treatise, often called simply Newton's *Principia*, presented his theory of motion, which included his three laws of motion and his law of universal gravitation. Together these laws explain most of what was then known about the motion of ordinary objects near the surface of the earth (terrestrial mechanics) as well as the motion of the planets around the sun (celestial mechanics). Along the way, Newton had to invent the mathematical techniques that we call calculus.

Newton's theory of mechanics described in the *Principia* was an incredible intellectual achievement that revolutionized both science and philosophy. The revolution did not begin with Newton, though. The true rebel was the Italian scientist, Galileo Galilei, who died just a few months after Newton was born in 1642. Galileo championed the sun-centered view of the solar system proposed a hundred years earlier by Nicolaus Copernicus and stood trial under the Inquisition for his pains. Galileo also challenged the conventional wisdom, based on Aristotle's teachings, about the motion of ordinary objects. In the process, he developed many of the principles of terrestrial mechanics that Newton later incorporated into his theory.

Although Newton's theory of motion does not accurately describe the motion of very fast objects (which are now described using Einstein's theory of relativity) and very small objects (where quantum mechanics must be used), it is still used extensively in physics and engineering to explain motion and to analyze structures. Newton's theory has had enormous influence over the last three hundred years in realms of thought that extend well beyond the natural sciences and deserves to be understood by anyone claiming to be well educated.

Central to Newton's theory is his second law of motion. It states that the acceleration of an object is proportional to the net force acting on the object and inversely proportional to the mass of the object. Push an object and that object accelerates in the direction of the applied force. Contrary to intuition and to Aristotle's teachings, acceleration, not velocity, is proportional to the applied force. To understand this idea, we will thoroughly examine acceleration, which involves a *change* in the motion of an object.

Rather than plunging into Newton's theory, we begin this unit by studying Galileo's insights into motion and free fall. This provides the necessary foundation to tackle Newton's ideas. To see well, we need to stand on the shoulders of these giants.

Describing Motion

chapter

2

chapter overview

The main purpose of this chapter is to provide clear definitions and illustrations of the terms used in physics to describe motion, such as the motion of the car described in this chapter's opening example. Speed, velocity, and acceleration are crucial concepts for the analysis of motion in later chapters. Precise description is the first step to understanding. Without it, we remain awash in vague ideas that are not defined well enough to test our explanations.

Each numbered topic in this chapter builds on the previous section, so it is important to obtain a clear understanding of each topic before going on. The distinctions between speed and velocity and velocity and acceleration are particularly important.

chapter outline

1 **Average and instantaneous speed.** How do we describe how fast an object is moving? How does instantaneous speed differ from average speed?

2 **Velocity.** How do we introduce direction into descriptions of motion? What is the distinction between speed and velocity?

3 **Acceleration.** How do we describe changes in motion? What is the relationship between velocity and acceleration?

4 **Graphing motion.** How can graphs be used to describe motion? How can the use of graphs help us gain a clearer understanding of speed, velocity, and acceleration?

5 **Uniform acceleration.** What happens when an object accelerates at a steady rate? How do the velocity and distance traveled vary with time when an object is uniformly accelerating?

Imagine that you are in your car stopped at an intersection. After waiting for cross traffic, you pull away from the stop sign, accelerating eventually to a speed of 56 kilometers per hour (35 miles per hour). You maintain that speed until a dog runs in front of your car and you hit the brakes, reducing your speed rapidly to 10 km/h (fig. 2.1). Having missed the dog, you speed up again to 56 km/h. After another block, you come to another stop sign and reduce your speed gradually to zero.

We can all relate to this description. Measuring speed in miles per hour (MPH) may be more familiar than the use of kilometers per hour (km/h), but speedometers in cars now show both. The use of the term *acceleration* to describe an increase in speed is also common. In physics, however, these concepts take on more precise and specialized meanings that make them even more useful in describing exactly what is happening. These meanings are sometimes different from those in everyday use. The term *acceleration,* for example, is used by physicists to describe any situation in which velocity is changing, even when the speed may be decreasing or the direction of the motion may be changing.

How would you define the term *speed* if you were explaining the idea to a younger brother or sister? Does *velocity* mean the same thing? What about *acceleration*—is the notion vague or does it have a precise meaning? Is it the same thing as velocity? Clear definitions are essential to developing clear explanations. The language used by physicists differs from our everyday language, even though the ideas are related and the same words are used. What are the exact meanings that physicists attach to these concepts, and how can they help us to understand motion?

figure **2.1**   As the car brakes for the dog, there is a sudden change in speed.

study hint

Science has always relied on pictures and charts to get points across. Throughout the book, a number of concepts will be introduced and illustrated. In the illustrations, the same color will be used for certain phenomena.

→   Blue arrows are velocity vectors.

→   Green arrows depict acceleration vectors.

→   Red arrows depict force vectors.

→   Purple arrows show momentum, a concept we will explore in chapter 7.

2.1 Average and Instantaneous Speed

Since driving or riding in cars is a common activity in our daily lives, we are familiar with the concept of speed. Most of us have had experience in reading a speedometer (or perhaps failing to read it carefully enough to avoid the attention of law enforcement). If you describe how fast something is moving, as we did in our example in the introduction, you are talking about **speed.**

How is average speed defined?

What does it mean to say that we are traveling at a speed of 55 MPH? It means that we would cover a distance of 55 miles in a time of 1 hour if we traveled steadily at that speed. Carefully note the structure of this description: there is a number, 55, and some units or dimensions, miles per hour. Numbers and units are both essential parts of a description of speed.

The term *miles per hour* implies that miles are divided by hours in arriving at the speed. This is exactly how we would compute the **average speed** for a trip: suppose, for example, that we travel a distance of 260 miles in a time of 5 hours, as shown on the road map of figure 2.2. The average speed is then 260 miles divided by 5 hours, which is

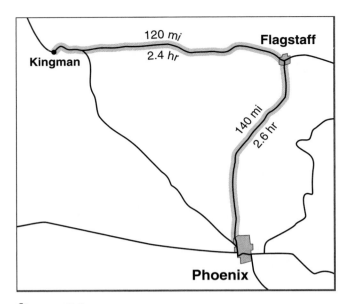

figure **2.2**   A road map showing a trip of 260 miles, with driving times for the two legs of the trip.

equal to 52 MPH. This type of computation is familiar to most of us.

We can also express the definition of average speed in a word equation as

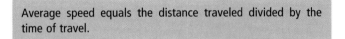

Average speed equals the distance traveled divided by the time of travel.

or

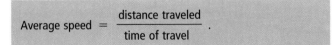

$$\text{Average speed} = \frac{\text{distance traveled}}{\text{time of travel}}.$$

We can represent this same definition with symbols by writing

$$s = \frac{d}{t},$$

where the letter s represents the speed, d represents distance, and t represents the time. As noted in chapter 1, letters or symbols are a compact way of saying what could be said with a little more effort and space with words. Judge for yourself which is the more efficient way of expressing this definition of average speed. Most people find the symbolic expression easier to remember and use.

The average speed that we have just defined is the **rate** at which distance is covered over time. Rates always represent one quantity divided by another. Gallons per minute, pesos per dollar, and points per game are all examples of rates. If we are considering time rates, the quantity that we divide by is *time,* which is the case with average speed.

Many other quantities that we will be considering involve time rates.

What are the units of speed?

Units are an essential part of the description of speed. Suppose you say that you were doing 70—without stating the units. In the United States, that would probably be understood as 70 MPH, since that is the unit most frequently used. In Europe, on the other hand, people would probably assume that you are talking about the considerably slower speed of 70 km/h. If you do not state the units, you will not communicate effectively.

It is easy to convert from one unit to another if the conversion factors are known. For example, if we want to convert kilometers per hour to miles per hour, we need to know the relationship between miles and kilometers. A kilometer is roughly 6/10 of a mile (0.6214, to be more precise). As shown in Try This Box 2.1, 70 km/h is equal to 43.5 MPH. The process involves multiplication or division by the appropriate conversion factor.

Units of speed will always be a distance divided by a time. In the metric system, the fundamental unit of speed is meters per second (m/s). Try This Box 2.1 also shows the conversion of kilometers per hour to meters per second, done as a two-step process. As you can see, 70 km/h can also be expressed as 19.4 m/s or roughly 20 m/s. This is a convenient size for discussing the speeds of ordinary objects. (As shown in Try This Box 2.2, the convenient unit for measuring the growth of grass has a very different size.) Table 2.1 shows some familiar speeds expressed in miles per hour, kilometers per hour, and meters per second to give you a sense of their relationships.

What is instantaneous speed?

If we travel a distance of 260 miles in 5 hours, as in our earlier example, is it likely that the entire trip takes place

try this box 2.1

Unit Conversions   1 km = 0.6214 miles
                          1 mile = 1.609 km

Convert 70 kilometers per hour to miles per hour.

   70 ~~km~~/h × 0.6214 miles/~~km~~ = **43.5 MPH**

Convert 70 kilometers per hour to meters per second.

   70 ~~km~~/h × 1000 m/~~km~~ = 70 000 m/h

   But 1 h = 60 ~~min~~ × 60 s/~~min~~ = 3600 s

   $$\frac{70\,000 \text{ m/}\cancel{h}}{3600 \text{ s/}\cancel{h}} = \textbf{19.4 m/s}$$

Lines drawn through the units indicate cancellation.

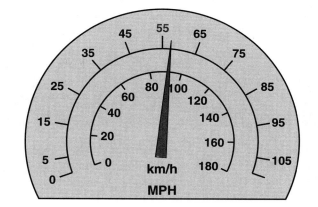

figure 2.3 A speedometer with two scales for measuring instantaneous speed, MPH and km/h.

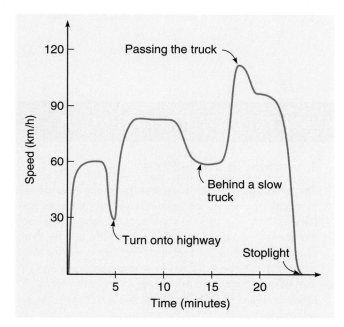

figure 2.4 Variations in instantaneous speed for a portion of a trip on a local highway.

at a speed of 52 MPH? Of course not; the speed goes up and down as the road goes up and down, when we overtake slower vehicles, when rest breaks occur, or when the highway patrol looms on the horizon. If we want to know how fast we are going at a given instant in time, we read the speedometer, which displays the **instantaneous speed** (fig. 2.3).

How does instantaneous speed differ from average speed? The instantaneous speed tells us how fast we are going at a given instant but tells us little about how long it will take to travel several miles, unless the speed is held constant. The average speed, on the other hand, allows us to compute how long a trip might take but says little about the variation in speed during the trip. A more complete description of how the speed of a car varies during a portion of a trip could be provided by a graph such as that shown in figure 2.4. Each point on this graph represents the instantaneous speed at the time indicated on the horizontal axis.

Even though we all have some intuitive sense of what instantaneous speed means from our experience in reading speedometers, computing this quantity presents some problems that we did not encounter in defining average speed. We could say that instantaneous speed is the rate that distance is being covered at a given instant in time, but how

do we compute this rate? What time interval should we use? What is an instant in time?

Our solution to this problem is simply to choose a very short interval of time during which a very short distance is covered and the speed does not change drastically. If we know, for example, that in 1 second a distance of 20 meters was covered, dividing 20 meters by 1 second to obtain a speed of 20 m/s would give us a good estimate of the instantaneous speed, provided that the speed did not change much during that single second. If the speed was changing rapidly, we would have to choose an even shorter interval of time. In principle, we can choose time intervals as small as we wish, but in practice, it can be hard to measure such small quantities.

table 2.1

Familiar Speeds in Different Units

20 MPH	=	32 km/h	=	9 m/s
40 MPH	=	64 km/h	=	18 m/s
60 MPH	=	97 km/h	=	27 m/s
80 MPH	=	130 km/h	=	36 m/s
100 MPH	=	160 km/h	=	45 m/s

If we put these ideas into a word definition of instantaneous speed, we could state it as

> Instantaneous speed is the rate at which distance is being covered at a given instant in time. It is found by computing the average speed for a very short time interval in which the speed does not change appreciably.

Instantaneous speed is closely related to the concept of average speed but involves very short time intervals. When discussing traffic flow, average speed is the critical issue, as shown in Everyday Phenomenon Box 2.1.

We find an average speed by dividing the distance traveled by the time required to cover that distance. Average speed is therefore the average rate at which distance is being covered. Instantaneous speed is the rate that distance is being covered at a given instant in time and is found by considering very small time intervals or by reading a speedometer. Average speed is useful for estimating how long a trip will take, but instantaneous speed is of more interest to the highway patrol.

2.2 Velocity

Do the words *speed* and *velocity* mean the same thing? They are often used interchangeably in everyday language, but physicists make an important distinction between the two terms. The distinction has to do with direction: which way is the object moving? This distinction turns out to be essential to understanding Newton's theory of motion (introduced in chapter 4), so it is not just a matter of whim or jargon.

Everyday Phenomenon
box 2.1

Transitions in Traffic Flow

The Situation. Jennifer commutes into the city on a freeway every day for work. As she approaches the city, the same patterns in traffic flow seem to show up in the same places each day. She will be moving with the flow of traffic at a speed of approximately 60 MPH when suddenly things will come to a screeching halt. The traffic will be stop and go briefly and then will settle into a wavelike mode with speeds varying between 10 and 30 MPH. Unless there is an accident, this will continue for the rest of the way into the city.

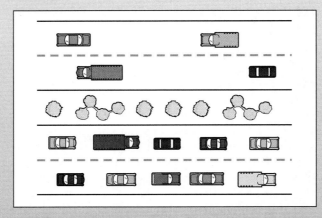

The traffic in the upper lanes is flowing freely with adequate spacing to allow higher speeds. The higher-density traffic in the lower lanes moves much more slowly.

What causes these patterns? Why does the traffic stop when there is no apparent reason such as an accident? Why do metered ramp lights seem to help the situation? Questions like these are the concern of the growing field of traffic engineering.

The Analysis. Although a full analysis of traffic flow is complex, there are some simple ideas that can explain many of the patterns that Jennifer observes. The density of vehicles, measured in vehicles per mile, is a key factor. Adding vehicles at entrance ramps increases the density. The spacing between vehicles varies with speed so that speed and density are interrelated.

When Jennifer and other commuters are traveling at 60 MPH, they need to keep a spacing of several car lengths between vehicles. Most drivers do this without thinking about it, although there are always some who follow too closely or *tailgate*. Tailgating runs the risk of rear-end collisions when the traffic suddenly slows.

When more vehicles are added at an entrance ramp, the density must increase, reducing the distance between vehicles. As the distance between vehicles decreases, drivers should reduce their speed to maintain a safe stopping distance. If this occurred uniformly, there would be a gradual decrease in the average speed of the traffic to accommodate the greater density. This is not what usually happens, however.

(continued)

What is the difference between speed and velocity?

Imagine that you are driving a car around a curve (as illustrated in figure 2.5) and that you maintain a constant speed of 60 km/h. Is your velocity also constant in this case? The answer is no, because **velocity** involves the direction of motion as well as how fast the object is going. The direction of motion is changing as the car goes around the curve.

To simply state this distinction, speed as we have defined it tells us how fast an object is moving but says nothing about the direction of the motion. Velocity includes the idea of direction. To specify a velocity, we must give both its size or **magnitude** (how fast) and its direction (north, south, east, up, down, or somewhere in between). If you tell me that an object is moving 15 m/s, you have told me its *speed*. If you tell me that it is moving due west at 15 m/s, you have told me its *velocity*.

At point A on the diagram in figure 2.5, the car is traveling due north at 60 km/h. At point B, because the road curves, the car is traveling northwest at 60 km/h. Its velocity at point B is different from its velocity at point A (because the directions are different). The speeds at point A and B are the same. Direction is irrelevant in specifying the speed of the object. It has no effect on the reading on your speedometer.

Changes in velocity are produced by forces acting upon the car, as we will discuss further in chapter 4. The most important force involved in changing the velocity of a car is the frictional force exerted on the tires of the car by the road surface. A force is required to change either the size or the direction of the velocity. If no net force were acting on the car, it would continue to move at constant speed in a straight line. This happens sometimes when there is ice or oil on the road surface, which can reduce the frictional force to almost zero.

A significant proportion of drivers will attempt to maintain their speed of 50 to 60 MPH even when densities have increased beyond the point where this is advisable. This creates an unstable situation. At some point, usually near an entrance ramp, the density becomes too large to sustain these speeds. At this point there is a sudden drop in average speed and a large increase in the local density. As shown in the drawing, cars can be separated by less than a car length when they are stopped or moving very slowly.

Once the average speed of a few vehicles has slowed to less than 10 MPH, vehicles moving at 50 to 60 MPH begin to pile up behind this slower moving jam. Because this does not happen smoothly, some vehicles must come to a complete stop, further slowing the flow. At the front end of the jam, on the other hand, the density is reduced due to the slower flow behind. Cars can then start moving at a speed consistent with the new density, perhaps around 30 MPH. If every vehicle moved with the appropriate speed, flow would be smooth and the increased density could be safely accommodated. More often, however, overanxious drivers exceed the appropriate speed, causing fluctuations in the average speed as vehicles begin to pile up again.

Notice that we are using average speed with two different meanings in this discussion. One is the average speed of an individual vehicle as its instantaneous speed increases and decreases. The other is the average speed of the overall traffic flow involving many vehicles. When the traffic is flowing freely, the average speed of different vehicles may differ. When the traffic is in a slowly moving jam, the average speeds of different vehicles are essentially the same, at least within a given lane.

Traffic lights at entrance ramps that permit vehicles to enter one-at-a-time at appropriate intervals can help to smoothly integrate the added vehicles to the existing flow. This reduces the sudden changes in speed caused by a rapid increase in density. Once the density increases beyond the certain level, however, a slowing of traffic is inevitable. The abrupt change from low-density, high-speed flow to higher-density, slow flow is analogous to a phase transition from a gas to a liquid. (See chapter 10.) Traffic engineers have used this analogy to better understand the process.

If we could automatically control and coordinate the speeds of all the vehicles on the highway, the highway might carry a much greater volume of traffic at a smooth rate of flow. Speeds could be adjusted to accommodate changes in density and smaller vehicle separations could be maintained at higher speeds because the vehicles would all be moving in a synchronized fashion. Better technology may someday achieve this dream.

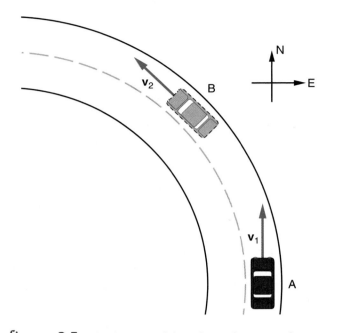

figure 2.5 The direction of the velocity changes as the car moves around the curve, so that the velocity v_2 is not the same as the velocity v_1 even though the speed has not changed.

What is a vector?

Velocity is a quantity for which both the size and direction are important. We call such quantities **vectors.** To describe these quantities fully, we need to state *both the size and the direction.* Velocity is a vector that describes how fast an object is moving and in what direction it is moving. Many of the quantities used in describing motion (and in physics more generally) are **vector quantities.** These include velocity, acceleration, force, and momentum, to name a few.

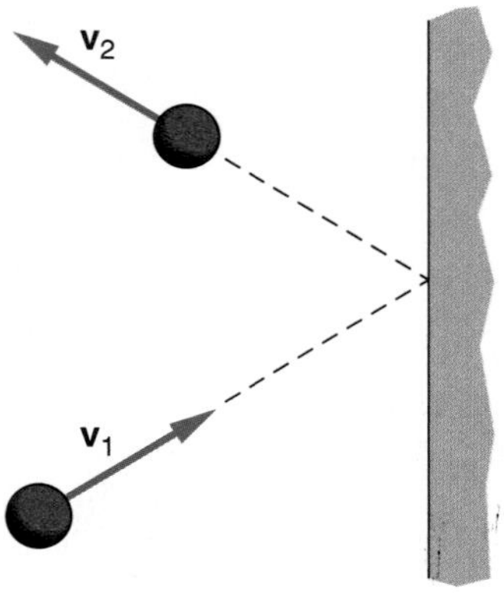

figure 2.6 The direction of the velocity changes when a ball bounces from a wall. The wall exerts a force on the ball in order to produce this change.

Think about what happens when you throw a rubber ball against a wall, as shown in figure 2.6. The speed of the ball may be about the same after the collision with the wall as it was before the ball hit the wall. The velocity has clearly changed in the process, though, because the ball is moving in a different direction after the collision. Something has happened to the motion of the ball. A strong force had to be exerted on the ball by the wall to produce this change in velocity.

The velocity vectors in figures 2.5 and 2.6 are represented by arrows. This is a natural choice for depicting vectors, since the direction of the arrow clearly shows the direction of the vector, and the length can be drawn proportional to the size. In other words, the larger the velocity, the longer the arrow (fig. 2.7). In the text, we will represent vectors by printing their symbols in boldface and larger than other symbols: **v** is thus the symbol for velocity. A fuller description of vectors can be found in appendix C.

How do we define instantaneous velocity?

In considering automobile trips, *average* speed is the most useful quantity. We do not really care about the direction of motion in this case. *Instantaneous* speed is the quantity of interest to the highway patrol. **Instantaneous velocity,** however, is most useful in considering physical theories of motion. We can define instantaneous velocity by drawing on our earlier definition of instantaneous speed.

> Instantaneous velocity is a vector quantity having a size equal to the instantaneous speed at a given instant in time and having a direction corresponding to that of the object's motion at that instant.

Instantaneous velocity and instantaneous speed are closely related, but velocity includes direction as well as size. It is *changes* in instantaneous velocity that require the intervention of forces. These changes will be emphasized when we explore Newton's theory of mechanics in chapter 4. We can also define the concept of average velocity, but that is a much less useful quantity for our purposes than either instantaneous velocity or average speed.

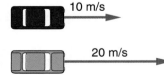

figure 2.7 The length of the arrow shows the size of the velocity vector.

To specify the velocity of an object, we need to state both how fast and in what direction the object is moving; velocity is a vector quantity. Instantaneous velocity has a magnitude equal to the instantaneous speed and points in the direction that the object is moving. Changes in instantaneous velocity are where the action is, so to speak, and we will consider these in more detail when we discuss acceleration in section 2.3.

2.3 Acceleration

Acceleration is a familiar idea. We use the term in speaking of the acceleration of a car away from a stop sign or the acceleration of a running back in football. We feel the effects of acceleration on our bodies when a car's velocity changes rapidly and even more strikingly when an elevator lurches downward, leaving our stomachs slightly behind (fig. 2.8). These are all accelerations. You can think of your stomach as an acceleration detector—a roller-coaster gives it a real workout!

Understanding acceleration is crucial to our study of motion. **Acceleration** is the rate at which velocity *changes.* (Note that we said velocity, not speed.) It plays a central role in Newton's theory of motion. How do we go about finding a value of an acceleration, though? As with speed, it is convenient to start with a definition of average acceleration and then extend it to the idea of instantaneous acceleration.

How is average acceleration defined?

How would we go about providing a quantitative description of an acceleration? Suppose that your car, pointing due east, starts from a full stop at a stop sign, and its velocity increases from zero to 20 m/s as shown in figure 2.9. The change in velocity is found simply by subtracting the initial velocity from the final velocity (20 m/s − 0 m/s = 20 m/s). To find its *rate of change,* however, we also need to know the time needed to produce this change. If it took just 5 seconds for the velocity to change, the rate of change would be larger than if it took 30 seconds.

Suppose that a time of 5 seconds was required to produce this change in velocity. The rate of change in velocity could then be found by dividing the size of the change in velocity by the time required to produce that change. Thus the size of the **average acceleration,** *a,* is found by dividing the change in velocity of 20 m/s by the time of 5 seconds,

$$a = \frac{20 \text{ m/s}}{5 \text{ s}} = 4 \text{ m/s/s}.$$

The unit m/s/s is usually written m/s^2 and is read as *meters per second squared.* It is easier to understand it, however, as *meters per second per second.* The car's velocity (measured in m/s) is changing at a rate of 4 m/s every second. Other units could be used for acceleration, but they will all have this same form: distance per unit of time per unit of time. In discussing the acceleration of a car on a drag strip, for example, the unit *miles per hour per second* is sometimes used.

The quantity that we have just computed is the size of the average acceleration of the car. The average acceleration is found by dividing the total change in velocity for some time interval by that time interval, ignoring possible differences in the rate of change of velocity that might be occurring within the time interval. Its definition can be stated in words as

> Average acceleration is the change in velocity divided by the time required to produce that change.

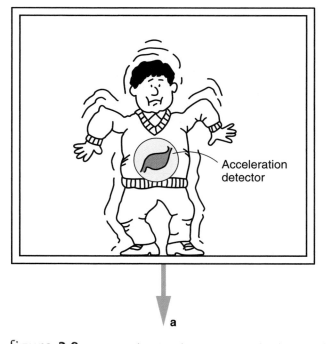

figure **2.8**   Your acceleration detector senses the downward acceleration of the elevator.

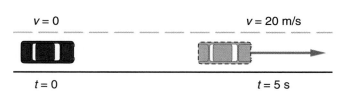

figure **2.9**   A car, starting from rest, accelerates to a velocity of 20 m/s due east in a time of 5 s.

We can restate it in symbols as

$$\text{Acceleration} = \frac{\text{change in velocity}}{\text{elapsed time}}$$

or

$$\mathbf{a} = \frac{\Delta\mathbf{v}}{t}.$$

Because change is so important in this definition, we have used the special symbol Δ (the Greek letter delta) to mean a change in a quantity. Thus $\Delta\mathbf{v}$ is a compact way of writing *the change in velocity,* which otherwise would be expressed as $\mathbf{v}_2 - \mathbf{v}_1$, since a change is the difference between two quantities. Because the concept of change is critical, this notation will appear often.

The idea of change is all-important. Acceleration is *not* velocity over time. It is the *change* in velocity divided by time. It is common for people to associate large accelerations with large velocities, when in fact the opposite is often true. The acceleration of a car may be largest, for example, when it is just starting up and its velocity is near zero. The rate of change of velocity is greatest then. On the other hand, a car can be traveling at 100 MPH but still have a zero acceleration if its velocity is not changing.

What is instantaneous acceleration?

Instantaneous acceleration is similar to average acceleration with an important exception. Just as with instantaneous speed or velocity, we are now concerned with the rate of change at a given instant in time. It is *instantaneous* acceleration that our stomachs respond to. It can be defined as

> Instantaneous acceleration is the rate at which velocity is changing at a given instant in time. It is computed by finding the average acceleration for a very short time interval during which the acceleration does not change appreciably.

If the acceleration is changing with time, choosing a very short time interval guarantees that the acceleration computed for that time interval will not differ too much from the instantaneous acceleration at any time within the interval. This is the same idea used in finding an instantaneous speed or instantaneous velocity.

What is the direction of an acceleration?

Like velocity, acceleration is a vector quantity. Its direction is important. The direction of the acceleration vector is that of the *change* in velocity $\Delta\mathbf{v}$. If, for example, a car is moving in a straight line and its velocity is increasing, the change in velocity is in the same direction as the velocity itself, as shown in figure 2.10. The change in velocity $\Delta\mathbf{v}$

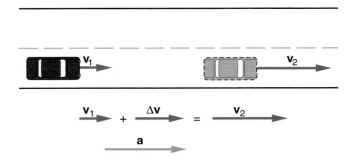

figure 2.10   The acceleration vector is in the same direction as the velocity vectors when the velocity is increasing.

must be added to the initial velocity $\mathbf{v}_1$ to obtain the final velocity $\mathbf{v}_2$. All three vectors point forward. The process of adding vectors can be readily seen when we represent the vectors as arrows on a graph. (More information on vector addition can be found in appendix C.)

If the velocity is decreasing, however, the change in velocity $\Delta\mathbf{v}$ points in the opposite direction to the two velocity vectors, as shown in figure 2.11. Because the initial velocity $\mathbf{v}_1$ is larger than the final velocity $\mathbf{v}_2$, the change in velocity must point in the opposite direction to produce a shorter $\mathbf{v}_2$ arrow. The acceleration is also in the opposite direction to the velocity, since it is in the direction of the *change* in velocity. In Newton's theory of motion, the force required to produce this acceleration would also be opposite in direction to the velocity. It must push backward on the car to slow it down.

The term *acceleration* describes the rate of *any* change in an object's velocity. The change could be an increase (as in our initial example), a decrease, or a change in direction. The term applies even to decreases in velocity (*decelerations*). To a physicist these are simply accelerations with a direction opposite that of the velocity. If a car is braking while traveling in a straight line, its velocity is decreasing and its acceleration is negative if the velocity is positive. This situation is illustrated in the sample exercise in Try This Box 2.3.

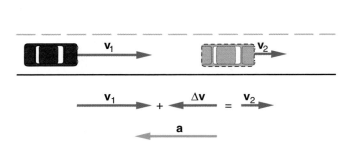

figure 2.11   The velocity and acceleration vectors for decreasing velocity: $\Delta\mathbf{v}$ and $\mathbf{a}$ are now opposite in direction to the velocity. The acceleration $\mathbf{a}$ is proportional to $\Delta\mathbf{v}$.

Sample Exercise: Negative Accelerations

The driver of a car steps on the brakes, and the velocity drops from 20 m/s due east to 10 m/s due east in a time of 2.0 seconds. What is the acceleration?

$\mathbf{v}_1 = 20$ m/s due east

$\mathbf{v}_2 = 10$ m/s due east

$t = 2.0$ s

$\mathbf{a} = ?$

$$a = \frac{\Delta v}{t} = \frac{v_2 - v_1}{t}$$

$$= \frac{10 \text{ m/s} - 20 \text{ m/s}}{2.0 \text{ s}}$$

$$= \frac{-10 \text{ m/s}}{2.0 \text{ s}}$$

$$= -5 \text{ m/s}^2$$

$$\mathbf{a} = 5.0 \text{ m/s}^2 \text{ due west}$$

Notice that when we are dealing just with the magnitude of a vector quantity, we do not use the boldface notation. The sign can indicate direction, however, in a problem involving straight-line motion.

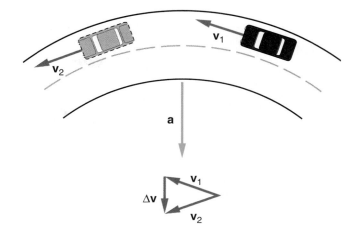

figure 2.12 A change in the direction of the velocity vector also involves an acceleration, even though the speed may be constant.

The minus sign is an important part of the result in the example in Try This Box 2.3 because it indicates that the change in velocity is negative. The velocity is getting smaller. We can call it a deceleration if we like, but it is the same thing as a negative acceleration. One word, *acceleration*, covers all situations in which the velocity is changing.

Acceleration is the rate of change of velocity and is found by dividing the change in the velocity by the time required to produce that change. Any change in velocity involves an acceleration, whether an increase, a decrease, or a change in direction. Acceleration is a vector having a direction corresponding to the direction of the change in velocity, which is not necessarily the same direction as the instantaneous velocity itself. The concept of change is crucial. The graphical representations in section 2.4 will help you visualize changes in velocity as well as in other quantities.

Can a car be accelerating when its speed is constant?

What happens when a car goes around a curve at constant speed? Is it accelerating? The answer is yes, because the direction of its velocity is changing. If the direction of the velocity vector is changing, the velocity is changing. This means that there must be an acceleration.

This situation is illustrated in figure 2.12. The arrows in this drawing show the direction of the velocity vector at different points in the motion. The change in velocity $\Delta \mathbf{v}$ is the vector that must be added to the initial velocity $\mathbf{v}_1$ to obtain the final velocity $\mathbf{v}_2$. The vector representing the change in velocity points toward the center of the curve, and therefore, the acceleration vector also points in that direction. The size of the change is represented by the length of the arrow $\Delta \mathbf{v}$. From this we can find the acceleration.

Acceleration is involved whenever there is a change in velocity, regardless of the nature of that change. Cases like figure 2.12 will be considered more fully in chapter 5 where circular motion is discussed.

2.4 Graphing Motion

It is often said that a picture is worth a thousand words, and the same can be said of graphs. Imagine trying to describe the motion depicted in figure 2.4 precisely in words and numbers. The graph provides a quick overview of what took place. A description in words would be much less efficient. In this section, we will show how graphs can also help us to understand velocity and acceleration.

What can a graph tell us?

How can we produce and use graphs to help us describe motion? Imagine that you are watching a battery-powered toy car moving along a meter stick (fig. 2.13). If the car is moving slowly enough, you could record the car's position while also recording the elapsed time using a digital watch. At regular time intervals (say, every 5 seconds), you would note the value of the position of the front of the car on the meter stick and write these values down. The results might be something like those shown in table 2.2.

figure **2.13** A toy car moving along a meter stick. Its position can be recorded at different times.

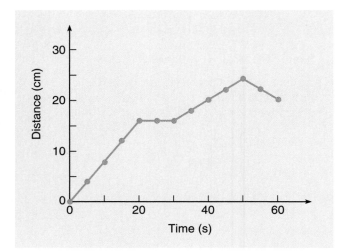

figure **2.14** Distance plotted against time for the motion of the toy car. The data points are those listed in table 2.2.

How do we graph these data? First, we create evenly spaced intervals on each of two perpendicular axes, one for distance traveled (or position) and the other for time. To show how distance varies with time, we usually put time on the horizontal axis and distance on the vertical axis. Such a graph is shown in figure 2.14, where each data point from table 2.2 is plotted and a line is drawn through the points. To make sure that you understand this process, choose different points from table 2.2 and find where they are located on the graph. Where would the point go if the car was at 21 centimeters at 25 seconds?

table **2.2**	
Position of the Toy Car along the Meter Stick at Different Times	
Time	**Position**
0 s	0 cm
5 s	4.1 cm
10 s	7.9 cm
15 s	12.1 cm
20 s	16.0 cm
25 s	16.0 cm
30 s	16.0 cm
35 s	18.0 cm
40 s	20.1 cm
45 s	21.9 cm
50 s	24.0 cm
55 s	22.1 cm
60 s	20.0 cm

The graph summarizes the information presented in the table in a visual format that makes it easier to grasp at a glance. The graph also contains information on the velocity and acceleration of the car, although that is less obvious. For example, what can we say about the average velocity of the car between 20 and 30 seconds? Is the car moving during this time? A glance at the graph shows us that the distance is not changing during that time interval, so the car is *not* moving. The velocity is zero during that time, which is represented by a horizontal line on our graph of distance versus time.

What about the velocity at other points in the motion? The car is moving more rapidly between 0 and 20 seconds than it is between 30 and 50 seconds. The distance curve is rising more rapidly between 0 and 20 seconds than between 30 and 50 seconds. Since more distance is covered in the same time, the car must be moving faster there. A steeper slope to the curve is associated with a larger speed.

In fact, the **slope** of the distance-versus-time curve at any point on the graph is equal to the *instantaneous velocity* of the car.* The slope indicates how rapidly the distance is changing with time at any instant in time. The rate of change of distance with time is the instantaneous speed according to the definition given in section 2.1. Since the motion takes place along a straight line, we can then represent the direction of the velocity with plus or minus signs. There are only two possibilities, forward or backward. We then have the instantaneous velocity, which includes both the size (speed) and direction of the motion.

---

*Since the mathematical definition of slope is the change in the vertical coordinate Δd divided by the change in the horizontal coordinate Δt, the slope, $\Delta d/\Delta t$, is equal to the instantaneous velocity, provided that Δt is sufficiently small. It is possible to grasp the concept of slope, however, without appealing to the mathematical definition.

When the car travels backward, its distance from the starting point decreases. The curve goes down, as it does between 50 and 60 seconds. We refer to this downward-sloping portion of the curve as having a *negative slope* and also say that the velocity is negative during this portion of the motion. A large upward slope represents a large instantaneous velocity, a zero slope (horizontal line) a zero velocity, and a downward slope a negative (backward) velocity. Looking at the slope of the graph tells us all we need to know about the velocity of the car.

Velocity and acceleration graphs

These ideas about velocity can be best summarized by plotting a graph of velocity against time for the car (fig. 2.15). The velocity is constant wherever the slope of the distance-versus-time graph of figure 2.14 is constant. Any straight-line segment of a graph has a constant slope, so the velocity changes only where the slope of the graph in figure 2.14 changes. If you compare the graph in figure 2.15 to the graph in figure 2.14 carefully, these ideas should become clear.

What can we say about the acceleration from these graphs? Since acceleration is the rate of change of velocity with time, the velocity graph (fig. 2.15) also provides information about the acceleration. In fact, the instantaneous acceleration is equal to the slope of the velocity-versus-time graph. A steep slope represents a rapid change in velocity and thus a large acceleration. A horizontal line has zero slope and represents zero acceleration. The acceleration turns out to be zero for most of the motion described by our data. The velocity changes at only a few points in the motion. The acceleration would be large at these points and zero everywhere else.

Since our data do not indicate how rapidly the changes in velocity actually occur, we do not have enough information to say just how large the acceleration is at those few points where it is not zero. We would need measurements of distance or velocity every tenth of a second or so to get a clear idea of how rapid these changes are. As we will see in chapter 4, we know that these changes in velocity cannot occur instantly. Some time is required. So we can sketch an approximate graph of acceleration versus time, as shown in figure 2.16.

The spikes in figure 2.16 occur when the velocity is changing. At 20 seconds, there is a rapid decrease in the velocity represented by a downward spike or negative acceleration. At 30 seconds, the velocity increases rapidly from zero to a constant value, and this is represented by an upward spike or positive acceleration. At 50 seconds, there is another negative acceleration as the velocity changes from a positive to a negative value. If you could put yourself inside the toy car, you would definitely feel these accelerations. (Everyday Phenomenon Box 2.2 provides another example of how a graph is useful for analyzing motion.)

Can we find the distance traveled from the velocity graph?

What other information can be gleaned from the velocity-versus-time graph of figure 2.15? Think for a moment about how you would go about finding the distance traveled if you knew the velocity. For a constant velocity, you can get the distance simply by multiplying the velocity by the time, $d = vt$. In the first 20 seconds of the motion, for example, the velocity is 0.8 cm/s and the distance traveled is 0.8 cm/s times 20 seconds, which is 16 cm. This is just the reverse of what we used in determining the velocity in the first place. We found the velocity by dividing the distance traveled by the time.

How would this distance be represented on the velocity graph? If you recall formulas for computing areas, you

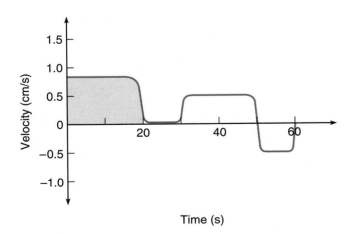

figure 2.15 Instantaneous velocity plotted against time for the motion of the toy car. The velocity is greatest when distance traveled is changing most rapidly.

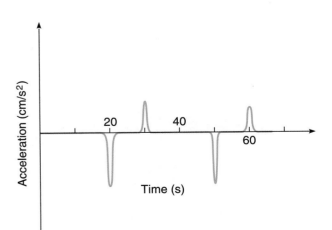

figure 2.16 An approximate sketch of acceleration plotted against time for the toy-car data. The acceleration is non-zero only when the velocity is changing.

Everyday Phenomenon

box 2.2

The 100-m Dash

The Situation. A world-class sprinter can run 100 m in a time of a little under 10 s. The race begins with the runners in a crouched position in the starting blocks, waiting for the sound of the starter's pistol. The race ends with the runners lunging across the finish line, where their times are recorded by stopwatches or automatic timers.

Runners in the starting blocks, waiting for the starter's pistol to fire.

What happens between the start and finish of the race? How do the velocity and acceleration of the runners vary during the race? Can we make reasonable assumptions about what the velocity-versus-time graph looks like for a typical runner? Can we estimate the maximum velocity of a good sprinter? Most importantly for improving performance, what factors affect the success of a runner in the dash?

The Analysis. Let's assume that the runner covers the 100-m distance in a time of exactly 10 s. We can compute the average speed of the runner from the definition $s = d/t$:

$$s = \frac{100 \text{ m}}{10 \text{ s}} = 10 \text{ m/s}.$$

Clearly, this is not the runner's instantaneous speed throughout the course of the race, since the runner's speed at the beginning of the race is zero and it takes some time to accelerate to the maximum speed.

The objective in the race is to reach a maximum speed as quickly as possible and to sustain that speed for the rest of the race. Success is determined by two things: how quickly the runner can accelerate to this maximum speed and the value of this maximum speed. A smaller runner often has better acceleration but a smaller maximum speed, while a larger runner sometimes takes longer to reach top speed but has a larger maximum speed.

The typical runner does not reach top speed before traveling at least 10 to 20 m. If the average speed is 10 m/s, the runner's maximum speed must be somewhat larger than this value, since we know that the instantaneous speed will be less than 10 m/s while the runner is accelerating. These ideas

are easiest to visualize by sketching a graph of velocity plotted against time, as shown. Since the runner travels in a straight line, the magnitude of the instantaneous velocity is equal to the instantaneous speed. The runner reaches top speed at approximately 2 s into the race.

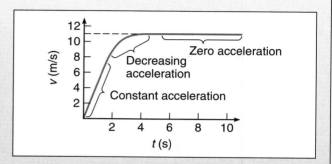

A graph of velocity versus time for a hypothetical runner in the 100-m dash.

The average speed (or velocity) during the time that the runner is accelerating is approximately half of its maximum value if the runner's acceleration is more or less constant during the first 2 s. If we assume that the runner's average speed during this time is about 5.5 m/s (half of 11 m/s), then the speed through the remainder of the race would have to be about 11.1 m/s to give an average speed of 10 m/s for the entire race. This can be seen by computing the distance from these values:

$$d = (5.5 \text{ m/s})(2 \text{ s}) + (11.1 \text{ m/s})(8 \text{ s})$$

$$= 11 \text{ m} + 89 \text{ m} = 100 \text{ m}.$$

What we have done here is to make some reasonable guesses for these values that will make the average speed come out to 10 m/s; we then checked these guesses by computing the total distance. This suggests that the maximum speed of a good sprinter must be about 11 m/s (25 MPH). For sake of comparison, a distance runner who can run a 4-min mile has an average speed of about 15 MPH, or 6.7 m/s.

The runner's strategy should be to get a good jump out of the blocks, keeping the body low initially and leaning forward to minimize air resistance and maximize leg drive. To maintain top speed during the remainder of the race, the runner needs good endurance. A runner who fades near the end needs more conditioning drills. For a given runner with a fixed maximum speed, the average speed depends on how quickly the runner can reach top speed. This ability to accelerate rapidly depends upon leg strength (which can be improved by working with weights and other training exercises) and natural quickness.

may recognize that the distance *d* is the area of the shaded rectangle on figure 2.15. The area of a rectangle is found by multiplying the height times the width, just what we have done here. The velocity, 0.8 cm/s, is the height and the time, 20 seconds, is the width of this rectangle on the graph.

It turns out that we can find the distance this way even when the areas involved on the graph are not rectangles, although the process is more difficult when the curves are more complicated. The general rule is that the distance traveled is equal to the area under the velocity-versus-time curve. When the velocity is negative (below the time axis on the graph), the object is traveling backward and its distance from the starting point is decreasing.

Even without computing the area precisely, it is possible to get a rough idea of the distance traveled by studying the velocity graph. A large area represents a large distance. Quick visual comparisons give a good picture of what is happening without the need for lengthy calculations. This is the beauty of a graph.

A good graph can present a picture of motion that is rich in insight. Distance traveled plotted against time tells us not only where the object is at any time, but its slope also indicates how fast it was moving. The graph of velocity plotted against time also contains information on acceleration and on the distance traveled. Producing and studying such graphs can give us a more general picture of the motion and the relationships between distance, velocity, and acceleration.

2.5 Uniform Acceleration

If you drop a rock, it falls toward the ground with a constant acceleration, as we will see in the next chapter. An unchanging or **uniform acceleration** is the simplest form of accelerated motion. It occurs whenever there is a constant force acting on an object, which is the case for a falling rock as well as for many other situations.

How do we describe the resulting motion? The importance of this question was first recognized by Galileo, who studied the motion of balls rolling down inclined planes as well as objects in free fall. In his famous work, *Dialogues Concerning Two New Sciences,* published in 1638 near the end of his life, Galileo developed the graphs and formulas that are introduced in this section and that have been studied by students of physics ever since. His work provided the foundation for much of Newton's thinking a few decades later.

How does velocity vary in uniform acceleration?

Suppose a car is moving along a straight road and accelerating at a constant rate. We have plotted the acceleration

against time for this situation in figure 2.17. The graph is very simple, but it illustrates what we mean by uniform acceleration. A uniform acceleration is one that does not change as the motion proceeds. It has the same value at any time, which produces a horizontal-line graph.

The graph of velocity plotted against time for this same situation tells a more interesting story. From our discussion in section 2.4, we know that the slope of a velocity-versus-time graph is equal to the acceleration. For a uniform positive acceleration, the velocity graph should have a constant upward slope; the velocity increases at a steady rate. A constant slope produces a straight line, which slopes upward if the acceleration is positive as shown in figure 2.18. In plotting this graph, we assumed that the initial velocity is zero.

This graph can also be represented by a formula. The velocity at any time *t* is equal to the original velocity *plus* the velocity that has been gained because the car is accelerating. The change in velocity Δv is equal to the acceleration times the time, $\Delta v = at$ since acceleration is defined as $\Delta v/t$. These ideas result in the relationship

$$v = v_0 + at.$$

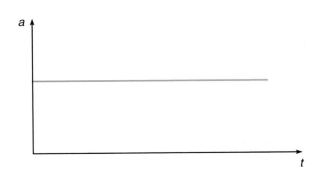

figure **2.17** The acceleration graph for uniform acceleration is a horizontal line. The acceleration does not change with time.

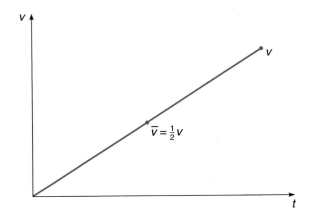

figure **2.18** Velocity plotted against time for uniform acceleration, starting from rest. The average velocity is equal to one-half the final velocity.

The first term on the right, v_0, is the original velocity (assumed to be zero in figure 2.18), and the second term, *at*, represents the change in velocity due to the acceleration. Adding these two terms together yields the velocity at any later time *t*.

A numerical example applying these ideas to an accelerating car is found in Try This Box 2.4. The car could not keep on accelerating indefinitely at a constant rate because the velocity would reach incredible values before long. Not only is this dangerous, but physical limits imposed by air resistance and other factors prevent this from happening.

What happens if the acceleration is negative? Velocity would decrease rather than increase, and the slope of the velocity graph would slope downward rather than upward. Because the acceleration is then negative, the second term in the formula for *v* would subtract from the first term, causing the velocity to decrease from its initial value. The velocity then decreases at a steady rate.

How does distance traveled vary with time?

If the velocity is increasing at a steady rate, what effect does this have on the distance traveled? As the car moves faster and faster, the distance covered grows more and more rapidly. Galileo showed how to find the distance for this situation.

We find distance by multiplying velocity by time, but in this case we must use an average velocity since the velocity is changing. By appealing to the graph in figure 2.18, we can see that the average velocity should be just half the final velocity, *v*. If the initial velocity is zero, the final

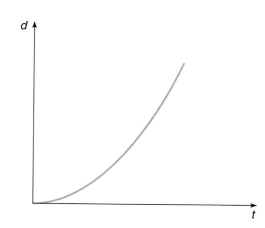

figure 2.19 As the car accelerates uniformly, the distance covered grows more and more rapidly with time because the velocity is increasing.

velocity is *at*, so multiplying the average velocity by the time yields

$$d = \tfrac{1}{2}at^2.$$

The time *t* enters twice, once in finding the average velocity and then again when we multiply the velocity by time to find the distance.*

The graph in figure 2.19 illustrates this relationship; the distance curve slopes upward at an ever-increasing rate. This formula and graph are only valid if the object starts from rest as shown in figure 2.18. Since distance traveled is equal to the area under the velocity-versus-time curve (as discussed in section 2.4), this expression for distance can also be thought of as the area under the triangle in figure 2.18. The area of a triangle is equal to one-half its base times its height, which produces the same result.

If the car is already moving before it begins to accelerate, the velocity graph can be redrawn as pictured in figure 2.20. The total area under the velocity curve can then be split in two pieces, a triangle and a rectangle, as shown. The total distance traveled is the sum of these two areas,

$$d = v_0 t + \tfrac{1}{2}at^2.$$

The first term in this formula represents the distance the object would travel if it moved with constant velocity v_0, and the second term is the additional distance traveled because the object is accelerating (the area of the triangle in figure 2.20). If the acceleration is negative, meaning that the object is slowing down, this second term will subtract from the first.

---

*Expressing this argument in symbolic form, it becomes

The average velocity $\bar{v} = \tfrac{1}{2}v = \tfrac{1}{2}at$

$$d = \bar{v}t = \left(\tfrac{1}{2}at\right)t = \tfrac{1}{2}at^2.$$

---

try this box 2.4

Sample Exercise: Uniform Acceleration

A car traveling due east with an initial velocity of 10 m/s accelerates for 6 seconds at a constant rate of 4 m/s².
 a. What is its velocity at the end of this time?
 b. How far does it travel during this time?

a. v_0 = 10 m/s $v = v_0 + at$
  a = 4 m/s²   = 10 m/s + (4 m/s²)(6 s)
  t = 6 s   = 10 m/s + 24 m/s
  v = ?   = **34 m/s**

        v = 34 m/s due east

b. $d = v_0 t + \tfrac{1}{2}at^2$

  = (10 m/s)(6 s) + $\tfrac{1}{2}$(4 m/s²)(6 s)²

  = 60 m + (2 m/s²)(36 s²)

  = 60 m + 72 m = **132 m**

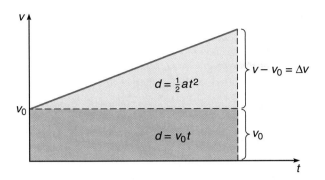

figure **2.20** The velocity-versus-time graph redrawn for an initial velocity different from zero. The area under the curve is divided into two portions, a rectangle and a triangle.

This more general expression for distance may seem complex, but the trick to understanding it is to break it down into its parts, as just suggested. We are merely adding two terms representing different contributions to the total distance. Each one can be computed in a straightforward manner, and it is not difficult to add them together. The two portions of the graph in figure 2.20 represent these two contributions.

The sample exercise in Try This Box 2.4 provides a numerical example of these ideas. The car in this example accelerates uniformly from an initial velocity of 10 m/s due east to a final velocity of 34 m/s due east and covers a distance of 132 meters while this acceleration is taking place. Had it not been accelerating, it would have gone only 60 meters in the same time. The additional 72 meters comes from the acceleration of the car.

Acceleration involves change, and uniform acceleration involves a steady rate of change. It therefore represents the simplest kind of accelerated motion that we can imagine. Uniform acceleration is essential to an understanding of free fall, discussed in chapter 3, as well as to many other phenomena. Such motion can be represented by either the graphs or the formulas introduced in this section. Looking at both and seeing how they are related will reinforce these ideas.

---

summary

The main purpose of this chapter is to introduce concepts that are crucial to a precise description of motion. To understand acceleration, you must first grasp the concept of velocity, which in turn builds on the idea of speed. The distinctions between speed and velocity, and between velocity and acceleration, are particularly important.

1   Average and instantaneous speed.   Average speed is defined as the distance traveled divided by the time. It is the average rate at which distance is covered. Instantaneous speed is the rate at which distance is being covered at a given instant in time and requires that we use very short time intervals for computation.

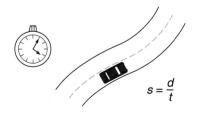

2   Velocity.   The instantaneous velocity of an object is a vector quantity that includes both direction and size. The size of the velocity vector is equal to the instantaneous speed, and the direction is that of the object's motion.

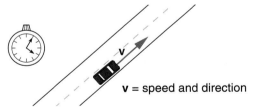

$\mathbf{v}$ = speed and direction

3   Acceleration.   Acceleration is defined as the time rate of change of velocity and is found by dividing the *change* in velocity by the time. Acceleration is also a vector quantity. It can be computed as either an average or an instantaneous value. A change in the direction of the velocity can be as important as a change in magnitude. Both involve acceleration.

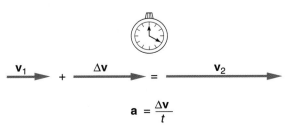

$$\mathbf{a} = \frac{\Delta \mathbf{v}}{t}$$

4   **Graphing motion.**   Graphs of distance, speed, velocity, and acceleration plotted against time can illustrate relationships between these quantities. Instantaneous velocity is equal to the slope of the distance-time graph. Instantaneous acceleration is equal to the slope of the velocity-time graph. The distance traveled is equal to the area under the velocity-time graph.

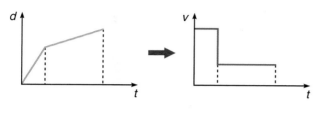

5   **Uniform acceleration.**   When an object accelerates at a constant rate producing a constant-slope graph of velocity versus time, we say that it is uniformly accelerated. Graphs help us to understand the two formulas describing how velocity and distance traveled vary with time for this important special case.

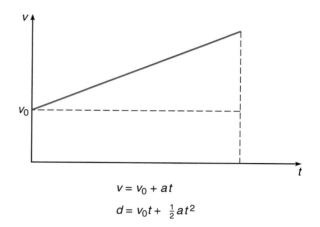

$$v = v_0 + at$$

$$d = v_0 t + \frac{1}{2} a t^2$$

key terms

Speed, 19
Average speed, 19
Rate, 20
Instantaneous speed, 21
Velocity, 23

Magnitude, 23
Vector, 24
Vector quantity, 24
Instantaneous velocity, 24
Acceleration, 25

Average acceleration, 25
Instantaneous acceleration, 26
Slope, 28
Uniform acceleration, 31

questions

*Questions identified with an asterisk are more open-ended than the others. They call for lengthier responses and are more suitable for group discussion.

Q1. Suppose that critters are discovered on Mars who measure distance in *boogles* and time in *bops*.
  a. What would the units of speed be in this system? Explain.
  b. What would the units of velocity be? Explain.
  c. What would the units of acceleration be? Explain.

Q2. Suppose that we choose inches as our basic unit of distance and days as our basic unit of time.
  a. What would the units of velocity and acceleration be in this system? Explain.
  b. Would this be a good choice of units for measuring the acceleration of an automobile? Explain.

Q3. What units would have an appropriate size for measuring the rate at which fingernails grow? Explain.

Q4. A tortoise and a hare cover the same distance in a race. The hare goes very fast for brief intervals, but stops frequently, whereas the tortoise plods along steadily and finishes the race ahead of the hare.
  a. Which of the two racers has the greater average speed over the duration of the race? Explain.
  b. Which of the two racers is likely to reach the greatest instantaneous speed during the race? Explain.

Q5. A driver states that she was doing 80 when stopped by the police. Is that a clear statement? Would this be interpreted differently in England than it would be in the United States? Explain.

Q6. Does the speedometer on a car measure average speed or instantaneous speed? Explain.

*Q7. The highway patrol sometimes uses radar guns to identify possible speeders and at other times uses associates in airplanes who note the time taken for a car to pass between two marks some distance apart on the highway. What do each of these methods measure, average speed or instantaneous speed? Can you think of situations in which either one of these methods might unfairly penalize a driver? Explain.

Q8. A car traveling around a circular track moves with constant speed. Is this car moving with constant velocity? Explain.

Q9. A ball is thrown against a wall and bounces back toward the thrower with the same speed as it had before hitting the wall. Does the velocity of the ball change in this process? Explain.

Q10. A ball attached to a string is whirled in a horizontal circle such that it moves with constant speed.
a. Does the velocity of the ball change in this process? Explain.
b. Is the acceleration of the ball equal to zero? Explain.

*Q11. A ball tied to a string fastened at the other end to a rigid support forms a pendulum. If we pull the ball to one side and release it, the ball moves back and forth along an arc determined by the string length.
a. Is the velocity constant in this process? Explain.
b. Is the speed likely to be constant in this process? What happens to the speed when the ball reverses direction?

Q12. A dropped ball gains speed as it falls. Can the velocity of the ball be constant in this process? Explain.

Q13. A driver of a car steps on the brakes, causing the velocity of the car to decrease. According to the definition of acceleration provided in this chapter, does the car accelerate in this process? Explain.

Q14. At a given instant in time, two cars are traveling at different velocities, one twice as large as the other. Based upon this information is it possible to say which of these two cars has the larger acceleration at this instant in time? Explain.

Q15. A car just starting up from a stop sign has zero velocity at the instant that it starts. Must the acceleration of the car also be zero at this instant? Explain.

Q16. A car traveling with constant speed rounds a curve in the highway. Is the acceleration of the car equal to zero in this situation? Explain.

Q17. A racing sports car traveling with a constant velocity of 100 MPH due west startles a turtle by the side of the road who begins to move out of the way. Which of these two objects is likely to have the larger acceleration at that instant? Explain.

Q18. In the graph shown here, velocity is plotted as a function of time for an object traveling in a straight line.
a. Is the velocity constant for any time interval shown? Explain.
b. During which time interval shown does the object have the greatest acceleration? Explain.

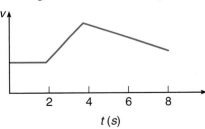

Q19. A car moves along a straight line so that its position (distance from some starting point) varies with time as described by the graph shown here.
a. Does the car ever go backward? Explain.
b. Is the instantaneous velocity at point A greater or less than that at point B? Explain.

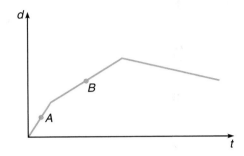

Q20. For the car whose distance is plotted against time in Q19, is the velocity constant during any time interval shown in the graph? Explain.

Q21. A car moves along a straight section of road so that its velocity varies with time as shown in the graph.
a. Does the car ever go backward? Explain.
b. At which of the labeled points on the graph, A, B, or C, is the magnitude of the acceleration the greatest? Explain.

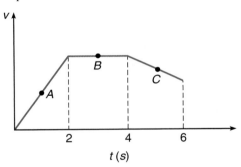

Q22. For the car whose velocity is plotted in Q21, in which of the equal time segments 0–2 seconds, 2–4 seconds, or 4–6 seconds, is the distance traveled by the car the greatest? Explain.

Q23. Look again at the velocity-versus-time graph for the toy car shown in figure 2.15.
a. Is the instantaneous speed greater at any time during this motion than the average speed for the entire trip? Explain.
b. Is the car accelerated when the direction of the car is reversed at $t = 50$ s? Explain.

Q24. Suppose that the acceleration of a car increases with time. Could we use the relationship $v = v_0 + at$ in this situation? Explain.

Q25. When a car accelerates uniformly from rest, which of these quantities increases with time: acceleration, velocity, and/or distance traveled? Explain.

Q26. The velocity-versus-time graph of an object curves as shown in the diagram. Is the acceleration of the object constant? Explain.

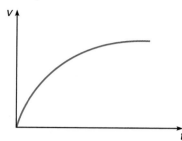

Q27. For a uniformly accelerated car, is the average acceleration equal to the instantaneous acceleration? Explain.

Q28. A car traveling in the forward direction experiences a *negative* uniform acceleration for 10 seconds. Is the distance covered during the first 5 seconds equal to, greater than, or less than the distance covered during the second 5 seconds? Explain.

Q29. A car starts from rest, accelerates uniformly for 5 seconds, travels at constant velocity for 5 seconds, and finally decelerates uniformly for 5 seconds. Sketch graphs of velocity versus time and acceleration versus time for this situation.

Q30. Suppose that two runners run a 100-meter dash, but the first runner reaches maximum speed more quickly than the second runner. Both runners maintain constant speed once they have reached their maximum speed and cross the finish line at the same time. Which runner has the larger maximum speed? Explain.

Q31. Sketch a graph showing velocity-versus-time curves for the two runners described in Q30. (Sketch both curves on the same graph, so that the differences are apparent.)

*Q32. A physics instructor walks with increasing speed across the front of the room then suddenly reverses direction and walks backward with constant speed. Sketch graphs of velocity and acceleration consistent with this description.

exercises

E1. A traveler covers a distance of 460 miles in a time of 8 hours. What is the average speed for this trip?

E2. A walker covers a distance of 1.8 km in a time of 30 minutes. What is the average speed of the walker for this distance in km/h?

E3. Grass clippings are found to have an average length of 4.8 cm when a lawn is mowed 12 days after the previous mowing. What is the average speed of growth of this grass in cm/day?

E4. A driver drives for 3.5 hours at an average speed of 58 MPH. What distance does she travel in this time?

E5. A woman walks a distance of 240 m with an average speed of 1.2 m/s. What time was required to walk this distance?

E6. A person in a hurry averages 62 MPH on a trip covering a distance of 300 miles. What time was required to travel that distance?

E7. A hiker walks with an average speed of 1.2 m/s. What distance in kilometers does the hiker travel in a time of one hour?

E8. A car travels with an average speed of 25 m/s.
   a. What is this speed in km/s?
   b. What is this speed in km/h?

E9. A car travels with an average speed of 58 MPH. What is this speed in km/h? (See Try This Box 2.1.)

E10. Starting from rest and moving in a straight line, a runner achieves a velocity of 7 m/s in a time of 2 s. What is the average acceleration of the runner?

E11. Starting from rest, a car accelerates at a rate of 4.2 m/s^2 for a time of 5 seconds. What is its velocity at the end of this time?

E12. The velocity of a car decreases from 30 m/s to 18 m/s in a time of 4 seconds. What is the average acceleration of the car in this process?

E13. A car traveling with an initial velocity of 12 m/s accelerates at a constant rate of 2.5 m/s^2 for a time of 2 seconds.
   a. What is its velocity at the end of this time?
   b. What distance does the car travel during this process?

E14. A runner traveling with an initial velocity of 5 m/s accelerates at a constant rate of 1.2 m/s^2 for a time of 2 seconds.
   a. What is his velocity at the end of this time?
   b. What distance does the runner cover during this process?

E15. A car moving with an initial velocity of 30 m/s slows down at a constant rate of -3 m/s^2.
   a. What is its velocity after 3 seconds of deceleration?
   b. What distance does the car cover in this time?

E16. A runner moving with an initial velocity of 9.0 m/s slows down at a constant rate of -1.5 m/s^2 over a period of 2 seconds.
   a. What is her velocity at the end of this time?
   b. What distance does she travel during this process?

E17. If a world-class sprinter ran a distance of 100 meters starting at his top speed of 11 m/s and running with constant speed throughout, how long would it take him to cover the distance?

E18. Starting from rest, a car accelerates at a constant rate of 3.0 m/s^2 for a time of 5 seconds.
   a. Compute the velocity of the car at 1 s, 2 s, 3 d, 4 d, and 5 d and plot these velocity values against time.
   b. Compute the distance traveled by the car for these same times and plot the distance values against time.

challenge problems

CP1. A railroad engine moves forward along a straight section of track for a distance of 80 m due west at a constant speed of 5 m/s. It then reverses its direction and travels 20 m due east at a constant speed of 4 m/s. The time required for this deceleration and reversal is very short due to the small speeds involved.
   a. What is the time required for the entire process?
   b. Sketch a graph of average speed versus time for this process. Show the deceleration and reacceleration upon reversal as occurring over a very short time interval.
   c. Using negative values of velocity to represent reversed motion, sketch a graph of velocity versus time for the engine.
   d. Sketch a graph of acceleration versus time for the engine.

CP2. The velocity of a car increases with time as shown in the graph.
   a. What is the average acceleration between 0 seconds and 4 seconds?
   b. What is the average acceleration between 4 seconds and 8 seconds?
   c. What is the average acceleration between 0 seconds and 8 seconds?
   d. Is the result in part (c) equal to the average of the two values in parts (a) and (b)? Compare and explain.

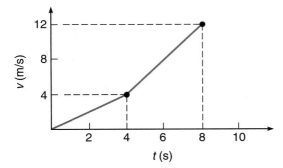

CP3. A car traveling due west on a straight road accelerates at a constant rate for 10 seconds increasing its velocity from zero to 24 m/s. It then travels at constant speed for 10 seconds and then decelerates at a steady rate for the next 5 seconds to a velocity of 10 m/s. It travels at this velocity for 5 seconds and then decelerates rapidly to a stop in a time of 2 seconds.
   a. Sketch a graph of the car's velocity versus time for the entire motion just described. Label the axes of your graph with the appropriate velocities and times.
   b. Sketch a graph of acceleration versus time for the car.
   c. Does the distance traveled by the car continually increase in the motion described? Explain.

CP4. A car traveling in a straight line with an initial velocity of 14 m/s accelerates at a rate of 2.0 m/s^2 to a velocity of 24 m/s.
   a. How much time does it take for the car to reach the velocity of 24 m/s?
   b. What is the distance covered by the car in this process?
   c. Compute values of the distance traveled at 1-second intervals and carefully draw a graph of distance plotted against time for this motion.

CP5. Just as car A is starting up, it is passed by car B. Car B travels with a constant velocity of 10 m/s, while car A accelerates with a constant acceleration of 4.5 m/s^2, starting from rest.
   a. Compute the distance traveled by each car for times of 1 s, 2 s, 3 s, and 4 s.
   b. At what time, approximately, does car A overtake car B?
   c. How might you go about finding this time exactly? Explain.

home experiments and observations

HE1. How fast do you normally walk? Using a meter stick or a string of known length, lay out a straight course of 40 or 50 meters. Then use a watch with a second hand or a stopwatch to determine:
   a. Your normal walking speed in m/s.
   b. Your walking speed for a brisk walk.
   c. Your jogging speed for this same distance.
   d. Your sprinting speed for this distance.

   Record and compare the results for these different cases. Is your sprinting speed more than twice your speed for a brisk walk?

HE2. The speed with which hair or fingernails grow provides some interesting measurement challenges. Using a milli-meter rule, estimate the speed of growth for one or more of: fingernails, toenails, facial hair if you shave regularly, or hair near your face (such as sideburns) that will provide an easy reference point. Measure the average size of clippings or of growth at regular time intervals.

   a. What is the average speed of growth? What units are most appropriate for describing this speed?

   b. Does the speed appear to be constant with time? Does the speed appear to be the same for different nails (thumb versus fingers, fingernails versus toenails), or in the case of hair, for different positions on your face?

Falling Objects and Projectile Motion

chapter 3

chapter overview

Our main purpose in this chapter is to explore how objects move under the influence of the gravitational acceleration near the surface of the earth. Uniform acceleration, introduced in the previous chapter, plays a prominent role. We begin by considering carefully the acceleration of a dropped object, and then we will extend these ideas to thrown objects or objects projected at an angle to the ground.

chapter outline

1 **Acceleration due to gravity.** How does a dropped object move under the influence of the earth's gravitational pull? How is its acceleration measured, and in what sense is it constant?

2 **Tracking a falling object.** How do velocity and distance traveled vary with time for a falling object? How can we quickly estimate these values knowing the gravitational acceleration?

3 **Beyond Free Fall: Throwing a ball upward.** What changes when a ball is thrown upward rather than being dropped? Why does the ball appear to hover near the top of its flight?

4 **Projectile motion.** What determines the motion of an object that is fired horizontally? How do the velocity and position of the object change with time in this case?

5 **Hitting a target.** What factors determine the trajectory of a rifle bullet or football that has been launched at some angle to the horizontal to hit a target?

Have you ever watched a leaf or a ball fall to the ground? At times during your first few years of life, you probably amused yourself by dropping an object repeatedly and watching it fall. As we grow older, that experience becomes so common that we usually do not stop to think about it or to ask why objects fall as they do. Yet this question has intrigued scientists and philosophers for centuries.

To understand nature, we must first carefully observe it. If we control the conditions under which we make our observations, we are doing an experiment. The observations of falling objects that you performed as a young child were a simple form of experiment, and we would like to rekindle that interest in experimentation here. Progress in science has depended on carefully controlled experiments, and your own progress in understanding nature will depend on your active testing of ideas through experiments. You may be amazed at what you discover.

Look around for some small, compact objects. A short pencil, a rubber eraser, a paper clip, or a small ball will all do nicely. Holding two objects at arm's length, release them simultaneously and watch them fall to the floor (fig. 3.1). Be careful to release them from the same height above the floor without giving either one an upward or downward push.

How would you describe the motion of these falling objects? Is their motion accelerated? Do they reach the floor at the same time? Does the motion depend on the shape and composition of the object? To explore this last question, you might take a small piece of paper and drop it at the same time as an eraser or a ball. First,

figure 3.1 An experimenter dropping objects of different mass. Do they reach the ground at the same time?

drop the paper unfolded. Then, try folding it or crumpling it into a ball. What difference does this make?

From these simple experiments, we can draw some general conclusions about the motion of falling objects. We can also try throwing or projecting objects at different angles to study the motion of a projectile. We will find that a constant downward gravitational acceleration is involved in all of these cases. This acceleration affects virtually everything that we do when we move or play on the surface of this earth.

3.1 Acceleration Due to Gravity

If you dropped a few objects as suggested in the introduction, you already know the answer to one of the questions posed there. Are the falling objects accelerated? Think for a moment about whether the velocity is changing. Before you release an object, its velocity is zero, but an instant after the object is released, the velocity has some value different from zero. There has been a change in velocity. If the velocity is changing, there is an acceleration.

Things happen so rapidly that it is difficult, just from watching the fall, to say much about the acceleration. It does appear to be large, because the velocity increases rapidly. Does the object reach a large velocity instantly, or does the acceleration occur more uniformly? To answer this question, we must slow the motion down somehow so that our eyes and brains can keep up with what is happening.

How can we measure the gravitational acceleration?

There are several ways to slow down the action. One was pioneered by the Italian scientist, Galileo Galilei (1564–1642), who was the first to accurately describe the *acceleration due to gravity*. Galileo's method was to roll or slide objects down a slightly inclined plane. This allows only a small portion of the gravitational acceleration to come into play, just that part in the direction of motion along the plane. Thus a smaller acceleration results. Other methods (not available to Galileo) use time-lapse photography, spark timers, or video recording to locate the position of the falling object at different times.

If you happen to have a grooved ruler and a small ball or marble handy, you can make an inclined plane yourself. Lift one end of the ruler slightly by placing a pencil under one end, and let the ball or marble roll down the ruler under the influence of gravity (fig. 3.2). Can you see it gradually pick up speed as it rolls? Is it clearly moving faster at the bottom of the incline than it was halfway down?

Galileo was handicapped by a lack of accurate timing devices. He often had to use his own pulse as a timer. Despite this limitation, he was able to establish that the acceleration was uniform, or constant, with time and to estimate its value using inclined planes. We are more fortunate. We have devices that allow us to study the motion

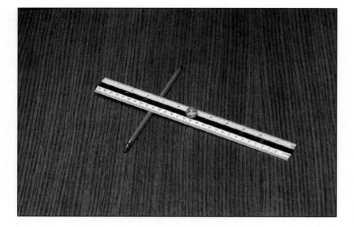

figure **3.2**   A marble rolling down a ruler serving as an inclined plane. Does the velocity of the marble increase as it rolls down the incline?

of a falling object more directly. One such device is a stroboscope, a rapidly blinking light whose flashes occur at regular intervals in time. Figure 3.3 is a photograph taken using a stroboscope to illuminate an object as it falls. The position of the object is pinpointed every time the light flashes.

If you look closely at figure 3.3, you will notice that the distance covered in successive time intervals increases regularly. The time intervals between successive positions of the ball are all equal. (If the stroboscope light flashes every $\frac{1}{20}$ of a second, you are seeing the position of the ball every $\frac{1}{20}$ of a second.) Since the distance covered by the ball in equal time intervals is increasing, the velocity must be increasing. Figure 3.3 shows a ball whose velocity is steadily increasing in the downward direction.

Computing values of the average velocity for each time interval will make this even clearer. The computation can be done if we know the time interval between flashes and can measure the position of the ball from the photograph, knowing the distance between the grid marks. Table 3.1 displays data obtained in this manner. It shows the position of a ball at intervals of $\frac{1}{20}$ of a second (0.05 second).

To see that the velocity is indeed increasing, we compute the average velocity for each successive time interval. For example, between the second and third flashes, the ball traveled a distance of 3.6 centimeters, which is found by subtracting 1.2 centimeters from 4.8 centimeters. Dividing this distance by the time interval of 0.05 second yields the average size of the velocity:

$$v = \frac{3.6 \text{ cm}}{0.05 \text{ s}} = 72 \text{ cm/s}.$$

You could verify the other values shown in the third column of table 3.1 by doing similar computations.

It is clear in table 3.1 that the velocity values steadily increase. To see that velocity is increasing at a constant rate, we can plot velocity against time (fig. 3.4). Notice that each velocity data point is plotted at the midpoint between the two times (or flashes) from which it was computed. This is because these values represent the average velocity for the short time intervals between flashes. For constant acceleration, the average velocity for any time interval is equal to the instantaneous velocity at the midpoint of that interval.

Did you notice that the slope of the line is constant in figure 3.4? The velocity values all fall approximately on a constant-slope straight line. Since acceleration is the slope of the velocity-versus-time graph, the acceleration must also be constant. The velocity increases uniformly with time.

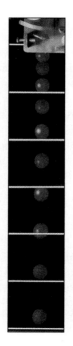

figure **3.3**   A falling ball is illuminated by a rapidly blinking stroboscope. The stroboscope blinks at regular time intervals.

table **3.1**		
Distance and Velocity Values for a Falling Ball		
Time	Distance	Velocity
0	0	
		24 cm/s
0.05 s	1.2 cm	
		72 cm/s
0.10 s	4.8 cm	
		124 cm/s
0.15 s	11.0 cm	
		174 cm/s
0.20 s	19.7 cm	
		218 cm/s
0.25 s	30.6 cm	
		268 cm/s
0.30 s	44.0 cm	
		320 cm/s
0.35 s	60.0 cm	
		368 cm/s
0.40 s	78.4 cm	
		416 cm/s
0.45 s	99.2 cm	
		464 cm/s
0.50 s	122.4 cm	

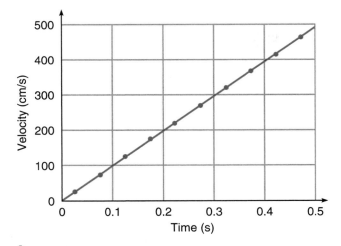

figure **3.4**   Velocity plotted against time for the falling ball. The velocity values are those shown in table 3.1.

To find the value of the acceleration, we choose two velocity values that lie on the straight line and calculate how rapidly the velocity is changing. For example, the last velocity value, 464 cm/s, and the second value, 72 cm/s, are separated by a time interval corresponding to 8 flashes or 0.40 second. The increase in velocity Δv is found by subtracting 72 cm/s from 464 cm/s, obtaining 392 cm/s. To find the acceleration, we divide this change in velocity by the time interval ($a = \Delta v/t$),

$$a = \frac{392 \text{ cm/s}}{0.4 \text{ s}} = 980 \text{ cm/s}^2 = 9.8 \text{ m/s}^2.$$

This result gives us the **acceleration due to gravity** for objects falling near the earth's surface. Its value actually varies slightly from point to point on the earth's surface because of differences in altitude and other effects. This acceleration is used so often that it is given its own symbol g where

$$g = 9.8 \text{ m/s}^2.$$

Called the *gravitational acceleration* or *acceleration due to gravity,* it is valid only near the surface of the earth and thus is not a fundamental constant.

How did Galileo's ideas on falling objects differ from Aristotle's?

There is another sense in which the gravitational acceleration is constant, which takes us back to the experiments suggested in the chapter opener, p. 39. When you drop objects of different sizes and weights, do they reach the floor at the same time? Except for an unfolded piece of paper, it is likely that all of the objects that you test, regardless of their weight, reach the floor at the same time when released simultaneously. This finding suggests that the gravitational acceleration does not depend on the weight of the object.

Galileo used similar experiments to prove this point. His experiments contradicted Aristotle's view that heavier objects fall more rapidly. How could Aristotle's idea have been accepted for so long when simple experiments can disprove it? Experimentation was not part of the intellectual outlook of Aristotle and his followers; they valued pure thought and logic more highly. Galileo and other scientists of his time broke new ground by using experiments as an aid to thinking. A new tradition was emerging.

On the other hand, Aristotle's view agrees with our intuition that heavy objects do fall more rapidly than some lighter objects. If, for example, we drop a brick together with a feather or unfolded piece of paper (fig. 3.5), the brick will reach the floor first. The paper or feather will not fall in a straight line but instead will flutter to the floor much as a leaf falls from a tree. What is happening here?

You will probably recognize that the effects of **air resistance** impede the fall of the feather or paper much more than the fall of the brick, a steel ball, or a paper clip. When we crumple the piece of paper into a ball and drop it simultaneously with a brick or other heavy object, the two objects reach the floor at approximately the same time. We live at the bottom of a sea of air, and the effects of air resistance can be substantial for objects like leaves, feathers, or pieces of paper. These effects produce a slower and less regular flight for light objects that have a large surface area.

If we drop a feather and a brick simultaneously in a vacuum or in the very thin atmosphere of the moon, they do reach the ground at the same time. Moonlike conditions are not part of our everyday experience, however, so we are used to seeing feathers fall more slowly than rocks or bricks. Galileo's insight was that the gravitational acceleration is the same for all objects, regardless of their weight, provided that the effects of air resistance are not significant. Aristotle did not separate the effect of air resistance from that of gravity in his observations.

figure **3.5**   The brick reaches the floor first when a brick and a feather are dropped at the same time.

The gravitational acceleration for objects near the surface of the earth is uniform and has the value of 9.8 m/s². It can be measured by using stroboscopes or similar techniques to record the position of a falling object at regular, very small time intervals. This acceleration is constant in time. Contrary to Aristotle's belief, it also has the same value for objects of different weight.

3.2 Tracking a Falling Object

Imagine yourself dropping a ball from a sixth-story window, as in figure 3.6. How long does it take for the ball to reach the ground below? How fast is it traveling when it gets there? Things happen quickly, so the answers to these questions are not obvious.

If we assume that air-resistance effects are small for the object we are tracking, we know that it accelerates toward the ground at the constant rate of 9.8 m/s². Let's make some quick estimates of how these values change with time without doing detailed computations.

How does the velocity vary with time?

In making estimates of velocity and distance for a falling object, we often take advantage of the fact that the gravitational-acceleration value of 9.8 m/s² is almost 10 m/s² and round it up. This makes the numerical values easier to calculate without sacrificing much in accuracy. Multiplying by 10 is quicker than multiplying by 9.8.

How fast is our dropped ball moving after 1 second? An acceleration of 10 m/s² means that the velocity is increasing by 10 m/s each second. If its original velocity is zero, then after 1 second its velocity has increased to 10 m/s, in 2 seconds to 20 m/s, and in 3 seconds to 30 m/s. For each additional second, the ball gains 10 m/s in velocity.*

To help you appreciate these values, look back at table 2.1, which shows unit comparisons for familiar speeds. A velocity of 30 m/s is roughly 70 MPH, so after 3 seconds the ball is moving quickly. After just 1 second, it is moving with a downward velocity of 10 m/s, which is over 20 MPH. The ball gains velocity at a faster rate than is possible for a high-powered automobile on a level surface.

How far does the ball fall in different times?

The high velocities are more meaningful if we examine how far the ball falls during these times. As the ball falls, it gains speed, so it travels farther in each successive time interval, as in the photograph in figure 3.3. Because of uniform

figure **3.6**   A ball is dropped from a sixth-story window. How long does it take to reach the ground?

acceleration, the distance increases at an ever-increasing rate.

During the first second of motion, the velocity of the ball increases from zero to 10 m/s. Its average velocity during that first second is 5 m/s, and it travels a distance of 5 meters in that second. This can also be found by using the relationship between distance, acceleration, and time in section 2.5. If the starting velocity is zero, we found that $d = \frac{1}{2}at^2$. After 1 second, the ball has fallen a distance

$$d = \frac{1}{2}(10 \text{ m/s}^2)(1 \text{ s})^2 = 5 \text{ m}.$$

Since the height of a typical story of a multistory building is less than 4 meters, the ball falls more than one story in just a second.

---

*In section 2.5, we noted that the velocity of an object moving with uniform acceleration is $v = v_0 + at$, where v_0 is the original velocity and the second term is the change in velocity, $\Delta v = at$. When a ball is dropped, $v_0 = 0$, so v is just at, the change in velocity.

During the next second of motion, the velocity increases from 10 m/s to 20 m/s, yielding an average velocity of 15 m/s for that interval. The ball travels 15 meters in that second, which, when added to the 5 meters covered in the first second, yields a total of 20 meters. After 2 seconds, the distance fallen is four times as large as the 5 meters traveled after 1 second.* Since 20 meters is roughly five stories in height, the ball dropped from the sixth story will be near the ground after 2 seconds.

Figure 3.7 gives the velocity and distance fallen at half-second time intervals for a ball dropped from a six-story building. Notice that in just half a second, the ball falls 1.25 meters. An object dropped to the floor from an out-stretched arm therefore hits the floor in roughly half a second. This makes it difficult to time with a stopwatch. (See Try This Box 3.1.)

The change in velocity is proportional to the size of the time interval selected. In 1 second the change in velocity is 10 m/s, so in half a second the change in velocity is 5 m/s. In each half-second the ball gains approximately 5 m/s in velocity, illustrated in figure 3.7. As the velocity gets larger, the arrows representing the velocity vectors grow. If we plotted these velocity values against time, we would get a simple upward-sloping straight-line graph as in figure 3.4.

What does the graph of the distance values look like? The distance values increase in proportion to the square of the time, which means that they increase more and more rapidly as time elapses. Instead of being a straight-line graph, the graph of the distance values curves upward as in figure 3.8. The rate of change of distance with time is itself increasing with time.

Throwing a ball downward

Suppose that instead of just dropping the ball, we throw it straight down, giving it a starting velocity v_0 different from zero. How does this affect the results? Will the ball reach the ground more rapidly and with a larger velocity? You would probably guess correctly that the answer is yes.

In the case of the velocity values, the effect of the starting velocity is not difficult to see. The ball is still being accelerated by gravity so that the change in velocity for each second of motion is still $\Delta v = 10$ m/s, or for a half-second, 5 m/s. If the initial downward velocity is 20 m/s, after half a second, the velocity is 25 m/s, and after 1 second, it is 30 m/s. We simply add the change in velocity to the initial velocity as indicated by the formula $v = v_0 + at$.

In the case of distance, however, the values increase more rapidly. The full expression for distance traveled by a uniformly accelerated object (introduced in section 2.5) is

$$d = v_0 t + \frac{1}{2}at^2.$$

*This is a result of the time being squared in the formula for distance. Putting 2 s in place of 1 s in the formula $d = \frac{1}{2}at^2$ multiplies the result by a factor of 4 ($2^2 = 4$), yielding a distance of 20 m.

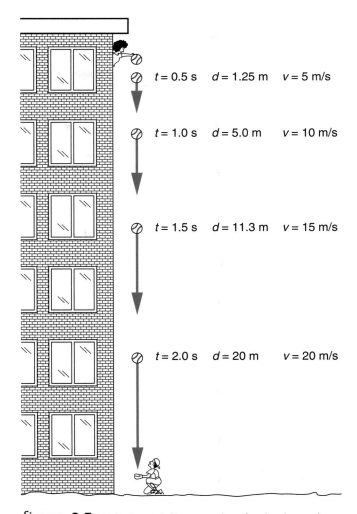

$t = 0.5$ s	$d = 1.25$ m	$v = 5$ m/s
$t = 1.0$ s	$d = 5.0$ m	$v = 10$ m/s
$t = 1.5$ s	$d = 11.3$ m	$v = 15$ m/s
$t = 2.0$ s	$d = 20$ m	$v = 20$ m/s

figure **3.7** Velocity and distance values for the dropped ball shown at half-second time intervals.

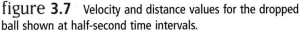

try this box **3.1**

Sample Question: Using a Pulse Rate to Time a Falling Object

Question: Suppose that Galileo's resting pulse rate was 60 beats per minute. Would his pulse be a useful timer for getting position-versus-time data for an object dropped from the height of 2 to 3 meters?

Answer: A pulse rate of 60 beats per minute corresponds to 1 beat per second. In the time of 1 second, a dropped object falls a distance of approximately 5 m. (It falls 1.22 m in just half a second as seen in table 3.1.) Thus this pulse rate (or most pulse rates) would not be an adequate timer for an object dropped from a height of a few meters. It could be slightly more effective for an object dropped from a tower several stories in height.

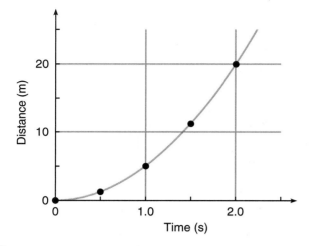

figure **3.8**   A plot of distance versus time for the dropped ball.

The first term is the distance that the ball would travel if it continued to move with just its original velocity. This distance also increases with time. The second term is due to the acceleration and has the same values as shown in figures 3.7 and 3.8.

In the sample exercise in Try This Box 3.2, we calculate velocity and distance traveled during the first 2 seconds of motion for a ball thrown downward. Notice that after 2 seconds the ball has traveled a distance of 60 meters, much larger than the 20 meters when the ball is simply dropped.

---

try this box 3.2

Sample Exercise: Throwing a Ball Downward

A ball is thrown downward with an initial velocity of 20 m/s. Using the value 10 m/s^2 for the gravitational acceleration, find (a) the velocity and (b) the distance traveled at 1-s time intervals for the first 2 s of motion.

a. $v_0 = 20$ m/s $v = v_0 + at$
    $a = 10$ m/s^2 for $t = 1$ s
    $v = ?$ $v = 20$ m/s $+ (10$ m/s$^2)(1$ s$)$
                   $= 20$ m/s $+ 10$ m/s
                   $= \textbf{30 m/s}$

    $t = 2$ s $v = 20$ m/s $+ (10$ m/s$^2)(2$ s$)$
                   $= 20$ m/s $+ 20$ m/s $= \textbf{40 m/s}$

b. $d = ?$ $d = v_0 t + \frac{1}{2} at^2$
    $t = 1$ s $d = (20$ m/s$)(1$ s$) + \frac{1}{2}(10$ m/s$^2)(1$ s$)^2$
                   $= 20$ m $+ 5$ m $= \textbf{25 m}$
    $t = 2$ s $d = (20$ m/s$)(2$ s$) + \frac{1}{2}(10$ m/s$^2)(2$ s$)^2$
                   $= 40$ m $+ 20$ m $= \textbf{60 m}$

After just 1 second the ball has already traveled 25 meters, which means that it would be near the ground if thrown from our sixth-story window.

Keep in mind, though, that we have ignored the effects of air resistance in arriving at these results. For a compact object falling just a few meters, the effects of air resistance are very small. These effects increase as the velocity increases, however, so that the farther the object falls, the greater the effects of air resistance. In chapter 4, we will discuss the role of air resistance in more depth in the context of sky diving.

When an object is dropped, its velocity increases by approximately 10 m/s every second due to the gravitational acceleration. The distance traveled increases at an ever-increasing rate because the velocity is increasing. In just a few seconds, the object is moving very rapidly and has fallen a large distance. In section 3.3, we will explore the effects of gravitational acceleration on an object thrown upward.

3.3 Beyond Free Fall: Throwing a Ball Upward

In section 3.2, we discussed what happens when a ball is dropped or thrown downward. In both of these cases, the ball gains velocity as it falls due to the gravitation acceleration. What if the ball is thrown upward instead, as in figure 3.9? How does gravitational acceleration affect the ball's motion? What goes up must come down—but when and how fast are interesting questions with everyday applications.

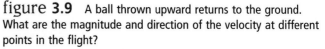

figure **3.9**   A ball thrown upward returns to the ground. What are the magnitude and direction of the velocity at different points in the flight?

The directions of the acceleration and velocity vectors merit our close attention. The gravitational acceleration is always directed downward toward the center of the earth, because that is the direction of the gravitational force that produces this acceleration. This means that the acceleration is in the opposite direction to the original upward velocity.

How does the ball's velocity change?

Suppose that we throw a ball straight up with an original velocity of 20 m/s. Many of us can throw a ball at this velocity: it is approximately 45 MPH. This is a lot less than a 90-MPH fastball, but throwing a ball upward with good velocity is harder than throwing it horizontally.

Once the ball leaves our hand, the primary force acting on it is gravity, which produces a downward acceleration of 9.8 m/s^2 or approximately 10 m/s^2. Every second, there is a change in velocity of 10 m/s. This change in velocity is directed downward, however, opposite to the direction of the original velocity. It subtracts from the original velocity rather than adding to it.

Once you are aware of how important direction is in observing the ball thrown upward, finding the velocity at different times is not hard. After 1 second, the velocity of the ball has decreased by 10 m/s, so if it started at +20 m/s (*choosing the positive direction to be upward in this case*), it is now moving upward with a velocity of just +10 m/s. After 2 seconds, it loses another 10 m/s, so its velocity is then zero. It does not stop there, of course. In another second (3 seconds from the start), its velocity decreases by another 10 m/s, and it is then moving downward at −10 m/s. *The sign of the velocity indicates its direction.* All of these values can be found from the relationship $v = v_0 + at$, where $v_0 = +20$ m/s and $a = -10$ m/s^2.

Clearly, the ball has changed direction, as you might expect. Just as before, the velocity changes steadily at −10 m/s each second, due to the constant downward acceleration. After 4 seconds, the ball is moving downward with a velocity of −20 m/s and is back at its starting position. These results are illustrated in figure 3.10. The high point in the motion occurs at a time 2 seconds after the ball is thrown, where the velocity is zero. If the velocity is zero, the ball is moving neither upward nor downward, so this is the turnaround point.

An interesting question, a favorite on physics tests (and often missed by students), asks for the value of acceleration at the high point in the motion. If the velocity is zero at this point, what is the value of the acceleration? The quick, but incorrect, response given by many people is that the acceleration must also be zero at that point. The correct answer is that the acceleration is still −10 m/s^2. The gravitational acceleration is constant and does not change. The velocity of the ball is still changing at that instant, from a positive to a negative value, even though the instantaneous velocity is zero. Acceleration is the *rate of change* of velocity and is *unrelated* to the size of the velocity.

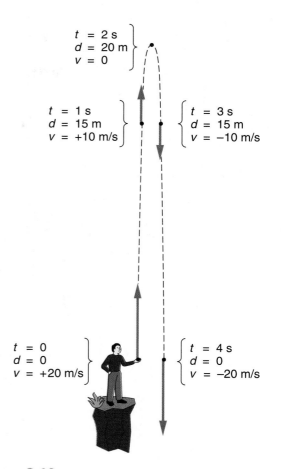

figure 3.10 The velocity vectors at different points in the flight of a ball thrown upward with a starting velocity of +20 m/s.

What would a graph of velocity plotted against time look like for the motion just described? If we make the upward direction of motion positive, the velocity starts with a value of +20 m/s and changes at a steady rate, decreasing by −10 m/s each second. This is a straight-line graph, sloping downward as in figure 3.11. The positive values of velocity represent upward motion, where the size of the velocity is decreasing, and the negative values of velocity represent downward motion. If the ball did not hit the ground, but was thrown from the edge of a cliff, it would continue to gain negative velocity as it moved downward.

How high does the ball go?

The position or height of the ball at different times can be computed using the methods in section 3.2. These distance computations involve the formula for uniform acceleration developed in section 2.5. In the sample exercise in Try This Box 3.3, we compute the height or distance traveled at 1-second intervals for the ball thrown upward at +20 m/s, using −10 m/s^2 for the gravitational acceleration.

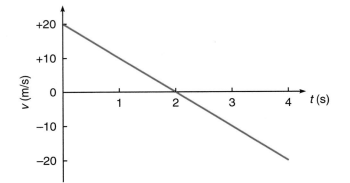

figure **3.11** A plot of the velocity versus time for a ball thrown upward with an initial velocity of +20 m/s. The negative values of velocity represent downward motion.

What should you notice about these results? First, the high point of the motion is 20 meters above the starting point. The high point is reached when the velocity is zero, and we determined earlier that this occurs at a time of 2 seconds. This time depends on how fast the ball is thrown initially. The larger the original velocity, the greater the time to reach the high point. Knowing this time, we can use the distance formula to find the height.

You should also notice that after just 1 second, the ball has reached a height of 15 meters. It covers just five additional meters in the next second of motion, and then falls back to 15 meters in the following second. The ball spends a full 2 seconds above the height of 15 meters, even though it only reaches a height of 20 meters. The ball is moving more slowly near the top of its flight than it is at lower points—this is why the ball appears to "hang" near the top of its flight.

Finally, the time taken for the ball to fall back to its starting point from the high point is equal to the time taken for the ball to reach the high point in the first place. It takes 2 seconds to reach the high point and another 2 seconds for it to return to the starting point. The total time of flight is just twice the time needed to reach the high point, in this case, 4 seconds. A larger starting velocity would produce a higher turnaround point and a greater "hang time" for the ball.

A ball thrown upward is slowed by the downward gravitational acceleration until its velocity is reduced to zero at the high point. The ball then falls from that high point accelerating downward at the same constant rate as when it was rising. The ball travels more slowly near the top of its flight, so it appears to "hang" there. It spends more time in the top few meters than it does in the rest of the flight. We will find that these features are also present when a ball is projected at an angle to the horizontal, as discussed in section 3.5.

3.4 Projectile Motion

Suppose that instead of throwing a ball straight up or down, you throw it horizontally from some distance above the ground. What happens? Does the ball go straight out until it loses all of its horizontal velocity and then starts to fall like the perplexed coyote in the *Roadrunner* cartoons (fig. 3.12)? What does the real path, or *trajectory,* look like?

Cartoons give us a misleading impression. In fact, two different things are happening at the same time: (1) the ball is accelerating downward under the influence of gravity, and (2) the ball is moving sideways with an approximately constant horizontal component of velocity. Combining these two motions gives the overall **trajectory** or path.

try this box 3.3

Sample Exercise: Throwing a Ball Upward

A ball is thrown upward with an initial velocity of 20 m/s. Find its height at 1-s intervals for the first 4 s of its flight.

$$d = ? \quad d = v_0 t + \frac{1}{2}at^2$$
$$t = 1\text{ s} \quad = (20\text{ m/s})(1\text{ s}) + \frac{1}{2}(-10\text{ m/s}^2)(1\text{ s})^2$$
$$= 20\text{ m} - 5\text{ m} = \textbf{15 m}$$

$$t = 2\text{ s} \quad d = (20\text{ m/s})(2\text{ s}) + \frac{1}{2}(-10\text{ m/s}^2)(2\text{ s})^2$$
$$= 40\text{ m} - 20\text{ m} = \textbf{20 m}$$

$$t = 3\text{ s} \quad d = (20\text{ m/s})(3\text{ s}) + \frac{1}{2}(-10\text{ m/s}^2)(3\text{ s})^2$$
$$= 60\text{ m} - 45\text{ m} = \textbf{15 m}$$

$$t = 4\text{ s} \quad d = (20\text{ m/s})(4\text{ s}) + \frac{1}{2}(-10\text{ m/s}^2)(4\text{ s})^2$$
$$= 80\text{ m} - 80\text{ m} = \textbf{0 m}$$

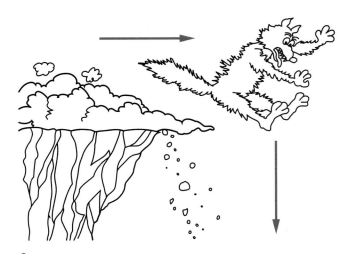

figure **3.12** A cartoon coyote falling off a cliff. Is this a realistic picture of what happens?

What does the trajectory look like?

You can perform a simple experiment to help you visualize the path that the projectile follows. Take a marble or small ball, roll it along the top of a desk or table, and let it roll off the edge. What does the path of the ball look like as it travels through the air to the floor? Is it like the coyote in figure 3.12? Roll the ball at different velocities and see how the path changes. Try to sketch the path after making these observations.

How do we go about analyzing this motion? The key lies in thinking about the horizontal and vertical components of the motion separately and then combining them to get the actual path (fig. 3.13).

The acceleration of the horizontal motion is zero, provided that air resistance is small enough to be ignored. This implies that the ball moves with a constant horizontal velocity once it has rolled off the table or has left the hand. The ball travels equal horizontal distances in equal time intervals, as shown across the top of figure 3.13. In constructing this diagram, we assumed an initial horizontal velocity of 2 m/s for the ball. Every tenth of a second, then, the ball travels a horizontal distance of 0.2 meter.

At the same time that the ball travels with constant horizontal velocity, it accelerates downward with the constant gravitational acceleration g. Its vertical velocity increases exactly like that of the falling ball photographed for figure 3.3. This motion is depicted along the left side of figure 3.13. In each successive time interval, the ball falls a greater distance than in the time interval before, because the vertical velocity increases with time.

Combining the horizontal and vertical motions, we get the path shown curving downward in figure 3.13. For each time shown, we draw a horizontal dashed line locating the vertical position of the ball, and a vertical dashed line for the horizontal position. The position of the ball at any time is the point where these lines intersect. The resulting trajectory (the solid curve) should look familiar if you have performed the simple experiments suggested in the first paragraph on this page.

If you understand how we obtained the path of the ball, you are well on your way to understanding **projectile motion**. The total velocity of the ball at each position pictured is in the direction of the path at that point, since this is the actual direction of the ball's motion. This total velocity is a vector sum of the horizontal and vertical components of the velocity (fig. 3.14). (See appendix C.) The horizontal velocity remains constant, because there is no acceleration in that direction. The downward (vertical) velocity gets larger and larger.

The actual shape of the path followed by the ball depends on the original horizontal velocity given the ball by throwing it or rolling it from the tabletop. If this initial horizontal velocity is small, the ball does not travel very far horizontally. Its trajectory will then be like the smallest starting velocity v_1 in figure 3.15.

The three trajectories shown in figure 3.15 have three different starting velocities. As you would expect, the ball travels greater horizontal distances when projected with a larger initial horizontal velocity.

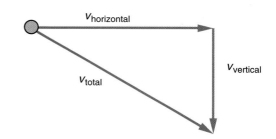

figure 3.14 The total velocity at any point is found by adding the vertical component of the velocity to the horizontal component.

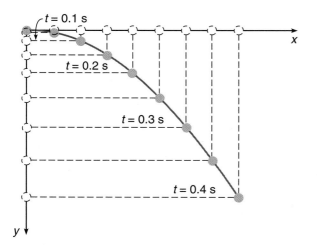

figure 3.13 The horizontal and vertical motions combine to produce the trajectory of the projected ball.

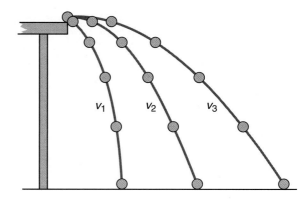

figure 3.15 Trajectories for different initial velocities of a ball rolling off a table: v_3 is larger than v_2, which in turn is larger than v_1.

What determines the time of flight?

Which of the three balls in figure 3.15 would hit the floor first if all three left the tabletop at the same time? Does the time taken for the ball to hit the floor depend on its horizontal velocity? There is a natural tendency to think that the ball that travels farther takes a longer time to reach the floor.

In fact, the three balls should all reach the floor at the same time. The reason is that they are all accelerating downward at the same rate of 9.8 m/s². This downward acceleration is not affected by how fast the ball travels horizontally. The time taken to reach the floor for the three balls in figure 3.15 is determined strictly by how high above the floor the tabletop is. The vertical motion is *independent* of the horizontal velocity.

This fact often surprises people. It contradicts our intuitive sense of what is going on but can be confirmed by doing simple experiments using two similar balls (fig. 3.16). If you throw one ball horizontally at the same time that you simply drop the second ball from the same height, the two balls should reach the floor at roughly the same time. They may fail to hit at the same time, most likely because it is hard to throw the first ball completely horizontally and to release both balls at the same time. A special spring gun, often used in demonstrations, will do this more precisely.

If we know how far the ball falls, we can compute the time of flight. This can then be used to determine the horizontal distance that the ball will travel, if we know the ini-

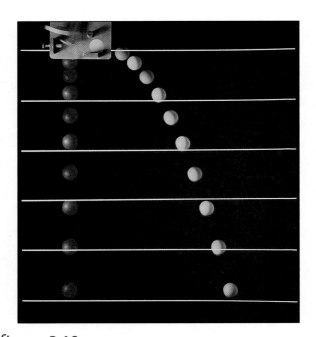

figure 3.16   A ball is dropped at the same time that a second ball is projected horizontally from the same height. Which ball reaches the floor first?

tial horizontal velocity. The sample exercise in Try This Box 3.4 shows this type of analysis. Notice that the horizontal distance traveled is determined by two factors: the time of flight and the initial velocity.

Treating the vertical motion independently of the horizontal motion and then combining them to find the trajectory is the secret to understanding projectile motion. A horizontal glide combines with a vertical plunge to produce a graceful curve. The downward gravitational acceleration behaves the same as for any falling object, but there is no acceleration in the horizontal direction if air resistance can be ignored. The projectile moves with constant horizontal velocity while it is accelerating downward.

3.5 Hitting a Target

As long as humans have been hunters or warriors, they have wanted to predict where a projectile such as a cannonball will land after it is fired. Being able to hit a target

Does the bullet fall when a rifle is fired?

Imagine that you are firing a rifle at a small target some distance away, with the rifle and target at exactly the same distance above the ground (fig. 3.17). If the rifle is fired directly at the target in a horizontal direction, will the bullet hit the center of the target? If you think of the ball rolling off the table in section 3.4, you should conclude that the bullet will strike the target slightly below the center. Why? The bullet will be accelerated downward by the gravitational pull of the earth and will fall slightly as it travels to the target.

Since the time of flight is small, the bullet does not fall very far, but it falls far enough to miss the center of the target. How do you compensate for the fall of the bullet? You aim a little high. You correct your aim either through trial and error or by adjusting your rifle sight so that your aim is automatically a little above center. Rifle sights are often adjusted for some average distance to the target. For longer distances you must aim high, for shorter distances a little low.

If you aim a little high, the bullet no longer starts out in a completely horizontal direction. The bullet travels up slightly during the first part of its flight and then comes down to meet the target. This also happens when you fire a cannon or throw a ball at a distant target. The rise and fall is more obvious for the trajectory of a football, though, than it is for a bullet.

The flight of a football

Whenever you throw a ball such as a football at a somewhat distant target, the ball must be launched at an angle above the horizontal so that the ball does not fall to the ground too soon. A good athlete does this automatically as a result of practice. The harder you throw, the less you need to direct the ball upward, because a larger initial velocity causes the ball to reach the target more quickly, giving it less time to fall.

Figure 3.18 shows the flight of a football thrown at an angle of 30° above the horizontal. The vertical position of the ball is plotted on the left side of the diagram, as in figure 3.13 for the horizontally projected ball. The horizontal position of the ball is shown across the bottom of the diagram. We have assumed that air resistance is small, so the ball travels with a constant horizontal velocity. Combining these two motions yields the overall path.

As the football climbs, the vertical component of its velocity decreases because of the constant downward gravitational acceleration. At the high point, this vertical component of the velocity is zero, just as it is for a ball thrown straight upward. The velocity of the ball is completely horizontal at this high point. The ball then begins to fall, gaining downward velocity as it accelerates. Unlike the ball thrown straight upward, however, there is a constant horizontal component to the velocity throughout the flight. We need to add this horizontal motion to the up-and-down motion that we described in section 3.3.

In throwing a ball, you can vary two quantities to help you hit your target. One is the initial velocity, which is determined by how hard you throw the ball. The other is the launch angle, which can be varied to fit the circumstances.

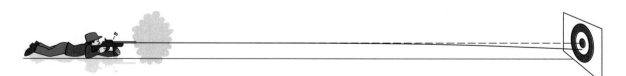

figure 3.17 A target shooter fires at a distant target. The bullet falls as it travels to the target.

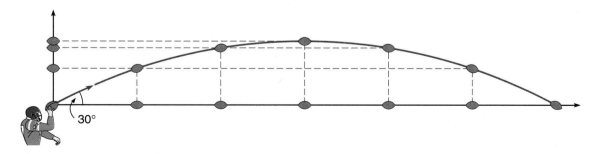

figure 3.18 The flight of a football launched at an angle of 30° to the horizontal. The vertical and horizontal positions of the ball are shown at regular time intervals.

A ball thrown with a large initial velocity does not have to be aimed as high and will reach the target more quickly. It may not clear the on-rushing linemen, however, and it might be difficult to catch because of its large velocity.

There is no time like the present to test these ideas. Take a page of scrap paper and crumple it into a compact ball. Then take your wastebasket and put it on your chair or desk. Throwing underhand, experiment with different throwing speeds and launch angles to see which is most effective in making a basket. Try to get a sense of how the launch angle and throwing speed interact to produce a successful shot. A low, flat-trajectory shot should require a greater throwing speed than a higher, arching shot. The flatter shot must also be aimed more accurately, since the effective area of the opening in the basket is smaller when the ball approaches at a flat angle. The ball "sees" a smaller opening. (This effect is discussed in Everyday Phenomenon Box 3.1.)

How can we achieve maximum distance?

In firing a rifle or cannon, the initial velocity of the projectile is usually set by the amount of gunpowder in the shell. The launch angle is then the only variable we can change in attempting to hit a target. Figure 3.19 shows three possible paths, or trajectories, for a cannonball fired at different launch angles for the same initial speed. For different launch angles, we tilt the cannon barrel by different amounts from the position shown.

Note that the greatest distance is achieved using an intermediate angle, an angle of 45° if the effects of air resistance are negligible. The same considerations are involved in the shot put in track-and-field events. The launch angle is very important and, for the greatest distance, will be near 45°. Air resistance and the fact that the shot hits the ground below the launch point are also factors, so the most effective angle is somewhat less than 45° in the shot put.

Everyday Phenomenon

box 3.1

Shooting a Basketball

The Situation. Whenever you shoot a basketball, you unconsciously select a trajectory for the ball that you believe will have the greatest likelihood of getting the ball to pass through the basket. Your target is above the launch point (with the exception of dunk shots and sky hooks), but the ball must be on the way down for the basket to count.

What factors determine the best trajectory? When is a high, arching shot desirable, and when might a flatter trajectory be more effective? Will these factors be different for a free throw than for a shot taken when you are guarded by another player? How can our understanding of projectile motion help us to answer these questions?

The Analysis. The diameter of the basketball and the diameter of the basket opening limit the angle at which the basketball can pass cleanly through the hoop. The second drawing shows the range of possible paths for a ball coming straight down and for one coming in at a 45° angle to the basket. The shaded area in each case shows how much the center of the ball can vary from the center line if the ball is to pass through the hoop. As you can see, a wider range of paths is available when the ball is coming straight down. The diameter of the basketball is a little more than half the diameter of the basket.

This second drawing illustrates the advantage of an arched shot. There is a larger margin of error in the path that the ball can take and still pass through the hoop cleanly.

For the dimensions of a regulation basketball and basket, the angle must be at least 32° for a clean shot. As the angle gets larger, the range of possible paths increases. At smaller angles, appropriate spin on the basketball will sometimes cause the ball to rattle through, but the smaller the angle, the less the likelihood of that happening.

Different possible trajectories for a basketball free throw. Which has the greatest chance of success?

(continued)

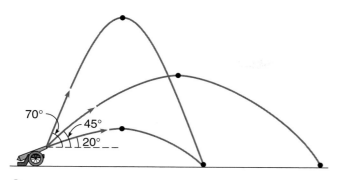

figure **3.19** Cannonball paths for different launch angles but the same initial launch speed.

Thinking about what happens to the horizontal and vertical components of the initial velocity at different launch angles will show us why the angle for maximum distance is approximately 45°. (See figure 3.20.) Velocity is a vector, and its horizontal and vertical components can be found by drawing the vector to scale and adding dashed lines to the horizontal and vertical directions (fig. 3.20). This process is described more fully in appendix C.

For the lowest launch angle 20°, we see that the horizontal component of the velocity is much larger than the vertical. Since the initial upward velocity is small, the ball does not go very high. Its time of flight is short, and it hits the ground sooner than in the other two cases shown. The ball gets there quickly because of its large horizontal velocity and short travel time, but it does not travel very far before hitting the ground.

The high launch angle of 70° produces a vertical component much larger than the horizontal component. The ball thus travels much higher and stays in the air for a longer time than at 20°. It does not travel very far horizontally,

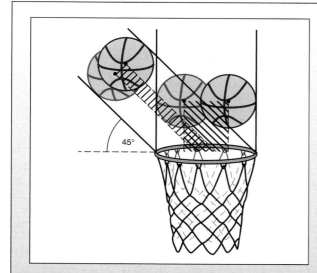

Possible paths for a basketball coming straight down and for one coming in at a 45° angle. The ball coming straight down has a wider range of possible paths.

The disadvantage of the arched shot is less obvious. As you get farther away from the basket, launching conditions for an arched shot must be more precise for the ball to travel the horizontal distance to the basket. If an arched shot is launched from 30 ft, it must travel a much higher path than a shot launched at the same angle closer to the basket, as shown in the third drawing. Since the ball stays in the air for a longer time, small variations in either the release speed or angle can cause large errors in the distance traveled. This distance depends on both the time of flight and the horizontal component of the velocity.

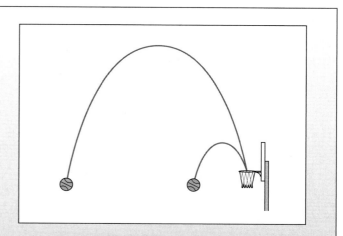

An arched shot launched from a large distance stays in the air longer than one launched at the same angle from much closer to the basket.

A highly arched shot is more effective when you are close to the basket. You can then take advantage of the greater range of paths available to the arched shot without suffering much from the uncertainty in the horizontal distance. Away from the basket, the desirable trajectories gradually become flatter, permitting more accurate control of the shot. An arched shot is sometimes necessary from anywhere on the court, however, to avoid having the shot blocked.

The spin of the basketball, the height of the release, and other factors all play a role in the success of a shot. A fuller analysis can be found in an article by Peter J. Brancazio in the *American Journal of Physics* (April 1981) entitled "Physics of Basketball." A good understanding of projectile motion might improve the game of even an experienced player.

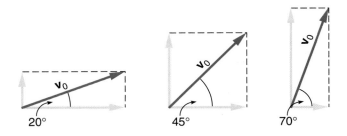

figure 3.20 Vector diagrams showing the horizontal and vertical components of the initial velocity for the three cases illustrated in figure 3.19.

however, because of its small horizontal velocity. The ball travels about the same horizontal distance as for the 20° launch, but it takes longer getting there. (If we shot it straight up, the horizontal distance covered would be zero, of course.)

The intermediate angle of 45° splits the initial velocity into equal-sized horizontal and vertical components. The ball therefore stays in the air longer than in the low-angle launch but also travels with a greater horizontal velocity than in the high-angle launch. In other words, with relatively large values for both the vertical and horizontal

components of velocity, the vertical motion keeps the ball in the air long enough for the horizontal velocity to be effective. This produces the greatest distance of travel.

The time of flight and the horizontal distance traveled can be found if the launch angle and the size of the initial velocity are known. It is first necessary to find the horizontal and vertical components of the velocity to do these computations, however, and this makes the problem more complex than those discussed earlier. The ideas can be understood without doing the computations. The key is to think about the vertical and horizontal motions separately and then combine them.

For a projectile launched at an angle, the initial velocity can be broken down into vertical and horizontal components. The vertical component determines how high the object will go and how long it stays in the air, while the horizontal component determines how far it will go in that time. The launch angle and the initial speed interact to dictate where the object will land. Through the entire flight, the constant downward gravitational acceleration is at work, but it changes only the vertical component of the velocity. Producing or viewing such trajectories is a common part of our everyday experience.

---

summary

The primary aim in this chapter has been to introduce you to the gravitational acceleration for objects near the surface of the earth and to show how that acceleration affects the motion of objects dropped or launched in various ways.

1   Acceleration due to gravity. To find the acceleration due to gravity, we use measurements of the position of a dropped object at different times. The gravitational acceleration is 9.8 m/s^2. It does not vary with time as the object falls, and it has the same value for different objects regardless of their weight.

2   Tracking a falling object. The velocity of a falling object increases by approximately 10 m/s every second of its fall. Distance traveled increases in proportion to the square of the time, so that it increases at an ever-increasing rate. In just 1 second, a dropped ball is moving with a velocity of 10 m/s and has traveled 5 meters.

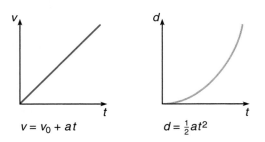

$$v = v_0 + at$$ $$d = \tfrac{1}{2}at^2$$

3   Beyond free fall: Throwing a ball upward. The speed of an object thrown upward first decreases due to the downward gravitational acceleration, passes through zero at the high point, and then increases as the object falls. The object spends more time near the top of its flight because it is moving more slowly there.

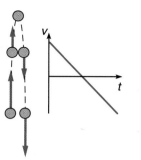

4   Projectile motion. If an object is launched horizontally, it moves with a constant horizontal velocity at the same time that it accelerates downward due to gravity. These two motions combine to produce the object's curved trajectory.

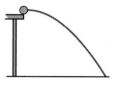

5   Hitting a target. There are two factors, the launch speed and the launch angle, that can be varied to determine the path of an object launched at an angle to the horizontal. Once again, the horizontal and vertical motions combine to produce the overall motion as the projectile moves toward a target.

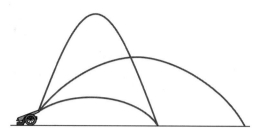

key terms

questions

*Questions identified with an asterisk are more open-ended than the others. They call for lengthier responses and are more suitable for group discussion.

Q1. A small piece of paper is dropped and flutters to the floor. Is the piece of paper accelerating at any time during this motion? Explain.

Q2. The diagram shows the positions at intervals of 0.10 seconds of a ball moving from left to right (as in a photograph taken with a stroboscope that flashes every tenth of a second). Is the ball accelerated? Explain.

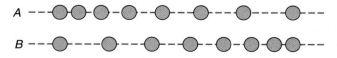

Q3. The diagram shows the positions at intervals of 0.05 seconds of two balls moving from left to right. Are either or both of these balls accelerated? Explain.

Q4. A lead ball and an aluminum ball, each 1 in. in diameter, are released simultaneously and allowed to fall to the ground. Due to its greater density, the lead ball has a substantially larger mass than the aluminum ball. Which of these balls, if either, has the greater acceleration due to gravity? Explain.

Q5. Two identical pieces of paper, one crumpled into a ball and the other left uncrumpled, are released simultaneously from the same height above the floor. Which one, if either, do you expect to reach the floor first? Explain.

*Q6. Aristotle stated that heavier objects fall faster than lighter objects. Was Aristotle wrong? In what sense could Aristotle's view be considered correct?

Q7. Two identical pieces of paper, one crumpled into a ball and the other left uncrumpled, are released simultaneously from inside the top of a large evacuated tube. Which one, if either, do you expect will reach the bottom of the tube first? Explain.

Q8. A rock is dropped from the top of a diving platform into the swimming pool below. Will the distance traveled by the rock in a 0.1-second interval near the top of its flight be the same as the distance covered in a 0.1-second interval just before it hits the water? Explain.

Q9. The graph shows the velocity plotted against time for a certain falling object. Is the acceleration of this object constant? Explain.

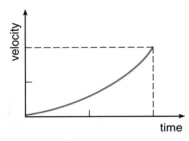

Q10. A ball is thrown downward with a large starting velocity.
a. Will this ball reach the ground sooner than one that is just dropped at the same time from the same height? Explain.
b. Will this ball accelerate more rapidly than one that is dropped with no initial velocity? Explain.

Q11. A ball thrown straight upward moves initially with a decreasing upward velocity. What are the directions of the velocity and acceleration vectors during this part of the motion? Does the acceleration decrease also? Explain.

Q12. A rock is thrown straight upward reaching a height of 20 meters. On its way up, does the rock spend more time in the top 5 meters of its flight than in its first 5 meters of its flight? Explain.

Q13. A ball is thrown straight upward and then returns to the earth. Choosing the positive direction to be upward, sketch a graph of the velocity of this ball against time. Where does the velocity change direction? Explain.

Q14. A ball is thrown straight upward. At the very top of its flight, the velocity of the ball is zero. Is its acceleration at this point also zero? Explain.

*Q15. A ball rolls up an inclined plane, slows to a stop, and then rolls back down. Do you expect the acceleration to be constant during this process? Is the velocity constant? Is

the acceleration equal to zero at any point during this motion? Explain.

Q16. A ball is thrown straight upward and then returns to the earth. Does the acceleration change direction during this motion? Explain.

Q17. A ball rolling rapidly along a tabletop rolls off the edge and falls to the floor. At the exact instant that the first ball rolls off the edge, a second ball is dropped from the same height. Which ball, if either, reaches the floor first? Explain.

Q18. For the two balls in question 17, which, if either, has the larger total velocity when it hits the floor? Explain.

Q19. Is it possible for an object to have a horizontal component of velocity that is constant at the same time that the object is accelerating in the vertical direction? Explain by giving an example, if possible.

Q20. A ball rolls off a table with a large horizontal velocity. Does the direction of the velocity vector change as the ball moves through the air? Explain.

Q21. A ball rolls off a table with a horizontal velocity of 5 m/s. Is this velocity an important factor in determining the time that it takes for the ball to hit the floor? Explain.

Q22. An expert marksman aims a high-speed rifle directly at the center of a nearby target. Assuming that the rifle sight has been accurately adjusted for more distant targets, will the bullet hit the near target above or below the center? Explain.

Q23. In the diagram, two different trajectories are shown for a ball thrown by a center fielder to home plate in a baseball game. Which of the two trajectories (if either), the higher one or the lower one, will result in a longer time for the ball to reach home plate? Explain.

Q24. For either of the trajectories shown in the diagram for question 23, is the velocity of the ball equal to zero at the high point in the trajectory? Explain.

Q25. Assuming that the two trajectories in the diagram for question 23 represent throws by two different center fielders, which of the two is likely to have been thrown by the player with the stronger arm? Explain.

Q26. A cannonball fired at an angle of 70° to the horizontal stays in the air longer than one fired at 45° from the same cannon. Will the 70° shot travel a greater horizontal distance than the 45° shot? Explain.

Q27. Will a shot fired from a cannon at a 20° launch angle travel a longer horizontal distance than a 45° shot? Explain.

Q28. The diagram shows a wastebasket placed behind a chair. Three different directions are indicated for the velocity of a ball thrown by the kneeling woman. Which of the three directions—A, B, or C—is most likely to result in the ball landing in the basket? Explain.

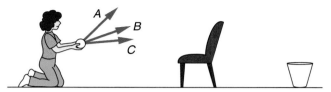

Q29. In the situation pictured in question 28, is the magnitude of the velocity important to the success of the shot? Explain.

Q30. In shooting a free throw in basketball, what is the primary advantage that a high, arching shot has over one with a flatter trajectory? Explain.

Q31. In shooting a basketball from greater than free-throw range, what is the primary disadvantage of a high, arching shot? Explain.

*Q32. A football quarterback must hit a moving target while eluding onrushing linemen. Discuss the advantages and disadvantages of a hard low-trajectory throw to a higher-lofted throw.

exercises

E1. A steel ball is dropped from a diving platform (with an initial velocity of zero). Using the approximate value of $g = 10$ m/s^2,
a. What is the velocity of the ball 0.8 seconds after its release?
b. What is its velocity 1.6 seconds after its release?

E2. For the ball in exercise 1:
a. Through what distance does the ball fall in the first 0.8 seconds of its flight? (Assume $g = 10$ m/s^2.)
b. How far does it fall in the first 1.6 seconds of its flight?

E3. A large rock is dropped from the top of a high cliff. Assuming that air resistance can be ignored and that the acceleration has the constant value of 10 m/s^2, how fast would the rock be traveling 5 seconds after it is dropped? What is this speed in MPH? (See inside front cover for conversion factors.)

E4. Suppose Galileo's pulse rate was 75 beats per minute.
a. What is the time in seconds between consecutive pulse beats?
b. How far does an object fall in this time when dropped from rest?

E5. A ball is thrown downward with an initial velocity of 12 m/s. Using the approximate value of $g = 10$ m/s^2, what is the velocity of the ball 1.0 seconds after it is released?

E6. A ball is dropped from a high building. Using the approximate value of $g = 10$ m/s^2, find the *change* in velocity between the first and fourth second of its flight.

E7. A ball is thrown upward with an initial velocity of 15 m/s. Using the approximate value of $g = 10$ m/s^2, what are the magnitude and direction of the ball's velocity:
a. 1 second after it is thrown?
b. 2 seconds after it is thrown?

E8. How high above the ground is the ball in exercise 7:
a. 1 second after it is thrown?
b. 2 seconds after it is thrown?

E9. At what time does the ball in exercise 7 reach the high point in its flight? (Use the approximate value of $g = 10$ m/s^2, and remember that the velocity is equal to zero at the high point.)

E10. Suppose that the gravitational acceleration on a certain planet is only 2.0 m/s^2. A space explorer standing on this planet throws a ball straight upward with an initial velocity of 18 m/s.
a. What is the velocity of the ball 4 seconds after it is thrown?
b. How much time elapses before the ball reaches the high point in its flight?

E11. A bullet is fired horizontally with an initial velocity of 900 m/s at a target located 150 m from the rifle.
a. How much time is required for the bullet to reach the target?
b. Using the approximate value of $g = 10$m/s^2, how far does the bullet fall in this time?

E12. A ball rolls off a shelf with a horizontal velocity of 6 m/s. At what horizontal distance from the shelf does the ball land if it takes 0.4 s to reach the floor?

E13. A ball rolls off a table with a horizontal velocity of 4 m/s. If it takes 0.5 seconds for the ball to reach the floor, how high above the floor is the tabletop? (Use $g = 10$ m/s^2.)

E14. A ball rolls off a table with a horizontal velocity of 5 m/s. If it takes 0.4 seconds for it to reach the floor:
a. What is the vertical component of the ball's velocity just before it hits the floor? (Use $g = 10$ m/s^2.)
b. What is the horizontal component of the ball's velocity just before it hits the floor?

E15. A ball rolls off a platform that is 5 meters above the ground. The ball's horizontal velocity as it leaves the platform is 6 m/s.
  a. How much time does it take for the ball to hit the ground? (See Try This Box 3.3, use $g = 10$ m/s^2.)
  b. How far from the base of the platform does the ball hit the ground?

E16. A projectile is fired at an angle such that the vertical component of its velocity and the horizontal component of its velocity are both equal to 30 m/s.
  a. Using the approximate value of $g = 10$m/s^2, how long does it take for the ball to reach its high point?
  b. What horizontal distance does the ball travel in this time?

challenge problems

CP1. A ball is thrown straight upward with an initial velocity of 16 m/s. Use $g = 10$ m/s^2 for computations listed here.
  a. What is its velocity at the high point in its motion?
  b. How much time is required to reach the high point?
  c. How high above its starting point is the ball at its high point?
  d. How high above its starting point is the ball 2 seconds after it is released?
  e. Is the ball moving up or down 2 seconds after it is released?

CP2. Two balls are released simultaneously from the top of a tall building. Ball A is simply dropped with no initial velocity, and ball B is thrown downward with an initial velocity of 12 m/s.
  a. What are the velocities of the two balls 1.5 seconds after they are released?
  b. How far has each ball dropped in 1.5 seconds?
  c. Does the difference in the velocities of the two balls change at any time after their release? Explain.

CP3. Two balls are rolled off a tabletop that is 0.8 m above the floor. Ball A has a horizontal velocity of 3 m/s and that of ball B is 5 m/s.
  a. Assuming $g = 10$ m/s^2, how long does it take each ball to reach the floor after it rolls off the edge?
  b. How far does each ball travel horizontally before hitting the floor?

  c. If the two balls started rolling at the same time at a point 1.2 m behind the edge of the table, will they reach the floor at the same time? Explain.

CP4. A cannon is fired over level ground at an angle of 30° to the horizontal. The initial velocity of the cannonball is 400 m/s, but because the cannon is fired at an angle, the vertical component of the velocity is 200 m/s and the horizontal component is 346 m/s.
  a. How long is the cannonball in the air? (Use $g = 10$ m/s^2 and the fact that the total time of flight is twice the time required to reach the high point.)
  b. How far does the cannonball travel horizontally?
  c. Repeat these calculations, assuming that the cannon was fired at a 60° angle to the horizontal, resulting in a vertical component of velocity of 346 m/s and a horizontal component of 200 m/s. How does the distance traveled compare to the earlier result?

CP5. A good pitcher can throw a baseball at a speed of 90 MPH. The pitcher's mound is approximately 60 ft from home plate.
  a. What is the speed in m/s?
  b. What is the distance to home plate in m?
  c. How much time is required for the ball to reach home plate?
  d. If the ball is launched horizontally, how far does the ball drop in this time, ignoring the effects of spin?

home experiments and observations

HE1. Gather numerous small objects and drop them from equal heights, two at a time. Record which objects fall significantly more slowly than a compact dense object such as a marble or similar object. Rank order these slower objects by their time of descent. What factors seem to be important in determining this time?

HE2. Try dropping a ball from one hand at the same time that you throw a second ball with your other hand. At first, try to throw the second ball horizontally, with no upward or downward component to its initial velocity. (It may take some practice.)
  a. Do the balls reach the floor at the same time? (It helps to enlist a friend for making this judgment.)

  b. If the second ball is thrown slightly upward from the horizontal, which ball reaches the ground first?
  c. If the second ball is thrown slightly downward from the horizontal, which ball reaches the ground first?

HE3. Take a ball outside and throw it straight up in the air as hard as you can. By counting seconds yourself, or by enlisting a friend with a watch, estimate the time that the ball remains in the air. From this information, can you find the initial velocity that you gave to the ball? (The time required for the ball to reach the high point is just half the total time of flight.)

HE4. Take a stopwatch to a football game and estimate the hang time of several punts. Also note how far (in yards) each punt travels horizontally. Do the highest punts have the longest hang times? Do they travel the greatest distances horizontally?

HE5. Using rubber bands and a plastic rule or other suitable support, design and build a marble launcher. By pulling the rubber band back by the same amount each time, you should be able to launch the marble with approximately the same speed each time.

a. Produce a careful drawing of your launcher and note the design features that you used. (Prizes may be available for the best design.)

b. Placing your launcher at a number of different angles to the horizontal, launch marbles over a level surface and measure the distance that they travel from the point of launch. Which angle yields the greatest distance?

c. Fire the marbles at different angles from the edge of a desk or table. Which angle yields the greatest horizontal distance?

HE6. Try throwing a ball or a wadded piece of paper into a wastebasket placed a few meters from your launch point.

a. Which is most effective, an overhanded or underhanded throw? (Five practice shots followed by ten attempts for each might produce a fair test.)

b. Repeat this process with a barrier such as a chair placed near the wastebasket.

Newton's Laws: Explaining Motion

chapter overview

The primary purpose of this chapter is to explain Newton's three laws of motion and how they apply in familiar situations. We begin with a historical sketch of their development and then proceed to a careful discussion of each law. The concepts of force, mass, and weight play critical roles in this discussion. We conclude the chapter by applying Newton's theory to several familiar examples.

chapter outline

1 **A brief history.** Where do our ideas and theories about motion come from? What roles were played by Aristotle, Galileo, and Newton?

2 **Newton's first and second laws.** How do forces affect the motion of an object? What do Newton's first and second laws of motion tell us, and how are they related to one another?

3 **Mass and weight.** How can we define mass? What is the distinction between mass and weight?

4 **Newton's third law.** Where do forces come from? How does Newton's third law of motion help us to define force, and how is it applied?

5 **Applications of Newton's laws.** How can Newton's laws be applied in different situations such as pushing a chair, sky diving, throwing a ball, and pulling two connected carts across the floor?

chapter

4

A large person gives you a shove, and you move in the direction of that push. A child pulls a toy wagon with a string, and the wagon lurches along. An athlete kicks a football or a soccer ball, and the ball is launched toward the goal. These are familiar examples involving forces in the form of pushes or pulls that cause changes in motion.

To pick a less complex example, imagine yourself pushing a chair across a wood or tile floor (fig. 4.1). Why does the chair move? Will it continue its motion if you stop pushing? What factors determine the velocity of the chair? If you push harder, will the chair's velocity increase? Up to this point, we have introduced ideas useful in describing motion, but we have not talked much about what causes changes in motion. Explaining motion is more challenging than describing it.

You already have some intuitive notions about what causes the chair to move. Certainly, the push that you exert on the chair has something to do with it. But is the strength of that push more directly related to the velocity of the chair or to its acceleration? At this point, intuition often serves us poorly.

Over two thousand years ago, the Greek philosopher Aristotle (384–322 B.C.) attempted to provide answers to some of these questions. Many of us would find that his explanations match our intuition for the case of the moving chair, but they are less satisfactory in the case of a thrown object where the push is not sustained. Aristotle's ideas were widely accepted until they were replaced by a theory introduced by Isaac Newton in the seventeenth century. Newton's theory of motion has proved to be a much more complete and satisfactory explanation of motion, and it permits quantitative predictions that were largely lacking in Aristotle's ideas.

Newton's three laws of motion form the foundation of his theory. What are these laws and how are they used in explaining motion? How do Newton's ideas differ from those of Aristotle, and why do Aristotle's ideas often seem to fit our commonsense notions of what is happening? A good understanding of Newton's laws will permit you to analyze and explain almost any simple motion. This understanding will provide you with insights useful in driving a car, moving heavy objects, and many other everyday activities.

figure **4.1**   Moving a chair. Will the chair continue to move when the person stops pushing?

4.1 A Brief History

Did some genius, sitting under an apple tree, concoct a full-blown theory of motion in a sudden, blinding flash of inspiration? Not quite. The story of how theories are developed and gain acceptance involves many players over long periods of time.

Let's highlight the roles of a few key people whose insights produced major advances. A glimpse of this history can help you appreciate the physical concepts we will discuss by showing when and how the theories emerged. It is important, for example, to know whether a theory was just proposed yesterday or has been tried and tested over a long time. Not all theories carry equal weight in their acceptance and use by scientists. Aristotle, Galileo, and Newton were major players in shaping our views of the causes of motion.

Aristotle's view of the cause of motion

Questions about the causes of motion and changes in motion had perplexed philosophers and other observers of nature for centuries. For over a thousand years, Aristotle's views prevailed. Aristotle was an astute and careful observer of nature. Aristotle investigated an incredible range of subjects, and he (or perhaps his students) produced extensive writings on topics such as logic, metaphysics, politics, literary criticism, rhetoric, psychology, biology, and physics.

In his discussions of motion, Aristotle conceived of force much as we have talked about it to this point: as a push or pull acting on an object. He believed that a force had to act for an object to move and that the velocity of the object was proportional to the strength of the force. A heavy object would fall more quickly toward the earth than a lighter object, because there was a larger force pulling the object to the earth. The strength of this force could be appreciated simply by holding the object in your hand.

Aristotle was also aware of the resistance that a medium offers the motion of an object. A rock falls more rapidly through air than through water. Water provides greater resistance to motion than air, as you surely know from trying to walk through waist-deep water at the beach. Aristotle thus saw the velocity of the object as being proportional to the force acting on it and inversely related to the resistance, but he never defined the concept of resistance quantitatively. He did not distinguish acceleration from velocity, and he spoke of velocity by stating the time required to cover a fixed distance.

Aristotle was an observer of nature rather than an experimenter. He did not make quantitative predictions that he checked by experiment. Even without such tests, however, some problems with his basic ideas of motion troubled Aristotle himself, as well as later thinkers. For example, in the case of a thrown ball or rock, the force that initially propels the object no longer acts once the ball leaves the hand. What keeps the ball moving?

Since the ball does keep moving for some time after leaving the hand that throws it, a force was necessary, according to Aristotle's theory. He suggested that the force that maintains the motion once the ball leaves the hand is provided by air rushing around to fill the vacuum in the spot where the ball has just been (fig. 4.2). This flow of air then pushes the ball from behind. Does this seem reasonable?

Following the decline of the Roman Empire, only fragments of Aristotle's writings were known to European thinkers for several centuries. His complete works, which had been preserved by Arab scholars, did not resurface in Europe until the twelfth century. Along with the work of other Greek thinkers, Aristotle's works were translated into Latin during the twelfth and thirteenth centuries.

How did Galileo challenge Aristotle's views?

By the time that the Italian scientist Galileo Galilei (1564–1642) came on the scene, Aristotle's ideas were well established at European universities, including the universities of Pisa and Padua where Galileo studied and taught. In fact, education at the universities was organized around the disciplines defined by Aristotle, and much of Aristotle's natural philosophy had been incorporated into the teaching of the Roman Catholic Church. The Italian theologian Thomas Aquinas had carefully interwoven Aristotle's thinking with the theology of the church.

To challenge Aristotle was equivalent to challenging the authority of the church and could carry heavy consequences. Galileo was not alone in questioning Aristotle's ideas on motion; others had noted that dropped objects of similar form but radically different weights fall at virtually the same rate, contrary to Aristotle's theory. Although Galileo may never have dropped objects from the Leaning Tower of Pisa, he did perform careful experiments with dropped objects and actively publicized his results.

Galileo's primary problems with the church came from advocating the ideas of Copernicus. Copernicus had proposed a sun-centered (*heliocentric*) model of the solar system (discussed in chapter 5), which opposed the prevailing earth-centered models of Aristotle and others. Galileo was an activist on several fronts in challenging Aristotle and the traditional thinking. This placed him in conflict with many of his university colleagues and with members of the church hierarchy. He was eventually tried by the Inquisition and found guilty of heresy. He was placed under house arrest and forced to retract some of his teachings.

In addition to his work on falling objects, Galileo developed new ideas on motion that contradicted Aristotle's theory. Galileo argued that the natural tendency of a moving object is to continue moving: no force is required to maintain this motion. (Think about the pushed chair again. Does this statement make sense in that situation?) Building on the work of others, Galileo also developed a mathematical description of motion that included acceleration. The relationship $d = \frac{1}{2} at^2$ for the distance covered by a uniformly accelerating object was carefully demonstrated by Galileo. He published many of these ideas near the end of his life in his famous *Dialogues Concerning Two New Sciences*.

What did Newton accomplish?

Isaac Newton (1642–1727; fig. 4.3) was born in England the same year that Galileo died in Italy. Building on the work of Galileo, he proposed a theory of the causes of motion that could explain the motion of any object—the motion of ordinary objects such as a ball or chair as well as the motion of heavenly bodies such as the moon and the planets. In the Greek tradition, celestial motions were thought of as an entirely different realm from earthbound motions, thus requiring different explanations. Newton abolished this distinction by explaining both terrestrial and celestial mechanics with one theory.

The central ideas in Newton's theory are his three laws of motion (discussed in sections 4.2 and 4.4) and his law of universal gravitation (discussed in chapter 5). Newton's theory provided successful explanations of aspects of motion already known and offered a framework for many new studies in physics and astronomy. Some of these studies led to predictions of phenomena not previously observed. For example, calculations applying Newton's theory to irregularities in the orbits of the known planets led to the

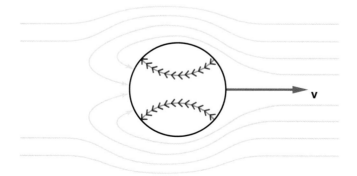

figure 4.2 Aristotle pictured air rushing around a thrown object to continue pushing the object forward. Does this picture seem reasonable?

figure **4.3** A portrait of Isaac Newton.

prediction of the existence of Neptune, which was quickly confirmed by observation. Confirmed predictions are one of the marks of a successful theory. Newton's theory served as the basic theory of mechanics for over two hundred years and is still used extensively in physics and engineering.

Newton developed the basic ideas of his theory around 1665, when he was still a young man. To avoid the plague, he had returned to his family's farm in the countryside where he had time to engage in serious thought with little interruption. He may have spent some time sitting under apple trees. The story has it that seeing an apple fall led to his insight that the moon also falls toward the earth and that the force of gravity is involved in both cases. (See chapter 5.) Flashes of insight or inspiration were surely a part of the process.

Although Newton developed much of his theory and its details in 1665, he did not formally publish his ideas until 1687. One reason for this delay was his need to develop some of the mathematical techniques required to calculate the effects of the proposed gravitational force on objects such as planets. (He is generally credited with being the coinventor of what we now call *calculus*.) The English title of Newton's 1687 treatise is *The Mathematical Principles of Natural Philosophy (Philosophiae Naturalis Principia Mathematica* in Latin), which is often referred to as Newton's *Principia*.

Scientific theories like Newton's do not just emerge in an intellectual vacuum. They are products of their time and the state of knowledge and worldview current then. They usually replace earlier and often cruder theories. The accepted theory of motion in Newton's day was still that of Aristotle, although it had come under attack by Galileo and others. Its shortcomings were generally recognized. Newton provided the capstone for a revolution in thought that was already well under way.

Although Aristotle's ideas on motion are now considered unsatisfactory and are worthless for making quantitative predictions, they do have an intuitive appeal much like our own untrained thinking about motion. For this reason, we often speak of the need to replace Aristotelian ideas about motion with Newtonian concepts as we learn mechanics. Even though our own naive ideas about motion are not usually as fully developed as those of Aristotle, you may find that some of your commonsense notions will require modification.

Newton's theory, in turn, has been partially superseded by more sophisticated theories that provide more accurate descriptions of motion. These include Einstein's theory of relativity as well as the theory of quantum mechanics, both of which arose early in the twentieth century. Although the predictions of these theories differ substantially from Newton's theory in the realm of the very fast (in the case of relativity) and the very small (quantum mechanics), they differ insignificantly for the motion of ordinary objects traveling at speeds much less than that of light. Newton's theory was a tremendous step forward and is still used extensively to analyze motion of ordinary objects.

Aristotle's ideas on motion, although not capable of making quantitative predictions, provided explanations that were widely accepted for many centuries and that fit well with some of our own commonsense thinking. Galileo challenged Aristotle's ideas on free fall as well as his general assumption that a force was required to keep an object in motion. Building on Galileo's work, Newton developed a more comprehensive theory of motion that replaced Aristotle's ideas. Newton's theory is still widely used to explain ordinary motions.

4.2 Newton's First and Second Laws

If we push a chair across the floor, what causes the chair to move or to stop moving? Newton's first two laws of motion address these questions and, in the process, provide part of a definition of **force.** The first law tells us what happens in the absence of a force, and the second describes the effects of applying a force to an object.

We discuss the first and second laws of motion together because the first law is actually a special case of the more general second law. Newton felt the need to state the first law separately, however, to counter strongly held Aristotelian ideas about motion. In doing so, Newton was following the lead of Galileo, who had stated a principle similar to Newton's first law several years earlier.

Newton's first law of motion

In language not too different from his own, **Newton's first law of motion** can be stated as

> An object remains at rest, or in uniform motion in a straight line, unless it is compelled to change by an externally imposed force.

In other words, unless there is a force acting on the object, its *velocity* will not change. If it is initially at rest, it will remain at rest; if it is moving, it will continue to do so with constant velocity (fig 4.4).

Notice that, in paraphrasing Newton's first law, we have used the term *velocity* rather than the term *speed*. Constant velocity implies that neither the direction nor the magnitude of the velocity changes. When the object is at rest, its velocity is zero, and that value remains constant in the absence of a force. If there is no force, the acceleration of the object is zero. The velocity does not change.

Although this law seems simple enough, it directly contradicts Aristotle's ideas (and perhaps your own intuition as well). Aristotle believed that a force is required to keep an object moving. His views make intuitive sense if we are talking about moving a heavy object such as the chair mentioned in our introduction. If you stop pushing, the chair stops moving. This view encounters problems, however, if we consider the motion of a thrown ball, or even a chair moving on a slippery surface. These objects continue to move after the initial push. Newton (and Galileo) made the

If **F** = 0

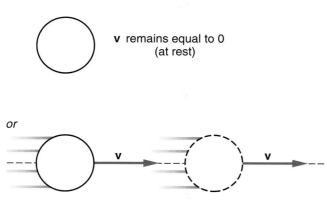

or

figure 4.4   Newton's first law: In the absence of a force, an object remains at rest or moves with constant velocity.

strong statement that no force is needed to keep an object moving.

How can Aristotle's ideas be so different from those of Newton and Galileo and yet seem so reasonable in some situations? The key to answering that question involves the existence of resistive or **frictional forces.** The chair does not move far after you stop pushing because the frictional forces of the floor acting on the chair cause the velocity to quickly decrease to zero. A thrown ball would eventually stop moving, even if it did not fall to the ground, because the force of air resistance is pushing against it. It is really quite difficult to find a situation in which there are *no* forces acting upon an object. Aristotle recognized the presence of air resistance and similar effects but did not treat them as forces in his theory.

How is force related to acceleration?

Newton's second law of motion is a more complete statement about the effect of an imposed force on the motion of an object. Stated in terms of acceleration, it says

> The acceleration of an object is directly proportional to the magnitude of the imposed force and inversely proportional to the mass of the object. The acceleration is in the same direction as that of the imposed force.

This statement is most easily grasped in symbolic form. By choosing appropriate units for force, we can state the proportionality of Newton's second law as the equation:

$$\mathbf{a} = \frac{\mathbf{F}}{m},$$

where **a** is the acceleration, **F** is the total force acting on the object, and m is the mass of the object. Since the acceleration is directly proportional to the imposed force, if we double the force acting on the object, we double the acceleration of the object. The same force acting on an object with a larger mass, however, will produce a smaller acceleration (fig. 4.5).

Note that the *acceleration* is directly related to the imposed force, not the velocity. Aristotle did not make a clear distinction between acceleration and velocity. Many of us also fail to make the distinction when we think informally about motion. In Newton's theory, this distinction is critical.

Newton's second law is *the* central idea of his theory of motion. According to this law, the acceleration of an object is determined by two quantities: the total force acting on the object and the mass of the object. In fact, the concepts of force and mass are, in part, defined by the second law. The total force acting on the object is the cause of its acceleration, and the magnitude of the force is defined by the size of the acceleration that it produces. Newton's third law, discussed in section 4.4, completes the definition of force by

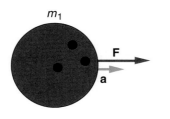

figure **4.5** The smaller-mass object experiences a larger acceleration than the larger-mass object when identical forces are applied to the two objects.

noting that forces result from interaction of the object with other objects.

The **mass** of an object is a quantity that tells us how much resistance an object has to a change in its motion, as indicated by the second law. We call this resistance to a change in motion **inertia,** following Galileo. (See Everyday Phenomenon Box 4.1.) We can define mass as

> Mass is a measure of an object's inertia, the property that causes it to resist a change in its motion.

The standard metric unit for mass is the kilogram (kg). We will say more about the determination of mass and its relationship to the weight of an object shortly (section 4.3).

Units of force can also be derived from Newton's second law. If we multiply both sides of the second-law equation by the mass, it can be expressed as

$$\mathbf{F} = m\mathbf{a}.$$

The appropriate unit for force must therefore be the product of a unit of mass and a unit of acceleration, or in the metric system, kilograms times meters per second squared. This frequently used unit is called the **newton** (N). Accordingly,

$$1 \text{ newton} = 1 \text{ N} = 1 \text{ kg·m/s}^2.$$

How do forces add?

Our version of the second law implies that the imposed force is the *total* or **net force** acting on the object. Force is a vector quantity whose direction is clearly important. If there is more than one force acting on an object, as there often is, we must then add these forces as vectors, taking into account their directions.

This process is illustrated in figure 4.6 and the sample exercise in Try This Box 4.1. A block is being pulled across a table by a force of 10 N applied through a string

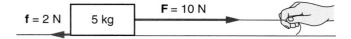

figure **4.6** A block being pulled across a table. Two horizontal forces are involved.

---

try this box 4.1

Sample Exercise: Finding the Net Force

A block with a mass of 5 kg is being pulled across a tabletop by a force of 10 N applied by a string tied to the front end of the block (fig 4.6). The table exerts a 2-N frictional force on the block. What is the acceleration of the block?

$\mathbf{F}_{\text{string}} = 10 \text{ N (to the right)}$ $\quad F = F_{\text{string}} - f_{\text{table}}$

$\mathbf{f}_{\text{table}} = 2 \text{ N (to the left)}$ $\quad\quad\quad = 10 \text{ N} - 2 \text{ N} = 8 \text{ N}$

$m = 5 \text{ kg}$ $\quad\quad\quad\quad\quad\quad \mathbf{F} = 8 \text{ N (to the right)}$

$\mathbf{a} = ?$

$$a = \frac{F}{m}$$

$$= \frac{8 \text{ N}}{5 \text{ kg}}$$

$$= \mathbf{1.6 \text{ m/s}^2}$$

$$(\mathbf{a} = \mathbf{1.6 \text{ m/s}^2 \text{ to the right}})$$

---

attached to the block. A frictional force of 2 N acts on the block, a result of contact with the table. What is the total force acting on the block?

Is the total force the numerical sum of the two forces, 10 N plus 2 N or 12 N? Looking at the diagram in figure 4.6 should convince you that this cannot be true. The two forces oppose one another. Because the forces are in opposite directions, the total force is found by subtracting the frictional force from the force applied by the string, resulting in a net force of 8 N. We cannot ignore the directions of the forces involved.

That forces are vectors whose directions must be taken into account when finding the net force is an important aspect of the second law. For forces restricted to one dimension, as in Try This Box 4.1, finding the total force is not difficult. In problems involving forces in two or three dimensions, addition is more complex but can be accomplished using techniques described in appendix C. In this chapter we will only consider one-dimensional cases.

A final point about Newton's first and second laws bears repeating: the first law is a special case of the second law. This can be seen by asking what happens, according to the second law, when the total force acting on an object is zero. In this case, the acceleration $\mathbf{a} = \mathbf{F}/m$ must also be zero. If the acceleration is zero, the velocity must be constant. The first law tells us that if the total force is zero, the object moves with constant velocity (or remains at rest). Newton's first law addresses the special case of the second law in which the total force acting on an object is zero.

The central principle in Newton's theory of motion is his second law of motion. This law states that the acceleration of an object is proportional to total force applied to the object and inversely proportional to the mass of the object. The mass of an object is its inertia or resistance to change in motion. Newton's first law is a special case of the second law when the total force acting on the object is zero. To find the total force acting on the object, we take into account the directions of the individual forces and add them as vectors.

4.3 Mass and Weight

What exactly is weight? Is your *weight* the same as your mass, or is there a difference in the meaning of these two terms? Clearly, mass plays an important role in Newton's second law. *Weight* is a familiar term often used interchangeably with *mass* in everyday language. Here again, physicists make a distinction between mass and weight that is important to Newton's theory.

How can masses be compared?

From the role that mass plays in Newton's second law, we can devise experimental methods of comparing masses. Mass is defined as the property of matter that determines how much an object resists a change in its motion. The greater the mass, the greater the *inertia* or resistance to change, and the smaller the acceleration provided by a given force. Imagine, for example, trying to decelerate a bowling

---

Everyday Phenomenon

box 4.1

The Tablecloth Trick

The Situation. When he was a child, Ricky Mendez saw a magician do the tablecloth trick. A full dinner place setting including a filled wineglass sat on a tablecloth covering a small table. The magician, with appropriate fanfare, pulled the tablecloth from the table without disturbing the dinnerware. Ricky ended up in the doghouse, however, when he tried this at home with disastrous results.

More recently Ricky saw his physics instructor do a similar trick with a simpler place setting. The students were told that the demonstration had something to do with inertia. Why does the trick work, and how is inertia involved? Why did the trick not work when Ricky tried it at home as a child?

The Analysis. The magician's trick, which is frequently used as a physics demonstration, is indeed an illustration of the effects of inertia. Since the nature of frictional forces also plays a role, the choice of a smooth material for the tablecloth is important. (Butcher paper is sometimes substituted in physics demonstrations.) Some practice is usually essential to the successful execution of the trick.

The performer, be it a magician, instructor, or student, must pull the cloth or paper very quickly, giving it a large initial acceleration. Pulling slightly downward across the

edge of the table helps to assure that there is no upward component to the acceleration and that the acceleration is reasonably uniform across the width of the tablecloth. As the tablecloth accelerates, it exerts a frictional force upon the tableware. If we pulled slowly, this frictional force would pull the dishes and glasses along with the tablecloth.

Inertia is the tendency of an object (related to its mass) to resist a change in its motion. When an object is at rest, it remains at rest unless a force is applied. There *is* a force acting on the plates and glasses, however—the frictional force exerted by the tablecloth. If the tablecloth is pulled quickly enough, the frictional force is in effect for only a very short time so the acceleration of the objects is very brief. The objects will accelerate slightly, but not nearly as much as the tablecloth.

There are two aspects of the frictional force that are important to our understanding of what happens. One is that the force of static friction (in effect when the surfaces are not sliding relative to one another) has a maximum value that is determined by the nature of the contacting surfaces and by the force pushing the surfaces together. The second is that once the objects start to slide, kinetic or sliding friction comes into play. The force of kinetic friction is usually smaller than that of static friction.

(continued)

ball and a ping-pong ball that are moving initially with equal velocities (fig. 4.7). A much greater force is required to decelerate the bowling ball than the ping-pong ball because of the difference in mass. According to the second law, the force required is proportional to the mass.

In effect, we are using Newton's second law to define mass. If we used the same force to accelerate different masses, the different accelerations could be used to compare the masses involved. If we choose one mass as a standard, any other mass can be measured against the standard mass by comparing the accelerations produced by equal forces. We could, in principle, determine the mass of any object this way.

How do we define weight?

In practice, the method just described is not convenient for comparing masses because of the difficulty of measuring acceleration. The more common method of comparing masses

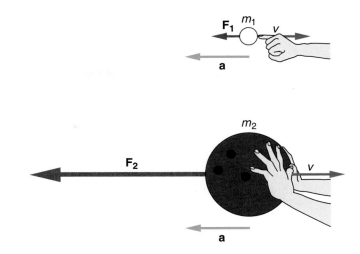

figure 4.7 Stopping a bowling ball and a ping-pong ball. A much larger force is required to produce the same *rate* of change in velocity for the larger mass.

When the tablecloth is given a large lateral acceleration, the force needed to also accelerate the tableware ($F = ma$) exceeds the maximum force of static friction between the dish or glass and the tablecloth. The tablecloth then begins to slide underneath the dish, reducing the size of the frictional force. If the surfaces are smooth, the frictional force is never large enough to produce an acceleration of the dish or glass that is anywhere near the size of the acceleration of the tablecloth. In the fraction of a second that this force acts, it does not have a chance to increase the velocity very much or to move the object very far. (See challenge problem 3.)

You can test these ideas yourself with a pencil, cup, or similar object (preferably nonbreakable) and a sheet of smooth tablet paper. Place the paper on a smooth desk or table surface with the end of the paper extending over the edge. Grasping the paper with both hands near the corners, as shown in the drawing, pull it downward across the edge of the desk or table. Notice that a slow pull brings the object along with the paper, but a very rapid pull leaves the object essentially in place. (The objects will usually move slightly in the direction of the pull.)

Before you graduate to tablecloths and full dinner place settings, a few cautions are in order. Objects that can tip, like filled wineglasses, are more difficult to work with. The bottom may start to move while the top portion (with its greater inertia) remains in place causing the glass to tip and spill the

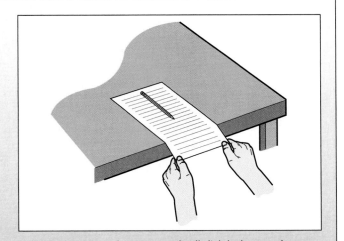

Grasp the paper near the corners and pull slightly downward across the edge of the table. A quick pull will leave the pencil near its initial position.

wine or water. Also, the larger the tablecloth, the more difficult it is to pull it clear of the table—your hands must move very rapidly through a large distance in the pull. Practice is essential, which is the case for most of the tricks that magicians (and physics instructors) perform.

is to "weigh" the objects on a balance or scale (fig. 4.8). What we actually do in weighing is to compare the gravitational force acting on the mass we wish to measure with that acting on some standard mass. The gravitational force acting on an object is the **weight** of the object. As a force, weight has different units (newtons) than mass (kilograms).

How is weight related to mass? From our discussion of gravitational acceleration in chapter 3, we know that objects of different mass experience the same gravitational acceleration near the earth's surface ($g = 9.8$ m/s²). This acceleration is caused by the gravitational force exerted by the earth on the object, which is the weight of the object. By Newton's second law, the force (the weight) is equal to the mass times the acceleration or

$$\mathbf{W} = m\mathbf{g}.$$

The symbol **W** represents the weight. It is a vector whose direction is straight down toward the center of the earth.

If we know the mass of an object, we can then compute its weight. An example is provided in Try This Box 4.2, where we show that a woman with a mass of 50 kg has a weight of 490 N. Since we are more used to expressing weights in the English system, we also convert her weight in newtons to pounds (lb), which yields a weight of 110 lb. The pound is most commonly used as a unit of *force*, not mass, in the English system. A mass of 1 kg weighs approximately 2.2 lb near the surface of the earth.

Although weight is proportional to mass, it also depends on the gravitational acceleration g. Since g varies slightly from place to place on the surface of the earth—and has a much smaller value on the moon or the smaller planets—the weight of an object clearly depends on where that object is. On the other hand, the mass of an object is a property of the object related to the quantity of matter making up that object and does not depend on the location of the object.

The gravitational acceleration on the moon is approximately one-sixth that on the surface of the earth. If we transported the woman whose weight we have just determined to the moon, her weight would decrease to about 18 lb (or 82 N), one-sixth her weight on earth. The woman's mass would still be 50 kg, provided that the trip did not take too much out of her. The mass of an object changes only if we add or subtract matter from it.

Why is the gravitational acceleration independent of mass?

The distinction between weight and mass can provide insight into why the gravitational acceleration is independent of mass. Let's turn to the case of a falling object and consider its motion using Newton's second law. Reversing the argument that we used in defining weight, we use the gravitational force (the weight) to determine the acceleration. By Newton's second law, the acceleration can be found by dividing the force ($\mathbf{W} = m\mathbf{g}$) by the mass:

$$\mathbf{a} = \frac{m\mathbf{g}}{m} = \mathbf{g}.$$

Mass cancels out of the equation when we compute the acceleration for a falling object. The gravitational force is proportional to the mass, but by Newton's second law, the acceleration is inversely proportional to the mass: these two effects cancel one another. This only holds true for falling objects. In most other cases, the net force does not depend directly on the mass.

Force and acceleration are *not* the same, although they are closely related by Newton's second law. A heavy object experiences a larger gravitational force (its weight) than a

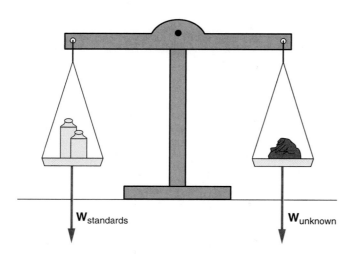

figure 4.8 Comparing an unknown mass to standard masses on a balance.

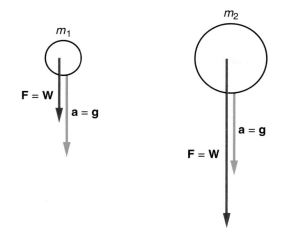

figure 4.9 Different gravitational forces (weights) act on falling objects of different masses, but because acceleration is inversely proportional to mass, the objects have the same acceleration.

lighter object, but the two objects will have the same gravitational acceleration (fig. 4.9). Because the gravitational force is proportional to mass, we find the same acceleration for different masses. The gravitational force will be discussed further in chapter 5 when we take up Newton's law of gravitation, a critical piece of his overall theory of motion.

Weight and mass are not the same. Weight is the gravitational force acting on an object, and mass is an inherent property related to the amount of matter in the object. Near the surface of the earth, weight is equal to the mass multiplied by the gravitational acceleration ($W = mg$), but the weight would change if we took the object to another planet where g has a different value. The reason that all objects experience the same gravitational acceleration near the earth's surface is that the gravitational force is proportional to the mass of the object, but acceleration is equal to the force divided by the mass.

4.4 Newton's Third Law

Where do forces come from? If you push on a chair to move it across the floor, does the chair also push back on you? If so, how does that push affect your own motion? Questions like these are important to what we mean by *force*. Newton's third law provides some answers.

Newton's third law of motion is an important part of his definition of force. It is an essential tool for analyzing the motion or lack of motion of real objects, but it is often misunderstood. For this reason, it is good to take a careful look at the statement and use of the third law.

How does the third law help us to define force?

If you push with your hand against a large chair or any large object, such as the wall of your room, you will feel the object push back against your hand. A force is acting on your hand that you can sense as it compresses your hand. Your hand is interacting with the chair or wall, and that object pushes back against your hand as you push against the object.

Newton's third law contains the idea that forces are caused by such interactions of two objects, each exerting a force on the other. It can be stated as

> If object A exerts a force on object B, object B exerts a force on object A that is equal in magnitude but opposite in direction to the force exerted on B.

The third law is sometimes referred to as the **action/ reaction principle**—for every action there is an equal but opposite reaction. Note that the two forces always act on two *different* objects, never on the same object. Newton's definition of force includes the idea of an *interaction* between objects. The forces represent that interaction.

If you exert a force $\mathbf{F}_1$ on the chair with your hand, the chair pushes back on your hand with a force $\mathbf{F}_2$ that is equal in size, but opposite in direction (fig. 4.10). Using this notation, Newton's third law can be stated in symbolic form as

$$\mathbf{F}_2 = -\mathbf{F}_1.$$

The minus sign indicates that the two forces have opposite directions. The force $\mathbf{F}_2$ acts on your hand and partly determines your own motion, but it has nothing to do with the motion of the chair. Of this pair of forces, the only one that

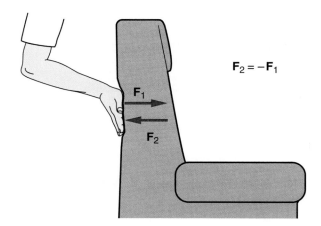

figure 4.10 The chair pushes back on the hand with a force $\mathbf{F}_2$ that is equal in size but opposite in direction to the force $\mathbf{F}_1$ exerted by the hand on the chair.

affects the motion of the chair is the one acting on the chair, $\mathbf{F}_1$.

Our definition of force is now complete. Newton's second law tells us how the motion of an object is affected by a force, and his third law tells where forces come from. They come from interactions with other objects. With a suitable definition of mass, which also depends upon the second law, we know how to measure the size of forces by determining the acceleration that they produce ($\mathbf{F} = m\mathbf{a}$). Both the second and third laws are necessary to define what we mean by *force*.

How can we use the third law to identify forces?

How do we identify the forces that act on an object to analyze how that object will move? First, we identify other objects that interact with the object of interest. Consider a book lying on a table (fig. 4.11). What objects are interacting with the book? Since it is in direct contact with the table, the book must be interacting with the table, but it also interacts with the earth through the gravitational attraction.

The downward pull of gravity that the earth exerts on the book is the book's weight $\mathbf{W}$. The object interacting with the book to produce this force is the earth itself. The book and the earth are attracted to one another (through gravity) with equal and opposite forces that form a third-law pair. The earth pulls down on the book with the force $\mathbf{W}$, and the book pulls upward on the earth with the force $-\mathbf{W}$. Because of the enormous mass of the earth, the effect of this upward force on the earth is extremely small.

The second force acting on the book is an upward force exerted on the book by the table. This force is often called the **normal force,** where the word *normal* means "perpendicular" rather than "usual." The normal force $\mathbf{N}$ is always perpendicular to the surfaces of contact. The book, in turn, exerts an equal but oppositely directed downward force $-\mathbf{N}$ on the table. These two forces, $\mathbf{N}$ and $-\mathbf{N}$, constitute another third-law pair. They result from the mutual compression of the book and table as they come into contact with one another. You could think of the table as a large and very stiff spring that compresses ever so slightly when the book is placed on it (fig. 4.12).

The two forces acting on the book, the force of gravity and the force exerted by the table, also happen to be equal in size and opposite to one another, but this is *not* due to the third law. How do we know that they must be equal? Since the book's velocity is not changing, its acceleration must be zero. According to Newton's *second* law, the total force F acting on the book must then be zero, since $\mathbf{F} = m\mathbf{a}$ and the acceleration $\mathbf{a}$ is zero. The only way that the total force can be zero is for the two contributing forces, $\mathbf{W}$ and $\mathbf{N}$, to cancel one another. They must be equal in magnitude and opposite in direction for their sum to be zero.

Even though equal in size and opposite in direction, these two forces do not constitute a third-law action/reaction pair. They both act on the *same* object, the book, and the third law always deals with interactions between *different* objects. So, $\mathbf{W}$ and $\mathbf{N}$ are equal in size and opposite in direction in this case as a consequence of the second law rather than the third law. If they did not cancel one another, the book would accelerate away from the tabletop. (Both the second and third laws are critical to the analysis of the elevator example in Everyday Phenomenon Box 4.2.)

Can a mule accelerate a cart?

Consider the story of the stubborn mule who, having had a brief exposure to physics, argued to his handler that there was no point in pulling on the cart to which he was connected. According to Newton's third law, the mule argued, the harder he pulls on the cart, the harder the cart pulls back on him (fig. 4.13). The net result is, therefore, nothing. Is he right, or is there a fallacy in his argument?

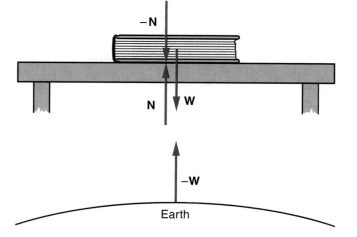

figure 4.11   Two forces, $\mathbf{N}$ and $\mathbf{W}$, act on a book resting on a table. The third-law reaction forces $-\mathbf{N}$ and $-\mathbf{W}$ act on different objects, the table and the earth.

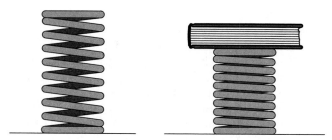

figure 4.12   An uncompressed spring and the same spring supporting a book. The compressed spring exerts an upward force on the book.

Everyday Phenomenon

box 4.2

Riding an Elevator

The Situation. We have all had the experience of riding an elevator and feeling sensations of heaviness or lightness as the elevator accelerates up or down. The feeling of lightness as the elevator accelerates downward is generally more striking, particularly if the acceleration is not smooth.

Do we really weigh more or less than usual in these situations? If you took a bathroom scale into the elevator, would it read your true weight when the elevator is accelerating? How can we apply Newton's laws of motion to explore these questions?

A woman standing on a bathroom scale inside an accelerating elevator. Will she read her true weight on the scale?

The Analysis. The first step in analyzing any situation using Newton's laws is to isolate the body of interest and carefully identify the forces that act on just that body. Different choices are possible for which objects to isolate, but some choices will be more productive than others. In this case, it makes sense to isolate the person standing on the scale, since her weight is the focus of our questions. The second drawing shows a **free-body diagram** of the woman indicating just those forces that act on her.

In this case, just two other objects interact with the woman, resulting in two forces. The earth pulls downward on the woman through the force of gravity **W**. The scale pushes upward on her feet with a force **N**, the normal force. The vector sum of these two forces determines her acceleration. If the elevator is accelerating upward with an acceleration **a**, the woman must also be accelerating upward at that rate. The total force must also be upward, which implies that the normal force **N** is larger than the gravitational force **W**. Using signs to indicate direction, and letting the positive direction be upward, Newton's second law requires that

$$F = N - W = ma.$$

What about the scale reading? By Newton's third law, the woman exerts a downward force on the scale equal in size to the normal force **N**, but opposite in direction. Since this is the force pushing down on the scale, the scale should read the value *N*, the magnitude of the normal force. The woman's true weight has not changed, but her apparent weight as measured by the scale has increased by an amount equal to *ma*. (Rearranging the second-law equation yields **N** = **W** + *m***a**.)

What happens when the elevator is accelerating downward? In that case, the total force acting upon the woman must be downward, and the normal force must be less than her weight. The scale reading *N* will then be less than the woman's true weight by the amount *ma*, perhaps producing a smile rather than a scowl.

If the elevator cable breaks, we have a particularly interesting special case. Both the woman and the elevator will accelerate downward with the gravitational acceleration **g**. Since the woman's weight is all that is required to give her that acceleration, the normal force acting on her feet must then be zero. The scale reading will likewise be zero, and the woman is apparently weightless!

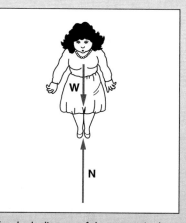

A free-body diagram of the woman in the elevator. Why is the normal force **N** larger than the weight **W**?

The sensation of our own weight is produced in part by the pressure on our feet and forces in our leg muscles needed to maintain our posture. The woman will feel weightless in this situation even though her true weight (the gravitational force acting on her) has not changed. In fact, she would be able to float around in the elevator as the astronauts do in the orbiting space shuttle. (The space shuttle is also falling toward the earth as it moves laterally in its orbit.) This happy scenario will come to a crashing halt for the woman, however, when the elevator reaches the bottom of the shaft.

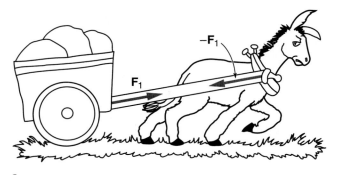

figure **4.13**   A mule and a cart. Does Newton's third law prevent the mule from moving the cart?

figure **4.14**   The car pushes against the road, and the road, in turn, pushes against the car.

The fallacy is simple but perhaps not obvious. The motion of the cart is affected by only one of the two forces that the mule is talking about, namely, the force that acts on the cart. The other force in this third-law pair acts on the mule and must be considered in conjunction with other forces that act on the mule to determine how he will move. The cart will accelerate if the force exerted by the mule on the cart is larger than the frictional forces acting on the cart. Try placing yourself in the role of the handler and explain the fallacy to the mule.

What force causes a car to accelerate?

As with the mule, the **reaction force** to a push or pull exerted by an object is often extremely important in describing the motion of the object itself. Consider the acceleration of a car. The engine cannot push the car because it is part of the car. The engine drives either the rear or front axle of the car, which causes the tires to rotate. The tires in turn push against the road surface through the force of friction **f** between the tires and the road (fig. 4.14).

According to Newton's third law, the road must then push against the tires with an equal but oppositely directed force $-$**f**. This external force causes the car to accelerate. Obviously, friction is desirable in this case. Without friction, the tires would spin, and the car would go nowhere. The case of the mule is similar. The frictional force exerted by the ground on his hooves causes him to accelerate forward. This frictional force is the reaction to his pushing against the ground.

Think about this next time you find yourself walking. What external force causes you to accelerate as you start out? What is your role and that of friction in producing this force? How would you walk on an icy or slippery surface?

To figure out what forces are acting on any object, we need first to identify the other objects with which it is interacting. Some of these will be obvious. Any object in direct contact with the object of interest will presumably contribute a force. Interactions producing other forces, such as air resistance or gravity, may be less obvious but still

recognizable with a little thought. The third law is the principle we use to identify any of these forces.

Newton's third law of motion completes his definition of force. The third law notes that forces arise from interactions between different objects. If object A exerts a force on object B, object B exerts an equal-size but oppositely directed force on A. We use the third law to identify the external forces that act on an object in order to apply the second law of motion.

4.5 Applications of Newton's Laws

We have now introduced Newton's laws of motion and discussed the definitions of force and mass within these laws. To appreciate their usefulness, however, we must be able to apply them to some familiar examples such as pushing a chair or throwing a ball. How do Newton's laws help us make sense of these motions? Do they provide a satisfactory picture of what is going on?

What forces are involved in moving a chair?

We have returned from time to time to the example of a chair being pushed but have not yet analyzed how and why it moves. As we indicated in the previous section, the first step in any analysis is to identify the forces that act on the chair. As shown in figure 4.15, four forces act on the chair from four separate interactions:

1. The force of gravity (the weight) **W** due to interaction with the earth.
2. The upward (normal) force **N** exerted by the floor due to compression of the floor.
3. The force exerted by the hand of the person pushing, **P.**
4. The frictional force **f** exerted by the floor.

Two of these forces, the normal force **N** and the frictional force **f,** are actually due to interactions with a single object, the floor. Since they are due to different effects and

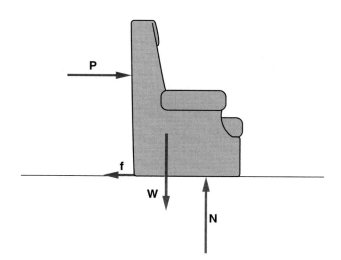

figure 4.15 Four forces act on a chair being pushed across the floor, the weight **W**, the normal force **N**, the force **P** exerted by the person pushing, and the frictional force **f**.

are perpendicular to one another, they are usually treated separately.

The effects of the two vertical forces acting on the chair, the weight **W** and the normal force **N,** cancel one another. Like the book on the table in section 4.4, this results because there is no acceleration of the chair in the vertical direction. By Newton's second law, the sum of the vertical forces must then be zero, which implies that the weight **W** and the normal force **N** are equal in size but opposite in direction. They play no direct role in the horizontal motion of the chair.

The other two forces, the push of the hand **P** and the frictional force **f,** do not necessarily cancel. These two forces together determine the horizontal acceleration of the chair. The push **P** must be larger than the frictional force **f** for the chair to accelerate. In the most likely scenario for moving the chair, you first give a push with your hand that is larger than the frictional force. This produces a total force, with magnitude $P - f$, in the forward direction, causing the chair to accelerate.

Once you have accelerated the chair to a reasonable velocity, you reduce the strength of your push **P** so that it is equal in size to the frictional force. The net horizontal force becomes equal to zero, and the horizontal acceleration is also zero by Newton's second law. If you sustain the push at this level, the chair moves across the floor with constant velocity.

Finally, you remove your hand and its push **P**, and the chair quickly decelerates to zero velocity under the influence of the frictional force **f**. If you happen to have a chair and a smooth floor handy, try to produce the motion that we have just been describing. See if you can feel differences in the force that you are exerting with your hand at various points in the motion. The force should be largest at the beginning of the motion.

The size of the force, needed to keep the chair moving with constant velocity is determined by the strength of the frictional force, which, in turn, is influenced by the weight of the chair and the condition of the floor surface. If you fail to recognize the importance of the frictional force, you may be led, like Aristotle, to think that a force is always needed to keep an object moving. Frictional forces are almost always present, but they are not as obvious as the forces applied directly.

Does a sky diver continue to accelerate?

In chapter 3, we considered the fact that an object falls with constant acceleration **g** if air resistance is not a significant factor. What about objects such as sky divers who fall for large distances? Do they continue to accelerate at this rate gaining larger and larger downward velocities? Any person with experience in sky diving knows that this does not happen. Why not?

If air resistance were not a factor, a falling object would experience only the gravitational force (its weight) and would indeed continue to accelerate. In sky diving, air resistance is an important factor, and its effects get larger as the velocity of the sky diver (or any object) increases. The sky diver has an initial acceleration of **g,** but as her velocity increases, the force of air resistance becomes significant. Her acceleration decreases (fig. 4.16).

For small velocities, the air-resistive force **R** is small, and the weight is the dominant force. As the velocity increases, the air-resistive force gets larger, causing the total magnitude of the downward force, $W - R$, to decrease. Since the total force is responsible for the acceleration, the acceleration will also decrease. Ultimately, as the velocity continues to increase, the air-resistive force reaches a value equal in size to the gravitational force. The net force is then zero, and the sky diver stops accelerating. We say that she has reached **terminal velocity,** and from there on, she moves downward with constant velocity. This terminal velocity is usually between 100 and 120 MPH.

Frictional or resistive forces play a critical role in analyzing the motion. Aristotle did not have the opportunity to try sky diving (nor have many of us), so this example was not a part of his experience. He did observe the terminal velocity, however, of very light objects such as feathers or leaves. The weight of such objects is small and the surface area is large relative to the weight, so the air-resistive force **R** becomes equal in size to the weight much sooner than for a heavier object.

Try tearing a small corner from a piece of paper and watching it fall. Does it appear to reach a constant (terminal) velocity? It will flutter as it falls, but it does not seem to accelerate much for most of its downward motion. You can see why Aristotle concluded that heavier objects fall faster than lighter objects. Dropping heavier objects through water can also show the terminal velocity. Water exerts a larger resistive force at lower velocities than air.

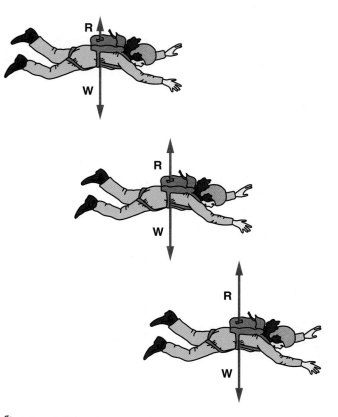

figure **4.16**   The force of air resistance **R** acting on a sky diver increases as the velocity increases.

What happens when a ball is thrown?

Aristotle had trouble explaining the motion of a thrown object such as a ball, once it had left the thrower's hand. Let's reconsider this example from a Newtonian perspective. Do we need a force to keep the ball moving? Not according to Newton's first law. Three forces, however, are involved in the flight of the ball: the initial push by the thrower, the downward pull of gravity, and (once again) air resistance (fig 4.17).

To highlight Newton's approach, it is best to break the motion down into two different spans of time. The first is the process of throwing, when the hand is in contact with the ball. During this interval, the force **P** exerted by the hand dominates the motion. The combined effects of the other forces (gravity and air resistance) must be smaller than the force **P** if the ball is to accelerate. Thus **P** acceler-

ates the ball to a velocity that we often refer to as the *initial* velocity. The magnitude and direction of the initial velocity are determined by the strength and direction of the force **P** and the length of time that it acts on the ball. Since this force usually varies with time, a full analysis of the process of throwing gets quite complex.

Once the ball leaves the hand, however, we are in the second time period, where **P** is no longer a consideration. During this interval, the gravitational force **W** and the air-resistive force **R** produce changes in the ball's velocity. From this point on, the problem becomes one of projectile motion (section 3.4). The gravitational force accelerates the ball downward, and the air-resistive force acts in a direction opposite to the velocity, gradually reducing the ball's velocity.

Contrary to Aristotle's view, no forces are needed to keep the ball moving once it has been thrown. In fact, if an object is thrown in deep space, where air resistance is negligible and gravitational forces are very weak, it would keep moving with constant velocity, as stated in Newton's first law. So, be careful with your tools when you are working in space outside of your spacecraft.

Because the air-resistive force or the push exerted by a person throwing a ball varies with time, we have avoided working out numerical examples for these situations. Just identifying the forces involved and their causes due to third-law interactions with other objects provides a useful description of what is happening.

How do we analyze the motion of connected objects?

Verification of Newton's laws of motion came initially from simpler examples that can be easily set up in the laboratory. One example not difficult to picture and set up in a physics laboratory (or even at home if suitable toys are available) is two connected carts accelerated by the pull of a string (fig. 4.18). To keep things simple, we will assume that the carts have excellent wheel bearings, so that they roll with very little friction. We will also assume that a scale is available to determine the masses of the carts and their contents.

To measure the magnitude of the force applied by the string, we would have to insert a small spring balance somewhere between the hand and the carts. The trickiest part of the entire experiment is applying a steady force with this arrangement while the carts are accelerating.

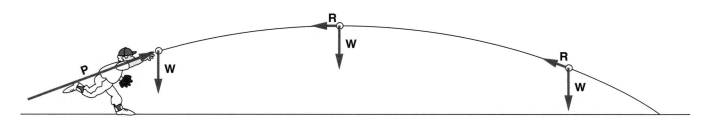

figure **4.17**   Three forces act on a thrown ball, the initial push **P**, the weight **W**, and air resistance **R**.

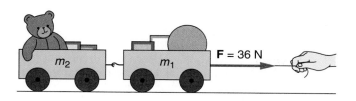

figure 4.18 Two connected carts being accelerated by a force **F** applied by a string.

If we know the masses of the carts and their contents, and the magnitude of the force applied by the string, we should be able to predict the value of the acceleration of the system from Newton's second law. (See Try This Box 4.3.) For the masses given, and an applied force of 36 N, we find an acceleration of 2.0 m/s^2 for the two carts. The acceleration could be verified experimentally by measuring the time required for the carts to travel a fixed distance and using the equations developed for constant acceleration in chapter 2 to calculate an experimentally determined value.

In Try This Box 4.3, we first treated the two carts as a single system to find the acceleration. Suppose, however, that we wanted to know the magnitude of the force exerted by the hooks connecting the two carts. In this case, it makes sense to treat the motion of the individual carts separately. Once we know the acceleration, we again apply Newton's second law to find the total force acting on each cart. This computation is done in the second part of Try This Box 4.3 and is illustrated in figure 4.19.

For the second cart, a force of 16 N is required to produce the acceleration of 2 m/s^2. By Newton's third law, there should then be a force of 16 N pulling back on the first cart. Combined with the forward force of 36 N applied by the string, this results in a total force of 20 N acting on the first cart (36 N − 16 N). This is exactly the value required to give the first cart an acceleration of 2 m/s^2.

From this example, we see that Newton's laws provide a completely consistent picture of the forces and accelerations of the different parts of the connected-cart system. This is a necessary condition for us to accept the laws as valid. Obviously, another condition is that any predictions be confirmed by experimental measurements. This has been done many times over by experiments similar to the one we have dealt with here.

We could try many variations on this experiment in the laboratory to see if the results agree with predictions derived from Newton's laws. Even with careful experimental technique using accurate stopwatches and balances, however, our results are unlikely to agree exactly with our predictions. It is impossible to eliminate the effects of friction completely, and none of our measurements can be made with infinite precision. The art of the experimentalist is to reduce these inaccuracies to a minimum as well as to predict how they affect our results.

try this box 4.3

Sample Exercise: Connected Objects

Two connected carts are pulled across the floor under the influence of a force of 36 N applied by a string (fig. 4.18). The forward cart and its contents have a mass of 10 kg, and the second cart and contents have a mass of 8 kg. Assuming that frictional forces are negligible:
    a. What is the acceleration of the two carts?
    b. What is the total force acting on each cart?

a. $m_1 = 10$ kg $F = ma$
    $m_2 = 8$ kg
    $F = 36$ N   or: $a = \dfrac{F}{m} = \dfrac{36 \text{ N}}{10 \text{ kg} + 8 \text{ kg}}$
    a = ?

$$= \frac{36 \text{ N}}{18 \text{ kg}} = \mathbf{2.0 \text{ m/s}^2}$$

    a = 2.0 m/s^2 in the forward direction

b. $F = ?$ first cart
   (for each cart) $F = ma$
                           $= (10 \text{ kg})(2 \text{ m/s}^2)$
                           $= \mathbf{20 \text{ N}}$

               second cart
                 $F = ma$
                           $= (8 \text{ kg})(2 \text{ m/s}^2)$
                           $= \mathbf{16 \text{ N}}$

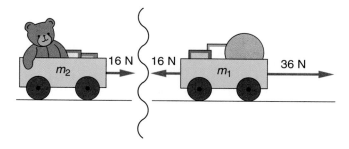

figure 4.19 The interaction between the two carts illustrates Newton's third law.

Newton's laws of motion provide both qualitative and quantitative explanations of any familiar motion. First, we identify the forces acting on the object by examining interactions with other objects. The relative sizes of these forces, when added together, give the acceleration of the object. The acceleration may change as the forces change with time, as in the case of a sky diver. Newton's laws have been verified many times over by experimental tests of their quantitative predictions. They are a much more consistent theory of the causes of motion than the older Aristotelian view.

summary

In 1685, Newton published his *Principia,* in which he introduced three laws of motion as the foundation of his theory of mechanics. These laws continue to serve as an extremely useful model for explaining the causes of motion and for predicting how objects will move in many familiar situations.

1   A brief history. Newton's theory was constructed on groundwork laid by Galileo and replaced a much earlier and less quantitative model developed by Aristotle to explain motion. Newton's theory had much greater predictive power than Aristotle's ideas. Although we now recognize its limitations, Newton's theory is still used extensively to explain the motion of ordinary objects.

2   Newton's first and second laws. Newton's second law states that the acceleration of an object is proportional to the total external force acting on that object and inversely proportional to the mass of the object. The first law, a *special case* of the second law, describes what happens when the total force is zero. The acceleration must then be zero, and the object moves with constant velocity.

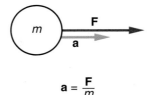

$$a = \frac{F}{m}$$

3   Mass and weight. Newton's second law defines the inertial mass of an object as the property that causes the object to resist a change in its motion. The weight of an object is the gravitational force acting on the object and is equal to the mass multiplied by the gravitational acceleration **g.** The weight of an object may vary as **g** varies, but mass is an inherent property of the object related to its quantity of matter.

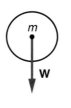

$$W = m\mathbf{g}$$

4   Newton's third law. Newton's third law completes the definition of force by showing that forces result from interactions between objects. If object A exerts a force on object B, then object B exerts an equal-size but oppositely directed force on object A.

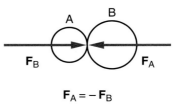

$$\mathbf{F}_A = -\mathbf{F}_B$$

5   Applications of Newton's laws. In analyzing the motion of an object using Newton's laws, the first step is to identify the forces that act on the object due to interactions with other objects. The strength and direction of the total force then determine how the object's motion will change.

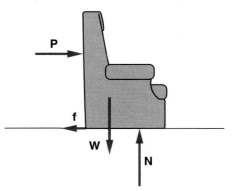

key terms

Force, 61
Newton's first law of motion, 62
Frictional force, 62
Newton's second law of motion, 62
Mass, 63

Inertia, 63
Newton (unit of force), 63
Net force, 63
Weight, 66
Newton's third law of motion, 67

Action/reaction principle, 67
Normal force, 68
Free-body diagram, 69
Reaction force, 70
Terminal velocity, 71

questions

*Questions identified with an asterisk are more open-ended than the others. They call for lengthier responses and are more suitable for group discussion.

Q1. Did Galileo's work on motion precede in time that of Aristotle or Newton? Explain.

Q2. Why did Aristotle believe that heavier objects fall faster than lighter objects? Explain.

Q3. Aristotle believed that a force was necessary to keep an object moving. Where, in his view, did this force come from in the case of a ball moving through the air? Explain.

Q4. Was Galileo aware of Aristotle's basic theory of motion? Explain.

*Q5. How did Aristotle explain the continued motion of a thrown object. Does this explanation seem reasonable to you? Explain.

Q6. Did Galileo develop a more complete theory of motion than that of Newton? Explain.

Q7. Two equal forces act on two different objects, one of which has a mass ten times as large as the other. Will the more massive object have a larger acceleration, an equal acceleration, or a smaller acceleration than the less massive object? Explain.

Q8. A 3-kg block is observed to accelerate at a rate twice that of a 6-kg block. Is the net force acting on the 3-kg block therefore twice as large as that acting on the 6-kg block? Explain.

Q9. Two equal-magnitude horizontal forces act on a box as shown in the diagram. Is the object accelerated horizontally? Explain.

Q10. Is it possible that the object pictured in question 9 is moving, given the fact that the two forces acting on it are equal in size but opposite in direction? Explain.

Q11. Suppose that a bullet is fired from a rifle in outer space where there are no appreciable forces due to gravity or air resistance acting on the bullet. Will the bullet slow down as it travels away from the rifle? Explain.

Q12. Two equal forces act on an object in the directions pictured in the diagram below. If these are the only forces involved, will the object be accelerated? Explain, using a diagram.

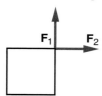

Q13. An object moving horizontally across a table is observed to slow down. Is there a non-zero total force acting on the object? Explain.

Q14. A car goes around a curve traveling at constant speed.
a. Is the acceleration of the car zero in this process? Explain.
b. Is there a non-zero total force acting on the car? Explain.

*Q15. Is Newton's first law of motion explained by the second law? Explain. Why did Newton state the first law as a separate law of motion?

Q16. Is the mass of an object the same thing as its weight? Explain.

Q17. The gravitational force acting on a lead ball is much larger than that acting on a wooden ball of the same size. When both are dropped, does the lead ball accelerate at the same rate as the wooden ball? Explain, using Newton's second law of motion.

Q18. The acceleration due to gravity on the moon is approximately one-sixth the gravitational acceleration near the earth's surface. If a rock is transported from the earth to the moon, will either its mass or its weight change in the process? Explain.

Q19. Is mass a force? Explain.

Q20. Two identical cans, one filled with lead shot and the other with feathers, are dropped from the same height by a student standing on a chair.
a. Which can, if either, experiences the greater force due to the gravitational attraction of the earth? Explain.
b. Which can, if either, experiences the greater acceleration due to gravity? Explain.

Q21. A boy sits at rest on the floor. What two vertical forces act upon the boy? Do these two forces constitute an action/reaction pair as defined by Newton's third law of motion? Explain.

Q22. The engine of a car is part of the car and cannot push directly on the car in order to accelerate it. What external force acting on the car is responsible for the acceleration of the car on a level road surface? Explain.

Q23. It is difficult to stop a car on an icy road surface. Is it also difficult to accelerate a car on this same icy road? Explain.

Q24. A ball hangs from a string attached to the ceiling, as shown in the diagram.
a. What forces act on the ball? How many are there?
b. What is the total (net) force acting on the ball? Explain.
c. For each force identified in part (a), what is the reaction force described by Newton's third law of motion?

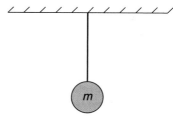

*Q25. When a magician performs the tablecloth trick, the objects on the table do not move very far. Is there a horizontal force acting on these objects while the tablecloth is being pulled off the table? Why do the objects not move very far? Explain.

Q26. A sprinter accelerates at the beginning of a 100-meter race and then tries to maintain maximum speed throughout the rest of the race.
   a. What external force is responsible for accelerating the runner at the beginning of the race? Explain carefully how this force is produced.
   b. Once the runner reaches her maximum velocity, is it necessary to continue pushing against the track in order to maintain that velocity? Explain.

Q27. A mule is attempting to move a cart loaded with rock. Since the cart pulls back on the mule with a force equal in size to the force that the mule exerts on the cart (according to Newton's third law), is it possible for the mule to accelerate the cart? Explain.

Q28. A toy battery-powered tractor pushes a book across a table. Draw separate diagrams of the book and the tractor identifying all of the forces that act upon each object. What is the reaction force described by Newton's third law of motion for each of the forces that you have drawn?

Q29. The upward normal force exerted by the floor on a chair is equal in size but opposite in direction to the weight of the chair. Is this equality an illustration of Newton's third law of motion? Explain.

Q30. Two masses, m_1 and m_2, connected by a string, are placed upon a fixed frictionless pulley as shown in the diagram. If m_2 is larger than m_1, will the two masses accelerate? Explain.

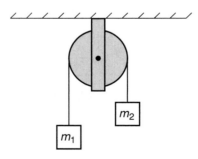

Q31. Two blocks with the same mass are connected by a string and are pulled across a frictionless surface by a constant force, **F,** exerted by a string (see diagram).
   a. Will the two blocks move with constant velocity? Explain.
   b. Will the tension in the connecting string be greater than, less than, or equal to the force **F**? Explain.

*Q32. Suppose that a sky diver wears a specially lubricated suit that reduces air resistance to a small constant force that does not increase as the diver's velocity increases. Will the sky diver ever reach a terminal velocity before opening her parachute? Explain.

Q33. If you get into an elevator on the top floor of a large building and the elevator begins to accelerate downward, will the normal force pushing up on your feet be greater than, equal to, or less than the force of gravity pulling downward on you? Explain.

Q34. If the elevator cable breaks and you find yourself in a condition of apparent weightlessness as the elevator falls, is the gravitational force acting upon you equal to zero? Explain.

exercises

E1. A single force of 40 N acts upon a 5-kg block. What is the magnitude of the acceleration of the block?

E2. A ball with a mass of 2.5 kg is observed to accelerate at a rate of 6.0 m/s². What is the size of the net force acting on this ball?

E3. A net force of 20 N acting on a wooden block produces an acceleration of 4.0 m/s² for the block. What is the mass of the block?

E4. A 2.5-kg block being pulled across a table by a horizontal force of 80 N also experiences a frictional force of 5 N. What is the acceleration of the block?

E5. A pulled tablecloth exerts a frictional force of 0.6 N on a plate with a mass of 0.4 kg. What is the acceleration of the plate?

E6. A 6-kg block being pushed across a table by a force **P** has an acceleration of 3.0 m/s^2.
   a. What is the net force acting upon the block?
   b. If the magnitude of **P** is 20 N, what is the magnitude of the frictional force acting upon the block?

E7. Two forces, one of 50 N and the other of 30 N, act in opposite directions on a box as shown in the diagram. What is the mass of the box if its acceleration is 4.0 m/s^2?

E8. A 5-kg block is acted upon by three horizontal forces as shown in the diagram.
   a. What is the net horizontal force acting on the block?
   b. What is the horizontal acceleration of the block?

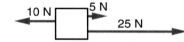

E9. What is the weight of a 40-kg mass?

E10. What is the mass of a 540-N weight?

E11. Jennifer has a weight of 110 lb.
   a. What is her weight in newtons? (1 lb = 4.45 N)
   b. What is her mass in kilograms?

E12. The author of this text has a weight of 600 N.
   a. What is his mass in kilograms?
   b. What is his weight in pounds? (1 lb = 4.45 N)

E13. Who has the larger mass, a man weighing 145 lb or one weighing 735 N?

E14. At a given instant in time, a 4-kg rock that has been dropped from a high cliff experiences a force of air resistance of 15 N. What are the magnitude and direction of the acceleration of the rock? (Do not forget the gravitational force!)

E15. At a given instant in time, a 5-kg rock is observed to be falling with an acceleration of 7.0 m/s^2. What is the magnitude of the force of air resistance acting upon the rock at this instant?

E16. A 0.4-kg book rests on a table. A downward force of 6 N is exerted on the top of the book by a hand pushing down on the book.
   a. What is the magnitude of the gravitational force acting upon the book?
   b. What is the magnitude of the upward (normal) force exerted by the table on the book? (Is the book accelerated?)

E17. An upward force of 18 N is applied via a string to lift a ball with a mass of 1.5 kg.
   a. What is the net force acting upon the ball?
   b. What is the acceleration of the ball?

E18. A 60-kg woman in an elevator is accelerating upward at a rate of 1.2 m/s^2.
   a. What is the net force acting upon the woman?
   b. What is the gravitational force acting upon the woman?
   c. What is the normal force pushing upward on the woman's feet?

challenge problems

CP1. A constant horizontal force of 30 N is exerted by a string attached to a 5-kg block being pulled across a tabletop. The block also experiences a frictional force of 5 N due to contact with the table.
   a. What is the horizontal acceleration of the block?
   b. If the block starts from rest, what will its velocity be after 3 seconds?
   c. How far will it travel in these 3 seconds?

CP2. A rope exerts a constant horizontal force of 250 N to pull a 60-kg crate across the floor. The velocity of the crate is observed to increase from 1 m/s to 3 m/s in a time of 2 seconds under the influence of this force and the frictional force exerted by the floor on the crate.
   a. What is the acceleration of the crate?
   b. What is the total force acting upon the crate?
   c. What is the magnitude of the frictional force acting on the crate?
   d. What force would have to be applied to the crate by the rope in order for the crate to move with constant velocity? Explain.

CP3. A dish with a mass 0.4 kg has a force of kinetic friction of 0.15 N exerted on it by a moving tablecloth for a time of 0.2 s.
   a. What is the acceleration of the dish?
   b. What velocity does it reach in this time, starting from rest?
   c. How far (in cm) does the dish move in this time?

CP4. A 60-kg crate is lowered from a loading dock to the floor using a rope passing over a fixed support. The rope exerts a constant upward force on the crate of 500 N.
   a. Will the crate accelerate? Explain.
   b. What are the magnitude and direction of the acceleration of the crate?
   c. How long will it take for the crate to reach the floor if the height of the loading dock is 1.4 m above the floor?
   d. How fast is the crate traveling when it hits the floor?

CP5. Two blocks tied together by a horizontal string are being pulled across the table by a horizontal force of 30 N as shown. The 2-kg block has a 6-N frictional force exerted on it by the table, and the 4-kg block has an 8-N frictional force acting on it.
  a. What is the net force acting on the entire two-block system?
  b. What is the acceleration of this system?
  c. What force is exerted on the 2-kg block by the connecting string? (Consider only the forces acting on this block. Its acceleration is the same as that of the entire system.)
  d. Find the net force acting on the 4-kg block and calculate its acceleration. How does this value compare to that found in part b?

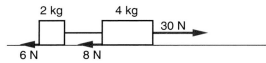

CP6. A 60-kg man is in an elevator that is accelerating downward at the rate of 1.4 m/s^2.
  a. What is the true weight of the man in newtons?
  b. What is the net force acting on the man required to produce the acceleration?
  c. What is the force exerted on the man's feet by the floor of the elevator?
  d. What is the apparent weight of the man in newtons? (This is the weight that would be read on the scale dial if the man were standing on a bathroom scale in the accelerating elevator.)
  e. How would your answers to parts (b) through (d) change if the elevator were accelerating upward with an acceleration of 1.4 m/s^2?

CP7. A sky diver has a weight of 750 N. Suppose that the air-resistance force acting on the diver increases in direct proportion to his velocity such that for every 10 m/s that the diver's velocity increases, the force of air resistance increases by 100 N.
  a. What is the net force acting on the sky diver when his velocity is 40 m/s?
  b. What is the acceleration of the diver at this velocity?
  c. What is the terminal velocity of the sky diver?
  d. What would happen to the velocity of the sky diver if for some reason (perhaps a brief down draft) his velocity exceeded the terminal velocity? Explain.

home experiments and observations

HE1. Collect a variety of small objects such as coins, pencils, keys, and bottle caps. Ice cubes, if they are available, also make excellent test objects. Try sliding these objects across a smooth surface such as a tabletop or floor, being as consistent as possible in the initial velocity that you give to them.
  a. Do the objects slide the same distance after they leave your hand? What differences are apparent, and how are they related to the nature of the surface and size of the objects? Which objects come closest to demonstrating Newton's first law of motion?
  b. What factors seem to be important in reducing the frictional force between the objects and the surface upon which they are sliding? If you see some general principle at work, test this idea by finding other objects that would support your hypothesis.

HE2. Place a sheet of paper under a medium-sized book lying on a smooth tabletop or desktop.
  a. Try to accelerate the book smoothly by exerting a constant pull on the sheet of paper. What happens if you try to accelerate the book too rapidly? Can you pull the paper cleanly from underneath the book without moving the book? Explain your observations in terms of Newton's laws of motion.
  b. Repeat these observations with a few books in a stack. How does increasing the mass of the books affect the results?
  c. Try other objects. Which objects move the least when the paper is pulled rapidly?

HE3. Falling objects whose surface area is large relative to their weight will reach terminal velocity more readily than a ball or a rock. Test several objects, such as a balloon, small pieces of paper, plant parts (leaves, flowers, or seeds), or whatever you think might work. Do these objects reach a terminal velocity? How far does each object fall before reaching constant velocity? How does the rate of fall differ for different objects when dropped at the same time? Which of the objects tested produces the clearest demonstration of terminal velocity, showing first a brief acceleration followed by a constant velocity?

HE4. Using elevators in your dormitory or other campus buildings, observe the effects of the elevator's acceleration. Most elevators accelerate briefly as they start and again as they stop (deceleration). Express elevators in high-rise buildings are best for observing the effects of acceleration.
  a. If you have a bathroom scale, see how much your apparent weight differs from your true weight when the elevator is stopping or starting. Can you estimate the rate of acceleration from this information? (See Everyday Phenomenon Box 4.2 and challenge problem 6.)
  b. Try holding your arm away from your body and maintaining it in this position as the elevator accelerates. How difficult is this to do for different conditions during the motion of the elevator? Explain your observations.

Circular Motion, the Planets, and Gravity

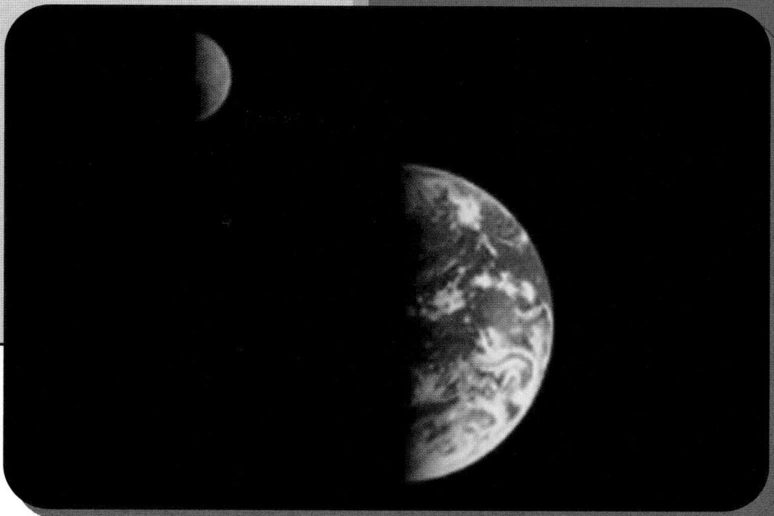

chapter overview

Using the example of a ball on a string, we first examine the acceleration involved in changing the direction of the velocity in circular motion (centripetal acceleration). Then we consider the forces involved in producing a centripetal acceleration in different cases, including that of a car rounding a curve. Kepler's laws of planetary motion will then be examined. Newton's law of universal gravitation will be introduced to explain the motion of the planets. We will also show how this gravitational force relates to the weight of an object and the gravitational acceleration near the earth's surface.

chapter outline

1 **Centripetal acceleration.** How can we describe the acceleration involved in changing the *direction* of an object's velocity? How does this acceleration depend on the object's speed?

2 **Centripetal forces.** What types of forces are involved in producing centripetal accelerations in different situations? What forces are involved for a car rounding a curve?

3 **Planetary motion.** How do the planets move around the sun? How has our understanding of planetary motion changed historically? What are Kepler's laws of planetary motion?

4 **Newton's law of universal gravitation.** What is the fundamental nature of the gravitational force, according to Newton? How does this force help to explain planetary motion?

5 **The moon and other satellites.** How does the moon orbit the earth? How do the orbits of artificial satellites differ from the moon and from each other?

chapter 5

unit one

"The car failed to negotiate the curve." How many times have you seen a phrase like that in an accident report in the newspapers? Either the road surface was slippery or the driver was driving too fast for the sharpness of the curve. In either case, poor judgment and probably a poor sense of the physics of the situation were at work (fig. 5.1).

When a car goes around a curve, the direction of its velocity changes. A change in velocity means acceleration, and by Newton's second law, an acceleration requires a force. The situation has much in common with a ball being twirled in a circle at the end of a string and other examples of circular motion.

What forces keep a car moving around a curve? How does the force required depend on the speed of the car and the sharpness of the curve? What other factors are involved? Finally, what does the car rounding a curve have in common with the ball on a string and the motions of the planets around the sun?

The motions of the planets around the sun and the moon around the earth played important roles in the development of Newton's theory of mechanics. Newton's law of universal gravitation was a crucial part of that theory. The gravitational force explains the behavior of objects falling near the surface of the earth, but it also

figure 5.1   The car failed to negotiate the curve. Newton's first law at work.

explains why the planets move in curved paths about the sun. Circular motion is a very important special case of motion in two dimensions, both in the history of physics and in our everyday experience.

5.1 Centripetal Acceleration

Suppose that we attach a ball to a string and twirl the ball in a horizontal circle (fig. 5.2). With a little practice it is not hard to keep the ball moving with a constant speed, but the direction of its velocity changes continually. A change in velocity implies an acceleration, but what is the nature of this acceleration?

The key to this situation involves taking a careful look at what happens to the velocity vector as the ball moves in a circle. How does this vector change as the path of the ball changes direction?

Can we evaluate the size of this change and how it is related to the speed of the ball or the radius of the curve? To define the concept of *centripetal acceleration,* we need to answer these questions.

What is a centripetal acceleration?

What do we have to do to get the ball on the string to change its direction? If you try twirling a ball as pictured in figure 5.2, you will feel a tension in the string. In other words, you have to apply a force by pulling on the string to cause the change in direction of the ball's velocity.

What would happen if this force were not present? According to Newton's first law of motion, an object will continue moving in a straight line with constant speed if there is no net force acting on the object. If the string breaks, or if we let go of the string, this is exactly what will hap-

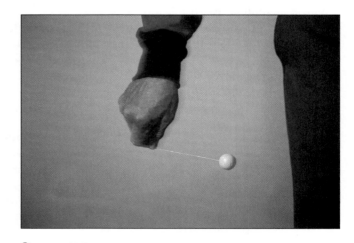

figure 5.2   A ball being twirled in a horizontal circle. Is the ball accelerated?

pen. The ball will fly off in the direction that it was traveling when the string broke (fig. 5.3). Without the pull of the string, the ball will move in a straight line. It will also fall, of course, as it is pulled down by the gravitational force.

According to Newton's second law of motion, if there is a force, there must be an acceleration ($\mathbf{F} = m\mathbf{a}$). This acceleration is associated with the change in the *direction* of the velocity vector. In the case of the ball on the string, the string pulls the ball toward the center of the circle causing

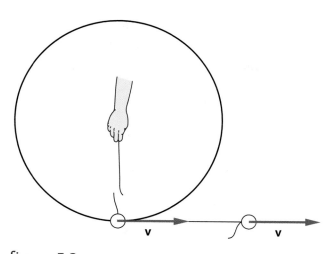

figure 5.3 If the string breaks, the ball flies off in a straight-line path in the direction it was traveling at the instant the string broke.

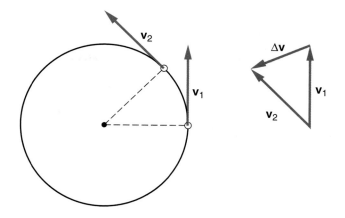

figure 5.4 The velocity vectors for two positions of a ball moving in a horizontal circle. The change in velocity, Δv, adds to v_1 to yield v_2.

the direction of the velocity vector to change continually. The direction of the force, and of the acceleration that it produces, is *toward the center* of the circle. We call this acceleration the **centripetal acceleration:**

Centripetal acceleration is the rate of change in velocity of an object that is associated with the change in *direction* of the velocity. Centripetal acceleration is always perpendicular to the velocity vector itself and toward the center of the curve.

To find the size of the centripetal acceleration, we need to determine how fast the velocity is changing. You might guess that this depends on how rapidly you are twirling the ball, but it also depends on the radius of the curve—the size of the circle.

How do we find the change in velocity Δv?

Figure 5.4 shows the ball and string as seen from above. The ball is moving in a horizontal circle. Velocity vectors are drawn on the circle at two positions separated by a short time interval. The velocity v_2 occurs a short time after the velocity v_1, as the ball moves counterclockwise around the circle. These two vectors are drawn with the same length, indicating that the speed of the ball is unchanged.

The change in velocity, Δv, is the difference between the initial velocity and the final velocity for a given time interval. In other words, the change in velocity is a vector that is added to the initial velocity to produce the final velocity. Adding Δv to v_1 produces v_2. This vector addition is shown in the vector triangle to the right of the circle in figure 5.4. (See appendix C for a discussion of vector addition by graphical methods.)

Note that the vector Δv has a direction different from either of the velocity vectors. If we choose a short enough time interval between the two positions, the direction of the *change* in velocity points toward the center of the circle, the direction of the instantaneous acceleration of the ball. (Acceleration always has the same direction as the change in velocity.) The ball is being accelerated toward the center of the circle, the direction of the tension in the string.

What is the size of the centripetal acceleration?

But how large is this centripetal acceleration, and how does it depend on the speed of the ball and the radius of the curve? The triangle illustrating the vector addition in figure 5.4 can be used to explore these questions. There are three effects to consider:

1. As the speed of the ball increases, the velocity vectors become longer, which makes Δv longer. The triangle in figure 5.4 becomes larger.
2. The greater the speed of the ball, the more rapidly the direction of the velocity vector changes, because the ball reaches the second position in figure 5.4 more quickly.
3. As the radius of the curve decreases, the rate of change in velocity increases because the direction of the ball changes more rapidly. A tight curve (small radius) produces a large change, but a gentle curve (large radius) produces a small change.

The first two effects both indicate that the rate of change in velocity will increase with an increase in the speed of the ball. Combining these two effects suggests that the centripetal acceleration should be proportional to the square of the speed. We need to multiply by the speed twice. The third effect suggests that the rate of change of velocity is inversely proportional to the radius of the curve. The larger the radius, the smaller the rate of change. Taken together, these effects produce the expression

$$a_c = \frac{v^2}{r}$$

for the size of the centripetal acceleration, a_c. It is proportional to the square of the speed and inversely proportional to the radius, r, of the curve. The direction of the centripetal-acceleration vector a_c is always toward the center of the curve, the direction of the change in velocity Δv.

The ball moving in a circle is accelerated, even though its speed remains constant. To change the direction of the velocity vector is to change the velocity, and an acceleration is involved. People often resist this idea: we use the term *acceleration* in everyday language to describe increases in speed without taking into account changes in direction.

What force produces the centripetal acceleration?

Since an object moving in a circle is accelerated, a force must be acting to produce that acceleration, according to Newton's second law. For the ball on the string, the tension in the string pulling on the ball provides the centripetal acceleration. A closer look shows that this tension has both horizontal and vertical components, since the string is not completely within the horizontal plane. As shown in figure 5.5, the horizontal component of the tension pulls the ball toward the center of the horizontal circle and produces the centripetal acceleration.

The total tension in the string is determined by both the horizontal and the vertical components of the tension. The vertical component is equal to the weight of the ball, since the net force in the vertical direction should be zero.

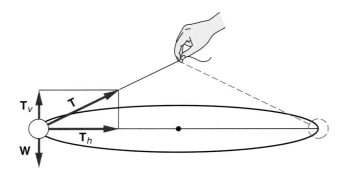

figure 5.5   The horizontal component of the tension is the force that produces the centripetal acceleration. The vertical component of the tension is equal to the weight of the ball.

The ball stays in the horizontal plane of the circle and is not accelerated in the vertical direction. In the example in Try This Box 5.1, the weight of the ball is approximately 0.50 N ($W = mg$), so that becomes the value of the vertical component of the tension.

The ball in Try This Box 5.1 has a slow speed. Even at this low speed, the horizontal component of the tension is larger than the vertical component. As the ball twirls at a faster rate, the centripetal acceleration increases even more rapidly, since it is proportional to the square of the speed of the ball. The horizontal component of the tension then becomes much larger than the vertical component, which remains equal to the weight of the ball (fig. 5.6). These effects can be readily observed with your own ball and string. Give it a try. You will feel the tension increase with increasing speed.

Centripetal acceleration involves the rate of change in the direction of the velocity vector. Its size is equal to the square of the speed of the object divided by the radius of the curve ($a_c = v^2/r$). Its direction is toward the center of the curve. Just as with any acceleration, there must be a force acting on the object to produce the centripetal acceleration. For a ball on a string, that force is the horizontal component of the tension in the string.

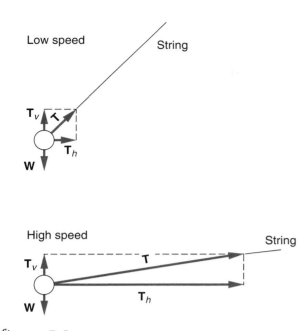

figure 5.6 At higher speeds, the string comes closer to lying in the horizontal plane because a large horizontal component of the tension is needed to provide the required centripetal force.

5.2 Centripetal Forces

For a ball twirled at the end of a string, the string pulls inward on the ball, providing the force that causes the centripetal acceleration. For a car rounding a curve, however, there is no string attached. Different forces must be at work to provide the centripetal acceleration. A person riding on a Ferris wheel also experiences circular motion. What forces produce centripetal acceleration in these situations?

The *total* force that produces a centripetal acceleration is often referred to as the **centripetal force.** This term is sometimes a source of confusion, because it implies that a special force is somehow involved. In fact, centripetal forces are *any* force, or combination of forces, that acts on an object in certain situations to produce the centripetal acceleration. Almost any force can play this role: pulls from strings, pushes from contact with other objects, friction, gravity, and so on. We need to analyze each situation separately to identify the forces and determine their effects.

What force helps a car negotiate a flat curve?

What forces are involved in producing the centripetal acceleration for a car rounding a curve? It depends on whether or not the curve is banked. The easiest situation to analyze is when a curve is not banked, so that we deal with a flat road surface.

For a flat road surface, friction alone produces the necessary centripetal acceleration. The tendency of the car to move in a straight line causes the tires to pull against the pavement as the car turns. By Newton's third law, the pavement then pulls in the opposite direction on the tires (fig. 5.7). The frictional force acting on the tires points toward the center of the curve. If this force were not present, the car could not turn.

The size of a frictional force depends on whether or not there is motion along the surfaces of contact producing the friction. If there is no motion in the direction of the force, we call it the **static force of friction.** If the object is sliding, as it might on a wet or icy surface, the **kinetic force of friction** is involved. Usually, the kinetic force of friction is smaller than the maximum possible static force of friction, so whether or not the car is skidding becomes an important factor.

Unless the car has already begun to skid, the *static* force of friction produces the centripetal acceleration for the car rounding the curve. The part of the tire in contact with the road is momentarily at rest on the road; it does not slide along the road. If the tires do not move in the direction of the frictional force, the static force is in effect.

How large is the required frictional force? It depends on the speed of the car and the radius of the curve. From Newton's second law, we know that the magnitude of the required force is $F = ma_c$, where the centripetal acceleration a_c is equal to v^2/r. Putting these two ideas together, we see that the frictional force f must be equal to mv^2/r, since it is the only force operating to produce the centripetal acceleration. The speed of the car is a critical factor in determining how large a force is needed, which is why we often slow down in approaching a curve.

figure 5.7 The centripetal acceleration of a car rounding a level curve is produced by frictional forces exerted on the tires by the road surface.

If the mass times the centripetal acceleration is greater than the maximum possible frictional force, we are in trouble. Because the square of the speed is involved in this relationship, doubling the speed would require a frictional force four times as large as that for the lower speed. A sharper curve with a smaller radius *r* also requires either a larger frictional force or a lower speed. Both the speed and the radius must be considered in making driving judgments.

What happens if the required centripetal force is larger than the maximum possible frictional force? The frictional force cannot produce the necessary centripetal acceleration, and the car begins to skid. Once it is skidding, kinetic friction comes into play rather than static friction. Since the force of kinetic friction is generally smaller than that of static friction, the frictional force decreases and the skid gets worse. The car, like the ball on the broken string, follows its natural tendency to move in a straight line.

The maximum possible value of the frictional force is dictated by the road and tire conditions. Any factor that reduces the force of static friction will cause problems. Wet or icy road surfaces are the usual culprits. In the case of ice, the force of friction may diminish almost to zero, and an extremely slow speed will be necessary to negotiate a curve. There is nothing like driving on an icy road to give you an appreciation of the value of friction. Newton's first law is illustrated vividly. (See also Everyday Phenomenon Box 5.1.)

What happens if the curve is banked?

If the road surface is properly banked, we are no longer totally dependent on friction to produce the centripetal acceleration. For the banked curve, the normal force between the car's tires and the road surface can also be helpful (fig. 5.8).

Everyday Phenomenon

box 5.1

Seat Belts, Air Bags, and Accident Dynamics

The Situation. In automobile accidents, serious or fatal injuries are often the result of riders being thrown from the vehicle. Since the 1960s, federal regulations have required that cars be equipped with seat belts. More recently, front-seat air bags have also been required in an effort to reduce the carnage. Still, we often read of people being thrown from their vehicle in accident reports.

How do air bags and seat belts help? If your car is equipped with air bags, as most now are, is it still necessary to wear your seat belt? In what situations are air bags most effective and when are seat belts essential?

In a head-on collision, the air bag inflates rapidly to prevent the rider from moving forward and colliding with the windshield or steering column.

The Analysis. Except in high-speed collisions where the passenger compartment of the vehicle is crushed, most injuries and fatalities are caused by motion of the rider within, and outside of, the vehicle. The vehicle stops or turns suddenly due to the collision and the rider continues to move in a straight line, following Newton's first law of motion.

In a head-on collision, the car stops while the rider continues to move forward unless constrained. In the absence of either seat belts or air bags, front-seat riders hit the windshield or the steering column, resulting in serious head or chest injuries. Seat belts can prevent this when used properly, but air bags are also designed to protect against these injuries. As the rider begins to move forward relative to the vehicle, the air bag inflates rapidly, providing a cushion between the rider and other objects in the car. The rider decelerates more gradually involving a smaller force and less trauma. (This idea is best understood in terms of the concept of impulse discussed in chapter 7.) Air bag usage has resulted in a significant reduction in serious head and chest injuries in head-on collisions with other vehicles or with fixed objects.

Head-on collisions are not the most frequent type of serious accident, however. Rollover accidents involving single vehicles are common, and vehicles can also collide in intersections, providing impacts to the side of the car. In the latter case, the struck vehicle will often go into a spin. In both of these cases, the vehicle undergoes rotational motion while the rider moves forward in a straight line.

(continued)

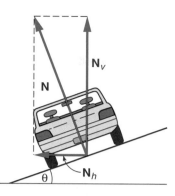

figure 5.8 The horizontal component of the normal force **N**$_h$ exerted by the road on the car can help to produce the centripetal acceleration when the curve is banked.

The normal force **N** is always perpendicular to the surfaces involved, so it points in the direction shown in the diagram. The total normal force acting on the car (indicated in the diagram) is the sum of those for each of the four tires.

Since the car is not accelerated vertically, the net force in the vertical direction must be zero. The vertical component of the normal force **N**$_v$ must be equal in magnitude to the weight of the car to yield a net vertical force of zero. This fact determines how large the normal force will be. Only the horizontal component of the normal force **N**$_h$ is in the appropriate direction to produce the centripetal acceleration.

The angle of the banking and the weight of the car determine the size of the normal force. They also determine the size of its horizontal component. At the appropriate speed, this horizontal component pushing on the tires of the car is all that is needed to provide the centripetal acceleration.

These are the accidents in which the rider is likely to be thrown from the vehicle.

Will air bags help in these situations? Air bags are most effective in head-on collisions and do not provide much protection against sideways motion of the rider. (A few vehicles do come equipped with air bags in the front-seat doors, which can protect against sideways movement, but air bags are seldom provided for the rear seats.) In a rollover accident, the vehicle goes into a spin about an axis through its long dimension. The doors will sometimes open or the windows will shatter during the first roll, providing openings for the rider to fly through as he or she continues to move forward while the vehicle turns. In some cases, the rider is thrown from the vehicle and the vehicle then rolls over the victim.

Seat belts can make a big difference. Because the vehicle is turning rapidly in a rollover accident, a centripetal force acting on the rider is necessary to hold the rider against the seat rather than moving forward in a straight line. In the absence of such a force, the rider is thrown outward against the sides of the vehicle. Attempts by riders to brace themselves are usually totally inadequate to provide the required centripetal force. The seat belt and shoulder harness, on the other hand, can provide the force necessary to hold the rider in place.

Statistics on accident fatalities are compelling. In rollover accidents, riders who are wearing their seat belts generally survive, while those who are not using their belts and shoulder harnesses are frequently killed or seriously injured. Often those killed are thrown from the vehicle, but even when they

As the vehicle rolls, a rear-seat passenger is thrown against the side of the vehicle (viewed from the back). A properly adjusted seat belt and shoulder harness can prevent this.

remain inside the vehicle, trauma from being thrown around inside the vehicle can be fatal. Statistics in a recent national study indicate that 63% of the deaths in rollover accidents involve riders ejected from the vehicle.

Newton's first law of motion is vividly illustrated in automobile accidents. An object keeps moving in a straight line with constant speed unless acted upon by an external force. Air bags and seat belts can provide that force, but seat belts provide better protection for all passengers in rollover accidents.

The higher the speed, the steeper the required banking angle because a steeper angle produces a larger horizontal component for the normal force. Fortunately, since both the normal force and the required centripetal force are proportional to the mass of the car, the same banking angle will work for vehicles of different mass.

A banked curve is designed for a particular speed. Since friction is also usually present, the curve can be negotiated at a range of speeds above and below the intended speed. Friction and the normal force combine to produce the required centripetal acceleration.

If the road is icy and there is no friction, the curve can still be negotiated at the intended speed. Speeds higher than that speed will cause the car to fly off the road, just as on a flat road surface. Speeds too low, on the other hand, will cause the car to slide down the icy banked incline toward the center of the curve.

What forces are involved in riding a Ferris wheel?

Riding a Ferris wheel is another example of circular motion that many of us have experienced. On a Ferris wheel, the circular motion is vertical, unlike the horizontal circles of our previous examples.

Figure 5.9 shows the forces exerted on the rider at the bottom of the circle as the Ferris wheel turns. At this point in the ride, the normal force acts upward and the weight downward. Since the centripetal acceleration of the rider is directed upward, toward the center of the circle, the total force acting on the rider must also be upward. In other

words, the normal force of the seat pushing on the rider must be larger than the weight of the rider.

By Newton's second law, the total force must be equal to the mass times the centripetal acceleration. In this case, the centripetal force is the difference of two forces, the upward normal force and the downward weight of the rider, so

$$N - W = ma_c.$$

Since the normal force is larger than her weight, she feels heavy in this position. The situation is similar to that in an upward accelerating elevator (Everyday Phenomenon Box 4.2).

As the rider moves up or down along the sides of the circle, a horizontal component of the normal force is needed to provide the centripetal acceleration. This horizontal component may be provided by the frictional force exerted by the seat on the rider, by the seat back pushing on the rider on the left side of the cycle, or by a seat belt or hand bar on the right side of the cycle. The latter case is more exciting.

At the top of the cycle, the weight of the rider is the only force (other than a possible seat-belt force) in the appropriate direction to produce the centripetal acceleration. Again, from Newton's second law, the total force must equal the mass of the rider times the centripetal acceleration, which is now directed downward. This yields the relationship

$$W - N = ma_c.$$

As the speed gets larger and the centripetal acceleration, $a_c = v^2/r$, increases, the normal force must get smaller to increase the total force. Usually, the top speed of the Ferris wheel is adjusted so that the normal force is small when the rider is at the top of the cycle. Since the force exerted by the seat on the rider is small, the rider feels light, part of the thrill of the ride.

If there is one nearby, take a break and go ride a Ferris wheel. There is nothing like direct experience to bring home the ideas we have just described. As you ride, try to sense the direction and magnitude of the normal force. The light feeling at the top and the sense of plunging outward in the downward portion of the cycle are what the price of the ride is all about.

A centripetal force is any force or combination of forces that produces the centripetal acceleration for an object moving around a curve. In the case of a car moving on a flat road surface, the centripetal force is provided by friction. If the road surface is banked, the normal force of the road pushing on the tires of the car also helps. In the case of a Ferris wheel, the weight of the rider and the normal force exerted by the seat on the rider combine to provide the centripetal force. We use Newton's laws of motion to identify the forces and analyze each situation.

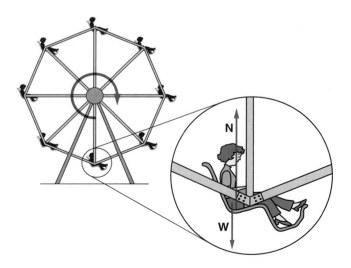

figure **5.9**   At the bottom of the cycle, the weight of the rider and the normal force exerted by the seat combine to produce the centripetal acceleration for a rider on a Ferris wheel.

5.3 Planetary Motion

From a historical perspective, the most important examples of centripetal acceleration are the heavenly bodies—the sun, the moon, and the planets. These objects are a part of our everyday experience, yet many of us are surprisingly unaware of their motions. How do the planets move? How can we make sense of these motions?

How did the Greeks explain the motions of the heavens?

Observing the heavens was probably a more popular pastime when there were fewer roofs over our heads. If you have ever spent a night in a sleeping bag under the stars, you probably experienced a sense of wonder and amazement at all of those bright objects out there. If you spent night after night under the stars, you might notice, as the ancients did, that some of the brightest objects move relative to the other stars.

These wanderers are the planets. The so-called fixed stars always maintain the same relative position to one another as they move across the sky. The Big Dipper never seems to change its shape, but the planets roam about with respect to the fixed stars in a regular but curious fashion. Their motions excited the curiosity of ancient observers of the heavens.

Suppose you were an early philosopher-scientist trying to make sense of these motions. What kind of model might you develop? Some features seem simple and regular. The sun, for example, moves across the sky each day, from east to west, as if it were at the end of an enormously long and invisible rope tethered at the center of the earth. The stars follow a similar pattern. Their apparent motion as seen from earth could be explained by picturing them as lying on a giant sphere that revolves around the earth. This earth-centered or *geocentric* view of the universe seemed natural and reasonable.

The moon also moves across the sky in an apparently circular orbit around the earth. Unlike the stars, the moon does not reappear in the same position each night. Instead, it goes through a series of regular changes in position and phase in a cycle of approximately 30 days. How many of us can provide a clear explanation of the phases of the moon? The motion of the moon will be considered more fully in the final section of this chapter.

Early models of the motions of the heavenly bodies developed by Greek philosophers involved a series of concentric spheres centered on the earth (fig. 5.10). Plato and others of his time viewed spheres and circles as ideal shapes that would reflect the beauty of the heavens. The sun, the moon, and the five planets known then each had its own sphere. The fixed stars were on the outermost sphere. These spheres were thought to revolve around the earth in ways that explained the positions of the heavenly bodies.

figure **5.10**   The heavenly spheres: an earth-centered view of the universe showing earth, water, air, and fire at the center. The sun is in the fourth sphere beyond the fire sphere.

figure **5.11**   The retrograde motion of Mars relative to the background of fixed stars. These changes take place over a period of several months.

Unfortunately, the planets do not behave as though they are on a continuously revolving sphere. Mars, Jupiter, and Saturn sometimes appear to move backward to their normal direction of motion, relative to the fixed stars. We call this **retrograde motion.** It takes a few months for Mars to trace one of these retrograde patterns (fig. 5.11).

To explain the apparent retrograde motion of these planets, Ptolemy (Claudius Ptolemaeus), working in the second century A.D., devised a more sophisticated model than the one used by earlier Greek philosophers. Ptolemy's model

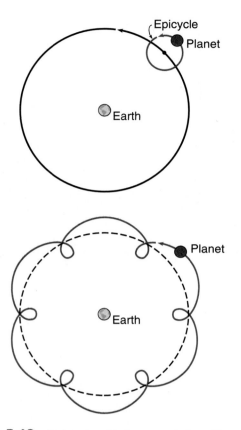

figure 5.12   Ptolemy's epicycles were circles rolling along the circular orbits of the planets. This model explained the retrograde motion observed for the outer planets.

used circular orbits rather than spheres but was still geocentric. He invented the idea of **epicycles,** circles that rolled along the larger basic orbit of the planet around the earth (fig. 5.12). The epicycles accounted for the retrograde motion and could also be used to explain other irregularities in planetary orbits.

Ptolemy's model accurately predicted where to find the planets at any given time of any year. As more accurate observations became available, refinements were needed, however, to improve the predictions. In some cases, this meant adding epicycles to epicycles, but the basic scheme of circles was retained. Ptolemy's system became part of the accepted knowledge during the Middle Ages and was incorporated, along with many of Aristotle's works, into the teachings of the Roman Catholic Church and the emerging European universities.

How did the Copernican model differ from Ptolemy's conception?

Ptolemy's model is not the one that you were introduced to in elementary school. It has been superseded. During the sixteenth century, a Polish astronomer, Nicolaus Copernicus (1473–1543) put forth a sun-centered or **heliocentric**

view, later championed by Galileo. Copernicus was not the first to suggest such a model, but earlier heliocentric versions had not taken hold. Copernicus spent many years working out the details of his model, but he did not publish it until within a year of his death.

Galileo was an early advocate of the Copernican model and promoted it more vigorously than Copernicus himself. In 1610, hearing of the invention of the telescope, Galileo built his own improved version and turned it to the heavens. He discovered that the moon has mountains, that Jupiter has moons, and that Venus goes through phases like our moon. He showed that the phases of Venus could be explained better by the Copernican model than by a geocentric model. Galileo became famous throughout Europe for his discoveries and ended up in trouble with church authorities, a problem not to be taken lightly in his day. People had been burned at the stake for similar offenses.

Copernicus placed the sun at the center of the circular orbits of the planets and demoted the earth to the status of the other planets. Also, the Copernican model requires that the earth *rotate* on an axis through its center—thus explaining the daily motions of the sun and the other heavenly bodies (including the fixed stars). This idea was revolutionary at the time. Why are we not blown away by the enormous winds that rotation would produce?

The advantage of the Copernican view is that it does not require complicated epicycles to explain retrograde motion, although epicycles were still used to make other adjustments to planetary orbits. Retrograde motion comes about because the earth is orbiting the sun along with the other planets. The position of Mars appears to change as *both* Mars and earth move in the same direction against the background of the fixed stars (fig. 5.13). As the more rapidly moving earth passes Mars, Mars slips behind and briefly appears to move backward.

Accepting the Copernican model meant giving up the earth-centered view of the universe to endorse what seemed to some to be an absurd proposition: that the earth rotates, with a frequency of one cycle per day. Since an approximation of the radius of the earth was known (6400 km), rotation implied that we must be moving at roughly 1680 km/h (or just over 1000 MPH) if we are standing near the equator on the earth! We certainly do not feel that motion.

Because Copernicus assumed the planets' orbits to be circular, the accuracy of his model for predicting was no better than Ptolemy's model. In fact, it required some adjustments (for which Copernicus used epicycles) just to make it agree with already known astronomical data. Settling the controversy generated by the competition between the two models called for more accurate observations, a project undertaken by a Danish astronomer, Tycho Brahe (1546–1601).

Tycho was the last great naked-eye astronomer. He developed a large quadrant (fig. 5.14) that he used to make very accurate sightings of the positions of the planets and stars. It was capable of measuring these positions to an accuracy of $\frac{1}{60}$ of a degree, considerably better than previously

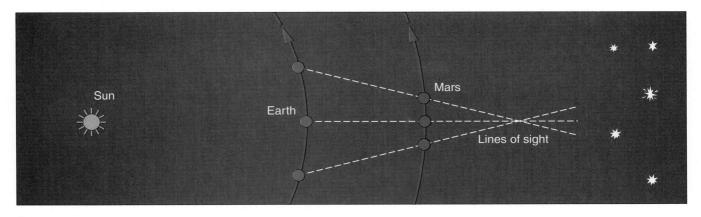

figure **5.13**   As the earth passes the more slowly moving Mars, Mars appears to move backward as seen against the background of the fixed stars. (Not drawn to scale.)

figure **5.14**   Tycho Brahe's large quadrant permitted accurate measurement of the positions of the planets and other heavenly bodies.

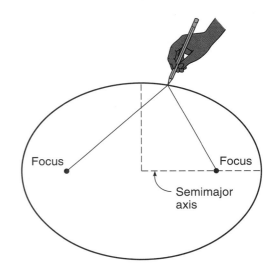

figure **5.15**   An ellipse can be drawn by fixing a string at two points (foci) and moving a pencil around the path permitted by the string.

available data. Tycho spent several years painstakingly collecting data on the precise positions of the planets and other bodies—all without the benefit of a telescope.

Kepler's laws of planetary motion

Analyzing the data collected by Tycho fell to his assistant, Johannes Kepler (1571–1630), after Tycho's death. It was an enormous task requiring the transformation of the data

to coordinates around the sun and then numerical trial and error to find regular planetary orbits. It was already known that these orbits were not perfect circles. Kepler was able to show that the orbits of the planets around the sun were ellipses, with the sun at one focus.

An **ellipse** can be drawn by attaching a string between two fixed foci and then moving a pencil around the perimeter of the path allowed by the string (fig. 5.15). A circle is a special case of an ellipse in which the two foci coincide. The orbits of most of the planets are close to being circles, but Tycho's data were so precise that they showed a difference between a perfect circle, on one hand, and an ellipse with two closely spaced foci. Kepler's first law of planetary motion states that the orbits of the planets are ellipses.

Kepler's other two laws of planetary motion came after even more laborious numerical trial and error with Tycho's

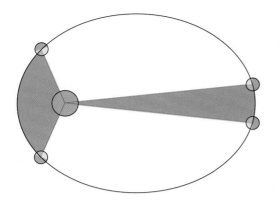

figure **5.16** Because planets move faster when nearer to the sun, the radius line for each planet sweeps out equal areas in equal times (Kepler's second law). In other words, the two blue sections each cover the same span of time and have the same area.

data. Kepler's second law describes how the planets move faster when they are nearer to the sun, so that the radius vector sweeps out equal areas in equal times regardless of where it is in its orbit (fig. 5.16). The first two laws were published in 1609.

The third law (published in 1619) states a relationship between the average radius of the orbit and the time taken for one complete cycle around the sun (the **period** of the orbit). Kepler found his third law after trying many other possible relationships between the periods, T, and the average radii of the planetary orbits, r. To a high degree of accuracy, he found that the ratio of the square of the period to the cube of the radius (T^2/r^3) was the same for all of the known planets. The behavior of the planets is surprisingly regular. Kepler published his findings in papers that also contained elaborate speculations on numerical mysticism and musical harmonies associated with the planets. Some of these ideas must have seemed strange to Galileo and others who admired Kepler's work.

Kepler's laws added to the accuracy with which we can predict the positions of the planets as they appear to wander among the fixed stars. Like the Copernican model, Kepler's

---

Kepler's Laws of Planetary Motion

1. The planets all move in elliptical orbits about the sun, with the sun at one focus of the ellipse.
2. The radius vector drawn from the sun to the planet sweeps out equal areas in equal times.
3. The cube of the average radius about the sun for each planet is proportional to the square of the period of the orbit.

---

model was heliocentric (sun-centered), so it supported Galileo's efforts to overthrow the geocentric (earth-centered) model of Ptolemy. More importantly, however, Kepler's laws described a new set of precisely stated relationships that called for explanation. The stage was set for Isaac Newton to incorporate these relationships into a grand theory that explains both celestial mechanics (the motion of the heavenly bodies) and the more mundane motion of everyday objects near the earth's surface.

Many of the early models for describing the motion of the planets were geocentric (earth-centered). Ptolemy's model included epicycles to explain the apparent retrograde motion of the planets. Copernicus introduced a heliocentric (sun-centered) model, which explained retrograde motion more simply. This model was championed by Galileo. Galileo was one of the first scientists to use a telescope systematically, and he made significant discoveries supporting the heliocentric view. Kepler refined the heliocentric model by showing that planetary orbits are ellipses with some surprising regularities.

5.4 Newton's Law of Universal Gravitation

Planetary motion and centripetal acceleration lead us to the next question. If the planets are moving in curved paths around the sun, what force must be present to produce the centripetal acceleration? You are probably aware that gravity is involved, but that involvement was not at all obvious when Newton began his work. How did Newton put it all together?

What was Newton's breakthrough?

Newton realized that there is a similarity between the motion of a projectile launched near the earth's surface and the orbit of the moon. A famous drawing in Newton's *Principia* depicts this similarity (fig. 5.17).

The idea is simple but earthshaking. Imagine, as Newton did, a projectile being launched horizontally from an incredibly high mountain. The larger the launch velocity, the farther away from the base of the mountain the projectile will land. At very large launch velocities, the curvature of the earth becomes a significant factor. In fact, if the launch velocity is large enough, the projectile would never reach the surface of the earth. It keeps falling, but the curvature of the earth falls away, too. The projectile goes into a circular orbit around the earth.

Newton's insight was that the moon, under the influence of gravity, is actually falling, just as a projectile does. The moon, of course, is at a distance from the earth much greater than the height of any mountain. The same force that accounts for the acceleration of objects near the earth's surface, as described by Galileo, explains the orbit of the moon.

Newton's law of universal gravitation

From Galileo's work, Newton knew that near the earth's surface the gravitational force is proportional to the mass of the object, **F** = m**g**. Mass, then, should be involved in any more general expression for the gravitational force.

Does the gravitational force vary with distance, though, and, if so, how? The idea that a force could influence two masses separated by a large distance was hard to accept in Newton's day (and, in some ways, even now). If such a force exists, we would expect that this force "acting at a distance" would decrease in strength as the distance increases. Using geometrical reasoning (fig. 5.18), other scientists had speculated that the force might be inversely proportional to the square of the distance r between the masses, but they could not prove it.

At this point, Kepler's laws of planetary motion and the concept of centripetal acceleration came into play. Newton was able to prove mathematically that Kepler's first and third laws of planetary motion could be derived from the assumption that the gravitational force between the planets and the sun falls off with the inverse square of the distance. The proof involved setting the assumed $1/r^2$ force equal to the required centripetal force in Newton's second law of motion. All of Kepler's laws are consistent with this assumption.

The proof that Kepler's laws could be explained by a gravitational force proportional to the masses of two interacting objects, and inversely proportional to the square of the distance between the objects, led to **Newton's law of universal gravitation.** This law and Newton's three laws of motion are the fundamental postulates of his theory of mechanics. The law of gravitation can be stated as

> The gravitational force between two objects is proportional to the mass of each object and inversely proportional to the square of the distance between the centers of the masses:
>
> $$F = \frac{Gm_1m_2}{r^2},$$
>
> where G is a constant. The direction of the force is attractive and lies along the line joining the centers of the two masses (fig. 5.19).

figure **5.17** In a diagram similar to this in his *Principia*, Newton imagined a projectile fired from an incredibly high mountain. If fired with a large enough horizontal velocity, the projectile falls toward the earth but never gets there.

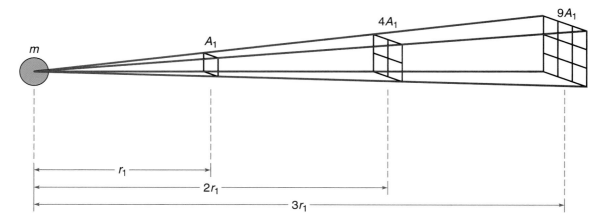

figure **5.18** If lines are drawn radiating outward from a point mass, the areas intersected by these lines increase in proportion to r^2. Does this suggest that the force exerted by the mass on a second mass might become weaker in proportion to $1/r^2$?

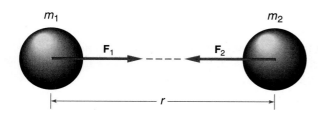

figure 5.19 The gravitational force is attractive and acts along the line joining the center of the two masses. It obeys Newton's third law of motion.

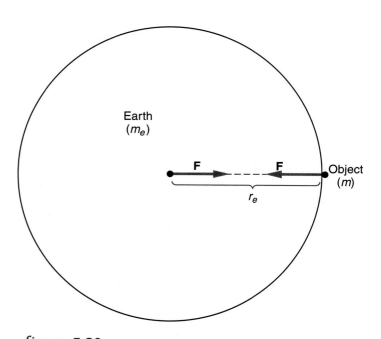

figure 5.20 For the earth and an object near the earth's surface, the distance between the centers of the two objects is equal to the radius of the earth.

For this statement to be completely valid, the masses in question must be either point masses or perfect spheres.

In Newton's law of gravitation, G is the **universal gravitational constant.** It has the same value for any two objects. Newton did not actually know the value of this constant, because he did not know the masses of the earth, the sun, and the other planets. Its value was determined more than a hundred years later in an experiment done by Henry Cavendish (1731–1810) in England. Cavendish measured the very weak gravitational force between two massive lead balls for different distances of separation. In metric units, the value of G is

$$G = 6.67 \times 10^{-11} \text{ N·m}^2/\text{kg}^2.$$

The power-of-10 notation (see appendix B) is useful here because G is a very small number. The power -11 means that the decimal point is located eleven places to the left of where it is shown. If we did not use power-of-10 notation, the number would appear as

$$G = 0.000\ 000\ 000\ 066\ 7 \text{ N·m}^2/\text{kg}^2.$$

Because of the small size of this constant, the gravitational force between two ordinary-sized objects, such as people, is extremely small and not usually noticeable. Cavendish's experiment required real ingenuity to measure such a weak force.

How is weight related to the law of gravitation?

Suppose that one of the masses is a planet or other very large object. The force of gravity then can be quite large because one of the masses is very large. Consider the force exerted on a person standing on the surface of the earth. As figure 5.20 illustrates, the distance between the centers of the two objects, the person and the earth, is essentially the radius of the earth, r_e.

From Newton's law of gravitation, the force on the person must be $F = Gmm_e/r_e^2$, where m is the mass of the person and m_e is the mass of the earth. Since this gravitational force is the weight of the person, we can also express the force as $F = W = mg$. For these two expressions for F to be the same, g, the gravitational acceleration, must be related to the universal gravitational constant G by $g = Gm_e/r_e^2$.

The gravitational acceleration near the earth's surface g is therefore *not* a universal constant. It will be different on different planets and even slightly different at different points on the earth because of variations in the radius of the earth and other factors. The constant G *is* a universal constant of nature that can be used to find the gravitational acceleration for any planet if we know the radius and mass of the planet.

If we know the gravitational acceleration near the surface of the earth, it is easier to use the expression **F** = m**g** to compute a weight than to use the law of universal gravitation. This computation is done both ways in the sample exercise in Try This Box 5.2. Either way, we get the same result. The weight of the 50-kg person is approximately 490 N. The mass of the earth, 5.98×10^{24} kg, is a very large number that was first determined by Cavendish when he measured the universal constant G. Cavendish weighed the earth by making that measurement.

If we wanted to know the gravitational force exerted on a 50-kg person in a space capsule several hundred kilometers above the earth, we would have to use the more general expression in Newton's law of gravitation. Likewise, if we wanted to know the weight of this person when standing on the moon, we would need to use the mass and radius of the moon in place of those of the earth in our calculation. The weight of a 50-kg person on the moon is

try this box 5.2

Sample Exercise: Gravity, Your Weight, and the Weight of the Earth

The mass of the earth is 5.98×10^{24} kg, and its average radius is 6370 km. Find the gravitational force (the weight) of a 50-kg person standing on the surface of the earth
    a. by using the gravitational acceleration.
    b. by using Newton's law of gravitation.

a. $m = 50$ kg $F = W = mg$
   $g = 9.8$ m/s^2 $= (50$ kg$)(9.8$ m/s$^2)$
   $F = ?$ $= \mathbf{490\ N}$

b. $m_e = 5.98 \times 10^{24}$ kg
   $r_e = 6.37 \times 10^6$ m
   $F = W = Gmm_e/r_e^2$

$$= \frac{(6.67 \times 10^{-11}\ \text{N·m}^2/\text{kg}^2)(50\ \text{kg})(5.98 \times 10^{24}\ \text{kg})}{(6.37 \times 10^6\ \text{m})^2}$$

$$= \mathbf{490\ N}$$

Most scientific calculators will handle the scientific notation directly. The powers add for multiplication and subtract in division.

only about ⅙ the value of 490 N that we computed for the same person standing on the earth. The expression $F = mg$ is valid *only* near the surface of the earth.

The weaker gravitational force and acceleration of the moon are explained by the moon's smaller mass. Since our muscles are adapted to conditions on earth, we would find that our smaller weight on the moon makes some amazing leaps and bounds possible. The smaller gravitational force on objects near the moon's surface also explains why the moon has no atmosphere. Gas molecules escape the gravitational pull of the moon much more readily than they can from the earth.

Newton recognized that the moon is falling toward the earth much like projectiles moving near the earth's surface. He proposed that the gravitational force that explains projectile motion is also involved in the motions of the planets around the sun and of the moon around the earth. Newton's law of universal gravitation states that the gravitational force between two masses is proportional to the product of the masses and inversely proportional to the square of the distance between them. Using this law and his laws of motion, Newton was able to explain Kepler's laws of planetary motion as well as the motion of ordinary objects near the surface of the earth.

5.5 The Moon and Other Satellites

The moon has fascinated people as long as humanity has existed and wondered about nature. In the twentieth century, we have actually visited the moon for the first time and brought back samples from its surface. That visit has not dulled the romance that the moon holds for us, but it may have reduced its mystery.

How are the *phases of the moon* associated with changes in its position? Are Kepler's laws of planetary motion valid for the moon? How are the orbits of other satellites of earth similar to the moon's?

How do we explain the phases of the moon?

The moon was the only earth satellite available to Newton and his predecessors to study. The moon played a pivotal role in Newton's thinking and in the development of his law of gravitation. Observations of the moon and its phases, however, go back much farther than Newton's day. The moon figures in many early religions and rituals. Its course must have been carefully followed even in prehistoric times.

How do we explain the **phases of the moon**? Is the time that the moon rises in the evening related to whether it will be a full moon or not? Moonlight is reflected sunlight. So, to understand the moon's phases, we have to take into account the positions of the sun, the moon, and the observer (fig. 5.21). When the moon is full, it is on the opposite side of the earth from the sun, and we see the side that is fully illuminated by the sun. The full moon rises in the east about the same time that the sun sets in the west. These events are determined by the earth's rotation.

Because the earth and the moon are both small compared to the distances between the earth, the moon, and the sun, they do not usually get in the way of light coming from the sun. When they do, however, there is an **eclipse.** During a *lunar eclipse,* the earth casts a full or partial shadow

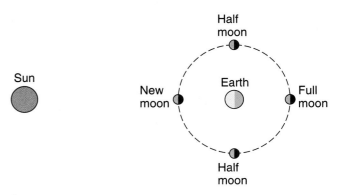

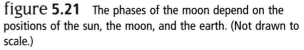

figure 5.21 The phases of the moon depend on the positions of the sun, the moon, and the earth. (Not drawn to scale.)

on the moon. From figure 5.21, we can see that a lunar eclipse can only occur during a full moon. A *solar eclipse* happens when the moon is in the right position to cast a shadow on the earth. During what phase of the moon will this occur?

At other times during the moon's 27.3-day revolution around the earth, we do not see all of the illuminated side of the moon; we see a crescent or a half-moon or some shape in between (fig. 5.22). The new moon occurs when the moon is on the same side of the earth as the sun and is more or less invisible. When we are a few days on either side of the new moon, we see the familiar crescent.

When the moon is between full moon and new moon, it can often be seen during daylight. In particular, when it is near half-moon, it rises around noon and sets around midnight (or vice versa, depending on where it is in its cycle). Under just the right conditions near sunset or sunrise, we can sometimes see the dark portion of the crescent moon illuminated by earthshine.

The next time you see the moon, think about where it is in the sky, when it will rise and set, and how this is related to its phase. Better yet, try explaining this to a friend. You too can be the wizard who predicts the motions of the heavens.

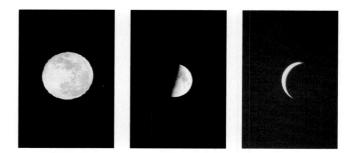

figure **5.22**   Photographs of different phases of the moon. When during the day will each rise and set?

Does the moon obey Kepler's laws?

The moon's orbit around the earth is more complicated than those of the planets because two bodies, the earth and the sun, exert strong forces on the moon, rather than just one (fig. 5.23). The earth is much closer to the moon than the sun is, but the sun has a much larger mass than the earth, so the sun's effect is still appreciable. First, let's consider just the effects of the earth on the moon's motion.

Everyday Phenomenon

box 5.2

Explaining the Tides

The Situation. Anyone who has lived near the ocean is familiar with the regular variation of the tides. Roughly twice a day the tides go in and go out again. The actual cycle of two high tides and two low tides is closer to 25 hours. Sometimes high tide is higher and low tide is lower than at other times—these times correspond to the full moon or the new moon.

The times when high tides and low tides happen shift from day to day because of their 25-hour cycle, but the pattern repeats monthly. How do we explain this behavior?

The Analysis. The monthly cycle and the correlations of the highest tides with the phase of the moon suggest a lunar influence. Both the moon and the sun exert gravitational forces on the earth. The sun exerts the stronger force because of its much larger mass, but the moon is much closer, and variations in its distance from the earth may be significant. The gravitational force depends on $1/r^2$, so its strength will vary as the distance r varies, as indicated in the drawing on page 95.

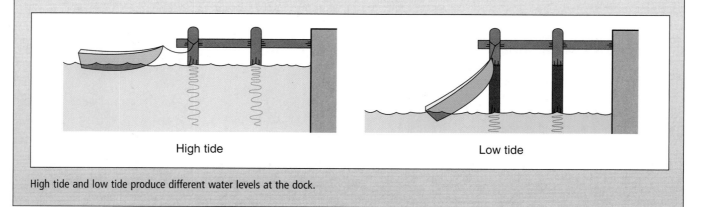

High tide and low tide produce different water levels at the dock.

(continued)

The physics of the situation is the same as for the orbits of the planets around the sun. The gravitational attraction between the moon and the earth provides the centripetal acceleration to keep the moon moving in its roughly circular orbit. By Newton's law of gravitation, the gravitational force acting on the moon is proportional to $1/r^2$, where r is the distance between the center of the moon and the center of the earth. The tides can be explained by this dependence of the gravitational force on distance. (See Everyday Phenomenon Box 5.2.)

Like the planets, the moon's orbit is an ellipse but with the earth at one focus of the ellipse rather than the sun. The sun also exerts a force on the moon that distorts the ellipse, causing the moon's orbit around the earth to oscillate about a true elliptical path as the moon and earth orbit together around the sun. Calculating these oscillations was a problem that kept mathematical physicists busy for many years.

Kepler's first and second laws of planetary motion are approximately true for the moon, provided that we substitute the earth for the sun in the statement of these laws.

Kepler's third law shows some differences between the moon and the planets. When Newton derived the expression for the ratio in Kepler's third law, he arrived at the expression

$$\frac{T^2}{r^3} = \frac{4\pi^2}{Gm_s}$$

where m_s is the mass of the sun. For the moon, we would replace the mass of the sun with the mass of the earth. We get a different ratio for the moon's orbit around the earth than for the orbits of the planets around the sun.

Orbits of artificial satellites

Any satellite orbiting the earth must have the same value for the ratio T^2/r^3 as the moon. Kepler's third law holds for any satellite of earth, then, as long as we keep in mind that the ratio will not have the same value as it does for the orbits of the planets. The value of this ratio for earth satellites is calculated either from the earth's mass or from the

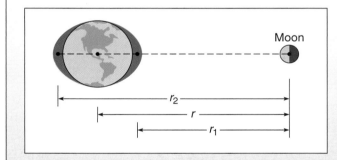

Because it depends on distance, the gravitational force per unit mass exerted by the moon on different parts of the earth (and water in the oceans) gets weaker as we move from the side nearer the moon to the far side. (The bulges here are greatly exaggerated.)

Since water is a fluid (except when frozen), the water that makes up the oceans moves over the more rigid crust of the earth. The primary force acting on the water is the gravitational attraction of the earth that holds the water to the earth's surface. The gravitational force exerted by the moon on the water is also significant, however, and its strength per unit mass is greatest on the side of the earth closest to the moon and weakest on the opposite side of the earth because of the difference in distance.

This difference in strength of the moon's pull produces a bulge in the water surface on both sides of the earth. The bulge on the side nearest the moon results from the water being pulled toward the moon by a stronger force per unit

mass than the force per unit mass exerted on the rest of the earth. This produces a high tide. The water will flow nearer to the top of the dock.

On the opposite side of the earth, the *earth* is being pulled by the moon with a stronger force per unit mass than the water. Since the earth is pulled away (slightly) from the water, this also produces a high tide. The forces exerted by the moon are small compared to the force that the water and the earth exert on each other but are still large enough to produce the tides.

When the sun and the moon both line up with the earth during the new moon or full moon, the sun also contributes to this difference in forces and produces bulges on either side of the earth, adding to those produced by the moon. The highest tides occur during a full moon or a new moon because of this combination of the moon and sun.

Why is the cycle 25 hours rather than 24 hours? The high-tide bulges occur on either side of the earth along the line joining the moon and the earth. The earth rotates underneath these bulges with a period of 24 hours, but in this time, the moon also moves, since it orbits the earth with a period of 27.3 days. In one day, therefore, the moon has moved through roughly $1/27$ of its orbital cycle, causing the time when the moon again lines up with a given point on the earth to be a little longer than 1 day. This additional time is approximately $1/27$ of 24 hours, or a little less than an hour.

This model was conceived by Newton and accounts neatly for the major features of the tides. The variation of the gravitational force with distance is the key to the explanation.

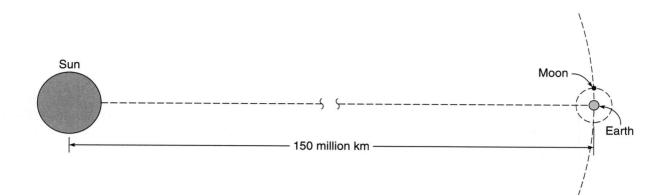

figure 5.23 The moon is influenced by gravitational attraction to both the earth and the sun. (Distances and sizes not drawn to scale.)

values of the period and average distance of the moon's orbit.

Any artificial satellite of earth must have the same value for this ratio. If its distance from the center of the earth r is smaller than the moon's distance, its orbital period T must also be smaller to keep the ratio T^2/r^3 the same. Using this ratio, we can calculate the appropriate distance from the earth for any satellite if we know its orbital period. For example, a satellite with a **synchronous orbit** has a period of 24 hours, which keeps it above the same point on the earth as the earth rotates. From the third-law ratio, we find a distance r of 42 000 km for such a satellite (measured from the center of the earth). Since the radius of the earth is 6370 km, this is roughly seven times the radius of the earth. Quite a ways up, but not nearly as high as the moon.

Most artificial satellites are even closer to the earth. The original Russian satellite, Sputnik, for example, had a period of about 90 minutes or 1.5 h. Using the third-law ratio, this yields an average distance from the center of the earth of 6640 km. Subtracting the radius of the earth, 6370 km, indicates that this distance is only 270 km above the surface of the earth. The shorter the period, the closer the satellite is to the earth. The orbital period cannot be much shorter than Sputnik's before atmospheric drag becomes too large for motion to be sustained. Obviously, the orbit cannot have a radius smaller than the earth's radius.

The orbits of different satellites are planned to meet different objectives. Some are close to circular, others much more elongated ellipses (fig. 5.24). The plane of the orbit can pass through the poles of the earth (polar orbit) or take any orientation between the poles and the equator. It all depends on the mission of the satellite.

Artificial satellites have become a routine feature of today's world that did not exist before 1958 when Sputnik was launched. Their uses are many, including communications, surveillance, weather observations, and various mili-

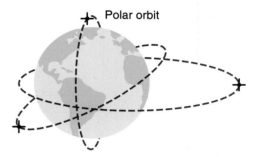

figure 5.24 The orbits of different artificial satellites can have different orientations and elliptical shapes.

tary applications. The basic physics of their behavior is accounted for by Newton's theory. If Newton could return, he might be amazed at the developments, but for him, the analysis would be routine.

The motion of the moon around the earth is governed by the same principles as that of the planets around the sun. The gravitational force provides the centripetal acceleration that keeps the moon in an approximately elliptical orbit. The moon is illuminated primarily by the sun, and the phases of the moon can be explained by the moon's position with regard to the sun and the earth. The full moon occurs when the sun and moon are on opposite sides of the earth. Other satellites of earth are governed by these principles, but Kepler's third-law ratio has a different value for satellites of earth (including the moon) than it does for the planets. The moon is no longer alone; it has been joined by many much smaller objects buzzing around the earth in lower orbits.

summary

Objects moving in circular paths are accelerated because the direction of the velocity vector continually changes. The forces involved in producing this centripetal acceleration were examined for the motion of a ball on a string, cars rounding curves, a rider on a Ferris wheel, and finally the planets moving around the sun. The force providing the centripetal acceleration for planetary motion is described by Newton's law of gravitation.

1 Centripetal acceleration. Centripetal acceleration is the acceleration involved in changing the direction of the velocity vector. It is proportional to the square of the speed of the object and inversely proportional to the radius of the curve.

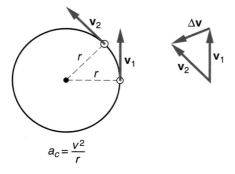

$$a_c = \frac{v^2}{r}$$

2 Centripetal forces. A centripetal force is any force or combination of forces that acts on a body to produce the centripetal acceleration, including friction, normal forces, tension in a string, or gravity. The total force is related to the centripetal acceleration by Newton's second law.

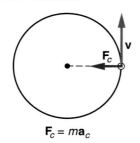

$$F_c = m\mathbf{a}_c$$

3 Planetary motion. Kepler's three laws of planetary motion describe the orbits of the planets around the sun. The orbits are ellipses that sweep out equal areas in equal times (the first and second laws). The third law states a relationship between the period of the orbit and the distance of the planet from the sun.

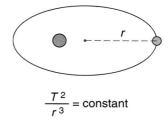

$$\frac{T^2}{r^3} = \text{constant}$$

4 Newton's law of universal gravitation. Newton's law of universal gravitation states that the gravitational force between two masses is proportional to each of the masses and inversely proportional to the square of the distance between the masses. Using this law with his laws of motion, Newton could derive Kepler's laws of planetary motion.

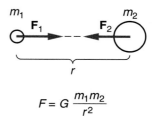

$$F = G \frac{m_1 m_2}{r^2}$$

5 The moon and other satellites. The moon's orbit around the earth can also be described by Kepler's laws, provided we substitute the mass of the earth for the sun in the expression for the period. Artificial satellites have the same ratio T^2/r^3 as that for the moon.

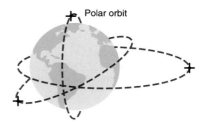

key terms

questions

*Questions identified with an asterisk are more open-ended than the others. They call for lengthier responses and are more suitable for group discussion.

Q1. Suppose that the speed of a ball moving in a horizontal circle is increasing at a steady rate. Is this increase in speed produced by the centripetal acceleration? Explain.

Q2. A car travels around a curve with constant speed.
   a. Does the velocity of the car change in this process? Explain.
   b. Is the car accelerated? Explain.

Q3. Two cars travel around the same curve, one at twice the speed of the other. After traveling the same distance, which car, if either, has experienced the larger change in velocity? Explain.

Q4. A car travels the same distance at constant speed around two curves, one with twice the radius of curvature of the other. For which of these curves is the change in velocity of the car greater? Explain.

*Q5. The centripetal acceleration depends upon the square of the speed rather than just being proportional to the speed. Why does the speed enter twice? Explain.

Q6. A ball on the end of a string is whirled with constant speed in a counterclockwise horizontal circle. At point A in the circle, the string breaks. Which of the curves sketched below most accurately represents the path that the ball will take after the string breaks (as seen from above)? Explain.

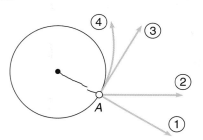

Q7. Before the string breaks in question 6, is there a net force acting upon the ball? If so, what is its direction? Explain.

Q8. For a ball being twirled in a horizontal circle at the end of a string, does the vertical component of the force exerted by the string produce the centripetal acceleration of the ball? Explain.

Q9. A car travels around a flat (nonbanked) curve with constant speed.
   a. Sketch a diagram showing all of the forces acting on the car.
   b. What is the direction of the net force acting on the car? Explain.

Q10. Is there a maximum speed at which the car in question 9 will be able to negotiate the curve? If so, what factors determine this maximum speed? Explain.

Q11. If a curve is banked, is it possible for a car to negotiate the curve even when the frictional force is zero due to very slick ice? Explain.

*Q12. If a ball is whirled in a vertical circle with constant speed, at what point in the circle, if any, is the tension in the string the greatest? Explain. (Hint: Compare this situation to the Ferris wheel described in section 5.2.)

Q13. Sketch the forces acting upon a rider on a Ferris wheel when the rider is at the top of the cycle, labeling each force clearly. Which force is largest at this point, and what is the direction of the total force? Explain.

*Q14. In what way did the heliocentric view of the solar system proposed by Copernicus provide a simpler explanation of planetary motion than the geocentric view of Ptolemy? Explain.

Q15. Did Ptolemy's view of the solar system require motion of the earth, rotational or otherwise? Explain.

*Q16. Heliocentric models of the solar system (Copernican or Keplerian) require that the earth rotate on its axis producing surface speeds of roughly 1000 MPH. If this is the case, why do we not feel this tremendous speed? Explain.

Q17. How did Kepler's view of the solar system differ from that of Copernicus? Explain.

Q18. Consider the method of drawing an ellipse pictured in figure 5.15. How would we modify this process to make the ellipse into a circle, which is a special case of an ellipse? Explain.

Q19. Does a planet moving in an elliptical orbit about the sun move fastest when it is farthest from the sun or when it is nearest to the sun? Explain by referring to one of Kepler's laws.

Q20. Does the sun exert a larger force on the earth than that exerted on the sun by the earth? Explain.

Q21. Is there a net force acting on the planet earth? Explain.

Q22. Three equal masses are located as shown in the diagram. What is the direction of the total force acting upon m_2? Explain.

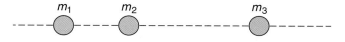

Q23. Two masses are separated by a distance r. If this distance is doubled, is the force of interaction between the two masses doubled, halved, or changed by some other amount? Explain.

Q24. A painter depicts a portion of the night sky as shown in the diagram below, showing the stars and a crescent moon. Is this view possible? Explain.

Q25. At what times during the day or night would you expect the new moon to rise and set? Explain.

Q26. At what times of the day or night does the half moon rise or set? Explain.

Q27. Are we normally able to see the new moon? Explain.

Q28. During what phase of the moon can a solar eclipse occur? Explain.

Q29. Is Kepler's third law valid for artificial satellites orbiting about the earth? Explain.

*Q30. A synchronous satellite is one that does not move relative to the surface of the earth; it is always above the same location. Why does such a satellite not just fall straight down to the earth? Explain.

Q31. Since the earth rotates on its axis once every 24 hours, why don't high tides occur exactly twice every 24 hours? Explain.

Q32. Would tides exist if the gravitational force did not depend upon the distance between objects? Explain.

Q33. Why is there a high tide rather than a low tide when the moon is on the opposite side of the earth from the ocean and the gravitational pull of the moon on the water is the weakest? Explain.

exercises

E1. A ball is traveling at a constant speed of 5 m/s in a circle with a radius of 0.8 m. What is the centripetal acceleration of the ball?

E2. A car rounds a curve with a radius of 25 m at a speed of 20 m/s. What is the centripetal acceleration of the car?

E3. A ball traveling in a circle with a constant speed of 3 m/s has a centripetal acceleration of 9 m/s². What is the radius of the circle?

E4. How much larger is the required centripetal acceleration for a car rounding a curve at 60 MPH than for one rounding the same curve at 30 MPH?

E5. A 0.25-kg ball moving in a circle at the end of a string has a centripetal acceleration of 4 m/s². What is the magnitude of the centripetal force exerted by the string on the ball to produce this acceleration?

E6. A car with a mass of 1200 kg is moving around a curve with a radius of 50 m at a constant speed of 20 m/s (about 45 MPH).
a. What is the centripetal acceleration of the car?
b. What is the magnitude of the force required to produce this centripetal acceleration?

E7. A car with a mass of 1000 kg travels around a banked curve with a constant speed of 27 m/s (about 60 MPH). The radius of curvature of the curve is 40 m.
a. What is the centripetal acceleration of the car?
b. What is the magnitude of the horizontal component of the normal force that would be required to produce this centripetal acceleration in the absence of any friction?

E8. A Ferris wheel at a carnival has a radius of 12 m and turns so that the speed of the riders is 8 m/s.
a. What is the magnitude of the centripetal acceleration of the riders?
b. What is the magnitude of the net force required to produce this centripetal acceleration for a rider with a mass of 70 kg?

E9. What is the ratio of the earth's orbital period about the sun to the earth's period of rotation about its own axis?

E10. Joe has a weight of 800 N (about 180 lb) when he is standing on the surface of the earth. What would his weight (the gravitational force due to the earth) be if he doubled his distance from the center of the earth by flying in a spacecraft?

E11. Two masses are attracted by a gravitational force of 0.36 N. What will the force of attraction be if the distance between the two masses is tripled?

E12. Two 200-kg masses (440 lb) are separated by a distance of 1 m. Using Newton's law of gravitation, find the magnitude of the gravitational force exerted by one mass on the other.

E13. Two masses are attracted by a gravitational force of 0.14 N. What will the force of attraction be if the distance between these two masses is halved?

E14. The acceleration of gravity at the surface of the moon is approximately ⅙ that at the surface of the earth (9.8 m/s²). What is the weight of an astronaut standing on the moon whose weight on earth is 180 lb?

E15. The acceleration of gravity on the surface of Jupiter is 26.7 m/s². What is the weight on Jupiter of a woman whose weight on earth is 110 lb?

E16. The time separating high tides is 12 hours and 25 minutes. If high tide occurs at 3:30 P.M. one afternoon:
a. At what time will high tide occur the next afternoon?
b. When would you expect low tides to occur the next day?

challenge problems

CP1. A 0.20-kg ball is twirled at the end of a string in a horizontal circle with a radius of 0.60 m. The ball travels with a constant speed of 4.0 m/s.
  a. What is the centripetal acceleration of the ball?
  b. What is the magnitude of the horizontal component of the tension in the string required to produce this centripetal acceleration?
  c. What is the magnitude of the vertical component of the tension required to support the weight of the ball?
  d. Draw to scale a vector diagram showing these two components of the tension and estimate the magnitude of the total tension from your diagram. (See appendix C.)

CP2. A Ferris wheel with a radius of 12 m makes one complete rotation every 8 seconds.
  a. Using the fact that the distance traveled by a rider in one rotation is $2\pi r$, the circumference of the wheel, find the speed with which the riders are moving.
  b. What is the magnitude of their centripetal acceleration?
  c. For a rider with a mass of 40 kg, what is the magnitude of the centripetal force required to keep that rider moving in a circle? Is the weight of the rider large enough to provide this centripetal force at the top of the cycle?
  d. What is the magnitude of the normal force exerted by the seat on the rider at the top of the cycle?
  e. What would happen if the Ferris wheel is going so fast that the weight of the rider is not sufficient to provide the centripetal force at the top of the cycle?

CP3. A car with a mass of 900 kg is traveling around a curve with a radius of 60 m at a constant speed of 25 m/s (56 MPH). The curve is banked at an angle of 15 degrees.
  a. What is the magnitude of the centripetal acceleration of the car?
  b. What is the magnitude of the centripetal force required to produce this acceleration?
  c. What is the magnitude of the vertical component of the normal force acting upon the car to counter the weight of the car?
  d. Draw a diagram of the car (as in fig. 5.8) on the banked curve. Draw to scale the vertical component of the normal force. Using this diagram, find the magnitude of the total normal force, which is perpendicular to the surface of the road.
  e. Using your diagram, estimate the magnitude of the horizontal component of the normal force. Is this component sufficient to provide the centripetal force?

CP4. Assume that a passenger in a rollover accident must turn through a radius of 3.0 m to remain in the seat of the vehicle. Assume also that the vehicle makes a complete turn in 1 second.
  a. Using the fact that the circumference of a circle is $2\pi r$, what is the speed of the passenger?
  b. What is the centripetal acceleration? How does it compare to the acceleration due to gravity?
  c. If the passenger has a mass of 60 kg, what is the centripetal force required to produce this acceleration? How does it compare to the passenger's weight?

CP5. The sun's mass is 1.99×10^{30} kg, the earth's mass is 5.98×10^{24} kg, and the moon's mass is 7.36×10^{22} kg. The average distance between the moon and the earth is 3.82×10^{8} m, and the average distance between the earth and the sun is 1.50×10^{11} m.
  a. Using Newton's law of gravitation, find the average force exerted on the earth by the sun.
  b. Find the average force exerted on the earth by the moon.
  c. What is the ratio of the force exerted on the earth by the sun to that exerted by the moon? Will the moon have much of an impact on the earth's orbit about the sun?
  d. Using the distance between the earth and the sun as the average distance between the moon and the sun, find the average force exerted on the moon by the sun. Will the sun have much impact on the orbit of the moon about the earth?

CP6. The period of the moon's orbit about the earth is 27.3 days, but the average time between full moons is approximately 29.3 days. The difference is due to the motion of the earth about the sun.
  a. Through what fraction of its total orbital period does the earth move in one period of the moon's orbit?
  b. Draw a sketch of the sun, the earth, and the moon with the moon in the full moon condition. Then, 27.3 days later sketch the moon's position again for the new position of the earth. If the moon is in the same position relative to the earth as it was 27.3 days earlier, is this a full moon?
  c. How much farther would the moon have to go to reach the full moon condition? Show that this represents approximately an extra two days.

home experiments and observations

HE1. Tape a string half a meter or so in length securely to a small rubber ball. Practice whirling the ball in both horizontal and vertical circles and make these observations:
  a. For horizontal motion of the ball, how does the angle that the string makes with the horizontal vary with the speed of the ball?
  b. If you let go of the string at a certain point in the circle, what path does the ball follow after release?
  c. Can you feel differences in tension in the string for different speeds of the ball? How does the tension vary with speed?
  d. For a vertical circle, how does the tension in the string vary for different points in the circle? Is it greater at the bottom than at the top when the ball moves with constant speed?

HE2. Tie a small paper cup to a string, attaching it at two points near the rim as shown in the diagram. Take a marble or other small object and place it in the cup.
  a. Whirl the cup in a horizontal circle. Does the marble stay in the cup?
  b. Whirl the cup in a vertical circle. Does the marble stay in the cup? What keeps the marble in the cup at the top of the circle?
  c. Try slowing the cup down. Does the marble stay in the cup?
  d. If you are brave, try replacing the marble with water. Under what conditions does the water stay in the cup?

HE3. Observe the position and phase of the moon on several days in succession and at regularly chosen times during the day and evening. (It is probably best to choose a point near the first quarter of the moon's cycle when the moon is visible in the afternoon and evening.)
  a. Sketch the shape of the moon on each successive day. Does this shape change for different times in the same day?
  b. Can you devise a method for accurately noting changes in the position of the moon at a set time, say, 10 P.M., on successive days? A fixed sighting point, a meter stick, and a protractor may be useful. Describe your technique.
  c. By how much does the position of the moon change from one day to the next at your regular chosen time?

HE4. Consult your instructor or other sources to find out what planets are observable in the evening during the current month. Venus, Jupiter, or Mars are usually the best candidates.
  a. Locate the planet visually and observe it with binoculars if possible. How does the planet differ in appearance from that of nearby stars?
  b. Sketch the position of the planet relative to nearby stars for several nights. How does this position change?

Energy and Oscillations

chapter

6

chapter overview

We usually approach energy by first considering how it is added to a system. This involves the concept of work, which has a specialized meaning in physics. If a force does work on a system, the energy of the system increases. Work is a means of transferring energy.

We begin by defining work and showing how to find it in simple cases. In different circumstances, work done on a system increases either the kinetic energy or the potential energy of the system. Finally, we will tie these ideas together by introducing the principle of conservation of energy and applying it to practical situations, including oscillations.

chapter outline

Have you ever watched a ball on the end of a string swing back and forth? A pendant on the end of a chain (fig. 6.1), a swing in the park, and the pendulum on a grandfather clock all display the same hypnotic motion. Galileo (it is said) amused himself during boring sermons in church by watching the chandeliers sway slowly back and forth at the end of their chains.

What intrigued Galileo is the way a pendulum always seems to return to the same position at the end of each swing. It may fall a little short of the earlier position in successive swings, but the motion goes on for a long time before coming to a complete stop. On the other hand, the velocity is continually changing, from zero at the end points of the swing to a maximum at the low point in the path. How can the pendulum go through such changes in velocity and yet always return to its starting point?

Evidently, something is being saved or *conserved.* The quantity that remains constant (and is conserved) turns out to be what we now call *energy.* Energy did not play a role in Newton's theory of mechanics. It was not until the nineteenth century that energy and energy transformations were elevated to the central position that they now hold in our understanding of the physical world.

The motion of a pendulum and other types of oscillation can be understood using the principle of conservation of mechanical energy. The potential energy that the pendulum has at its end points is converted to kinetic energy at the low point—and then back to potential

figure 6.1 A pendant swinging at the end of a chain. Why does it return to approximately the same point after each swing?

energy. What is energy, though, and how does it get into the system in the first place? Why does energy now play a central role in physics and all of science?

Energy is the basic currency of the physical world. To spend energy wisely, we must understand it.

6.1 Simple Machines, Work, and Power

If you make a pendulum by fastening a ball to the end of a string (fig. 6.2), what do you do to start it swinging? In other words, how do you get energy into the system? Usually, you would start by pulling the ball away from the center position directly below the point from which the string is suspended. To do so, you must apply a force to the ball with your hand and move the ball some distance.

To a physicist, applying a force to move an object some distance involves doing *work,* even though the actual exertion may be slight. Doing work on a system increases the energy of the system, and this energy can then be used in the motion of the pendulum. How do we define work, though, and how can simple machines demonstrate the usefulness of the idea?

What are simple machines?

An early application of work was the analysis of the devices such as levers, pulley systems, or inclined planes that we call *simple machines.* A **simple machine** is any mechanical device that multiplies the effect of an applied force. A lever is one example of a simple machine. By ap-

figure 6.2 The force applied does work to move the ball from its original position directly below the point of suspension.

plying a small force at one end of a lever, a larger force can be exerted on the rock at the opposite end (fig. 6.3).

What price do you pay for this multiplying effect of the applied force? To move the rock a small distance, the other end of the lever must move through a larger distance.

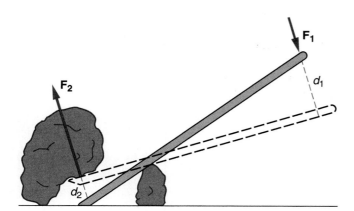

figure 6.3   A lever is used to lift a rock. A small force F_1 generates a larger force F_2 to lift the rock, but F_1 acts through a larger distance d_1 than does F_2.

Generally, with simple machines, we get by with a small force if we are willing to apply that force over a large distance. The output force at the other end may be large, but it acts only over a short distance.

The pulley system shown in figure 6.4 is another simple machine that achieves a similar result. In this system, the tension in the rope pulls up on either side of the pulley supporting the weight. If the system is in equilibrium, the tension in the rope is only half the weight being lifted, since there are, in effect, two ropes pulling up on the pulley. But to lift the pulley and its load a certain height, the person must move the rope *twice* the distance that the load moves. (Both rope segments on either side of the pulley must decrease in length by an amount equal to the increase in the height of the load.)

The net result of using the pulley system illustrated in figure 6.4 is that you can lift a weight a certain height by applying a force equal to only *half* the weight being lifted. However, we must pull the rope *twice* the distance the weight is lifted. This way the product of the force and the distance moved will be the same for the input force applied by the person to the rope as for the output force exerted on the load. The quantity *force times distance* is thus conserved (if frictional losses are small). We call this product *work,* and the result for an ideal simple machine is

work output = work input.

The ratio of the output force to the input force is called the **mechanical advantage** of the simple machine. For our pulley system, the mechanical advantage is 2. The output force that lifts the load is twice the input force exerted by the person pulling on the rope.

How is work defined?

Our discussion of simple machines shows that the quantity force times distance has a special significance. Suppose

figure 6.4   A simple pulley system is used to lift a weight. The tension in the rope pulls up on either side of the lower pulley, so the tension is only half the size of the supported weight.

that you apply a constant horizontal force to a heavy crate to move it across a concrete floor, as illustrated in figure 6.5. You would agree that you have done work to move the crate and that the farther you move it, the more work you will do.

The amount of work that you do also depends on how hard you have to push to keep the crate moving. These are the basic ideas that we use in defining work: work depends both on the *strength* of the applied force and the *distance* that the crate is moved. If the force and the distance moved are in the same direction, then **work** is the applied force multiplied by the distance that the crate moves under the influence of this force, or

$$\text{work} = \text{force} \times \text{distance}$$
$$W = Fd,$$

where W is the work and d is the distance moved. The units of work will be units of force multiplied by units of distance, or newton-meters (N·m) in the metric system. We call this unit a *joule* (J). The joule is the basic metric unit of energy. (1 J = 1 N·m)

The first part of the sample exercise in Try This Box 6.1 shows how we find the work done in a simple case. A horizontal force of 50 N is used to pull a crate a distance of 4 m, resulting in 200 J of work done on the crate by the applied force. In doing this work, we transfer 200 J of energy to the crate and its surroundings from the person applying the force. The person loses energy; the crate and its surroundings gain energy.

figure 6.5 A crate is moved a distance *d* across a concrete floor under the influence of a constant horizontal force **F**.

try this box 6.1

Sample Exercise: How Much Work?

A crate is pulled a distance of 4 m across the floor under the influence of a 50-N force applied by a rope to the crate. What is the work done on the crate by the 50-N force if
   a. the rope is horizontal, parallel to the floor?
   b. the rope pulls at an angle to the floor, so that the horizontal component of the 50-N force is 30 N (fig. 6.6)?

a. $F = 50$ N $W = Fd$
   $d = 4$ m $= (50$ N$)(4$ m$)$
   $W = ?$ $= \mathbf{200\ J}$

b. $F_h = 30$ N $W = Fd$
   $d = 4$ m $= (30$ N$)(4$ m$)$
   $W = ?$ $= \mathbf{120\ J}$

rected upward, rather than parallel to the floor. The box does not move in the direction of the force. Picture the force as having two components, one parallel to the floor and the other perpendicular to the floor. Only the component of the force in the direction of motion is used in computing the work. The component perpendicular to the motion does no work.

By taking direction into account, we can complete the definition of work:

> The work done by a given force is the product of the component of the force along the line of motion of the object multiplied by the distance that the object moves under the influence of the force.

How is power related to work?

When a car accelerates, energy is transferred from the fuel in the engine to the motion of the car. Work is done to move the car, but often we are more concerned with *how*

Does any force do work?

In our initial example, the force acting on the crate was in the same direction as the motion produced. What about other forces acting on the crate—do they do work? The normal force of the floor pushes upward on the crate, for example, but the normal force has no direct effect in producing the motion because it is perpendicular to the direction of the motion. Forces *perpendicular* to the motion, such as the normal force or the gravitational force acting on the crate, *do no work* when the crate moves horizontally.

What if the force acting on an object is neither perpendicular nor parallel to the direction of the object's motion? In this case, we do not use the total force in computing work. Instead, we use only that portion or component of the force in the direction of the motion. This idea is illustrated in figure 6.6 and in the second part of Try This Box 6.1.

In figure 6.6, the rope used to pull the crate is at an angle to the floor, so that part of the applied force is di-

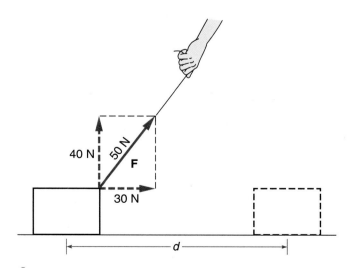

figure 6.6 A rope is used to pull a box across the floor. Only the portion of the force that is parallel to the floor is used in computing the work.

fast this work is accomplished. The rate at which this work can be done depends on the **power** of the engine. The shorter the time, the greater the power. Power can be defined as

---

Power is the rate of doing work; it is found by dividing the amount of work done by the time required.

$$\text{power} = \frac{\text{work}}{\text{time}}$$

$$P = \frac{W}{t}$$

In the first part of the example in Try This Box 6.1, we computed a work value of 200 J for moving a crate 4 m across the floor using a force of 50 N. If the crate is in motion for 10 seconds, the power is found by dividing 200 J by 10 seconds, yielding a power of 20 J/s. A joule per second (J/s) is called a *watt* (W), the metric unit of power. We use watts commonly in discussing electric power, but watts are also used more generally for any situation involving the rate of transfer of energy.

Another unit of power still used to describe the power of automobile engines is horsepower (hp). One horsepower is equal to 746 watts or 0.746 kilowatt (kW). The day may come when we routinely compare the power of different engines in kilowatts rather than in horsepower, but we are not there yet. The relationship of horsepower to the typical horse is dubious, but comparing the iron horse to the flesh-and-blood kind still has a certain appeal.

Work is the applied force times the distance moved, provided that the force acts along the line of motion of the object. In simple machines, work output can be no greater than work input, even though the output force is larger than the input force. Power is the rate of doing work: the faster the work is done, the greater the power. Doing work on an object increases the energy of the object or system, as in our initial example of pulling the pendulum bob away from equilibrium.

6.2 Kinetic Energy

Suppose that the force applied to move a crate is the only force acting on the crate in the direction of motion. What happens to the crate then? According to Newton's second law, the crate will accelerate, and its velocity will increase. Doing work on an object increases its energy. We call the energy associated with the motion of the object **kinetic energy.**

Since work involves the transfer of energy, the amount of kinetic energy gained by the crate should be equal to the amount of work done. How can we define kinetic energy

so that this is indeed the case? Work serves as the starting point.

How do we define kinetic energy?

Imagine that you are pushing a crate across the floor (fig. 6.5). If you place the crate on rollers with good bearings, the frictional forces may be small enough to be ignored. The force that you apply will then accelerate the crate. If you knew the mass of the crate, you could find its acceleration from Newton's second law of motion.

As the crate gains speed, you will have to move faster to keep applying a constant force. For equal time intervals, the crate would move larger distances as its speed increases, and you would find yourself doing work more rapidly. For constant acceleration, the distance traveled is proportional to the square of the final speed. The work done is therefore also proportional to the square of the speed.

Since the work done should equal the increase in kinetic energy, the kinetic energy must increase with the square of the speed. If the crate begins from rest, the exact relationship is

$$\text{work done} = \text{change in kinetic energy} = \tfrac{1}{2}mv^2.$$

We often use the abbreviation *KE* to represent kinetic energy.

Kinetic energy is the energy of an object associated with its motion and is equal to one-half the mass of the object times the square of its speed.

$$KE = \tfrac{1}{2}mv^2$$

Figure 6.7 illustrates the process. If the crate is initially at rest, its kinetic energy is equal to zero. After being accelerated over a distance d, it has a final kinetic energy of $\tfrac{1}{2}mv^2$, which is equal to the work done on the crate. The work done is actually equal to the *change* in kinetic energy. If the crate was already moving when you began pushing, its increase in kinetic energy would equal the work done.

In Try This Box 6.2, we highlight these ideas by calculating the energy gained by the crate in two different ways.

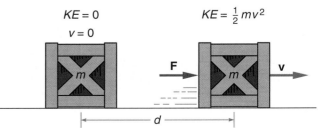

figure **6.7**   The work done on an object by the net force acting on the object results in an increase in the object's kinetic energy.

try this box 6.2

Sample Exercise: Work and Kinetic Energy

Starting from rest on a frictionless floor, you move a 100-kg crate by applying a net force of 50 N for a time of 4 s. This results in a final speed of 2 m/s after the crate moves a distance of 4 m (see challenge problem 2). Find
  a. the work done on the crate.
  b. the final kinetic energy of the crate.

a. $F = 50$ N $W = Fd$
   $d = 4$ m $= (50$ N$)(4$ m$)$
   $W = ?$ $= \textbf{200 J}$

b. $m = 100$ kg $KE = \frac{1}{2} mv^2$
   $v = 2.0$ m/s
   $KE = ?$ $= \frac{1}{2} (100$ kg$)(2$ m/s$)^2$

                                         $= \textbf{200 J}$

In the first method, we use the definition of work. In the second, we use the definition of kinetic energy. We find that 200 J of work done on the crate results in an increase in kinetic energy of 200 J. It is no accident that these values are equal. Our definition of kinetic energy guarantees this to be true.

What is negative work?

If work done on an object increases its kinetic energy, can work also decrease the energy of an object? Forces can decelerate objects as well as accelerate them. Suppose, for example, that we apply the brakes to a rapidly moving car, and the car skids to a stop. Does the frictional force exerted by the road surface on the tires of the car do work?

When the car skids to a stop, it loses kinetic energy. A decrease in kinetic energy can be thought of as a negative change in kinetic energy. If the change in kinetic energy is negative, the work done on the car should also be negative.

Note that the frictional force exerted on the car acts in the *opposite* direction to the motion of the car shown in figure 6.8. When this is so, we say that the work done on the car by the force is **negative work,** removing energy from the system (the car) rather than increasing its energy. For a frictional force of magnitude f, the work done is $W = -fd$, if the car moves a distance d while decelerating.

Stopping distance for a moving car

The kinetic energy of the car is not proportional to the speed but rather to the *square* of the speed. If we double the speed, the kinetic energy *quadruples*. Four times as much work must be done to reach the doubled speed as was done to reach the original speed. Likewise, if we stop the car, four times as much energy must be removed.

A practical application is the stopping distances of cars traveling at different speeds. The amount of negative work required to stop the car is equal to the kinetic energy of the car before the brakes are applied. This amount of energy must be removed from the system. Since kinetic energy is proportional to the square of the speed, the work required (and the stopping distance) increases rapidly with the speed of the car. For example, the kinetic energy is four times as large for a car traveling at 60 MPH as for one traveling at 30 MPH. Doubling the speed requires four times as much negative work to remove the kinetic energy. The stopping distance at 60 MPH will be four times that required at 30 MPH, since the work done is proportional to the distance (assuming the frictional force is constant).

In fact, the frictional force varies with the speed of the car. If you look at the stopping distances in driver-training manuals, you will see that they do indeed increase rapidly with speed, although not exactly in proportion to the square of the speed. The more kinetic energy present initially, the more negative work is required to reduce this energy to zero, and the greater the stopping distance.

> Kinetic energy is the energy associated with an object's motion, and it is equal to one-half the mass of the object times the square of its speed. The kinetic energy gained or lost by an object is equal to the work done by the net force accelerating or decelerating the object.

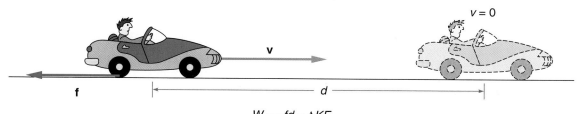

$$W = -fd = \Delta KE$$

figure **6.8** Frictional forces exerted on the car's tires by the road surface do negative work in stopping the car, resulting in a decrease in kinetic energy.

6.3 Potential Energy

Suppose that we lift a crate to a higher position on a loading dock, as in figure 6.9. Work is done in this process, but no kinetic energy is gained if the crate ends up just sitting on the dock. Has the energy of the crate increased? What happens to the work done by the lifting force?

Drawing back a bowstring or compressing a spring are similar. Work is done, but no kinetic energy is gained: instead the **potential energy** of the system increases. How does potential energy differ from kinetic energy?

Gravitational potential energy

To lift the crate in figure 6.9, we need to apply a force that pulls or pushes upward on the crate. The applied force will not be the only force acting on the crate. The gravitational attraction of the earth (the weight of the crate) pulls down on the crate. If we lift the crate with a force exactly equal to the force of gravity but opposite in direction, the net force acting on the crate will be zero, and the crate will not accelerate. We actually accelerate the crate a little bit at the start of the motion and decelerate it at the end of the motion, moving it with constant velocity during most of the motion.

The work done by the lifting force increases the **gravitational potential energy** of the crate. The lifting force and the gravitational force are equal in magnitude and opposite in direction, so the net force is zero and there is no acceleration. The lifting force does work by moving the object against the gravitational pull. If we let go of the rope, the crate will accelerate downward, gaining kinetic energy.

How much gravitational potential energy is gained? The work done by the lifting force is equal to the size of the force times the distance moved. The applied force is equal to the weight of the crate mg. If the crate is moved a height h, the work done is mg times h or mgh. The gravitational potential energy is equal to the work done,

$$PE = mgh,$$

where we use the abbreviation PE to represent potential energy.

The height h is the distance that the crate moves above some reference level or position. In Try This Box 6.3, we have chosen the original position of the crate on the ground to be our reference level. We usually choose the lowest point in the probable motion of the object as the reference level to avoid negative values of potential energy. The *changes* in potential energy are what is important, however, so the choice of reference level does not affect the physics of the situation.

The essence of potential energy

The term *potential energy* implies storing energy to use later on for other purposes. Certainly, this feature is present in the situation just described. The crate could be left indefinitely higher up on the loading dock. If we push it off the dock, though, it would rapidly gain kinetic energy as it fell. The kinetic energy, in turn, could be used to compress objects lying underneath, drive pilings into the ground, or for other useful mayhem (fig. 6.10). Kinetic energy also has this feature, however, so storing energy is not what distinguishes potential energy.

Potential energy involves *changing the position* of the object that is being acted on by a specific force. In the case of gravitational potential energy, that force is the gravitational attraction of the earth. The farther we move the object from the center of the earth, the greater the gravitational potential energy. Other kinds of potential energy involve different forces.

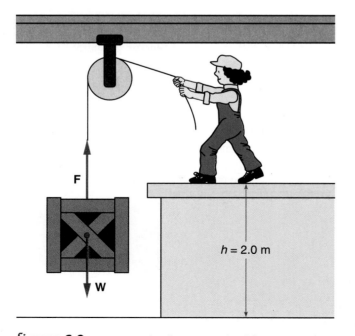

figure 6.9 A rope and pulley are used to lift a crate to a higher position on the loading dock, resulting in an increase in potential energy.

try this box 6.3

Sample Exercise: Potential Energy

A crate with a mass of 100 kg is lifted onto a loading dock 2 m above ground level. How much potential energy has been gained?

$m = 100$ kg $PE = mgh$
$h = 2$ m $= (100 \text{ kg})(9.8 \text{ m/s}^2)(2 \text{ m})$
                                       $= (980 \text{ N})(2 \text{ m})$
                                       $= \textbf{1960 J}$

figure **6.10** The potential energy of the raised crate can be converted to kinetic energy and used for other purposes.

What is elastic potential energy?

What happens if we pull on a bowstring or stretch a spring? In these examples, work is done by an applied force against an opposing **elastic force,** a force that results from stretching or compressing an object. Imagine a spring attached to a post, as in figure 6.11, with a wooden block or similar object attached to the other end of the spring. If we pull the block from the original position where the spring was unstretched, the system gains *elastic potential energy.* If we let go, the block would fly back.

Since a force must be applied over some distance to move the block, work is done in pulling against the force exerted by the spring. Most springs exert a force proportional to the distance the spring is stretched. The more the spring is stretched, the greater the force. This can be stated in an equation by defining the **spring constant** k that describes the stiffness of the spring. A stiff spring has a large spring constant. The force exerted by the spring is given by the spring constant multiplied by the distance stretched or $F = -kx$, where x is the distance that the spring is stretched, measured from its original unstretched position. This is often called Hooke's Law, named after Robert Hooke (1635–1703). The minus sign indicates that the force exerted by the spring pulls back on the object as the object moves away from its equilibrium position. Thus, if the mass is moved to the right, the spring pulls back to the left. If the spring is compressed, it pushes back to the right.

How do we find the increase in potential energy of such a system? As before, we need to find the work done by the force involved in changing the position of the object. We want the block to move without acceleration so the net force acting on the block is zero. The applied force must be adjusted so that it is always equal in magnitude but opposite in direction to the force exerted by the spring. This means that the applied force must increase as the distance x increases (fig. 6.12).

The increase in **elastic potential energy** is equal to the work done by the average force needed to stretch the spring. Figure 6.12 suggests that the average force is one-half the magnitude of the final force kx. The work done is the average force $\frac{1}{2}kx$ times the distance x, so

$$PE = \frac{1}{2}kx^2.$$

The potential energy of the stretched-spring system is one-half the spring constant times the square of the distance stretched. The same expression is valid when the spring is compressed. The distance x is then the distance that the spring is compressed from its original relaxed position.

The potential energy stored in the spring can be converted to other forms and put to various uses. If we let go of the block when the spring is either stretched or compressed,

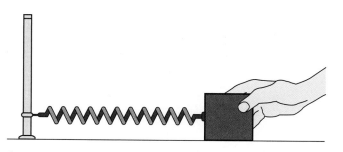

figure **6.11** A wooden block is attached to a spring tied to a fixed support at the opposite end. Stretching the spring increases the elastic potential energy of the system.

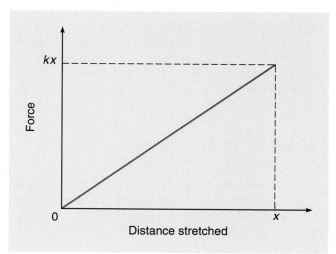

figure **6.12** The applied force used to stretch the spring varies with the distance stretched, going from an initial value of zero to a final value of kx.

the block will gain kinetic energy. Cocking a bow and arrow, squeezing a rubber ball, or stretching a rubber band are all familiar examples in which we generate elastic potential energy similar to the spring.

What are conservative forces?

Potential energy can result from work done against a variety of different forces besides gravity and springs. Work done against frictional forces, however, does not result in an increase in the potential energy of the system. Instead, heat is generated, which either transfers energy out of the system or increases the internal energy of the system at the atomic level. As discussed in chapter 11, this internal energy cannot be completely recovered to do useful work.

Forces such as gravity or elastic forces that lead to potential energy relationships are referred to as **conservative forces.** When work is done against conservative forces, the energy gained by the system is completely recoverable for use in other forms.

Potential energy is an object's energy by virtue of its position along the line of action of some conservative force (such as gravity or the spring force). Potential energy is stored energy associated with the position of the object rather than the object's motion. We find the potential energy by computing the work done to move the object against the conservative force. The system is poised to release that energy, converting it to kinetic energy or work done on some other system.

6.4 Conservation of Energy

The concepts of work, kinetic energy, and potential energy are now available to us. How can they help explain what is happening in systems like a pendulum?

Conservation of energy is the key. The total energy, the sum of the kinetic and potential energies, is a quantity that remains constant (is conserved) in many situations. We can describe the motion of a pendulum by tracking the energy transformations. What can this tell us about the system?

Energy changes in the swing of a pendulum

Imagine a pendulum consisting of a ball initially hanging motionless at the end of a string attached to a rigid support. You pull the ball to the side and release it to start it swinging. What happens to the energy of the system?

In the first step, work is done on the ball by your hand. The net effect of this work is to increase the potential energy of the ball, since the height of the ball above the ground increases as the ball is pulled to the side. The work

done transfers energy from the person doing the pulling to the system consisting of the pendulum and the earth. It becomes gravitational potential energy, $PE = mgh,$ where h is the height of the ball above its initial position (fig. 6.13).

When you release the ball, this potential energy begins to change to kinetic energy as the ball begins its swing. At the bottom of the swing (the initial position of the ball when it was just hanging), the potential energy is zero, and the kinetic energy reaches its maximum value. The ball does not stop at the low point; its motion continues to a point opposite the release point. During this part of the swing, the kinetic energy decreases, and the potential energy increases until it reaches the point where the kinetic energy is zero and the potential energy is equal to its initial value before release. The ball then swings back, repeating the transformation of potential energy to kinetic energy and back to potential energy (fig. 6.13).

What does it mean to say that energy is conserved?

As the pendulum swings, there is a continuing change of potential energy to kinetic energy and back again. The total mechanical energy of the system (the sum of the potential and kinetic energies) remains constant because there is no work being done on the system to increase or decrease its energy. The swing of the pendulum demonstrates the principle of **conservation of energy:**

If there are no forces doing work on a system, the total mechanical energy of the system (the sum of its kinetic energy and potential energy) remains constant.

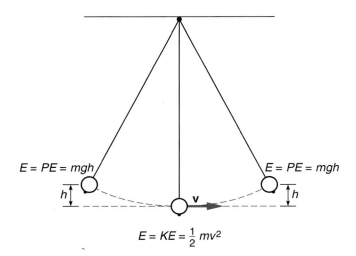

$$E = KE = \frac{1}{2} mv^2$$

figure **6.13** Potential energy is converted to kinetic energy and then back to potential energy as the pendulum swings back and forth.

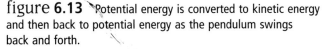

Work is pivotal. If no energy is added or removed by forces doing work, the total energy should not change. In symbols, this statement takes the form:

If $W = 0$, $E = PE + KE$ = constant,

where E is the symbol commonly used to represent the total energy.

We applied conservation of energy in describing the motion of the pendulum. Some points deserve close attention: for example, why do we *not* consider the work done by gravity on the pendulum? The answer is that the gravitational force becomes part of the system by including the gravitational potential energy of the ball in our description. Gravity is a conservative force already accounted for by potential energy.

What other forces act on the ball? The tension of the string acts in a direction perpendicular to the motion of the ball (fig. 6.14). This force does no work, because it has no component in the direction of the motion. The only other force that need concern us is air resistance. This force does negative work on the ball, slowly decreasing the total mechanical energy of the system. The total energy of the system is not completely constant in this situation. It would be constant only if air resistance were negligible. The air-resistive effects are often small, however, and can be ignored.

Why do we use the concept of energy?

What are the advantages of using the principle of conservation of energy? Imagine trying to describe the motion of the pendulum by direct application of Newton's laws of motion. You would have to deal with forces that vary continually in direction and magnitude as the pendulum moves. A full description using Newton's laws is quite complex.

Using energy considerations, however, we can make predictions about the behavior of a system much more easily than by applying Newton's laws. To the extent that we can ignore frictional effects, for example, we can predict that the ball will reach the same height at either end of its swing. The kinetic energy is zero at the end points of the swing where the ball momentarily stops, and at these points, the total energy equals potential energy. If no energy has been lost, the potential energy has the same value that it had at the point of release, which implies that the same height is reached ($PE = mgh$).

A demonstration sometimes performed in physics lecture rooms illustrates this idea dramatically by using a bowling ball as the pendulum bob. The bowling ball is suspended from a support near the ceiling so that, when pulled to one side, the ball is near the chin of the physics instructor. The instructor pulls the ball to this position, releases it to allow it to swing across the room, and stands without flinching as the ball returns and stops just a few inches from his or her chin (fig. 6.15). Not flinching requires some faith in the principle of conservation of energy! The success of this demonstration depends on the ball not be given any initial velocity when released—what happens if it is pushed?

We can also use the principle of conservation of energy to predict what the speed will be at any point in the swing. The speed is zero at the end points and has its maximum value at the low point of the swing. If we place our reference level for measuring potential energy at this low point,

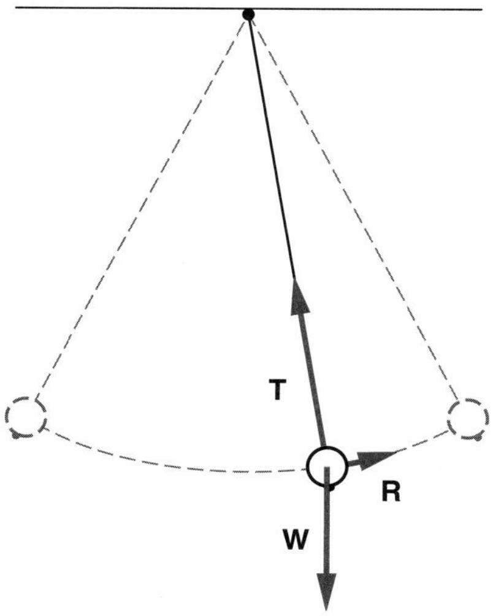

figure **6.14** Of the three forces acting on the ball, only the force of air resistance does work on the system to change its total energy. The tension does no work, and the work done by gravity is already included in the potential energy.

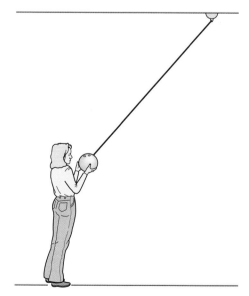

figure **6.15** A bowling ball at the end of a cable suspended from the ceiling is released and allowed to swing across the room and back, stopping just in time.

the potential energy will be zero there because the height is zero. All of the initial potential energy has been converted to kinetic energy. Knowing the kinetic energy at the low point allows us to compute the speed, as shown in Try This Box 6.4.

We could find the speed at any other point in the swing by setting the total energy at any point equal to the initial energy. Different values of the height h above the low point yield different values of the potential energy. The remaining energy must be kinetic energy. The system has only so much energy, either potential or kinetic energy or some of both, but it cannot exceed the initial value.

How is energy analysis like accounting?

A sled on a hill and a roller coaster illustrate the principle of conservation of energy. Conservation of energy can be used to make predictions about the speed of the sled or roller coaster that would be hard to make by direct application of Newton's laws. An energy accounting provides a better overview. The pole-vaulting example in Everyday Phenomenon Box 6.1 can also be analyzed in this way.

Consider the sled on the hill pictured in figure 6.16. A parent pulls the sled to the top of a hill, doing work on the sled and rider that increases their potential energy. At the top of the hill, the parent may do more work by giving the sled a push, providing it with some initial kinetic energy. The total work done by the parent is the energy input to the system and equals the sum of the potential and kinetic energies shown in table 6.1.

---

try this box 6.4

Sample Exercise: The Swing of a Pendulum

A pendulum bob with a mass of 0.50 kg is released from a position in which the bob is 12 cm above the low point in its swing. What is the speed of the bob as it passes through the low point in its swing?

$m = 0.5$ kg

$h = 12$ cm

$v = ?$
(at the low point)

The initial energy is

$E = PE = mgh$

$\quad = (0.5 \text{ kg})(9.8 \text{ m/s}^2)(0.12 \text{ m})$

$\quad = 0.588$ J

At the low point, the potential energy is zero, so

$$E = KE = 0.588 \text{ J}$$

$$\tfrac{1}{2} mv^2 = 0.588 \text{ J}$$

Dividing both sides by $\tfrac{1}{2}m$:

$$v^2 = \frac{KE}{\tfrac{1}{2} m}$$

$$\quad = \frac{(0.588 \text{ J})}{\tfrac{1}{2}(0.5 \text{ kg})}$$

$$\quad = 2.35 \text{ m}^2/\text{s}^2$$

Taking the square root of both sides:

$$v = \textbf{1.53 m/s}$$

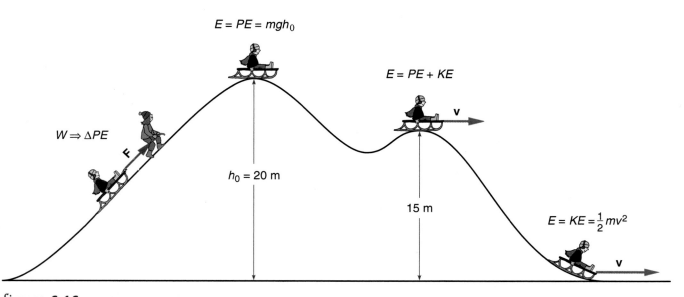

$E = PE = mgh_0$

$E = PE + KE$

$W \Rightarrow \Delta PE$

$h_0 = 20$ m

15 m

$E = KE = \tfrac{1}{2}mv^2$

figure **6.16**   Work done in pulling the sled up the hill produces an increase in potential energy of the sled and rider. This initial energy is then converted to kinetic energy as they slide down the hill.

Everyday Phenomenon

box 6.1

Energy and the Pole Vault

The Situation. Ben Lopez goes out for track. He specializes in the pole vault and helps out sometimes with the sprint relays where his speed can be used to good advantage. His coach, aware that Ben is also taking an introductory physics course, suggests that Ben try to understand the physics of the pole vault. What factors determine the height reached? How can he optimize these factors?

The coach knows that energy considerations are important in the pole vault. What type of energy transformations are involved? Could understanding these effects help Ben's performance?

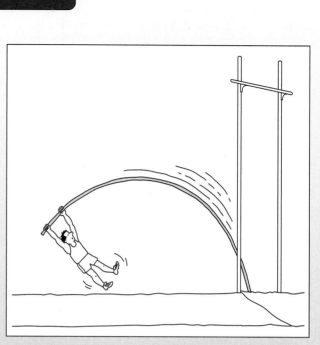

The flexibility of the pole and the point at which the vaulter grasps the pole are important to the success of the vault.

A pole-vaulter on the way up. What energy transformations are taking place?

The Analysis. It was not difficult for Ben to describe the energy transformations that take place in the pole vault: the vaulter begins by running down a path to the vaulting standard and pit. During this phase, he is accelerating and increasing his kinetic energy at the expense of chemical energy stored in his muscles. When he reaches the standard, he plants the end of the pole in a notch in the ground. At this point, some of his kinetic energy is stored in the elastic potential energy of the bent pole, which acts like a spring. The rest is converted to gravitational potential energy as he begins to rise over the standard.

Near the top of the vault, the elastic potential energy in the bent pole converts to gravitational potential energy as the pole straightens out. The vaulter does some additional work with his arm and upper-body muscles to provide an extra boost. At the very top of his flight, his kinetic energy should be zero, with only a minimal horizontal velocity left to carry him over the standard. Too large a kinetic energy at this

point would indicate that he had not optimized his jump by converting as much energy as possible into gravitational potential energy.

What can Ben learn from his analysis? First, the importance of speed. The more kinetic energy he generates during his approach, the more energy is available for conversion to gravitational potential energy (mgh), which will largely determine the height of his vault. Successful pole-vaulters are usually good sprinters.

The characteristics of the pole and Ben's grip on it are also important factors. If the pole is too stiff, or if he has gripped it too close to the bottom, he will experience a jarring impact in which little useful potential energy is stored in the pole, and some of his initial kinetic energy will be lost in the collision. If the pole is too limber, or if Ben's grip is too far from the bottom, it will not spring back soon enough to provide useful energy at the top of his vault.

Finally, upper-body strength is important in clearing the standard. Good upper-body conditioning should improve Ben's pole-vaulting. Timing and technique are also critical and can be improved only through practice. As far as his coach is concerned, that may be the most important message.

table **6.1**
Energy Balance Sheet for the Sled

A parent pulls a sled and rider with a combined weight of 50 kg to the top of a hill 20 m high and then gives the sled a push, providing an initial velocity of 4 m/s. Frictional forces acting on the sled do 2000 J of negative work as the sled moves down the hill.

Energy input

Potential energy gained by work done in pulling sled up the hill:	
$PE = mgh = (50 \text{ kg})(9.8 \text{ m/s}^2)(20 \text{ m})$	9 800 J
Kinetic energy gained by work done in pushing the sled at the top:	
$KE = \frac{1}{2}mv^2 = \frac{1}{2}(50 \text{ kg})(4 \text{ m/s})^2$	400 J
Total initial energy:	10 200 J

Energy expenditures

Work done against friction as the sled slides down the hill:	
$W = -fd$	−2 000 J
Energy balance:	8 200 J

Where does this initial energy come from? It came from the body of the parent doing the pulling and pushing. Muscle groups were activated, releasing chemical potential energy stored in the body. That energy came from food, which in turn involved solar energy stored by plants. A parent who does not eat a good breakfast, or attempts too many trips up the hill, may not have enough energy to get to the top.

If the sled and rider slide down the hill with negligible friction and air resistance, energy is conserved, and the total energy at any point during the motion should equal the initial energy. It is more realistic to assume that there is some friction as the sled slides down the hill (fig. 6.17).

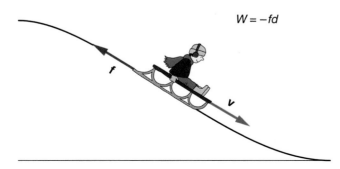

$W = -fd$

figure 6.17   The work done by frictional forces is negative, and it removes mechanical energy from the system.

Although it is difficult to predict the amount of work done against friction precisely, we can make an estimate if we know the total distance traveled and make some assumptions about the size of the average frictional force. In the energy accounting done in table 6.1, we assume that 2000 J of work has been done against friction by the time the sled reaches the bottom of the hill.

The work done against friction removes energy from the system and shows up as an expenditure on the account sheet. The energy balance at the bottom of the hill is 8200 J, rather than 10 200 J. This will lead to a smaller, more realistic value for the speed of the sled and rider at the bottom of the hill than if we ignored friction. Although precise calculations are not always possible, energy accounting sets limits on what is likely and helps us understand the behavior of systems such as the sled on the hill.

Energy is the currency of the physical world; an understanding of energy accounting is relevant to both science and economics. Doing work on a system puts energy in the bank. Total energy is then conserved, provided that only conservative forces are at work. Many aspects of the motion of the system can be predicted from a careful energy accounting.

6.5 Springs and Simple Harmonic Motion

If conservation of energy explains the motion of a pendulum, what about other systems that oscillate? Many systems involve springs or elastic bands that move back and forth, with potential energy being converted to kinetic energy and then back to potential energy repeatedly. What do such systems have in common? What makes them tick?

A mass on the end of a spring is one of the simplest oscillating systems. This system, and the simple pendulum described in section 6.4, are examples of *simple harmonic motion*.

Oscillation of a mass attached to a spring

If we attach a block to the end of a spring, as in figure 6.18, what happens when we pull it to one side of its equilibrium position? The equilibrium position is where the spring is neither stretched nor compressed. Doing work to pull the mass against the opposing force of the spring increases the potential energy of the spring-mass system. The potential energy in this case is elastic potential energy, $\frac{1}{2}kx^2$, rather than the gravitational potential energy associated with the pendulum. Increasing the potential energy of the mass on the spring is similar to cocking a bow and arrow or slingshot.

Once the mass is released, potential energy is converted to kinetic energy. Like the pendulum, the motion of the

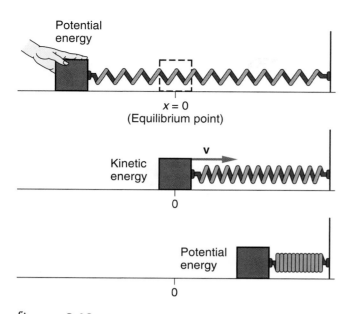

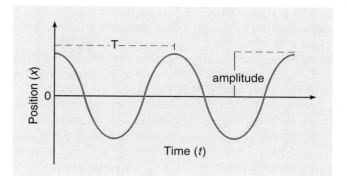

figure **6.19** The horizontal position x of the mass on the spring is plotted against time as the mass moves back and forth. The resulting curve is a harmonic function.

figure **6.18** Energy added by doing work to stretch the spring is then transformed back and forth between the potential energy of the spring and kinetic energy of the mass.

mass carries it beyond the equilibrium position, and the spring is compressed, gaining potential energy again. When the kinetic energy is completely reconverted to potential energy, the mass stops and reverses, and the whole process repeats (fig. 6.18). The energy of the system changes continuously from potential energy to kinetic energy and back again. If frictional effects can be ignored, the total energy of the system remains constant while the mass oscillates back and forth.

Using a video camera or other tracking techniques, it is possible to measure and plot the position of a pendulum bob or mass on a spring as it varies with time. If we plot the position of the mass against time, the resulting curve takes the form shown in figure 6.19. The mathematical functions that describe such curves are called "harmonic" functions, and the motion is called **simple harmonic motion,** * a term probably borrowed from musical descriptions of sounds produced by vibrating strings, reeds, and air columns. (See chapter 15.)

The line at zero on the graph in figure 6.19 is the equilibrium position for the mass on a spring. Points above this line represent positions on one side of the equilibrium point, and those below the line represent positions on the other side. The motion starts at the point of release, where the distance of the mass from equilibrium is a maximum. As the mass moves toward the equilibrium position ($x = 0$ on the graph) it gains speed, indicated by the increasing

slope of the curve. The object's position changes most rapidly when it is near the equilibrium point, where the kinetic energy and speed are the greatest.

As the mass passes through the equilibrium position, it starts to move away from equilibrium in the direction opposite to its initial position. The force exerted by the spring is now in the direction opposite to the velocity and is decelerating the mass. When the mass reaches the point farthest from its release point, the speed and kinetic energy are again zero, and the potential energy has returned to its maximum value. The slope of the curve is zero at this point, indicating that the mass is momentarily stopped (its velocity is zero). The mass continually gains or loses speed as it moves back and forth.

What are the period and the frequency?

If you look at the graph in figure 6.19, you will notice that the curve repeats itself regularly. The **period** T is the repeat time, or the time taken for one complete cycle. It is usually measured in seconds. You can think of the period as the time between adjacent peaks or valleys on the curve. A slowly oscillating system has a long period, and a rapidly oscillating system has a short period.

Suppose that the period of oscillation for a certain spring and mass is half a second. There are then two oscillations each second, which is the **frequency** of oscillation. The frequency f is the number of cycles per unit time, and it is found by taking the reciprocal of the period, $f = 1/T$. A rapidly oscillating system has a very short period and thus a high frequency. The unit commonly used for frequency is the *hertz,* which is defined as one cycle per second.

What determines the frequency of the spring-mass system? Intuitively, we expect a loose spring to have a low frequency of oscillation and a stiff spring to have a high frequency. This is indeed the case. The mass attached to the spring also has an effect. Larger masses offer greater resistance to a change in motion, producing lower frequencies.

*If you have studied trigonometry, you may know that the curve plotted in figure 6.19 is a cosine function. Sines and cosines are collectively referred to as harmonic functions.

The period and frequency of oscillation of a pendulum depend primarily on its length, measured from the pivot point to the center of the bob. To measure the period, you usually measure the time required for several complete swings and then divide by the number of swings to get the time for one swing.

Simple experiments with a ball on a string will give you an idea of how the period and frequency change with length. Try it and see if you can find a trend (see home experiment 1). The motion is regular—you can keep time by the swing of a pendulum or the motion of a mass on a spring.

Will any restoring force produce simple harmonic motion?

When a mass attached to a spring is moved to either side of equilibrium, the spring exerts a force that pulls or pushes the mass back toward the center. We call such a force the **restoring force.** In this case, it is the elastic force exerted by the spring. In any oscillation, there must be some such restoring force.

As discussed in section 6.3, the spring force is directly proportional to the distance x of the mass from its equilibrium position ($F = -kx$). The spring constant k has units of newtons per meter (N/m). Simple harmonic motion results whenever the restoring force has this simple dependence on distance. If the force varies in a more complicated way with distance, we may get an oscillation but not simple harmonic motion, and it will not produce a simple harmonic curve (fig. 6.19).

It is generally easiest to set up a spring-mass system by suspending the spring from a vertical support and hanging a mass on the end of the spring, as in figure 6.20. This arrangement avoids the frictional forces of the tabletop in a horizontal arrangement. In the vertical setup, when the mass is pulled down and released, the system oscillates up and down rather than horizontally. Two forces then act on the mass, the spring force pulling upward and the gravitational force pulling downward.

Since the gravitational force in the vertical setup is constant, it simply moves the equilibrium point lower. The equilibrium point is where the net force is zero—the downward pull of gravity is balanced by the upward pull of the spring. The restoring force is still provided by variations in the spring force, which are proportional to the distance from equilibrium. This system also meets the condition for simple harmonic motion. The potential energy involved, however, is the sum of the gravitational and elastic potential energies.

Gravity is the restoring force for the simple pendulum. When the pendulum bob is pulled to one side of its equilibrium position, the gravitational force acting on the bob

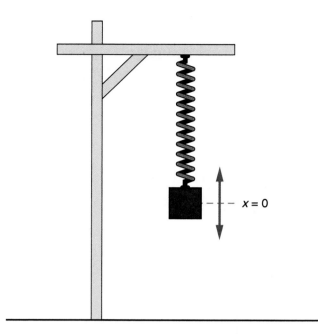

figure 6.20 A mass hanging from a spring will oscillate up and down with the same period as a horizontal mass-spring system using the same spring and mass.

pulls it back toward the center. The part of the gravitational force in the direction of motion is proportional to the displacement, if the displacement from equilibrium is not too large. Thus, for small *amplitudes* of swing, the simple pendulum also displays simple harmonic motion. **Amplitude** is the maximum distance from the equilibrium point.

Look around for systems that oscillate. There are many examples, ranging from a springy piece of metal to a ball rolling in a depression of some kind. What force pulls back toward the equilibrium position in each case? Is the motion likely to be simple harmonic motion, or a more complicated oscillation? What kind of potential energy is involved? The analysis of vibrations such as these forms an important subfield of physics that plays a role in music, communications, analysis of structures, and other areas.

Any oscillation involves a continuing interchange of potential and kinetic energies. If there are no frictional forces removing energy from the system, the oscillation will go on indefinitely. A restoring force that increases in direct proportion to the distance from the equilibrium position results in simple harmonic motion, with a simple curve (a harmonic function) describing the position of the object over time.

---

summary

The concept of work is central to this chapter. Energy is transferred into a system by doing work on the system, which can result in an increase in either the kinetic energy or the potential energy of the system. If no additional work is done on the system, the total energy of the system remains constant. This principle of conservation of energy allows us to explain many features of the behavior of the system.

1 Simple machines, work, and power. Work is defined as force times the distance involved in moving an object. Only the portion of the force in the direction of the motion is used. In simple machines, work output cannot exceed work input. Power is the rate of doing work.

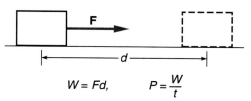

$$W = Fd, \qquad P = \frac{W}{t}$$

2 Kinetic energy. The work done by the net force acting on an object is used to accelerate the object, and the object gains kinetic energy. Kinetic energy is equal to one-half the mass of the object times the square of its speed. Negative work removes kinetic energy.

$$KE = \frac{1}{2}mv^2$$

3 Potential energy. If work done on an object moves the object against an opposing conservative force, the potential energy of the object is increased. Two types of potential energy were considered, gravitational potential energy and elastic potential energy.

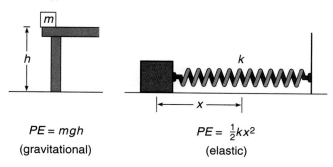

$$PE = mgh \qquad\qquad PE = \frac{1}{2}kx^2$$
(gravitational)   (elastic)

4 Conservation of energy. If no work is done on a system, the total mechanical energy (kinetic plus potential) remains constant. This principle of conservation of energy explains the behavior of many systems that involve exchanges of kinetic and potential energy. The system can be analyzed by energy accounting.

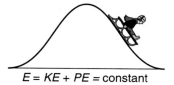

$$E = KE + PE = \text{constant}$$

5 Springs and simple harmonic motion. The motions of a simple pendulum and of a mass on a spring both illustrate the principle of conservation of energy, but they involve different kinds of potential energy. They are also examples of simple harmonic motion, which results whenever the restoring force is proportional to the distance of the object from its equilibrium position.

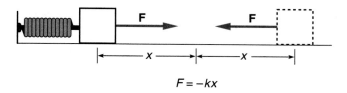

$$F = -kx$$

key terms

questions

*Questions identified with an asterisk are more open-ended than the others. They call for lengthier responses and are more suitable for group discussion.

Q1. Equal forces are used to move blocks A and B across the floor. Block A has twice the mass of block B, but block B moves twice the distance moved by block A. Which block, if either, has the greater amount of work done on it? Explain.

Q2. A man pushes very hard for several seconds upon a heavy rock, but the rock does not budge. Has the man done any work on the rock? Explain.

Q3. A string is used to pull a wooden block across the floor without accelerating the block. The string makes an angle to the horizontal as shown in the diagram.
   a. Does the force applied via the string do work on the block? Explain.
   b. Is the total force involved in doing work or just a portion of the force? Explain.

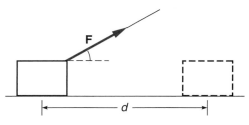

Q4. In the situation pictured in question 3, if there is a frictional force opposing the motion of the block, does this frictional force do work on the block? Explain.

Q5. In the situation pictured in question 3, does the normal force of the floor pushing upward on the block do any work? Explain.

Q6. A ball is being twirled in a circle at the end of a string. The string provides the centripetal force needed to keep the ball moving in the circle at constant speed. Does the force exerted by the string on the ball do work on the ball in this situation? Explain.

*Q7. A man walks across the room. What forces act on the man during this process? Which, if any, of these forces do work on the man? Explain.

Q8. A woman uses a pulley arrangement to lift a heavy crate. She applies a force that is one-fourth the weight of the crate, but moves the rope a distance four times the height that the crate is lifted. Is the work done by the woman greater than, equal to, or less than the work done by the rope on the crate? Explain.

Q9. A lever is used to lift a rock as shown in the diagram. Will the work done by the person on the lever be greater than, equal to, or less than the work done by the lever on the rock? Explain.

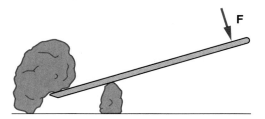

*Q10. A crate on rollers is pushed up an inclined plane into a truck. The pushing force required is less than half the force that would be needed to lift the crate straight up into the truck. Does the inclined plane serve as a simple machine in this situation? Explain.

Q11. A boy pushes his friend across a skating rink. Since the frictional forces are very small, the force exerted by the boy on his friend's back is the only significant force acting on the friend in the horizontal direction. Is the change in kinetic energy of the friend greater than, equal to, or less than the work done by the force exerted by the boy? Explain.

Q12. A child pulls a block across the floor with force applied by a horizontally held string. A smaller frictional force also acts upon the block, yielding a net force on the block that is smaller than the force applied by the string. Does the work done by the force applied by the string equal the change in kinetic energy in this situation? Explain.

Q13. If there is just one force acting on an object, does its work necessarily result in an increase in kinetic energy? Explain.

Q14. Two balls of the same mass are accelerated by different net forces such that one ball gains a velocity twice that of the other ball in the process. Is the work done by the net force acting on the faster-moving ball twice that done on the slower-moving ball? Explain.

Q15. A box is moved from the floor up to a tabletop but gains no speed in the process. Is there work done on the box, and if so, what has happened to the energy added to the system?

Q16. When work is done to increase the potential energy of an object without increasing its kinetic energy, is the *net* force acting on the object greater than zero? Explain.

Q17. Is it possible for a system to have energy if nothing is moving in the system? Explain.

Q18. Suppose that work is done on a large crate to tilt the crate so that it is balanced on one edge, as shown in the diagram, rather than sitting squarely on the floor as it was at first. Has the potential energy of the crate increased in this process? Explain.

*Q19. Which has the greater potential energy: a ball that is 10 feet above the ground, or one with the same mass that is 20 feet above the bottom of a nearby 50-foot-deep well? Explain.

Q20. When a bow and arrow are cocked, a force is applied to the string in order to pull it back. Is the energy of the system increased? Explain.

Q21. Suppose that the physics instructor pictured in figure 6.15 gives the bowling ball a push as she releases it. Will the ball return to the same point or will her chin be in danger? Explain.

Q22. A pendulum is pulled back from its equilibrium (center) position and then released.
a. What form of energy is added to the system prior to its release? Explain.
b. At what points in the motion of the pendulum after release is its kinetic energy the greatest? Explain.
c. At what points is the potential energy the greatest? Explain.

Q23. Is the total mechanical energy conserved in the motion of a pendulum? Will it keep swinging forever? Explain.

*Q24. A bird grabs a clam, carries it in its beak to a considerable height, and then drops it on a rock below, breaking the clam shell. Describe the energy conversions that take place in this process.

Q25. A mass attached to a spring, which in turn is attached to a wall, is free to move on a friction-free horizontal surface. The mass is pulled back and then released.
a. What form of energy is added to the system prior to the release of the mass? Explain.
b. At what points in the motion of the mass after its release is its potential energy the greatest? Explain.
c. At what points is the kinetic energy the greatest? Explain.

Q26. Suppose that the mass in question 25 is halfway between one of the extreme points of its motion and the center point. In this position, is the energy of the system kinetic energy, potential energy, or a combination of these forms? Explain.

*Q27. A spring gun is loaded with a rubber dart, the gun is cocked, and then fired at a target on the ceiling. Describe the energy transformations that take place in this process.

Q28. Suppose that a mass is hanging vertically at the end of a spring. The mass is pulled downward and released to set it into oscillation. Is the potential energy of the system increased or decreased when the mass is lowered? Explain.

*Q29. A sled is given a push at the top of a hill. Is it possible for the sled to cross a hump in the hill that is higher than its starting point under these circumstances? Explain.

*Q30. Can work done by a frictional force ever increase the total mechanical energy of a system? (Hint: Consider the acceleration of an automobile.) Explain.

Q31. Suppose that a pulley system is used to lift a heavy crate, but the pulleys have rusted and there are frictional forces acting on the pulleys. Will the useful work output of this system be greater than, equal to, or less than the work input? Explain.

Q32. Is the elastic potential energy stored in the pole the only type of potential energy involved in pole-vaulting? Explain.

Q33. If one pole-vaulter can run faster than another, will the faster runner have an advantage in the pole vault? Explain.

exercises

E1. A horizontally directed force of 40 N is used to pull a box a distance of 2.5 m across a tabletop. How much work is done by the 40-N force?

E2. A woman does 160 J of work to move a table 4 m across the floor. What is the magnitude of the force that the woman applied to the table if this force is applied in the horizontal direction?

E3. A force of 60 N used to push a chair across a room does 300 J of work. How far does the chair move in this process?

E4. A rope applies a horizontal force of 180 N to pull a crate a distance of 2 m across the floor. A frictional force of 120 N opposes this motion.
a. What is the work done by the force applied by the rope?
b. What is the work done by the frictional force?
c. What is the total work done on the crate?

E5. A force of 50 N is used to drag a crate 4 m across a floor. The force is directed at an angle upward from the crate so that the vertical component of the force is 30 N and the horizontal component is 40 N as shown in the diagram.
   a. What is the work done by the horizontal component of the force?
   b. What is the work done by the vertical component of the force?
   c. What is the total work done by the 50-N force?

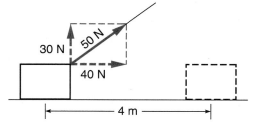

E6. A net force of 60 N accelerates a 4-kg mass over a distance of 10 m.
   a. What is the work done by this net force?
   b. What is the increase in kinetic energy of the mass?

E7. A 0.4-kg ball has a velocity of 20 m/s.
   a. What is the kinetic energy of the ball?
   b. How much work would be required to stop the ball?

E8. A box with a mass of 5.0 kg is lifted (without acceleration) through a height of 2.0 m, in order to place it upon the shelf of a closet.
   a. What is the increase in potential energy of the box?
   b. How much work was required to lift the box to this position?

E9. A spring with a spring constant k of 40 N/m is stretched a distance of 20 cm (0.20 m) from its original unstretched position. What is the increase in potential energy of the spring?

E10. To stretch a spring a distance of 0.20 m, 40 J of work is done.
   a. What is the increase in potential energy of the spring?
   b. What is the value of the spring constant k of the spring?

E11. Which requires more work: lifting a 2-kg rock to a height of 4 m without acceleration, or accelerating the same rock horizontally from rest to a speed of 10 m/s?

E12. At the low point in its swing, a pendulum bob with a mass of 0.2 kg has a velocity of 5 m/s.
   a. What is its kinetic energy at the low point?
   b. Ignoring air resistance, how high will the bob swing above the low point before reversing direction?

E13. A 0.20-kg mass attached to a spring is pulled back horizontally across a table so that the potential energy of the system is increased from zero to 120 J. Ignoring friction, what is the kinetic energy of the system after the mass is released and has moved to a point where the potential energy has decreased to 80 J?

E14. A sled and rider with a combined mass of 50 kg are at the top of a hill a height of 15 m above the level ground below. The sled is given a push providing an initial kinetic energy at the top of the hill of 1600 J.
   a. Choosing a reference level at the bottom of the hill, what is the potential energy of the sled and rider at the top of the hill?
   b. After the push, what is the total mechanical energy of the sled and rider at the top of the hill?
   c. If friction can be ignored, what will be the kinetic energy of the sled and rider at the bottom of the hill?

E15. A roller-coaster car has a potential energy of 450 000 J and a kinetic energy of 120 000 J at point A in its travel. At the low point of the ride, the potential energy is zero, and 50 000 J of work have been done against friction since it left point A. What is the kinetic energy of the roller coaster at this low point?

E16. A roller-coaster car with a mass of 800 kg starts at rest from a point 20 m above the ground. At point B, it is 12 m above the ground.
   a. What is the initial potential energy of the car?
   b. What is the potential energy at point B?
   c. If the initial kinetic energy was zero and the work done against friction between the starting point and point B is 30 000 J, what is the kinetic energy of the car at point B?

E17. The time required for one complete cycle of a mass oscillating at the end of a spring is 0.25 s. What is the frequency of oscillation?

E18. The frequency of oscillation of a pendulum is 8 cycles/s. What is the period of oscillation?

challenge problems

CP1. Suppose that two horizontal forces are acting upon a 0.25-kg wooden block as it moves across a laboratory table: a 5-N force pulling the block and a 2-N frictional force opposing the motion. The block moves a distance of 1.5 m across the table.
   a. What is the work done by the 5-N force?
   b. What is the work done by the net force acting upon the block?
   c. Which of these two values should you use to find the increase in kinetic energy of the block? Explain.

   d. What happens to the energy added to the system via the work done by the 5-N force? Can it all be accounted for? Explain.
   e. If the block started from rest, what are its kinetic energy and velocity at the end of the 1.5-m motion?

CP2. As described in Try This Box 6.2, a 100-kg crate is accelerated by a net force of 50 N applied for 4 s.
   a. What is the acceleration of the crate from Newton's second law?
   b. If it starts from rest, how far does it travel in the time of 4 s? (See section 2.5 in chapter 2.)

c. How much work is done by the 50-N net force?

d. What is the velocity of the crate at the end of 4 s?

e. What is the kinetic energy of the crate at this time? How does this value compare to the work computed in part c?

CP3. A slingshot consists of a rubber strap attached to a Y-shaped frame, with a small pouch at the center of the strap to hold a small rock or other projectile. The rubber strap behaves much like a spring. Suppose that for a particular slingshot a spring constant of 600 N/m is measured for the rubber strap. The strap is pulled back approximately 40 cm (0.4 m) prior to being released.

a. What is the potential energy of the system prior to release?

b. What is the maximum possible kinetic energy that can be gained by the rock after release?

c. If the rock has a mass of 50 g (0.050 kg), what is its maximum possible velocity after release?

d. Will the rock actually reach these maximum values of kinetic energy and velocity? Does the rubber strap gain kinetic energy? Explain.

CP4. Suppose that a 200-g mass (0.20 kg) is oscillating at the end of a spring upon a horizontal surface that is essentially friction-free. The spring can be both stretched and compressed and has a spring constant of 240 N/m. It was originally stretched a distance of 12 cm (0.12 m) from its equilibrium (unstretched) position prior to release.

a. What is its initial potential energy?

b. What is the maximum velocity that the mass will reach in its oscillation? Where in the motion is this maximum reached?

c. Ignoring friction, what are the values of the potential energy, kinetic energy, and velocity of the mass when the mass is 6 cm from the equilibrium position?

d. How does the value of velocity computed in part c compare to that computed in part b? (What is the ratio of the values?)

CP5. A sled and rider with a total mass of 40 kg are perched at the top of a hill as pictured in the diagram. The top of this hill is 40 m above the low point in the path of the sled. A second hump in the hill is 30 m above this low point. Suppose that we also know that approximately 2000 J of work is done against friction as the sled travels between these two points.

a. Will the sled make it to the top of the second hump if no kinetic energy is given to the sled at the start of its motion? Explain.

b. What is the maximum height that the second hump could be in order for the sled to reach the top, assuming that the same work against friction will be involved and that no initial push is provided? Explain.

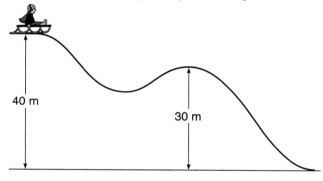

40 m

30 m

home experiments and observations

HE1. You can construct a simple pendulum easily by attaching a ball to a string (with tape or a staple) and fixing the other end of the string to a rigid support. (A pencil taped firmly to the end of a desk or table will do nicely.)

a. The frequency of oscillation can be measured by timing the swings. The usual method is to use a watch to measure the time required for ten or more complete swings. The period T (the time required for one swing) is then the total time divided by the number of swings counted and the frequency f is just $1/T$.

b. How does the frequency change if you vary the length of your pendulum? (Try at least three different lengths.)

HE2. A ramp for a marble or small steel ball can be made by bending a long strip of cardboard into a V-shaped groove. Two such ramps can be placed end to end, as pictured in the diagram, to produce a track in which the marble will oscillate.

a. Can you measure a frequency of oscillation for this system? Does this frequency depend upon how high up the ramp the marble starts?

b. How high up the second ramp does the marble go? Is more energy lost per cycle in this system than for a pendulum?

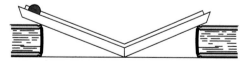

HE3. The height to which a ball bounces after being dropped provides a measure of how much energy is lost in the collision with the floor or other surface. A small portion of the energy is lost to air resistance as the ball is moving, but most is lost in the collision.

a. Trying a number of different balls that you may have available, test the height of the bounce using the same height of release for all of the balls tested. Which ball loses the most energy and which the least?

b. Can you explain why many balls return to a higher height than a marble will? What characteristics of the balls tested give the best bounce?

c. For a ball that bounces several times, does the period (time between bounces) change with each bounce? Does the bouncing ball undergo simple harmonic motion?

Momentum and Impulse

chapter 7

chapter overview

In this chapter, we explore momentum and impulse and examine the use of these concepts in analyzing events such as a collision between a fullback and defensive back. The principle of conservation of momentum is introduced and its limits explained. A number of examples will shed light on how these ideas are used, particularly conservation of momentum. Momentum is central to all of these topics—it is a powerful tool for understanding a lot of life's sudden changes.

chapter outline

The word *momentum* is overused by sports announcers to mean changes in the flow of a game. The "old mo" that announcers talk about bears only a metaphorical relationship to the physical concept of momentum. There are plenty of real examples of changes in momentum for us to consider in both the world of sports and the world more generally.

Take the collision between a hard-charging fullback and a defensive back on the football field (fig. 7.1). If they meet head-on, the velocity of the fullback is sharply reduced, although the two players might continue moving briefly in the original direction of the fullback's velocity. If the defensive back is moving before the collision, his velocity also changes abruptly. There must be strong forces at work to produce these accelerations, but these forces act for only an instant. How do we use Newton's laws to analyze this event?

Momentum, impulse, and conservation of momentum figure in any discussion of collisions. The total momentum of the fullback and defensive back is involved in predicting what will happen after the collision. How is momentum defined, and what does conservation of momentum have to do with Newton's laws? How is conservation of momentum useful in predicting what happens in collisions? These questions will be addressed as we examine a variety of collisions and other high-impact events.

figure **7.1**   A collision between a fullback and a defensive back. How will the two players move after the collision?

7.1 Momentum and Impulse

Imagine a baseball heading toward the catcher's mitt when its flight is rudely interrupted by the impact of a bat. In a very short time, the velocity of the ball changes direction and is accelerated in the direction opposite its original flight. Similar changes happen when a tennis racket hits a ball or when a ball bounces off a wall or the floor. In many everyday situations, a brief impact causes a rapid change in an object's velocity.

The forces responsible for such rapid changes in motion can be large, but they act for very short times and are difficult to measure. Not only are they brief, but they may change rapidly *during* the collision.

What happens when a ball bounces?

Consider the seemingly simple example of dropping a tennis ball. The ball is initially accelerated downward by the gravitational force. When it reaches the floor, its velocity quickly changes in direction, and the ball heads back up toward you (fig. 7.2). There must be a strong force exerted on the ball by the floor during the short time that they are in contact. This force provides the upward acceleration necessary to change the direction of the ball's velocity.

If we used a high-speed camera to catch the action during the time the ball is in contact with the floor, we would see that the ball's shape is distorted (fig. 7.3). The ball behaves like a spring, first compressing as it moves downward, then

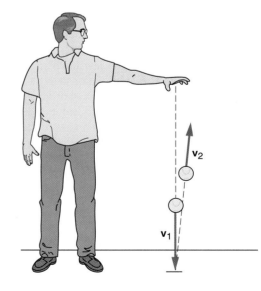

figure **7.2**   A tennis ball bouncing off the floor. There is a rapid change in the direction of the velocity when the ball hits the floor.

expanding (springing back) as it begins to move upward. A quick test (squeezing the ball with your hands) will persuade you that a strong force is required to distort the ball.

What we have, then, is *a strong force acting for a very brief time* producing a rapid acceleration that quickly changes the ball's velocity from a downward direction to an upward

figure 7.3 A high-speed photograph of a ball hitting the floor. The ball is compressed like a spring.

one. The magnitude of the ball's velocity decreases rapidly to zero and then increases equally rapidly in the opposite direction. This all happens in a time so short that we would miss it if we blinked.

How can we analyze such rapid changes?

We have described the collision of the ball with the floor using force and acceleration, and we could also use Newton's second law to predict how the velocity actually changes. The problem with this approach is that the time of the interaction is very short, and the force itself varies during this short time, so it is hard to describe the collision accurately. It is more productive to look at the *total change in motion* in this brief interaction.

We introduced Newton's second law in chapter 4 using the expression **F** = *m***a**. The acceleration **a** is the *rate of change* in velocity, which can be expressed as the change in velocity $\Delta\mathbf{v}$ divided by the time interval Δt required to produce that change. The time interval is important: the shorter the time for a given change in velocity, the larger the acceleration and the force needed to produce this change.

We can restate Newton's second law as

$$\mathbf{F} = m\left(\frac{\Delta\mathbf{v}}{\Delta t}\right),$$

expressing the acceleration in terms of the change in velocity. Multiplying both sides of this equation by the time interval Δt recasts the second law as

$$\mathbf{F}\Delta t = m\Delta\mathbf{v}.$$

While this is still Newton's second law, rewriting it offers us a different way of looking at events. This new view is more convenient for describing the overall change in motion.

What are impulse and momentum?

Impulse shows up as the quantity on the left side of the recast second law, **F**Δt. **Impulse** is the force acting on an object multiplied by the time interval over which the force

acts. If the force varies during this time interval, and it often does, we must use the *average* value of the force over this time interval.

> Impulse is the average force multiplied by its time interval of action:
>
> impulse = **F**Δt.

Since force is a vector quantity, impulse is also a vector in the direction of the average force.

How a force changes the motion of an object depends on both the size of the force and how long the force acts. The stronger the force, the larger the effect, and the longer the force acts, the greater its effect. Multiplying the two factors together to get the impulse shows the overall effect of the force.

On the right side of our recast second law, $m\Delta\mathbf{v}$ is the mass of the object multiplied by the change in velocity produced by the impulse. This product is the change in the *quantity of motion,* to use Newton's own term. We now call this product the *change in momentum* of the object, where **momentum** is defined as

> Momentum is the product of the mass of an object and its velocity, or
>
> **p** = *m***v**.

The symbol **p** is often used for momentum. If the mass of the object is constant, the change in momentum is the mass times the change in velocity or $\Delta\mathbf{p} = m\Delta\mathbf{v}$.

Like velocity, momentum is a vector and has the same direction as the velocity vector. Two different objects traveling in the same direction can have different masses and velocities but still have the same momentum. For example, a 7-kg bowling ball moving with the relatively slow speed of 2 m/s would have a momentum of 14 kg·m/s. On the other hand, a tennis ball with a mass of just 0.07 kg, moving with the much larger velocity of 200 m/s, has the same momentum as the bowling ball, 14 kg·m/s (fig. 7.4).

Using these definitions of impulse and momentum, we can state our recast form of Newton's second law as

impulse = change in momentum

$$= \Delta\mathbf{p}.$$

This statement of the second law is sometimes called the **impulse/momentum principle:**

> The impulse acting on an object produces a change in momentum of the object that is equal in both magnitude and direction to the impulse.

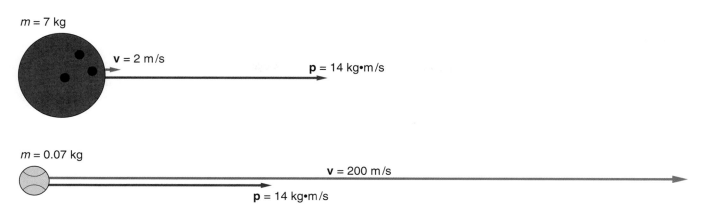

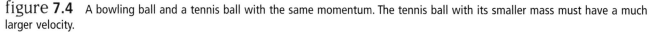

figure **7.4**   A bowling ball and a tennis ball with the same momentum. The tennis ball with its smaller mass must have a much larger velocity.

This principle is not a new law but another way of expressing Newton's second law of motion. It is particularly useful for looking at collisions.

How do we apply the impulse/momentum principle?

The impulse/momentum principle applies to almost any collision. Whacking a golf ball with a golf club is a good example (fig. 7.5). The impulse delivered by the golf club produces a change in the golf ball's momentum, also described in Try This Box 7.1. Note that the units of impulse (force multiplied by time, or N·s) must equal those of momentum (mass times velocity, or kg·m/s).

Does the momentum of the bouncing tennis ball we discussed earlier change when it hits the floor? Even if the ball loses no energy in its collision with the floor and bounces back with the same speed and kinetic energy it had just before hitting the floor, the momentum changes because its direction changes. The momentum decreases to

figure **7.5**   An impulse is delivered to the golf ball by the head of the club. If the initial momentum of the ball is zero, the final momentum is equal to the impulse delivered.

---

try this box 7.1

Sample Exercise: The Momentum and Impulse of Golf

A golf club exerts an average force of 500 N on a 0.1-kg golf ball, but the club is in contact with the ball for only a hundredth of a second.
   a. What is the magnitude of the impulse delivered by the club?
   b. What is the change in velocity of the golf ball?

a. $F = 500$ N   impulse $= F\Delta t$
   $\Delta t = 0.01$ s $= (500 \text{ N})(0.01 \text{ s})$
   impulse $= ?$ $= \textbf{5 N·s}$

b. $m = 0.1$ kg   impulse $= \Delta \mathbf{p} = m\Delta \mathbf{v}$
   $\Delta v = ?$
                    $\Delta v = \dfrac{\text{impulse}}{m}$

                     $= \dfrac{5 \text{ N·s}}{0.1 \text{ kg}}$

                     $= \textbf{50 m/s}$

Since the golf ball started at rest, this change in velocity equals the velocity of the ball as it leaves the face of the club. The direction of this velocity is the same as the impulse of the force exerted by the club.

---

zero as the ball comes to a momentary halt, and it changes again as the ball gains momentum in the opposite direction (fig. 7.6). The total change in momentum is larger than the change that would happen if the tennis ball stopped and did not bounce.

When the tennis ball bounces back with the same speed, the total change in momentum is *twice* the value of the momentum just before the ball hits the floor. Its final momentum is $m\mathbf{v}$, where the direction of $\mathbf{v}$ is upward, but

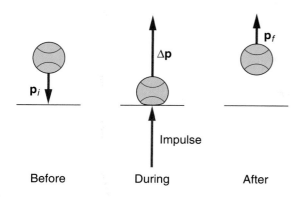

figure 7.6 The impulse exerted by the floor on the tennis ball produces a change in its momentum.

its *initial* momentum was $-m\mathbf{v}$ because the initial velocity was directed downward. We find the change in momentum by subtracting the initial value from the final value: $m\mathbf{v} - (-m\mathbf{v}) = 2m\mathbf{v}$. The impulse required to produce this change in momentum is twice as large as what is needed simply to stop the ball.

There are many practical lessons involving impulse and change in momentum. Why does it help to pull your hand back as you catch a hard-thrown ball? When you pull your hand back, you lengthen the time interval Δt. This reduces the average force that your hand must exert on the ball, since impulse is the product of the force and the time interval ($\mathbf{F}\Delta t$). If the time interval is longer, the force can be smaller yet still produce the same impulse and change in momentum. It hurts less this way! A padded dashboard or an air bag similarly lessens injury to passengers by increasing the time interval required to bring them to a halt in a collision.

Everything that we have done in this section is just another way of working with Newton's second law of motion. In fact, by dividing both sides of the impulse/momentum principle by the time interval Δt, Newton's second law can be expressed in the form that most nearly captures the meaning of Newton's original statement of the second law, $\mathbf{F} = \Delta\mathbf{p}/\Delta t$. In words, this form of the second law says that the net force acting on an object is equal to the rate of change in momentum of the object. This form covers a wider range of situations than the more familiar $\mathbf{F} = m\mathbf{a}$.

> Momentum and impulse are most useful for evaluating events such as collisions, where powerful forces act briefly to produce striking changes in the motion of objects. The impulse/momentum principle states that the change in momentum is equal to the impulse. This is a different way of stating Newton's second law. The impulse, the product of the average force and the time interval that it is applied, allows us to predict the change in momentum of the object. Large impulses yield large changes in momentum.

7.2 Conservation of Momentum

How do impulse and momentum help to explain the collision between the defensive back and fullback mentioned in the chapter introduction? The conditions described in section 7.1 are certainly present. The defensive back exerts a sizable but brief force on the fullback (fig. 7.7), and the momentum of both players changes rapidly in the collision.

The principle of **conservation of momentum** provides the key to understanding such a collision. This principle arises when we apply Newton's *third law* to impulse and changes in momentum. Conservation of momentum allows us to predict many features of collisions without requiring detailed knowledge of the forces of impact.

Why and when is momentum conserved?

Let's take a more detailed look at the head-on collision between the hard-charging fullback and the defensive back. To simplify the situation, we assume that the two players meet in midair and that after the collision they move together, with the fullback held in the tackle of the defensive back (fig. 7.7). What happens when they collide?

During the collision, the defensive back exerts a strong force on the fullback, and by Newton's third law, the fullback exerts a force equal in magnitude but opposite in direction on the defensive back. Since the time interval of action Δt is the same for both forces, the impulses $\mathbf{F}\Delta t$ must also be equal in magnitude but opposite in direction. From the impulse/momentum principle (Newton's second law), changes in the momentum $\Delta\mathbf{p}$ for each player are also equal in magnitude but opposite in direction.

If the two players experience equal but oppositely directed momentum changes, the *total change* in momentum of the two players together is zero. We look at the overall *system* and define the total momentum of that system as the sum of the momentum values of the two players. There

figure 7.7 Two football players colliding. The impulses acting on the two players are equal in magnitude but opposite in direction.

is no change in the momentum of this system because the changes in the momentums of the parts cancel one another. The total momentum of the system is *conserved*.

To reach this conclusion, we ignored external forces (produced by other objects) acting on the two players and assumed that the only significant forces were their own forces of interaction. The forces that they exert on one another are *internal* to the system consisting of both players. The principle of conservation of momentum can therefore be stated as

> If the net external force acting on a system of objects is zero, the total momentum of the system is conserved.

The forces of interaction between the objects in a system are internal forces whose effects on the total momentum cancel one another, because of Newton's third law of motion. Different portions of the system can exchange momentum without affecting the total momentum of the system. If there is a net *external* force acting on the system because of interaction with some object that is not part of the system, the entire system will be accelerated—and the momentum of the system will change.

Conservation of momentum and collisions

Using the principle of conservation of momentum, what information can we obtain about the results of a collision (like the one between two football players)? If we know the masses of the players and their initial velocities, we can find how fast and in what direction the players will move after they collide. We do not need to know anything at all about the details of the strong forces involved in the collision itself.

The sample exercise in Try This Box 7.2 treats a head-on collision between a fullback and a defensive back using realistic numerical values. The fullback has a mass of 100 kg (equivalent to a weight of about 220 lb) and is moving straight downfield with a velocity of 5 m/s through the hole created by his linemen. The somewhat smaller defensive back charges up to meet him with a velocity in the opposite direction of −4 m/s (fig. 7.8). The minus sign indicates direction: we have chosen the fullback's direction of motion to be positive.

The total momentum of the system before the collision in Try This Box 7.2 is found by adding the initial momentum of the fullback to the momentum of the defensive back, taking into account the difference in sign. If we assume that both players' feet leave the ground just before the collision (so that there are no frictional forces between their feet and the ground), momentum should be conserved in the collision. The total momentum of the two players moving together after the collision has the same value as it did immediately before the collision (fig. 7.9).

---

try this box 7.2

Sample Exercise: A Head-on Collision

A 100-kg fullback moving straight downfield with a velocity of 5 m/s collides head-on with a 75-kg defensive back moving in the opposite direction with a velocity of −4 m/s. The defensive back hangs on to the fullback, and the two players move together after the collision.

  a. What is the initial momentum of each player?
  b. What is the total momentum of the system?
  c. What is the velocity of the two players immediately after the collision?

a. *fullback:* $\qquad\qquad p = mv$

$\quad m = 100$ kg $\qquad = (100$ kg$)(5$ m/s$)$

$\quad v = 5$ m/s $\qquad\quad = \textbf{500 kg·m/s}$

$\quad p = ?$

defensive back: $\qquad p = mv$

$\quad m = 75$ kg $\qquad\quad = (75$ kg$)(-4$ m/s$)$

$\quad v = -4$ m/s $\qquad\quad = \textbf{−300 kg·m/s}$

b. $P_{total} = ? \qquad P_{total} = p_{fullback} + p_{defensive\ back}$

$\qquad\qquad\qquad = 500$ kg·m/s $+ (-300$ kg·m/s$)$

$\qquad\qquad\qquad = \textbf{200 kg·m/s}$

c. $v = ?$ (for both players after the collision)

$\quad m = 100$ kg $+ 75$ kg $\qquad p = mv$

$\qquad = 175$ kg

$$v = \frac{P_{total}}{m}$$

$$= \frac{200 \text{ kg·m/s}}{175 \text{ kg}}$$

$$= \textbf{1.14 m/s}$$

The positive value of the momentum after the collision means that the two players are traveling in the direction of the fullback's initial motion. The fullback had a larger initial momentum than the defensive back, so his direction of motion prevails when the two values are added. The defensive back will be carried backward briefly before the two players hit the turf.

Conservation of momentum results when the changes in momentum of different parts of a system cancel each other by Newton's third law. If there are no external forces acting on the system, its total momentum is conserved. The principle applies to all sorts of situations involving collisions and explosions or other forms of brief but forceful interaction between objects.

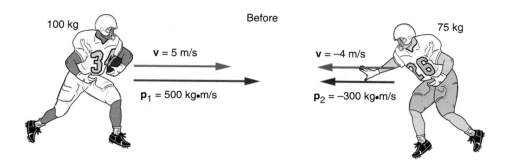

100 kg

Before

75 kg

v = 5 m/s

v = −4 m/s

p₁ = 500 kg•m/s

p₂ = −300 kg•m/s

figure **7.8**   The two players before the collision, with velocity and momentum vectors for each.

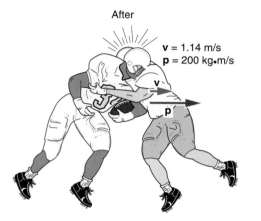

After

v = 1.14 m/s
p = 200 kg•m/s

v

p

figure **7.9**   The two players after the collision, with velocity and momentum vectors indicated.

the initial total momentum of the system is zero. If momentum is conserved, the total momentum of the system after the push-off will also be zero. How can the total momentum be zero when at least one of the skaters is moving? Both skaters must move with momentum values equal in magnitude but opposite in direction $\mathbf{p}_2 = -\mathbf{p}_1$. The momentum of the second skater $\mathbf{p}_2$ must be opposite that of the first skater $\mathbf{p}_1$. When added together to find the total momentum of the system, these individual values will cancel each other to produce a total momentum of zero.

After the push-off, the two skaters move in opposite directions with momentum vectors equal in magnitude (fig. 7.11), but their velocities are not of equal magnitude. Since momentum is mass times velocity ($\mathbf{p} = m\mathbf{v}$), the skater with the smaller mass must have the larger velocity to yield

7.3 Recoil

Why does a shotgun slam against your shoulder when fired, sometimes with painful consequences? How can a rocket accelerate in empty space when there is nothing there to push against but itself? These are examples of the phenomenon of recoil, a common part of everyday experience. Conservation of momentum is the key to understanding recoil.

What is recoil?

Imagine two ice skaters facing one another and pushing against each other with their hands (fig. 7.10). The frictional forces between their skates and the ice are presumably very small, so we can neglect them. The upward normal force and the downward force of gravity cancel one another, too, since we know that there is no acceleration in the vertical direction. The net external force acting on the system of the two skaters is effectively zero, and conservation of momentum should apply.

How do we apply conservation of momentum in this situation? Since neither skater is moving before the push-off,

figure **7.10**   Two skaters of different masses prepare to push off against one another. Which one will gain the larger velocity?

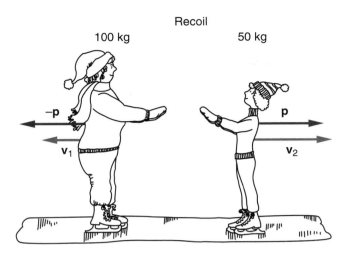

Recoil
100 kg   50 kg

figure **7.11** The two skaters after pushing off, with the velocity and momentum vectors indicated.

figure **7.12** The shot and the shotgun have equal but oppositely directed momentums after the gun is fired.

the same magnitude of momentum as the larger skater. Suppose that the smaller skater's mass is just half the mass of the larger skater. The smaller skater's velocity will then be twice as large as the larger skater after pushing off.

The ice skaters illustrate the basics of **recoil.** A brief force between two objects causes the objects to move in opposite directions. The lighter object attains the larger velocity to equalize the magnitudes of the momentums of the two objects. The total momentum for the system after the push-off equals zero, the value of the momentum of the system before the push-off if the objects were initially at rest. The total momentum of the system is conserved and does not change.

Recoil of a shotgun

If you have ever fired a shotgun without holding it firmly against your shoulder, you have probably had a painful experience of recoil. What happened? The explosion of the powder in the shotgun causes the shot to move very rapidly in the direction of the gun's aim. If the gun is free to move, it will recoil in the opposite direction with a momentum equal in magnitude to the momentum of the shot (fig. 7.12).

Even though the mass of the shot is considerably less than the shotgun, the momentum of the shot is quite large as a result of its large velocity. If the external forces acting on the system can be ignored, the shotgun recoils with a momentum equal in magnitude to the momentum of the shot. The recoil velocity of the shotgun will be smaller than the shot's velocity because of the larger mass of the gun, but it is still sizable. As the gun slams back against your shoulder, you will know that it has recoiled.

How can you avoid a bruised shoulder? The trick is to hold the gun firmly against your shoulder. (See Try This

try this box 7.3

Sample Question: Is Momentum Conserved When Shooting a Shotgun?

Question: When a shotgun is held firmly against your shoulder, is the momentum of the system conserved?

Answer: It depends on how you define the system. If the system is defined as just the shotgun and the pellets, there is then a strong external force exerted on the system by the shoulder of the shooter. Since the condition for conservation of momentum is that the net external force acting on the system be zero, the momentum of this system is not conserved.

If we included the shooter and the earth in our system, then momentum would be conserved because all of the forces would be internal to this system. The change in the momentum of the earth would be imperceptible, however.

Box 7.3.) Your own mass then becomes part of the system. This increased mass will produce a smaller recoil velocity, even if you happen to be standing on ice with no frictional forces between your feet and the earth. More important, the shotgun will not move against your shoulder.

How does a rocket work?

The firing of a rocket is another example of recoil. The exhaust gases rushing out of the tail of the rocket have both mass and velocity and, therefore, momentum. If we ignore external forces, the momentum gained by the rocket in the forward direction will equal in magnitude the momentum of the exhaust gases in the opposite direction (fig. 7.13). Momentum is conserved, just as in our other examples of recoil. The rocket and the exhaust gases push against one another, and Newton's third law applies.

The difference between a rocket and our earlier examples of the skaters and the shotgun is that firing a rocket is usually a continuous process. The rocket gains momentum gradually rather than in a single short blast. The mass

figure **7.13** If a short blast is fired, the rocket gains momentum equal in magnitude but opposite in direction to the momentum of the exhaust gases.

of the rocket also changes as fuel is consumed and gases exhaust from the rocket engines. Computation of the final velocity becomes more difficult than for the skaters. For a brief blast of the rocket, though, the same analysis can be used.

Recoil works in empty outer space: the two objects need only push against each other, just as with the skaters and the shotgun. Rocket engines can be used for space travel, unlike the propeller engines or jet engines used on airplanes. Airplane engines depend on the presence of the atmosphere, both as a source of oxygen used in burning fuel and as something to push against. An airplane propeller pushes against the air, and the air, by Newton's third law, pushes against the propeller. This interaction accelerates the airplane. A rocket, on the other hand, is self-contained. It pushes against its own exhaust gases.

During recoil, objects push against one another, moving in opposite directions. If external forces can be neglected, momentum is conserved. The total momentum before and after the interaction equals zero. After the interaction, the two objects move away with equal but oppositely directed momentum vectors that cancel one another. Recoil is one of many kinds of brief interaction to which conservation of momentum applies.

7.4 Elastic and Inelastic Collisions

As the example involving football players showed, collisions are one of the most fruitful areas for applying conservation of momentum. Collisions involve large forces of interaction acting for very brief times, and they produce dramatic changes in the motion of the colliding objects. Because the forces of interaction are so large, any external forces acting on the system usually are unimportant by comparison: momentum is conserved.

Different kinds of collisions produce different results. Sometimes the objects stick together and sometimes they bounce apart. What distinguishes these different cases, and what do the terms *elastic, inelastic,* and *perfectly inelastic* mean when applied to collisions? Is energy conserved as well as momentum? Railroad cars, bouncing balls, and pool balls help illustrate the differences.

What is a perfectly inelastic collision?

The easiest type of collision to analyze is one where two objects collide head-on and stick together after the collision, like the two football players discussed earlier. Because they stuck together and moved as one object after the collision, we had just one final velocity to contend with.

A collision in which the objects stick together after collision is called a **perfectly inelastic collision.** The objects do not bounce at all. If we know the total momentum of the system before the collision (and external forces are ignored), we can readily compute the final momentum and velocity of the now-joined objects.

Coupling railroad cars are another example of this type of collision. Try This Box 7.4 uses conservation of momentum to predict the final momentum and velocity of coupled railroad cars from knowledge of the momentum of the system before the collision. The process is much the same as the one used to predict the final velocity of the football players in section 7.2. In both cases, the separate objects move as one following the collision.

In Try This Box 7.4, the total mass of the coupled cars after the collision is five times that of car 5, so the final velocity of the coupled cars must be one-fifth that of car 5 to conserve momentum. The momentum of the system immediately after the collision is equal to that just before the collision, but the velocities have changed. The "final" velocity that we calculated is valid immediately after the collision. As the cars continue to move following the collision, frictional forces will gradually decelerate them until they come to rest.

Is energy conserved in collisions?

Is the kinetic energy after the railroad cars collide equal to the original kinetic energy of car 5 in the example in Try This Box 7.4? Using the relationship $KE = \frac{1}{2}mv^2$ introduced in chapter 6, we can compute the kinetic energy before and after the collision. The original kinetic energy of car 5 is 2250 kJ. (A kilojoule, kJ, is a thousand joules.) Immediately after the collision, the kinetic energy of the five cars moving together is just 450 kJ. (You can check these values.) A portion of the original kinetic energy is lost in any perfectly inelastic collision.

If we put a large spring on the front of the moving railroad car and allowed it to bounce off the other four cars rather than coupling, we will find that a greater portion of the kinetic energy is retained in the collision. When the objects bounce, the collision is either *elastic* or only *partially inelastic* rather than perfectly inelastic. The distinction is based on energy. An **elastic collision** is one in which *no* energy is lost. A **partially inelastic collision** is one in which *some* energy is lost, but the objects do not stick together. The greatest portion of energy is lost in the *perfectly inelastic collision,* when the objects stick.

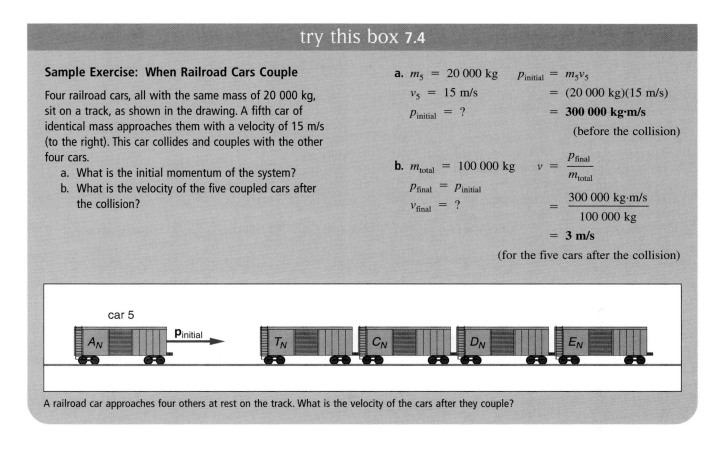

try this box 7.4

Sample Exercise: When Railroad Cars Couple

Four railroad cars, all with the same mass of 20 000 kg, sit on a track, as shown in the drawing. A fifth car of identical mass approaches them with a velocity of 15 m/s (to the right). This car collides and couples with the other four cars.

    a. What is the initial momentum of the system?

    b. What is the velocity of the five coupled cars after the collision?

a. $m_5 = 20\,000$ kg $p_{initial} = m_5 v_5$

     $v_5 = 15$ m/s $= (20\,000$ kg$)(15$ m/s$)$

     $p_{initial} = ?$ $= \mathbf{300\,000}$ **kg·m/s**

                      (before the collision)

b. $m_{total} = 100\,000$ kg $v = \dfrac{p_{final}}{m_{total}}$

     $p_{final} = p_{initial}$

     $v_{final} = ?$ $= \dfrac{300\,000 \text{ kg·m/s}}{100\,000 \text{ kg}}$

                $= \mathbf{3}$ **m/s**

             (for the five cars after the collision)

A railroad car approaches four others at rest on the track. What is the velocity of the cars after they couple?

In most collisions, some kinetic energy is lost because the collisions are not perfectly elastic. Heat is generated, the objects may be deformed, and sound waves are created, all of which involve conversions of the kinetic energy of the objects to other forms of energy. Even if the objects bounce, we cannot assume that the collision is elastic. More likely, the collision will be partially inelastic, implying that some of the initial kinetic energy has been lost.

A ball bouncing off a floor or wall with no decrease in the magnitude of its velocity is an example of an elastic collision. Since the magnitude of the velocity does not change (only the direction changes), the kinetic energy does not decrease. No energy has been lost. More likely, of course, some energy will be lost in such a collision, and the magnitude of the ball's velocity after the collision will be a little smaller than before.

The opposite extreme to an elastic collision of a ball with the wall would be a perfectly inelastic collision in which the ball sticks to the wall. In this case, the velocity of the ball after the collision is zero. So is its kinetic energy. All of the kinetic energy is lost (fig. 7.14).

What happens when pool balls bounce?

Very little energy is lost when pool balls collide with one another. (Time spent playing pool can be justified as a form of experimental physics. Your intuition about elastic

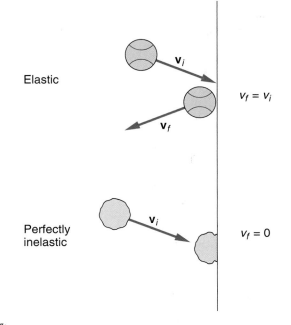

figure 7.14 An elastic collision and a perfectly inelastic collision of a ball with a wall. The ball sticks to the wall in the perfectly inelastic collision.

collisions can be improved in the process!) The collisions are basically elastic. Both momentum and kinetic energy are conserved in most collisions of pool balls.

When colliding objects such as pool balls bounce off one another, we must deal with two final velocities rather than one. We can readily compute the total momentum of the system before and after the collision from our knowledge of the initial momentum values of the objects. More information is needed to determine the individual velocities of the objects after the collision, however, since one value is not enough to determine two unknown velocities. (This is why the case of the perfectly inelastic collision, where objects stick together, is particularly easy to analyze.) In an elastic collision, conservation of energy provides the additional information.

For pool balls, the simplest case is the white cue ball colliding head-on with a second ball that is not moving before it is hit (the eleven ball in fig. 7.15). What happens? If spin is a minor factor in the collision, the cue ball stops dead on impact, and the eleven ball moves forward with a velocity equal to that of the cue ball before the collision. If the eleven ball acquires the same velocity that the cue ball had before the collision, it also has the same momentum $m\mathbf{v}$ as the initial momentum of the cue ball because both balls have the same mass. Momentum is conserved.

Kinetic energy is also conserved. The cue ball had a kinetic energy of $\frac{1}{2}mv^2$ before the collision. After the collision, the velocity and kinetic energy of the cue ball are zero, but the eleven ball now has a kinetic energy of $\frac{1}{2}mv^2$, since its mass and speed are the same as the cue ball before the collision. Given the equal masses of the two balls, the only way that both momentum and kinetic energy can be conserved is for the cue ball to stop and the eleven ball to move forward with the same momentum and kinetic energy that the cue ball had before the collision. This effect is familiar to any pool player.

The same phenomenon is involved in the familiar swinging-ball demonstration often seen as a decorative toy on mantels or desktops (fig. 7.16). A row of steel balls hangs

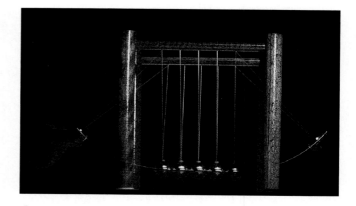

figure 7.16   The swinging-ball apparatus provides an example of collisions that are approximately elastic.

by threads from a metal or wooden frame. If one ball is pulled back and released, the collision with the other balls results in a single ball from the other end of the chain flying off with the same velocity as the first ball just before the collision. Both momentum and kinetic energy are conserved.

If two balls on one side are pulled back and released, two balls fly off from the opposite side of the row of balls after the collision. Again, both momentum and kinetic energy are conserved by this result. You can explore a variety of other combinations. It can be entertaining as well as addictive.

Collisions between hard spheres, such as pool balls or the steel balls in the swinging-ball apparatus, will generally be more or less elastic. Most collisions involving everyday objects, though, are inelastic to some degree. Some kinetic energy is lost. Momentum will be conserved, however, as long as our concern is with the values of momentum and velocity immediately before and after the collision.

Conservation of momentum is the primary tool used in understanding collisions. External forces can be ignored for the brief time of the collision, when the collision forces are dominant, and the law of conservation of momentum applies. Kinetic energy can also be conserved if the collision is elastic, as it is approximately for pool balls or other hard spheres. Most collisions involving familiar objects are partially inelastic and involve some loss of energy. The greatest proportion of energy is lost in perfectly inelastic collisions where objects stick together.

7.5 Collisions at an Angle

What happens when objects such as pool balls or automobiles collide at an angle, rather than head-on? Some interesting applications of conservation of momentum arise

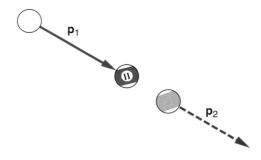
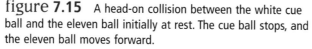

figure 7.15   A head-on collision between the white cue ball and the eleven ball initially at rest. The cue ball stops, and the eleven ball moves forward.

when motion is not confined to a straight line. It becomes more apparent that momentum is a vector when objects are free to move in two dimensions. Balls on a pool table, cars colliding at an intersection, or football players tackling all provide interesting examples.

An inelastic two-dimensional collision

Two football players, originally traveling at right angles to one another, collide and stick together, as in figure 7.17. What will be their direction of motion after the collision? How do we apply conservation of momentum in this two-dimensional case? In figure 7.17, we assume that the two players have the same masses and initial speeds as in our earlier example (section 7.2), but we no longer have a head-on collision. The momentum of the defensive back is now directed across the field as the fullback heads downfield.

Because momentum is a vector, we need to add the individual momentum vectors of the fullback and the defensive back to get the total momentum of the system before the collision. This can be done most readily by using a vector diagram with the vectors drawn to scale. The vectors can then be added graphically as we have done before. As shown in figure 7.18, the total momentum of the system before the collision is the hypotenuse of the right triangle formed by adding the other two momentum vectors.

If momentum is conserved in the collision, the total momentum of the two players after the collision will equal the total momentum before the collision. Since the two players move together after the collision, they will travel in the

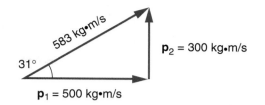

figure **7.18** The total momentum of the two football players prior to the collision is the vector sum of their individual momentums.

direction of the total momentum vector shown in figure 7.18. The direction of motion of both players changes as a result of the collision. The larger momentum of the fullback before the collision dictates that the final direction of motion is more downfield than across the field, but it is some of both. This result makes intuitive sense if you imagine yourself as one of the players.

The final direction of motion of the two football players after the collision depends on their momentum values before the collision. If the defensive back were bigger or moving faster than we assumed initially, he would have a larger momentum, and his tackle would cause a more impressive change in the direction of the fullback's motion. On the other hand, if the defensive back is small and moving slowly, his effect on the fullback's direction will be small. Adjusting the length of the momentum arrow $\mathbf{p}_2$, the momentum of the defensive back in figure 7.18, will illustrate these changes.

Everyday Phenomenon Box 7.1 describes a similar situation. Two cars approaching an intersection at right angles to one another collide and stick together. Working backward from information about the final direction of travel, the investigator can draw conclusions about the initial velocities of the two cars. Conservation of momentum is extremely important in accident analysis.

What happens in elastic two-dimensional collisions?

When pool balls collide, they do not stick together after the collision: when objects bounce, we have to contend with two final velocities with different directions. Although many real collisions are like this, analysis is more complicated for them than for the perfectly inelastic examples we have discussed. More information is needed to predict the final velocities. If we know that the collision is elastic, however, conservation of kinetic energy can provide that additional information.

Experimental physics on the pool table comes through again with an interesting example (and one of practical value to any pool player). Suppose that the cue ball strikes the

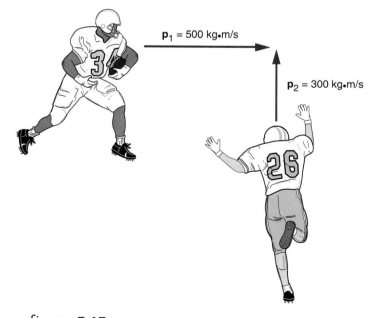

figure **7.17** The fullback and the defensive back approaching each other at a right angle.

Everyday Phenomenon

box 7.1

An Automobile Collision

The Situation. Officer Jones is investigating an automobile collision at the intersection of Main Street and Nineteenth Avenue. Driver A was traveling east on 19th Avenue when she was struck from the side by driver B who was traveling north on Main Street. The two cars stuck together after the collision and ended up against the lamppost on the northeast corner of the intersection.

Both drivers claim they started up just as the light in their direction changed to green and then collided with the other driver, who was running a red light and speeding. There are no other witnesses. Which driver is telling the truth?

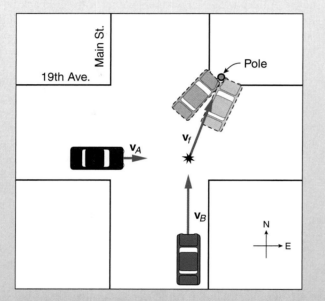

The collision at Main Street and 19th Avenue.

The Analysis. Officer Jones, having taken a physics course during college and being trained in the art of accident investigation, makes these observations:

1. The point of impact is well marked by the shards of glass from the headlights of B's car and other debris. Officer Jones indicates this point on the diagram in her accident report form.
2. The direction the two cars are traveling after the impact is also obvious. (She indicated this by a line drawn from the point of impact to the cars' final resting spot.)

3. Both cars have about the same mass (both are compacts of roughly the same vintage and size).
4. Conservation of momentum should determine the direction of the momentum vector after the collision.

After sketching the diagram and noting the direction of the final momentum vector, Officer Jones concludes that B is lying. Why? The final momentum vector must be equal to the sum of the initial momentum vectors of the two cars before the collision. Since the cars were traveling at right angles to one another, the two initial momentum vectors form the sides of a right triangle whose hypotenuse is the total momentum of the system. The diagram clearly shows that the momentum of B's car must have been considerably larger than the momentum of A's car.

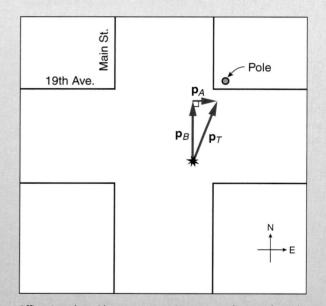

Officer Jones's accident report contains a vector diagram derived from conservation of momentum.

Since both drivers claimed to have just started from a complete stop after the traffic signal changed, the driver with the larger velocity before the collision is not telling it like it was. Driver B is the one who had the larger velocity, and so was presumably speeding through the red light. Driver B is thus cited by Officer Jones.

stationary eleven ball at an angle (off-center), as in figure 7.19. What happens to the two balls after the collision? The combined effects of conservation of momentum and conservation of kinetic energy lead to a unique result well-known to serious pool players.

The initial momentum of the system is simply that of the cue ball, the only one moving. Its direction is indicated by the arrow labeled $\mathbf{p}_i$ in figure 7.19 and in the drawings in figure 7.20. The force of interaction (and the impulse) between the two balls is along a line joining the centers of the balls at the point of impact. The eleven ball moves off along this line because the force of contact pushes it in that direction.

The total momentum of the system after the collision must still be in the direction of the initial momentum because momentum is conserved in the collision. Conservation of momentum also restricts the possible momentum and direction of the cue ball's motion after the collision (fig. 7.20). The momentum vectors of the two balls after the collision are added here to give the total momentum of the system, $\mathbf{p}_{total}$, which must be equal in both magnitude and direction to the initial momentum of the system.

Since the collision is elastic, the initial kinetic energy of the cue ball, $\frac{1}{2}mv^2$, must also equal the sum of the kinetic energies of the two balls after the collision. Since the masses of the two balls are equal, conservation of kinetic energy in the collision requires that*

$$v^2 = (v_1)^2 + (v_2)^2$$

where v is the speed of the cue ball before the collision, and v_1 and v_2 are the speeds of the two balls afterward. The velocity vectors form a triangle like the one formed by the momentum vectors in figure 7.20. If the sum of the squares of the two sides equals the square of the third side of the triangle, this triangle must be a right triangle according to the Pythagorean theorem from plane geometry. If the velocity vectors form a right triangle, so do the momentum vectors, which have the same directions as the corresponding velocity vectors.

Conservation of momentum requires that the momentum vectors add to form a triangle, but conservation of kinetic energy dictates that it be a *right* triangle. The cue ball will move off at a right angle (90°) to the direction of motion of the eleven ball after the collision. In playing pool, this is an important piece of intelligence if you are planning your next shot. Conservation of momentum *and* conservation of kinetic energy determine the shot.

If you have a pool table handy, test these conclusions using a variety of impact angles. (Marbles or steel balls are

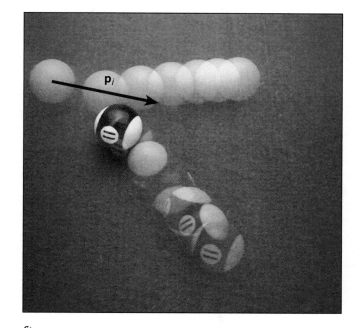

figure **7.19** The cue ball is aimed at a point off-center on the second ball to produce an angular collision.

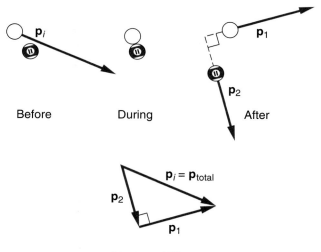

Vector addition

figure **7.20** The momentum vectors of the two balls after the collision add to give the total (initial) momentum of the system. The paths of the two balls are approximately at right angles after the collision.

a suitable substitute.) You may not get perfect 90° angles after the collision: the collision is not perfectly elastic, and spin can sometimes be a factor. The angle between the two final velocities, though, will usually be within a few degrees of a right angle. Seeing is believing—give it a try.

---

* Conservation of kinetic energy requires that $\frac{1}{2}mv^2 = \frac{1}{2}m(v_1)^2 + \frac{1}{2}m(v_2)^2$, but the mass and the factor of $\frac{1}{2}$ can be divided out of the equation.

Conservation of momentum requires that the direction of the momentum vector be conserved as well as its size. When collisions occur at an angle, this requirement restricts the directions and velocities of the resulting motions. If the collision is elastic, as with pool balls, conservation of energy adds another restriction. If you can

imagine the direction and magnitude of the original momentum vector, you will have some sense of the outcome. These conservation laws are powerful predictors of what happens when people, pool balls, cars, and even subatomic particles or stars collide.

summary

In this chapter, we recast Newton's second law in terms of impulse and momentum to describe interactions between objects, such as collisions, that involve strong interaction forces acting over brief time intervals. The principle of conservation of momentum, which follows from Newton's second and third laws, plays a central role.

1   Momentum and impulse.   Newton's second law can be recast in terms of momentum and impulse, yielding the statement that the impulse acting on an object equals the change in momentum of the object. Impulse is defined as the average force acting on an object multiplied by the time interval during which the force acts. Momentum is defined as the mass of an object times its velocity.

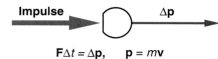

$$\mathbf{F}\Delta t = \Delta \mathbf{p}, \quad \mathbf{p} = m\mathbf{v}$$

2   Conservation of momentum.   Newton's second and third laws combine to yield the principle of conservation of momentum: if the net external force acting on a system is zero, the total momentum of the system is a constant.

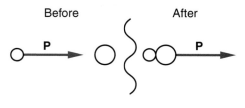

If $\mathbf{F}_{external} = 0$

$\mathbf{P}_{total} = $ constant

3   Recoil.   If an explosion or push occurs between two objects initially at rest, conservation of momentum dictates that the total momentum after the event must still be zero if there is no net external force. The final momentum vectors of the two objects are equal in size but opposite in direction.

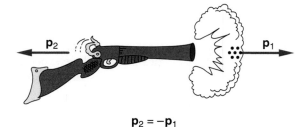

$$\mathbf{p}_2 = -\mathbf{p}_1$$

4   Elastic and inelastic collisions.   A perfectly inelastic collision is one in which the objects stick together after the collision. If external forces can be ignored, the total momentum is conserved. An elastic collision is one in which the total kinetic energy is also conserved.

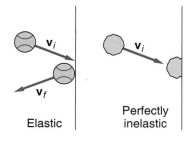

Elastic   Perfectly inelastic

5   Collisions at an angle.   Conservation of momentum is not restricted to one-dimensional motion. When objects collide at an angle, the total momentum of the system before and after the collision is found by adding the momentum vectors of the individual objects.

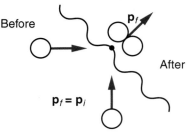

$$\mathbf{p}_f = \mathbf{p}_i$$

key terms

Impulse, 124
Momentum, 124
Impulse/momentum principle, 124

Conservation of momentum, 126
Recoil, 129
Perfectly inelastic collision, 130

Elastic collision, 130
Partially inelastic collision, 130

study hint

Except for the examples involving impulse, most of the situations described in this chapter highlight the principle of conservation of momentum. The basic ideas used in applying conservation of momentum are:

1. External forces are assumed to be much smaller than the very strong forces of interaction in a collision or other brief event. If external forces acting on the system can be ignored, momentum is conserved.
2. The total momentum of the system before the collision or other brief interaction $p_{initial}$ is equal to the momentum after the event p_{final}. Momentum is conserved and does not change.

3. Equality of momentum before and after the event can be used to obtain other information about the motion of the objects.

For review, look back at how these three points are used in each of the examples in this chapter. The total momentum of the system before and after the event is always found by adding the momentum values of the individual objects as *vectors*. You should be able to describe the magnitude and direction of this total momentum for each of the examples.

questions

*Questions identified with an asterisk are more open-ended than the others. They call for lengthier responses and are more suitable for group discussion.

Q1. Does the length of time that a force acts on an object have any effect on the strength of the impulse produced? Explain.

Q2. Two forces produce equal impulses, but the second force acts for a time twice that of the first force. Which force, if either, is larger? Explain.

Q3. Is it possible for a baseball to have as large a momentum as a much more massive bowling ball? Explain.

Q4. Are impulse and force the same thing? Explain.

Q5. Are impulse and momentum the same thing? Explain.

Q6. If a ball bounces off a wall so that its velocity coming back has the same magnitude that it had prior to bouncing:
    a. Is there a change in the momentum of the ball? Explain.
    b. Is there an impulse acting on the ball during its collision with the wall? Explain.

Q7. Is there an advantage to following through when hitting a baseball with a bat, thereby maintaining a longer contact between the bat and the ball? Explain.

Q8. What is the advantage of a padded dashboard compared to a rigid dashboard in reducing injuries during collisions? Explain using momentum and impulse ideas.

Q9. What is the advantage of an air bag in reducing injuries during collisions? Explain using impulse and momentum ideas.

*Q10. If an air bag inflates too rapidly and firmly during a collision, it can sometimes do more harm than good in low-velocity collisions. Explain using impulse and momentum ideas.

Q11. If you catch a baseball or softball with your bare hand, will the force exerted on your hand by the ball be reduced if you pull your arm back during the catch? Explain.

Q12. A truck and a bicycle are moving side by side with the same velocity. Which, if either, will require the larger impulse to bring it to a halt? Explain.

Q13. Is the principle of conservation of momentum always valid, or are there special conditions necessary for it to be valid? Explain.

Q14. A ball is accelerated down a fixed inclined plane under the influence of the force of gravity. Is the momentum of the ball conserved in this process? Explain.

Q15. Two objects collide under conditions where momentum is conserved. Is the momentum of each object conserved in the collision? Explain.

Q16. Which of Newton's laws of motion are involved in justifying the principle of conservation of momentum? Explain.

*Q17. A compact car and a large truck have a head-on collision. During the collision, which vehicle, if either, experiences:
   a. the greater force of impact? Explain.
   b. the greater impulse? Explain.
   c. the greater change in momentum? Explain.
   d. the greater acceleration? Explain.

Q18. A fullback collides midair and head-on with a lighter defensive back. If the two players move together following the collision, is it possible that the fullback will be carried backward? Explain.

Q19. Two ice skaters, initially at rest, push off one another. What is the total momentum of the system after they push off? Explain.

Q20. Two shotguns are identical in every respect (including the size of shell fired) except that one has twice the mass of the other. Which gun will tend to recoil with greater velocity when fired? Explain.

*Q21. When a cannon rigidly mounted on a large boat is fired, is momentum conserved? Explain, being careful to clearly define the system being considered.

Q22. Is it possible for a rocket to function in empty space (in a vacuum) where there is nothing to push against except itself? Explain.

Q23. Suppose that you are standing on a surface that is so slick that you can get no traction at all in order to begin moving across this surface. Fortunately, you are carrying a bag of oranges. Explain how you can get yourself moving.

Q24. A railroad car collides and couples with a second railroad car that is standing still. If external forces acting on the system are ignored, is the velocity of the system after the collision equal to, greater than, or less than that of the first car before the collision? Explain.

Q25. Is the collision in question 24 elastic, partially inelastic, or perfectly inelastic? Explain.

Q26. If momentum is conserved in a collision, does this indicate conclusively that the collision is elastic? Explain.

Q27. A ball bounces off a wall with a velocity whose magnitude is less than it was before hitting the wall. Is the collision elastic? Explain.

*Q28. A ball bounces off a wall that is rigidly attached to the earth.
   a. Is the momentum of the ball conserved in this process? Explain.
   b. Is the momentum of the entire system conserved? Explain, being careful to clarify how you are defining the system.

Q29. A cue ball strikes an eight ball of equal mass, initially at rest. The cue ball stops and the eight ball moves forward with a velocity equal to the initial velocity of the cue ball. Is the collision elastic? Explain.

Q30. Two lumps of clay traveling through the air in opposite directions collide and stick together. Their momentum vectors prior to the collision are shown in the diagram. Sketch the momentum vector of the combined lump of clay after the collision, making the length and direction appropriate to the situation. Explain your result.

Q31. Two lumps of clay, of equal mass, are traveling through the air at right angles to each other with velocities of equal magnitude. They collide and stick together. Is it possible that their velocity vector after the collision is in the direction shown in the diagram? Explain.

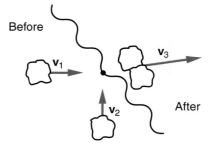

Q32. Two cars of equal mass collide at right angles to one another in an intersection. Their direction of motion after the collision is as shown in the diagram. Which car had the greater velocity before the collision? Explain.

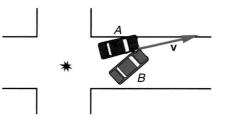

Q33. A car and a small truck traveling at right angles to one another with the same speed collide and stick together. The truck's mass is roughly twice the car's mass. Sketch the direction of their momentum vector immediately after the collision. Explain your result.

*Q34. A cue ball strikes a glancing blow against a second pool ball initially at rest. Sketch the situation indicating the magnitudes and directions of the momentum vectors of each ball before and after the collision.

*Q35. Suppose that on a perfectly still day, a sailboat enthusiast decides to bring along a powerful battery-operated fan in order to provide an air current for his sail, as shown in the diagram.

a. What are the directions of the change in momentum of the air at the fan and at the sail?

b. What are the directions of the forces acting on the fan and on the sail due to these changes in momentum?

c. Would the sailor in this picture be better off with the sail furled (down) or unfurled (up)? Explain.

exercises

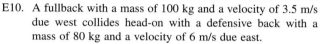

E1. An average force of 300 N acts for a time interval of 0.04 s on a golf ball.

a. What is the magnitude of the impulse acting on the golf ball?

b. What is the change in the golf ball's momentum?

E2. What is the momentum of a 1200-kg car traveling with a speed of 27 m/s (60 MPH)?

E3. A bowling ball has a mass of 6 kg and a speed of 1.5 m/s. A baseball has a mass of 0.12 kg and a speed of 40 m/s. Which ball has the larger momentum?

E4. A force of 35 N acts on a ball for 0.2 s. If the ball is initially at rest:

a. What is the impulse on the ball?

b. What is the final momentum of the ball?

E5. A 0.12-kg ball traveling with a speed of 40 m/s is brought to rest in a catcher's mitt. What is the size of the impulse exerted by the mitt on the ball?

E6. A ball experiences a change in momentum of 9.0 kg·m/s.

a. What is the impulse acting on the ball?

b. If the time of interaction is 0.15 s, what is the magnitude of the average force acting on the ball?

E7. A 60-kg front-seat passenger in a car moving initially with a speed of 18 m/s (40 MPH) is brought to rest by an air bag in a time of 0.4 s.

a. What is the impulse acting on the passenger?

b. What is the average force acting on the passenger in this process?

E8. A ball traveling with an initial momentum of 2.5 kg·m/s bounces off a wall and comes back in the opposite direction with a momentum of −2.5 kg·m/s.

a. What is the change in momentum of the ball?

b. What impulse would be required to produce this change?

E9. A ball traveling with an initial momentum of 4.0 kg·m/s bounces off a wall and comes back in the opposite direction with a momentum of −3.5 kg·m/s.

a. What is the change in momentum of the ball?

b. What impulse is required to produce this change?

E10. A fullback with a mass of 100 kg and a velocity of 3.5 m/s due west collides head-on with a defensive back with a mass of 80 kg and a velocity of 6 m/s due east.

a. What is the initial momentum of each player?

b. What is the total momentum of the system before the collision?

c. If they stick together and external forces can be ignored, what direction will they be traveling immediately after they collide?

E11. An ice skater with a mass of 80 kg pushes off against a second skater with a mass of 32 kg. Both skaters are initially at rest.

a. What is the total momentum of the system after they push off?

b. If the larger skater moves off with a speed of 3 m/s, what is the corresponding speed of the smaller skater?

E12. A rifle with a mass of 2.2 kg fires a bullet with a mass of 5.0 g (0.005 kg). The bullet moves with a muzzle velocity of 600 m/s after the rifle is fired.

a. What is the momentum of the bullet after the rifle is fired?

b. If external forces acting on the rifle can be ignored, what is the recoil velocity of the rifle?

E13. A rocket ship at rest in space gives a short blast of its engine, firing 50 kg of exhaust gas out the back end with an average velocity of 400 m/s. What is the change in momentum of the rocket during this blast?

E14. A railroad car with a mass of 12 000 kg collides and couples with a second car of mass 18 000 kg that is initially at rest. The first car is moving with a speed of 12 m/s prior to the collision.

a. What is the initial momentum of the first car?

b. If external forces can be ignored, what is the final velocity of the two railroad cars after they couple?

E15. For the railroad cars in Try This Box 7.4:

a. What is the kinetic energy of car 5 before the collision?

b. What is the kinetic energy of all five cars just after the collision?

c. Is energy conserved in this collision?

E16. A 4000-kg truck traveling with a velocity of 10 m/s due north collides head-on with a 1200-kg car traveling with a velocity of 20 m/s due south. The two vehicles stick together after the collision.
   a. What is the momentum of each vehicle prior to the collision?
   b. What are the size and direction of the total momentum of the two vehicles after they collide?

E17. For the two vehicles in exercise 16:
   a. Sketch to scale the momentum vectors of the two vehicles prior to the collision.
   b. Add the two vectors on your sketch graphically.

E18. A truck with a mass of 4000 kg traveling with a speed of 10 m/s collides at right angles with a car with a mass of 1500 kg traveling with a speed of 20 m/s.
   a. Sketch to proper scale and direction the momentum vectors of each vehicle prior to the collision.
   b. Using the graphical method of vector addition, add the momentum vectors to get the total momentum of the system prior to the collision.

challenge problems

CP1. A fast ball thrown with a velocity of 40 m/s (approximately 90 MPH) is struck by a baseball bat, and a line drive comes back toward the pitcher with a velocity of 60 m/s. The ball is in contact with the bat for a time of just 0.04 s. The baseball has a mass of 120 g (0.120 kg).
   a. What is the change in momentum of the baseball during this process?
   b. Is the change in momentum greater than the final momentum? Explain.
   c. What is the magnitude of the impulse required to produce this change in momentum?
   d. What is the magnitude of the average force that acts on the baseball to produce this impulse?

CP2. A bullet is fired into a block of wood sitting on a block of ice. The bullet has an initial velocity of 500 m/s and a mass of 0.005 kg. The wooden block has a mass of 1.2 kg and is initially at rest. The bullet remains embedded in the block of wood afterward.
   a. Assuming that momentum is conserved, find the velocity of the block of wood and bullet after the collision.
   b. What is the magnitude of the impulse that acts on the block of wood in this process?
   c. Does the change in momentum of the bullet equal that of the block of wood? Explain.

CP3. Consider two cases in which the same ball is thrown against a wall with the same initial velocity. In case A, the ball sticks to the wall and does not bounce. In case B, the ball bounces back with the same speed that it came in with.
   a. In which of these two cases is the change in momentum the largest?
   b. Assuming that the time during which the momentum change takes place is approximately the same for these two cases, in which case is the larger average force involved?
   c. Is momentum conserved in this collision? Explain.

CP4. A car traveling at a speed of 18 m/s (approximately 40 MPH) crashes into a solid concrete wall. The driver has a mass of 90 kg.
   a. What is the change in momentum of the driver as he comes to a stop?
   b. What impulse is required in order to produce this change in momentum?
   c. How does the application and magnitude of this force differ in two cases: the first, in which the driver is wearing a seat belt, and the second, in which he is not wearing a seat belt and is stopped instead by contact with the windshield and steering column? Will the time of action of the stopping force change? Explain.

CP5. A 1500-kg car traveling due north with a speed of 25 m/s collides head-on with a 4500-kg truck traveling due south with a speed of 15 m/s. The two vehicles stick together after the collision.
   a. What is the total momentum of the system prior to the collision?
   b. What is the velocity of the two vehicles just after the collision?
   c. What is the total kinetic energy of the system before the collision?
   d. What is the total kinetic energy just after the collision?
   e. Is the collision elastic? Explain.

home experiments and observations

HE1. Take two marbles or steel balls of the same size and practice shooting one into the other. Make these observations:
a. If you produce a head-on collision with the second marble initially at rest, does the first marble come to a complete stop after the collision?
b. If the collision with a second marble occurs at an angle, is the angle between the paths of the two marbles after the collision a right angle (90°)?
c. If marbles of different sizes and masses are used, how do the results of parts a and b differ from those obtained with marbles of the same mass?

HE2. If you have access to a pool table, try parts a and b of the observations in home experiment 1 on the pool table. What effect does putting spin on the first ball have on the collisions?

HE3. If you have both a basketball and a tennis ball, try dropping the two of them onto a floor with a hard surface, first individually and then with the tennis ball placed on top of the basketball before the two are dropped together.
a. Compare the height of the bounce of each ball in these different cases. The case where the two are dropped together may surprise you.

b. Can you devise an explanation for these results using impulse and Newton's third law? (Consider the force between the basketball and the floor as well as that between the tennis ball and the basketball for the case where they are dropped together.)

HE4. Place a cardboard box on a smooth tile or wood floor. Practice rolling a basketball or soccer ball at different speeds and allowing the ball to collide with the box. Observe the motion of both the box and the ball just after the collision.
a. How do the results of the collision vary for different speeds of the ball (slow, medium, fast)?
b. If we increase the weight of the box by placing books inside, how do the results of the collision change for the cases in part a?
c. Can you explain your results using conservation of momentum?

Rotational Motion of Solid Objects

chapter overview

Starting with a merry-go-round—and making use of the analogy between linear and rotational motion—we first consider what concepts are needed to describe rotational motion. We then turn to the causes of rotational motion, which involves a modified form of Newton's second law. Torque, rotational inertia, and angular momentum will be introduced as we proceed. Our goal is to develop a clear picture of both the description and causes of rotational motion. After studying this chapter, you should be able to predict what will happen in many common examples of spinning or rotating objects, such as ice skaters and divers. The world of sports is rich in examples of rotational motion.

chapter outline

In the park next door to the author's house, there is a child-propelled merry-go-round (fig. 8.1). It consists of a circular steel platform mounted on an excellent bearing so that it rotates without much frictional resistance. Once set in rotation, it will continue to rotate. Even a child can start it moving and then jump on (and sometimes fall off). Along with the swings, the slide, and the little animals mounted on heavy-duty springs, the merry-go-round is a popular center of activity in the park.

The motion of this merry-go-round bears both similarities and differences to motions we have already considered. A child sitting on the merry-go-round experiences circular motion, so some of the ideas discussed in chapter 5 will come into play. What about the merry-go-round itself, though? It certainly moves, but it goes nowhere. How do we describe its motion?

Rotational motion of solid objects such as the merry-go-round is common: the rotating earth, a spinning skater, a top, and a turning wheel all exhibit this type of motion. For Newton's theory of motion to be broadly useful, it should explain what is happening in rotational motion as well as in **linear motion** (where an object moves from one point to another in a straight line). What causes rotational motion? Can Newton's second law be used to explain such motion?

figure **8.1** The merry-go-round in the park is an example of rotational motion. How do we describe and explain this motion?

We will find that there is a useful analogy between the linear motion of objects and rotational motion. The questions just posed can be answered best by making full use of this analogy. Taking advantage of the similarities between rotational motion and linear motion saves space in our mental computers, thus making the learning process more efficient.

8.1 What Is Rotational Motion?

A child begins to rotate the merry-go-round described in the introduction. She does so by holding on to one of the bars on the edge of the merry-go-round (fig. 8.1) as she stands beside it. She begins to push the merry-go-round, accelerating as she goes, until eventually she is running, and the merry-go-round is rotating quite rapidly.

How do we describe the rotational motion of the merry-go-round or that of a spinning ice skater? What quantities would we use to describe how fast they are rotating or how far they have rotated?

Rotational displacement and rotational velocity

How would you measure how fast the merry-go-round is rotating? If you stood to one side and watched the child pass your position, you could count the number of revolutions that the child makes in a given time, measured with your watch. Dividing the number of revolutions by the time in minutes yields the average rotational speed in revolutions per minute (rpm), a commonly used unit for describing the rate of rotation of motors, Ferris wheels, and other rotating objects.

If you say that the merry-go-round rotates at a rate of 15 rpm, you have described how fast an object is turning. The rate is analogous to speed or velocity, quantities used

to describe how fast an object is moving in the case of linear motion. We usually use the term **rotational velocity** to describe this rate of rotation. Revolutions per minute is just one of several units used to measure this quantity.

In measuring the rotational velocity of the merry-go-round, we describe how far it rotates in revolutions or complete cycles. Suppose that an object rotates less than one complete revolution. We could then use a fraction of a revolution to describe how far it has turned, but we might also use an angle measured in degrees. Since there are 360° in one complete revolution or circle, revolutions can be converted to degrees by multiplying by 360°/rev.

The quantity measuring how far an object has turned or rotated is an angle, often called the **rotational displacement.** It can be measured in revolutions, degrees, or a simple but less familiar unit used in mathematics and physics called the **radian.*** The three units commonly used to describe rotational displacement are summarized in figure 8.2.

*The radian is defined by dividing the arc length through which the point travels by the radius of the circle on which it is moving. Thus, in figure 8.2, if the point on the merry-go-round moves along the arc length a distance s, the number of radians involved is s/r where r is the radius. Since we are dividing one distance by another, the radian itself has no dimensions. Also, since the arc length s is proportional to the radius r, it does not matter how large a radius we choose. The ratio of s to r will be the same for a given angle. By definition of the radian, 1 revolution (rev) = 360° = 2π radians, and 1 radian (rad) = 57.3°.

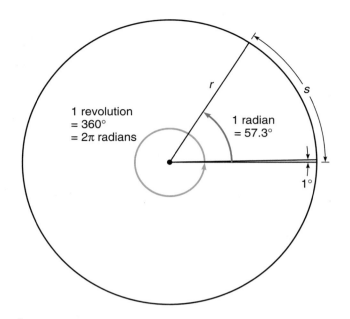

figure **8.2** Revolutions, degrees, or radians are different units for describing the rotational displacement of the merry-go-round.

Rotational displacement is analogous to the distance traveled by an object in linear motion. If we include the direction of travel, this distance is sometimes called the **linear displacement.**

The symbols used to describe rotational quantities mainly come from the Greek alphabet. Greek letters are used to avoid confusion with other quantities represented by letters of our ordinary Roman alphabet. The Greek letter theta (θ) is commonly used to represent angles (rotational displacements), and the Greek letter omega (ω) is used to represent rotational velocities.

The quantities that we have just introduced for describing the motion of an object such as the merry-go-round can be summarized as

Rotational displacement θ is an angle showing how far an object has rotated.

and

Rotational velocity ω is the rate of change of rotational displacement. It is found by dividing the rotational displacement by the time taken for this displacement to happen

$$\omega = \frac{\theta}{t}.$$

In describing rotational velocity, we usually use either revolutions or radians as the measure of rotational displacement. Degrees are less commonly used.

What is rotational acceleration?

In our original description of the child pushing the merry-go-round, the rate of rotation increased as she ran alongside. This involves a *change* in the rotational velocity, which suggests the concept of **rotational acceleration.** The Greek letter alpha (α) is the symbol used for rotational acceleration. It is the first letter in the Greek alphabet and corresponds to the letter *a* used to represent linear acceleration.

Rotational acceleration can be defined similarly to linear acceleration (see chapter 2):

Rotational acceleration is the rate of change in rotational velocity. It is found by dividing the change in rotational velocity by the time required for this change to occur,

$$\alpha = \frac{\Delta\omega}{t}.$$

The units of rotational acceleration are rev/s^2 or rad/s^2.

These definitions for both rotational velocity and rotational acceleration actually yield the *average* values of these quantities. To get *instantaneous* values, the time interval t must be made very small, as in the linear-motion definitions of instantaneous velocity and instantaneous acceleration (see sections 2.2 and 2.3). This then yields the rate of change of either displacement or velocity at a given instant in time.

You will remember these definitions of rotational displacement, velocity, and acceleration better if you keep in mind the complete analogy that exists between linear and rotational motion. This analogy is summarized in figure 8.3. In one dimension, distance *d* represents the change in position or *linear* displacement, which corresponds to *rotational* displacement θ. Average velocity and acceleration for linear motion are defined as before, with the corresponding definitions of rotational velocity and acceleration shown on the right side of the diagram in figure 8.3.

Constant rotational acceleration

In chapter 2, we introduced equations for the special case of constant linear acceleration because of its many important applications. By comparing linear and rotational quantities, we can write similar equations for constant rotational acceleration by substituting the rotational quantities for the corresponding linear quantities in the equations developed for linear motion. Table 8.1 shows the results beside corresponding equations for linear motion. Try This Box 8.1 is an application of the equations for constant rotational acceleration.

The merry-go-round in Try This Box 8.1 starts from rest and rotates through nine complete revolutions in 1 minute, a good effort by the person pushing. It is unlikely that this rate of acceleration could be sustained much longer than 1 minute. The rotational velocity reached in this time is a

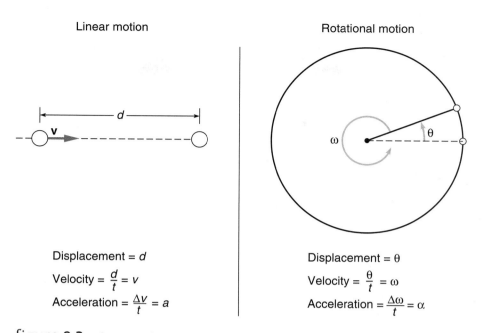

figure 8.3 There is a close resemblance between quantities used to describe linear motion and those used to describe rotational motion.

little less than a third of a revolution per second, a fast rotational velocity for such a merry-go-round.

How are linear and rotational velocity related?

How fast is the rider going when the merry-go-round in Try This Box 8.1 is rotating with a velocity of 0.30 rev/s? The answer to this question depends on where the rider is sitting. He or she will move faster when seated near the edge of the merry-go-round than near the center. The question involves a relationship between the linear speed of the rider and the rotational velocity of the merry-go-round.

Figure 8.4 shows two circles on the merry-go-round with different radii representing different positions of riders. The rider seated at the greater distance from the center travels a larger distance in 1 revolution than the rider near the center because the circumference of his circle is greater. The outside rider is therefore moving with a greater linear speed than the rider near the center.

The farther the rider is from the center, the farther he travels in 1 revolution, and the faster he is moving. The circumference of the circle on which the rider is traveling increases in proportion to the radius of the circle r, the distance of the rider from the center. If we express the rotational velocity in radians per second (rad/s), the relationship for the linear speed of the rider takes the form

$$v = r\omega.$$

try this box 8.1

Sample Exercise: Rotating a Merry-Go-Round

Suppose that a merry-go-round is accelerated at a constant rate of 0.005 rev/s^2, starting from rest.
  a. What is its rotational velocity at the end of 1 min?
  b. How many revolutions does the merry-go-round make in this time?

a. $\alpha = 0.005$ rev/s^2 $\quad \omega = \omega_0 + \alpha t$
  $\omega_0 = 0$ $\qquad\qquad = 0 + (0.005 \text{ rev/s}^2)(60 \text{ s})$
  $t = 60$ s $\qquad\qquad = \mathbf{0.30 \text{ rev/s}}$

b. $\theta = ?$ $\qquad$ Since ω_0 is equal to zero,

$$\theta = \tfrac{1}{2}\alpha t^2$$

$$= \tfrac{1}{2}(0.005 \text{ rev/s}^2)(60 \text{ s})^2$$

$$= \mathbf{9 \text{ rev}}$$

table 8.1

Constant Acceleration Equations for Linear and Rotational Motion

Linear motion	Rotational motion
$v = v_0 + at$	$\omega = \omega_0 + \alpha t$
$d = v_0 t + \tfrac{1}{2}at^2$	$\theta = \omega_0 t + \tfrac{1}{2}\alpha t^2$

figure **8.4** The rider near the edge travels a greater distance in 1 revolution than one near the center.

The linear speed *v* of a rider seated a distance *r* from the center of a merry-go-round is equal to *r* times the rotational velocity ω of the merry-go-round. (For this simple relationship to be valid, however, the rotational velocity must be expressed in radians per second rather than revolutions or degrees per second.)

The rate at which the merry-go-round or other object turns will affect how fast a point on the rotating object will move—in other words, its linear speed. Linear speed will depend on the distance from the axis of rotation. A child out at the edge of the merry-go-round will get a bigger thrill from the ride than one more timidly parked near the middle.

> Rotational displacement, rotational velocity, and rotational acceleration are the quantities that we need to fully describe the motion of a rotating object. They describe how far the object has rotated (rotational displacement), how fast it is rotating (rotational velocity), and the rate at which the rotation may be changing (rotational acceleration). These definitions are analogous to similar quantities used to describe linear motion. They tell us how the object is rotating, but not why. Causes of rotation are considered next.

8.2 Torque and Balance

What causes the merry-go-round to rotate in the first place? To get it started, a child has to push it, which involves applying a force. The direction and point of application of force are important to the success of the effort. If the child

pushes straight in toward the center, nothing happens. How do we apply a force to produce the best effect?

Unbalanced torques cause objects to rotate. What are *torques,* though, and how are they related to forces? A look at a simple scale or *balance* can help us get at the idea.

When is a balance balanced?

Consider a balance made of a thin but rigid beam supported by a **fulcrum** or pivot point, as in figure 8.5. If the beam is balanced before we place weights on it, and if we put equal weights at equal distances from the fulcrum, we expect that the beam will still be balanced. By *balanced,* we mean that it will *not* tend to rotate about the fulcrum.

Suppose that we wish to balance unequal weights on the beam. To balance a weight twice as large as a smaller weight, would we place the two weights at equal distances from the fulcrum? Intuition suggests that the smaller weight needs to be placed farther from the fulcrum than the larger weight for the system to be balanced, but it may not tell you how much farther (fig. 8.6). Trial and error with a simple balance will show that the smaller weight must be placed twice as far from the fulcrum as the larger (twice as large) weight.

Try it yourself using a ruler for the beam and a pencil for the fulcrum. Coins can be used as the weights. Experiments will show that both the weight and the distance from the fulcrum are important. The farther a weight sits from the fulcrum, the more effective it will be in balancing larger weights on the other side of the fulcrum. *Weight times distance from the fulcrum* determines the effect. If this product is the same for weights placed on either side of the fulcrum, the balance will not rotate.

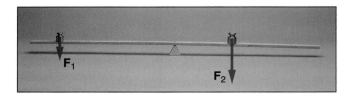

figure **8.5** A simple balance with equal weights placed at equal distances from the fulcrum.

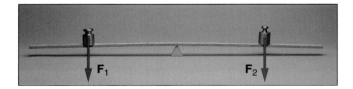

figure **8.6** A simple balance with unequal weights placed at different distances from the fulcrum. What determines whether the system will be balanced?

What is a torque?

This product of the force and the distance from the fulcrum—which describes the tendency of a weight to produce a rotation, is called the **torque.** More generally:

> The torque, τ, about a given axis or fulcrum is equal to the product of the applied force and the lever arm, *l;*
>
> $\tau = Fl.$
>
> The **lever arm** is the perpendicular distance from the axis of rotation to the line of action of the applied force.

The symbol τ is the Greek letter tau and is commonly used for torque.

The length *l* is the distance from the fulcrum to the point of application of the force and must be measured in a direction perpendicular to the line of action of the force. This distance is called the *lever arm* or *moment arm* of the force in question. The strength of the torque depends directly on both the size of the force and the length of its lever arm. If the torques produced by weights on either side of the fulcrum of our balance are equal in magnitude, the scale is balanced. It will not rotate.

Most of us have tried to turn a nut with a wrench at some time. We exert the force at the end of the wrench, in a direction perpendicular to the handle (fig. 8.7). The handle is the lever, and its length determines the lever arm. A longer handle is more effective than a shorter one because the resulting torque is greater.

As the term suggests, *lever arm* comes from our use of levers to move objects. Moving a large rock with a crowbar, for example, involves leverage. The applied force is most effective if it is applied at the end of the bar *and* perpendicular to the bar. The lever arm *l* is then just the distance from the fulcrum to the end of the bar. If the force is applied in some other direction, as in figure 8.8, the lever arm is shorter than it would be if the force is applied perpendicular to the bar. The lever arm is found by drawing the perpendicular line from the fulcrum to the line of action of the force, as indicated in figure 8.8.

How do torques add?

The direction of rotation associated with a torque is also important. Some torques tend to produce clockwise rotations and others counterclockwise rotations about a particular axis. For example, the torque due to the heavier weight on the right side of the fulcrum in figure 8.6 will produce a clockwise rotation about the fulcrum if it acts by itself. This is opposed by the equal-magnitude torque of the weight on the left side of the fulcrum, which would produce a counterclockwise rotation. The two torques cancel one another when the system is balanced.

Since torques can have opposing effects, we assign opposite signs to torques that produce rotations in opposite directions. If, for example, we chose to call torques that produce a counterclockwise rotation positive, torques producing clockwise rotations would be negative. (This is the conventional choice—it is unimportant which direction is chosen as positive as long as you are consistent in a given situation.) Identifying the sign of the torque indicates whether it will add or subtract from other torques.

In the case of the balance beam, the net torque will be zero when the beam is balanced, because the two torques

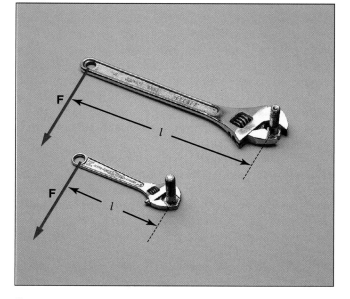

figure **8.7**   A wrench with a long handle is more effective than one with a short handle because of the longer lever arm for the longer wrench.

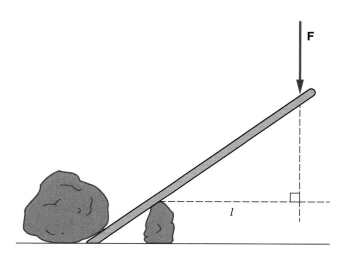

figure **8.8**   When the applied force is not perpendicular to the crowbar, the lever arm is found by drawing a perpendicular line from the fulcrum to the line of action of the force.

are equal but have opposite signs. The condition for balance or equilibrium is that the net torque acting on the system be zero. Either no torques act or the sum of the positive torques equals the sum of the negative torques, canceling one another by adding up to zero.

In Try This Box 8.2, we find the distance that a 3-N weight must be placed from a fulcrum to balance a 5-N weight producing a net torque of zero. (Since **W** = m**g**, a 5-N weight has a mass of approximately 0.5 kg or 500 g.) The units of torque are those of force times distance, newton-meters (N·m) in the metric system.

What is the center of gravity of an object?

Often, the weight of an object is itself an important factor in whether the object will rotate. How far, for example, could the child in figure 8.9 walk out on the plank without

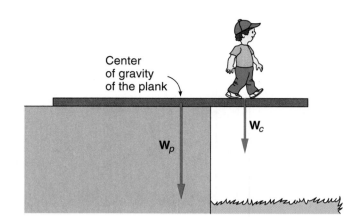

figure 8.9 How far can the child walk without tipping the plank? The entire weight of the plank can be treated as though it is located at the center of gravity.

the plank tipping? The weight of the plank is important in this case, and the concept of *center of gravity* is useful.

The **center of gravity** is the point about which the weight of the object itself exerts no net torque. If we suspend the object from its center of gravity, there would be no net torque at the suspension or support point. The object would be balanced. We can locate the center of gravity of a rodlike object by finding the point where it balances on your finger or other suitable fulcrum. For a more complex two-dimensional (planar) object, you can locate the center of gravity by suspending the object from two different points, drawing a line straight down from the point of suspension in each case, and locating the point of intersection of the two lines, as figure 8.10 illustrates.

In the case of the plank (fig. 8.9), the center of gravity is at the geometric center of the plank, provided that the plank is uniform in density and cut. The pivot point will be the edge of the supporting platform, the point to consider when computing torques. The plank will not tip as long as the counterclockwise torque produced by the weight of the plank about the pivot point is larger than the clockwise torque produced by the weight of the child. The weight of the plank is treated as though it is concentrated at the center of gravity of the plank.

The plank will verge on tipping when the torque of the child about the edge equals the torque of the plank in magnitude. This determines how far the child can walk on the plank before it tips. As long as the torque of the plank about the edge of the platform is larger than the torque of the child, the child is safe. The platform keeps the plank from rotating counterclockwise.

The location of the center of gravity is important in any effort at balancing. If the center of gravity lies below the pivot point, as in the balancing toy in figure 8.11, the toy will automatically regain its balance when disturbed. The center of gravity returns to the position directly below the pivot point, where the weight of the toy produces no torque.

---

try this box 8.2

Sample Exercise: Balancing a System

Suppose we have a 3-N weight that we want to balance against a 5-N weight on a beam, which is balanced when no masses are in place. The 5-N weight is placed 20 cm to the right of the fulcrum.

a. What is the torque produced by the 5-N weight?
b. How far would we have to place the 3-N weight from the fulcrum to balance the system?

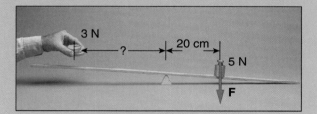

Where should the 3-N weight be placed on the beam to balance the system?

a. $F = 5$ N $\tau = -Fl$
   $l = 20$ cm $= 0.2$ m $= -(5 \text{ N})(0.2 \text{ m})$
   $\tau = ?$ $= -1$ **N·m**

The minus sign indicates that this torque would produce a clockwise rotation.

b. $F = 3$ N $\tau = Fl$

   $l = ?$ $l = \dfrac{\tau}{F}$

   $= \dfrac{+1 \text{ N·m}}{3 \text{ N}}$

   $= \mathbf{0.33 \ m}$ **(33 cm)**

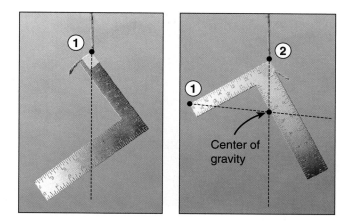

figure **8.10** Locating the center of gravity of a planar object. The center of gravity does not necessarily lie within the object.

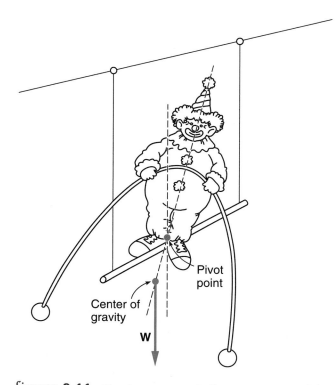

figure **8.11** The clown automatically returns to an upright position because the center of gravity is below the pivot point.

In this position, the lever arm for the weight of the clown and bar is zero.

Similarly, the location of your center of gravity is important in performing various maneuvers, athletic or otherwise. Try, for example, touching your toes with your back and heels against a wall. Why is this apparently simple trick impossible for most people to do? Where is your center of gravity relative to the pivot point determined by your feet? Center of gravity and torque are at work here.

Torques determine whether or not something will rotate. A torque is found by multiplying a force by its lever arm (the perpendicular distance from the axis of rotation to the line of action of the force). If the torque tending to produce a clockwise rotation equals the torque tending to produce a counterclockwise rotation, there is no rotation. If one of these torques is larger than the other, the torque will be unbalanced and the system will rotate.

8.3 Rotational Inertia and Newton's Second Law

When a child runs beside a merry-go-round, starting it to rotate, the force exerted by the child produces a torque about the axle. From our discussion in section 8.2, we know that the net torque acting on an object determines whether or not it will begin to rotate. Can we predict the rate of rotation by knowing the torque?

In linear motion, net force and mass determine the acceleration of an object, according to Newton's second law of motion. How do we adapt Newton's second law to cases of rotational motion? In this case, *torque* determines the rotational acceleration. A new quantity, the rotational inertia, takes the place of mass.

What is rotational inertia?

Let's return to the merry-go-round. The propulsion system (one energetic child or tired parent) applies a force at the edge of the merry-go-round. The torque about the axle is found by multiplying this force by the lever arm, in this case the radius of the merry-go-round (fig 8.12). If the frictional torque at the axle is small enough to be ignored, the torque produced by the child is the only one acting on the system. This torque produces the rotational acceleration of the merry-go-round.

How would we find this rotational acceleration? To find the linear acceleration produced by a force acting on an object, we use Newton's second law, $\mathbf{F} = m\mathbf{a}$. By analogy, we can develop a similar expression for rotational motion, where the torque τ replaces the force and the rotational acceleration α replaces the linear acceleration. But what quantity should we use in place of the *mass* of the merry-go-round?

In linear motion, mass represents the inertia or resistance to a change in motion. For rotational motion, a new concept is needed, **rotational inertia,** also referred to as the **moment of inertia.** The rotational inertia is the resistance of an object to change in its rotational motion. Rotational inertia is related to the mass of the object but also depends on how that mass is distributed about the axis of rotation.

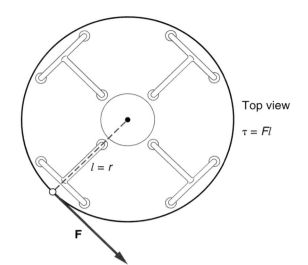

Top view

$\tau = Fl$

$l = r$

F

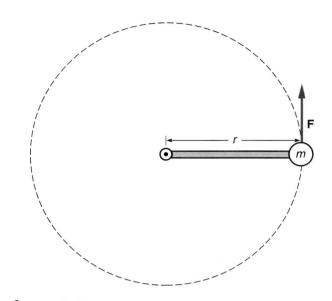

figure 8.13   A single concentrated mass at the end of a very light rod is set into rotation by the applied force **F**. Use Newton's second law to find the acceleration.

figure 8.12   The child exerts a force at the rim of the merry-go-round that produces a torque about the axle.

the size of the mass, the rotational inertia of a concentrated mass is

> rotational inertia = mass × square of distance
> from axis

$$I = mr^2,$$

where I is the symbol commonly used for rotational inertia, and r is the distance of the mass m from the axis of rotation. The total rotational inertia of an object like the merry-go-round can be found by adding the contributions of different parts of the object lying at different distances from the axis.

Newton's second law modified for rotational motion

By analogy to Newton's second law, **F** = m**a,** we can state the second law for rotational motion as

> The net torque acting on an object about a given axis is equal to the rotational inertia of the object about that axis times the rotational acceleration of the object, or
>
> $\tau = I\alpha.$

To put it differently, the rotational acceleration produced is equal to the torque divided by the rotational inertia, $\alpha = \tau/I$. The larger the torque, the larger the rotational acceleration, *but* the larger the rotational inertia, the smaller the rotational acceleration. Rotational inertia dictates how hard it is to change the rotational velocity of the object.

To get a feeling for a concept, physicists often use the trick of considering the simplest possible situation. For rotational motion, the simplest case is a single concentrated mass at the end of a very light rod, as in figure 8.13. If a force is applied to this mass in a direction perpendicular to the rod, the rod and mass will begin to rotate about the fixed axis at the other end of the rod.

For the rod and mass to undergo a rotational acceleration, the mass itself must have a linear acceleration. Like riders on a merry-go-round, however, the farther the mass is from the axis, the faster it moves for a given rotational velocity ($v = r\omega$). To produce the same rotational acceleration, a mass at the end of the rod must receive a larger linear acceleration than one nearer the axis. It is harder to get the system rotating when the mass is at the end of the rod than when it is nearer to the axis.

Applying Newton's second law to this situation, we find that the resistance to a change in rotational motion depends on the *square* of the distance of this mass from the axis of rotation. Since the resistance to change also depends on

To get a feel for these ideas, consider a simple object such as a twirler's baton. A baton consists of two masses at the end of a rod (fig. 8.14). If the rod itself is light, most of the baton's rotational inertia comes from the masses at either end. If you hold the baton at the center, a torque can be applied with your hand, producing a rotational acceleration and starting the baton to rotate.

Suppose that we could move these masses along the rod. If we moved the masses toward the center of the rod so that the distance from the center is half the original distance, what happens to the rotational inertia? The rotational inertia decreases to one-fourth of its initial value, ignoring the contribution of the rod. Rotational inertia depends on the *square* of the distance of the mass from the axis. Doubling the distance quadruples the rotational inertia. Halving the distance divides the rotational inertia by four.

The baton will be four times as hard to get to rotate when the masses are at the ends as when they are halfway from the ends. In other words, the torque needed to produce a rotational acceleration will be four times as large when the masses are at the ends as when they are at the intermediate positions. If you had a rod with adjustable masses, you could feel the difference in the amount of torque needed to start it rotating. Try a pencil with lumps of clay for the masses as a substitute.

Finding the rotational inertia of the merry-go-round

Finding the rotational inertia of an object like a merry-go-round is more difficult than just multiplying the mass by the square of the radius. Not all of the mass of the merry-go-round is at the outer edge—some of it is closer to the

axis and will make a smaller contribution to the rotational inertia. Imagine breaking the merry-go-round down into several pieces, finding the rotational inertia of each piece, and adding the rotational inertias for each piece together to get the total.

Results of this process for a few simple shapes are shown in figure 8.15. The equations illustrate the ideas we have discussed. For example, a solid disk has a smaller rotational inertia than a ring of the same mass and radius, because the mass of the disk is, on average, closer to the axis. The location of the axis is also important. A rod has a larger rotational inertia about an axis through one end than about an axis through the middle. When the axis of rotation is at the end of the rod, there is more mass at greater distances from the axis.

Depending on how it is constructed, the merry-go-round might be like a solid disk. A child sitting on the merry-go-round will also affect the rotational inertia. If several children all sit near the edge of the merry-go-round, their rotational inertia makes it more difficult to get the merry-go-round moving. If the children cluster near the center, they provide less additional rotational inertia. If you are feeling tired, have the children sit near the middle. You will save some effort.

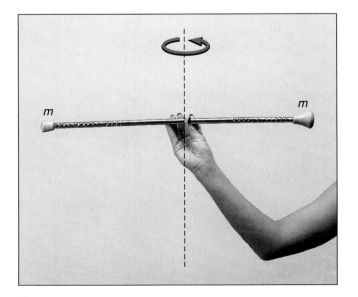

figure 8.14 The rotational inertia of a baton is determined largely by the masses at either end.

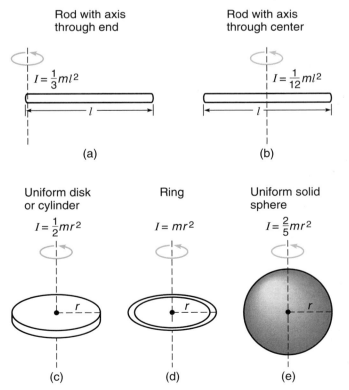

figure 8.15 Expressions for the rotational inertia of several objects, each with a uniform distribution of mass over its volume. The letter *m* is used to represent the total mass of the object.

try this box 8.3

Sample Exercise: Turning a Merry-Go-Round and a Rider

A simple merry-go-round has a rotational inertia of 800 kg·m² and a radius of 2 m. A child with a mass of 40 kg sits near the edge of the merry-go-round.

    a. What is the total rotational inertia of the merry-go-round and the child about the axis of the merry-go-round?

    b. What torque is required to give the merry-go-round a rotational acceleration of 0.05 rad/s²?

a. $I_{\text{merry-go-round}} = 800 \text{ kg·m}^2$ $I_{\text{child}} = mr^2$

    $m_{\text{child}} = 40 \text{ kg}$ $= (40 \text{ kg})(2 \text{ m})^2$

    $r = 2 \text{ m}$ $= 160 \text{ kg·m}^2$

The total rotational inertia is

$$I_{\text{total}} = I_{\text{merry-go-round}} + I_{\text{child}}$$
$$= 800 \text{ kg·m}^2 + 160 \text{ kg·m}^2$$
$$= \mathbf{960 \text{ kg·m}^2}$$

b. $\alpha = 0.05 \text{ rad/s}^2$ $\tau = I\alpha$

    $\tau = ?$ $= (960 \text{ kg·m}^2)(0.05 \text{ rad/s}^2)$

                              $= \mathbf{48 \text{ N·m}}$

Try This Box 8.3 attaches some numbers to these quantities. A child sitting on a merry-go-round is being accelerated by a push at a rate of 0.05 rad/s².* A torque of 48 N·m is needed to produce this rotational acceleration. A force of 24 N applied at the edge would have a lever arm of 2 m and produce the necessary torque of 48 N·m, a reasonable force for a child to generate if the child is not too small.

Rotational inertia is the resistance to change in rotational motion. It depends on both the mass of the object and the distribution of that mass about the axis of rotation. The rotational form of Newton's second law, $\tau = I\alpha$, shows the quantitative relationship between torque, rotational inertia, and rotational acceleration. Torque takes the place of force, rotational inertia replaces mass, and rotational acceleration replaces linear acceleration.

---

* To use Newton's second law for rotational motion, the rotational acceleration must be stated in radians per second squared. If the rotational acceleration is provided in rev/s² or some other angular unit, we convert it to rad/s² before proceeding.

8.4 Conservation of Angular Momentum

Have you ever watched an ice skater go into a spin? She starts the spin with her arms and one leg extended, then brings them in toward her body. As she brings her arms in, the rate of the spin increases; as she extends her arms again, her rotational velocity decreases (fig. 8.16).

The concept of angular or rotational momentum is useful in situations like this. The principle of conservation of angular momentum explains a variety of phenomena like the ice skater, including tumbling divers or gymnasts as well as the motion of planets around the sun. How can we use the analogy between linear and rotational motion to understand these ideas?

What is angular momentum?

If you were asked to invent the idea of angular (rotational) momentum, how might you go about it? Linear momentum is the mass (the inertia) times the linear velocity of an object ($\mathbf{p} = m\mathbf{v}$). An increase in either the mass or the velocity increases the momentum. Since it is a measure both of how much is moving and how fast it is moving, Newton called momentum the *quantity of motion*.

What is momentum's rotational equivalent? In comparing rotational and linear motion, rotational inertia plays the role of mass and rotational velocity replaces linear velocity. By analogy, we can define **angular momentum** as

Angular momentum is the product of the rotational inertia and the rotational velocity, or

$$L = I\omega,$$

where L is the symbol used for angular momentum.

The term *angular momentum* is more common than *rotational momentum*, but either can be used.

figure **8.16** The rotational velocity of the skater increases as she pulls her arms and leg in toward her body.

Like linear momentum, angular momentum is the product of two quantities, an inertia and a velocity. A bowling ball spinning slowly might have the same angular momentum as a baseball spinning much more rapidly, because of the larger rotational inertia I of the bowling ball. With its enormous rotational inertia, the earth has a huge angular momentum associated with its daily turn about its axis, even though the rotational velocity is small.

When is angular momentum conserved?

We have used the analogy between linear and rotational motion to introduce angular momentum. Can we also use it to state the principle of conservation of angular momentum? In chapter 7, we found that linear momentum is conserved when there is no net external force acting on a system. When would angular momentum be conserved?

Since torque takes the role of force for rotational motion, we can state the principle of **conservation of angular momentum** as

> If the net torque acting on a system is zero, the total angular momentum of the system is conserved.

Torque replaces force, and angular momentum replaces ordinary or linear momentum. Table 8.2 lists some important parallels between linear and rotational motion.

Changes in the ice skater's rate of spin

Conservation of angular momentum is the key to understanding what happens when the spinning ice skater increases her rotational velocity by pulling in her arms. The external torque acting on the skater about her axis of rotation is very small, so the condition for conservation of angular momentum exists. Why does her rotational velocity increase?

When the skater's arms and one leg are extended, they contribute a relatively large portion to her total rotational inertia—their average distance from her axis of rotation is much larger than for other portions of her body. Rotational inertia depends on the square of the distance of various portions of her mass from the axis ($I = mr^2$). The effect of this distance is substantial, even though her arms and one leg are only a small part of the total mass of the skater. When the skater pulls her arms and leg in toward her body, their contribution to her rotational inertia decreases, and therefore, her total rotational inertia decreases.

Conservation of angular momentum requires that her angular momentum remain constant. Since angular momentum is the product of the rotational inertia and rotational velocity, $L = I\omega$, if I decreases, ω must increase for angular momentum to stay constant. She can slow her rate of spin by extending her arms and one leg again, which she does at the end of the spin. This increases her rotational inertia and decreases her rotational velocity: angular momentum is conserved. These ideas are illustrated by the example in Try This Box 8.4.

This phenomenon can be explored using a rotating platform or stool with good bearings to keep the frictional torques small (fig. 8.17). In these demonstrations, we often have the students hold masses in their hands, which increase the changes in rotational inertia that happen as the arms are drawn in toward the body. A striking increase in rotational velocity can be achieved!

A similar effect is at work when a diver pulls into a tuck position to produce a spin. In this case, the diver starts with her body extended and a slow rate of rotation about an axis through her body's center of gravity (fig. 8.18). As she goes into a tuck, the rotational inertia about this axis is reduced, and rotational velocity increases. As her dive nears

table 8.2

Corresponding Concepts of Linear and Rotational Motion

Concept	Linear motion	Rotational motion
Inertia	m	I
Newton's second law	$\mathbf{F} = m\mathbf{a}$	$\tau = I\alpha$
Momentum	$\mathbf{p} = m\mathbf{v}$	$L = I\omega$
Conservation of momentum	If $\mathbf{F} = 0$, $\mathbf{p} = $ constant	If $\tau = 0$, $L = $ constant
Kinetic energy	$KE = \frac{1}{2}mv^2$	$KE = \frac{1}{2}I\omega^2$

try this box 8.4

Sample Exercise: Some Physics of Figure Skating

An ice skater has a rotational inertia of 1.2 kg·m² when her arms are extended and a rotational inertia of 0.5 kg·m² when her arms are pulled in close to her body. If she goes into a spin with her arms extended and has an initial rotational velocity of 1 rev/s, what is her rotational velocity when she pulls her arms in close to her body?

$I_1 = 1.2$ kg·m²   Since angular momentum is conserved:

$I_2 = 0.5$ kg·m² $L_{final} = L_{initial}$

$\omega_1 = 1$ rev/s $I_2\omega_2 = I_1\omega_1$

$\omega_2 = ?$

Dividing both sides by I_2,

$$\omega_2 = (I_1/I_2)\omega_1$$
$$= (1.2 \text{ kg·m}^2/0.5 \text{ kg·m}^2)(1 \text{ rev/s})$$
$$= (2.4)(1 \text{ rev/s})$$
$$= \textbf{2.4 rev/s}$$

figure **8.17**   A student holding masses in each hand while sitting on a rotating stool can achieve a large increase in rotational velocity by bringing his arms in toward his body.

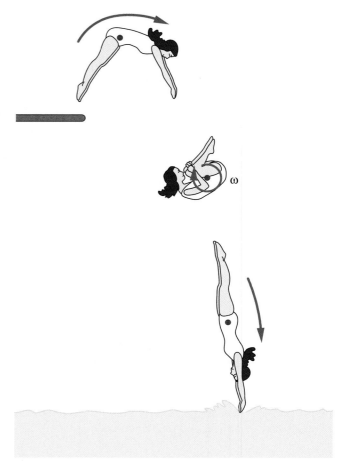

figure **8.18**   The diver increases her rotational velocity by pulling into a tuck position, thus reducing her rotational inertia about her center of gravity.

completion, she comes out of the tuck, increasing the rotational inertia and decreasing the rotational velocity. (The torque about the center of gravity due to the gravitational force acting on the diver is zero.)

There are many examples of varying the rotational velocity by changing the rotational inertia. It is much easier to produce a change in the rotational inertia of a body than to change the mass of the body. We simply change the distance of various portions of the mass from the axis of rotation. Conservation of angular momentum provides a quick explanation for these phenomena.

Kepler's second law

Conservation of angular momentum also plays a role in the orbit of a planet about the sun, and in fact, it can be used to explain Kepler's second law of planetary motion (see section 5.3). Kepler's second law says that the radius line from the sun to the planet sweeps out equal areas in equal times. The planet moves faster in its elliptical orbit when it is nearer to the sun than when it is farther from the sun.

The gravitational force acting on the planet produces no torque about the sun, because its line of action passes directly through the sun (fig. 8.19). The lever arm for this force is zero, and the resulting torque must also be zero. Angular momentum, therefore, is conserved.

When the planet moves nearer to the sun, its rotational inertia I about the sun decreases. To conserve angular momentum, the rotational velocity of the planet about the sun

(and thus its linear velocity*) must increase to keep the product $L = I\omega$ constant. This requirement results in equal areas being swept out by the radius line in equal times. The velocity of the planet must be larger when the radius gets smaller to keep the area being swept out the same.

You can observe a related effect in a simple experiment with a ball on a string. If you let the string wrap around your finger as it rotates, which produces a smaller radius of rotation, the ball will increase its rotational velocity about your finger. The rotational velocity ω increases as the rotational inertia I decreases because of the decreased radius. Angular momentum is conserved. Try it!

Everyday Phenomenon Box 8.1 provides an example in which angular momentum is conserved at some points in the motion of a yo-yo. At other points the angular momentum changes under the influence of torques.

---

* For a compact mass rotating about some axis, the definition of angular momentum reduces to $L = mvr$, where mv is the linear momentum and r is the perpendicular distance from the axis of rotation to the line along which the object is moving at that instant. If r decreases, v must increase to conserve angular momentum.

Everyday Phenomenon

box 8.1

Achieving the State of Yo

The Situation. A physics professor noticed that one of his students often carried a yo-yo to class and was proficient at putting the yo-yo through its paces. The professor challenged the student to explain the behavior of the yo-yo using the principles of torque and angular momentum.

In particular, the professor asked the student to explain why the yo-yo sometimes comes back but sometimes can be made to "sleep," or continue to rotate, at the end of the string. What are the differences in these two situations?

A yo-yo will come back to your hand, or with sufficient skill, you can make it "sleep" at the end of its string.

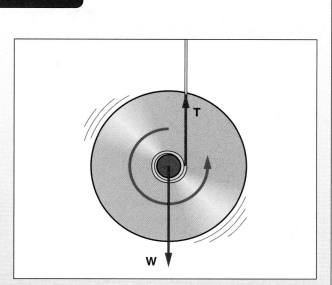

A cut-away diagram showing the forces acting on the yo-yo when it is falling. Its weight and the tension in the string are the only significant forces.

The Analysis. The student carefully examined the yo-yo's construction and how the string is attached. He noticed that the string is not tied tightly to the axle of the yo-yo, but ends in a loose loop around the axle instead. When the yo-yo is at the end of its string, the string can slip on the axle. When wound around the axle, on the other hand, the string is less likely to slip.

Usually, the yo-yo is started with the string wound around the axle and looped around the middle finger. When the yo-yo is released from the hand, the string unwinds, and the yo-yo gains rotational velocity and angular momentum. The student reasoned that a torque must be at work, and he drew a force diagram for the yo-yo that looked like the one shown here. Two forces act on the yo-yo, its weight acting downward and the tension in the string acting upward.

Since the yo-yo is accelerated downward, the weight must be greater than the tension to produce a downward net force. The weight does not produce a torque about the center of gravity of the yo-yo, though, because its line of action passes through the center of gravity, and the lever arm is zero. The tension acts along a line that is off-center and produces a torque that will cause a counterclockwise rotation about the center of gravity, as in the drawing.

The torque due to the tension in the string produces a rotational acceleration, and the yo-yo gains rotational velocity

and angular momentum as it falls. The yo-yo has a sizable angular momentum when it reaches the bottom of the string, and in the absence of external torques to change this angular momentum, it will be conserved. This is what happens when the yo-yo "sleeps" at the bottom of the string: the only torque acting is the frictional torque of the string slipping on the axle, and this will be small if the axle is smooth.

What happens, however, when the yo-yo returns to the student's hand? The yo-yo artist (yo-yoist?) jerks lightly on the string at the instant that the yo-yo reaches the bottom of the string. This jerk provides a brief impulse and upward acceleration of the yo-yo. Since it is already spinning, the yo-yo continues spinning in the same direction and the string rewinds itself around the axle of the yo-yo. The line of action of the tension in the string is now on the opposite side of the axle, though, and its torque causes the rotational velocity and angular momentum to decrease. The rotation should stop when the yo-yo slips back into the student's hand.

When the yo-yo is rising, the net force acting on the yo-yo is still downward, and the linear velocity of the yo-yo decreases along with its rotational velocity. The only time that a net force acts upward is when the upward impulse is delivered by jerking on the string. The situation is similar to a ball bouncing on the floor—the net force is downward except during the very brief time of contact with the floor. Our ability to affect the nature and timing of the impulse through the string causes the yo-yo either to sleep or return. This is what the "art of yo" is all about.

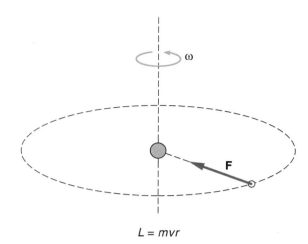

$$L = mvr$$

figure **8.19**   The gravitational force acting on the planet produces no torque about an axis through the sun because the lever arm is zero for this force.

By analogy to linear momentum, angular momentum is the product of the rotational inertia and the rotational velocity. Angular momentum is conserved when the net external torque acting on a system is zero. Decreases in rotational inertia lead to increases in rotational velocity, as demonstrated by the spinning ice skater. A spinning diver, a ball rotating at the end of a string, and a planet orbiting the sun are other examples of this effect.

8.5 Riding a Bicycle and Other Amazing Feats

Have you ever wondered why a bicycle remains upright when it is moving but promptly falls over when not moving? Angular momentum is involved, but some additional wrinkles are needed in the explanation. The direction of angular momentum is an important consideration. How can angular momentum have direction, and how is this direction involved in explaining the behavior of a bicycle, a spinning top, or other phenomena?

Is angular momentum a vector?

Linear momentum is a vector, and the direction of the momentum **p** is the same as for the velocity **v** of the object. Since angular momentum is associated with a rotational velocity, the question comes down to whether rotational velocities have direction. How would we define the direction of a rotational velocity?

If a merry-go-round (or just a disk) is rotating in a counterclockwise direction, as in figure 8.20, how might we indicate that direction with an arrow? The term *counterclockwise* indicates the direction of rotation as seen from a

certain perspective, but it does not define a unique direction. To complete our description, we would also have to specify the axis of rotation and our perspective or viewpoint. An object seen rotating counterclockwise when viewed from above is seen rotating clockwise when viewed from below. We could draw an axis of rotation and a curved arrow around it, as we often do, but it would be more desirable to specify direction with a simple straight arrow.

The usual solution to this problem is to define the direction of the rotational-velocity vector as being along the axis of rotation and in the upward direction for the counterclockwise rotation in figure 8.20. A rule for whether the vector should point up or down along the axis can be defined with the help of your right hand. If you hold your right hand with the fingers curling around the axis of rotation in the direction of the rotation, your thumb points in the direction of the rotational-velocity vector. If the merry-go-round were rotating clockwise (instead of counterclockwise), your thumb would point down, the direction of the rotational-velocity vector.

The direction of the angular-momentum vector is the same as the rotational velocity, since **L** = *I*ω. Conservation of angular momentum requires that the *direction* of the angular-momentum vector remain constant, as well as its magnitude.

Angular momentum and bicycles

Most of us have had some experience with riding a bicycle. The wheels of a bicycle acquire angular momentum when the bicycle is moving. Torque is applied to the rear wheel by the pedals and chain to produce a rotational acceleration. If the bicycle is moving in a straight line, the direction of the angular-momentum vector is the same for both wheels and is horizontal (fig. 8.21).

To tip the bike over, the direction of the angular-momentum vector must change, and that requires a torque. This torque would normally come from the gravitational

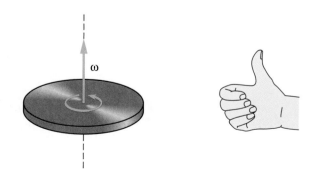

figure **8.20**   The direction of the rotational-velocity vector for the counterclockwise rotation is defined to be upward along the axis of rotation, as indicated by the thumb on the right hand with the fingers curled in the direction of rotation.

figure **8.21** The angular-momentum vector for each wheel is horizontal when the bicycle is upright.

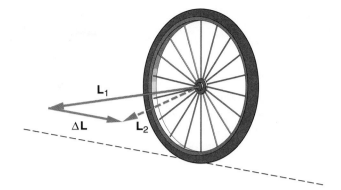

figure **8.22** The change in angular momentum associated with a leftward tilt points straight back, parallel to the line of contact of the tires with the road.

force acting on the rider and the bicycle through their center of gravity. When the bicycle is exactly upright, this force acts straight downward and passes through the axis of rotation for the falling bike. This axis of rotation is the line along which the tires contact the road. The torque about this axis will be zero, because the line of action of the force passes through the axis of rotation and the lever arm is zero. The direction and magnitude of the original angular momentum are conserved.

If the bike is not perfectly upright, a gravitational torque acts about the line of contact of the tires with the road. As the bike begins to fall, it acquires a rotational velocity and angular momentum about this axis. By our "right-hand rule," the direction of that angular-momentum vector is along the axis and points forward or backward depending on the direction of tilt. If the bike tilts to the left as seen from behind, the change in angular momentum associated with this torque points straight back, as in figure 8.22.

If the bike is standing still, that is all there is to it—the gravitational torque causes the bike to fall. When the bike is moving, however, the change in angular momentum ΔL produced by the gravitational torque adds to the angular momentum already present (L_1) from the rotating tires. As shown in figure 8.22, this causes a change in the direction of the total angular-momentum vector (L_2). This change in direction can be accommodated simply by turning the wheel

of the bicycle rather than letting the bike fall. We compensate for the effects of the gravitational torque by turning the bicycle towards the direction of the impending fall. The larger the initial angular momentum, the smaller the turn required. The angular momentum of the wheels is a major factor in stabilizing the bicycle.

This result may be surprising—yet all of us who have ridden bicycles take advantage of it routinely. When the bike is moving slowly, sharp turns of the wheel can keep it from falling while you shift your weight. Smaller adjustments suffice when the bike is moving more rapidly. By leaning into a curve, you use the gravitational torque to change the direction of angular momentum, helping to round the curve. Likewise, if you roll a coin along a tabletop, you will see it curve as it begins to fall. The path curves in the direction that the coin is tilting.

You can also observe this effect of torque in changing the direction of an angular-momentum vector by holding a bicycle upright on its rear wheel and having a friend spin the front wheel. It is harder to change the direction of this wheel when it is spinning rapidly than when it is spinning slowly or not at all. You will also get the feeling that the wheel has a mind of its own. As you try to tilt the wheel, it will tend to turn in a direction perpendicular to the tilt.

A bicycle tire mounted on a hand-held axle is even more effective for sensing the effects of torques applied to the axle. This is a common demonstration apparatus, but usually the tire is filled with steel cable rather than air. The steel cable gives the wheel a larger rotational inertia and a larger angular momentum for a given rate of spin. If you hold the axle on either side while the wheel is spinning in a vertical plane and then try to tip the wheel, you get a sense of what happens when you are riding a bicycle. It also demonstrates how hard it is to change the direction of the angular momentum of a rapidly spinning wheel. Everyday Phenomenon Box 8.2 discusses how torques are involved in the gear system of a bicycle.

Everyday Phenomenon

box 8.2

Bicycle Gears

The Situation. Most modern bicycles come equipped with the ability to change gears. When we are climbing a hill, we shift into a low gear making pedaling easier. When we are on level ground or a downgrade, we shift into higher gears, allowing us to cover more ground per turn of the pedal crank.

How do these gears work? How does the torque exerted on the rear wheel change when we change gears? What is the advantage of having many different gears? Adding a rotational twist to the concept of simple machines can help in understanding how gears work. Similar ideas apply to an automobile transmission.

The Analysis. The photograph shows the pedal wheel and the rear-wheel gear assembly for a 21-speed bicycle. There are

There are seven different sprockets on the rear-wheel gear assembly of a 21-speed bicycle. The two smaller sprockets on the pedal wheel lie behind the largest sprocket.

seven sprockets (toothed wheels) of different sizes on the rear wheel hub. There are also three different sprockets on the pedal wheel, only one of which is fully visible in the photograph. A pulley and lever mechanism (called a derailleur) allows us to move the chain from one sprocket to another. This is controlled by levers mounted on the handlebars that are linked to the derailleurs by cables.

When we pedal the bicycle we apply a torque to the pedal sprocket by pushing on the pedals. If our feet push perpendicularly to the pedal shaft, then the lever arm is just the length of the shaft. That will be the case when the pedal is in the forward position where we get maximum torque as shown in the drawing. This maximum-torque position alternates from the left foot to the right foot as the crank turns.

The torque exerted on the pedal chain ring causes the sprocket to accelerate rotationally provided that this torque is larger than the opposing torque exerted by the tension in the chain pulling back on the sprocket. This tension, in turn, produces a torque on the rear wheel via the rear-wheel sprocket. The size of the torque transmitted to the rear wheel depends on which of the several sprockets is engaged with the chain. A larger sprocket radius yields a larger torque because of the greater lever arm. (The lever arm is equal to the radius of the sprocket engaged.) As with an automobile, the rear wheel pushes against the road surface via friction, and by Newton's third law, the frictional force pushes forward on the bicycle.

(continued)

Rotating stools and tops

The hand-held bicycle wheel is good for other demonstrations that highlight angular momentum as a vector. If a student holds the wheel with its axle in the vertical direction while sitting on a rotating stool, conservation of angular momentum produces striking results. It is best to start the wheel spinning while holding the stool so that the stool does not rotate initially. We then have the student turn the wheel over, as in figure 8.23, reversing the direction of the angular-momentum vector of the wheel.

Can you imagine what happens then? To conserve angular momentum, the original direction of the angular-momentum vector must be maintained. The only way this can happen is for the stool with the student volunteer to begin to rotate in the same direction that the wheel was rotating *initially*. The sum of the angular-momentum vector of the wheel and the vector of the student and stool add to yield the original angular momentum (fig. 8.24). This will

figure **8.23**   A student holds a spinning bicycle wheel while sitting on a stool that is free to rotate. What happens if the wheel is turned upside down?

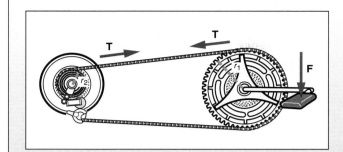

A force applied to the pedal produces a torque on the pedal wheel. This torque produces a tension in the chain that exerts a smaller torque on the rear-wheel sprocket due to its smaller radius.

How are simple-machine ideas involved? Suppose that the chain is engaged with the smallest sprocket on the rear-wheel assembly. The torque exerted on the rear wheel by the chain is then relatively small because of the small lever arm. The wheel turns several times, however, for each turn of the pedal sprocket. If, for example, the radius of the pedal sprocket is five times that of the rear-wheel sprocket, then the rear wheel turns five times for each turn of the pedal crank. (The circumference of each sprocket ($2\pi r$) is proportional to the radius, and the circumference determines how far the chain must travel for each turn.)

In this case, then, a small torque turns the rear wheel through a large angle while a larger input torque turns the pedal sprocket through a smaller angle. By analogy to linear work (force times distance moved), rotational work can be defined as the torque times the angle through which the sprocket turns ($\tau\,\theta$). As for any simple machine, the work output equals the work input, ignoring frictional torques at the wheel axles.

Would the situation we have just described represent a high gear or a low gear? Since we are getting several turns of the rear wheel for each turn of the pedal crank, this is a high gear. For a lower gear, we need to move the chain to a larger sprocket on the rear wheel (or a smaller sprocket on the pedal wheel). This will transmit a larger torque to the rear wheel at the expense of turning the wheel through a smaller angle and moving the bicycle through a smaller distance for each turn of the crank. When we are going uphill, we need this larger torque to overcome the pull of gravity.

For a 21-speed bike, there are three sprocket sizes on the pedal wheel and seven on the rear wheel allowing twenty-one (3 × 7) different ratios between the two sprockets. The advantage of having all these choices is that we can adjust the mechanical advantage of our gear system to the conditions we encounter, thus adjusting the force that we need to apply to the pedals to achieve the desired torque. If this force is too large, we will quickly tire. When it is small, however, we may not be taking maximum advantage of the easier riding conditions. We can go faster in a higher gear.

be true if the angular momentum gained by the student and stool is exactly twice the original angular momentum of the wheel. The student can stop the rotation of the stool by flipping the wheel axis back to its original direction.

The direction of angular momentum and its conservation are important in many other situations. The angular momentum of the helicopter's rotors, for example, is an extremely important factor in helicopter design. The motion of a top also shows fascinating effects. If you have a top, observe what happens to the direction of the angular-momentum vector as the top slows down. As the top begins to totter, the change in direction of the angular-momentum vector causes the rotation axis of the top to rotate (precess) about a vertical line. Does this remind you of what happens with a bicycle wheel?

Angular momentum and its direction are also central to atomic and nuclear physics. The particles that make up atoms have spins, and these spins imply angular momentum. The ways these angular-momentum vectors add are used to explain a variety of atomic phenomena. While the size scales differ enormously, it is useful to recognize the common ground that atoms and nuclei share with bicycle wheels and the solar system.

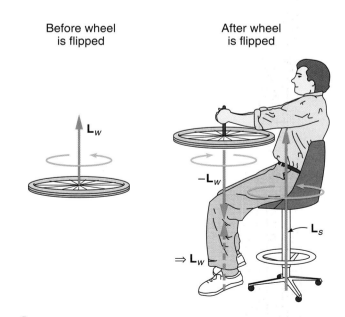

Before wheel is flipped

After wheel is flipped

figure 8.24 The angular momentum of the student and stool, $\mathbf{L}_s$, adds to that of the wheel, $-\mathbf{L}_w$, to yield the direction and magnitude of the original angular momentum, $\mathbf{L}_w$.

Like linear momentum, angular momentum is a vector. Its direction is the same as that of the rotational-velocity vector, which is along the axis of rotation, with the "right-hand rule" specifying which way it points along that axis. Conservation of angular momentum requires that both the magnitude and direction of the angular-momentum vector be constant (if there are no external torques). Many interesting phenomena can be explained using these ideas, including the stability of a moving bicycle, the motion of a spinning top, and the behavior of atoms and galaxies.

summary

We have considered the rotational motion of a solid object and what causes changes in rotational motion. We have used an analogy between linear motion and rotational motion to develop many of the concepts. The key ideas are summarized here:

1 What is rotational motion? Rotational displacement is described by an angle. Rotational velocity is the rate of change of that angle with time. Rotational acceleration is the rate of change of rotational velocity with time.

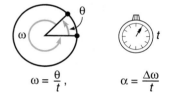

$$\omega = \frac{\theta}{t}, \qquad \alpha = \frac{\Delta\omega}{t}$$

2 Torque and balance. A torque is what causes an object to rotate. It is defined as a force times the lever arm of the force, which is the perpendicular distance from the line of action of the force to the axis of rotation. If the net torque acting on an object is zero, the object will not change its state of rotation.

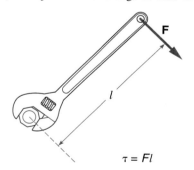

$$\tau = Fl$$

3 Rotational inertia and Newton's second law. In the form of Newton's second law that relates to rotation, torque takes the place of force, rotational acceleration replaces ordinary linear acceleration, and rotational inertia replaces mass. Rotational inertia depends on the distribution of mass about the axis of rotation.

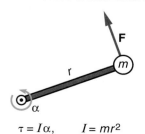

$$\tau = I\alpha, \qquad I = mr^2$$

4 Conservation of angular momentum. By analogy to linear momentum, angular momentum is defined as the rotational inertia times the rotational velocity. It is conserved when no net external torque acts on the system.

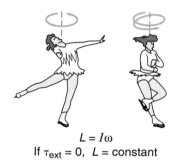

$$L = I\omega$$

If $\tau_{ext} = 0$, $L = $ constant

5 Riding a bicycle and other amazing feats (angular momentum as a vector). The direction of the rotational-velocity and angular-momentum vectors are defined by the right-hand rule. These vectors explain the stability of a moving bicycle and other phenomena. If there are no external torques, the direction of the angular momentum is conserved, as well as its magnitude.

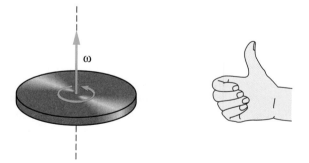

key terms

questions

*Questions identified with an asterisk are more open-ended than the others. They call for lengthier responses and are more suitable for group discussion.

Q1. Which units would not be appropriate for describing a rotational velocity: rad/min², rev/s, rev/h, m/s? Explain.

Q2. Which units would not be appropriate for describing a rotational acceleration: rad/s, rev/s², rev/m², degrees/s²? Explain.

Q3. A coin rolls down an inclined plane gaining speed as it rolls. Does the coin have a rotational acceleration? Explain.

Q4. The rate of rotation of an object is gradually slowing down. Does this object have a rotational acceleration? Explain.

Q5. Is the rotational velocity of a child sitting near the center of a rotating merry-go-round the same as that of another child sitting near the edge of the same merry-go-round? Explain.

Q6. Is the linear speed of a child sitting near the center of a rotating merry-go-round the same as that of another child sitting near the edge of the same merry-go-round? Explain.

Q7. If an object has a constant rotational acceleration, is its rotational velocity also constant? Explain.

*Q8. A ball rolls down an inclined plane gaining speed as it goes. Does the ball experience both linear and rotational acceleration? How far does the ball travel in one revolution? How is the linear velocity of the ball related to its rotational velocity? Explain.

Q9. Which, if either, will produce the greater torque: a force applied at the end of a wrench handle (perpendicular to the handle) or an equal force applied in the same direction near the middle of the handle? Explain.

Q10. Which of the forces pictured as acting upon the rod in the diagram will produce a torque about an axis perpendicular to the plane of the diagram at the left end of the rod? Explain.

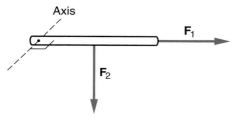

Q11. The two forces in the diagram have the same magnitude. Which orientation will produce the greater torque on the wheel? Explain.

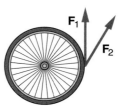

Q12. Is it possible to balance two objects of different weight on a simple balance beam resting upon a fulcrum? Explain.

*Q13. Is it possible for the net force acting on an object to be zero, but the net torque to be greater than zero? Explain. (Hint: The forces contributing to the net force may not lie along the same line.)

Q14. You are trying to move a large rock using a steel rod as a lever. Will it be more effective to place the fulcrum nearer to your hands or nearer to the rock? Explain.

Q15. A pencil is balanced on a fulcrum located two-thirds of the distance from one end. Is the center of gravity of this pencil located at its center point? Explain.

Q16. A solid plank with a uniform distribution of mass along its length rests on a platform with one end of the plank protruding over the edge. How far out can we push the plank before it tips? Explain.

*Q17. A uniform metal wire is bent into the shape of an L. Will the center of gravity for the wire lie on the wire itself? Explain.

*Q18. A tall crate has a higher center of gravity than a shorter crate. Which will have the greater tendency to tip over if we push near the top of the crate? Explain with a force diagram. Where is the probable axis of rotation?

Q19. An object is rotating with a constant rotational velocity. Can there be a net torque acting on the object? Explain.

Q20. Two objects have the same total mass, but object A has its mass concentrated closer to the axis of rotation than object B. Which object will be easier to set into rotational motion? Explain.

Q21. Is it possible for two objects with the same mass to have different rotational inertias? Explain.

Q22. Can you change your rotational inertia about a vertical axis through the center of your body without changing your total weight? Explain.

Q23. A solid sphere and a hollow sphere made from different materials have the same mass and the same radius. Which of these two objects, if either, will have the greater rotational inertia about an axis through its center? Explain.

Q24. Is angular momentum always conserved? Explain.

Q25. A metal rod is rotated first about an axis through its center and then about an axis passing through one end. If the rotational velocity is the same in both cases, is the angular momentum also the same? Explain.

Q26. A child on a freely rotating merry-go-round moves from near the center to the edge. Will the rotational velocity of the merry-go-round increase, decrease, or not change at all? Explain.

*Q27. Moving straight inward, a large kid jumps onto a freely rotating merry-go-round. What effect will this have on the rotational velocity of the merry-go-round? Explain.

Q28. Is it possible for an ice skater to change his rotational velocity without involving any external torque? Explain.

Q29. Suppose you are rotating a ball attached to a string in a circle. If you allow the string to wrap around your finger, does the rotational velocity of the ball change as the string shortens? Explain.

Q30. Does the direction of the angular-momentum vector of the wheels change when a bicycle goes around a corner? Explain.

Q31. An ice skater is spinning counterclockwise about a vertical axis when viewed from above. What is the direction of her angular-momentum vector? Explain.

Q32. A pencil, balanced vertically on its eraser, falls to the right.
   a. What is the direction of its angular-momentum vector as it falls?
   b. Is its angular momentum conserved during the fall? Explain.

*Q33. A top falls over quickly if it is not spinning, but will stay approximately upright for some time when it is spinning. Explain why this is so.

Q34. Can a yo-yo be made to "sleep" if the string is tied tightly to the axle? Explain.

exercises

E1. Suppose that a merry-go-round is rotating at the rate of 10 rev/min.
   a. Express this rotational velocity in rev/s.
   b. Express this rotational velocity in rad/s.

E2. When the author was a teenager, the rate of rotation for popular music records on a record player was 45 RPM.
   a. Express this rotational velocity in rev/s.
   b. Through how many revolutions does the record turn in a time of 5 s?

E3. Suppose that a disk rotates through three revolutions in 4 seconds.
   a. What is its displacement in radians in this time?
   b. What is its average rotational velocity in rad/s?

E4. The rotational velocity of a merry-go-round increases at a constant rate from 1.0 rad/s to 1.8 rad/s in a time of 4 s. What is the rotational acceleration of the merry-go-round?

E5. A bicycle wheel is rotationally accelerated at the constant rate of 1.2 rev/s^2.
   a. If it starts from rest, what is its rotational velocity after 4 s?
   b. Through how many revolutions does it turn in this time?

E6. The rotational velocity of a spinning disk decreases from 6 rev/s to 3 rev/s in a time of 12 s. What is the rotational acceleration of the disk?

E7. Starting from rest a merry-go-round accelerates at a constant rate of 0.2 rev/s^2.
   a. What is its rotational velocity after 5 s?
   b. How many revolutions occur during this time?

E8. A force of 50 N is applied at the end of a wrench handle that is 24 cm long. The force is applied in a direction perpendicular to the handle as in the diagram.
   a. What is the torque applied to the nut by the wrench?
   b. What would the torque be if the force were applied half way up the handle instead of at the end?

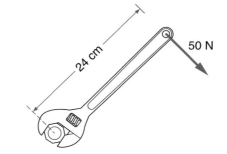

E9. A weight of 30 N is located a distance of 10 cm from the fulcrum of a simple balance beam. At what distance from the fulcrum should a weight of 20 N be placed on the opposite side in order to balance the system?

E10. A weight of 5 N is located 10 cm from the fulcrum of a simple balance beam. What weight should be placed at a point 4 cm from the fulcrum on the opposite side in order to balance the system?

E11. Two forces are applied to a merry-go-round with a radius of 1.2 m as shown in the diagram below. One force has a magnitude of 80 N and the other a magnitude of 50 N.
   a. What is the torque about the axle of the merry-go-round due to the 80-N force?
   b. What is the torque about the axle due to the 50-N force?
   c. What is the net torque acting on the merry-go-round?

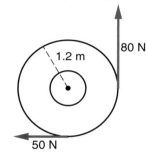

E12. A net torque of 60 N·m is applied to a disk with a rotational inertia of 12.0 kg·m². What is the rotational acceleration of the disk?

E13. A wheel with a rotational inertia of 4.5 kg·m² accelerates at a rate of 3.0 rad/s². What net torque is needed to produce this acceleration?

E14. A torque of 50 N·m producing a counterclockwise rotation is applied to a wheel about its axle. A frictional torque of 10 N·m acts at the axle.
   a. What is the net torque about the axle of the wheel?
   b. If the wheel is observed to accelerate at the rate of 2 rad/s² under the influence of these torques, what is the rotational inertia of the wheel?

E15. Two 0.2-kg masses are located at either end of a 1-m long, very light and rigid rod as in the diagram. What is the rotational inertia of this system about an axis through the center of the rod?

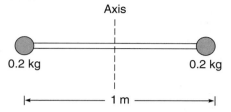

E16. A mass of 0.8 kg is located at the end of a very light and rigid rod 50 cm in length. The rod is rotating about an axis at its opposite end with a rotational velocity of 3 rad/s.
   a. What is the rotational inertia of the system?
   b. What is the angular momentum of the system?

E17. A uniform disk with a mass of 4 kg and a radius of 0.2 m is rotating with a rotational velocity of 20 rad/s.
   a. What is the rotational inertia of the disk? (See fig. 8.15.)
   b. What is the angular momentum of the disk?

E18. A student, sitting on a stool rotating at a rate of 20 RPM, holds masses in each hand. When his arms are extended, the total rotational inertia of the system is 4.5 kg·m². He pulls his arms in close to his body, reducing the total rotational inertia to 1.5 kg·m². If there are no external torques, what is the new rotational velocity of the system?

challenge problems

CP1. A merry-go-round in the park has a radius of 1.8 m and a rotational inertia of 900 kg·m². A child pushes the merry-go-round with a constant force of 80 N applied at the edge and parallel to the edge. A frictional torque of 12 N·m acts at the axle of the merry-go-round.
   a. What is the net torque acting on the merry-go-round about its axle?
   b. What is the rotational acceleration of the merry-go-round?
   c. At this rate, what will the rotational velocity of the merry-go-round be after 15 s if it starts from rest?
   d. What is the rotational acceleration of the merry-go-round if the child stops pushing after 15 s? How long will it take for the merry-go-round to stop turning?

CP2. A 4-m long plank with a weight of 80 N is placed on a dock with 1 m of its length extended over the water, as in the diagram. The plank is uniform in density so that the center of gravity of the plank is located at the center of the plank. A boy with a weight of 150 N is standing on the plank and moving out slowly from the edge of the dock.
   a. What is the torque exerted by the weight of the plank about the pivot point at the edge of the dock? (Treat all the weight as acting through the center of gravity of the plank.)
   b. How far from the edge of the dock can the boy move until the plank is just on the verge of tipping?
   c. How can the boy test this conclusion without falling in the water? Explain.

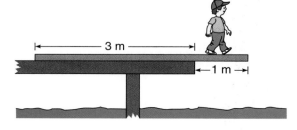

CP3. In the park, several children with a total mass of 240 kg are riding on a merry-go-round that has a rotational inertia of 1500 kg·m² and a radius of 2.2 m. The average distance of the children from the axle of the merry-go-round is 2.0 m initially, since they are all riding near the edge.

a. What is the rotational inertia of the children about the axle of the merry-go-round? What is the total rotational inertia of the children and the merry-go-round?

b. The children now move inward toward the center of the merry-go-round so that their average distance from the axle is 0.5 m. What is the new rotational inertia for the system?

c. If the initial rotational velocity of the merry-go-round was 1.2 rad/s, what is the rotational velocity after the children move in toward the center, assuming that the frictional torque can be ignored? (Use conservation of angular momentum.)

d. Is the merry-go-round rotationally accelerated during this process? If so, where does the accelerating torque come from?

CP4. A student sitting on a stool that is free to rotate but is initially at rest, holds a bicycle wheel. The wheel has a rotational velocity of 5 rev/s about a vertical axis, as shown in the diagram. The rotational inertia of the wheel is 2 kg·m² about its center and the rotational inertia of the student and wheel and platform about the rotational axis of the platform is 6 kg·m².

a. What are the magnitude and direction of the initial angular momentum of the system?

b. If the student flips the axis of the wheel, reversing the direction of its angular-momentum vector, what is the rotational velocity (magnitude and direction) of the student and the stool about their axis after the wheel is flipped? (Hint: See fig. 8.23.)

c. Where does the torque come from that accelerates the student and the stool? Explain.

home experiments and observations

HE1. If there is a park nearby containing a freely rotating child's merry-go-round, take some time with a friend to observe some of the phenomena discussed in this chapter. In particular, make these observations:

a. What is a typical rotational velocity that can be achieved with the merry-go-round? How would you go about measuring this?

b. How long does it take for the merry-go-round to come to rest after you stop pushing? Could you estimate the frictional torque from this information? What other information would you need?

c. If you or your friend are riding on the merry-go-round, what happens to the rotational velocity when you move inward or outward from the axis of the merry-go-round? How do you explain this?

HE2. Create a simple balance using a ruler as the balance beam and a pencil as the fulcrum. (A pencil or pen with a hexagonal cross section is easier to use than one with a round cross-section.)

a. Does the ruler balance exactly at its midpoint? What does this imply about the ruler?

b. Using a nickel as your standard, what are the ratios of the weights of pennies, dimes, and quarters to that of the nickel? Describe the process used to find these ratios.

c. Is the distance from the fulcrum necessary to balance two nickels on one side with a single nickel on the opposite side exactly half the distance for the single nickel? How would you account for any discrepancy?

HE3. You can make a simple top by cutting a circular piece of cardboard, poking a hole through the center, and using a short dull pencil for the post. A short wooden dowel with a rounded end works even better than a pencil.

a. Try building such a top and testing it. How far up the pencil should the cardboard disk sit for best stability?

b. Observe what happens to the axis of rotation of the top as it slows down. What is the direction of the angular-momentum vector, and how does it change?

c. What happens to the stability of your top if you tape two pennies near the edge on opposite sides of the cardboard disk?

HE4. Try spinning a quarter or other large coin about its edge on a smooth tabletop or other similar surface. Describe the motion that follows, paying particular attention to the direction of the angular-momentum vector.

unit TWO

Fluids and Heat

Although Newton's theory of mechanics was introduced in the seventeenth century, other areas of classical physics did not make great strides until the nineteenth century. Some of these developments, particularly in the fields of fluid mechanics and thermodynamics, were stimulated by the industrial revolution. The industrial revolution could not have happened without the invention of steam engines and means of using power from these engines to run factories. Understanding the behavior of fluids, particularly gases, is central to learning about such engines.

Thermodynamics is especially important because it describes the energy conversions that take place in heat engines and other systems. The study of fluid behavior and thermodynamics is an important part of the training of mechanical engineers who design these systems.

The history of the field of thermodynamics has taken some strange twists and turns. A French scientist and engineer, Sadi Carnot (1796–1832), developed a theory of heat engines around 1820, but Carnot's theory raised more questions than it answered. Carnot had some understanding of what we now call the second law of thermodynamics, but he lacked the crucial insights of the first law. The first law involves con-servation of energy, but that idea came into its own only around 1850, about thirty years later. A comprehensive theory had to wait until the middle of the nineteenth century for the statement of the first and second laws of thermodynamics.

Lord Kelvin (William Thompson) (1824–1907) is generally credited with introducing the first and second laws of thermodynamics. Many others contributed, including James Prescott Joule (1818–1889), who measured the heating effect of mechanical work, a key idea in the first law of thermodynamics. Together, the first and second laws of thermodynamics place limits on what can be accomplished with energy in the form of heat.

The laws of thermodynamics play an enormous role in any discussion of the use of energy resources. Energy is still a critical issue—our heavy dependence on fossil fuels cannot go on indefinitely. Concerns about global warming (associated with the greenhouse effect) and other environmental issues related to energy use have become hot political issues. Understanding the science underlying these issues is important for economists, politicians, environmentalists, and citizens in general. Chapters 9, 10, and 11 address these ideas.

The Behavior of Fluids

chapter

9

unit two

chapter overview

Our first objective in this chapter is to explore the meaning of pressure. We will then investigate atmospheric pressure and how pressure varies with depth in a fluid. Those ideas will prepare us to explore the behavior of floating objects as well as what happens when fluids are in motion. Moving fluids are described by Bernoulli's principle, which helps us to explain why a curveball curves and many other phenomena.

chapter outline

Boats hold a special fascination for many of us. As a child, you probably floated twigs or sticks in streams. A stick or toy boat will follow the current of a stream, sometimes moving swiftly and other times getting caught in an eddy or stranded near the bank. You were observing some characteristics of fluid flow.

You probably also noticed that some things float and some do not. Stones sink quickly to the bottom of the stream. As you grew older, you may have wondered why a steel boat floats, while a piece of metal dropped in the water quickly sinks (fig. 9.1). Does the shape of the material have something to do with whether it floats or not? Can you make a boat of concrete?

Things can float in air as well as in water. A balloon filled with helium pulls up on the string, but a balloon filled with air drifts down to the floor. What makes the difference?

The behavior of things that float (or sink) in water or air is one aspect of the behavior of fluids. Water and air are both examples of fluids, although one is a liquid and the other a gas. They both flow readily and conform to the shapes of their containers, unlike solids, which have shapes of their own. Although liquids are usually much denser than gases, many of the principles

figure 9.1   A steel boat floats, but a piece of metal dropped in the water quickly sinks. How do we explain this?

that apply to liquids also apply to gases, so it makes sense to consider them together under the common heading of fluids.

Pressure plays a central role in describing the behavior of fluids. We will explore pressure thoroughly in this chapter. Pressure is involved in Archimedes' principle, which explains how things float, but pressure is also important in other phenomena we will consider, including the flow of fluids.

9.1 Pressure and Pascal's Principle

A small woman wearing high-heel shoes sinks into soft ground, but a large man wearing size-13 shoes may walk across the same ground without difficulty (fig. 9.2). Why is this so? The man weighs much more than the woman, so he must exert a larger force on the ground. But the woman's high heels leave much deeper indentations in the ground.

Clearly weight alone is not the determining factor. How the force is distributed across the area of contact between the shoes and the earth is more important. The woman's shoes have a small area of contact, while the man's shoes have a much larger area of contact. The force exerted on the ground by the man's feet due to his weight is distributed over a larger area.

How is pressure defined?

What is happening when you stand or walk on soft ground? If you are not accelerating in the vertical direction, your weight must be balanced by the normal force exerted upward on your feet by the ground. By Newton's third law, you exert a downward normal force on the ground equal to your weight.

The quantity that determines whether the soil will yield, letting your shoes sink into the ground, is the **pressure** exerted on the soil by your shoes. The total normal force does not matter as much as the force per unit area:

> Pressure is the ratio of the force to the area over which it is applied:
>
> $$P = \frac{F}{A}$$

Pressure is measured in units of newtons per meter squared (N/m^2)—the metric unit of force divided by the metric unit of area. This unit is also called a pascal (1 Pa = 1 N/m^2).

The heel area of a woman's high-heel shoe can be as small as just 1 or 2 square centimeters (cm^2). When you walk, there are times when almost all of your weight is supported by your heel. The weight shifts from heel to toe as you move forward with your other foot off the ground. Take a few steps to test this statement. The woman's weight divided by the small area of her heel produces a large pressure on the ground.

The man's shoe, on the other hand, may have a heel area of as much as 100 cm^2 (for a size-13 shoe). Since his heel area may be 100 times larger than the woman's heel, he can weigh two or three times as much as the woman yet exert a pressure on the ground ($P = F/A$) that is a small fraction of the pressure exerted by the woman. His smaller pressure leads to a smaller indentation in the ground.

The area over which a force is distributed is the critical factor in pressure. The area of a surface increases more rapidly than the linear dimensions of the surface. A square

figure **9.3**   A square with sides 2 cm long has an area that is four times as large as a square with sides 1 cm long.

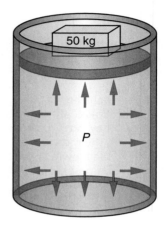

figure **9.2**   A woman's high heels sink into the soft ground, but the larger shoes of the much bigger man do not.

figure **9.4**   The pressure exerted on the piston extends uniformly throughout the fluid, causing it to push outward with equal force per unit area on the walls and bottom of the cylinder.

that is 1 cm on each side has an area of 1 cm², for example, but a square that is 2 cm on a side has an area of 4 cm² (2 cm × 2 cm), four times as large as the smaller square (fig. 9.3). For a circle, the area is equal to π times the square of the radius of the circle ($A = \pi r^2$). A circle with a radius of 10 cm has an area 100 times larger (10^2) than a circle with a radius of 1 cm.

What is Pascal's principle?

What happens inside a fluid when pressure is exerted on it? Does pressure have direction? Does it transmit a force to the walls or bottom of a container? These questions point to another important feature of pressure in fluids.

When we apply a force by pushing down on the piston in a cylinder, as in figure 9.4, the piston exerts a force on the fluid. By Newton's third law, the fluid also exerts a force (in the opposite direction) on the piston. The fluid inside the cylinder will be squeezed and may decrease in volume somewhat. It has been *compressed*. Like a compressed spring, the compressed fluid will push outward on the walls and bottom of the cylinder as well as on the piston.

Although the fluid behaves like a spring, it is an unusual spring. It pushes outward *uniformly* in all directions when compressed. Any increase in pressure is transmitted uniformly throughout the fluid, as figure 9.4 indicates. If

we ignore variations in pressure due to the weight of the fluid itself, the pressure that pushes upward on the piston is equal to the pressure that pushes outward on the walls and downward on the bottom of the cylinder.

The ability of a fluid to transmit the effects of pressure uniformly is the core of Pascal's principle and the basis of the operation of a hydraulic jack and other hydraulic devices. Blaise Pascal (1623–1662) was a French scientist and philosopher whose primary contributions were in the areas of fluid statics and probability theory. **Pascal's principle** is usually stated as

> Any change in the pressure of a fluid is transmitted uniformly in all directions throughout the fluid.

How does a hydraulic jack work?

Hydraulic systems are the most common applications of Pascal's principle. They depend on the uniform transmission of pressure, as well as on the relationship between pressure, force, and area stemming from the definition of pressure. The basic idea is illustrated in figure 9.5.

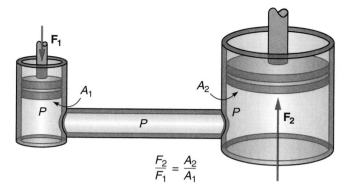

$$\frac{F_2}{F_1} = \frac{A_2}{A_1}.$$

figure **9.5** A small force **F₁** applied to a piston with a small area produces a much larger force **F₂** on the larger piston. This allows a hydraulic jack to lift heavy objects.

figure **9.6** A hydraulic jack can easily lift a car.

A force applied to a piston with a small area can produce a large increase in pressure in the fluid because of the small area of the piston. This increase in pressure is transmitted through the fluid to the piston on the right in figure 9.5, which has a much larger area. Since pressure is force per unit area, the force exerted on the larger piston by this pressure is proportional to the area of the piston ($F = PA$). Applying the same pressure to the much larger area of the second piston results in a much larger force on the second piston.

Since it is feasible to build a system in which the area of the second piston is more than 100 times larger than the first piston, we can produce a force on the second piston more than 100 times larger than the input force. The mechanical advantage, the ratio of the output force to the input force, of a hydraulic system will usually be large compared to a lever or other simple machine. (See section 6.1.) A small force applied to the input piston of a hydraulic jack can be multiplied to produce a force large enough to lift a car (fig. 9.6). These ideas are illustrated in Try This Box 9.1.

As discussed in chapter 6, we pay a price for getting a larger output force. The work done by the larger piston in lifting the car cannot be any greater than the work input to the handle of the jack. Since work is equal to force times distance ($W = Fd$), a large output force means that the larger piston moves just a small distance. Conservation of energy requires that if the output force is 50 times larger than the input force, the input piston must move 50 times farther than the output piston.

With a hand-pumped hydraulic jack, you have to move the smaller piston several times, letting the jack's chamber refill with fluid after each stroke. The total distance moved by the smaller piston is the sum of the distances moved on each stroke. The larger piston inches upward as you pump. A similar process occurs in larger hydraulic jacks.

Hydraulic systems and fluid are also used in the brake systems of cars and in many other applications. Oil is more

try this box **9.1**

Sample Exercise: Some Basics of Jacks

A force of 10 N is applied to a circular piston with an area of 2 cm² in a hydraulic jack. The output piston for the jack has an area of 100 cm².
- a. What is the pressure in the fluid?
- b. What is the force exerted on the output piston by the fluid?

a. $F_1 = 10$ N $\qquad P = \dfrac{F_1}{A_1}$

$\quad A_1 = 2$ cm²

$\qquad = 0.0002$ m² $\qquad = \dfrac{10 \text{ N}}{0.0002 \text{ m}^2}$

$\quad P = ?$

$\qquad\qquad\qquad = 50\,000$ N/m²

$\qquad\qquad\qquad = \mathbf{50\ kPa}$

A kilopascal (kPa) is 1000 Pa, or 1000 N/m².

b. $A_2 = 100$ cm² $\qquad F_2 = PA_2$

$\quad = 0.01$ m² $\qquad = (50\,000 \text{ N/m}^2)(0.01 \text{ m}^2)$

$\quad P = 50$ kPa $\qquad = \mathbf{500\ N}$

$\quad F_2 = ?$

The mechanical advantage of this jack is 500 N divided by 10 N, or 50: the output force is 50 times *larger* than the input force.

effective than water as a hydraulic fluid because it is not corrosive and can also lubricate, ensuring smooth operation of the system. Hydraulic systems take good advantage of the ability of fluids to transmit changes in pressure, as described by Pascal's principle. They also use the multiplying effect produced by pistons of different areas. Thinking about how a hydraulic system ·works is an example of the concept of pressure in action.

Pressure is the ratio of the force to the area over which it is applied. A force applied to a small area exerts a much larger pressure than the same force applied to a larger area. Changes in pressure are transmitted uniformly through a fluid, and the pressure pushes outward in all directions, according to Pascal's principle. These ideas explain the operation of a hydraulic jack and other hydraulic systems.

9.2 Atmospheric Pressure and the Behavior of Gases

Living on the surface of the earth, we are at the bottom of a sea of air. Except for smog or haze, air is usually invisible. We seldom give it a second thought. How do we know it is there? What measurable effects does it have?

We feel the presence of air, of course, when riding a bike or walking on a gusty day. Skiers, bicycle racers, and car designers are all conscious of the need to reduce resistance to the flow of air past themselves and their vehicles. The labored breathing of a mountain climber is partly due to the thinning of the atmosphere near the top of a high mountain. How do we measure **atmospheric pressure** and its variations with weather or altitude?

How is atmospheric pressure measured?

Atmospheric pressure was first measured during the seventeenth century. Galileo noticed that the water pumps he designed were capable of pumping water to a height of only 32 feet, but he never adequately explained why. His disciple, Evangelista Torricelli (1608–1647), invented the barometer as he attempted to answer this question.

Torricelli was interested in vacuums. He reasoned that Galileo's pumps created a partial vacuum and that the pressure of the air pushing down on the water at the pump intake was responsible for lifting the water. In thinking about how to test this hypothesis, he was struck by the idea of using a much denser fluid than water. Mercury or quicksilver was the logical choice, because it is a fluid at room temperature and has a density approximately 13 times that of water. **Density** is the mass of an object divided by its volume: a given volume of mercury has a mass (and weight) that is 13 times an equal volume of water.

In his early experiments, Torricelli used a glass tube about 1 meter in length, sealed at one end and open at the other. He filled the tube with mercury and then, holding his finger over the open end, inverted the tube and placed this end in the open container of mercury (fig. 9.7). The mercury flowed from the tube into the container until an equilibrium was reached, leaving a column of mercury in

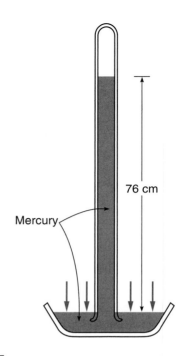

figure 9.7   Torricelli filled a tube with mercury and inverted it into an open container of mercury. Air pressure acting on the mercury in the dish can support a column of mercury 76 cm in height.

the tube approximately 76 cm (30 in.) high. The pressure of air pushing down on the surface of the mercury in the open container apparently was strong enough to support a column of mercury 76 cm high.

Torricelli was careful to demonstrate experimentally that there was a vacuum in the space at the top of the tube above the column of mercury. The reason that the mercury column does not fall is that the pressure at the top of the column is zero, while the pressure at the bottom of the column is equal to atmospheric pressure. We still often quote atmospheric pressure in either centimeters of mercury or inches of mercury (the commonly used unit in the United States). How are these units related to the pascal?

We can establish the relationship between these units if we know the density of mercury. From this, we can find the weight of the column of mercury being supported by the atmosphere. Dividing this weight by the cross-sectional area of the tube gives us force per unit area, or pressure. Using this reasoning, we find that a column of mercury 76 cm in height produces a pressure of 1.01×10^5 Pa, known as standard atmospheric pressure.

Atmospheric pressure is approximately 100 kilopascals (kPa) at sea level, or 14.7 pounds per square inch (psi). Living at the bottom of this sea of air, you have 14.7 pounds pushing on you for every square inch of your body. Why do you not notice this? Fluids permeate your body and push back out—the interior and exterior pressures are essentially equal.

A famous experiment designed to demonstrate the effects of air pressure was performed by Otto von Guericke (1602–1686). Von Guericke designed two bronze hemispheres that could be smoothly joined together at their rims. He then pumped the air out of the sphere formed from the two hemispheres, using a crude vacuum pump that he had invented. As shown in figure 9.8, two eight-horse teams were unable to pull the hemispheres apart. When the stopcock was opened to let air back into the evacuated sphere, the two hemispheres could easily be separated.

Variations in atmospheric pressure

If we live at the bottom of a sea of air, we might expect pressure to decrease as we go up in altitude. The pressure that we experience results from the weight of the air above us. As we go up from the earth's surface, there is less atmosphere above us, so the pressure should decrease.

Similar reasoning led Blaise Pascal to try to measure atmospheric pressure at different altitudes shortly after Torricelli's invention of the mercury barometer. Because Pascal was in poor health through most of his adult life, and not up to climbing mountains, he sent his brother-in-law to the top of the Puy-de-Dome mountain in central France with a barometer similar to Torricelli's. Pascal's brother-in-law found that the height of the mercury column supported by the atmosphere was about 7 cm lower at the top of the 1460-m (4800-ft) mountain than at the bottom.

Pascal also had his brother-in-law take a partially inflated balloon to the top of the mountain. As Pascal predicted, the balloon expanded as the climbers gained elevation, indicating a decrease in the external atmospheric pressure (fig. 9.9). Pascal was even able to show a decrease in pressure within the city of Clermont between the low

figure **9.9**   A balloon that was partially inflated near sea level expanded as the experimenters climbed the mountain.

point in town and the top of the cathedral tower. The decrease was small but measurable.

Using the newly invented barometer, Pascal also observed variations in pressure related to changes in the weather. Column heights were lower on stormy days than on clear days. This pressure variation has been used ever since to indicate changes in weather. Falling atmospheric pressure points to stormy weather ahead. Readings are usually corrected to sea level so that variations in altitude will not mask changes related to weather.

The weight of a column of air

Can we calculate the variation in atmospheric pressure with altitude the same way that we compute the pressure at the bottom of a column of mercury? We would need to know the weight of the column of air above us. Even though mercury and air are both fluids, there is a significant difference (besides the difference in density) between the behavior of a column of mercury and a column of air.

Mercury, like most liquids, is not readily compressible. In other words, increasing the pressure on a given amount of mercury does not change the volume of the mercury much. The density of the mercury (mass per unit volume) is the same near the bottom of the column of mercury as near the top. A gas like air, on the other hand, is easy to compress. As the pressure changes, the volume changes, and so does the density. Therefore, we cannot use a single

figure **9.8**   Two teams of eight horses were unable to separate von Guericke's evacuated metal sphere. What force pushes the two hemispheres together?

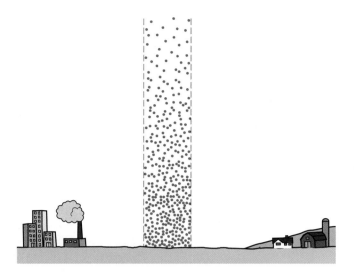

figure **9.10**   The density of a column of air decreases as altitude increases because air expands as pressure decreases.

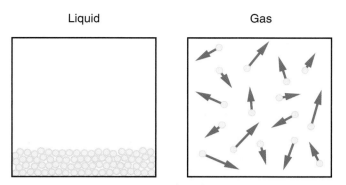

figure **9.11**   The atoms in a liquid are closely packed while those in a gas are separated by much larger distances.

value for the density to compute the weight of a column of air. Its density decreases as we rise in the atmosphere (fig. 9.10).

This major difference between gases and liquids comes from differences in their atomic or molecular "packing." Except at very high pressures, the atoms or molecules of a gas are separated by large distances compared to the size of the atoms themselves, as in figure 9.11. The atoms in a liquid, on the other hand, are closely packed, much like those in a solid. They cannot be easily squeezed.

Gases are springy. They can readily be compressed to a small fraction of their initial volume. They can also expand if pressure is reduced, like the gas in Pascal's balloon. Changes in temperature are likely to affect the volume or pressure of a gas much more than they affect a liquid. If the temperature is held constant, however, the volume changes with changes in pressure.

How does the volume of a gas change with pressure?

Variations in the volume and density of a gas that accompany changes in pressure were studied by Robert Boyle (1627–1691) in England, as well as by Edme Mariotte (1620–1684) in France. Boyle's results were first reported in 1660 but went unnoticed on the European continent where Mariotte published his similar conclusions in 1676. Both were interested in the springiness, or compressibility, of air.

Both experimenters used a bent glass tube sealed on one end and open on the other (fig. 9.12). In Boyle's experiment, the tube was partially filled with mercury, so that air was trapped in the closed portion of the tube. He allowed air to pass back and forth initially so that the pres-

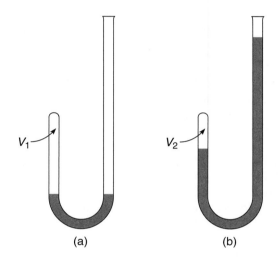

figure **9.12**   In Boyle's experiment, adding mercury to the open side of the bent tube caused a decrease in the volume of the trapped air in the closed side.

sure in the closed side of the tube was equal to atmospheric pressure and the column of mercury on either side was at the same height.

As Boyle added mercury to the open end of the tube, the volume of the air trapped on the closed side decreased. When he added enough mercury to increase the pressure to twice atmospheric pressure, the height of the air column on the closed side decreased by one-half. In other words, doubling the pressure caused the volume of air to decrease by half. Boyle discovered that the volume of a gas is inversely proportional to the pressure.

We can express **Boyle's law** in symbols:

$$PV = \text{constant},$$

where P is the pressure in the gas, and V is the volume of the gas. If the pressure increases, the volume decreases in inverse proportion to keep the product of pressure and volume constant. We often write Boyle's law (also known as Mariotte's law in continental Europe) as

$$P_1 V_1 = P_2 V_2,$$

where P_1 and V_1 are the initial pressure and volume, and P_2 and V_2 are the final pressure and volume (see Try This Box 9.2). For Boyle's law to be valid, a fixed mass or quantity of gas must be kept at a constant temperature while the pressure and volume change.

As we gain altitude, atmospheric pressure decreases, and the volume of a given mass of air increases. Since density is the ratio of the mass to the volume, the density of the air must be decreasing as the volume increases. In computing the weight of a column of air, we have to take the change in density into account, and the computation becomes more complex than for the column of mercury. The density of a gas also depends on temperature, which usually decreases as we gain altitude, a further complication.

We live at the bottom of a sea of air, whose pressure we can measure with barometers. The earliest barometers were a column of mercury in a closed glass tube. The height of the column supported by atmospheric pressure is a measure of the pressure. Atmospheric pressure decreases with increasing altitude, because it is determined by the weight of the column of air above. Determining the weight of a column of air is harder than determining the weight of a column of mercury, however, because air is compressible, and its density varies with altitude. Boyle's law describes how the volume of a gas changes with pressure: increasing the pressure decreases the volume in inverse proportion.

---

try this box 9.2

Sample Exercise: How Does a Gas Change with Pressure?

A fixed quantity of gas is held in a cylinder capped at one end by a movable piston. The pressure of the gas is initially 1 atmosphere (101 kPa) and the volume is initially 0.3 m³. What is the final volume of the gas if the pressure is increased to 3 atmospheres at constant temperature?

$P_1 = 1$ atm

$V_1 = 0.3$ m³

$P_2 = 3$ atm

$V_2 = ?$

$P_1 V_1 = P_2 V_2 = $ constant

$$V_2 = \frac{P_1 V_1}{P_2}$$

$$= \frac{(1 \text{ atm})(0.3 \text{ m}^3)}{3 \text{ atm}}$$

$$= \frac{1}{3} (0.3 \text{ m}^3)$$

$$= \mathbf{0.1 \text{ m}^3}$$

9.3 Archimedes' Principle

Why do some things float while others do not? Is floating determined by the weight of the object? A large ocean liner floats, but a small pebble quickly sinks. Clearly, it is not a matter of the total weight of the object. The density of the object is the key. Objects that are denser than the fluid they are immersed in will sink—those less dense will float. The complete answer to the question of why things sink or float is found in Archimedes' principle, which describes the *buoyant force* exerted on any object fully or partly immersed in a fluid.

What is Archimedes' principle?

A block of wood floats, but a metal block of the same shape and size sinks. The metal block weighs more than the block of wood, even though it is the same size, which means that the metal is denser than wood. Density is the ratio of the mass to the volume (or mass per unit volume). The metal has a greater mass than wood for the same volume. Since weight is found by multiplying the mass by the gravitational acceleration g, the metal also has a greater weight for the same volume.

If we compared the densities of the metal and wood blocks to water, we would find that the metal block has a density greater than water and that the density of the wood block is less than water. The *average* density of an object compared to a fluid is what determines whether the object will sink or float in that fluid.

If we push down on a block of wood floating in a pool of water, we can clearly feel the water pushing back up on the block. In fact, it is hard to submerge a large block of wood or a rubber inner tube filled with air. They keep popping back up to the surface. The upward force that pushes such objects back toward the surface is called the **buoyant force.** If the block of wood is partially submerged at first and you push down to submerge it farther, the buoyant force gets larger as more of the block is underwater.

Legend has it that Archimedes was sitting in the public baths observing floating objects when he realized what determines the strength of the buoyant force. When an object is submerged, its volume takes up space occupied by water: it *displaces* the water, in other words. The more water displaced by the object as you push downward, the greater the upward buoyant force. **Archimedes' principle** can be stated as

---

The buoyant force acting on an object fully or partially submerged in a fluid is equal to the weight of the fluid displaced by the object.

If the fluid has a greater density than the object, the weight of the fluid displaced when the object is fully submerged will be greater than the weight of the object. By Archimedes' principle, the buoyant force will be greater than the weight of the object, and there will be a net upward force on the object that pushes it toward the surface.

What is the source of the buoyant force?

The source of the buoyant force described by Archimedes' principle is the increase in pressure that occurs with increasing depth in a fluid. When we swim to the bottom of the deep end of a swimming pool, we can feel pressure building on our ears. The pressure near the bottom of the pool is larger than near the surface for the same reason that the pressure of the atmosphere is greater near the surface of the earth than at higher altitudes. The weight of the fluid above us contributes to the pressure that we experience.

By Pascal's principle, the atmospheric pressure pushing on the surface of the pool extends uniformly throughout the fluid. To get the total pressure at some depth, we add the excess pressure resulting from the weight of the water to the atmospheric pressure. For many purposes, the excess pressure above atmospheric pressure is of most interest. Since our bodies have internal pressures equal to atmospheric pressure, our eardrums are sensitive to the increase in pressure *beyond* atmospheric pressure rather than to the total pressure.

To find the excess pressure at a certain depth in a liquid, we need to determine the weight of the liquid above that depth. The problem is similar to finding the pressure at the bottom of a column of mercury discussed in section 9.2. If we imagine a column of water in a swimming pool (fig. 9.13), the weight of the column depends on the volume of the column and the density of the water. The volume of the column is directly proportional to its height h and therefore so is its weight. The excess pressure increases in direct proportion to the depth h below the surface.

A large can filled with water demonstrates this variation of pressure with depth. If we punch holes in the can at different depths, the water shoots from a hole near the bottom of the can with a much greater horizontal velocity than from a hole punched near the top (fig. 9.14). Since the can is submerged in the atmosphere, the excess pressure above

figure **9.14**   Water emerging from a hole near the bottom of a can filled with water has a larger horizontal velocity than water emerging from a hole near the top.

atmospheric pressure is most significant. A larger excess pressure provides a larger accelerating force for the emerging water.

How does the fact that pressure increases with depth explain the buoyant force? Imagine a rectangular-shaped object, such as a steel block, submerged in water. If we suspend the block from a string, as in figure 9.15, the pressure of the water will push on the block from all sides. Because the pressure increases with depth, however, the pressure at the bottom of the block is greater than at the top. This greater pressure produces a larger force ($F = PA$) pushing up on the bottom of the block than that pushing down on the top. The difference in these two forces is the buoyant force. The buoyant force is proportional to both the height and the cross-sectional area of the block, and thus to its volume, Ah. The volume of the fluid displaced by the object is directly related to the weight of the fluid displaced, which leads to the statement of Archimedes' principle.

What forces act on a floating object?

If there are no strings attached, or other forces pushing or pulling on an object partially or fully submerged in a fluid, only the weight of the object and the buoyant force determine what happens. The weight is proportional to the density and volume of the object, and the buoyant force depends on the density of the fluid and the volume of fluid displaced by the object. The motion of the object is governed by the sizes of the buoyant force pushing upward and the weight pulling downward. There are just three possibilities:

1. The density of the object is greater than the density of the fluid. If the object has an average density greater than the fluid it is submerged in, the weight of the

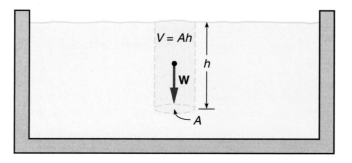

figure **9.13**   The weight of a column of water is proportional to the volume of the column. The volume V is equal to the area A times the height h.

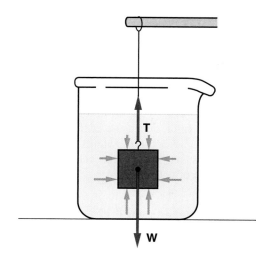

figure 9.15 The pressure acting on the bottom of the suspended metal block is greater than that acting on the top due to the increase of pressure with depth.

object will be greater than the weight of the fluid displaced by the fully submerged object, because the same volume is involved. Since the weight, which acts downward, is greater than the buoyant force, which acts upward, the net force acting on the object will be downward and the object will sink (unless it is supported by another force such as one exerted by a string tied to the block).

2. The density of the object is less than the density of the fluid. If the object's density is less than the fluid's, the buoyant force will be larger than the weight of the object when the object is fully submerged. The net force acting on the object will be upward, and the object will float to the top. When it reaches the surface of the fluid, just enough of the object remains submerged so that the weight of the fluid displaced by the submerged portion of the object (the buoyant force) will equal the weight of the object. The net force is zero, and the object is in equilibrium. It has no acceleration.

3. The density of the object equals the density of the fluid. The weight of the object then equals the weight of the fluid displaced when the object is submerged. The object floats when fully submerged, rising or sinking in the fluid by changing its average density slightly, which is what a fish or a submarine does. The average density of a submarine can be increased or decreased by taking in or releasing water.

Why does a boat made of steel float?

Steel is a lot denser than water, and a large steel boat is a very heavy object. Why does it float? The answer is that the boat is not made of solid steel all the way through. There are open spaces filled with air and other materials in the boat. A solid piece of steel will quickly sink, but if the

average density of the boat is less than the water it is displacing, a steel boat will certainly float. Because of the air spaces and other materials, the average density of the boat will be much less than steel.

According to Archimedes' principle, the buoyant force that acts on the boat must equal the weight of the water displaced by the hull of the boat. For the boat to be in equilibrium (with a net force of zero), the buoyant force must equal the weight of the boat. As we load the boat with cargo, the total weight of the boat increases. So must the buoyant force. The amount of water displaced by the hull must increase, so the boat sinks lower in the water. There is a limit to how much weight can be added to a boat (often expressed as tons of displacement). A fully loaded oil tanker will ride much lower in the water than an unloaded tanker (fig. 9.16).

Other important considerations in designing a boat are the shape of the hull and how the boat will be loaded. If the center of gravity of the boat is too high or if the boat is unevenly loaded, there is some danger that the boat will tip over. Wave action and winds add to this danger, so a margin of safety must be included in the design. Once water enters the boat, the overall weight of the boat and its average density increase. When the boat's average density becomes larger than the average density of water, down she goes.

figure 9.16 A fully loaded tanker rides much lower in the water than an empty tanker.

When will a balloon float?

Buoyant forces also act on objects submerged in a gas such as air. If a balloon is filled with a gas whose density is less than air, the average density of the balloon is less than air, and the balloon will rise. Helium and hydrogen are the two common gases with densities less than air, but helium is more commonly used in balloons even though its density is somewhat greater than hydrogen. Hydrogen can combine explosively with the oxygen in air, making its use dangerous.

The average density of the balloon is determined by the material the balloon is made of as well as by the density of the gas with which it is filled. Ideal materials for making balloons stretch very thin without losing strength. They should also remain impermeable to the flow of gas, so that the helium or other gas will not be lost rapidly through the skin of the balloon. Balloons made of mylar (which is often coated with aluminum) are much less permeable than ordinary latex balloons.

Hot-air balloons take advantage of the fact that any gas will expand when it is heated. If the volume of the gas increases, its density decreases. As long as the air inside the balloon is much hotter than the air surrounding the balloon, there will be an upward buoyant force. The beauty of a hot-air balloon is that we can readily adjust the density of the air within the balloon by turning the gas-powered heater on or off. This gives us some control over whether the balloon will ascend or descend.

Buoyant forces and Archimedes' principle are useful in applications besides boats and balloons. We can use Archimedes' principle to determine the density of objects or the fluid in which they are submerged. This, in fact, was Archimedes' original application. Archimedes is said to have used his idea to determine the density of the king's crown to ascertain whether it was truly pure gold. (The king suspected a goldsmith of fraud.) Gold is denser than baser metals that might have been substituted for gold.

Archimedes' principle states that the buoyant force acting on an object is equal to the weight of the fluid displaced by the object. If the average density of the object is greater than the density of the fluid being displaced, the weight of the object will exceed the buoyant force and the object will sink. The buoyant force results because the pressure on the bottom of the object is greater than on top since pressure increases with depth. Archimedes' principle can be used to understand the behavior of boats, balloons, or objects floating in a bathtub or stream.

9.4 Fluids in Motion

If we return to the bank of the stream mentioned in the introduction to this chapter, what else can we see? When a stick or toy boat floats down the stream, we might notice that the speed of the current varies from point to point in the stream. Where the stream is wide, the flow is slow. Where the stream narrows, the speed increases. Also, the speed is usually greater near the middle of the stream than close to the banks. Eddies and other features of turbulent flow may be observed.

These are all characteristics of the flow of fluids. The speed of flow is affected by the width of the stream and by the *viscosity* of the fluid, a measure of the frictional effects within the fluid. Some of these features are easy to understand, while others, particularly the behavior of turbulent flow, are still areas of active research.

Why does the water speed change?

One of the most obvious features of the stream's current is that the water speed increases where the stream narrows. Our stick or toy boat moves slowly through the wider portions of the stream but gains speed when it passes through a narrow spot or rapids.

As long as no tributaries add water to the stream and there are no significant losses through evaporation or seepage, the flow of the stream is continuous. In a given time, the same amount of water that enters the stream at some upper point leaves the stream at some lower point. We call this *continuity of flow*. If flow were not continuous, water would collect at some point or, perhaps, be lost somewhere within that segment of the stream, something that does not usually happen.

How would we describe the *rate* of flow of water through a stream or pipe? A flow rate is a volume divided by time, so many gallons per minute or, in the metric system, liters per second or cubic meters per second. As figure 9.17 shows, the volume of a portion of water of length L flowing past some point in a pipe is the product of the length times the cross-sectional area A, or LA. The speed with which this volume moves determines the rate of flow.

What is the rate of flow through the pipe? To find the rate, we divide the water's volume LA by the time interval t, which gives us LA/t. Since L/t is the speed v of the water, we get

rate of flow $= vA$.

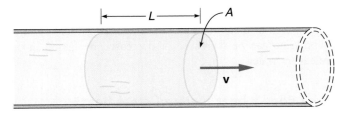

Rate of flow = vA

figure 9.17 The rate at which water moves through a pipe is defined by the volume divided by time. This is equal to the speed of the water times the cross-sectional area.

This expression is valid for any fluid and makes intuitive sense: the greater the speed, the greater the rate of flow, and the larger the cross section of the pipe or stream, the greater the rate of flow.

How does the rate of flow explain changes in the speed of the water? If the flow through the pipe is continuous, the rate of flow must be the same at any point along the pipe. The same number of gallons per minute flow past each point. If the cross-sectional area A decreases, the speed v increases to keep the rate vA constant. If the cross-sectional area increases, the speed decreases to maintain the same rate of flow.

The same principle applies to a stream. Where the stream is narrower, the cross-sectional area of the stream will generally be smaller than at a wider point in the stream. The stream may be deeper at the narrow places, but usually not enough to make the cross-sectional area as large as at the wider places. If the cross-sectional area decreases, the fluid speed must increase to maintain the rate of flow.

How does viscosity affect the flow?

Up to this point, we have ignored any variation in the fluid speed across its cross-sectional area. We mentioned that the water speed will usually be greatest near the middle of the stream. The reason is the frictional or viscous effects between layers of the fluid itself and between the fluid and the walls of the pipe or the banks of the stream.

Imagine the fluid as made up of layers, and you can see why the speed will be greatest near the center. Figure 9.18 shows different layers of a fluid moving through a trough. Since the bottom of the trough is not moving, it exerts a frictional force on the bottom layer of fluid, which moves more slowly than the layer immediately above it. This layer exerts a frictional drag, in turn, on the layer above it, which flows more slowly than the next one above it, and so on.

The **viscosity** of the fluid is the property of fluid that determines the strength of the frictional forces between the layers of the fluid—the larger the viscosity, the larger the frictional force. The magnitude of the frictional force also depends on the area of contact between layers and the rate at which the speed is changing across the layers. If these other factors are the same, a fluid with a high viscosity like molasses experiences a larger frictional force between layers than a fluid with a low viscosity such as water.

A thin layer of fluid that does not move at all is usually found next to the walls of the pipe or trough. The fluid speed increases as the distance from the wall increases. The exact variation with distance depends on the viscosity of the fluid and the overall rate of flow of the fluid through the pipe. For a fluid with a low viscosity, the transition to the maximum speed occurs over a short distance from the wall. For a fluid with a high viscosity, the transition takes place over a larger distance, and the speed may vary throughout the pipe or trough (fig. 9.19).

The viscosity of different fluids varies enormously. Molasses, thick oils, and syrup all have much larger viscosities than water or alcohol. Most liquids have much higher viscosities than gases. The viscosity of a given fluid also changes substantially if its temperature changes. An increase in temperature usually produces a decrease in viscosity. Heating a bottle of syrup, for example, makes it less viscous causing it to flow more readily.

Laminar and turbulent flow

One of the most fascinating questions about the flow of fluids is why the flow is smooth or *laminar* under some conditions but *turbulent* in others. Both kinds of flow can be observed in rivers or creeks. How do they differ—and what determines which type of flow prevails?

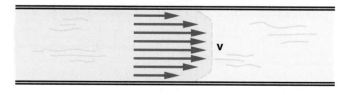

figure **9.19** The velocity increases rapidly from the wall inward for a low-viscosity fluid but more gradually for a high-viscosity fluid.

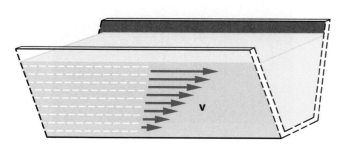

figure **9.18** Because of the frictional or viscous forces between layers, each layer of fluid flowing in the trough moves more slowly than the layer immediately above.

In sections of a stream where flow is smooth or laminar, there are no eddies or other similar disturbances. The flow of the stream can be described by *streamlines* that indicate the direction of flow at any point. The streamlines for laminar flow are roughly parallel, as in figure 9.20. The speeds of different layers may vary, but one layer moves smoothly past another.

As the stream narrows and the fluid speed increases, the simple laminar-flow pattern disappears. Ropelike twists in the streamlines appear, then whorls and eddies: the flow becomes *turbulent*. In most applications, **turbulent flow** is undesirable, because it greatly increases the fluid's resistance to flow through a pipe or past other surfaces. It *does* make river rafting much more exciting, though.

If the density of a fluid and the width of the pipe or stream do not vary, the transition from laminar to turbulent flow is predicted by two quantities, the average fluid speed and the viscosity. Higher speeds are more likely to produce turbulent flow, as we would expect. On the other hand, higher viscosities inhibit turbulent flow. Larger fluid densities and pipe widths are also more conducive to turbulent flow. From experiments, scientists have been able to use these quantities to predict with some accuracy the speed at which the transition to turbulent flow will begin.

You can observe the transition from laminar to turbulent flow in many common phenomena. The higher water speeds at the narrowing of a stream often produce turbulent flow. The transition can also be seen in the flow of water from a spigot. A small flow rate will usually produce laminar flow, but as the flow rate increases, the flow becomes turbulent. Flow may be smooth near the top of the water column but turbulent lower down, as the water is accelerated by gravity. Try it next time you are near a sink.

You can also see this phenomenon in the smoke rising from a cigarette or candle. Near the source, the upward flow of the smoke is usually laminar. As the smoke accelerates upward (due to the buoyant force), the column widens, and the flow becomes turbulent (fig. 9.21). The whorls and eddies are much like those you see in a stream.

The conditions that produce turbulence are well understood, but until recently, scientists could not explain why the flow patterns develop as they do. Some surprising patterns can be discerned in the seemingly chaotic behavior of turbulent flow in different situations. Recent theoretical advances in the study of chaos have produced a much better understanding of the reasons for these patterns.

The study of chaos and the regular behaviors that appear in turbulent flow have provided new insights into global weather patterns and other phenomena. Perhaps the most striking examples of atmospheric flow patterns are the photographs sent back from the Voyager flyby of the planet Jupiter. Whorls and eddies can be seen in the flow of the atmospheric gases on Jupiter. These include the famous red spot, which is now thought to be a giant and very stable atmospheric eddy (fig. 9.22).

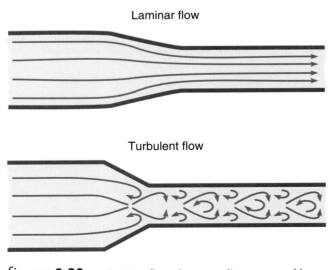

figure **9.20**   In laminar flow, the streamlines are roughly parallel to one another. In turbulent flow, the stream-flow patterns are much more complicated.

figure **9.21**   Smoke rising from a cigarette first exhibits laminar flow, and then as the speed increases and the column widens, turbulent flow appears.

figure **9.22** Whorls and eddies, including the giant red spot (lower left), can be seen in the atmosphere of the planet Jupiter.

The rate of flow of a stream is equal to the fluid speed multiplied by the cross-sectional area through which the stream is flowing. For the flow to be continuous, the speed must increase where the stream narrows to pass through a smaller cross-sectional area. The fluid speed of the stream also varies across its area because of viscosity: the fluid speed is largest at the center and smallest near the banks or pipe walls. Smooth or laminar flow gives way to turbulent flow with its eddies and whorls as the fluid speed increases or the viscosity decreases.

9.5 Bernoulli's Principle

Have you ever wondered how a large passenger jet gets off the ground? You know that those things fly, but somehow it still seems improbable. What forces keep it in the air?

Although the forces that act on an airplane wing, or **air foil,** can be analyzed in different ways, a partial explanation is found in a principle regarding the flow of fluids published by Daniel Bernoulli (1700–1782) in a treatise on hydrodynamics in 1738.

What is Bernoulli's principle?

What happens if we do work on a fluid, increasing its energy? This increase may show up as an increase in kinetic energy of the fluid, leading to an increase in the fluid's speed. It could also appear as an increase in potential energy if the fluid is squeezed (elastic potential energy) or if the fluid is raised in height (gravitational potential energy). Bernoulli considered all of these possibilities. Bernoulli's principle is a direct result of applying conservation of energy to the flow of fluids.

The most interesting examples of Bernoulli's principle involve changes in kinetic energy. If a fluid that is not compressible (not squeezable) is flowing in a level pipe or stream, any work done on this fluid will increase its kinetic energy. To accelerate the fluid and increase its kinetic energy, there must be a net force doing work on the fluid. This force is associated with a difference in pressure from one point to another within the fluid.

If there is a difference in pressure, the fluid accelerates from the region of higher pressure toward the region of lower pressure, because that is the direction of the force acting on the fluid. We expect to find higher fluid speeds in the regions of lower pressure. In the simple case of an incompressible fluid flowing in a level pipe or stream, work and energy considerations lead to a statement of **Bernoulli's principle:**

> The sum of the pressure plus the kinetic energy per unit volume of a flowing fluid must remain constant.

$$P + \tfrac{1}{2}dv^2 = \text{constant}$$

Here, P is the pressure, d is the density of the fluid, and v is the fluid's speed. The second term in this sum is the kinetic energy per unit volume (kinetic energy divided by volume) of the fluid, since density is mass divided by volume.

A fuller statement of Bernoulli's principle would include the effects of gravitational potential energy, allowing for changes in the height of the fluid. However, many of the most interesting effects can be investigated using the form just stated. In applying Bernoulli's principle, the association of lower pressures with higher fluid speeds is often the key point.

How does the pressure vary in pipes and hoses?

Consider a pipe with a constriction in its center section, as in figure 9.23. Would you expect the pressure of water flowing in the pipe to be greatest at the constriction or in the wider sections of the pipe? Intuition leads you to suspect that the pressure is greater in the constricted section, but this is *not* the case.

We know that the speed of the water will be greater in the constricted section (where the cross-sectional area is smaller) than in the wider portions of the pipe, because of the concept of continuous flow. What does Bernoulli's principle tell us? To keep the sum $P + \tfrac{1}{2}dv^2$ constant, the

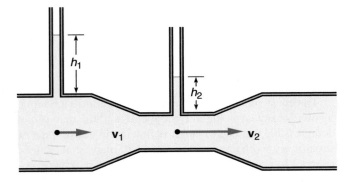

figure 9.23 Vertical open pipes can serve as pressure gauges. The height of the column of water is proportional to the pressure. The pressure of a moving fluid is greater where the fluid velocity is smaller.

pressure must be larger where the fluid speed is smaller. In other words, if the speed increases, the pressure must decrease.

Open tubes at different places in the pipe (fig. 9.23) can serve as simple pressure gauges. For the water in these open tubes to rise, fluid pressure must be greater than atmospheric pressure. The height that the fluid rises will depend on how much greater its pressure is. The water level in these tubes will reach a greater height above the main pipe where the pipe is wider rather than at the constriction, indicating that the pressure is higher in the wider portion of the pipe.

This result goes against our intuition because we tend, incorrectly, to associate higher pressure with higher speed. Another confirming example is the nozzle on a hose. The nozzle constricts the area of flow and increases the fluid's speed. By Bernoulli's principle, the pressure of the water is smaller at the narrow end of the nozzle than farther back in the hose, contrary to what we expect.

If you place your hand in front of the nozzle, you will feel a large force on your hand as the water strikes it. This force results from the change in the velocity and momentum of the water as it strikes your hand. By Newton's second law, a large force is required to produce this change in momentum, and by Newton's third law, the force exerted on the water by your hand is equal in magnitude to the force exerted on your hand by the water. That force is not directly associated with the fluid pressure in the hose—the pressure is actually greater farther back in the hose where the water is not moving very fast.

How does an airplane wing work?

Bernoulli's principle, as we have stated it, is only valid for fluids whose density does not change (noncompressible fluids), but we often extend it to investigate effects of the mo-

tion of air and other compressible gases. Even for compressible fluids, a larger fluid speed is usually associated with a smaller fluid pressure.

A simple demonstration will make a believer of you. Take half a sheet of tablet paper (or even a facial tissue) and hold it in front of your mouth, as in figure 9.24. The paper should hang down limply in front of your chin. If you blow across the top of it, the paper rises, and if you blow hard enough, the paper may even stand straight out horizontally. What is happening?

Blowing across the top of the paper makes the air flow across the top with a greater speed than underneath the paper. The air underneath presumably is not moving much at all. This greater speed leads to a reduction in pressure. Since the air pressure is then greater on the bottom of the paper than on the top, the upward force on the bottom of the paper is larger than the downward force on the top, so the paper rises. Contrary to intuition, blowing between two strips of paper makes them move closer together rather than farther apart, for the same reason. Try it.

Similar effects are at work on an airplane wing. The shape and tilt of the wing can cause the air to move faster across the top of the wing than across the bottom. By Bernoulli's principle, we expect this greater speed to be associated with a lower pressure on the top of the wing than on the bottom. This pressure difference produces a net upward force, or *lift,* acting on the wing. The lift force balances the weight of the airplane and thus enables it to fly.

figure 9.24 Blowing across the top of a limp piece of paper causes the paper to rise, demonstrating Bernoulli's principle.

Although it provides a simple explanation of the lift force, Bernoulli's principle cannot be used to accurately compute the lift force because air is a compressible fluid. It is also difficult to predict the difference in air speed of the airflow across the top and bottom of the wing. A more complete analysis must take into account the different directions and magnitudes of the forces acting at different points on the wing. As can be seen in figure 9.25, air is deflected downward by the wing, which requires a downward force exerted on the air by the wing. By Newton's third law, there must then be an equal but opposite (upward) force exerted on the wing by the air.

The design of airplane wings and the flow of air past wings have been extensively studied in wind tunnels, where a wing is kept stationary and air blown past it. The effects of the angle at which the wing is set and the use of flaps to change the curvature can be thoroughly explored in wind tunnels. Under some conditions, airflow over the wing becomes turbulent, an undesirable effect that reduces the lift force. Fluid-flow considerations are extremely important to the design and operation of all kinds of aircraft.

What keeps the department-store ball in the air?

Another demonstration of Bernoulli's principle that uses airflow is often seen in department stores when vacuum cleaners are being advertised. A ball can be suspended in an upward-moving column of air produced by a vacuum cleaner. As the air moves upward, the speed of airflow is greatest in the center of the stream and falls off to zero away from the center. Once again, Bernoulli's principle requires that the pressure will be smallest in the center, where the speed is greatest.

The pressure increases away from the center of the air column toward regions where the air is moving less rapidly. If the ball moves out of the center of the air column, a

figure **9.26**   A ball is suspended in an upward-moving column of air produced by a hair dryer. The air pressure is smallest in the center of the column where the air is moving with the greatest speed.

larger pressure and a larger force act on the outer side of the ball than on the side nearer the center of the column. The ball moves back into the center. The upward force of air hitting the bottom of the ball holds it up, while the low pressure in the center of the column keeps the ball near the center. You can produce the same effect with a small ball and a hair dryer (fig. 9.26).

The motion of a curveball discussed in Everyday Phenomenon Box 9.1 provides another example of Bernoulli's principle at work. In all of these phenomena, we see the effects of a reduction in fluid pressure associated with an increase in fluid speed, as predicted by Bernoulli's principle. We do not always expect this. We are tempted to think that blowing between two pieces of paper will push them apart, for example, but a simple experiment shows otherwise. Understanding such surprises is part of the fun of physics.

> Bernoulli's principle is derived from energy considerations and says that the sum of the pressure plus the kinetic energy per unit volume remains constant from one point in a fluid to another. This holds true if the fluid is not compressible and changes in height are not involved. Pressure is then lower where fluid speed is higher. This effect explains many surprising phenomena involving pipes, hoses, airplane wings, and the fact that blowing across the top of a piece of paper causes the paper to rise rather than drop.

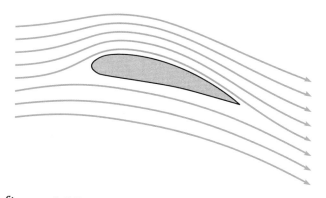

figure **9.25**   The shape of an airplane wing can cause an airflow pattern in which the air flowing over the wing moves with a greater speed than the air flowing underneath.

Everyday Phenomenon

box **9.1**

Throwing a Curveball

The Situation. Baseball players know that they can be badly fooled by a well-thrown curveball. The idea that a fast-moving curveball can be deflected by as much as a foot on its way to the plate is hard for some people to accept. Many people have insisted, over the years, that the curve is just an illusion. Does the path of a curveball really curve? If so, how can we explain it?

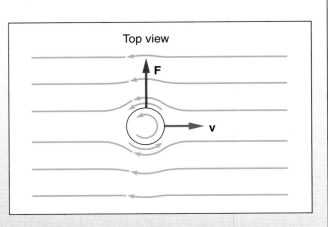

A batter is badly fooled by a curveball. Does the path of the ball really curve?

The whirlpool of air created by the spin of the ball causes the air to move more rapidly on one side of the ball than on the other. This produces a deflecting force, as predicted by Bernoulli's principle.

The Analysis. There is no secret to throwing a curveball. A right-handed pitcher throws a curveball with a counter-clockwise spin (as viewed from above), so that it curves away from a right-handed batter. The pitch is most effective when it starts out looking as if it is headed toward the inside of the plate and then curves over the plate and down and away from the batter.

Bernoulli's principle can explain the deflection of the ball's path. Because the surface of the spinning ball is rough, it drags a layer of air around with it, creating a whirlpool of air near the ball. The ball is also moving toward the plate, which produces an additional flow of air past the ball in the direction opposite to the velocity of the ball. The whirlpool created by the spin of the ball causes the air to move more swiftly on the side opposite the right-handed batter, where the two effects add, than on the side nearer the batter, as shown in the drawing.

By Bernoulli's principle, a greater speed of airflow is associated with a lower pressure: the air pressure is lower on the side of the ball opposite the batter than on the near side.

This difference in pressure produces a deflecting force on the ball, pushing it away from the right-handed batter.

Although Bernoulli's principle gives a good explanation of the direction of the deflecting force and the direction of the curve, it cannot be used to make accurate quantitative predictions. Air is a compressible fluid, and the usual form of Bernoulli's principle is valid only for noncompressible fluids such as water or other liquids. More accurate methods of treating the effects of airflow past the ball must be used for predicting the degree of curvature.

Both theoretical computations and experimental measurements have confirmed that there is a deflecting force on the ball and that the path of the ball does indeed curve. The degree of curvature depends on the rate of spin on the ball and the roughness of the surface of the ball, as you might expect from Bernoulli's principle. Some pitchers have been known to cheat by roughening the ball's surface with sandpaper hidden in their gloves. Debate continues about whether the orientation of the seams of the baseball also has an effect. The pitcher's grip on the ball is an important factor in how much spin he can put on the ball. Once the ball is released, however, experimental evidence suggests that the orientation of the seams is not important in the strength of the deflecting force.

A good discussion of the theory and the experimental evidence is found in an article by Robert Watts and Ricardo Ferrer in the January 1987 issue of the *American Journal of Physics,* pages 40–44. The curved motion of spinning balls is important in other sports, such as golf and soccer. A good athlete needs to be able to recognize and take advantage of the effects of these curves.

summary

Fluid pressure is central to understanding the behavior of fluids, which include both liquids and gases. This chapter has focused on the effects of pressure in determining the behavior of both stationary fluids and fluids in motion.

1 Pressure and Pascal's principle.
Pressure is defined as the ratio of force to area exerted on or by a fluid. According to Pascal's principle, pressure extends uniformly in all directions through a fluid, which explains the operation of hydraulic systems.

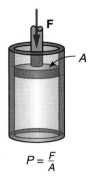

$$P = \frac{F}{A}$$

2 Atmospheric pressure and the behavior of gases.
We can measure atmospheric pressure by determining the height of a column of mercury supported by the atmosphere. Atmospheric pressure decreases with increasing altitude. The density of air also changes with altitude, because a lower pressure leads to a larger volume, according to Boyle's law.

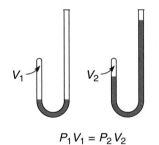

$$P_1 V_1 = P_2 V_2$$

3 Archimedes' principle.
In a fluid, pressure increases with depth, producing a buoyant force on objects submerged in the fluid. Archimedes' principle states that this force is equal to the weight of the fluid displaced by the object. If the buoyant force is less than the weight of the object, it will sink in the fluid. Otherwise, it will float.

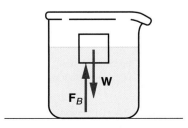

4 Fluids in motion.
The rate of flow of a fluid is equal to the speed times the cross-sectional area, vA. In continuous flow, the speed increases if the area decreases. Fluids with high viscosity have a greater resistance to flow than those with low viscosity. As the flow speed increases, the flow may change from laminar (smooth) flow to turbulent flow.

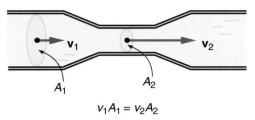

$$v_1 A_1 = v_2 A_2$$

5 Bernoulli's principle.
Energy considerations require that the kinetic energy per unit mass of a fluid increase as the pressure decreases. This is Bernoulli's principle for cases in which changes in height are not important. Higher fluid speeds are associated with lower pressures. These ideas can be used to explain the curve of a curveball and other phenomena.

$$P + \frac{1}{2} dv^2 = \text{constant}$$

key terms

Pressure, 167
Pascal's principle, 168
Atmospheric pressure, 170
Density, 170

Boyle's law, 172
Buoyant force, 173
Archimedes' principle, 173
Viscosity, 177

Turbulent flow, 178
Air foil, 179
Bernoulli's principle, 179

questions

*Questions identified with an asterisk are more open-ended than the others. They call for lengthier responses and are more suitable for group discussion.

Q1. Is it possible for a 100-lb woman to exert a greater pressure on the ground than a 250-lb man? Explain.

Q2. If we measure force in pounds (lb) and distance in feet (ft), what are the units of pressure in this system? Explain.

Q3. The same force is applied to two cylinders that contain air. One has a piston with a large area, and the other has a piston with a small area. In which cylinder will the pressure be greater? Explain.

Q4. A penny and a quarter are embedded in the concrete bottom of a swimming pool filled with water. Which of these coins experiences the greater downward force due to water pressure acting on it? Explain.

*Q5. Why are bicycle tires often inflated to a higher pressure than automobile tires, even though the automobile tires must support a much larger weight? Explain.

Q6. The fluid in a hydraulic system pushes against two pistons, one with a large area and the other with a small area.
   a. Which piston experiences the greater force due to fluid pressure acting on it? Explain.
   b. When the smaller piston moves, does the larger piston move through the same distance, a greater distance, or a smaller distance than the smaller piston? Explain.

Q7. If the output piston in a hydraulic pump exerts a greater force than one applied to the input piston, is the pressure at the output piston also larger than at the input piston? Explain.

Q8. When a mercury barometer is used to measure atmospheric pressure, does the closed end of the tube above the mercury column usually contain air? Explain.

*Q9. Could we use water instead of mercury to make a barometer? What advantages and disadvantages would be associated with the use of water? Explain.

Q10. If you climbed a mountain carrying a mercury barometer, would the level of the mercury column in the glass tube of the barometer increase or decrease (compared to the mercury reservoir) as you climb the mountain? Explain.

Q11. If you filled an airtight balloon at the top of a mountain, would the balloon expand or contract as you descend the mountain? Explain.

*Q12. When you go over a mountain pass in an automobile, your ears often "pop" both on the way up and on the way down. How can you explain this effect?

Q13. The plunger of a sealed hypodermic syringe containing air is slowly pulled out. Does the air pressure inside the syringe increase or decrease when this happens? Explain.

Q14. Helium is sealed inside a balloon impermeable to the flow of gas. If a storm suddenly comes up, would you expect the balloon to expand or contract? Explain. (Assume that there is no change in temperature.)

Q15. Is it possible for a solid metal ball to float in mercury? Explain.

Q16. A rectangular metal block is suspended by a string in a beaker of water so that the block is completely surrounded by water. Is the water pressure at the bottom of the block equal to, greater than, or less than the water pressure at the top of the block? Explain.

Q17. Is it possible for a boat made of concrete to float? Explain.

Q18. A block of wood is floating in a pool of water.
   a. Is the buoyant force acting on the block greater than, less than, or equal to the weight of the block? Explain.
   b. Is the volume of the fluid displaced by the block greater than, less than, or equal to the volume of the block? Explain.

Q19. A large bird lands on a rowboat that is floating in a swimming pool. Will the water level in the pool increase, decrease, or remain the same when the bird lands on the boat? Explain.

*Q20. A rowboat is floating in a swimming pool when the anchor is dropped over the side. When the anchor is dropped, will the water level in the swimming pool increase, decrease, or remain the same? Explain.

Q21. Is it possible that some objects might float in salt water but sink in fresh water? Explain.

Q22. If an object has the same density as water, will the object float to the top, sink to the bottom, or take neither course? Explain.

Q23. A steady stream of water flowing in a narrow pipe reaches a point where the pipe widens. Does the speed of the water increase, decrease, or remain the same when the pipe widens? Explain.

Q24. Why does the stream of water flowing from a faucet often get more narrow as the water falls? Explain.

Q25. Does a stream of liquid with a high viscosity flow more rapidly under the same conditions as a stream with low viscosity? Explain.

Q26. If the speed of flow in a stream decreases, is the flow likely to change from laminar to turbulent flow? Explain.

Q27. Why is the flow of smoke from a cigarette often laminar near the source but turbulent farther from the source? Explain.

Q28. If you blow between two limp pieces of paper held hanging down a few inches apart, will the pieces of paper come closer together or move farther apart? Explain.

*Q29. A wind gust blows sideways across an outward-swinging door that is slightly ajar. Will this cause the door to slam shut or swing open? Explain.

Q30. A hair dryer can be used to create a stream of air. Is the air pressure in the center of the stream greater than, less than, or equal to the air pressure at some distance from the center of the stream? Explain.

Q31. From the perspective of a right-handed batter, does a rising fastball spin clockwise or counterclockwise? Explain.

exercises

E1. A force of 40 N pushes down on the movable piston of a closed cylinder containing a gas. The piston's area is 0.5 m². What pressure does this produce in the gas?

E2. A 110-lb woman puts all of her weight on one heel of her high-heel shoes. The heel has an area of 0.4 in². What is the pressure that her heel exerts on the ground in pounds per square inch (psi)?

E3. A 250-lb man supports all of his weight on a showshoe with an area of 200 in². What pressure is exerted on the snow (in pounds per square inch)?

E4. The pressure of a gas contained in a cylinder with a movable piston is 300 Pa (300 N/m²). The area of the piston is 0.2 m². What is the magnitude of the force exerted on the piston by the gas?

E5. In a hydraulic system, a force of 400 N is exerted on a piston with an area of 0.001 m². The load-bearing piston in the system has an area of 0.2 m².
  a. What is the pressure in the hydraulic fluid?
  b. What is the magnitude of the force exerted on the load-bearing piston by the hydraulic fluid?

E6. The load-bearing piston in a certain hydraulic system has an area 50 times as large as the input piston. If the larger piston supports a load of 6000 N, how large a force must be applied to the input piston?

E7. A column of water in a vertical pipe has a cross-sectional area of 0.2 m² and a weight of 450 N. What is the increase in pressure (in Pa) from the top to the bottom of the pipe?

E8. With the temperature held constant, the pressure of a gas in a cylinder with a movable piston is increased from 10 kPa to 90 kPa. The initial volume of the gas in the cylinder is 0.6 m³. What is the final volume of the gas after the pressure is increased?

E9. With the temperature held constant, the piston of a cylinder containing a gas is pulled out so that the volume increases from 0.1 m³ to 0.3 m³. If the initial pressure of the gas was 80 kPa, what is the final pressure?

E10. A 0.5-kg block of wood is floating in water. What is the magnitude of the buoyant force acting on the block?

E11. A block of wood of uniform density floats so that exactly half of its volume is underwater. The density of water is 1000 kg/m³. What is the density of the block?

E12. A certain boat displaces a volume of 2.5 m³ of water. (The density of water is 1000 kg/m³.)
  a. What is the mass of the water displaced by the boat?
  b. What is the buoyant force acting on the boat?

E13. A rock with a volume of 0.2 m³ is fully submerged in water having a density of 1000 kg/m³. What is the buoyant force acting on the rock?

E14. A stream moving with a speed of 0.5 m/s reaches a point where the cross-sectional area of the stream decreases to one-fourth of the original area. What is the water speed in this narrowed portion of the stream?

E15. Water emerges from a faucet at a speed of 1.5 m/s. After falling a short distance, its speed increases to 3 m/s as a result of the gravitational acceleration. By what number would you multiply the original cross-sectional area of the stream to find the area at the lower position?

E16. An airplane wing with an average cross-sectional area 10 m² experiences a lift force of 60 000 N. What is the difference in air pressure, on the average, between the bottom and top of the wing?

challenge problems

CP1. Suppose that the input piston of a hydraulic jack has a diameter of 2 cm and the load piston a diameter of 25 cm. The jack is being used to lift a car with a mass of 1400 kg.
   a. What are the areas of the input and load pistons in square centimeters? ($A = \pi r^2$)
   b. What is the ratio of the area of the load piston to the area of the input piston?
   c. What is the weight of the car in newtons? ($W = mg$)
   d. What force must be applied to the input piston to support the car?

CP2. Water has a density of 1000 kg/m^3. The depth of a swimming pool at the deep end is about 3 m.
   a. What is the volume of a column of water 3 m deep and 0.5 m^2 in cross-sectional area?
   b. What is the mass of this column of water?
   c. What is the weight of this column of water in newtons?
   d. What is the excess pressure (above atmospheric pressure) exerted by this column of water on the bottom of the pool?
   e. How does this value compare to atmospheric pressure?

CP3. A steel block with a density of 7800 kg/m^3 is suspended from a string in a beaker of water so that the block is completely submerged but not resting on the bottom. The block is a cube with sides of 3 cm (0.03 m).
   a. What is the volume of the block in cubic meters?
   b. What is the mass of the block?
   c. What is the weight of the block?
   d. What is the buoyant force acting on the block?
   e. What tension in the string is needed to hold the block in place?

CP4. A flat-bottomed wooden box is 3 m long and 1.5 m wide, with sides 1 m high. The box serves as a boat carrying five people as it floats on a pond. The total mass of the boat and the people is 1200 kg.
   a. What is the total weight of the boat and the people in newtons?
   b. What is the buoyant force required to keep the boat and its load afloat?
   c. What volume of water must be displaced to support the boat and its load? (The density of water is 1000 kg/m^3.)
   d. How much of the boat is underwater? (What height of the sides must be submerged to produce a volume equal to that of part c?)

CP5. A pipe with a circular cross section has a diameter of 8 cm. It narrows at one point to a diameter of 5 cm. The pipe is carrying a steady stream of water that completely fills the pipe and is moving with a speed of 1.5 m/s in the wider portion.
   a. What are the cross-sectional areas of the wide and narrow portions of the pipe? ($A = \pi r^2$, and the radius is half the diameter.)
   b. What is the speed of the water in the narrow portion of the pipe?
   c. Is the pressure in the narrow portion of the pipe greater than, less than, or equal to the pressure in the wider portion? Explain.

home experiments and observations

HE1. Using an awl or drill, punch three similar holes in an empty plastic milk jug at three different heights. Plug the holes and fill the jug with water.
   a. Placing the jug at the edge of the sink, unplug the holes and let the water flow out. Which hole produces the greatest initial speed for the emerging water?
   b. Which stream travels the greatest horizontal distance? What factors determine how far the water will go horizontally?
   c. Observe the shape of the water streams as they fall into the sink. Do the streams narrow as the water falls? If so, how can this be explained?

HE2. If you have some modeling clay available, try building a boat from the clay.
   a. Does the clay sink when it is rolled into a ball? What does this indicate about the density of the clay?
   b. Is a flat boat more effective than a canoe in carrying the maximum load of steel washers or other weights? Try them both. What are the problems with each?

HE3. If you have a hair dryer handy, try supporting a ping-pong ball in a vertical column of air coming from the dryer.
   a. Can you get the ping-pong ball to stay in place? How far from the center can the ball wander and still return?
   b. Will a tennis ball or a small balloon work? What differences do you note in each of these cases?

HE4. Using paper clips and a pen or pencil as a support rod, suspend two full-sized pieces of paper a few inches apart so that they hang vertically.
   a. Blow downward between the two pieces and observe the effect. How does increasing the spacing affect the results?
   b. Try blowing at different rates. How does this affect the results?

Temperature and Heat

chapter overview

The first two sections of this chapter explore the ideas of temperature, heat, and the relationship between these two distinct concepts. We then introduce the first law of thermodynamics, which helps us explain why a drill bit gets hot as it works, as well as many aspects of gas behavior. Finally, we discuss the ways heat is transferred from one object to another.

chapter outline

1 **Temperature and its measurement.** What is temperature? How do we go about measuring it? Where should we place the zero of a temperature scale?

2 **Heat and specific heat capacity.** What is heat? How does it differ from temperature? Does adding heat always change the temperature of a substance? How is heat involved in changes of phase?

3 **Joule's experiment and the first law of thermodynamics.** Are there other ways of changing the temperature of an object besides adding heat? What does the first law of thermodynamics say and mean?

4 **Gas behavior and the first law.** How can the behavior of gases be explained using the first law of thermodynamics? What is an ideal gas?

5 **The flow of heat.** What are the different ways that heat can be transferred from one object to another? How do these ideas apply to heating or cooling a house?

chapter

10

unit two

Have you ever touched a drill bit right after drilling a hole in a piece of wood or metal (fig. 10.1)? Chances are that you removed your hand quickly, since the bit was probably hot, particularly if you had been drilling in metal. The same phenomenon is at work when brakes get hot on your bicycle or car, or in any process where one surface is rubbing on another.

We could also make the drill bit hot by placing it in a pot of boiling water or the flame of a blowtorch. In either case, when the bit gets hotter, we say that its temperature has increased. What is temperature, though, and how do we go about comparing one temperature to another? Is the final condition of the drill bit any different if it has been warmed by drilling rather than by being placed in contact with hot water?

Questions such as these lie in the realm of **thermodynamics,** the study of heat and its effects on matter. Thermodynamics is ultimately about energy but in a broader context than mechanical energy, which we discussed in chapter 6. The first law of thermodynamics, introduced in this chapter, extends the principle of conservation of energy to include the effects of heat.

What is the difference between heat and temperature? Our everyday language often blends these two ideas. A true appreciation for this distinction did not emerge until about the middle of the nineteenth century when the laws of thermodynamics were developed. Heat, temperature, and the distinction between them

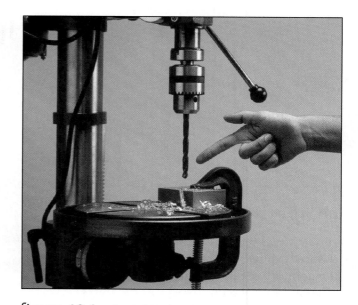

figure **10.1**   The drill bit feels hot after drilling a hole in a piece of metal. What causes the increase in its temperature?

are critical, however, to understanding why things get hot or cold and why some remain that way longer than others. The same ideas involved in cooling a drink are important to understanding global weather patterns.

10.1 Temperature and Its Measurement

Suppose that you are not feeling well and are looking for an excuse to skip class. You think you might have a fever, and your roommate, by placing a hand on your forehead, agrees that you feel a little hot. To confirm your hunch, however, you find a thermometer and take your temperature (fig. 10.2). The thermometer reads 101.3°.

What exactly does this reading tell you? It indicates that your temperature is above 98.6° Fahrenheit, generally considered normal, and that you have a fever. You are probably justified in spending the day in bed. The thermometer provides a quantitative measure of how hot or cold things are and a basis for comparing your current temperature to your normal temperature. This quantitative measure usually carries more credibility with the boss or a professor than the vaguer statement that you felt like you had a fever.

How do we measure temperature?

Taking a temperature or reading a thermometer is a common experience. Besides the obvious function of comparing the "hotness" or "coldness" of different objects, however, do the numbers have any more fundamental meaning? To ask a basic question, how do hot objects differ from cold objects? What is *temperature*?

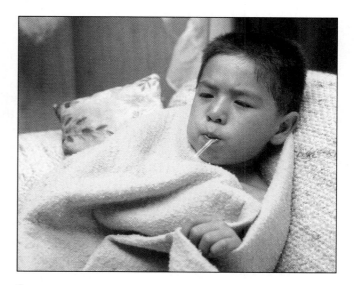

figure **10.2**   Taking a temperature. What does the thermometer tell us?

Although we all have an intuitive feel for what hot and cold mean, putting that impression into words can be hard. Try it before you read on: how would you define the term *hot*? Even our senses can mislead us. The pain that we feel when we touch something very hot can be hard to distinguish from the pain that we feel when we touch a very cold object. A metal block feels colder to the touch than a wooden block, even though the two blocks may have the same temperature. In the end, temperature measurement is a comparison, and the comparative terms *hotter* and *colder* have more meaning than *hot* or *cold* themselves.

A closer look at what happens when we use a thermometer is helpful. The traditional clinical thermometer was a sealed glass tube partially filled with some fluid, usually mercury. (These have now been largely replaced by digital thermometers.) The inside diameter of the tube was very small, but it widened at the bottom into a reservoir containing most of the mercury (fig. 10.3). Usually you placed the thermometer in your mouth under your tongue and waited a few minutes until the thermometer reached the same temperature as the inside of your mouth. At first, the mercury rose in the thin tube. When the mercury level in the tube was no longer changing, you assumed that it had reached the same temperature as your mouth.

Why does the mercury rise? Most materials expand as they get warmer, and mercury and many other liquids expand at a greater rate than glass. As it expands, the mercury in the reservoir must go somewhere, so it rises in the narrow tube. We use the physical property of the thermal expansion of mercury to give an indication of temperature; by placing marks along the tube, we create a temperature scale. Any other physical property that changes with temperature could, in principle, be used. Such properties include changes in electrical resistance, thermal expansion of metals, and even changes in color.

In this whole process, we assume that if two objects are in contact with one another long enough so that the physical properties (such as volume) of the objects are no longer changing, the two objects have the same temperature. When we take a temperature with the clinical thermometer, we wait until the mercury stops rising before reading the

thermometer. This procedure gives us part of the definition of temperature by defining when two or more objects have the *same* temperature. When the physical properties are no longer changing, the objects are said to be in **thermal equilibrium.** Two or more objects in thermal equilibrium have the same temperature. This assumption is sometimes referred to as the **zeroth law of thermodynamics,** because it underlies the definition of temperature and the process of temperature measurement.

How were temperature scales developed?

What do the numbers on a thermometer mean? When the first crude thermometer was made, the numbers marking the divisions on the scale were arbitrary. They were useful for comparing temperatures only if the *same* thermometer was used. To compare a temperature measured in Germany with one thermometer to another measured in England with a different thermometer, a standard temperature scale was needed.

The first widely used temperature scale was devised by Gabriel Fahrenheit (1686–1736) in the early 1700s. Anders Celsius (1701–1744) invented another widely used scale somewhat later, in 1743. Both of these scales use the freezing point and boiling point of water to anchor the scales. The modern Celsius scale uses the triple point of water, where ice, water, and water vapor are all in equilibrium. This point varies only slightly from the freezing point at normal atmospheric pressure. Fahrenheit set the freezing point of water at 32° and the boiling point at 212° on his scale. These two points are set at 0° and 100° on the Celsius scale (fig. 10.4).

As figure 10.4 shows, the Celsius degree is larger than the Fahrenheit degree. Only 100 Celsius degrees span the temperature range between the freezing point and boiling point of water, while 180 Fahrenheit degrees (212° − 32°) are needed to span the same range. The ratio of Fahrenheit degrees to Celsius degrees is therefore 180/100, or 9/5. The Fahrenheit degree is 5/9 the size of the Celsius degree, so more Fahrenheit degrees are needed to cover the same range in temperature.

The Celsius scale is used in science and in most of the world. Because Fahrenheit temperatures are still more familiar in the United States, we commonly have to convert from one scale to the other. Since the zero points and the size of the degree differ, we take both factors into account in the conversion. For example, an ordinary room temperature of 72° Fahrenheit is 40 degrees above the 32°F freezing point of water. Because it takes only 5/9 as many Celsius degrees to span the same range, this temperature would be (5/9)(40°) or 22 Celsius degrees above the freezing point of water. The Celsius scale has its zero at the freezing point of water, so 72°F is equal to 22°C.

Let's recap. First, we found how many Fahrenheit degrees this temperature is above the freezing point of water

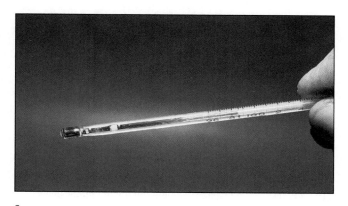

figure **10.3** The traditional clinical thermometer contained mercury in a thin glass tube with a wider reservoir of mercury at the bottom.

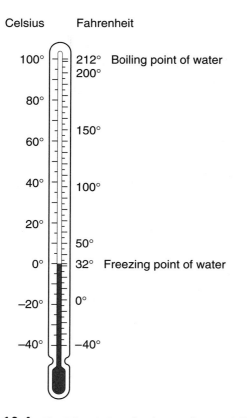

Celsius   Fahrenheit

100°   212°   Boiling point of water
              200°

80°

              150°
60°

40°   100°

20°

              50°
0°   32°   Freezing point of water

−20°   0°

−40°   −40°

figure 10.4 The Fahrenheit and Celsius scales use different numerical values for the freezing and boiling points of water. The Celsius degree is larger than the Fahrenheit degree.

Is there an absolute zero?

Do the zero points on either scale have any special significance? Although the zero point on the Celsius scale is the freezing point of water, there is no special significance to these points. They were selected arbitrarily. In fact, the temperature goes below zero on either scale, as happens frequently in places like Minnesota and Alaska in the winter. Fahrenheit's zero point was based on the temperature of a mixture of salt and ice in a salt solution.

Although the zero points of both the Fahrenheit and Celsius scales were selected arbitrarily, there is an absolute zero with more fundamental significance. Absolute zero was not proposed until 100 years or so after these scales were originally developed.

The first indication of an absolute zero for temperature measurements came from studying changes in pressure and volume that take place in gases when the temperature is changed. If we hold the volume of a gas constant while increasing the temperature, the pressure of the gas will increase. Pressure is another physical property that changes with temperature, just like the volume of mercury in the case of a mercury thermometer. (See chapter 9 for a discussion of pressure.) Because the pressure of a gas of constant volume increases with temperature, we can use this property as a means of measuring temperature (fig. 10.5).

If we plot the pressure of a gas as a function of the temperature measured on the Celsius scale, we get a graph like figure 10.6. If the temperatures and pressures are not too high, a striking feature emerges from such graphs. The

by subtracting 32. Then we multiplied by the factor $\frac{5}{9}$, which is the ratio of the size of the Fahrenheit degree to the Celsius degree. We know it takes fewer Celsius degrees to span this range, since the Celsius degree is larger than the Fahrenheit degree. We can express this conversion in an equation:

$$T_C = \frac{5}{9}(T_F - 32).$$

Using the same logic to perform the reverse conversion from Celsius to Fahrenheit, or simply rearranging the equation, it becomes

$$T_F = \frac{9}{5}T_C + 32.$$

Multiplying the temperature in Celsius by $\frac{9}{5}$ tells us how many Fahrenheit degrees above the freezing point it is. This value then is added to 32°F, the freezing point on the Fahrenheit scale. Using this relationship, you can confirm that the boiling point of water ($T_C = 100°C$) equals 212°F. Normal body temperature (98.6°F) is equivalent to 36°C. A body temperature of 38°C would be enough to keep you home from school or work.

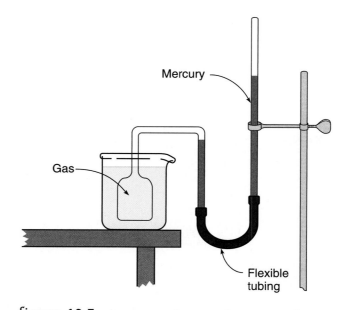

figure 10.5 A constant-volume gas thermometer allows the pressure to change with temperature while the volume is held constant. The difference in height of the two mercury columns is proportional to the pressure.

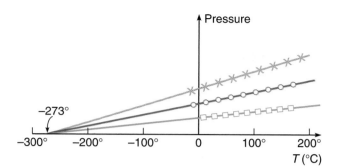

figure **10.6** The pressure plotted as a function of temperature for different amounts of different gases. When extended backward, the lines all intersect the temperature axis at the same point.

curves for different gases or amounts all are straight lines, and when these lines are extended backward to zero pressure, they all intersect the temperature axis at the same point. This is true regardless of the gas used (oxygen, nitrogen, helium, and so on) or the quantity of gas.

The curves all intersect the axis at $-273.2°C$. Since a negative pressure has no meaning, this intersection suggests that the temperature cannot get any lower than $-273.2°C$. It is important to keep in mind that most gases will condense to liquids and then solidify well before this point is reached.

We refer to this temperature, $-273.2°C$, as **absolute zero.** The Kelvin, or absolute, temperature scale has its zero at this point and uses intervals the same size as the Celsius degree. To convert Celsius temperatures to kelvins, we simply add 273.2 to the Celsius temperature, or

$$T_K = T_C + 273.2.$$

Room temperature of 22°C is an **absolute temperature** of approximately 295 K. (The term *degree* and the degree symbol are not used in expressing absolute temperatures; they are simply called *kelvins.*) The absolute temperature scale currently in use is basically the same as the one originally suggested by observing the behavior of gases.

Absolute zero is a temperature that we can approach but never really reach. It represents a limiting value. Speaking of anything getting any colder than absolute zero is not meaningful.

Temperature measurements are based on physical properties that change with temperature, such as the volume of a liquid or the pressure of a gas. The Fahrenheit and Celsius scales have different zero points and different-sized degrees. The Kelvin or absolute scale starts at absolute zero and uses intervals the same size as the Celsius degree.

10.2 Heat and Specific Heat Capacity

You are already late for class or work, but your morning cup of coffee is too hot to drink. What can you do to cool it to the point where you won't scald your tongue? You can blow on it, but blowing is not a very efficient way of cooling coffee quickly. If you take milk with your coffee, you can pour a little cold milk into the cup, and the temperature of the coffee will be lowered (fig. 10.7).

What happens when objects or fluids at different temperatures come in contact with one another? The temperature of the colder object increases, and the temperature of the warmer one decreases, with both objects eventually coming to the same intermediate temperature. Something is flowing from the hotter to the colder body (or vice versa). But what is flowing?

What is specific heat capacity?

Early attempts at explaining phenomena like our coffee-cooling example involved the idea that an invisible fluid called *caloric* flowed from the hotter to the cooler object. The amount of caloric transferred dictated the extent of the temperature change. Different materials were thought to store different amounts of caloric in the same amount of mass, which helped to explain some observations. Although the caloric model successfully explained many simple

figure **10.7** Pouring cold milk into a hot cup of coffee lowers the temperature of the coffee.

phenomena involving temperature changes, there were also problems with it, which we will discuss in section 10.3.

We now use the term **heat** for the quantity that flows from one object to another when two objects having different temperatures come in contact. As we will see in section 10.3, heat flow is a form of energy transfer between objects.

The idea of exchange of heat sheds light on a large variety of phenomena. If you drop a 100-gram steel ball, initially at room temperature, into your cup of coffee, you will find that it is less effective in cooling your coffee than 100 grams of room-temperature milk or water. Steel has a lower **specific heat capacity** than milk or water. (Milk is mostly water, as is coffee.) Less heat is needed to change the temperature of 100 grams of steel than to change the temperature of an equal mass of water by the same amount.

> The specific heat capacity of a material is the quantity of heat needed to change a unit mass of the material by a unit amount in temperature (for example, to change 1 gram by 1 Celsius degree). It is a property of the material.

The specific heat capacity of a material is the relative amount of heat needed to raise its temperature. This number is determined experimentally for each substance. For example, the specific heat capacity of water is 1 cal/g·C°—that is, it takes 1 calorie of heat to raise the temperature of 1 gram of water 1°C. The *calorie* is a commonly used unit of heat. It is defined as the quantity of heat required to raise the temperature of 1 gram of water 1 Celsius degree. Likewise, if 1 calorie is removed from 1 gram of water, its temperature decreases 1°C. Water happens to have an unusually large specific heat capacity. Therefore, we use water as a reference material for measuring the specific heat capacities of other materials, a few of which are listed in table 10.1.

The specific heat capacity of steel is approximately 0.11 cal/g·C°, much less than water. Dropping 100 grams of room-temperature steel shot into our cup of coffee is much less effective in cooling the coffee than pouring 100 grams of room-temperature water into the cup (fig. 10.8). The steel absorbs less heat from the coffee for each degree of temperature that it changes than the water does.

When the cold steel is dropped into the hot water, heat flows from the water into the steel shot. From the definition of specific heat capacity, we can find how much heat must be absorbed by the steel to change its temperature by a given amount. Since specific heat capacity is the amount of heat per unit mass per unit temperature change, the total heat required is

$$Q = mc\Delta T,$$

where Q is the standard symbol for a quantity of heat, m is the mass, c is the symbol for specific heat capacity, and ΔT is the change in temperature. Because the specific heat capacity c is much larger for water than for steel, a larger quantity of heat is needed to warm 100 grams of water to a given temperature than 100 grams of steel. Since this heat comes from the hot coffee, water is more effective than steel in cooling the coffee.

The large specific heat capacity of water is partly responsible for its moderating effect on temperatures near the coast of an ocean or large lake. A large body of water, with its large specific heat capacity, requires the addition or removal of a large quantity of heat to change its temperature. The body of water will have a moderating influence on the air temperature in its vicinity. Nights will be warmer and days cooler than at some distance inland, because it is difficult to change the temperature of the water.

What is the distinction between heat and temperature?

When two objects at different temperatures are placed in contact, heat will flow from the object with the higher temperature to the one with the lower temperature (fig 10.9).

table **10.1**	
Specific Heat Capacities of Some Common Substances	
Substance	Specific heat capacity in cal/g·C°
water	1.0
ice	0.49
steam	0.48
ethyl alcohol	0.58
glass	0.20
granite	0.19
steel	0.11
aluminum	0.215
lead	0.0305

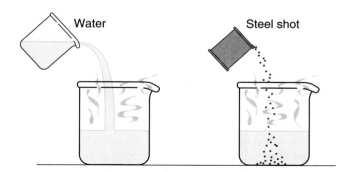

figure **10.8**   One-hundred grams of room-temperature water (20°C) is more effective than 100 grams of room-temperature steel shot in cooling a hot cup of water.

$T_1 > T_2$

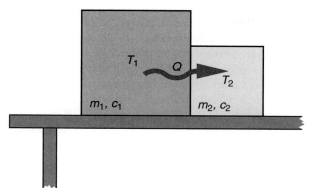

figure **10.9** Heat flows from the hotter object to the colder one when two objects at different temperature are placed in contact. The changes in temperature that result depend on the quantities of material and the specific heat capacity of each object.

Heat added increases the temperature, and heat removed decreases the temperature—heat and temperature are not the same. The quantity of heat added or removed to produce a given change in temperature depends on the amount of material involved (its mass) and the specific heat capacity of the material.

Temperature is a quantity that tells us which direction the heat will flow. If two objects are at the same temperature, no heat will flow. If they are at different temperatures, the direction of heat flow is from the higher to the lower temperature. The quantity of heat transferred depends on the temperature difference between the two materials as well as on their masses and specific heat capacities.

Heat and temperature are closely related concepts but play different roles in our explanation of heating and cooling processes:

Heat is energy that flows from one object to another when there is a difference in temperature between the objects.

Temperature is the quantity that indicates whether or not, and in which direction, heat will flow. Objects at the same temperature are in thermal equilibrium, and no heat will flow from one to the other.

How is heat involved in melting or freezing?

Can heat be added to or removed from a substance without changing the temperature at all? This does indeed occur when substances go through a **change of phase** or state. The melting of ice and the boiling of water are the most familiar examples of changes in phase. Whenever we cool a

drink using ice or cook an egg in boiling water we produce changes of phase. What happens?

Ice and liquid water are different *phases* of the same substance, water. If the temperature of water is cooled to 0°C, and we continue to remove heat, the water will freeze into ice. Likewise, if we warm ice to 0°C and continue to add heat, the ice will melt (fig. 10.10). As we add heat, the temperature of the ice and water remains at 0°C while the melting is taking place. Heat is being added, but no change in temperature occurs. Apparently, adding or removing heat produces changes other than just temperature changes.

Careful measurements have shown that it takes approximately 80 calories of heat to melt 1 gram of ice. This ratio, 80 cal/g, is referred to as the *latent heat of fusion* of water and is often denoted by the symbol L_f. **Latent heat** changes the *phase* of water without changing its temperature. Likewise, approximately 540 calories of heat is required to turn 1 gram of water at 100°C into steam. This ratio, 540 cal/g, is called the *latent heat of vaporization*, L_v. These values are valid only for water. Other substances have their own particular latent heats of fusion and vaporization.

What happens when we cool a glass of water by adding ice? At first, the ice and water are at different temperatures, the ice somewhere below 0°C and the water somewhere

figure **10.10** Adding heat to a mixture of water and ice at 0°C melts the ice without changing the temperature of the mixture.

above this temperature. Heat flows from the water to the ice until the ice reaches a temperature of 0°C. At this point, the ice begins to melt as heat continues to flow from the water to the ice. If enough ice is present, this flow of heat will proceed until the water also reaches a temperature of 0°C. Try This Box 10.1 illustrates these ideas.

If the glass were well insulated so that heat could not flow into the glass from the warmer surroundings, the ice and water mixture would remain at 0°C after both the water and the ice have reached this temperature, and no more ice would melt. Usually, though, heat flows into the system from the surroundings, and the ice continues to melt more slowly. Once the ice is gone, the drink will begin to warm and will eventually reach room temperature as heat flows in from the surrounding air.

Once it reaches thermal equilibrium, an ice-water mixture, well stirred, remains at 0°C. Small quantities of heat may flow into or out of the mixture from the surroundings, and small quantities of ice may melt or freeze, but the temperature will not change. This is one reason why the temperature of 0°C is useful as a reference point for our temperature scales. It is a stable and reproducible temperature.

When we cook food by boiling, we take advantage of the fact that water boils at 100°C—and that we can continue to add heat *without* changing the temperature. Adding heat from the burner causes the liquid water to change to water vapor while the temperature remains at 100°C. Since the temperature is constant, the cooking time required

to boil an egg or a potato is also fairly constant, depending on the size of the egg or potato.

Cooks should be aware, however, of the effects of altitude on the boiling temperature. In cities like Denver or Albuquerque, both located approximately a mile above sea level, water boils at a temperature of about 96°C because of the lower atmospheric pressure at this altitude. It takes longer to boil an egg or potato in these places than near sea level, where the boiling point is 100°C. If you do not make the adjustment in cooking time, your potatoes will be crunchy in the center.

Temperature changes can be caused by the flow of energy as heat from one object to another. The amount of heat required to produce a given change in temperature depends on the mass and specific heat capacity of the object as well as the temperature change. Temperature is a quantity that tells us when and in which direction heat will flow: heat flows from the hotter to the cooler body. When a substance changes phase, as from ice to water or water to steam, heat is added or removed without changing the temperature. The amount of heat needed per unit mass to produce a phase change is called the latent heat.

10.3 Joule's Experiment and the First Law of Thermodynamics

Can the temperature of an object be increased without placing the object in contact with a warmer object? What is going on when things get warmer from rubbing? These questions were a subject of scientific debate during the first half of the nineteenth century. Their resolution came about through the experimental work of James Prescott Joule (1818–1889), which was followed by the statement of the first law of thermodynamics around mid-century.

One of the first persons to raise these questions forcefully was the colorful American-born scientist and adventurer, Benjamin Thompson (1753–1814), also known as Count Rumford. Thompson, finding himself on the losing side of the American Revolutionary War, migrated to Europe, where he served the king of Bavaria as a consultant on weaponry. In Bavaria, he received the title Count Rumford. From his experience in supervising the boring of cannon barrels, Rumford knew firsthand that the barrels and drill bits became quite hot during drilling (fig. 10.11). He once demonstrated that water could be boiled by placing it in contact with the cannon barrel as it was being drilled.

What was the source of the heat that caused the increase in temperature? No hotter body was present that heat might flow from. Horses were used to turn the drill, but their temperature, although warm, was certainly below the boiling point of water. The horses performed mechanical work on the drilling engine. Was this work capable of generating heat?

---

try this box 10.1

Sample Exercise: Changing Ice to Water

If the specific heat capacity of ice is 0.5 cal/g·C°, how much heat would have to be added to 200 g of ice, initially at a temperature of −10°C, to:
   a. raise the ice to the melting point?
   b. completely melt the ice?

a. Heat required to raise the temperature:

$m = 200$ g $Q = mc\Delta T$
$c = 0.5$ cal/g·C° $= (200 \text{ g})(0.5 \text{ cal/g·C°})(10°C)$
$T = -10°C$ $= \textbf{1000 cal}$
$Q_{\text{raise}} = ?$

b. Heat required to melt the ice:

$L_f = 80$ cal/g $Q = mL_f$
$Q_{\text{melt}} = ?$ $= (200 \text{ g})(80 \text{ cal/g})$
                  $= \textbf{16 000 cal}$

The total heat required to raise the ice to 0°C *and* to melt it is

1000 cal + 16 000 cal = **17 000 cal** or 17 kcal.

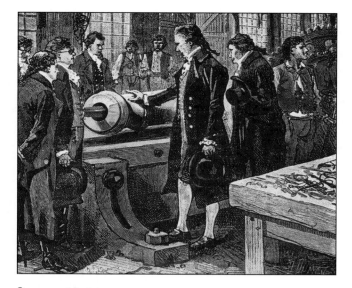

figure **10.11** Rumford's cannon-drilling apparatus. What is the source of heat that raises the temperature of the drill and the cannon barrel?

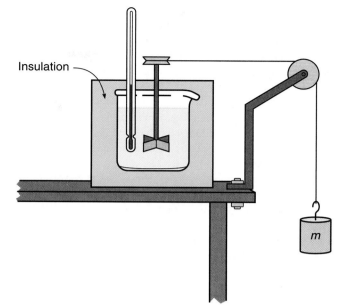

figure **10.12** A falling mass turns a paddle in an insulated beaker of water in this schematic representation of Joule's apparatus for measuring the temperature increase produced by doing mechanical work on a system.

What did Joule's experiments show?

Rumford's demonstration took place in 1798 and was discussed by many scientists during the early part of the nineteenth century. Quantitative research on Rumford's observations was not done until the 1840s, when Joule performed a famous series of experiments. Joule showed that various ways of performing mechanical work on a system had a consistent and predictable effect in raising the temperature of the system.

In the most dramatic of Joule's experiments, he turned a simple paddle wheel in an insulated beaker of water and measured the increase in temperature (fig. 10.12). A system of weights and pulleys turned the paddle wheel, transferring energy from the weights to the water. As the weights fell, losing gravitational potential energy, the paddle wheel did work against the resisting viscous forces of the water. This work was equal to the loss in potential energy of the weights.

A thermometer placed in the water measured the increase in temperature. Joule found that 4.19 J of work was required to raise the temperature of 1 gram of water by 1 C°. Since we can also raise the temperature of 1 gram of water 1 C° by adding 1 calorie of heat, this implies that 4.19 J of work is equivalent to 1 calorie of heat. (The use of the joule (J) as an energy unit did not come about until well after Joule's work. He stated his results in more archaic energy units. See chapter 6 for the definition of the joule.)

Joule used several ways of performing work in his experiments, as well as several kinds of systems. The results were always the same: 4.19 J of work produced the same temperature increase as 1 calorie of heat. The final state of the system provided no clue as to whether the temperature was increased by adding heat or by doing mechanical work.

The first law of thermodynamics

At the time of Joule's work, several people had already suggested that the transfer of heat was a transfer of kinetic energy between the atoms and molecules of the systems involved. Joule's experiments supported this view and led directly to the statement of the first law of thermodynamics by William Thomson (Lord Kelvin). The idea underlying the first law is that both work and heat represent transfers of energy into or out of a system.

If energy is added to a system either as work or as heat, the *internal energy* of the system increases accordingly. The change in internal energy is equal to the net amount of heat and work transferred into the system. An increase in temperature is one way an increase in internal energy might show up. The **first law of thermodynamics** summarizes these ideas:

> The increase in the internal energy of a system is equal to the amount of heat added to a system minus the amount of work done *by* the system.

In symbols, we often write the first law of thermodynamics as

$$\Delta U = Q - W,$$

where U represents the internal energy of the system, Q represents the amount of heat added to the system, and W represents the amount of work done *by* the system (fig. 10.13). The minus sign in this statement is a direct result of the convention of choosing work done *by* the system to be positive rather than work done *on* the system. Work done on the system adds energy to the system, but work done by the system on its surroundings removes energy from the system and reduces the internal energy. This sign convention is convenient for discussing heat engines, the main focus of chapter 11.

On its surface, the first law seems like a statement of conservation of energy, and indeed it is. Its apparent simplicity obscures the important insights that it contains. First and foremost among these insights is that heat flow is a transfer of energy, an idea strongly reinforced by Joule's experiments. Before 1850, the concept of energy had been restricted to mechanics. In fact, it was only then coming into its own in mechanics and did not play an important role in Newton's original theory. The first law of thermodynamics extended energy into new areas.

What is internal energy?

The first law also introduced the concept of the **internal energy** of a system. An increase in internal energy can show up in a variety of ways: one is an increase in temperature, as in Joule's experiment with the paddle wheel. A temperature increase is related to an increase in the average kinetic energy of the atoms or molecules making up the system.

A change in phase, such as melting or vaporization, is another way that an increase in the internal energy of a system affects the system. In a phase change, the average *potential* energy of the atoms and molecules is increased as the atoms and molecules are pulled farther away from each other. No temperature change occurs, but there is still an increase in internal energy.

An increase in internal energy, then, can show up as either an increase in the kinetic energy or potential energy (or both) of the atoms or molecules making up the system:

> The internal energy of the system is the sum of the kinetic and potential energies of the atoms and molecules making up the system.

Internal energy is a property of the system uniquely determined by the state of the system. If we know that the system is in a certain phase, and the temperature, pressure, and quantity of material are specified, there is only one possible value for the internal energy. The amount of heat or work that have been transferred to the system, on the other hand, are not uniquely determined by its state. As Joule's experiment showed, *either* heat or work could be transferred to the beaker of water to produce the same increase in temperature.

The system under study can be anything we wish it to be—a beaker of water, a steam engine, or an elephant. It is best, however, to consider a system that has distinct boundaries, so that we can easily define where it begins and ends. Choosing the system is similar to the process of isolating an object in mechanics to apply Newton's second law. In mechanics, forces define the interaction of the chosen object with other objects. In thermodynamics, the transfer of heat or work defines the interaction of a system with other systems or the surroundings.

Try This Box 10.2 applies the first law of thermodynamics. The system is a beaker containing water and ice (fig. 10.14). The internal energy of the system is increased by both heating on a hot plate and doing work by stirring. Notice that the work done by stirring is a negative quantity, since it represents work done *on* the system. The increase in internal energy results in the melting of the ice.

Counting food calories

A final note on energy units is worth mentioning. When we eat food, we add energy to our system by taking in material that can release potential energy in chemical reactions. We often measure that energy in a unit called a *Calorie*. This unit is *actually* a kilocalorie, or 1000 calories. The calorie counting that is done for food uses the unit:

$$1 \text{ Cal} = 1 \text{ kilocalorie} = 1000 \text{ cal}$$

The capital C used in the abbreviation indicates that this is a larger unit than the ordinary calorie.

Caloric values given for different types of food are the amount of energy these foods release when they are digested and metabolized. This energy may be stored in various ways within the body, including fat cells. Our bodies are constantly converting energy stored in muscles to other forms of energy. When we do physical work, for example,

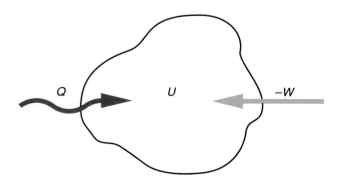

figure **10.13**   The internal energy U of a system is increased by the transfer of either heat or work into the system.

try this box 10.2

Sample Exercise: Applying the First Law of Thermodynamics

A hot plate is used to transfer 400 cal of heat to a beaker containing ice and water. 500 J of work are also done on the contents of the beaker by stirring.
  a. What is the increase in internal energy of the ice-water mixture?
  b. How much ice melts in this process?

a. We first convert the heat units to joules:

$Q = 400$ cal $Q = (400 \text{ cal})(4.19 \text{ J/cal})$

$W = -500$ J $= 1680$ J

$\Delta U = ?$ $\Delta U = Q - W$

                     $= 1680 \text{ J} - (-500 \text{ J})$

                     $= \mathbf{2180 \ J}$

b. $L_f = 80$ cal/g $\Delta U = mL_f$

    $= 335$ J/g

$m = ?$ $m = \dfrac{\Delta U}{L_f}$

                $= \dfrac{2180 \text{ J}}{335 \text{ J/g}}$

                $= \mathbf{6.5 \ g}$
                (amount of ice melted)

Since the internal energy was expressed in joules, we converted the latent heat to units of J/g by multiplying by 4.19 J/cal.

figure 10.14 The ice in the beaker can be melted either by adding heat from a hot plate or by doing work stirring the ice and water mixture.

we transfer mechanical energy to other systems and warm the surroundings by heat flow from our bodies. When you count calories, you are counting energy intake. If your output is less than your intake, you should expect to get larger.

Joule discovered that doing 4.19 joules of work on a system increased the temperature of the system by the same amount as adding 1 calorie of heat. This discovery convinced scientists that the flow of heat was a form of energy transfer and led to the statement of the first law of thermodynamics. The first law says that the change in the internal energy of a system is equal to the net amount of heat and work transferred into the system. The internal energy is the sum of the kinetic and potential energies of the atoms making up the system. These ideas extend the principle of conservation of energy.

10.4 Gas Behavior and the First Law

What happens when we compress air in a cylinder? How does a hot-air balloon work? The behavior of gases in these and other situations can be investigated using the first law of thermodynamics, with the help of some additional facts about the nature of gases. Our atmosphere is another interesting system for exploring the laws of thermodynamics.

What happens when we compress a gas?

Suppose that a gas is contained in a cylinder with a movable piston, as in figure 10.15. If the piston is pushed inward by an external force, work is done by the piston *on* the gas in the cylinder. What effect does this have on the pressure or temperature of the gas?

From the first law of thermodynamics, we know that doing work on a system adds energy to the system. By itself, work would increase the internal energy of the gas, but the change in internal energy also depends on the heat flow into or out of the gas. Knowing how much work is done on the gas provides only part of the picture.

From the definition of work (section 6.1), we know that the work done on the gas equals the force exerted by the piston times the distance that the piston moves. Since pressure is the force per unit area, the force exerted on the piston by the gas equals the pressure of the gas times the area of the piston ($F = PA$). If we move the piston without accelerating it, the net force acting on the piston is zero, and the external force applied to the piston equals the force exerted by the gas on the piston.

Putting these ideas together, the work done on the gas is the force times the distance d that the piston moves, or $W = Fd = (PA)d$. The motion of the piston produces a change in the volume of the gas, and this change in volume equals the area of the piston times the distance that it

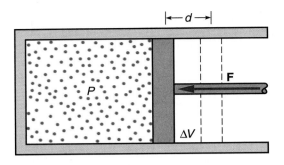

figure 10.15 A movable piston compresses a gas in a cylinder. The work done is $W = Fd = P\Delta V$.

moves, $\Delta V = Ad$ (see figure 10.15). The work done on the gas is therefore $W = P\Delta V$.

If the gas is being compressed, the volume of the gas is decreasing, and the change in volume ΔV will be negative, indicating that the work done on the gas also is negative. Work is negative when it is done on the system. From the first law, $\Delta U = Q - W$, negative work increases the energy of the system, since the minus signs cancel. We do work on the gas in compressing it, and that work increases the internal energy of the gas. If the gas is expanding, positive work is done by the gas on its surroundings, and the internal energy of the gas decreases.

Compressing a gas causes its internal energy to increase in proportion to the change in volume of the gas, provided that there is no heat flow across the boundaries of the container. What effect does this have on the gas? How does a change in internal energy relate to the properties of the gas?

How is internal energy related to temperature?

Many natural processes involving gases are **adiabatic,** meaning that no heat flows into or out of the gas during the process. If we compress a gas in a cylinder, for example, the compression can occur quickly enough that there is no time for a significant amount of heat to be transferred. The increase in internal energy is equal to the work done on the gas. What happens to the gas?

In general, internal energy is made up of both the kinetic and potential energies of the atoms and molecules in the system. In a gas, the internal energy is almost exclusively kinetic energy, because the molecules are so far apart (on average) that the potential energy of interaction between them is negligible. An **ideal gas** is one whose forces between atoms (and the associated potential energy) are small enough to be completely ignored. This is a good approximation for gases in many situations.

For an ideal gas, the internal energy of the gas is exclusively kinetic energy: increasing the internal energy increases the kinetic energy of the gas molecules. Absolute temperature is directly related to the average kinetic energy

of the molecules of a system. If the internal energy of an ideal gas increases, the temperature also increases in direct proportion. A plot of internal energy as a function of absolute temperature for an ideal gas looks like the graph in figure 10.16.

The temperature of a gas will therefore increase in an adiabatic compression. According to the first law of thermodynamics, the internal energy of the gas increases by an amount equal to the work done on the gas. The temperature of an ideal gas is directly related to its internal energy, so the temperature increases in direct proportion to the amount of work done in the compression.

The reverse is also true. If we allow the gas to expand against the piston, doing work on its surroundings, the internal energy of the gas decreases. The temperature of the gas decreases in an adiabatic expansion—this is how a refrigerator works. A pressurized gas is allowed to expand, lowering its temperature. The cool gas then circulates through coils inside the refrigerator, removing heat from the contents of the refrigerator.

How can we keep the temperature of a gas from changing?

Is it possible for the temperature of a gas to remain constant during a compression or expansion? In an **isothermal** process, the temperature does not change. (*Iso* means equal, and *thermal* refers to warmth or temperature.) Because of the direct relationship between temperature and internal energy in an ideal gas, the internal energy must be constant if there is no change in temperature. The change in internal energy, ΔU, is zero for an isothermal process in an ideal gas.

If ΔU is zero, the first law of thermodynamics ($\Delta U = Q - W$) requires that $W = Q$. In other words, if an amount of heat Q is added to the gas, an equal amount of work W

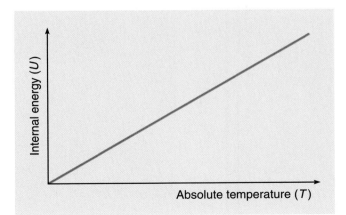

figure 10.16 The internal energy is plotted as a function of absolute temperature for an ideal gas. The internal energy increases in direct proportion to the temperature.

will be done by the gas on its surroundings if the temperature and the internal energy are to remain constant. It is possible to add heat to a gas without changing its temperature, as long as the gas does an equal amount of work on its surroundings by expanding.

Here again, the reverse is also true. If we compress a gas while holding the temperature constant, heat is removed from the gas, according to the first law of thermodynamics. For the internal energy and the temperature to remain constant, energy added to the gas in the form of work must result in an equal amount of energy in the form of heat being removed. Isothermal compressions and expansions are important processes in heat engines, which we will consider in chapter 11.

What happens to the gas in a hot-air balloon?

When the gas is heated in a hot-air balloon (fig. 10.17), the pressure, not the temperature, remains constant. The pressure of the gas inside the balloon cannot be significantly larger than the surrounding atmosphere. When pressure remains constant in a process, it is called **isobaric** (*baric* refers to pressure). The internal energy increases as the gas

figure **10.17**   Heat is added to the air in a hot-air balloon by the propane burner above the gondola. The heated air expands making it less dense than the surrounding air.

is heated, and so does the temperature. The gas also expands in this process, however, which removes some internal energy.

The gas expands in the isobaric heating process because of another property of ideal gases discovered near the beginning of the nineteenth century. A series of experiments showed that the pressure, volume, and absolute temperature of an ideal gas are related by the equation $PV = NkT$ where N is the number of molecules in the gas and k is a constant of nature called *Boltzmann's constant.* (Ludwig Boltzmann (1844–1906) was an Austrian physicist who developed a statistical theory of gas behavior.) This equation is called the **equation of state** of an ideal gas.

The temperature T used in the equation of state must be the absolute temperature. Recall that the concept of absolute temperature originally arose in the study of the behavior of gases (see section 10.1). The equation of state combines Boyle's law (section 9.2) with the effect of temperature on pressure discussed in the first section of this chapter. The equation of state shows that if the temperature increases while the pressure and number of molecules are held constant, the volume of the gas also must increase. In other words, the gas expands. Likewise, if the gas expands at constant pressure, the temperature must increase.

For an isobaric expansion, the first law of thermodynamics indicates that the heat added to the gas must be greater than the amount of work done by the gas in expanding. The internal energy, U, increases when the temperature increases. Since $\Delta U = Q - W$, Q must be larger than W (which is positive, since the gas is expanding) for the internal energy to increase. The first law of thermodynamics and the equation of state for an ideal gas can then be used to compute how much heat must be added to the system to produce a given amount of expansion.

Since the gas inside the balloon has expanded, the same number of molecules now occupy a larger volume, and the density of the gas has decreased. In other words, the balloon rises because the density of the gas inside the balloon is less than the density of the air outside the balloon. The buoyant force acting on the balloon is described by Archimedes' principle (section 9.3).

The same phenomenon occurs on a much larger scale in the atmosphere. A warm air mass will rise compared to cooler air masses for the same reason that a hot-air balloon rises. Air heated near the ground on a warm summer day expands, becomes less dense, and rises in the atmosphere. Rising air carries water vapor that provides the moisture for the formation of clouds at higher elevations. The rising columns of warm air also create air turbulence, which can become a problem for aircraft. These *thermals* are a real boon, however, to hang-glider enthusiasts and soaring eagles.

The behavior of gases and the implications of the first law of thermodynamics play important roles in our understanding of weather phenomena. Latent heat is involved in the evaporation and condensation of water as well as in the formation of ice crystals in snow or sleet. The source of energy for all of these processes is the sun: surface water and the atmosphere of the earth are a huge engine driven by the sun.

The first law of thermodynamics explains many processes involving gases. The absolute temperature of a gas is directly related to the internal energy of the gas: in the case of an ideal gas, the internal energy is the kinetic energy of the gas molecules. The amount of heat added and the work done by the gas determine its temperature, yielding different results for adiabatic, isothermal, and isobaric processes. The first law, together with the equation of state, can be used to predict the behavior of gases in heat engines, hot-air balloons, and air masses in the atmosphere.

10.5 The Flow of Heat

In most areas of the United States, we need to heat our homes and buildings in the winter. We warm our houses by burning wood or fossil fuels such as coal, oil, or natural gas in a furnace, or by electric heating. We pay for the fuel that we use and thus have an incentive to keep the heat it generates from escaping. How does heat flow away from a house or any other warm object? What mechanisms are involved, and how can we lessen the loss of heat by our awareness of these mechanisms?

Conduction, convection, and radiation are the three basic means of heat flow. All three are important in heat loss (or gain) from a house. An understanding of these mechanisms can help you achieve comfort while saving fuel in your house, residence-hall room, or workplace.

Heat flow by conduction

In **conduction,** heat flows *through* a material when objects at different temperatures are placed in contact with one another (fig 10.18). Heat flows directly from the hotter body to the cooler body at a rate that depends on the temperature difference between the bodies and a property of the materials called the **thermal conductivity.** Some materials have much larger thermal conductivities than others. Metals, for example, are much better heat conductors than wood or plastic.

Imagine that you pick up a block of metal in one hand and a block of wood in the other (fig. 10.19). The two blocks are at room temperature, but your body is typically 10 to 15 Celsius degrees warmer than room temperature, so heat will flow from your hands to the blocks. The metal block, however, will feel colder than the wood block. Since the two blocks are initially at the same temperature, the difference lies in the thermal conductivity, not in the temperature, of the two materials. Metal conducts heat more readily than wood does, so the heat flows more rapidly from your hand into the metal than into the wood. Since contact with

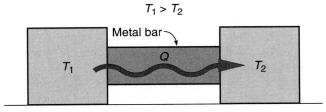

figure **10.18**   In thermal conduction, energy flows through a material when there is a temperature difference across the material.

figure **10.19**   When a metal block and a wooden block, both at room temperature, are picked up, the metal block feels cooler.

the metal cools your hand more rapidly than contact with the wood, the metal block feels colder.

Heat flow is a flow of energy, the actual transfer of kinetic energy of atoms and electrons. Because of the higher temperature of your hands, the average kinetic energy of the atoms in your hands is larger than the average kinetic energy of the atoms in the blocks. As the atoms bump into one another, kinetic energy is transferred, and the average kinetic energy of the atoms and electrons in the blocks increases at the expense of the atoms in your hands. Free electrons in metals play an important role in this process: metals are good conductors of both heat and electrical charge.

In discussing home insulation, we often use the concept of thermal resistance or *R values* to compare the effectiveness of different materials as insulators. The greater the conductivity, the lower the R value, since good thermal conductors are poor thermal insulators. The R value, in fact, relates both to the thickness of the material and to its thermal conductivity. The insulating effect of a material increases as its thickness increases, so R values increase with the thickness of the material.

Still air is a good thermal insulator. Porous materials with trapped pockets of air like rock wool or fiberglass insulation make excellent insulators and have high R values. Their low thermal conductivities result from the air pockets trapped in these materials rather than from the material itself. When we use such materials to insulate the walls, ceilings, and floors of a dwelling, we reduce their ability to conduct heat and lessen heat loss from the house.

What is convection?

If we heat a volume of air, then move the air by blowers or natural flow, we transfer heat by convection. In heating a house, we often move air, water, or steam through pipes or ducts to carry heat from a central furnace to different rooms. **Convection** transfers heat by the *motion of a fluid* containing thermal energy.

Convection, therefore, is the main way of heating a house. Warm air is less dense than cooler air, so it tends to rise from radiators or baseboard heaters toward the ceiling. As it does, the warm air sets up air currents within the room that distribute energy around the room. Warm air rises along the wall containing the heat source, and cooler air falls along the opposite wall (fig. 10.20).

Convection is also involved in heat loss from buildings. When warm air leaks out of a building or cooler air leaks in from outside, we say that we are losing heat to *infiltration,* a form of convection. Infiltration can be prevented, to some extent, by weatherstripping doors and windows and by sealing other cracks or holes. It is not desirable, however, to eliminate infiltration completely. A turnover of the air in a house is needed to keep the air fresh and free of odors.

What is radiation, and how does it transfer heat?

Radiation involves the flow of energy *by electromagnetic waves.* Wave motion and electromagnetic waves are discussed in chapters 15 and 16. The forms of electromagnetic waves primarily involved in the transfer of heat lie in the infrared portion of the electromagnetic wave spectrum. Infrared wavelengths are shorter than those of radio waves but longer than the wavelengths of visible light. (See fig. 16.5.)

While conduction requires a medium to travel through, and convection a medium to be carried along with, radiation can take place across a vacuum, such as the evacuated barrier in a thermos bottle (fig. 10.21). The radiation is reduced to a minimum by silvering the facing walls of the evacuated space. Silvering causes the electromagnetic waves to be reflected rather than absorbed and reduces the energy flow into or out of the container.

The same principle is involved in the use of foil-backed insulation in housing construction. Some of the heat flow across the dead-air space within a wall is due to radiation and

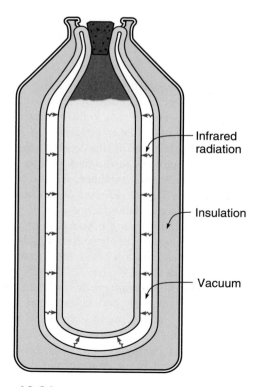

figure **10.21** Radiation is the only mechanism that can carry thermal energy across the evacuated space in a thermos bottle. Silvering the walls reduces the flow of radiation.

Infrared radiation

Insulation

Vacuum

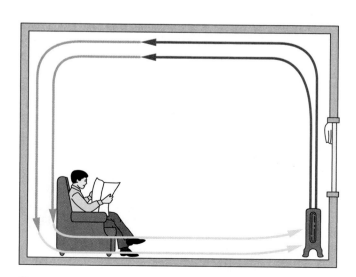

figure **10.20** In convection, thermal energy is transferred by the motion of a heated fluid. In a room, this occurs when heated air flows from a radiator or heater duct.

Everyday Phenomenon
box 10.1

Solar Collectors and the Greenhouse Effect

The Situation. Flat-plate solar collectors are often used for collecting solar energy to heat water or houses. The flat-plate collector consists of a metal plate with water-carrying tubes bonded to its surface. The plate and tubes are painted black and are held in a frame insulated below the plate, and with a glass or transparent plastic cover on top. How does the flat-plate collector work? What heat-flow processes are involved, and what relationship, if any, does the solar collector have to the much-discussed greenhouse effect?

The Analysis. The flat-plate collector receives energy from the sun as electromagnetic radiation, the only form of heat flow that can take place across the vacuum of space. Electromagnetic waves emitted by the sun are mainly in the visible part of the electromagnetic spectrum, the wavelengths to which our eyes are sensitive. These waves pass easily through the transparent cover plate of the collector.

The black surface of the collector plate absorbs most of the visible light that falls on it, reflecting very little. This absorbed energy raises the internal energy and temperature of the plate. Water traveling through the tubes bonded to the plate is heated by conduction. The plate must be at a higher temperature than the water for heat flow to take place. The insulation below the plate reduces heat flow to the surroundings.

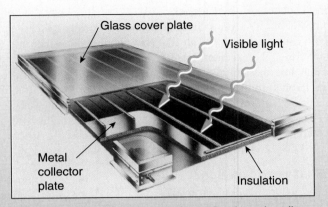

The black-surfaced metal collecting plate of a flat-plate solar collector is insulated below and covered with a glass or transparent plastic plate above.

Since it is warm, the metal plate will radiate heat in the form of infrared electromagnetic waves. Glass (or plastic) is opaque to infrared radiation, so these waves cannot pass through the glass cover. The glass cover also reduces heat loss from convection, since it prevents air currents from passing directly over the metal plate. The glass enclosure of a greenhouse serves the same purpose. Besides reducing convection

(continued)

some to conduction through the air. Using foil-backed insulation reduces heat flow from radiation. Thin sheets of foil-backed insulation may have *R* values equal to that of thicker sheets of insulation without the foil backing.

Radiation is also involved in the loss of energy from roofs or from road surfaces to the dark (and cold) night sky. Blacktopped road surfaces in particular and black surfaces in general are effective radiators of electromagnetic waves. The road surface radiates energy to the sky, cooling the surface more rapidly than the surrounding air. Even when the air temperature is still above freezing, the surface of a blacktopped road can drop below the freezing point, creating icy conditions.

Good absorbers of electromagnetic radiation are also good emitters: black materials both absorb more solar energy in the summer and lose more heat in the winter. A black roof may look nice, but it is not the best color for energy conservation. A lighter color is more effective both in keeping your home cool in the summer and warm in the winter.

Knowledge of the basic mechanisms of heat transfer is useful in designing a house. Conduction, convection, and

radiation are all important in understanding how heat flows into and out of a house, as well as how it circulates inside the house. Glass is a good conductor of heat, so large windows produce high heat loss, even when double-pane windows are used. If the window is on the south side of the house, however, this heat loss may be partially offset by the heat gain by radiation from the sun. Heat-flow mechanisms are also important in understanding how to make the best use of solar energy, discussed in Everyday Phenomenon Box 10.1.

The flow of heat can occur by three distinct means—conduction, convection, and radiation. In conduction, energy is transferred through a material: some materials are better thermal conductors than others. Convection involves transfer of heat by the flow of a fluid containing thermal energy. Radiation is the flow of energy through electromagnetic waves. All three of these processes are important in any application of the first law of thermodynamics, since the flow of heat is one way the internal energy of a system can change.

losses, the glass allows visible sunlight in while preventing heat loss from radiation at longer wavelengths. It is an energy trap. This one-way transmission of radiation at visible wavelengths, but not at longer infrared wavelengths, is the **greenhouse effect**. It is what causes your car to heat up on a sunny day when the windows are closed.

The earth also resembles a large greenhouse. Carbon dioxide gas, as well as certain other gases present in the atmosphere in smaller quantities, is transparent to visible radiation but not to infrared radiation. Carbon dioxide is a natural by-product of the burning of any carbon-based fuel such as oil, natural gas, coal, or wood. Carbon dioxide is taken up by plant life and converted to other carbon compounds, so that the global plant cover is also a factor in determining the amount of carbon dioxide in the atmosphere.

Our heavy use of fossil fuels such as oil, natural gas, and coal appears to be increasing the amount of carbon dioxide in the upper atmosphere. If this is so, and some measurements seem to support this conclusion, the greenhouse effect attributable to the carbon dioxide will let less heat escape by radiation, slowly increasing the earth's temperature. The loss of forests and other forms of plant life may also contribute to the carbon dioxide buildup.

If, indeed, the earth's temperature is slowly increasing, the polar ice caps will begin to melt. Thawing of the ice caps will

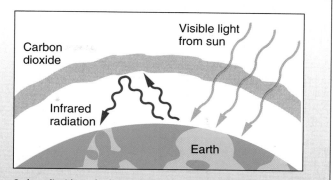

Carbon dioxide and other gases in the atmosphere of the earth play the same role as the cover plate on a solar collector. Visible light passes through, but infrared radiation does not.

cause the oceans to rise, and low-lying coastal regions may be flooded. Changes in global weather patterns could also affect crop production. The only solutions to this problem are to reduce our dependence on the burning of fossil fuels or somehow to increase the plant cover on the earth's surface. To develop rational environmental policies, we need to monitor the greenhouse effect and any other effects that can increase or decrease the earth's temperature.

summary

Heat, which plays an important role in the first law of thermodynamics, is the focus of this chapter. We used the first law of thermodynamics to explain changes in temperature or phase of a substance as well as the behavior of an ideal gas. The distinction between the concepts of heat and temperature is critical.

1   Temperature and its measurement.   Temperature scales have been devised to provide a consistent means of describing how hot or cold an object is. When objects are at the same temperature, no heat flows between them, and the physical properties remain constant. The idea of an absolute zero of temperature emerged from studies of gas behavior.

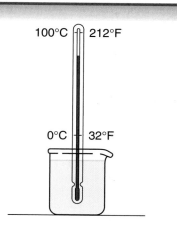

2   **Heat and specific heat capacity.**   The energy that flows between objects because they are at different temperatures is called *heat*. The amount of heat required to cause a one-degree change in temperature to one unit of mass is called the *specific heat capacity*. Latent heat is the amount of heat per unit mass required to change the phase of a substance without changing the temperature.

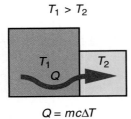

$$T_1 > T_2$$

$$Q = mc\Delta T$$

3   **Joule's experiment and the first law of thermodynamics.**   The first law of thermodynamics states that the internal energy of a substance or system can be increased either by adding heat or by doing work on the system. Increasing the internal energy increases the potential and kinetic energies of the atoms and molecules making up the system, and it may show up as an increase in temperature, a change of phase, or as changes in other properties of the system.

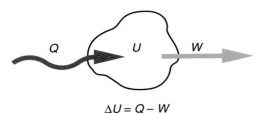

$$\Delta U = Q - W$$

4   **Gas behavior and the first law.**   The internal energy of an ideal gas equals the kinetic energy of the gas molecules. It varies directly with the absolute temperature of the gas. Work done by the gas is proportional to its change in volume, so the first law can be used to predict how the temperature or pressure will change when work is done or heat is added.

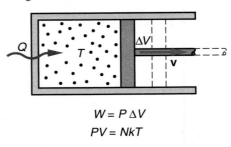

$$W = P\Delta V$$

$$PV = NkT$$

5   **The flow of heat.**   There are three basic mechanisms of heat flows from one system to another. Conduction is the direct flow of heat through materials. Convection is the flow of heat carried by a moving fluid. Radiation is the flow of heat by electromagnetic waves.

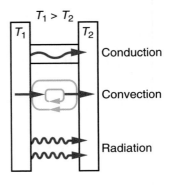

key terms

questions

*Questions identified with an asterisk are more open-ended than the others. They call for lengthier responses and are more suitable for group discussion.

Q1. Is an object that has a temperature of 0°C hotter than, colder than, or at the same temperature as one that has a temperature of 0°F? Explain.

Q2. Which spans a greater range in temperature, a change in temperature of 10 Fahrenheit degrees or a change of 10 Celsius degrees? Explain.

Q3. The volume of a gas held at constant pressure increases in a predictable way when its temperature increases. Could such a system be used as a thermometer? Explain.

*Q4. We sometimes attempt to determine whether another person has a fever by placing a hand on their forehead. Is this a reliable procedure? What assumptions do we make in this process?

Q5. Is it possible for a temperature to be lower than 0°C? Explain.

Q6. Is it possible for a temperature to be lower than 0 K on the Kelvin temperature scale? Explain.

Q7. Is an object with a temperature of 273.2 K hotter than, colder than, or at the same temperature as an object with a temperature of 0°C? Explain.

Q8. Two objects at different temperatures are placed in contact with one another but are insulated from the surroundings. Will the temperature of either object change? Explain.

Q9. Is it possible for the final temperature of the objects discussed in question 8 to be greater than the initial temperatures of both objects? Explain.

Q10. Two objects of the same mass, but made of different materials, are initially at the same temperature. Equal amounts of heat are added to each object. Will the final temperature of the two objects necessarily be the same? Explain.

Q11. Two cities, one near a large lake and the other in the desert, both reach the same high temperature during the day. Which city, if either, would you expect to cool down more rapidly once the sun has set? Explain.

Q12. Is it possible to add heat to a substance without changing its temperature? Explain.

Q13. What happens if we add heat to water that is at the temperature of 100°C? Does the temperature change? Explain.

Q14. What happens if we remove heat from water at 0°C? Does the temperature change? Explain.

Q15. Is it possible to change the temperature of a glass of water by stirring the water, even though the glass is insulated from its surroundings? Explain.

Q16. A hammer is used to pound a piece of soft metal into a new shape. If the metal is thermally insulated from its surroundings, will its temperature change due to the pounding? Explain.

Q17. Which represents the greater amount of energy, 1 J or 1 cal? Explain.

*Q18. Suppose that the internal energy of a system has been increased, resulting in an increase in the temperature of the system. Is it possible to tell from the final state of the system whether the change in internal energy was due to the addition of heat or of work to the system? Explain.

Q19. Is it possible for the internal energy of a system to be greater than the kinetic energy of the molecules and atoms making up the system? Explain.

*Q20. Based upon his experiments, Joule proposed that the water in a pool at the bottom of a waterfall should have a higher temperature than that at the top. Why might this be so? Explain.

Q21. An ideal gas is compressed without allowing any heat to flow into or out of the gas. Will the temperature of the gas increase, decrease, or remain the same in this process? Explain.

Q22. Is it possible to decrease the temperature of a gas without removing any heat from the gas? Explain.

Q23. Heat is added to an ideal gas, and the gas expands in the process. Is it possible for the temperature to remain constant in this situation? Explain.

Q24. Heat is added to an ideal gas maintained at constant volume. Is it possible for the temperature of the gas to remain constant in this process? Explain.

Q25. Heat is added to a hot-air balloon causing the air to expand. Will this increased volume of air cause the balloon to fall? Explain.

*Q26. Heat is added to ice causing the ice to melt but producing no change in temperature. Because water expands when it freezes into ice, the volume of the water obtained from melting the ice is less than the initial volume of ice. Does the internal energy of the ice-water system change in this process? Explain.

Q27. A block of wood and a block of metal have been sitting on a table for a long time. The block of metal feels colder to the touch than the block of wood. Does this mean that the metal is actually at a lower temperature than the wood? Explain.

Q28. Heat is sometimes lost from a house through cracks around windows and doors. What mechanism of heat transfer (conduction, convection, or radiation) is involved? Explain.

Q29. Is it possible for water on the surface of a road to freeze even though the temperature of the air just above the road is above 0°C? Explain.

Q30. What heat transfer mechanisms (conduction, convection, or radiation) are involved when heat flows through a glass windowpane? Explain.

Q31. Is it possible for heat to flow across a vacuum? Explain.

Q32. What property does glass share with carbon dioxide gas that makes them both effective in producing the greenhouse effect? Explain.

Q33. What mechanism of heat flow (conduction, convection, or radiation) is involved in transporting useful thermal energy away from a flat-plate solar collector? Explain.

Q34. Will a solar power plant (one that generates electricity from solar energy) have the same tendency to increase the greenhouse effect in the atmosphere as a coal-fired power plant? Explain.

exercises

E1. An object has a temperature of 45°C. What is its temperature in degrees Fahrenheit?

E2. The temperature on a winter day is 14°F. What is the temperature in degrees Celsius?

E3. The temperature in a residence-hall room is 24°C. What is the temperature of the room on the absolute (Kelvin) temperature scale?

E4. The temperature on a warm summer day is 86°F. What is this temperature
   a. in degrees Celsius?
   b. on the absolute (Kelvin) scale?

E5. The temperature of a beaker of water is 318.2 K. What is this temperature
   a. in degrees Celsius?
   b. in degrees Fahrenheit?

E6. How much heat is required to raise the temperature of 70 g of water from 20°C to 80°C?

E7. How much heat must be removed from a 200-g block of copper to lower its temperature from 150°C to 30°C? The specific heat capacity of copper is 0.093 cal/g·C°.

E8. How much heat must be added to 150 g of ice at 0°C to melt the ice completely?

E9. If 600 cal of heat are added to 50 g of water initially at a temperature of 10°C, what is the final temperature of the water?

E10. How much heat must be added to 120 g of water at an initial temperature of 60°C to
   a. heat it to the boiling point?
   b. completely convert the 100°C water to steam?

E11. If 200 cal of heat are added to a system, how much energy has been added in joules?

E12. If 600 J of heat are added to 50 g of water initially at 20°C,
   a. how much energy is this in calories?
   b. what is the final temperature of the water?

E13. While a gas does 300 J of work on its surroundings, 800 J of heat is added to the gas. What is the change in the internal energy of the gas?

E14. The volume of an ideal gas is increased from 1 m^3 to 3 m^3 while maintaining a constant pressure of 1000 Pa (1 Pa = 1 N/m^2).
   a. How much work is done by the gas in this expansion?
   b. If no heat has been added, what is the change in the internal energy of the gas?

E15. Work of 1500 J is done on an ideal gas, but the internal energy increases by only 800 J. What is the amount and direction of heat flow into or out of the system?

E16. If 500 cal of heat are added to a gas, and the gas expands doing 500 J of work on its surroundings, what is the change in the internal energy of the gas?

E17. Work of 600 J is done by stirring an insulated beaker containing 100 g of water.
   a. What is the change in the internal energy of the system?
   b. What is the change in the temperature of the water?

challenge problems

CP1. Heat is added to an object initially at 30°C, increasing its temperature to 80°C.
   a. What is the temperature change of the object in Fahrenheit degrees?
   b. What is the temperature change of the object in kelvins?
   c. Is there any difference in the numerical value of a heat capacity expressed in cal/g·C° from one expressed in cal/g·K? Explain.

CP2. A student constructs a thermometer and invents her own temperature scale with the ice point of water at 0°S (S for student) and the boiling point of water at 50°S. She measures the temperature of a beaker of water with her thermometer and finds it to be 15°S.
   a. What is the temperature of the water in degrees Celsius?
   b. What is the temperature of the water in degrees Fahrenheit?
   c. What is the temperature of the water in kelvins?
   d. Is the temperature range spanned by 1 Student degree larger or smaller than the range spanned by 1 Celsius degree? Explain.

CP3. The initial temperature of 150 g of ice is $-20°C$. The specific heat capacity of ice is 0.5 cal/g·C° and water's is 1 cal/g·C°. The latent heat of fusion of water is 80 cal/g.
   a. How much heat is required to raise the ice to 0°C and completely melt the ice?
   b. How much additional heat is required to heat the water (obtained by melting the ice) to 25°C?
   c. What is the total heat that must be added to convert the 80 g of ice at $-20°C$ to water at $+25°C$?
   d. Can we find this total heat simply by computing how much heat is required to melt the ice and adding the amount of heat required to raise the temperature of 80 g of water by 45°C? Explain.

CP4. A 150-g quantity of a certain metal, initially at 120°C, is dropped into an insulated beaker containing 100 g of water at 20°C. The final temperature of the metal and water in the beaker is measured as 35°C. Assume that the heat capacity of the beaker can be ignored.
   a. How much heat has been transferred from the metal to the water?
   b. Given the temperature change and mass of the metal, what is the specific heat capacity of the metal?
   c. If the final temperature of the water and this metal is 70°C instead of 35°C, what quantity of this metal (initially at 120°C) was dropped into the insulated beaker?

CP5. A beaker containing 400 g of water has 1200 J of work done on it by stirring and 200 cal of heat added to it from a hot plate.
   a. What is the change in the internal energy of the water in joules?
   b. What is the change in the internal energy of the water in calories?
   c. What is the temperature change of the water?
   d. Would your answers to the first three parts differ if there had been 200 J of work done and 1200 cal of heat added? Explain.

home experiments and observations

HE1. Look around your home to see what kinds of thermometers you can find. Indoor, outdoor, clinical, or cooking thermometers can be found in most homes. For each thermometer, note:
   a. What temperature scales are used?
   b. What is the temperature range of the thermometer?
   c. What is the smallest temperature change that can be read on each thermometer?
   d. What changing physical property is used to indicate change in temperature for each thermometer?

HE2. Take two Styrofoam cups, partially fill them with equal amounts of cold water, and drop equal amounts of ice into each cup to obtain a mixture of ice and water.
   a. Stir the water and ice mixture in one of the cups vigorously with a nonmetallic stirrer until all of the ice has been melted. Note the time that it takes to melt the ice.
   b. Set the second cup aside and observe it every 10 minutes or so until all the ice has melted. Note the time that it takes for the ice to melt.
   c. How do the times compare? Where is the energy coming from to melt the ice in each cup?

HE3. Fill a Styrofoam cup with very hot water (or coffee). Take objects made of different materials such as a metal spoon, a wooden pencil, a plastic pen, or a glass rod. Put one end of each object into the water and make these observations:
   a. How close to the surface of the water do you have to hold the object for it to feel noticeably warm to the touch?
   b. From your observations, which material would you judge to be the best conductor of heat? Which is the worst? Can you rank the materials according to their ability to conduct heat?

HE4. Conduct an energy survey of your residence-hall room or other similar living space. Note:
   a. What is the heat source (or sources) for the room and how does this heat flow into the room? What heat-flow mechanisms are involved?
   b. How is heat lost from the room? What heat-flow mechanisms are involved?
   c. What could be done to reduce heat loss or to otherwise improve the energy efficiency of the room?

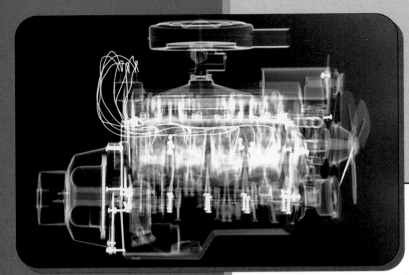

Heat Engines and the Second Law of Thermodynamics

chapter

11

chapter overview

After a discussion of heat engines, we will explore the second law of thermodynamics. The second law is central to understanding the efficiency of heat engines. Using both the first and second laws of thermodynamics, we then consider the use of energy resources in today's global economy. These issues have fundamental significance to the quality of our environment as well as to the health of our economy. The laws of thermodynamics are critical to making intelligent choices about energy policy.

chapter outline

1 **Heat engines.** What is a heat engine? What does the first law of thermodynamics tell us about how heat engines work?

2 **The second law of thermodynamics.** How would an ideal heat engine operate if we could build one? How is the concept of an ideal engine related to the second law of thermodynamics—what does the second law say and mean?

3 **Refrigerators, heat pumps, and entropy.** What does a refrigerator or heat pump do? What is entropy? How is it related to limitations on the use of heat energy?

4 **Thermal power plants and energy resources.** How can we generate power efficiently? What are the implications of the second law of thermodynamics in our use of energy resources such as fossil fuels and solar energy?

5 **Perpetual motion and energy frauds.** Is perpetual motion possible, according to the laws of thermodynamics? How can we judge the claims of inventors?

Many of us spend a good portion of our lives driving or riding in automobiles. We talk about how many miles per gallon we get or discuss the merits of different kinds of engines. We regularly drive into gas stations to buy unleaded gasoline, and we generally have a good idea of what this fuel costs at different places (fig. 11.1).

How many of us really understand what is going on inside that engine, though? As we turn the ignition switch and depress the gas pedal, the engine roars into action, consumes fuel, and powers the car at our command, without requiring the driver to have a detailed knowledge of its principles of operation. Ignorance may be bliss—until something goes wrong—and then some understanding of the engine may be useful. What does an automobile engine do, and how does it work?

The internal combustion engine used in most modern automobiles is a **heat engine**. The science of thermodynamics arose in an attempt to better understand the principles of operation of steam engines, the first practical versions of heat engines. Building more efficient engines was the primary objective. Although steam engines are usually external-combustion engines, whose fuel is burned outside of the engine rather than inside, the basic principles of operation of steam engines and automobile engines are the same.

figure **11.1**   Filling up the family chariot. How does the engine use this fuel to produce motion?

How do heat engines work? What factors determine the efficiency of a heat engine, and how can efficiency be maximized? The second law of thermodynamics plays a central role. Heat engines will lead us to an exploration of the second law and the related concept of entropy.

11.1 Heat Engines

Think for a moment about what the internal-combustion engine in your car does. We have said that both steam engines and gasoline engines are heat engines. What does this mean? We know that fuel is burned in the engine and that work is done to move the car. Somehow, work is generated from heat released in burning the fuel. Can we develop a model that describes the basic features of all heat engines?

What does a heat engine do?

Let's sketch a description of what the gasoline engine in your car does: Fuel in the form of gasoline is mixed with air and introduced into a can-shaped chamber in the engine called a *cylinder*. A spark produced by a spark plug ignites the gas-air mixture, which burns rapidly (fig. 11.2). Heat is released from the fuel as it burns, which causes the gases in the cylinder to expand, doing work on the piston.

Through mechanical connections, work done on the pistons is transferred to the drive shaft and to the wheels of the car. The wheels push against the road surface, and by Newton's third law, the road exerts a force on the tires that does the work to move the car.

Not all of the heat obtained from burning fuel is converted to work done in moving the car. The exhaust gases

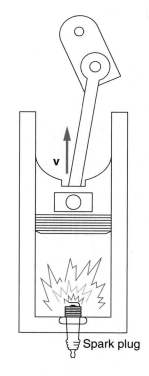

figure **11.2**   Heat released by burning gasoline in the cylinder of an automobile engine causes the piston to move, converting some of the heat to work.

emerging from the tailpipe of your car are hot, so some heat is released into the environment. Unused heat is a general feature of the operation of a heat engine.

This general description shows you the features common to all heat engines: Thermal energy (heat) is introduced into the engine. Some of this energy is converted to mechanical work. Some heat is released into the environment at a temperature lower than the input temperature. Figure 11.3 presents these ideas schematically. The circle represents the heat engine. The box at the top is the high-temperature source of heat, and the box at the bottom is the lower-temperature environment into which the *waste heat* is released.

Efficiency of a heat engine

How much useful mechanical work can an engine produce for a given energy input of heat? It is important to know how productive or efficient an engine is. **Efficiency** is defined as the *ratio* of the net work done by the engine to the amount of heat that must be supplied to accomplish this work. In symbols,

$$e = \frac{W}{Q_H},$$

where e is the efficiency, W is the net work done by the engine, and Q_H is the quantity of heat taken in by the engine

try this box 11.1

Sample Exercise: How Efficient Is This Engine?

A heat engine takes in 1200 J of heat from the high-temperature heat source in each cycle and does 400 J of work in each cycle.
  a. What is the efficiency of this engine?
  b. How much heat is released into the environment in each cycle?

a. $Q_H = 1200$ J $e = \dfrac{W}{Q_H}$
   $W = 400$ J
   $e = ?$ $= \dfrac{400 \text{ J}}{1200 \text{ J}}$

                     $= \dfrac{1}{3} = 0.33$

                     $= \mathbf{33\%}$

b. $Q_C = ?$ $W = Q_H - Q_C$
                  so $Q_C = Q_H - W$
                     $= 1200 \text{ J} - 400 \text{ J}$
                     $= \mathbf{800 \text{ J}}$

from the high-temperature source or *heat reservoir.* The work W is positive, since it represents work done *by* the engine on the surroundings.

In Try This Box 11.1, a heat engine takes in 1200 J of heat from a high-temperature source and does 400 J of work in each cycle, resulting in an efficiency of $\frac{1}{3}$. We usually state efficiency as a decimal fraction, 0.33 or 33% in this case. An efficiency of 33% is greater than that of most automobile engines but less than that of the steam turbines powered by coal or oil used in many electric power plants.

In computing this efficiency, we used heat and work values for one *complete* cycle. An engine usually functions in cycles where the engine repeats the same process over and over. It is necessary to use a complete cycle or an average of several complete cycles to compute efficiency, because heat and work exchanges occur at different points within the cycle.

What does the first law of thermodynamics tell us about heat engines?

The first law of thermodynamics places some limits on what a heat engine can do. Since the engine returns to its initial state at the end of each cycle, its internal energy at the end of the cycle has the same value as at the beginning.

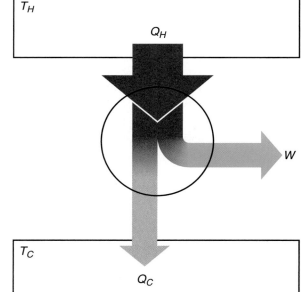

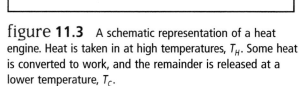

figure 11.3   A schematic representation of a heat engine. Heat is taken in at high temperatures, T_H. Some heat is converted to work, and the remainder is released at a lower temperature, T_C.

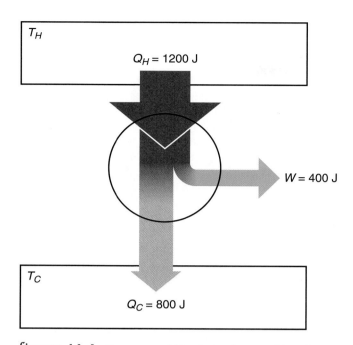

figure **11.4**   The arrow widths depict the quantities of energy in the sample exercise in Try This Box 11.1.

reservoir. The arrow representing the released heat is twice as wide as the arrow depicting the work, and the combined width of these two arrows equals the width of the arrow of the original heat input.

Automobile engines, diesel engines, jet engines, and the steam turbines used in power plants all are heat engines. You can create a simple steam turbine by placing a pinwheel in front of the spout of a tea kettle, as in figure 11.5. The input heat is supplied to the tea kettle by the stove or hot plate. Some of this heat is converted to work in turning the pinwheel. The rest of the heat is released into the room and does no work on the pinwheel, although it does warm the room slightly.

If a string or thread is attached to the shaft of the pinwheel, work done on the pinwheel could be used to lift a small weight. This pinwheel steam turbine would not be powerful or efficient. Designing better engines with both high power (the rate of doing work) and high efficiency (the fraction of input heat converted to work) has been the goal of scientists and mechanical engineers for the last 200 years. Everyday Phenomenon Box 11.1 discusses a recent development in this effort. Discovering the factors that have an impact on efficiency is an important part of this quest.

The change in the internal energy of the engine for one complete cycle is zero.

The first law states that the change in internal energy is the difference between the net heat added and the net work done by the engine ($\Delta U = Q - W$). Since the change in internal energy ΔU is zero for one complete cycle, the net heat Q flowing into or out of the engine per cycle must equal the work W done by the engine during the cycle. Energy is conserved.

Net heat is the difference between the heat input from the higher-temperature source Q_H and heat released to the lower-temperature environment Q_C. (The subscripts H and C stand for *hot* and *cold* here.) The first law of thermodynamics shows that the net work done by the heat engine is:

$$W = Q_H - Q_C,$$

since the work done equals the net heat.*

In the sample exercise in Try This Box 11.1, 800 J of heat are released to the environment. The width of the arrows in figure 11.4 corresponds to the quantities of heat and work involved in this example. Start with the wide arrow at the top: in each cycle, 1200 J of heat are taken in, 400 J ($\frac{1}{3}$ of the total) are converted to work, and 800 J of heat ($\frac{2}{3}$ of the total) are released to the lower-temperature

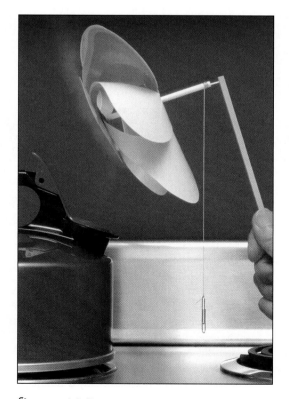

figure **11.5**   Steam issuing from the kettle makes the pinwheel turn in this simple steam turbine. Work could be done to lift a small weight with such an engine.

---

* We are using the absolute value of Q_C. Since Q_C flows *out* of the engine, it is often considered a negative quantity. The relationship is clearer if we make the minus sign explicit.

Everyday Phenomenon

box 11.1

Hybrid Automobile Engines

The Situation. Automobile companies have been working for years to develop better engines and drive systems. The goals are to achieve greater efficiency and lower exhaust emissions without excessive increases in cost. Federal and state environmental regulations provide incentive.

Electric cars have been one solution. Although electric motors produce essentially no exhaust emissions, electric cars suffer from limited range, weight issues associated with the storage batteries, and the need to recharge for several hours with relatively high-cost electric power. The introduction of hybrid systems that involve both an electric motor and a gasoline engine is a more recent development. How do these hybrid systems work? What are their advantages and disadvantages?

The Analysis. Although different arrangements are possible, most hybrid designs allow both the electric motor and the gasoline engine to turn the transmission, which transmits power to the drive wheels of the car. The gasoline engine is a heat engine, but the electric motor is not. The electric motor converts electrical energy stored in batteries to mechanical energy to drive the car. Either or both engines can power the car depending upon conditions. A sophisticated power-splitting transmission is needed to direct the energy flows.

In city-driving conditions, the electric motor is used to power the car when starting from a complete stop or accelerating at low speeds. This avoids the exhaust emissions associated with using a gasoline engine at speeds at which it is not very efficient. It also avoids the need for a large gasoline engine that

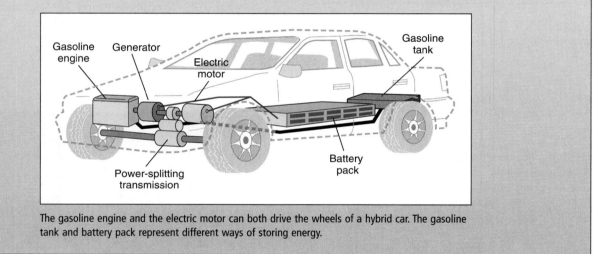

The gasoline engine and the electric motor can both drive the wheels of a hybrid car. The gasoline tank and battery pack represent different ways of storing energy.

(continued)

A heat engine takes in heat from a high-temperature source, converts part of this heat into work, and releases the remaining heat to the environment at a lower temperature. Gasoline engines in cars, jet engines, rocket engines, and steam turbines are all heat engines. The efficiency of an engine is the ratio of work done by the engine to the amount of heat taken in from the high-temperature source. Since the change in internal energy is zero for a complete cycle, the first law of thermodynamics requires that the work done in one cycle equal the net heat flow into and out of the engine.

11.2 The Second Law of Thermodynamics

If the efficiency of a typical automobile engine is less than 30%, we seem to be wasting a lot of energy. What is the best efficiency that we could achieve? What factors determine efficiency? These questions are as important today for automotive engineers or designers of modern power plants as they were for the early designers of steam engines.

What is a Carnot engine?

One of the earliest scientists to be intrigued by these questions was a young French engineer named Sadi Carnot

would otherwise be required for good acceleration. The gasoline engine can even be turned off in these situations.

In highway driving, the gasoline engine provides the primary power with the electric motor kicking in when extra acceleration is needed. At times when the full power of the gasoline engine is not needed to drive the car, it is also used to drive a generator to produce electricity to charge the batteries. (See chapter 14 for a discussion of electric motors and generators.) Electric power can also be generated by running the electric motor in reverse when the car is going downhill or braking for a stop. In this way we recapture and store energy in the batteries that would otherwise be lost as low-grade heat. The flow chart shows the different possible directions of energy flow in a hybrid vehicle.

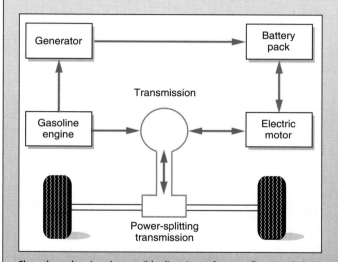

Flow chart showing the possible directions of energy flow in a hybrid vehicle. During braking, energy flows from the wheels to the battery pack via the electric motor.

The advantages of the hybrid system are:

1. Because we are not using the gasoline engine for the peak power needed for acceleration, we can get by with a small gasoline engine. The smaller the engine, the greater the fuel economy.
2. The batteries of the hybrid car are charged by the gasoline engine and by recapturing energy in the braking process. They do not need to be recharged overnight with high-cost electric power as the batteries for an all-electric car must be. Also, we do not need as large a battery pack for the hybrid vehicle as that required for decent range in an all-electric car.
3. Because it is used mainly for highway driving and running the generator, the gasoline engine can operate at its most efficient speed. This leads to lower exhaust emissions, which are greatest at low speeds and high accelerations.

The disadvantages of hybrid systems lie primarily in the extra cost of having both a gasoline engine and a relatively large electric motor, as well as a sophisticated transmission for directing the power from these two sources. (Standard gasoline-engine automobiles require an electric motor as the starter, but the electric motor used in hybrid cars must have higher power and greater size.) The hybrid car also has an expensive battery pack that ultimately will need to be replaced.

These disadvantages are reflected in higher cost for the hybrid vehicle than that of a standard gasoline vehicle of similar size and features. Despite this, hybrid cars have gained quick acceptance from people who are aware of the environmental benefits of better fuel economy and lower emissions. More stringent environmental regulations will make hybrid vehicles an even more attractive option.

(1796–1832). Carnot was inspired by the work of his father, also an engineer, who had studied and written about the design of water wheels, an important source of mechanical power at the time.

The workings of a water wheel provided Carnot with the basis for modeling an ideal heat engine. Carnot's father realized that bringing in all of the water at the highest point and releasing it at the lowest point maximized the efficiency of the water wheel. Carnot reasoned that the greatest efficiency of a heat engine would be obtained by taking in all of the input heat at a single high temperature and releasing all of the unused heat at a single low temperature.

This would maximize the effective temperature difference within the limits of the temperatures of the heat source and the environment.

Carnot imposed another requirement on his ideal engine: all of the processes had to occur without undue turbulence or departure from equilibrium. This requirement also paralleled his father's ideas on water wheels. In an ideal heat engine, the working fluid of the engine (steam, or whatever else might be used) should be roughly in equilibrium at all points in the cycle. This condition means that the engine is completely **reversible**—it can be turned around and run the other way at any point in the cycle, because it is always near equilibrium. This was Carnot's ideal—a real engine might depart considerably from these conditions.

When Carnot published his paper on the ideal heat engine in 1824, the energy aspects of heat and the first law of thermodynamics were not yet understood. Carnot pictured caloric flowing through his engine just as water flowed through a water wheel. It was not until after the development of the first law around 1850 that the full impact of Carnot's ideas became apparent. It then became clear that heat does not simply flow through the engine. Some of the heat is converted to mechanical work done by the engine.

What are the steps in a Carnot cycle?

The cycle devised by Carnot for an *ideal* heat engine is illustrated in figure 11.6. Imagine a gas or some other fluid contained in a cylinder with a movable piston. In step 1 of the cycle (the energy input), heat flows into the cylinder at a single high temperature T_H. The gas or fluid expands *isothermally* (at constant temperature) during this process and does work on the piston. In step 2, the fluid continues to expand, but no heat is allowed to flow between the cylinder and its surroundings. This expansion is *adiabatic* (with no heat flow).

Step 3 is an isothermal compression—work is done by the piston on the fluid to compress the fluid. This is the exhaust step, since heat Q_C flows out of the fluid at a single low temperature T_C during this compression. The final step, step 4, returns the fluid to its initial condition by an additional compression done adiabatically. All four steps must be done slowly so that the fluid is in approximate equilibrium at all times. The complete process is then *reversible,* a crucial feature of the **Carnot engine.**

When the fluid is expanding in steps 1 and 2, it is doing positive work on the piston that can be transmitted by mechanical links for another use. In steps 3 and 4, the fluid is being compressed, which requires that work be done on the engine by external forces. The work added in steps 3 and 4 is less than the amount done by the engine in steps 1 and 2, however, so that the engine does a net amount of work on the surroundings.

What is the efficiency of a Carnot engine?

The first law of thermodynamics made it possible to compute the efficiency of the *Carnot cycle* by assuming that the working fluid was an ideal gas. The process involves computing the work done on or by the gas in each step and the quantities of heat that must be added and removed in steps 1 and 3. Using the definition of efficiency from section 11.1, we obtain

$$e_c = \frac{T_H - T_C}{T_H},$$

where e_c is the **Carnot efficiency** and T_H and T_C are the *absolute* temperatures at which heat is taken in and released.

Try This Box 11.2 illustrates these ideas. For the temperatures given, a Carnot efficiency of approximately 42% is obtained. According to Carnot's ideas, this would be the *maximum efficiency possible* for any engine operating between these two temperatures. Any real engine operating between these same two temperatures has a somewhat lower efficiency because it is impossible to run a real engine in the completely reversible manner required for a Carnot engine.

The second law of thermodynamics

The absolute temperature scale was developed in the 1850s by Lord Kelvin in England. Kelvin was aware of Joule's experiments and had a hand in the statement of the first law of thermodynamics. He was in an excellent position to take a fresh look at Carnot's ideas on heat engines.

By combining Carnot's ideas with the new recognition that heat flow is a transfer of energy, Kelvin put forth a

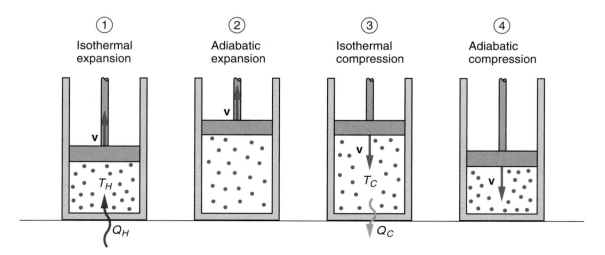

figure **11.6** The Carnot cycle. Step 1: isothermal expansion; step 2: adiabatic expansion; step 3: isothermal compression; step 4: adiabatic compression.

try this box 11.2

Sample Exercise: Carnot Efficiency

A steam turbine takes in steam at a temperature of 400°C and releases steam to the condenser at a temperature of 120°C.

    a. What is the Carnot efficiency for this engine?

    b. If the turbine takes in 500 kJ of heat in each cycle, what is the maximum amount of work that could be generated by the turbine in each cycle?

a. $T_H = 400°C$
$\qquad = 673\ K$

$T_C = 120°C$
$\qquad = 393\ K$

$e = ?$

$e_c = \dfrac{T_H - T_C}{T_H}$

$\qquad = \dfrac{673\ K - 393\ K}{673\ K}$

$\qquad = \dfrac{280\ K}{673\ K}$

$\qquad = \mathbf{0.416}\quad (41.6\%)$

b. $Q_H = 500\ kJ$

$W = ?$

$e = \dfrac{W}{Q_H}$, so $W = eQ_H$

$\qquad = (0.416)(500\ kJ)$

$\qquad = \mathbf{208\ kJ}$

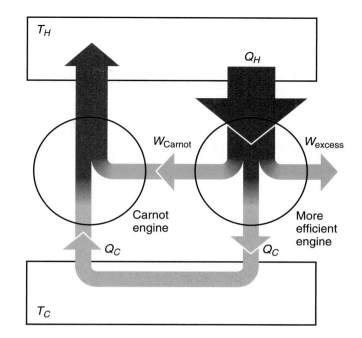

figure 11.7 A diagram showing that if some engine had a greater efficiency than a Carnot engine operating between the same two temperatures, the second law of thermodynamics would be violated.

general principle or law of nature. This principle, which we now call the **second law of thermodynamics,** is usually stated as

> No engine, working in a continuous cycle, can take heat from a reservoir at a single temperature and convert that heat completely to work.

In other words, it is not possible for any heat engine to have an efficiency of 1 or 100%.

Using the second law of thermodynamics, we can also show that no engine operating between two given temperatures can have a greater efficiency than a Carnot engine operating between the same two temperatures. This proof, which involves the reversible feature of the Carnot cycle, justifies Carnot's contention that the Carnot engine is the best that can be achieved. In fact, if some engine did have an efficiency greater than the Carnot efficiency, the second law of thermodynamics would be violated. This argument is illustrated in figure 11.7.

Here is how the argument goes. An engine with an efficiency greater than the Carnot engine would produce a greater amount of work than the Carnot engine for the same amount of heat input Q_H. Some of this work could be used to run the Carnot engine in reverse, returning the heat released by the first engine to the higher-temperature reser-

voir. To run the engine in reverse, we reverse the directions of the arrows without changing the quantities of heat and work.

The remaining work (W_{excess} in figure 11.7) would be available for external use, and no heat would end up in the lower-temperature reservoir. The two engines operating in tandem would take a small quantity of heat from the higher-temperature reservoir and convert it *completely* to work. This violates the second law of thermodynamics. Therefore, no engine can have a greater efficiency than a Carnot engine operating between these two reservoirs.

The Carnot efficiency is thus the maximum possible efficiency for any heat engine operating between these two temperatures. Since the proof follows from the second law of thermodynamics, the Carnot efficiency is sometimes referred to as the *second-law efficiency.*

The second law of thermodynamics cannot be proved. It is a law of nature, which, as near as we know, cannot be violated. It sets a limit on what can be achieved with heat energy. Time has shown it to be an accurate statement. The second law is consistent with what we know about heat transfer, heat engines, refrigerators, and many other phenomena.

Sadi Carnot developed the concept of a completely reversible, ideal heat engine. The Carnot engine takes in all of its heat at a single high temperature and releases the unused heat at a single low temperature; its efficiency depends on this temperature difference. The second law of thermodynamics as stated by Lord Kelvin says that no continuously operating engine can take in heat at a single temperature and convert that heat completely to work. Relying on the second law, we can show that no engine can have a greater efficiency than a Carnot engine operating between the same two temperatures.

11.3 Refrigerators, Heat Pumps, and Entropy

To a teenager, the automobile and the refrigerator just may be the two most important inventions. The gasoline engine in an automobile is a heat engine. What is a refrigerator? In section 11.2, we talked about running a heat engine in reverse, with the result that work was used to pump heat from a colder to a hotter reservoir. Is this what a refrigerator does? Is there a relationship between heat engines and refrigerators?

What do refrigerators and heat pumps do?

The term *refrigerator* requires little explanation. We are all familiar with refrigerators, even if we do not understand how they function. A refrigerator keeps food cold by pumping heat out of the cooler interior of the refrigerator into the warmer room (fig. 11.8). An electric motor or gas-powered engine does the necessary work. A refrigerator also warms the room, as you can tell by holding your hand near the coils on the back of the refrigerator when it is running.

Figure 11.9 shows a diagram of a heat engine run in reverse: It is the same diagram used for heat engines, but the directions of the arrows showing the flow of energy have been reversed. Work W is done on the engine, heat Q_C is removed from the lower-temperature reservoir, and a greater quantity of heat Q_H is released to the higher-temperature reservoir. A device that moves heat from a cooler reservoir to a warmer reservoir by means of work supplied from some external source is called a **heat pump** or a *refrigerator*.

The first law of thermodynamics requires that, for a complete cycle, the heat released at the higher temperature must equal the energy put into the engine in the form of both heat *and work*. As before, the engine returns to its initial condition at the end of each cycle—the internal energy of the engine does not change. More heat is released at the higher temperature than was taken in at the lower temperature (fig. 11.9).

While a refrigerator is a heat pump, the term *heat pump* usually refers to a device that heats a building by pumping heat from the colder outdoors to the warmer interior (fig. 11.10). An electric motor does the work needed to run the pump. The amount of heat energy available to heat the house is greater than the work supplied because Q_H is equal in magnitude to the sum of the work W and the heat Q_C removed from the outside air. We can get a larger amount of heat from the heat pump than by converting the electrical energy directly to heat, as is done in an electric furnace.

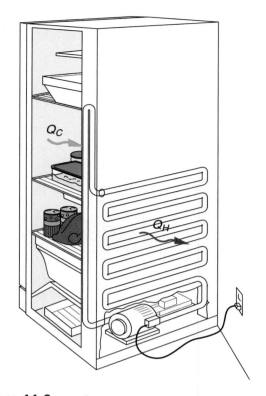

figure 11.8   A refrigerator pumps heat from the cooler interior of the refrigerator to the warmer room. The heat exchange coils that release heat to the room are usually on the back side of the refrigerator.

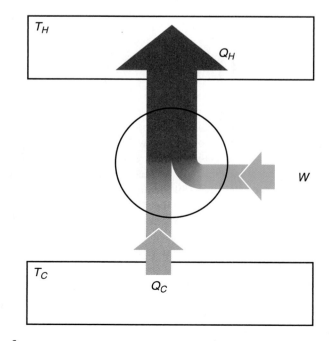

figure 11.9   A diagram of a heat engine run in reverse. In reverse, the heat engine becomes a heat pump or refrigerator.

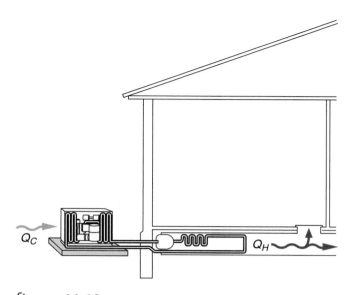

figure **11.10**   A heat pump removes heat from the outside air and pumps it into the warmer house.

figure **11.11**   The exterior heat-exchange coils for an air-to-air heat pump are usually located on a concrete pad behind the house or building.

Two sets of coils are used for heat exchange in the heat pump—one sits outside (fig. 11.11) and removes heat from the outside air, while the other releases heat to the air inside the building. Many heat pumps are designed to pump heat in either direction and can be used as air conditioners in the summer. Heat pumps are most effective for heating houses in climates where the difference between the outside temperature and inside temperature is not too large, such as the Southeast or the Pacific Northwest, where the winters are mild.

A heat pump can deliver a quantity of heat to the inside of the building that is often two to three times the amount of electric energy supplied as work. Again, the first law of thermodynamics is not violated, since the extra energy is being supplied from the outside air. The work applied to the heat pump allows us to move thermal energy in the direction opposite to its natural tendency, much as a water pump moves water uphill.

The Clausius statement of the second law of thermodynamics

Heat normally flows from hotter objects to colder objects. This natural tendency is the basis for another statement of the second law of thermodynamics. Often called the Clausius statement after its originator, Rudolf Clausius (1822–1888), it takes the form:

> Heat will not flow from a colder body to a hotter body unless some other process is also involved.

In the case of a heat pump, the other process is the work used to pump the heat against its usual direction of flow.

Although this statement of the second law sounds very different from the Kelvin statement, they both express the same fundamental law of nature. They both place limits on what can be done with heat, and the limits in each statement are equivalent. This can be shown by an argument similar to the one that confirmed that no engine can have a greater efficiency than a Carnot engine operating between the same two temperatures. Figure 11.12 illustrates the argument.

If it were possible for heat to flow from the colder to the hotter reservoir without any work, heat released by the heat engine on the right side of the diagram could flow back to the hotter reservoir. Some heat could then be removed from the hotter reservoir and converted *completely* to work. Heat would not be added to the cooler reservoir. This result violates the Kelvin statement of the second law of thermodynamics, which says that taking heat from a reservoir at a single temperature and converting it completely to work is impossible.

A violation of the Clausius statement of the second law is therefore a violation of the Kelvin statement. A similar argument will show that a violation of the Kelvin statement is a violation of the Clausius statement. (You may want to try your hand at developing this argument.) These two statements express the same fundamental law of nature in two different ways.

What is entropy?

Is there something inherent in heat that leads to the limitations described in the two statements of the second law of thermodynamics? Both statements of the second law describe processes that do not violate the first law of thermodynamics

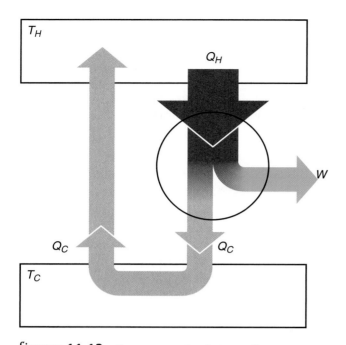

figure **11.12**   If we assume that heat can flow spontaneously from a colder to a hotter reservoir, the Kelvin statement of the second law of thermodynamics is violated.

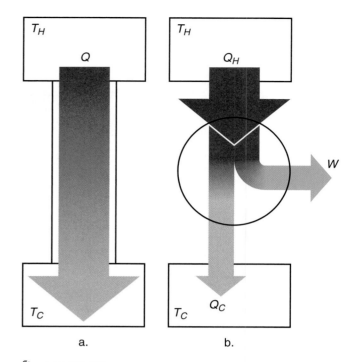

a.   b.

figure **11.13**   Heat can be removed from a high-temperature source either by direct flow through a conducting material or by being used to run a heat engine.

(conservation of energy)—but apparently are not possible. Certain things cannot be done with heat energy.

Suppose that we had a quantity of heat in a high-temperature reservoir (perhaps just a container of hot water). We could imagine two different ways to remove heat from the source. First, we could simply let the heat flow through a good heat conductor into the cooler environment, as shown in figure 11.13a.

Second, we could use this heat to run a heat engine and do some useful work (fig. 11.13b). If we were to use a Carnot engine, the process is completely reversible and we have obtained the maximum possible work from the available heat. The first process, heat simply flowing through a conductor to the lower-temperature reservoir, is *irreversible*. The system is not in equilibrium while this heat flow takes place, and the energy is not converted to useful work. In the irreversible process, *we lose some ability to do useful work*.

Entropy is the quantity that describes the extent of this loss. As entropy increases, we lose the ability to do work. Entropy is sometimes defined as a *measure of the disorder of the system*. The entropy of a system increases any time the disorder or randomness of the system increases. A system organized into two reservoirs at two distinct temperatures is, in this sense, more organized than having the energy all at a single intermediate temperature.

If we use the heat available in the hotter reservoir to run a completely reversible Carnot engine, there is no increase in the entropy of the system and its surroundings. We get the maximum useful work from the available energy. In iso-

lated systems (and in the universe as well), entropy remains constant in *reversible* processes but increases in *irreversible* processes where conditions are not in equilibrium. The entropy of a system decreases only if it interacts with some other system whose entropy is increased in the process. The entropy of the universe never decreases. This statement is yet another version of the second law of thermodynamics:

> The entropy of the universe or of an isolated system can only increase or remain constant. Its entropy cannot decrease.

The randomness of heat energy is responsible for the limitations in the second law of thermodynamics. The entropy of the universe would decrease if heat could flow by itself from a colder to hotter body—but the Clausius statement of the second law says that this cannot happen. Likewise, the entropy of the universe would decrease if heat at a single temperature could be converted completely to work, violating the Kelvin statement of the second law.

The thermal energy of a gas consists of the kinetic energy of the molecules. The velocities of these molecules are randomly directed, however, as in figure 11.14. Only some of them move in the proper direction to push the piston to produce work. If the molecules all moved in the same direction, we could convert their kinetic energy completely to work. This would be a lower-entropy (more organized) condition, but it does not represent the normal condition

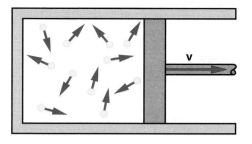

figure 11.14 The random directions of gas molecules prevent us from completely converting their kinetic energy to useful mechanical work.

for the thermal energy in a gas. The basic disorganization of heat energy is responsible for the limitations encompassed by the three statements of the second law.

Like internal energy, entropy can be computed for any state of a given system. The notion of disorder or randomness is at the heart of the matter. The state of my office or of a student's residence-hall room tends to become more disordered. This natural tendency of entropy to increase can only be countered by introducing energy in the form of work to straighten things up again.

A heat pump or refrigerator is a heat engine run in reverse: work is supplied to pump heat from a colder to a hotter body. The Clausius statement of the second law of thermodynamics, which is equivalent to the Kelvin statement, says that heat normally flows from a hotter to a colder body and cannot flow in the opposite direction without some other process being involved. The limitations on the use of heat expressed in the second law come about because of the disorganized nature of heat energy. Entropy is a measure of the disorder of a system. The entropy of the universe can only increase.

11.4 Thermal Power Plants and Energy Resources

We use electric power so routinely that most of us do not stop to think about where that energy comes from. Concerns about the greenhouse effect (see Everyday Phenomenon Box 10.1) and other environmental issues have recently pushed questions about how we generate electric power back into public debate. How is electric power generated in your area? If hydroelectric power is not a major contributor, most of your power comes from thermal power sources that use heat engines.

Thermodynamics plays an extremely important role in any discussion of the use of energy. What are the most efficient ways of using energy resources such as coal, oil, natural gas, nuclear energy, solar energy, or geothermal energy?

What bearing do the laws of thermodynamics and the efficiency of heat engines have on these questions?

How does a thermal power plant work?

The most common way of producing electric power in this country is a **thermal power plant** fueled by coal, oil, or natural gas (fossil fuels). The heart of such a plant is a heat engine. The fuel is burned to release heat that causes the temperature of the working fluid (usually water and steam) to increase. Hot steam is run through a turbine (fig. 11.15) that turns a shaft connected to an electric generator. Electricity is then transmitted through power lines to consumers (such as homes, offices, and factories).

Because the steam turbine is a heat engine, its efficiency is inherently limited by the second law of thermodynamics to the maximum given by the Carnot efficiency. The efficiency of a real heat engine is always less than this limit because it falls short of the ideal conditions of a Carnot engine. In any real engine, there are always irreversible processes taking place. A steam turbine generally comes closer to the ideal, however, than the internal-combustion engines used in automobiles. Rapid burning of gasoline-and-air mixtures in an automobile engine are highly turbulent and irreversible processes.

Since the maximum possible efficiency is dictated by the temperature difference between the hot and cold reservoirs, heating the steam to as high a temperature as the materials will permit is advantageous. The materials the boiler and turbine are made from set an upper limit on the temperatures that can be tolerated. If these materials begin to soften or melt, the equipment will obviously deteriorate. For most steam turbines, the upper temperature limit is around 600°C, well below the melting point of steel. In practice, most turbines operate at temperatures below this limit—550°C is typical.

If a steam turbine is operating between an input temperature of 600°C (873 K) and an exhaust temperature near the boiling point of water (100°C or 373 K) where the steam condenses to water, we can compute the maximum possible efficiency for this turbine:

$$e_c = \frac{T_H - T_C}{T_H}.$$

The difference between these two temperatures is 500 K. Dividing this by the input temperature of 873 K yields a Carnot efficiency of 0.57 or 57%. This is the ideal efficiency. The actual efficiency will be somewhat lower and usually runs between 40% and 50% for modern coal- or oil-fired power plants.

At best, we can convert about half of the thermal energy released in burning coal or oil to mechanical work or electrical energy. The rest must be released into the environment at temperatures too low for running heat engines or most other functions except space heating. The exhaust side of the turbines must be cooled to achieve maximum

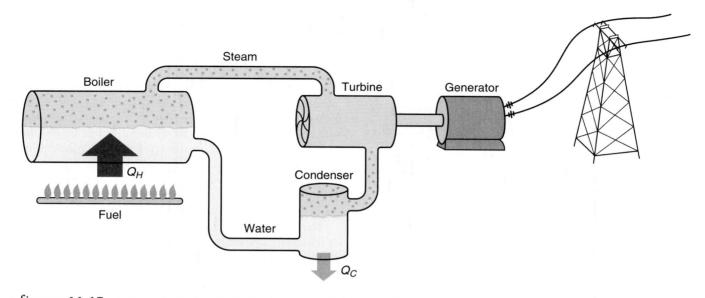

figure 11.15   A diagram showing the basic components of a thermal-electric power plant. Heat from the boiler generates steam, which turns the steam turbine. The turbine does work to generate electric power in the generator.

efficiency, and the cooling water is either returned to some body of water or run through cooling towers, where the heat is dissipated into the atmosphere (fig. 11.16). This is the *waste heat* often referred to in discussions of the environmental impact of power plants. If the waste heat is dumped into a river, the temperature of the river will rise, with undesirable effects on the fish and other wildlife.

Alternatives to fossil fuels

Nuclear power plants, discussed in chapter 19, also generate heat to run steam turbines. Because of the effects of radiation on materials, however, it is not feasible to run the turbines in a nuclear plant at temperatures as high as those in fossil-fuel plants. The thermal efficiencies for nuclear plants are somewhat lower, typically between 30% and 40%.

For equal amounts of energy generated, the amount of heat released into the environment is somewhat larger for a nuclear plant than for a fossil-fuel plant. On the other hand, nuclear power plants do *not* release carbon dioxide and other exhaust gases into the atmosphere and do not contribute to the greenhouse effect. Processing and disposal of nuclear wastes and concerns about nuclear accidents continue to be significant issues.

Heat is also available from other sources such as geothermal energy, which is heat that comes from the interior of the earth. Hot springs and geysers indicate the presence of hot water near the surface of the earth that might be used for power production, but usually the temperature of the water is not much greater than 200°C. In places where steam is available from geysers, low-temperature steam turbines can be run, as in northern California at the Geysers power plant (fig. 11.17).

figure 11.16   The cooling towers that are a common feature of many thermal power plants transfer heat from the cooling water to the atmosphere.

If the water temperature is below 200°C, steam turbines are not effective, and some other fluid with a lower boiling temperature than water is preferable for running a heat engine. Isobutane has been studied as a possible working fluid for low-temperature heat engines. The efficiency of such an engine would be quite low, however. If water were available at a temperature of 150°C (423 K), for example, and cooling water was available at 20°C (293 K), the Carnot

figure **11.17** The Geysers power plant in California uses geothermal energy to run steam turbines.

figure **11.18** At a solar-thermal power plant at Barstow, California, high temperatures are achieved from solar energy by focusing sunlight on a central boiler with an array of mirrors.

efficiency would be 31%. In practice, the efficiency will be even less, so it is common to find efficiencies of only 20% to 25% in proposed geothermal power plants.

Warm ocean currents are yet another source of heat. Power plants that take advantage of the temperature difference between warm water at the surface of an ocean and cooler water drawn from greater depths have been proposed and prototypes developed. Typically, the temperature difference is only 20°C or so, and the thermal efficiency is very low. For an input temperature of 25°C (298 K) and an output temperature of 5°C (278 K), the Carnot efficiency is only 6.7%. Although the efficiency is low, so is the cost of water warmed by the sun. It still may be economically feasible to produce power in this manner.

The sun is an energy source with enormous potential for development if the costs become competitive. The temperatures that can be achieved with solar power depend on the type of collection system used. The ordinary flat-plate collector only achieves relatively low temperatures of 50°C to 100°C, but concentrating collectors that use mirrors or lenses to focus the sunlight provide much higher temperatures. In a solar power plant near Barstow, California, an array of mirrors focuses sunlight on a boiler on a central tower (fig. 11.18). The temperatures generated are comparable to those of fossil-fuel plants, so similar steam turbines can be used.

High-grade and low-grade heat

The temperature makes a big difference in how much useful work can be extracted from heat. The second law of thermodynamics and the related Carnot efficiency define the limits. What effects do these factors have on our national energy policies and our day-to-day use of energy?

Clearly, heat at temperatures around 500°C or higher is much more useful for running heat engines and producing mechanical work or electrical energy than heat at lower temperatures. Heat at these high temperatures is sometimes

called **high-grade heat** because of its potential for producing work. Even then, only 50% or less of the heat can actually be converted to work.

Heat at lower temperatures can produce work but with considerably lower efficiency. Heat at temperatures around 100°C or lower is generally called **low-grade heat.** Low-grade heat is better used for purposes like heating homes or buildings (space heating). Space heating is the optimal use for heat collected by flat-plate solar collectors, and even from geothermal sources, if they are near enough to buildings that need heat. Geothermal heat is used for space heating in Klamath Falls, Oregon, as well as in certain other parts of the world where conditions are favorable.

Much low-grade heat, such as the low-temperature heat released from power plants, goes to waste simply because it is not economical to transport it to places where space heating may be needed. Nuclear power plants, for example, are not usually built in populated areas. There are other possible uses of low-grade heat, such as in agriculture or aquaculture, but we have not gone far in developing such uses.

The main advantage of electrical energy is that it can be easily transported through power lines to users far from the point of generation. Electrical energy can also be readily converted to mechanical work by electric motors. Electric motors operate at efficiencies of 90% or greater because they are *not* heat engines. The efficiency involved in producing the electrical power in the first place, of course, may have been considerably lower.

Electrical energy can also be converted back to heat if desired. Electricity is used for space heating in regions like the Pacific Northwest, where electric power has been relatively cheap because of the extensive hydroelectric resources on the Columbia River and its tributaries. The development of these resources has been subsidized by the federal government. Low prices encourage the use of a high-grade form of energy for purposes that might be just as well served by lower-grade sources.

Energy is still relatively cheap in this country and in many other parts of the world. Continued economic development and depletion of fossil-fuel resources will gradually change this picture. As scarcity occurs, questions of the optimal uses of energy resources will become critical. Wise decisions will depend on the participation of an informed, scientifically literate citizenry.

Thermal power plants use heat engines—steam turbines—to generate electric power. These engines are limited by the Carnot efficiency, which depends on input and output temperatures. The best that can be achieved is the 50% or so efficiency of fossil-fuel plants. The temperatures that can be obtained with other energy resources dictate the best way of using these resources. Lower temperatures are more suitable for space heating and similar uses than for power generation. The laws of thermodynamics set limits on power generation.

11.5 Perpetual Motion and Energy Frauds

The idea of perpetual motion has long fascinated inventors. The lure of inventing an engine that could run without fuel, or from some plentiful source like water, is like finding gold. If such an engine could be developed and patented, the inventor would become even richer than if he or she had discovered gold.

Is such an engine possible? The laws of thermodynamics impose some limits. Since these laws are consistent with everything that physicists know about energy and engines, any claim that violates the laws of thermodynamics should be suspect. We can analyze claims of perpetual motion or miracle engines by testing to see if they violate either the first or second laws of thermodynamics.

Perpetual-motion machines of the first kind

A proposed engine or machine that would violate the first law of thermodynamics is called a **perpetual-motion machine of the first kind.** Since the first law of thermodynamics involves conservation of energy, a perpetual-motion machine of the first kind is one that puts out more energy as work or heat than it takes in. If the machine or engine is operating in a continuous cycle, the internal energy must return to its initial value, and the energy output of the engine must equal its energy input, as we have already seen.

In figure 11.19, the total magnitude of the work and heat output is greater than the magnitude of the heat input (as represented by the width of the arrows). This could only happen if there was some source of energy, such as a battery, within the engine itself. If this were so, the energy

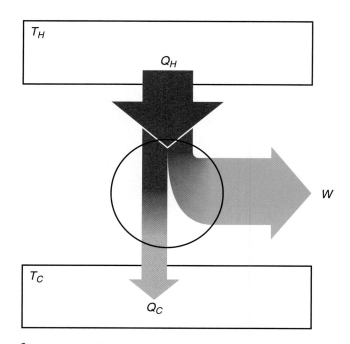

figure **11.19**   A perpetual-motion machine of the first kind. The energy output exceeds the input and therefore violates the first law of thermodynamics.

in the battery would gradually be depleted, and the internal energy of the engine would decrease. The engine could not run indefinitely.

The fact that physicists reject such an engine as impossible does not keep inventors from proposing them. From time to time, we see claims reported in newspapers and other popular media of engines that can run indefinitely fueled only by a gallon of water or a minuscule quantity of gasoline. Given the sometimes high price of gasoline, such claims have an obvious appeal and often attract investors and other interest. With your knowledge of physics, though, you could ask the inventor some simple questions: Where is the energy coming from? How can the machine put more energy out than went in? Hang on to your wallet—it looks like a poor investment.

What is a perpetual-motion machine of the second kind?

Engines that would violate the second law of thermodynamics without violating the first law often are a little more subtle. Their inventors have learned how to answer the questions in the previous paragraph. They have a source of energy—perhaps they plan to extract heat from the atmosphere or the ocean. These claims must be evaluated in terms of the second law of thermodynamics. If the second

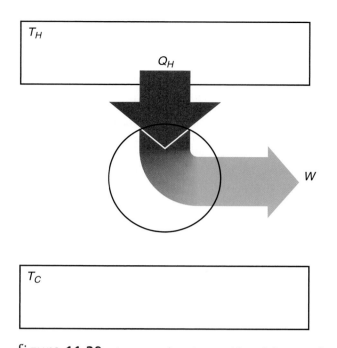

figure 11.20 A perpetual-motion machine of the second kind. Heat is extracted from a reservoir at a single temperature and converted completely to work, thus violating the second law of thermodynamics.

law is violated, the inventor has proposed a **perpetual-motion machine of the second kind** (fig. 11.20).

The second law states that it is not possible to take heat from a reservoir at a single temperature and convert it completely to work. We must have a lower-temperature reservoir available, and some heat must be released into that reservoir. In addition, even if we have a lower-temperature reservoir, the efficiency of any engine running between the two reservoirs will be quite low if the temperature difference between the reservoirs is not great. Any claim of efficiency greater than the Carnot efficiency (for the available temperature difference) also violates the second law of thermodynamics.

The laws of thermodynamics are useful in evaluating an inventor's claims, and they also guide our attempts to build better engines and to use energy more efficiently. Better engines can and will be developed, and their inventors may indeed be enriched in the process. A host of possibilities exist that involve relatively low-grade energy sources or other special circumstances. Engineers and scientists are also working on developing special materials and innovative designs for engines that might use even higher temperatures than are common in today's fossil-fuel plants. These efforts do not violate the laws of thermodynamics.

Most physics departments receive inquiries from local inventors seeking endorsement or help with their schemes for producing or using energy. The inventors are often sincere and sometimes quite well informed. Occasionally their ideas have merit, although they are often based on misunderstandings of the laws of thermodynamics (see Everyday Phenomenon Box 11.2). Unfortunately, it is often hard to persuade inventors that their ideas will not work, even when they involve clear violations of the laws of thermodynamics.

Other cases are more clearly the work of charlatans. Inventors who begin with good intentions sometimes discover that their ideas attract money from eager investors even when the invention fails to work. Some inventors have managed to raise several million dollars for the design and testing of prototype engines. Somehow, the engines are never quite finished, or tests are inconclusive and more money is needed to continue the work. The inventors, in the meantime, live quite well and find that promoting their inventions is more lucrative than actually building and testing them.

Investors beware. We know of no circumstances that violate the laws of thermodynamics. The failure of repeated attempts to violate them reinforces our belief in their validity. The laws cannot be proved, but their proven ability to describe experimental results successfully gives physicists confidence that they will continue to be useful indicators of what is possible.

A perpetual-motion machine of the first kind violates the first law of thermodynamics because more work is obtained than the energy input. A perpetual-motion machine of the second kind violates the second law of thermodynamics either by converting heat completely to work or by claiming an efficiency greater than the Carnot efficiency. Physicists do not believe that either of these options is possible, but inventors keep trying and investors keep wasting money on such schemes. The laws of thermodynamics guide our attempts to build better engines and place limits on what we can do.

Everyday Phenomenon

box 11.2

A Productive Pond

The Situation. A local farmer consulted with the author about an idea he had for generating electricity on his farm. The farmer had a pond that he thought could be used to run a water wheel, which could power an electric generator. He brought a sketch that looked something like the drawing here.

The sketch showed a drain pipe at the bottom of the pond. The farmer was aware that water would flow down such a drain with considerable velocity. His plan was to direct water through a pipe to the side of the pond and then up above the pond level, where it would flow onto a water wheel, powering the generator. The water would return to the pond after leaving the water wheel, so there would be no need to replenish the water supply, except to replace water lost to evaporation or seepage.

How would you advise the farmer? Will this plan work? Does it represent a perpetual-motion machine, and, if so, of what kind?

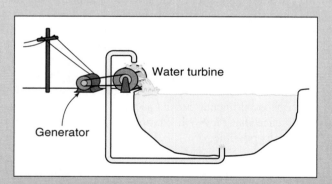

A sketch of the farmer's plan for obtaining electrical power from his pond. Which law of thermodynamics does he propose to violate?

The Analysis. Just by looking at the overall result, we see that work is being obtained to turn the generator without the input of any energy in the form of heat or work. Because the pond returns to its initial state (with the same internal energy), the first law of thermodynamics (or the principle of conservation of energy) is violated by the proposal. The farmer's plan is a perpetual-motion machine of the first kind.

If we look at the mechanics in more detail, we see that the water will indeed gain kinetic energy as it flows down the drain. This gain in kinetic energy comes at the expense of a loss in potential energy as the pond level is lowered. Directing the water upward causes it to regain potential energy at the expense of kinetic energy: it will slow down. If there are no losses due to friction, its velocity should reach zero at the point where it reaches the original level of the pond.

In an initial surge when the valve is opened, the water might overshoot the original pond level, but water cannot be raised above this level in a continuous process. Eventually the water in the vertical pipe will reach the same level as the pond, and no water will flow. The proposal will not work.

Although the author explained these ideas to the farmer carefully, and the farmer was an educated and intelligent person, the farmer still was not persuaded that his idea would not work. Because the farmer was not convinced by the theoretical arguments, the author encouraged him to make a small scale model before investing any money in plumbing his pond. Models or prototypes are a good way to test ideas and can sometimes be more convincing than theoretical arguments. (Whether or not the farmer actually tried out his plan was never reported to the author.)

summary

Attempts to build better steam engines led to a more general analysis of heat engines and, ultimately, to a statement of the second law of thermodynamics. The second law explains the workings of heat engines and refrigerators, and the concept of entropy. An understanding of the laws of thermodynamics is crucial to making wise energy choices and policy decisions.

1   Heat engines. A heat engine is any device that takes in heat from a high-temperature source and converts some of this heat to useful mechanical work. In the process, some heat is always released at lower temperatures into the environment. The efficiency of a heat engine is defined as the work output divided by the heat input.

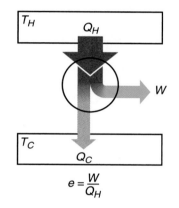

$$e = \frac{W}{Q_H}$$

2   The second law of thermodynamics. The Kelvin statement of the second law says that it is not possible for an engine, working in a continuous cycle, to take in heat at a single temperature and convert that heat completely to work. The maximum possible efficiency for any engine operating between two given temperatures is that of a Carnot engine. It is always less than 1 or 100%.

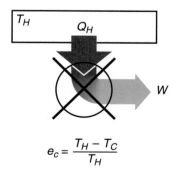

$$e_c = \frac{T_H - T_C}{T_H}$$

3   Refrigerators, heat pumps, and entropy. A heat pump or refrigerator is a heat engine run in reverse—it uses work to move heat from a low-temperature body to a hotter one. The Clausius statement of the second law says that heat will not flow spontaneously from a colder to a hotter body. Entropy can be

thought of as a measure of the disorder of a system. This disorder is responsible for the limitations expressed in the second law. Irreversible processes always increase the entropy of the universe.

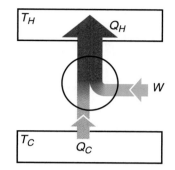

4   Thermal power plants and energy resources. Power plants that use heat generated from coal, oil, natural gas, nuclear fuels, geothermal sources, or the sun to produce electrical energy are all examples of thermal power plants. Their efficiency cannot be greater than the Carnot efficiency, so high input temperatures are desirable.

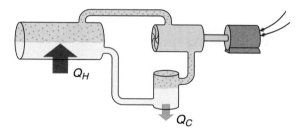

5   Perpetual motion and energy frauds. Any proposed device that would violate the first law of thermodynamics is called a perpetual-motion machine of the first kind. One that would violate the second law, but not the first, is called a perpetual-motion machine of the second kind. Neither will work.

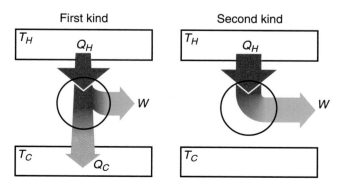

key terms

questions

*Questions identified with an asterisk are more open-ended than the others. They call for lengthier responses and are more suitable for group discussion.

Q1. Which of these types of motors or engines are heat engines?
   a. an automobile engine
   b. an electric motor
   c. a steam turbine
   Explain your reasons for classifying or not classifying each of these as a heat engine.

Q2. Could a simple machine such as a lever, a pulley system, or a hydraulic jack be considered a heat engine? Explain.

Q3. In applying the first law of thermodynamics to a heat engine, why is the change in the internal energy of the engine assumed to be zero? Explain.

Q4. Is the total amount of heat released by a heat engine to the low-temperature reservoir in one cycle ever greater than the amount of heat taken in from the high-temperature reservoir in one cycle? Explain.

Q5. From the perspective of the first law of thermodynamics, is it possible for a heat engine to have an efficiency greater than 1? Explain.

Q6. Which law of thermodynamics requires the work output of the engine to equal the difference in the quantities of heat taken in and released by the engine? Explain.

Q7. Is it possible for a heat engine to operate as shown in the following diagram? Explain, using the laws of thermodynamics.

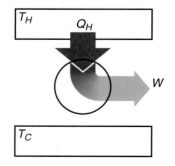

Q8. Is it possible for a heat engine to operate as shown in the following diagram? Explain, using the laws of thermodynamics.

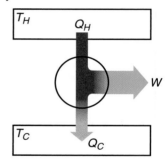

Q9. Is it possible for a heat engine to operate as shown in the following diagram? Explain, using the laws of thermodynamics.

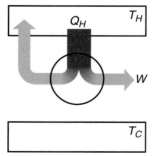

Q10. Is it possible for the efficiency of a heat engine to equal 1? Explain.

Q11. Can a Carnot engine operate in an irreversible manner? Explain.

Q12. Does an automobile engine operate in a reversible manner? Explain.

Q13. Which would have the greater efficiency—a Carnot engine operating between the temperatures of 400°C and 300°C, or one operating between the temperatures of 400 K and 300 K? Explain.

*Q14. If we want to increase the efficiency of a Carnot engine, would it be more effective to raise the temperature of the high-temperature reservoir by 50°C or lower the temperature of the low-temperature reservoir by 50°C? Explain.

Q15. Is a heat pump the same thing as a heat engine? Explain.

Q16. Is a heat pump essentially the same thing as a refrigerator? Explain.

Q17. When a heat pump is used to heat a building, where does the heat come from? Explain.

Q18. Is it possible to cool a closed room by leaving the door of a refrigerator open in the room? Explain.

Q19. Is it ever possible to move heat from a cooler to a warmer temperature? Explain.

Q20. Is it possible for a heat pump to operate as shown in the diagram? Explain, using the laws of thermodynamics.

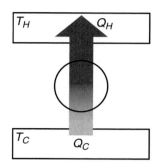

Q21. Is it possible for a heat pump to operate as shown in the diagram? Explain, using the laws of thermodynamics.

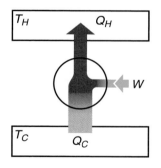

Q22. Is it possible for a heat pump to operate as shown in the diagram? Explain, using the laws of thermodynamics.

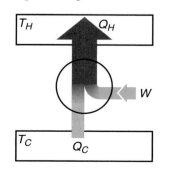

Q23. Which has the higher entropy, a deck of cards in which the cards are organized by suit or a shuffled deck of cards? Explain.

Q24. A hot cup of coffee is allowed to cool down, thus warming its surroundings. Does the entropy of the universe increase in this process? Explain.

*Q25. When a substance freezes, the molecules become more organized and the entropy decreases. Does this involve a violation of the entropy statement of the second law of thermodynamics? Explain.

Q26. Which would normally have the greater thermal efficiency, a coal-fired power plant or a geothermal power plant? Explain.

Q27. In what ways is a nuclear power plant similar to a coal-fired plant? Explain.

Q28. What is the distinction between high-grade heat and low-grade heat? Explain.

Q29. Is heat obtained from a flat-plate solar collector best used for running a heat engine or for space heating? Explain.

*Q30. Electric motors convert electric energy to mechanical work at a much higher efficiency than gasoline engines can convert heat to work. Why might it not make sense, then, to run all of our vehicles on electric power? Where does the electric power originate? Explain.

Q31. Is an automobile engine a perpetual-motion machine? Explain.

Q32. An engineer proposes a power plant that will extract heat from warm surface water in the ocean, convert some of it to work, and release the remaining heat into cooler water at greater depths. Is this a perpetual-motion machine and, if so, what kind? Explain.

Q33. An engineer proposes a device that will extract heat from the atmosphere, convert some of it to work, and release the remaining heat back into the atmosphere at the same temperature as the input heat. Is this a perpetual-motion machine and, if so, what kind? Explain.

exercises

E1. In one cycle, a heat engine takes in 1000 J of heat from a high-temperature reservoir, releases 600 J of heat to a lower-temperature reservoir, and does 400 J of work. What is its efficiency?

E2. A heat engine with an efficiency of 25% does 600 J of work in each cycle. How much heat must be supplied from the high-temperature source in each cycle?

E3. In one cycle, a heat engine takes in 900 J of heat from a high-temperature reservoir and releases 600 J of heat to a lower-temperature reservoir.
  a. How much work is done by the engine in each cycle?
  b. What is its efficiency?

E4. A heat engine with an efficiency of 40% takes in 600 J of heat from the high-temperature reservoir in each cycle.
  a. How much work does the engine do in each cycle?
  b. How much heat is released to the low-temperature reservoir?

E5. In one cycle, a heat engine does 400 J of work and releases 500 J of heat to a lower-temperature reservoir.
  a. How much heat does it take in from the higher-temperature reservoir?
  b. What is the efficiency of the engine?

E6. A Carnot engine takes in heat at a temperature of 650 K and releases heat to a reservoir at a temperature of 350 K. What is its efficiency?

E7. A Carnot engine takes in heat from a reservoir at 400°C and releases heat to a lower-temperature reservoir at 150°C. What is its efficiency?

E8. A Carnot engine operates between temperatures of 600 K and 400 K and does 200 J of work in each cycle.
  a. What is its efficiency?
  b. How much heat does it take in from the higher-temperature reservoir in each cycle?

E9. A heat pump takes in 300 J of heat from a low-temperature reservoir in each cycle and uses 150 J of work per cycle to move the heat to a higher-temperature reservoir. How much heat is released to the higher-temperature reservoir in each cycle?

E10. In each cycle of its operation, a refrigerator removes 18 J of heat from the inside of the refrigerator and releases 30 J of heat into the room. How much work per cycle is required to operate this refrigerator?

E11. A typical electric refrigerator has a power rating of 400 W, which is the rate (in J/s) at which electrical energy is supplied to do the work needed to remove heat from the refrigerator. If the refrigerator releases heat to the room at a rate of 900 W, at what rate (in watts) does it remove heat from the inside of the refrigerator?

E12. A typical nuclear power plant delivers heat from the reactor to the turbines at a temperature of 540°C. If the turbines release heat at a temperature of 200°C, what is the maximum possible efficiency of these turbines?

E13. An ocean thermal-energy power plant takes in warm surface water at a temperature of 22°C and releases heat at 10°C to cooler water drawn from deeper in the ocean. Is it possible for this power plant to operate at an efficiency of 8%? Justify your answer.

E14. An engineer designs a heat engine using flat-plate solar collectors. The collectors deliver heat at 70°C and the engine releases heat to the surroundings at 35°C. What is the maximum possible efficiency of this engine?

challenge problems

CP1. Suppose that a typical automobile engine operates at an efficiency of 25%. One gallon of gasoline releases approximately 150 MJ of heat when it is burned. (A megajoule, MJ, is a million joules.)
  a. Of the energy available in a gallon of gas, how much energy can be used to do useful work in moving the automobile and running its accessories?
  b. How much heat per gallon is released to the environment in the exhaust gases and via the radiator?
  c. If the car is moving at constant speed, how is the work output of the engine used?
  d. Would you expect the efficiency of the engine to be greater on a very hot day or on a cold day? Explain. (There may be competing effects at work.)

CP2. Suppose that a certain Carnot engine operates between the temperatures of 500°C and 150°C and produces 30 J of work in each complete cycle.
  a. What is the efficiency of this engine?
  b. How much heat does it take in from the 450°C reservoir in each cycle?
  c. How much heat is released to the 150°C reservoir in each cycle?
  d. What is the change, if any, in the internal energy of the engine in each cycle?

CP3. A Carnot engine operating in reverse as a heat pump moves heat from a cold reservoir at 5°C to a warmer one at 25°C.
a. What is the efficiency of a Carnot engine operating between these two temperatures?
b. If the Carnot heat pump releases 200 J of heat into the higher-temperature reservoir in each cycle, how much work must be provided in each cycle?
c. How much heat is removed from the 5°C reservoir in each cycle?
d. The performance of a refrigerator or heat pump is described by a "coefficient of performance" defined as $K = Q_c/W$. What is the coefficient of performance for our Carnot heat pump?
e. Are the temperatures used in this example appropriate to the application of a heat pump for home heating? Explain.

CP4. In section 11.3, we showed that a violation of the Clausius statement of the second law of thermodynamics is a violation of the Kelvin statement. Develop an argument to show that the reverse is also true: a violation of the Kelvin statement is a violation of the Clausius statement.

CP5. Suppose that an oil-fired power plant is designed to produce 100 MW (megawatts) of electrical power. The turbine operates between temperatures of 650°C and 240°C and has an efficiency that is 80% of the ideal Carnot efficiency for these temperatures.
a. What is the Carnot efficiency for these temperatures?
b. What is the efficiency of the actual oil-fired turbines?
c. How many kilowatt-hours (kW·h) of electrical energy does the plant generate in 1 h? (The kilowatt-hour is an energy unit equal to 1 kW of power multiplied by 1 h.)
d. How many kilowatt-hours of heat must be obtained from the oil in each hour?
e. If one barrel of oil yields 1700 kW·h of heat, how much oil is used by the plant each hour?

home experiments and observations

HE1. If there is a refrigerator handy in your residence hall or home, study the construction of the refrigerator to make these observations:
a. What energy source (gas or electricity) provides the work necessary to remove the heat? Can you find a power rating somewhere on the refrigerator?
b. Where are the heat-exchange coils that are used to release heat into the room? (They are usually on the back of the refrigerator, often near the bottom.)
c. Use a thermometer to determine the temperatures inside the refrigerator, inside the freezer compartment, near the heat-exchange coils outside, and in the room at some distance from the refrigerator. You should take measurements at different times to see how much these temperatures vary.

HE2. Find a local car dealer who sells hybrid automobiles. (Honda and Toyota are currently the best bets.) Collect literature describing the specifications for the hybrid vehicle as well as for other vehicles of similar size. (This information may also be obtained on the Internet.)
a. Compare the gasoline engine size, horsepower, and estimated gas mileage for the hybrid vehicle and the standard vehicle of similar size.
b. What are the voltage and current ratings for the batteries in hybrid and standard vehicles?
c. What tradeoffs do you see in purchasing the hybrid versus the standard vehicle of similar size?

HE3. What types of power plants are used to generate electricity in your area? Do they employ heat engines, and if so, what is the source of heat? Are there cooling towers or other cooling structures used to release waste heat into the atmosphere? (Your local utility should be able to provide such information.)

unit Three

Electricity and Magnetism

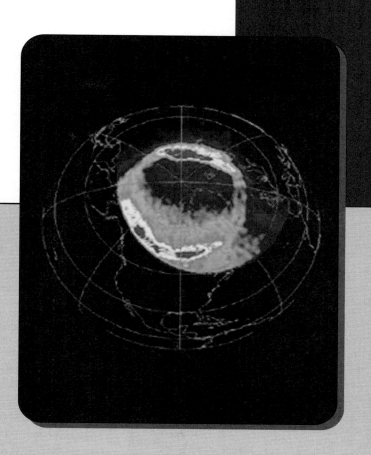

No area of physics has had a greater impact on the way we live than the study of electricity and magnetism. Using electricity and electronic devices has become second nature to us now but would have been hard to imagine 200 years ago. The invention and design of television sets, microwave ovens, laptop computers, and thousands of other familiar appliances and devices have all required an understanding of the basic principles of electricity and magnetism.

Although the effects of magnets and static electricity had been known for a long time, our basic knowledge of electricity and magnetism is mainly a product of the nineteenth century. A key invention at the turn of the century opened the door to these developments. In 1800, the Italian scientist Alessandro Volta (1745–1827) invented the battery. Volta's invention grew out of the work of an Italian physician, Luigi Galvani (1737–1798), who had discovered effects that he called *animal electricity*. Galvani found that he could produce electrical effects by probing a frog's leg with metal scalpels.

Volta discovered that the frog was not necessary. Two different kinds of metal separated by a suitable chemical solution were sufficient to produce many of the electrical effects observed by Galvani. Volta's *voltaic piles*, which consisted of alternating plates of copper and zinc separated by paper impregnated with a chemical solution, were capable of producing sustained electric currents. As often happens, this new device made many new experiments and investigations possible.

The invention of the battery led to the discovery of the magnetic effects of electric currents by Hans Christian Oersted (1777–1851) in 1820. Oersted's discovery made a formal connection between electricity and magnetism, leading to the modern term *electromagnetism*. Electromagnetism was the hot area of research for physicists during the 1820s and 1830s, and key advances were made by Ampère, Faraday, Ohm, and Weber. In 1865, the Scottish physicist James Clerk Maxwell (1831–1879) published a comprehensive theory of electric and magnetic fields that brought together the insights of many of these other scientists. Maxwell invented the concepts of electric and magnetic fields, ideas that proved to be tremendously productive.

Electromagnetism is still an active area of research. Its ramifications are important in radio and television, computers, communications, and other areas of technology. Despite this importance, the invisibility of the underlying phenomena makes the subject of electromagnetism seem abstract and mysterious to many people. The basic ideas are not difficult, though, and can be understood by carefully examining familiar phenomena.

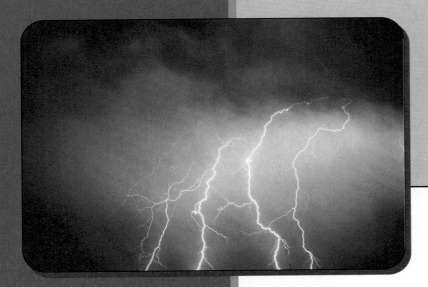

Electrostatic Phenomena

chapter 12

chapter overview

The main purpose of this chapter is to describe and explain the electrostatic force. We will also investigate concepts such as electric field and electric potential. These concepts will be made more vivid by describing and interpreting some simple experiments. Experimentation can lead to a better understanding of electric charge, the distinction between conductors and insulators, and many other ideas.

chapter outline

Most of us have had the experience of running a comb through our hair on a dry winter day and hearing a crackling sound or, if the room is dark enough, even seeing sparks. Depending on the length and looseness of the hair, strands of it may have stood on end, as in figure 12.1. This phenomenon is intriguing but also annoying. You often have to wet the comb to get your hair to behave.

What is happening? Some force seems to be at work, causing the individual hairs to repel one another. You would probably identify that force as an electrostatic force, or at least, you might say that static electricity is the cause of your unruly hair. To name a phenomenon is not the same as explaining it, though. Why does static electricity occur under these conditions, and what is the force in question?

Hair-and-comb fireworks are just one example of electrostatic phenomena that are part of our everyday experience. Why does a piece of plastic refuse to leave your hand after you peeled it off a package? Why do you get a slight shock after you walk across a carpeted floor and touch a light switch? The phenomena range from static cling and similar annoyances to dramatic displays of lightning in a big thunderstorm. They are familiar and yet, to many of us, mysterious.

figure **12.1**   Hair sometimes seems to have a mind of its own when combed on a dry winter day. What causes the hairs to repel one another?

Despite its familiarity, many of us are surprisingly unaware of what static electricity is. People are often put off by a fear of electric shock as well as by the abstract nature of the subject. It need not be so. Many of the phenomena can be explained with ideas accessible to all of us.

12.1 Effects of Electric Charge

What do the hair-combing example and the other phenomena mentioned in the chapter introduction have in common? The comb passing through your hair, your shoes passing over the rug, or plastic being peeled from a box all involve different materials rubbing against one another. Perhaps, this rubbing is the cause of the phenomena.

How can we test such an idea? One approach would be to collect different materials, rub them together in various combinations, and see what we generate in the way of sparks or other observable effects. Experiments like these could help to establish which combinations of materials are most effective in producing electrostatic crackles and pops. Of course, we would also need some consistent way of gauging the strength of these effects.

What can we learn from experiments with pith balls?

A common way of demonstrating electrostatic effects is to rub plastic or glass rods with different furs or fabrics. If we rub a plastic or hard rubber rod with a piece of cat fur, for example, we can see the hairs on the fur standing out and snapping at one another. Like the hair-combing fireworks, these effects are most striking on a dry winter day when little moisture is in the air. Another piece of equipment often used in these demonstrations is a small stand to which two pith balls are attached by threads (fig. 12.2).

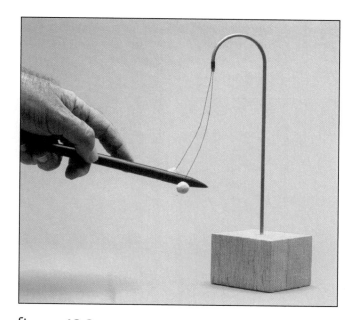

figure **12.2**   Pith balls suspended from a small stand are attracted to a charged plastic rod.

Pith balls are small wads of dry, paperlike material light enough to be strongly influenced by electrostatic forces.

An interesting sequence of events happens when a plastic rod, vigorously rubbed with cat fur on a dry day, is brought near the pith balls. At first, the pith balls are attracted to the rod like bits of iron to a magnet. After contacting the rod, and perhaps sticking to it for a few seconds, the pith balls dance away from the rod. They are repelled by the rod at this point and also by each other. The threads supporting the pith balls now form an angle to the vertical, as in figure 12.3, rather than hanging straight down.

How can we explain what has happened? A repulsive force must be acting between the two pith balls after they have been in contact with the rod. We might imagine that the balls have received something (call it **electric charge**) from the rod that is responsible for the force we observe. This charge, whatever it is, was somehow generated by rubbing the rod with the cat fur. The force that is exerted by one stationary charge on another is called the **electrostatic force.**

We can observe further changes if we rub a glass rod with a piece of synthetic fabric such as nylon. If the glass rod is brought near the pith balls that had been charged by the plastic rod, the pith balls are now attracted to the glass rod. They are still repelled by the plastic rod. If we allow the pith balls to touch the glass rod, they repeat the sequence of events that we observed earlier with the plastic rod. At first, they are attracted and stick briefly to the rod. Then, they dance away and are repelled by the rod and by each other. If we bring the plastic rod near, we now find that the balls are attracted to the plastic rod.

These observations complicate the picture somewhat. Apparently, there are at least two types of charge: one generated by rubbing a plastic rod with fur and another generated by rubbing a glass rod with nylon. Could there be more kinds of charge? Further experiments with different materials would indicate that these two types are all that we need to explain the effects. Other charged objects will cause either an attraction or repulsion with the two types of charge already identified and can be placed in one of these two categories.

What is an electroscope?

A simple *electroscope* consists of two leaves of thin metallic foil suspended face-to-face from a metal hook (fig. 12.4). At one time, gold foil was used, but aluminized mylar is now preferred. The leaves are connected by the hook to a metal ball that protrudes from the top of the instrument. The foil leaves are protected from air currents and other disturbances by a glass-walled container.

If the foil leaves are uncharged, they will hang straight down. If we bring a charged rod in contact with the metal ball on the top, however, the leaves immediately spread apart. They will remain spread apart even after the rod is taken away, presumably because the leaves are now charged.

If any charged object is now brought near the metal ball, the electroscope shows what type of charge is on the object and gives a rough indication of how much charge is present. If an object with the same type of charge as the original rod is brought near the ball, the leaves will spread farther apart. An object with the opposite charge will make the leaves come closer together. A larger charge produces a larger effect.

Although less dramatic than the pith balls, the electroscope offers several advantages for detecting the type and strength of charge. First, the foil leaves do not dance around

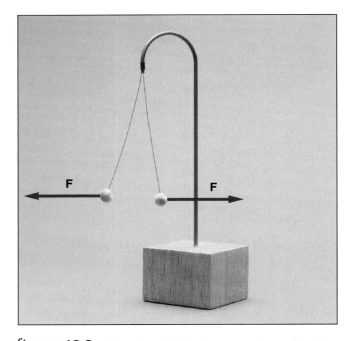

figure 12.3 Once the pith balls have moved away from the rod, they also repel each other, indicating the presence of a repulsive force.

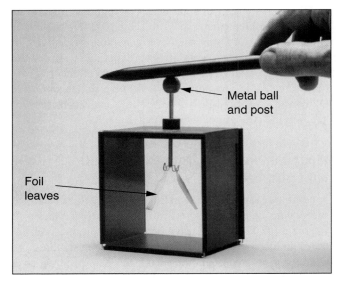

figure 12.4 A simple electroscope consists of two metallic-foil leaves suspended from a metal post inside a glass-walled container.

and become tangled together as pith balls often do. The metal ball on the electroscope offers a stationary point for testing other objects. Also, the distance the leaves move apart or come together when a charged object is brought near the ball gives a consistent indication of the strength of the charge.

An electroscope gives further information about the charging process. When we charge the plastic rod by rubbing it with the cat fur, the fur also is charged but with a charge of the opposite type. We can verify this by bringing the fur close to the ball on the electroscope after the electroscope has been charged by the rod. When the fur approaches the metal ball, the leaves move closer together. They move farther apart when the rod approaches. Similar experiments show that the glass rod and the nylon fabric also acquire opposite types of charge when rubbed together. Try This Box 12.1 provides another example of charging by rubbing.

Benjamin Franklin's single-fluid model

By the middle of the eighteenth century, it was well known that there were two types of charge that produced electrostatic forces between charged objects. The forces were either attractive or repulsive according to the simple rule:

Like charges repel each other, and unlike charges attract each other.

There was less agreement, however, on what to call these two types of charge. *Charge-produced-on-a-rubber-rod-when-rubbed-by-cat-fur* and *charge-produced-on-a-glass-rod-when-rubbed-with-silk* are unwieldy. *Positive* and *negative*, the names we now use, were introduced by the American statesman and scientist Benjamin Franklin (1706–1790) around 1750. During the 1740s, Franklin had performed a series of experiments on static electricity like those we have described.

Franklin proposed that the facts, as they were known, could be explained by the action of a single fluid that was transferred from one object to another during charging. The charge acquired when a surplus of this fluid was gained was *positive* (+) and the charge associated with a shortage of this fluid was *negative* (−) (fig. 12.5). Which of these was which was not clear, since the fluid itself was invisible and not otherwise detectable. Franklin arbitrarily proposed that the charge on a glass rod when rubbed with silk (there were no synthetic fabrics then) be called positive.

Besides simplifying the names of the two types of charge, Franklin's model offered a picture of what might be happening during charging. In his model, two objects become oppositely charged because some *neutral* or stable amount of the invisible fluid is present in all objects, and rubbing objects together transfers some of this fluid from one object to the other. During rubbing, one object gains

try this box 12.1

Sample Question: Carpet Fireworks

Question: Why do sparks fly when we shuffle across a carpet on a dry day and then touch a light switch?

Answer: Shuffling our feet on a carpet represents a rubbing process between two different materials. The soles of our shoes are often made of a rubberlike material and the carpet fibers are usually a synthetic fabric. The rubbing process creates a separation of charge similar to those we have been describing using rods and fabric. The charge that our shoes acquire can flow to other parts of our bodies, which are good conductors.

When we touch a light switch, the charge that we have acquired discharges producing the sparks. This is because, when properly wired, the switch is *grounded*. This means that the external parts of the switch are connected via wires and other conductors to the earth, which serves as a large sink for charge. The excess charge on our bodies is discharged to ground via the switch.

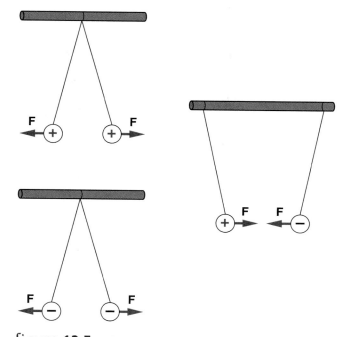

figure 12.5 Like charges repel and unlike charges attract. The plus and minus signs were introduced in Franklin's single-fluid model.

an excess of the fluid while the other experiences a short-age. Franklin's model was simpler than an earlier theory that had proposed two different substances for the two types of charge.

Franklin's model comes surprisingly close to our modern view of what takes place during charging. We now know that electrons are transferred between objects when they are rubbed together. **Electrons** are small, negatively charged particles present in all atoms and, therefore, in all materials. A negatively charged object has a surplus of electrons, and a positively charged object has a shortage of electrons. The atomic or chemical properties of materials dictate which way the electrons flow when objects are rubbed together.

Rubbing different materials together sometimes causes electric charge to move from one material to the other. This charge then produces attractive or repulsive electrostatic forces between objects. Like charges repel one another, and unlike charges attract one another. An electroscope consistently gauges the sign and strength of charge present. Using Benjamin Franklin's labels, there are two types of charge, positive and negative. The positive and negative labels originally meant a surplus or shortage of an invisible fluid in Franklin's model, but we now know that negatively charged electrons are transferred during rubbing.

12.2 Conductors and Insulators

The experiments described in section 12.1 gave us some basic information about the electrostatic force. Different properties of materials are also important, however, in understanding the range of electrostatic phenomena. Why were the pith balls initially attracted to the rod, for exam-ple, even when they were not charged? Why are the leaves in the electroscope made of metal? The distinction between insulators and conductors is a big piece of the puzzle.

How do insulators differ from conductors?

Suppose that you touch the electroscope with a charged plastic or glass rod. The leaves repel one another. What happens if you then touch the metal ball on top of the electroscope with your finger? The leaves of the electroscope immediately droop straight down. You have *discharged* the electroscope by touching the ball with your finger (fig. 12.6).

Suppose that you charge up the electroscope again. Now, touch the metal ball on top with an uncharged rod made of plastic or glass. There is no effect on the leaves of the electroscope. If, however, you touch the ball with a hand-held metal rod, the leaves immediately droop straight down. The electroscope is discharged.

How can we explain these observations? Apparently, both the metal rod and your finger let charge flow from the leaves of the electroscope to your body. Your body is a large neutral *sink* for charge. You can easily absorb the charge on the electroscope without much change in your overall charge. The plastic or glass rods, on the other hand, do not seem to permit the flow of charge from the electroscope to your body.

The metal rod and our bodies are examples of **conductors,** materials through which charge can flow readily. Plastic and glass are examples of **insulators,** materials that do not ordinarily permit the flow of charge. By performing similar experiments using the charged electroscope, we could test many other materials. We would discover that all metals are good conductors, while glass, plastic, and most nonmetallic materials are good insulators. Table 12.1 lists some examples of insulating and conducting materials.

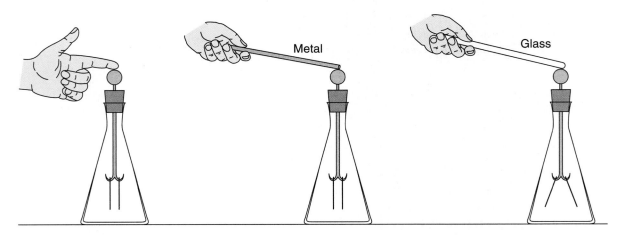

figure **12.6**   Touching the ball on top of a charged electroscope with either your finger or a metal rod causes the electroscope to discharge. Touching a charged electroscope with an uncharged glass rod produces no effect.

table **12.1**		
Some Common Conductors, Insulators, and Semiconductors		
Conductors	Insulators	Semiconductors
copper	glass	carbon
silver	plastic	silicon
iron	ceramics	germanium
gold	paper	
salt solution	oil	
acids		

The difference in the ability of good conductors and good insulators to conduct electric charge is amazingly large. Charge flows much more readily through several miles of copper wire than it does through the few inches of ceramic material used as an insulator on electric transmission lines.

Some materials, however, **semiconductors,** are intermediate between a good conductor and a good insulator. Carbon and silicon are probably the most familiar examples. A wooden rod behaves like a semiconductor in discharging the electroscope. With the appropriate amount of moisture in the wood, the rod will cause a slow discharge of the electroscope.

Although semiconductors are far less common than good conductors or good insulators, their importance to modern technology is enormous. The ability to control the level of conduction in these materials by mixing in small amounts of other substances has led to the development of miniaturized electronic devices like transistors and integrated circuits. The entire computer revolution has relied on the use of these materials, mostly silicon. Chapter 21 gives a closer look.

Charging a conductor by induction

Can you charge an object without actually touching it with another charged object? It turns out that you can. The process is called *charging by induction,* and it involves the conducting property of metals.

Suppose that you charge a plastic rod with cat fur and then bring the rod near a metal ball mounted on an insulating post, as in figure 12.7. The free electrons in the metal ball will be repelled by the negatively charged rod. Free to flow within the ball, they produce a negative charge (excess electrons) on the side opposite the rod and a positive charge (shortage of electrons) on the side near the rod. The overall charge of the metal ball is still zero.

To charge the ball by **induction,** you now touch the ball with your finger on the side opposite the rod, still holding the rod near, but not touching, the ball. The negative charge flows from the ball to your body, since it is still being re-

figure 12.7 The negatively charged rod is brought near a metal ball mounted on an insulating post, thus producing a separation of charge on the ball.

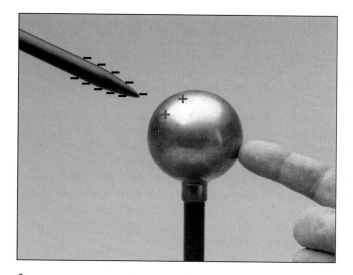

figure 12.8 Touching a finger to the opposite side of the metal ball draws off the negative charge, leaving the ball with a net positive charge.

pelled by the negative charge on the rod. If you now remove your finger and then the rod (in that order), a net positive charge will be left on the ball (fig. 12.8). You can easily test this fact by bringing the metal ball near an electroscope that has been given a negative charge from the plastic rod. The leaves will come closer together when the ball charged by induction approaches, indicating the presence of positive charge on this ball.

In this experiment, the proper sequence of events is important for the experiment to work. The charged rod must be held in place while the finger touches the ball on the opposite side and is removed. Only after the finger has been removed can the charged rod be moved away. The ball ends up with a charge opposite to the charge on the rod. This would also be true if we had used a positively charged glass rod. Then, the metal ball would end up with a negative charge.

The process of charging by induction illustrates the mobility of charges on a conducting object such as the metal ball. The process will not work with a glass ball. Charging by induction is an important process in machines used for generating electrostatic charges, and in many other practical devices. It also explains some of the phenomena associated with lightning storms. (See page 247.)

Why are insulators attracted to charged objects?

In our initial experiments with the pith balls, we noted that the pith balls were attracted to the charged rod before they had a chance to become charged themselves (fig. 12.2). How can we explain this phenomenon? What happens inside an insulating object when it is brought near a charged object?

Unlike the charges in the metal ball, the electrons in the pith ball or other insulating material are not free to migrate through the material. Instead, they are tied to the atoms or molecules of the material. Within an atom or molecule, however, charges have some freedom of movement. The distribution of charge within the atom or molecule can change.

Without delving into details of atomic structure, we can develop a rough picture of what happens to the charge in atoms when an insulating material is brought near a charged object. The basic idea is illustrated in figure 12.9, which magnifies the pith ball and greatly exaggerates the size of the atoms. Within each atom, a small distortion of the charge distribution takes place. The negative charge in the atom is attracted to the positively charged rod, and the positive charge is repelled.

Each atom has now become an **electric dipole,** in which the center, or average location, of the negative charge is separated by a slight distance from the center of the positive charge. The atom now has positive and negative *poles*—hence the term *dipole*—and we say that the material has become **polarized.** Overall, these atomic dipoles within the insulating material produce a slight negative charge on the surface of the pith ball near the positively charged rod and a slight positive charge on the opposite surface. The adjacent positive and negative charges within the material cancel one another.

The pith ball itself then becomes an electric dipole in the presence of the charged rod. Since the negatively charged surface is closer to the rod than the positively charged sur-

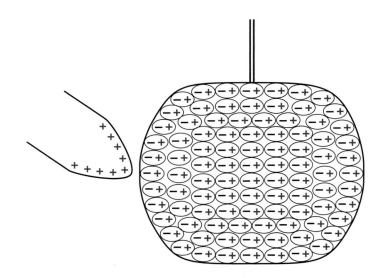

figure **12.9**   The negative charge in the atoms is attracted to the positively charged glass rod, while the positive charge is repelled. This produces a polarization of the charge in the atoms. The size of the atoms is grossly exaggerated.

face, it experiences a stronger electrostatic force, with the result that the pith ball is attracted to the charged rod. At this point, the overall charge on the pith ball is still zero. Once the ball comes in contact with the charged rod, however, some of the charge on the rod can be transferred to the pith ball, which becomes positively charged like the rod. The pith balls are then repelled by the rod, as we observed earlier.

The ability to become polarized is an important property of insulating materials. Polarization explains why small bits of paper or Styrofoam will be attracted to a charged object such as a synthetic-fabric sweater rubbed against some other material. Electrostatic precipitators used to remove particles from smoke in industrial smoke stacks take advantage of this property. Polarized particles are attracted to charged plates in the precipitator, removing them from the emitted gases.

Experiments in discharging an electroscope show that different materials vary widely in their ability to allow the flow of electric charge. Most metals are good conductors, but glass, plastic, and many other nonmetallic materials are poor conductors and, therefore, good insulators. A few materials, called semiconductors, have intermediate abilities to conduct charge. Conductors can be charged by induction without actually touching the conductor to another charged object. Insulators become polarized in the presence of charged objects, which explains why they are attracted to charged objects.

12.3 The Electrostatic Force: Coulomb's Law

Although we cannot see electric charge, we can see the effects of the forces exerted on charged objects. Pith balls and foil leaves refuse to hang straight down when they carry like charges. A repulsive force pushes them apart. Can we describe this force in a quantitative way? How does it vary with distance and quantity of charge? Is it similar in some way to the gravitational force?

Questions such as these were actively explored by scientists in the latter part of the eighteenth century. Like the gravitational force, the electrostatic force was apparently present even when objects were not in contact—it acted at a distance. Although others had already speculated about the form of the force law, the experiments of Charles Coulomb (1736–1806) during the 1780s settled the matter, establishing what we now know as Coulomb's law.

How did Coulomb measure the electrostatic force?

At first glance, measuring the strength of a force such as the electrostatic force might seem like a simple exercise. In reality, it is not an easy thing to do. Although much stronger than the gravitational force between ordinary-sized objects, the electrostatic force is still relatively weak, so Coulomb needed to develop techniques for measuring small forces. In addition, he was faced with the problem of defining how much charge was present, far from a trivial matter.

Coulomb's answer to the problem of measuring weak forces was to develop what we now call a *torsion balance,* shown in figure 12.10. Two small metal balls are balanced on an insulating rod suspended at the middle from a thin wire. The balls and wire are contained in a glass-walled enclosure to avoid disturbance from air currents. A force applied to either ball perpendicular to the rod produces a torque that causes the wire to twist. If we have previously determined how much torque is required to produce a given angle of twist, we have a method of measuring weak forces.

To measure the electrostatic force, we must somehow charge one of the balls. A third ball, also charged, can then be inserted on the end of an insulating rod (fig. 12.10). If it has the same type of charge as the ball on the end of the suspended rod, the two charges will repel one another, and the resulting torque will twist the wire. By adjusting the distance between the two charged balls, we can measure the strength of the repulsive force at different distances.

The problem of determining the amount of charge placed on the balls required even more ingenuity. A simple electroscope gave only a rough indication of the quantity of charge present, and no consistent set of units or procedures for specifying quantity of charge were in use at the

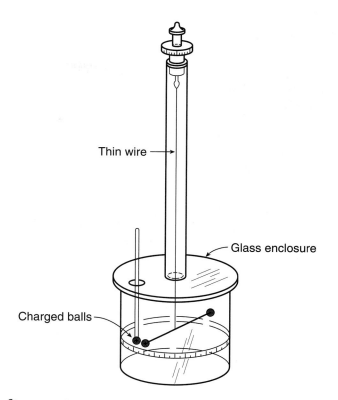

figure 12.10 A diagram of Coulomb's torsion balance. The degree of twist of the wire provides a measure of the repulsive force between the two charges.

time of Coulomb's work. His solution was to develop a system of charge division. He started with an unknown quantity of charge on a single metal ball mounted on an insulating stand, as in figure 12.11. He touched this ball to an identical ball mounted on a similar stand. He reasoned that the two balls would now contain equal quantities of charge (half of the original amount on the first ball), which was verified by bringing the balls near an electroscope.

If a third identical metal ball touched one of the first two, the charge would again be divided equally, and these two balls would each have one-half the charge of the ball that did not participate in this second exchange. If several more identical balls are available, splitting the charge can be continued. Although this process does not provide an absolute measure of charge, we can say with confidence that one ball contains twice as much charge, or perhaps four times as much charge, as another.

Coulomb used this process to test the effects of different amounts of charge on the electrostatic force. If the balls used in splitting the charge were identical in size to those in the torsion balance, touching one of the torsion-balance balls with one of the others was just one more division of charge. Using these procedures, Coulomb was able to determine how the strength of the electrostatic

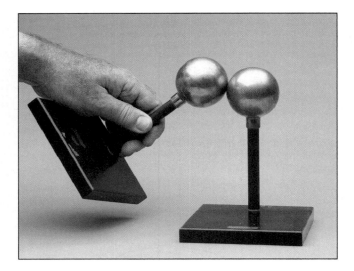

figure **12.11**   By bringing two identical metal balls into contact, one charged and the other one initially uncharged, equal quantities of charge are obtained on the two balls.

force varied with the quantity of charge on each object and the distance between the two charged objects.

What were the results of Coulomb's measurements?

The results of Coulomb's work can be stated in a relationship usually referred to as **Coulomb's law:**

> The electrostatic force between two charged objects is proportional to the quantity of each of the charges and inversely proportional to the square of the distance between the charges, or
>
> $$F = \frac{kq_1q_2}{r^2}.$$

Here the letter q represents the quantity of charge, k is a constant (called *Coulomb's constant*) whose value depends on the units used, and r is the distance between the centers of the two charges.

Although Coulomb himself used different units, we now usually express charge in units called *coulombs* (C). If distance is measured in meters, the value of Coulomb's constant turns out to be approximately $k = 9 \times 10^9$ N·m²/C². The coulomb itself is determined by measurements involving electric current discussed in chapter 14. A force computation using Coulomb's law is shown in Try This Box 12.2.

Figure 12.12 illustrates Coulomb's law. The forces obey Newton's third law—the two charges experience equal but oppositely directed forces. The two charges shown are both positive, so the force is repulsive, since like charges repel.

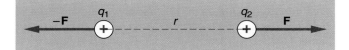

---

try this box 12.2

Sample Exercise: Calculating Electrostatic Force

Two positive charges, one 2 μC and the other 7 μC, are separated by a distance of 20 cm. What is the magnitude of the electrostatic force that each charge exerts upon the other?

$q_1 = 2$ μC   (1 μC = 10^{-6} C = 1 microcoulomb)

$q_2 = 7$ μC

$r = 20$ cm = 0.2 m

$F = ?$

$$F = \frac{kq_1q_2}{r^2}$$

$$= \frac{(9.0 \times 10^9 \text{ N·m}^2/\text{C}^2)(2 \times 10^{-6} \text{ C})(7 \times 10^{-6} \text{ C})}{(0.2 \text{ m})^2}$$

$$= \frac{0.126 \text{ N·m}^2}{0.04 \text{ m}^2}$$

$$= \textbf{3.15 N}$$

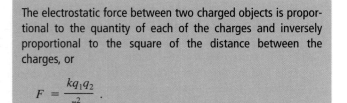

figure **12.12**   Two positive charges exert equal but oppositely directed forces upon one another, according to Coulomb's law and Newton's third law of motion. The force is inversely proportional to the square of the distance r between the two charges.

If one charge were negative and the other positive, the directions of both forces would be reversed.

Coulomb's law compared to Newton's law of gravity

The electrostatic force has the same inverse-square dependence on distance as Newton's law of gravitation (see chapter 5). If we double the distance between the two charges, the force between the two charges falls to only one-fourth what it was before the distance was doubled. Tripling the distance produces a force that is one-ninth the force at the original distance, and so forth. The strength of the interaction between the two charges falls off rapidly when the distance between them is increased.

Since the gravitational force and the electrostatic force are two of the fundamental forces of nature, it is interesting to compare them. How do they differ? Placing their symbolic expressions side by side highlights both the similarities and the differences:

$$F_g = \frac{Gm_1 m_2}{r^2} \quad \text{and} \quad F_e = \frac{kq_1 q_2}{r^2} \ .$$

An obvious difference is that the gravitational force depends on the product of the masses of the two objects, and the electrostatic force depends on the product of the charges of the two objects. Otherwise, the forms of these two force laws are similar.

A more subtle difference involves the direction. The gravitational force is always attractive. As far as we know, there is no such thing as negative mass. The electrostatic force, on the other hand, can be either attractive or repulsive depending on the signs of the two charges. The rule that like charges repel and unlike charges attract determines the direction.

Another difference has to do with the strengths of these two forces. For objects of ordinary size, and for subatomic particles, the gravitational force is much weaker than the electrostatic force. At least one of the objects must have an enormous mass (like the earth) to produce a significant gravitational force. For charged particles at the atomic or subatomic level, the electrostatic force is far more important than the much weaker gravitational force. It is the electrostatic force that holds atoms together and binds one atom to another in liquids and solids.

Although the basic forms of these two force laws have been known for more than 200 years, physicists are still trying to understand the underlying reasons for the relative strengths of these and other fundamental forces of nature. The search for a *unified field theory* that would explain the relationships between all of the fundamental forces is a major area of research in modern theoretical physics (see chapter 21). Some of today's students will undoubtedly play important roles in this search.

Charles Coulomb designed a torsion balance to measure the strength of the electrostatic force between two charges. He found that the force was proportional to the size of each charge and inversely proportional to the square of the distance between the two charges. Coulomb's force law has a form very similar to Newton's law of gravitation describing the force between two masses. The gravitational force is always attractive, however, and generally much weaker than the electrostatic force.

12.4 The Electric Field

Coulomb's law tells us how to find the force between any two charged objects if the objects are small compared to the distance between them. These forces act at a distance: the charges do not have to be in contact for the force to be exerted. Does the presence of electric charge somehow modify the space surrounding the charge? How can we describe the effects of a large distribution of charges on some other charge?

The concept of **electric field** describes the effect of such a distribution of charges on another individual charge. The usefulness of the idea of a field has made it a central concept in modern theoretical physics. To most of us, the word *field* suggests a field of wheat or a meadow filled with wild flowers. The concept of electric field in physics is somewhat more abstract. An example involving just a few charges can help to introduce the ideas.

Finding the force exerted by several charges

Using Coulomb's law, we can find the magnitude of the electrostatic force between any two charged objects. If the charged objects are small compared to the distance between them, we often call them *point charges*. If there are more than two point charges, we can compute the net force on any one of them by adding (as vectors) the forces due to each of the other charges.

In the first part of Try This Box 12.3, we find the force on a charge q_0 exerted by two other charges for the situation shown in figure 12.13, where the charge q_0 lies between the charges q_1 and q_2. The resulting forces are added as vectors to obtain the net force on q_0. The forces due to each of the other charges must be computed separately using Coulomb's law (see Try This Box 12.2 for details of this process).

Notice that the force $\mathbf{F}_1$ on the charge q_0 exerted by the charge q_1 is considerably larger than the force $\mathbf{F}_2$ exerted by q_2, mainly because q_2 is farther from q_0 than q_1 is. The electrostatic force described by Coulomb's law is inversely proportional to the square of the distance between the two charges. The two forces acting on q_0 are in opposite directions, yielding a net force of 9 N in the direction of the larger force.

What is an electric field?

Suppose that we wanted to know the force on some other charge placed at the same location as q_0. Would we have to use Coulomb's law again, find the forces due to each of the other charges, and add them as we did in Try This Box 12.3? There is an easier way that involves the concept of electric field.

Think of the charge q_0 as a test charge that has been inserted at this particular location to assess the strength of the electrostatic effect at that point. By the nature of Coulomb's law, the force on this test charge will be proportional to the magnitude of the charge selected. If we divide the net force by the magnitude of the test charge, we find the force per unit charge at this location. (See the second part of the sample exercise in Try This Box 12.3.) Knowing the force per unit charge permits us to compute the force on any other charge placed at that same point.

try this box 12.3

Sample Exercise: An Electric Field

Two point charges with charges $q_1 = 3\ \mu C$ and $q_2 = 2\ \mu C$ are separated by a distance of 30 cm (fig. 12.13). A third charge $q_0 = 4\ \mu C$ is placed between the initial two charges 10 cm from q_1. From Coulomb's law, the force exerted by q_1 on q_0 is 10.8 N, and the force exerted by q_2 on q_0 is 1.8 N.

   a. What is the net electrostatic force acting on the charge q_0?

   b. What is the electric field (force per unit charge) at the location of the charge q_0 due to the other two charges?

a. $F_1 = 10.8$ N to the right $F = F_1 - F_2$

   $F_2 = 1.8$ N to the left $= 10.8$ N $- 1.8$ N

   $F = ?$ $= 9$ N

$$F = 9 \text{ N to the right}$$

b. $E = ?$ $E = \dfrac{F}{q_0}$

                    $= \dfrac{9 \text{ N}}{4 \times 10^{-6} \text{ C}}$

                    $= 2.25 \times 10^6$ N/C

$$E = 2.25 \times 10^6 \text{ N/C to the right}$$

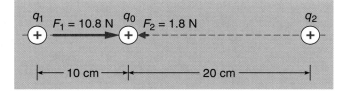

figure 12.13   The two forces acting upon q_0 are in opposite directions, yielding a net force of 9 N (see Try This Box 12.3).

is opposite to the direction of the field. The direction of the electric field at any point in space is the direction of the force that would be exerted on a *positive* charge placed at that point.

Keep in mind that the electric field and the electrostatic force are not the same. We can talk about the field at a point in space even if there is no charge at that point. The field tells us the magnitude and direction of the force that would be exerted on any charge placed at that point. There must be a charge if there is to be a force, but the field exists regardless of whether there is a charge at that point or not.

The electric field can exist even in a vacuum. When we use the field concept, we shift our focus from the interactions between particles or objects to the way a charged object affects the space surrounding it. The field concept is not restricted to electrostatics. We can also define a gravitational field or a magnetic field, as well as others.

How are electric field lines used?

The concept of electric field was formally introduced by James Clerk Maxwell (1831–1879) around 1865 as part of his highly successful theory of electromagnetism. The idea had already been used informally by Michael Faraday (1791–1867), who developed the concept of what we now call **field lines** as an aid in visualizing electric and magnetic effects. Faraday was not trained in mathematics but was a brilliant experimentalist who made good use of mental pictures.

To illustrate field lines, we can use a positive test charge to assess the field direction and strength at various points around a single positive charge. We will find that the test charge is repelled by the original positive charge wherever we place it around that charge. If we draw lines to indicate the direction of the force on the test charge (which is also the direction of the field), we obtain a drawing like figure 12.14.

Figure 12.14 is only a two-dimensional slice of a three-dimensional phenomena. The electric field lines associated with a single positive charge radiate in all directions from the charge. If we also adopt the convention that we always start the lines on positive charges and end them on negative charges, the density of the field lines is proportional to the strength of the field. The closer together the field lines are, the stronger the field.

Field lines are a means of visualizing both the direction and the strength of the field. Figure 12.15 shows a two-dimensional slice of the electric field lines associated with

Using the force per unit charge as a measure of the strength of the electrostatic effect at some point in space lies at the heart of the concept of electric field. In fact, we define *electric field* as

> The electric field at a given point in space is the electric force per unit positive charge that would be exerted on a charge if it were placed at that point.
>
> $$\mathbf{E} = \frac{\mathbf{F}}{q_0}$$
>
> It is a vector having the same direction as the force on a positive charge placed at that point.

The symbol $\mathbf{E}$ represents the electric field.

In other words, the ratio $\mathbf{F}/q_0$ that we computed in Try This Box 12.3 is the magnitude of the electric field at that point. We can then use the field to find the force on any other charge placed at that point by multiplying the charge by the electric field,

$$\mathbf{F} = q\mathbf{E}.$$

If the charge happens to be negative, the minus sign indicates that the direction of the force on a negative charge

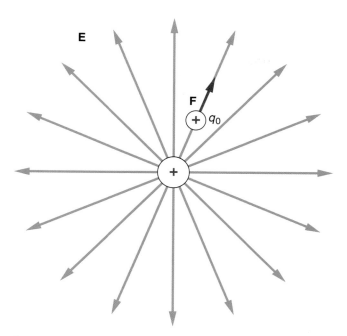

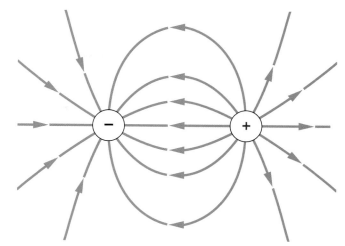

figure **12.16**   The electric field lines associated with two equal but opposite-sign charges (an electric dipole).

figure **12.14**   The direction of the electric field lines around a positive charge can be found by imagining a positive test charge q_0 placed at various points around the source charge. The field has the same direction as the force on a positive test charge.

a negative charge. Here the field lines end on the charge and must be directed inward to indicate the proper direction for the force on a positive test charge.

As a final example, consider the field lines associated with the electric dipole in figure 12.16. An electric dipole is two charges of equal magnitude but opposite sign, separated by a small distance. The field lines originate on the positive charge and end on the negative charge. Imagine a positive test charge placed at various points around the dipole. Do the field lines agree with your expectations for the correct direction of the force on the test charge? This is the acid test for any field-line diagram.

The concept of electric field focuses our attention on how the space surrounding an electric charge is affected by that charge. The field is defined as the force per unit charge that would be exerted on a test charge placed at some point in space in the vicinity of the source charges. The direction of the electric field is the same as the direction of the force on a positive test charge placed at that point in space. The field is a useful concept for treating the effects of many point charges on some other charge. Electric field lines can be used to visualize these effects.

12.5 Electric Potential

In chapter 6, we defined the potential energy associated with the gravitational force, as well as the potential energy of a spring. Can we define the potential energy of a charged particle being acted upon by an electrostatic force?

The electrostatic force is a conservative force, which means that we can define an electrostatic potential energy. This potential energy leads to the related concept of *electric potential,* often referred to simply as *voltage.* Voltage

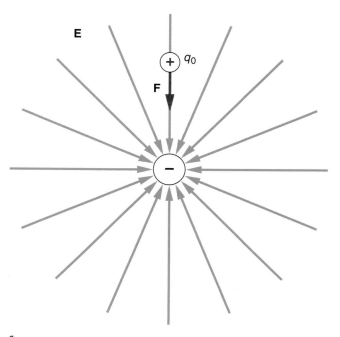

figure **12.15**   The electric field lines associated with a negative charge are directed inward, as indicated by the force on a positive test charge, q_0.

is familiar because we use it to discuss batteries and household circuits. What is voltage, and how is it related to electrostatic potential energy?

Finding the change in potential energy of a charge

To see how the potential energy of a charge changes with position, the easiest case to consider is a charged particle moving in a uniform electric field. In a uniform field, the electric field lines are parallel and evenly spaced. The field does not vary in direction or strength as we move from point to point within the region where the field exists: it is constant within this region.

How do we go about producing a uniform electric field? Two parallel metal plates with opposite charges, arranged as in figure 12.17, will do the trick. If one plate is positively charged and the other has an equal negative charge, the field lines are straight lines originating at the positive charges and terminating on the negative charges, as shown. You can test this conclusion yourself by thinking about the force on a positive test charge placed between the two plates, as we did in section 12.4.

Two conductors separated by an insulating material like air represent a useful means of storing charge. If the two conductors have charges of opposite sign, these charges are held in place by the attractive electrostatic force between the charges on the opposing plates. We call this arrangement a **capacitor,** which is basically a device to store charge. Capacitors have many applications, particularly in electric circuits.

Suppose that we now place a positive test charge in the uniform-field region between the two plates of the parallel-plate capacitor. This charge will experience an electrostatic force in the direction of the electric field. It will be attracted toward the negative charges on the bottom plate and repelled by the positive charges on the top plate. If we release the charge, it accelerates toward the bottom plate.

(This acceleration has nothing to do with gravity. We can assume that the gravitational force on the charged particle is very small compared to the electrostatic force.)

If we apply an external force to move the test charge in the opposite direction to the electric field, as in figure 12.18, this external force does work on the charge. This work increases the potential energy of the charge (see chapter 6). The process is similar to what happens when we lift an object against the gravitational force or pull on a bowstring doing work against the elastic force. Doing work against a conservative force increases the potential energy of the system.

To move the charge without accelerating it, the external force must be equal in magnitude but opposite in direction to the electrostatic force, so that the net force acting on the charge is zero. The electrostatic force acting on the charge is equal in magnitude to the charge times the electric field (qE), so the external force must also have this magnitude. The work done by the external force is qEd, the force times the distance, and this is equal to the increase in potential energy of the charge ($\Delta PE = qEd$).

What we are describing here is similar to using an external force to lift an object against the force of gravity, thus increasing its gravitational potential energy (fig. 12.19). The gravitational potential energy always increases when we lift an object, but the direction of increase for the electrostatic potential energy depends on the direction of the electric field. If we place the positively charged plate on

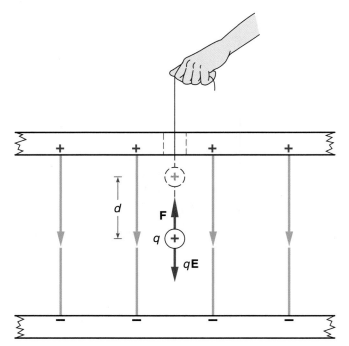

figure **12.18**   An external force **F**, equal in magnitude to the electrostatic force q**E**, is used to move the charge q a distance d in a uniform field.

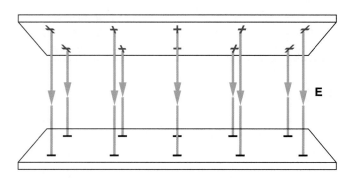

figure **12.17**   Two parallel metal plates containing equal but opposite-sign charges produce a uniform electric field in the region between the plates.

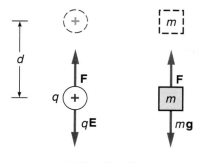

$$\Delta PE = W = Fd$$

figure **12.19** The increase in potential energy when a charge q is moved against the electrostatic force is analogous to what happens when a mass m is lifted against the gravitational force.

the bottom so that the field direction is upward, the potential energy of the positive charge increases when we move the charge downward.

Whenever work is done to move a charged particle against the direction that the electrostatic force would normally tend to move it, we increase its potential energy. If, for example, we move a negatively charged particle away from a positively charged particle, we increase the potential energy of the system. Like stretching a spring, the increase in potential energy can easily be converted to kinetic energy if we release the negative charge and allow it to accelerate toward the positive charge.

What is electric potential?

How does electric potential or voltage relate to potential energy? **Electric potential** is related to electrostatic potential energy in much the same way as electric field is related to the electrostatic force. We can regard the positive charge that is moved in figure 12.18 as a test charge used to determine how the potential energy varies with position. The change in electric potential can be defined as

> The change in electric potential is equal to the change in electrostatic potential energy per unit of positive test charge, or
>
> $$\Delta V = \frac{\Delta PE}{q}.$$

The symbol V represents the electric potential (voltage).

As you can see from this equation, the units of electric potential are those of energy per unit of charge. In the metric system, this unit is called a volt (V), which is defined as 1 joule per coulomb (1 J/C = 1 V). The unit and the term **voltage** suggest the symbol V commonly used for electric potential.

Like the electric field, no charge need be present to talk about the electric potential at some point in space. The

change in electric potential as we move from one point to another is equal to change in potential energy *per unit of positive charge* that would occur if a positive charge were moved between these two points. In other words, to find the change in potential energy for such a charge, we would multiply the change in electric potential by the magnitude of the charge, $\Delta PE = q\,\Delta V$.

Electric potential and potential energy are closely related, but they are not the same. If the charge q happens to be negative, its potential energy will *decrease* when it is moved in the direction of *increasing* electric potential.

As with gravitational potential energy, it is the *change* in electrostatic potential energy that is meaningful rather than a specific value of potential energy. To state a specific value of either electrostatic potential energy or electric potential, we define a reference point at which the potential is zero. Other values of potential energy are defined from that position. A numerical example involving the computation of electric potential is found in Try This Box 12.4.

Figure 12.20 illustrates the example in Try This Box 12.4. The potential energy of the charge increases by 0.15 J as the charge is moved from the bottom to the top plate of the parallel-plate capacitor. Since the charge is positive, it can serve as a test charge so we can compute the change in electric potential, leading to a change in potential of 30 V. If we chose a reference value of 0 V for the bottom plate,

---

try this box 12.4

Sample Exercise: Finding the Electric Potential

A uniform electric field of 1000 N/C is established between two oppositely charged metal plates. A particle with a charge of +0.005 C is moved from the bottom (negatively charged) plate to the top plate. (Imagine that a string is tied to the charge and is pulling it upward.) The distance between the plates is 3 cm.

   a. What is the change in potential energy of the charge?

   b. What is the change in electric potential from the bottom to the top plate?

a. $E = 1000$ N/C $\Delta PE = W = Fd$

    $q = 0.005$ C $= qEd$

    $d = 3$ cm $= (0.005\ \text{C})(1000\ \text{N/C})(0.03\ \text{m})$

    $\Delta PE = ?$ $= \mathbf{0.15\ J}$

b. $\Delta V = ?$ $\Delta V = \dfrac{\Delta PE}{q}$

                     $= \dfrac{0.15\ \text{J}}{0.005\ \text{C}}$

                     $= \mathbf{30\ V}$

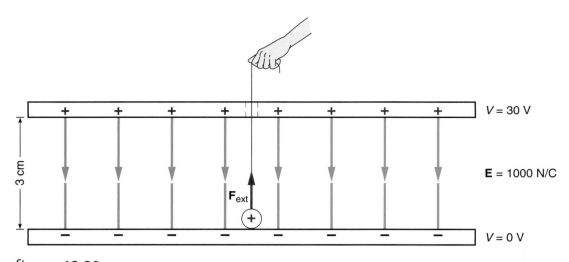

figure **12.20**   A positive charge is moved from the bottom plate to the top plate by an external force.

the top plate would have a potential of 30 V. Halfway between the two plates, the electric potential would be 15 V. The electric potential increases continuously from 0 V to 30 V from the bottom plate to the top plate.

How are electric potential and electric field related?

In the case of a uniform electric field, there is a simple relationship between the magnitude of the electric field and the change in electric potential. Since $\Delta PE = qEd$ for this case, and by definition the difference in electric potential is $\Delta V = \Delta PE/q$, dividing ΔPE by q yields:

$$\Delta V = Ed.$$

In figure 12.20, however, the increase in electric potential occurs in the direction opposite to the direction of the field, because the potential energy of a positive charge increases when we move it against the field.

This simple relationship between E and ΔV is valid only for a uniform field. If the field strength varies with position, the computation is more complex, and different relationships are found. In many practical situations, however, the field is more or less uniform. In fact, since $E = \Delta V/d$ in this case, we often express the value of the field strength in units of volts per meter (V/m), which is equal to newtons per coulomb (N/C).

The electric potential always increases most rapidly in the direction opposite to the electric field. For the field associated with a positive charge, for example, the electric potential increases as you move toward the charge, and the field lines radiate outward from the charge. Figure 12.21 depicts the electric field for a positive charge with a few values of electric potential indicated at different distances from the charge. The reference level in this case is defined by setting the zero value of electric potential at an infinite distance from the charge.

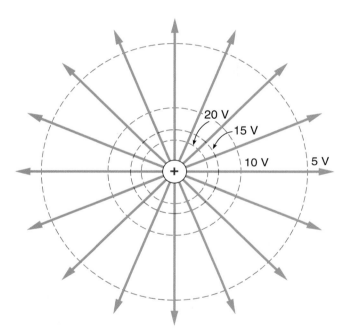

figure **12.21**   The electric potential (represented by the dashed lines of constant potential) increases as we move closer to a positive charge.

In any situation, we can always determine how the electric potential varies by thinking about what would happen to a positive test charge being moved within the electric field in question. The electric potential always increases as we move toward positive charges and away from negative charges, because the potential energy of a positive charge increases under those conditions. A positive charge moves toward regions of lower electric potential, decreasing its potential energy while gaining kinetic energy, much like a falling rock. A lightning storm (see Everyday Phenomenon Box 12.1) provides several striking examples of this process.

Everyday Phenomenon

box 12.1

Lightning

The Situation. We have all observed the awe-inspiring beauty and power of a good electrical storm. The flashes of lightning, followed at varying time intervals by claps of thunder, can be both fascinating and frightening. What is lightning? How are thunderclouds capable of producing the impressive electrical discharges that we see? What happens in an electrical storm?

Flashes of lightning illuminate the area. What is lightning, and how is it produced?

The charge distribution within a thundercloud induces a positive charge on objects on the earth directly below the cloud.

The Analysis. Most thunderclouds generate a separation of charge within the cloud that produces a net positive charge near the top of the cloud and a net negative charge near the bottom. Highly turbulent convection taking place in the cloud separates and transports the charge: thunderclouds consist of rapidly rising and falling columns of air and water, with cells of rising air often being found next to cells of falling air and water.

The charge separation within a thundercloud produces strong electric fields in the cloud as well as between the cloud and the earth. Since moist earth is a reasonably good conductor of electricity, a positive charge is induced on the surface of the earth below the cloud because of the negative charge on the bottom of the cloud.

The electric field generated by this charge distribution (pictured in the drawing) can be several thousand volts per meter. Since the base of the cloud usually floats several hundred meters above the surface of the earth, the potential difference between the cloud's base and the earth can easily be several million volts! (Even during fair weather, there is an electric field of a few hundred volts per meter in the atmosphere near the surface of the earth. This field is weaker, however, and in the opposite direction to what is usually generated between the cloud and the earth in an electrical storm.)

What happens when lightning strikes? Dry air is a good insulator, but moist air conducts electricity somewhat more readily. Any material will conduct, however, if the voltage across the material is large enough. The very large voltage between the cloud's base and the earth creates an initial flow of charge (called the *leader*) along a path that offers the best conducting properties over the shortest distance. This leader heats the air and **ionizes** (removes electrons from) some of the atoms along that path. Since the ionized atoms are charged, they enhance the air's ability to conduct along the path blazed by the leader, and a much greater flow of charge can then proceed.

The following strokes or discharges all take place along this same conducting path in very rapid succession, each one increasing the conductivity along the path. A very large discharge or flow of charge takes place between the earth and the cloud in a very short period of time. The heating and ionization of the air by the discharge produces the lightning that we see. The sound wave that we hear as thunder is produced at the same time, but takes longer to reach us because sound travels at a much slower speed than light.

A large tree or a person standing on top of a treeless hill provide favorable paths for a lightning discharge. Standing under an isolated tree during a lightning storm is therefore dangerous. Your best bet is to be inside a building or car, but if you are outdoors, find a place that is not near the highest conducting object in your immediate vicinity. Choose a low (and preferably dry) spot and hunker down.

The potential energy of a positive charge increases when we apply an external force to move the charge in a direction opposite the electric field. The work done is equal to the increase in potential energy. A change in electric potential (also called voltage) is defined as the change in potential

energy per unit of test charge. The electric potential of a positive charge increases when you move it closer to other positive charges or away from negative charges because this increases its potential energy. The normal tendency of positive charges is to move to regions of lower electric potential.

summary

Curiosity about simple electrostatic phenomena led us to the description of the electrostatic force and Coulomb's law. The difference between conductors and insulators, and the concepts of electric field and electric potential, are also important in explaining the variety of electrostatic phenomena that we observe.

1   Effects of electric charge. Rubbing different materials together separates a quantity that we call *electric charge,* which is capable of exerting a force on other charges. There are two types of charge called *positive* and *negative* following the single-fluid model introduced by Benjamin Franklin. Like charges repel and unlike charges attract.

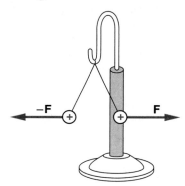

2   Conductors and insulators. Different materials vary widely in their ability to permit the flow of charge. These differences between conductors and insulators help to explain why charging by induction works and why uncharged bits of paper (or other insulators) are attracted to charged objects.

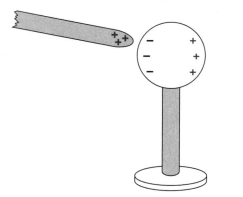

3   The electrostatic force: Coulomb's law. By careful experiments with his torsion balance, Coulomb was able to show that the electrostatic force that two charged objects exert on each other is proportional to the product of the two charges and inversely proportional to the square of the distance r between the charges.

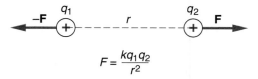

$$F = \frac{kq_1 q_2}{r^2}$$

4   The electric field. The electric field is defined as the electric force per unit of positive test charge that would be exerted if a charge were present at a point in space. Knowledge of the field at some point allows us to compute the force on any charge placed at that point. Field lines help us to visualize the electric field.

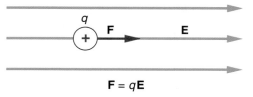

$$\mathbf{F} = q\mathbf{E}$$

5   Electric potential. Electric potential is defined as the potential energy per unit of positive charge that would exist at some point in space if a charge were present there. It is also called voltage, and its units are volts. The change in potential energy of a charge can be found by computing the work done to move the charge against the electrostatic force.

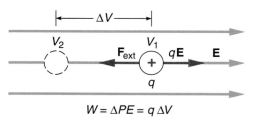

$$W = \Delta PE = q\,\Delta V$$

key terms

questions

*Questions identified with an asterisk are more open-ended than the others. They call for lengthier responses and are more suitable for group discussion.

Q1. When two different materials are rubbed together, do the two materials acquire the same type of charge or different types of charge? Explain how you could justify your answer with a simple experiment.

Q2. Two pith balls are both charged by contact with a plastic rod that has been rubbed by cat fur,
  a. What sign will the charges on the pith balls have? Explain.
  b. Will the two pith balls attract or repel one another? Explain.

Q3. Two pith balls are charged by touching one to a glass rod that has been rubbed with a nylon cloth and the other to the cloth itself,
  a. What sign will the charge on each pith ball have? Explain.
  b. Will the two pith balls attract or repel one another? Explain.

Q4. When a glass rod is rubbed by a nylon cloth, which of these two objects gains electrons? Explain.

Q5. Do the two metal-foil leaves of an electroscope gain charges of opposite sign when the electroscope is charged? Explain.

Q6. If you charge an electroscope with a plastic rod that has been rubbed with cat fur, will the metal leaves of the electroscope move farther apart or come closer together when you bring the cat fur near the ball of the electroscope? Explain.

Q7. When you comb your hair with a plastic comb, what will the sign be on the charge acquired by the comb? Explain. (Hint: Compare this process to rubbing a plastic rod with cat fur.)

*Q8. Describe how Benjamin Franklin's single-fluid model can explain what happens when we charge a glass rod by rubbing it with a nylon cloth. How do we get two types of charge from a single fluid? Explain.

Q9. If you touch the metal ball of a charged electroscope with an uncharged glass rod held in your hand, will the electroscope discharge completely? Explain.

Q10. If you touch the ball of a charged electroscope with your finger, will it discharge? What does this suggest about the conducting properties of people? Explain.

Q11. When a metal ball is charged by induction using a negatively charged plastic rod, what is the sign on the charge acquired by the ball? Explain.

*Q12. If, when charging by induction, you remove the charged rod from the vicinity of the metal ball before moving your finger from the ball, what will happen? Will the ball end up being charged? Explain.

Q13. Will bits of paper be attracted to a charged rod even if they have no net charge? Explain.

*Q14. Why are pith balls initially attracted to a charged rod and later repelled by the same rod, even though they have not touched any other charged object? Explain.

Q15. What is a torsion balance? Explain.

Q16. If you had several identical metal balls mounted on insulating stands, explain how you could obtain a quantity of charge on one ball that is four times as large as the quantity on another ball.

Q17. If the distance between two charged objects is doubled, will the electrostatic force that one object exerts on the other be cut in half? Explain.

Q18. If two charges are both doubled in magnitude without changing the distance between them, will the force that one charge exerts on the other also be doubled? Explain.

Q19. Can both the electrostatic force and the gravitational force be either attractive or repulsive? Explain.

Q20. Is it possible for an electric field to exist at some point in space at which there is no charge? Explain.

*Q21. Two charges, of equal magnitude but opposite sign, lie along a line as shown in the diagram. Using arrows, indicate the directions of the electric field at points A, B, C, and D shown on the diagram.

*Q22. If we change the negative charge in the diagram for question 21 to a positive charge of the same magnitude, what are the directions of the electric field at points A, B, C, and D? (Indicate with arrows.)

*Q23. Three equal positive charges are located at the corners of a square, as in the diagram. Using arrows, indicate the direction of the electric field at points A and B on the diagram.

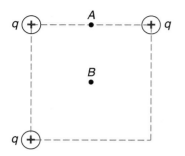

Q24. Is the electric field produced by a single positive charge a uniform field? Explain.

Q25. If we move a positive charge toward a negative charge, does the potential energy of the positive charge increase or decrease? Explain.

Q26. If we move a negative charge toward a second negative charge, does the potential energy of the first charge increase or decrease? Explain.

Q27. If a negative charge is moved in the same direction as the electric field lines in some region of space, does the potential energy of the negative charge increase or decrease? Explain.

Q28. Does the electric potential increase or decrease as we move toward a negative charge? Explain.

Q29. Is electric potential the same as electric potential energy? Explain.

*Q30. In the drawing for question 21, which point, B or C, will have the higher electric potential? Explain.

Q31. Will a negatively charged particle, initially at rest in an electric field, tend to move toward a region of lower electric potential if released? Explain.

Q32. Would you be more likely to be struck by lightning if you stood on a platform made from a good electrical insulator than if you stood on the ground? Explain.

exercises

E1. An electron has a charge of -1.6×10^{-19} C. How many electrons would be needed to produce a net charge of -4.8×10^{-6} C?

E2. Two identical brass balls mounted on wooden posts initially have different amounts of charge, one $+3$ μC and the other $+15$ μC. The balls are allowed to touch and then are separated again. What is the final charge on each ball?

E3. Two identical steel balls mounted on wooden posts initially have different amounts of charge, one $+12$ μC and the other -4 μC. The balls are allowed to touch and then are separated again. What is the final charge on each ball?

E4. Two charged particles exert an electrostatic force of 4 N on each other. What will the magnitude of the electrostatic force be if the distance between the two charges is reduced to one-half of the original distance?

E5. Two charged particles exert an electrostatic force of 27 N on each other. What will the magnitude of the force be if the distance between the two particles is increased to three times the original distance?

E6. Two positive charges, each of magnitude 4×10^{-6} C, are located a distance of 10 cm from each other.
a. What is the magnitude of the force exerted on each charge?
b. On a drawing, indicate the directions of the forces acting on each charge.

E7. A charge of $+2 \times 10^{-6}$ C is located 20 cm from a charge of -4×10^{-6} C.
a. What is the magnitude of the force exerted on each charge?
b. On a drawing, indicate the directions of the forces acting on each charge.

E8. An electron and a proton have charges of an equal magnitude but opposite sign of 1.6×10^{-19} C. If the electron and proton in a hydrogen atom are separated by a distance of 5×10^{-11} m, what are the magnitude and direction of the electrostatic force exerted on the electron by the proton?

E9. A uniform electric field is directed upward and has a magnitude of 20 N/C. What are the magnitude and direction of the force on a charge of -5 C placed in this field?

E10. A test charge of $+4 \times 10^{-6}$ C experiences a downward electrostatic force of 12 N when placed at a certain point in space. What are the magnitude and direction of the electric field at this point?

E11. A $+1.5 \times 10^{-6}$ C test charge experiences forces from two other nearby charges: a 12-N force due east and an 8-N force due west. What are the magnitude and direction of the electric field at the location of the test charge?

E12. A charge of -4×10^{-6} C is placed at a point in space where the electric field is directed toward the right and has a magnitude of 8.5×10^4 N/C. What are the magnitude and direction of the electrostatic force on this charge?

E13. A charge of +0.25 C is moved from a position where the electric potential is 10 V to a position where the electric potential is 60 V. What is the change in potential energy of the charge associated with this change in position?

E14. Three coulombs of charge flow from the +6 V positive terminal of a battery to the negative terminal at 0 V. What is the change in potential energy of the charge?

E15. The potential energy of a $+2 \times 10^{-6}$ C charge decreases from 0.06 J to 0.02 J when it is moved from point A to point B. What is the change in electric potential between these two points?

E16. The electric potential increases from 100 V to 500 V from the bottom plate to the top plate of a parallel-plate capacitor.
   a. What is the magnitude of the change in potential energy of a -5×10^{-4} C charge that is moved from the bottom plate to the top plate?
   b. Does the potential energy increase or decrease in this process?

challenge problems

CP1. Three positive charges are located along a line, as in the diagram. The 0.10-C charge at point A is 2 m to the left of the 0.02-C charge at point B, and the 0.04-C charge at point C is 1 m to the right of point B.
   a. What is the magnitude of the force exerted on the 0.02-C charge by the 0.10-C charge?
   b. What is the magnitude of the force exerted on the 0.02-C charge by the 0.04-C charge?
   c. What is the net force exerted on the 0.02-C charge by the other two charges?
   d. If we regard the 0.02-C charge as a test charge used to probe the strength of the electric field produced by the other two charges, what are the magnitude and direction of the electric field at point B?
   e. If the 0.02-C charge at point B is replaced by a −0.06-C charge, what are the magnitude and direction of the electrostatic force exerted on this new charge? (Use the electric field value to find this force.)

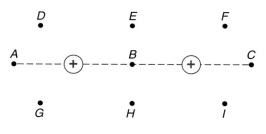

+0.10 C   2 m   +0.02 C   +0.04 C
     A   1 m   B   C

CP2. Suppose that two equal positive charges lie near one another, as shown in the diagram.
   a. Using small arrows, indicate the direction of the electric field at the labeled points on the diagram. Think about the direction of the force that would be exerted on a positive charge placed at each of these points.
   b. By drawing an equal number of field lines emerging from each charge, sketch the electric field lines for this distribution of charge. (See the diagrams in section 12.4.)

CP3. Suppose that one of the two charges in challenge problem 2 is twice as large as the other one. Use the procedures suggested in parts a and b of challenge problem 2 for this new situation. (When sketching the field, there should now be twice as many field lines emerging from the larger charge as from the smaller charge.)

CP4. Suppose that four equal positive charges are located at the corners of a square, as in the diagram.
   a. Using small arrows, indicate the direction of the electric field at each of the labeled points.
   b. Would the magnitude of the electric field be equal to zero at any of the labeled points? Explain.

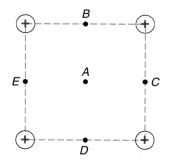

CP5. Suppose that the top plate of a parallel-plate capacitor has an electric potential of 0 V and the bottom plate has a potential of 400 V. There is a distance of 1.2 cm between the plates.
   a. What is the change in potential energy of a charge of $+3 \times 10^{-4}$ C that is moved from the bottom plate to the top plate?
   b. What is the direction of the electrostatic force exerted on this charge when it is between the plates?
   c. What is the direction of the electric field between the plates?
   d. What is the magnitude of the electric field between the plates?

home experiments and observations

HE1. The pith ball experiments described in section 12.1 can be done with homemade equipment. Small, lightly wadded pieces of paper or pieces of Styrofoam torn from a coffee cup can be used in place of the pith balls. These can be tied to pieces of thread and hung from any convenient support. Wooden pencils, plastic pens, or glass stirring rods can be used in place of the rods, and many different kinds of fabric are available in the form of clothing.

    a. Using available materials on a dry day, test to see which combinations of rod and fabric produce the best charge when rubbed.

    b. Repeat the experiments described in section 12.1. Can you get both types of charge with materials available?

HE2. To extend the observations of the previous experiment, a simple electroscope can also be constructed. Use light aluminum foil as the leaves, which can be suspended from a paper clip. One end of the clip can be straightened, poked through a piece of cardboard, and rebent. Place the piece of cardboard on top of a drinking glass or a glass jar (see the drawing), and your electroscope is complete.

    a. Test your electroscope with some of the materials suggested in the previous experiment.

    b. Test the conductivity of different materials, as in section 12.2.

    c. Try charging a metal spoon by induction. Wrap the handle of the spoon in a napkin, so that you will not discharge it when you handle it. Test the charge on the spoon with your electroscope.

Electric Circuits

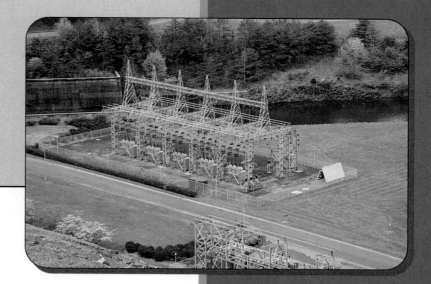

chapter overview

Investigating the concepts of electric circuits and electric current will be our main goal in this chapter. Together with electric potential (or voltage), introduced in the previous chapter, these ideas are central to understanding how simple electrical devices work. As the chapter outline indicates, we will begin with circuits and current and then explore how electric current is related to voltage (Ohm's law) and to energy and power. These ideas are then applied to household circuits.

chapter outline

253

Have you ever wondered how a flashlight works? The components are simple and familiar—a light bulb, a couple of batteries, and a cylindrical case with a switch. The operation of a flashlight is familiar, too. Pushing the switch turns the light on or off. The batteries run down and need to be replaced or recharged. Occasionally, the bulb burns out and must be replaced. But what is happening inside?

You turn electric switches on every day to produce light, heat, sound, or to run an electric motor. In fact, every time you start your car, you use an electric motor (the starting motor) powered by a battery. You know that electricity is involved in these situations, but *exactly* how things work may be murky.

Suppose that you are presented with the components of a flashlight—a bulb, a battery, and a single piece of metal wire (fig. 13.1). Your task is to get the bulb to light. How would you do it? What principles would guide you in producing a working arrangement? If you have these items handy, see if you can get the bulb to light.

This battery-and-bulb exercise presents quite a challenge to many people. Even those who quickly get the bulb to light may not be able to explain the principles involved. This simple example, though, is a good start to

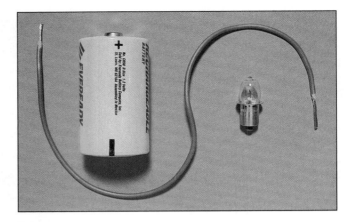

figure **13.1** A battery, a wire, and a flashlight bulb. Can you get the bulb to light?

a basic understanding of electric circuits. From the time your clock radio comes on in the morning to turning out the last light before going to bed, you are constantly using such circuits.

study suggestion

If you have the materials handy, you should try the battery-and-bulb experiment before reading further. The delight of figuring out how to get the bulb to light is something not to be spoiled by reading on prematurely. Once you get it to light (without, we hope, killing the battery), you may wish to experiment with other configurations and try to understand what distinguishes working arrangements from nonworking ones. Experimenting will help to make the concept of a circuit more vivid.

13.1 Electric Circuits and Electric Current

A flashlight, an electric toaster, and the starting motor in your car all involve electric circuits, and all use electric current to fulfill their purposes. The concepts of a circuit and current go hand in hand and are crucial to understanding how electrical devices operate. How can we use the battery-and-bulb exercise mentioned in the chapter introduction to get a handle on these ideas?

How do we get the bulb to light?

The battery-and-bulb exercise strips the flashlight down to its bare essentials—the bulb, the battery, and a single conductor. The rest of the flashlight is necessary only to hold

it together and provide a more convenient way of switching the light on and off. How do we get the bulb to light with just one wire?

What many people fail to recognize in trying this experiment is that the light bulb has two distinct connecting places that are electrically insulated from each other. You have to connect to both of them. You also have to complete a path from the light bulb to both ends of the battery. Such a *closed* or complete path is called a **circuit.** The word *circuit* itself implies a closed loop.

Figure 13.2 shows three possible arrangements, the one that works and two that do not. (Which is which?) The circuit shown in figure 13.2a is not complete. Nothing will happen. The bulb will not light, and the wire will not get warm. This case is an incomplete or *open* circuit. To be complete, a circuit must have a closed path of conducting elements joining the two ends of the battery. Without a complete path, nothing happens.

Figure 13.2b shows a complete circuit, but it does not pass through the bulb. In this arrangement, the bulb will not light but the wire will get warm. The battery will quickly die if the wire is left in place. In figure 13.2c, a wire runs from the bottom of the battery to the side of the light bulb, and the tip of the bulb rests on the other terminal of the battery. This is the arrangement that works. A complete circuit passes through the bulb and the battery.

The circuit in a flashlight is basically the same as the working arrangement in figure 13.2c. The bulb sits on top of a column of two or more batteries in direct contact with

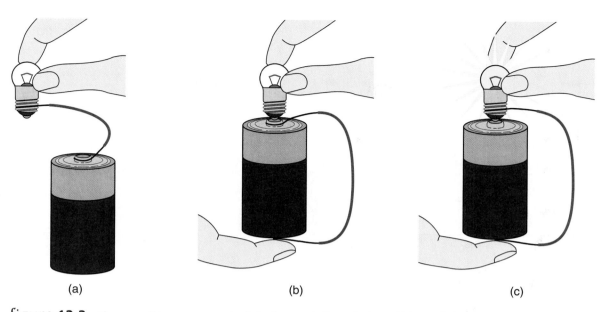

(a)   (b)   (c)

figure 13.2   Three possible arrangements of the battery, bulb, and wire. Which one lights the bulb, and why does it work?

the top battery. The side of the bulb is held within a metal sleeve, which is insulated from the rest of the flashlight canister. The switch connects this sleeve to the bottom of the battery, either through the canister itself or through a metal strip that runs to the bottom of the flashlight. Pushing the switch on closes the path. If you have a flashlight handy, take it apart and see if you can determine how this is done.

What is electric current?

Let's take a closer look at what is happening in the arrangement shown in figure 13.2c. The battery is the energy source for this circuit (as we will discuss in more detail in section 13.4). The battery uses energy from chemical reactions to separate positive and negative charges within the battery. The work done in this process produces an increase in the electrostatic potential energy of the charges, which leads to a voltage difference. Flashlight batteries typically generate a potential difference of 1.5 volts between the terminals.

Since there is an excess of positive charge at one end of the battery and an excess of negative charge at the other, these charges will tend to rejoin by flowing from one terminal to the other if we provide a suitable conducting path. These charges can flow only by an *external* conducting path, however, because of opposing forces associated with the chemical reactions inside the battery. If we simply connect a metal wire to the two terminals, charge will flow through the wire to the opposite terminal.

A flow of electric charge is an **electric current.** To be more precise, current is the *rate* of flow:

Electric current is the rate of flow of electric charge. In symbols,

$$I = \frac{q}{t},$$

where I is the symbol for electric current, q is the charge, and t is time. The direction of the current is defined as the direction of flow of positive charge.

The standard unit for electric current is the *ampere*, defined as 1 coulomb per second (1 A = 1 C/s). The coulomb is the unit of charge introduced in chapter 12. The ampere is named after the French mathematician and physicist André Marie Ampère (1775–1836), who made many contributions to the theory of electromagnetism. Some of these contributions are discussed in chapter 14. The ampere is often referred to informally as an "amp," but the correct abbreviation is A.

From the definition and its units of measurement, we see that the size of an electric current depends on how much charge flows in a given time. Figure 13.3 shows two views of charges flowing in a conductor. If the charge carriers were positively charged, their direction of motion, by definition, would be the direction of the conventional electric current (to the right, as in figure 13.3a). In reality, the charge carriers in a metal wire are electrons, which are negatively charged. Since removing negative charges increases the net

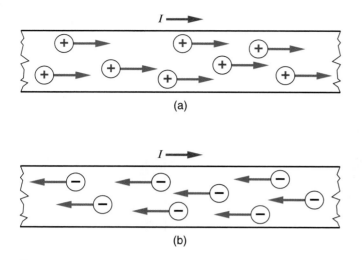

figure 13.3 Positive charges moving to the right have the same effect as negative charges moving to the left. By definition, the electric current is to the right in both cases.

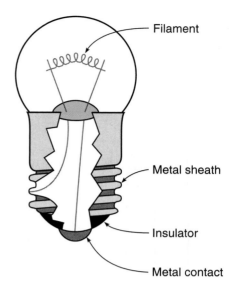

figure 13.4 A cutaway view of a flashlight bulb. The filament is connected to the metal post at the bottom and to the inside of the metal cylinder.

positive charge, negative charges flowing to the left produce the same effect as positive charges flowing to the right. The direction of the conventional electric current is to the right in figure 13.3b also—in the opposite direction to the motion of the electrons. Charge carriers are sometimes positive, such as positive ions in a chemical solution or the *holes* in a semiconductor (see chapter 21). The electric current is then in the same direction as the motion of the charge carriers.

What limits the flow of current?

If the battery in our simple circuit is fresh, the chemical reactions will continue to separate charge within the battery, and the current will continue to flow. A metal wire is an excellent conductor, however, so if we directly connect the two terminals of the battery with a wire, creating what is called a *short circuit,* a large current will flow. This will deplete the chemical reactants, and the battery will die after a short time. The wire will also become quite warm as a result of the large current. To avoid this problem, we need to place an element in our circuit (such as the light bulb) that will provide greater *resistance* to the flow of charge.

If you look closely at the light bulb, you will see that it consists of a very thin wire filament enclosed inside the glass bulb. This filament is connected to two points inside the bulb: one end is connected to the metal cylinder forming the lower sides of the bulb, and the other is connected to a metal post in the center of the bulb's base (fig. 13.4). These two points are electrically insulated from one another by a ceramic material surrounding the center post. The thin wire filament, although metal and a good conductor, restricts the flow of charge (the current) because of its very small cross-sectional area.

If we force the current to flow through the light bulb, as in figure 13.2c, a much smaller current will flow than when the wire is connected directly between the terminals of the battery. The bulb's thin wire filament restricts the amount of charge that can flow because the resistance of the filament is much greater than the resistance of the thicker connecting wires. The wire filament is the bottleneck in the circuit, and it gets very hot as we force charges through this constriction. Its high temperature makes it glow, and we have light.

An analogy to the flow of water

In discussing the battery-and-bulb circuit, we described an electric current as the flow of electric charge. Since charge is invisible, picturing the flow of charge takes a bit of imagination. An analogy can help us visualize what is happening here. Water flowing in a pipe is similar to electric current flowing in a circuit.

Figure 13.5a shows a pump that pumps water from a lower tank to a higher tank, increasing the gravitational potential energy of the water (as discussed in chapter 6). The water could remain in the higher tank indefinitely unless we provide a means for it to flow back to the lower tank. If we use a large pipe for this purpose, the water will flow back very quickly. Its rate of flow (the water current) can be measured in units of gallons per second or liters per second. The upper tank will empty quickly unless the pump runs continuously to replace the lost water. If we place a narrow pipe at some point in the path of the returning water, however, this constriction limits the amount of water that will flow. The narrow pipe provides a greater *resistance* to the flow of water.

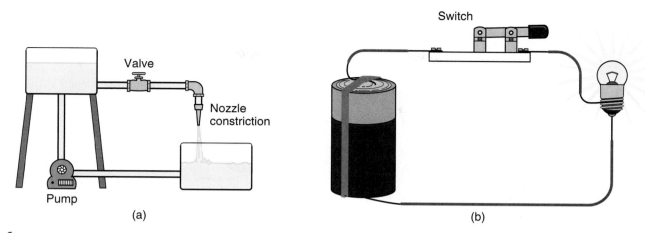

figure 13.5 Electric current is analogous to water flow. Which elements correspond in the two systems?

table **13.1**	
Corresponding Elements in the Analogy between a Water-Flow System and an Electric Circuit	
Water-flow system	**Electric circuit**
water	charge
pump	battery
pipes	wires
narrow pipe	wire filament
valve	switch

In figure 13.5b, the battery corresponds to the pump. Both produce an increase in potential energy—of water on one hand and of charge on the other. The thick pipe is like the connecting wires in the circuit, and the nozzle is analogous to the filament in the light bulb. In fact, the water does become warmer as it flows through the constriction in the pipe, like the warming of the wire filament. A valve placed at some point in the water system corresponds to a switch in an electric circuit. Table 13.1 summarizes the corresponding elements.

Understanding electric current may be easier if you keep in mind its resemblance to a water current. Both can be defined as rates of flow—of water on one hand, of charge on the other. In an electric circuit, there must be a continuous path of conductors (the circuit) for charge to flow. If we break the circuit at any point, the current stops. The water system also requires a complete loop (unless we are continually supplying water to the upper tank from an external source).

To get a bulb to light, or for any electric circuit to work, there must be a closed-loop conducting path from one terminal of the power source to the other. This closed path

is the circuit, and the rate of flow of charge around that loop is the electric current. The filament in a light bulb serves as a constriction in the conducting path that limits the flow of current. The filament gets hot as charge flows through it, thus producing light. An analogy to a water-flow system, in which a pump replaces the battery and water takes the place of charge, can help us visualize the concept of current.

13.2 Ohm's Law and Resistance

What determines the size of an electric current? In section 13.1, we noted that the large resistance of the wire filament restricted the flow of current. Is this the only effect, or does the voltage of the battery also play a role? Is it possible to predict how much current will flow in a given circuit?

How does electric current depend on voltage?

In a water-flow system, a high pressure difference between two points in the system will produce a large rate of water flow or current. High pressure can be produced by raising the storage tank to a considerable height above the point of use—this pressure is related to the gravitational potential energy of the water.

Likewise, a large difference in potential energy between the charges at the two ends of a battery is associated with a high voltage and a greater tendency for charge to flow. The size of the electric current is related to the voltage of the battery and to the resistance of the circuit elements. For most components in a circuit, there is a simple relationship between the current, the voltage, and the resistance of the component that can be used to predict the magnitude of the current. This relationship was discovered experimentally by the German physicist Georg Ohm (1789–1854) during the 1820s and is known as **Ohm's law:**

> The electric current flowing through a given portion of a circuit is equal to the voltage difference across that portion divided by the resistance. In symbols,
>
> $$I = \frac{\Delta V}{R},$$
>
> where R is the resistance and I and V stand for current and voltage, respectively.

In other words, the current is directly proportional to the voltage difference and inversely proportional to the resistance.

Ohm's law is a statement of the experimental fact that resistance is approximately constant for different values of current and voltage. In section 13.1, we used the term *resistance* in a qualitative sense as a property that restricts the flow of current. A quantitative definition of **resistance** can be obtained by using algebra to rearrange Ohm's law:

$$R = \frac{\Delta V}{I}.$$

Resistance R is the ratio of the voltage difference to the current for a given portion of a circuit. The unit of resistance is volts per ampere, which is called an *ohm* (1 ohm = 1 V/A). The ohm is often abbreviated as Ω, the Greek letter capital omega.

The resistance of a wire or other circuit element depends on several factors, including the **conductivity** of the material the wire is made of. A wire made of material with high electrical conductivity will have a smaller resistance than one made from a material of lower conductivity. The resistance also depends on the length of the wire and its cross-sectional area. The longer the wire, the greater its resistance—but a thicker wire will have a smaller resistance. Temperature also has an effect. When we heat the filament of a light bulb, its resistance increases. This is true for any metallic conductor.

If we know both the resistance of a given portion of a circuit and the applied voltage, the expected current can then be found. Suppose, for example, that a 1.5-volt battery is connected to a light bulb with a resistance of 20 ohms, as in figure 13.6. If the resistance of the battery itself is negligible, the current can be found by applying Ohm's law, $I = \Delta V/R$. Dividing 1.5 volts by the resistance of 20 ohms yields:

$$I = \frac{1.5 \text{ V}}{20 \ \Omega} = 0.075 \text{ A}.$$

This result could be expressed as 75 milliamperes (mA).

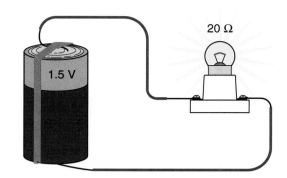

figure 13.6 A simple circuit connecting a 1.5-V battery to a 20-Ω light bulb. What is the magnitude of the electric current?

What is the electromotive force of a battery?

In finding the current through the 20-ohm light bulb, we ignored the resistance of the battery itself. We also ignored the very small resistances of the connecting wires. If the battery is fresh, its internal resistance is small and can often be neglected. The internal resistance gets larger, however, as the battery is used and its chemical reactants are depleted. In finding the current then, the total resistance of the circuit must be considered, including the resistance of the battery. This is done in the sample exercise in Try This Box 13.1, using the values shown in figure 13.7. A smaller current is obtained when the internal resistance of the battery is taken into account.

In Try This Box 13.1, we give the voltage of the battery the symbol $\mathcal{E}$, which stands for **electromotive force.** The term itself is misleading, since it is a potential difference, or voltage, *not* a force. The electromotive force is the increase in potential energy per unit charge provided by the chemical reactions in the battery. Its units are volts (joules/coulomb), and it is usually denoted by $\mathcal{E}$. (The concept of electromotive force is discussed in more detail in section 13.4, where we consider the energy aspects of circuits.)

If we measure the voltage difference across either the battery or the light bulb with a voltmeter while the bulb is operating, we obtain a value of 1.2 V, as in the second part of the sample exercise. If we break the circuit by disconnecting the light bulb and measure the voltage across the terminals of the battery again, we get a reading of about 1.5 V for the battery. This value of 1.5 V, measured when there is no current flowing through the battery, is the electromotive force of the battery. It is the voltage of the battery when there are no energy losses due to resistive effects within the battery.

To find the current in Try This Box 13.1, the electromotive force $\mathcal{E}$ of the battery is divided by the total resistance of the circuit ($I = \mathcal{E}/R$). This equation is similar to Ohm's law with $\mathcal{E}$ substituting for ΔV, but there is an important

try this box 13.1

Sample Exercise: Examining a Circuit's Current and Voltage

A 1.5-V battery with an internal resistance of 5 Ω is connected to a light bulb with a resistance of 20 Ω in a simple, single-loop circuit.
  a. What is the current flowing in this circuit?
  b. What is the voltage difference across the light bulb?

a. $\varepsilon = 1.5$ V   The total resistance of the circuit is:

$R_{\text{battery}} = 5\ \Omega$ $R = R_{\text{battery}} + R_{\text{bulb}}$

$R_{\text{bulb}} = 20\ \Omega$ $= 5\ \Omega + 20\ \Omega$

$I = ?$ $= 25\ \Omega$

$$I = \frac{\varepsilon}{R}$$

$$= \frac{1.5\ \text{V}}{25\ \Omega}$$

$$= 0.06\ \text{A} = 60\ \text{mA}$$

b. $\Delta V_{\text{bulb}} = ?$ $\Delta V = IR$

$$= (0.06\ \text{A})(20\ \Omega)$$

$$= 1.2\ \text{V}$$

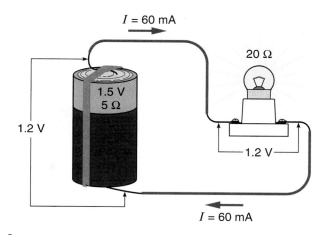

figure **13.7**   Voltage values for the battery-and-bulb circuit, assuming that the battery has an internal resistance of 5 Ω. The current is now 60 mA.

difference: Ohm's law can be applied to any portion of a circuit. The voltage difference ΔV across that portion is equal to the current times the resistance of that portion ($\Delta V = IR$). The equation involving the electromotive force, on the other hand, is applied to the entire circuit, or loop, and is sometimes referred to as the *loop equation.*

What happens when a battery dies?

As a battery gets older, its internal resistance gets larger and larger. The total resistance of the circuit increases and reduces the current flowing through the circuit, as predicted by the loop equation. As the current gets smaller, the bulb becomes dimmer until finally it glows no more. In a dead battery, the internal resistance has become so large that the battery can no longer produce a measurable current.

Surprisingly, if this dead battery is removed from the circuit and tested with a good voltmeter, it will still give a voltage reading of almost 1.5 volts. How can this be? The battery still has an electromotive force, but its internal resistance is so large that it is no longer able to deliver any appreciable current to an external element such as the light bulb. A good voltmeter does not draw much current, how-

ever, so it can still measure approximately the electromotive force of the battery. The condition of a battery is described by its internal resistance rather than by its electromotive force.

Ohm's law says that the electric current through any part of a circuit is proportional to the voltage difference across that segment and inversely proportional to the resistance of that segment. To find the current through a circuit, we divide the electromotive force of the battery by the total resistance of the circuit. This total resistance includes the internal resistance of the battery itself, which gets larger as the battery weakens.

13.3 Series and Parallel Circuits

An electric current, like a meandering stream, can split into different streams that rejoin later. Sometimes, there are advantages to arranging a circuit so that this happens rather than connecting all of the elements in a single loop. How do we describe and analyze these different ways of connecting circuits? Distinguishing between *series* and *parallel* connections is an important step of this process.

What is a series circuit?

The simple light-bulb circuit that we have been discussing is a single-loop, or *series,* circuit. In a **series circuit,** there are no points in the circuit where the current can branch into side streams or secondary loops. All of the elements line up on a single loop. The current that passes through one element must also pass through the others, since there is nowhere else for it to go. This can be seen in our diagram of this circuit in figure 13.6, but it is often easier to see in the diagrams (often called schematics) used to represent circuits.

In figure 13.8, the battery-and-bulb circuit is shown with its schematic alongside. The symbols used to represent various components in the schematic are standard. They will be recognized anywhere in the world where people study or use electric circuits. The light bulb, for example is shown as a resistance, for which the standard symbol is a zigzag line, $\Lambda\Lambda\Lambda$. In this case, it appears inside a circle that represents the glass bulb.

The lines in the schematic represent connecting wires, which are usually drawn as straight lines even though the wires are not likely to be straight in an actual circuit. We generally assume that the resistances of the connecting wires are small enough to be ignored (in comparison to the other resistances in the circuit). The schematic shows the components of the circuit and the connections between them in a way that is both clear and easy to draw.

What happens if we put more light bulbs in the circuit, *in series* with the others, as shown in both pictorial and schematic form in figure 13.9? In a series combination of resistances, each resistance contributes to restricting the flow of current around the loop. The total series resistance of the combination R_s, is the sum of the individual resistances:

$$R_{\text{series}} = R_1 + R_2 + R_3.$$

This is true no matter how many resistances are combined in series. We just add more terms to the sum if there are more resistances.

People often think that the current somehow gets used up in passing through the resistances in a series circuit. This is *not* the case. The same current must pass through each component much like the continuous flow of water in a pipe. *Voltage* is what changes as the current flows through the circuit. Voltage decreases by Ohm's law ($\Delta V = IR$) as the current passes through each resistor. The total voltage difference across the combination is the sum of these individual changes.

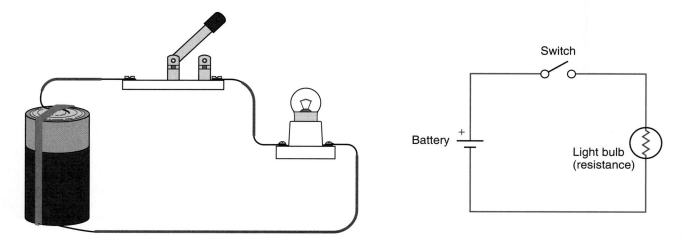

figure **13.8**   The battery-and-bulb circuit with its corresponding schematic.

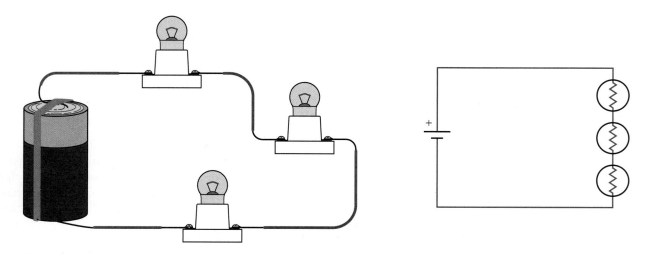

figure **13.9**   A series combination of three light bulbs and the corresponding schematic.

If two light bulbs are connected in series with a battery, the current will be less than with a single bulb, because the total series resistance is larger. The bulbs will glow less brightly in this arrangement. If we want brighter light, we would have to add more batteries in series to increase the voltage. This is what is done in many flashlights (though they usually have just one bulb).

Besides dimmer lights, there is another serious disadvantage to connecting bulbs in series. When a light bulb burns out, the filament breaks, which breaks the circuit. When one bulb in a series circuit burns out, the others also go out, because no current can flow around the broken loop. Usually, this is highly undesirable, particularly in a system like automobile lighting. It also makes it very hard to find the burned-out bulb, as it was in the old-fashioned strings of Christmas-tree lights that were connected in series.

What is a parallel circuit?

Another way of connecting the bulbs that avoids these problems is shown in figure 13.10. Here, the bulbs are not all in line around the loop. Instead, they are connected in *parallel* with one another. Note that there is now more than one loop in the circuit. In a **parallel circuit,** there are points at

which the current can *branch* or split up into different paths like the meandering stream mentioned at the beginning of this section. The flow divides and later rejoins.

Does the total effective resistance of the circuit increase or decrease if we add bulbs in parallel with one another? The water-flow analogy may help you here. Suppose that we add pipes in parallel in a water-flow system (fig. 13.10b). Does this increase or decrease the amount of water (the current) flowing in the system? What does this say about the resistance to the flow of water?

You should conclude that the flow of water is increased when we add pipes in parallel. In effect, we are increasing the total cross-sectional area the water flows through, therefore decreasing the resistance to flow. The same is true in an electric circuit when resistances are connected in parallel. Stated in terms of individual resistances, the effective parallel resistance R_p is

$$\frac{1}{R_{\text{parallel}}} = \frac{1}{R_1} + \frac{1}{R_2} + \frac{1}{R_3}.$$

To find R_p, we add the reciprocals of the individual resistances and then take the reciprocal of this sum. This always yields a value that is less than any of the individual resistances, as the example in Try This Box 13.2 illustrates.

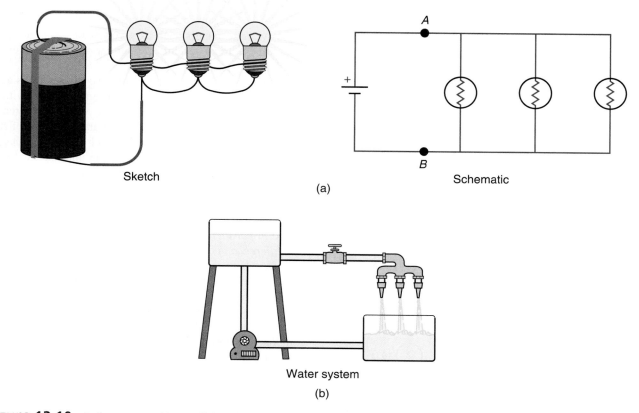

Sketch

Schematic

(a)

Water system

(b)

figure **13.10** Bulbs connected in parallel, with the schematic and the water-flow analogy also shown. The current now splits into three different paths.

try this box 13.2

Sample Exercise: Resistance and Current of Light Bulbs in Parallel

Two 10-Ω light bulbs are connected in parallel to one another, and this combination is connected to a 6-V battery, as shown in the schematic.

   a. What is the total current flowing around the loop?
   b. How much current passes through each light bulb?

a. $R_1 = R_2 = 10 \ \Omega$

   $\mathcal{E} = 6$ V

   $I = ?$

$$\frac{1}{R_p} = \frac{1}{R_1} + \frac{1}{R_2}$$

$$= \frac{1}{10 \ \Omega} + \frac{1}{10 \ \Omega}$$

$$= \frac{2}{10 \ \Omega}$$

$$R_p = \frac{10 \ \Omega}{2} = 5 \ \Omega$$

$$I = \frac{\mathcal{E}}{R}$$

$$= \frac{6 \text{ V}}{5 \ \Omega}$$

$$= \mathbf{1.2 \ A}$$

b. $I = ?$
(for either bulb)

$$I = \frac{\Delta V}{R}$$

$$= \frac{6 \text{ V}}{10 \ \Omega}$$

$$= \mathbf{0.6 \ A}$$

A current of 0.6 A flows through each bulb for a total of 1.2 A.

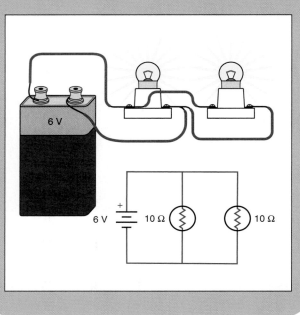

In a parallel combination of resistances, the voltage difference across each resistance is the same, since they are all connected between the same two points (points A and B in figure 13.10). The currents, on the other hand, can be different—they add to give the total current through the combination. The current splits up and recombines. A portion of the total current flows through each branch.

In Try This Box 13.2, the two bulbs have equal resistances—equal currents flow through each. If one bulb had a larger resistance than the other, the current would not divide equally. Instead, a larger portion of the current would flow through the bulb with the smaller resistance, as you might expect. The calculations of equivalent resistance and current would proceed just as shown in Try This Box 13.2.

As the example in Try This Box 13.2 shows, the resistance of the parallel combination is less than the resistance of either bulb by itself. Parallel combinations decrease the resistance and increase the amount of current that will flow. Increased current causes the bulbs to burn more brightly than in a series circuit but also depletes the batteries more quickly. There is no free lunch: we can choose dim light and a long lifetime or bright light and a short lifetime depending on how we connect the bulbs to the batteries. The energy available from the batteries is the same in either case.

Use of ammeters and voltmeters

How do we measure currents or voltage differences? You may have used meters such as a voltmeter or ammeter to check the electrical systems of your car or for other similar troubleshooting.

The *voltmeter* is the easier of the two to use and is also more commonly found in auto repair and other activities. Suppose that we want to measure the voltage difference across a light bulb in a circuit, for example. Many of the meters available are *multimeters,* which measure voltage, current, and resistance. You choose the appropriate function and range with a switch. Figure 13.11 shows both an *analog* multimeter that uses a needle and scale and a more modern *digital* multimeter that gives a digital readout.

To measure voltage, the leads of the voltmeter are placed *in parallel* with the bulb, as in figure 13.12. A voltage *difference* is the difference in potential energy per unit of charge between two points in a circuit. The voltmeter must, therefore, be connected between these two points, regardless of what other paths for current flow exist. Given the parallel connection, a voltmeter should have a large resistance, so that it does not divert much current from the component whose voltage is being measured.

Would this same type of connection work for measuring *current* with an *ammeter*? Think of water flow: to measure current, which is a rate of flow, you need to insert the gauge directly into the flow. If we were measuring water flow, we would have to cut the pipe and insert a flow gauge between the cut sections, so that the current being measured flows directly through the gauge. Likewise, in measuring electric current, we need to break the circuit and insert the meter *in series* with the other components, as in figure 13.13.

Since the ammeter is inserted in series in the circuit and must have some resistance, it will inevitably increase the total resistance of the circuit and decrease the current. An ammeter should have a small resistance, so that its effect on the current is small. Because of this small resistance, using an ammeter calls for more care than using a volt-meter. If you place an ammeter directly across the terminals of a battery, for example, a large current will flow. This could damage the meter and the battery.

With both ammeters and voltmeters, the positive terminal of the meter has to be inserted in the proper direction and connected as in figures 13.12 and 13.13. Tracing from the positive terminal of the battery or power supply, you should come first to the positive terminal of the meter. If you get it backwards, the needle of the analog meter will deflect in the wrong direction, which can damage the meter. The positive and negative terminals of a meter are usually clearly marked.

The components of a circuit can be connected either in series or in parallel, or in combinations of the two. A series connection has no branches. The current passes through each element in succession, and the resistance of the combination is the sum of the individual resistances. The current in a parallel combination splits into different branches. The effective resistance of the combination is less than any of the individual resistances. Voltmeters are connected in parallel and measure the voltage difference, but ammeters are inserted in series to measure current.

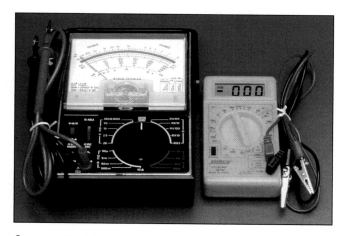

figure **13.11** An analog multimeter (with needle and separate scales) and a digital multimeter are commonly used measuring instruments.

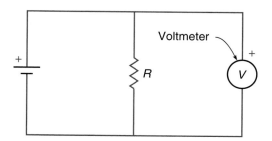

figure **13.12** A voltmeter is inserted in parallel with the element whose voltage difference is being measured.

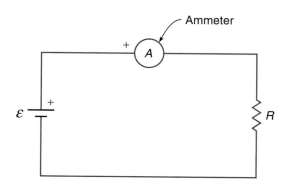

figure **13.13** An ammeter is inserted in series into the circuit whose current is being measured.

13.4 Electric Energy and Power

We have talked about batteries as sources of electrical energy in a circuit. We also use the term *power* in discussing our everyday applications of electricity. Energy or power can be used to gain more insight into the behavior of electric circuits, just as we used these ideas in chapter 6 to examine mechanical systems. What happens, for example, to the energy supplied by a battery?

What energy transformations take place in a circuit?

The analogy between electric circuits and water-flow systems can be used to good advantage here. Energy is supplied to the water-flow system by the pump, which receives its energy from some external source. Electricity, gasoline, or the wind are energy sources commonly used for pumping water. The pump increases the gravitational potential energy of the water by lifting it up to a higher tank (fig. 13.14). As the water flows down through pipes to a lower tank or reservoir, gravitational potential energy is transformed into kinetic energy of the moving water.

Once the water comes to rest in the lower tank, the kinetic energy is dissipated by frictional or *viscous* forces within the water or between the water and the inside surfaces of the pipes and tank. Frictional forces generate heat increasing the internal energy of the water and the surrounding pipes and air. This added internal energy shows up as an increase in the temperature of the water and its surroundings.

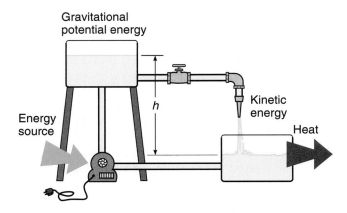

figure **13.14**   Energy transformations in the water-flow system. What happens to the energy introduced by the pump?

An analogy can be made for the electric circuit. The energy is supplied by the battery, which draws its energy from the potential energy stored in its chemical reactants. (Electric energy can also be generated from mechanical energy, as we will see in chapter 14.) The battery, much like the pump, increases the potential energy of electric charges as it moves positive charges toward the positive terminal and negative charges toward the negative terminal. When we complete the circuit by providing an external conducting path from the positive to the negative terminal, charge flows from points of higher potential energy to points of lower potential energy through the connecting wires and resistances.

As potential energy is lost, kinetic energy is gained by the moving electrons in the electric current. In the end, this kinetic energy is randomized by collisions with other electrons and atoms within the resistors in the circuit. This increases the internal energy of the resistors and connecting wires, which again shows up as an increase in temperature. Kinetic energy has been converted to heat through these collisions.

In both the water-flow system and the electric circuit, the following energy transformations take place:

energy source → potential energy → kinetic energy → heat

Pipes and resistances get warm as the current flows. In electric circuits, we often use the heat for some specific purpose, to light a lamp, to toast bread, or to warm our homes.

How is electric power related to current and voltage?

The potential difference produced by a battery when there is no current being drawn from the battery is often called the *electromotive force* and is represented by the symbol $\mathcal{E}$. It is simply the potential energy per unit charge supplied by the energy source, and its units are volts.

Since electromotive force (a voltage) represents a difference in potential energy per unit charge, multiplying electromotive force by charge yields energy. If we multiply by electric current rather than charge, we get energy per unit time, since current is the flow of charge per unit of time. This quantity is the power. Power is the rate of doing work or using energy and has units of energy per unit time. (See section 6.1.) The power supplied by any source of electrical energy is equal to the electromotive force times the current:

$$P = \mathcal{E}I.$$

A similar expression can be obtained for the power dissipated when current passes through a resistance. We replace the electromotive force $\mathcal{E}$ with the voltage difference ΔV across the resistance to give $P = \Delta VI$. Using Ohm's law, the voltage difference can be expressed as $\Delta V = IR$, so power takes the form:

$$P = (IR)(I) = I^2R.$$

The power dissipated in the resistance R is proportional to the square of the current I.

What is happening to the power in a simple circuit then? The power delivered by the battery must equal the power dissipated in the resistances in a steady-state situation. In symbols:

$$\mathcal{E}I = I^2R.$$

As long as the current remains constant, energy is supplied by the battery at a steady rate and is dissipated at the same rate in the resistances. There is no buildup of energy at any point in the circuit. The energy or power input equals the energy or power output (fig. 13.15): energy is conserved.

This principle of energy balance is the basis for the loop equation (introduced in section 13.2) used in circuit analysis. Dividing both sides of the energy-balance equation by the current I yields $\mathcal{E} = IR$, the loop equation. Conservation of energy underlies our process of analyzing circuits. Try This Box 13.3 illustrates these ideas.

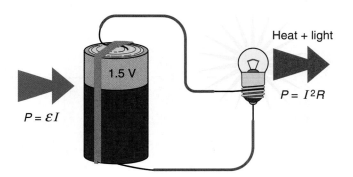

figure **13.15**   The power supplied by the battery in an electric circuit equals the power dissipated in the resistances as heat.

try this box 13.3

Sample Exercise: Analyzing a Circuit

What is the power dissipated in a 20-Ω light bulb powered by two 1.5-V batteries in series?

$\mathcal{E} = \mathcal{E}_1 + \mathcal{E}_2 = 3$ V

$R = 20\ \Omega$

$\mathcal{E} = IR$

$$I = \frac{\mathcal{E}}{R}$$

$$= \frac{3\text{ V}}{20\ \Omega}$$

$$= 0.15\text{ A}$$

$$P = I^2R$$

$$= (0.15\text{ A})^2(20\ \Omega)$$

$$= \mathbf{0.45\ W}$$

This can be checked by calculating the power delivered by the batteries:

$$P = \mathcal{E}I$$

$$= (3\text{ V})(0.15\text{ A})$$

$$= \mathbf{0.45\ W}$$

How do we distribute and use electric power?

Whenever you turn on an electric light or use an electric appliance, you are using electric power. Usually, that power does not come from a battery but from power lines (wires) that deliver power from a distant generating source (fig. 13.16). The American inventor Thomas Alva Edison (1847–1931) developed many electrical devices and played a major role in promoting the use of electrical power. Edison's in-

vention and refinement of the electric light bulb provided the original incentive for creating power-distribution systems. The ease with which electric power can be transmitted over considerable distances is one of its main advantages over other forms of energy.

What sources of energy are involved when you use electric power? The source might be the gravitational potential energy of water stored behind a dam. It could also be chemical potential energy stored in fossil fuels such as coal, oil, or natural gas, or the nuclear potential energy stored in uranium. Like the chemical fuel in a battery, fossil and nuclear fuels can all be depleted since they are mined from the earth. (What is the energy source involved in lifting the water stored behind a dam? Can it be depleted?)

In the case of chemical or nuclear fuels, the potential energy stored in the fuel is first converted to heat, then used to run a heat engine, usually a steam turbine (see chapter 11). In the case of hydroelectric power, the water stored behind the dam is run through water turbines to convert potential energy to kinetic energy. Whatever the source of energy, power plants all use electric generators that convert the mechanical kinetic energy produced by the turbines to electric energy. Electric generators are the source of the electromotive force for the power-distribution system. (How electric generators work is discussed in chapter 14 when Faraday's law of electromagnetic induction is introduced.)

When you pay your electric bill, you are paying for the amount of electric energy used during the previous month. The unit of energy used for this purpose is the kilowatt-hour, which is obtained by multiplying a unit of power (the kilowatt) by a unit of time (an hour). Since a kilowatt is 1000 watts and an hour is 3600 seconds, 1 kilowatt-hour equals 3.6 million joules. The kilowatt-hour is a much larger unit of energy than the joule (introduced in chapter 6). It is a convenient size, however, for the amounts of electrical energy typically used in a home.

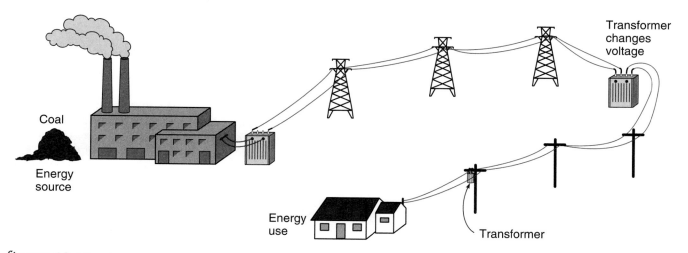

figure 13.16   Electric power is generated at a central power plant and distributed to users by power lines.

How much does it cost to light a 100-watt light bulb for one day? Electric rates vary across the country, from 4 to 5 cents per kilowatt-hour to as much as 15 cents per kilowatt-hour, but an average rate might be about 10 cents per kilowatt-hour. If the light burns for 24 hours, the energy used is found by multiplying the power by the time:

$$(100 \text{ W})(24 \text{ hr}) = 2400 \text{ Wh}$$

$$2400 \text{ Wh} = 2.4 \text{ kWh}.$$

At 10 cents a kilowatt-hour, the cost is approximately 24 cents. This may seem like a bargain, but as anyone who pays electric bills knows, the cost adds up quickly as the number of appliances multiply. Many appliances require larger amounts of power than a light bulb.

The first power-distribution systems were built late in the nineteenth century, so our growing dependence on electric power has been a twentieth-century phenomenon. The convenience of electric power for running appliances without the noise and exhaust associated with direct use of chemical fuels is the major reason for its wide use. A few places in the world still lack electric power, but it is difficult for most of us to imagine life without it.

The energy transformations that take place in an electric circuit resemble a water-flow system. The power source (pump or battery) increases the potential energy of the water or charges. This potential energy is converted to the kinetic energy of the current, which, in the end, is dissipated as heat. The rate of energy use in a circuit is equal to the electromotive force of the source times the current, which must also equal the rate at which energy is dissipated in the resistances. Various energy sources are used to generate the electric power used in our power-distribution systems.

13.5 Alternating Current and Household Circuits

What kinds of circuits are involved in our daily use of electric power? Are they similar to the simple circuits that use batteries and light bulbs? You probably know that the current we draw from a wall outlet is *alternating current* (ac) rather than *direct current* (dc). What is the distinction between these kinds of current? Can we use the same ideas employed in discussing simple battery-and-bulb circuits to describe household circuits that use alternating current?

How does alternating current differ from direct current?

Direct current implies that the current flows in a single direction from the positive terminal of a battery or power supply to the negative terminal. This is the normal result when a battery is used. **Alternating current,** on the other hand, continually reverses its direction—it flows first in one

direction, then in the other, and then back again. The alternating current used in North America goes through 60 back-and-forth cycles each second, so its frequency is 60 cycles per second or 60 hertz (Hz).

A *galvanometer* is an ammeter designed so that the needle will deflect in either direction depending on the direction of the current. If we place a galvanometer in an alternating-current circuit, the needle would fly back and forth rapidly—assuming that the meter could actually keep up with the changes. Since simple galvanometers cannot respond to such swift changes, the usual result is that the needle vibrates but remains centered on zero. The *average value* of an ordinary alternating current is zero.

An *oscilloscope* is an electronic instrument that plots electric voltage as it varies with time. We can use an oscilloscope to measure a current by displaying the voltage difference across a resistance in the circuit ($\Delta V = IR$). For an alternating current, the resulting graph would look like figure 13.17. On this graph, positive values of current represent one direction of flow, and negative values of current represent the opposite direction. This curve is called a **sinusoidal curve** because it is described by a trigonometric function, the sine. We met a similar curve earlier in our description of simple harmonic motion in chapter 6.

An alternating current's continually changing direction makes little difference for many electrical applications. In a light bulb, for example, the heating effect of the charges moving through the filament does not depend on the direction the charges are moving. Many electrical appliances make use of this heating effect (light bulbs, stoves, hair dryers, toasters, electric heaters, and so on). The operation of a toaster is described in Everyday Phenomenon Box 13.1.

Certain applications, such as electric motors, do depend on the kind of current used. We can design motors that operate on direct current, and we can also design (more easily) motors that operate on alternating current. A dc motor may not operate on alternating current, though, and an ac motor may not operate on direct current, depending upon the design. (See Everyday Phenomenon Box 14.1.)

What is the effective current or voltage?

Since the average value of an alternating current is zero, how can we characterize the size of an alternating current? Because the power dissipated in a resistance is proportional to the square of the current, it makes sense to use an average of the squared current as a measure. (Squaring a negative number yields a positive value.) To find an *effective current,* then, we first square the current, average this value over time, and take the square root of the result. For a sinusoidal variation with time, this process yields an effective current approximately seven-tenths (more accurately, 0.707) of the peak value of the current (fig. 13.17).

If the variation of the voltage across an electrical outlet is plotted against time, we get another sinusoidal curve (fig. 13.18). The effective value of this voltage, obtained in

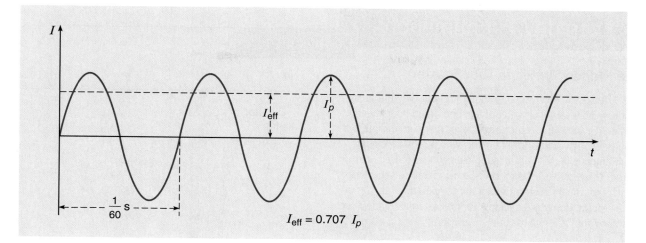

figure **13.17**   Electric current plotted as a function of time for an alternating current. The effective current I_{eff} is 0.707 times the peak current I_p.

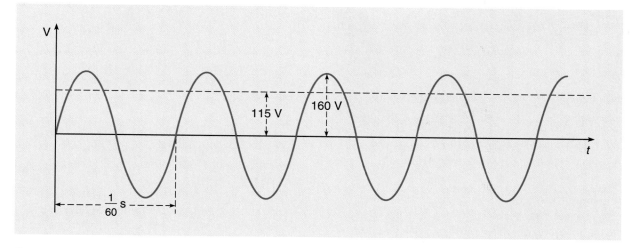

figure **13.18**   Household voltage plotted against time. The peak voltage is about 160 V if the effective voltage is 115 V.

the same manner as for the current, is typically between 110 and 120 volts in North America. The standard household power supplied in this country is 115 volts, 60 hertz ac. Other voltages and frequencies are used in other parts of the world, so electric appliances with electric motors (such as electric shavers or hair dryers) may require adapters to function properly in different places.

We can use these values of effective current and effective voltage to analyze alternating-current circuits and to compute values of electric power as we did for direct-current circuits. Try This Box 13.4 provides an example.

How are household circuits wired?

Suppose that you plug an appliance into an electrical outlet. Does this create a series circuit or a parallel circuit

with other appliances? Does the voltage available depend on what else is connected to the same circuit? These are practical questions with everyday implications.

Household circuits are always wired in parallel to assure that different appliances can be added to or removed from the circuit without affecting the voltage available to each (fig. 13.19). Usually, several outlets are wired as part of the same circuit. Each time you plug in another appliance, you create another loop in the parallel circuit.

As you add more appliances, the total current drawn from the circuit increases, because the total effective resistance of the circuit decreases when resistances are added in parallel. Too large a current could cause the wires to overheat. To protect against overheating, we add a fuse or circuit breaker in series with one leg of the circuit. If the current gets too large, the fuse will blow or the circuit breaker

Everyday Phenomenon

box 13.1

The Hidden Switch in Your Toaster

The Situation. Have you ever thought about why the toast suddenly pops up when you are toasting bread in an electric toaster? Or why an electric coffeemaker turns on and off to keep the coffee warm? Where is the switch in these circuits?

Many appliances with heating elements also use some kind of thermostat. A thermostat is a temperature-sensitive switch that breaks a circuit when the temperature reaches a certain point. How does a simple thermostat work, and where might we find it in the toaster?

Toasters, electric heaters, and coffeemakers are among the appliances that contain thermostats. How do their thermostats work?

The Analysis. If you take apart a toaster or coffeemaker (*making sure that it is unplugged first!*) and trace the wires inside the appliance, you will usually find a strip or band of metal that is part of the circuit. This strip is most often located near the entering wires, at a point where it can sense the heat generated by the heating coils of the appliance. Although simple in appearance, this special strip of metal and its connections make up the thermostat.

The special strip of metal is a *bimetallic strip* made up of two kinds of metal bonded together along their length. Because different metals expand at different rates as their temperature increases, the heating of the appliance will cause one side of the bimetallic strip to grow longer than the other side. Since the two metals are bonded together, this differential rate of expansion causes the strip to bend, as in the drawing. The metal with the larger rate of expansion lies on the outside of the curve, and the metal with the smaller rate of expansion lies on the inside, thus compensating for the greater length of the metal with the larger rate of expansion.

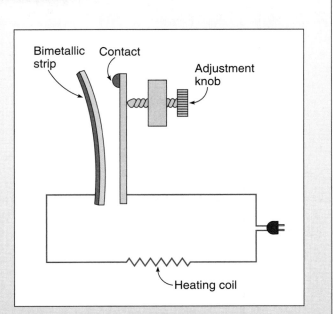

A bimetallic strip bends when heated because the two metals have different rates of thermal expansion. The bending of the strip makes or breaks a circuit.

It is not hard to see how such a device is used as a switch. If we make the strip part of the electric circuit, it can make contact with a metal tab in its unbent state and then pull away and break the circuit as it warms up and bends, as in the drawing. In a toaster, the strip is often part of a mechanical release system. When the toaster handle is pushed down, a ratchet holds it in place against the tension of a spring. The bending of the bimetallic strip releases the ratchet, causing the toast to pop up at the same time that the electric circuit is broken.

In devices like room thermostats, where greater sensitivity to small temperature changes is required, the bimetallic strip is usually bent into a coil. This allows for a much longer strip to be used in a small space. It also provides a convenient method for adjusting the set point of the thermostat. A knob or dial allows us to loosen or tighten the coil, resetting the temperature at which changes caused by heating will break the circuit.

This simple physical effect involving different rates of thermal expansion of different metals is used extensively in electric appliances. There are probably applications that have not yet been invented. Perhaps you could dream up a new one and obtain the patent.

figure 13.19 A typical household circuit may have several appliances connected in parallel with one another. A fuse or circuit breaker is in series with one leg of the circuit.

try this box 13.4

Sample Exercise: Light-Bulb Physics

A 60-W light bulb is designed to operate on 120 V ac.
   a. What is the effective current drawn by the bulb?
   b. What is the resistance of the bulb's filament?

a. $P = 60$ W $\qquad\qquad P = I\Delta V$

$\Delta V_{\text{effective}} = 120$ V $\qquad I = \dfrac{P}{\Delta V}$

$I_{\text{effective}} = ?$

$\qquad\qquad\qquad\qquad = \dfrac{60 \text{ W}}{120 \text{ V}}$

$\qquad\qquad\qquad\qquad = \mathbf{0.5\ A}$

b. $R = ?$ $\qquad$ From Ohm's law:

$\qquad\qquad\qquad \Delta V = IR$

$\qquad\qquad R = \dfrac{\Delta V_{\text{effective}}}{I_{\text{effective}}}$

$\qquad\qquad\quad = \dfrac{120 \text{ V}}{0.5 \text{ A}}$

$\qquad\qquad\quad = \mathbf{240\ \Omega}$

table 13.2

Power and Current Ratings of Some Common Appliances

Appliance	Power (W)	Current (A)
stove	6000 (220 V)	27
clothes dryer	5400 (220 V)	25
water heater	4500 (220 V)	20
clothes washer	1200	10
dishwasher	1200	10
electric iron	1100	9
coffeemaker	1000	8
toaster	850	7
hair dryer	650	5
food processor	500	4
large fan	240	2
color television	100	0.8
small fan	50	0.4
personal computer	45	0.4
clock radio	12	0.1

These values vary, depending on the size and design of a particular appliance.

will trip. The entire circuit is disrupted, and no current will flow to any of the attached appliances.

The current or power rating of an appliance indicates the maximum current normally used by that appliance. A 60-watt light bulb will draw just half an ampere, as in the sample exercise in Try This Box 13.4. Since a typical household circuit is fused at 15 to 20 A, several 60-watt bulbs could be turned on without blowing the fuse.

A toaster or a hair dryer, on the other hand, needs much more current than the typical light bulb. A toaster will often draw as much as 5 to 10 amperes by itself and can cause problems if connected to the same circuit as another appliance with a heating element. You will find current or power ratings printed somewhere on the appliance. Table 13.2 gives some typical values. Except where otherwise indicated, an effective line voltage of 120 V is used to compute the current from the power ratings in this table. Appliances with larger power requirements like stoves, clothes dryers, and water heaters are usually connected to a separate 220-V line.

As you can see from table 13.2, appliances with heating elements require more power than appliances, such as fans or food processors, that use power mainly to run an electric motor. Electronic devices such as televisions or radios require even less power. It is a good idea to check the power or current rating of appliances and to be aware of what other appliances are present when you plug a toaster or similar appliance into a new location.

You do not have to be an electrician to grasp the basics of household circuits. The fuse or breaker box in your house should indicate where each circuit is located in the house and the value of the current limit of the fuse or circuit breaker. Replacing a fuse or resetting a breaker is a routine operation that all of us may need to do. Under-

standing the ideas presented in this chapter will help us use household appliances safely.

Direct current flows in a single direction, but alternating current flows back and forth, changing its direction continually. Because the average values of alternating current and voltage are zero, we describe the magnitudes of these quantities by using the effective current or voltage. Household circuits operate on 115 V, 60 Hz ac in North America. When an appliance is plugged into an outlet, it connects in parallel with other appliances on that circuit. A fuse or circuit breaker in one leg of the circuit limits the total current that can flow in that circuit. You should be aware of the current requirements when several appliances are operating on the same circuit, particularly if the appliances have heating elements.

---

summary

Starting with the simple example of lighting a flashlight bulb, we introduced the concepts of electric circuit, electric current, and electric resistance. Ohm's law and the relationship between power and current were also used to analyze series and parallel connections and some basic features of household circuits.

1   Electric circuits and electric current.   An electric circuit is a closed conducting path around which charge can flow. The rate of flow of charge is called *electric current*. It has units of charge per unit of time, or amperes.

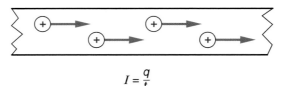

$$I = \frac{q}{t}$$

2   Ohm's law and resistance.   Resistance is the property of a circuit element that opposes the flow of current. Ohm's law states that the current that flows through an element is proportional to the voltage difference across that element and inversely proportional to the resistance.

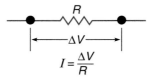

$$I = \frac{\Delta V}{R}$$

3   Series and parallel circuits.   When elements are connected in line so that the current that flows through one must also flow through the others, the elements are connected in *series*. When the current can branch into different paths, the elements are connected in *parallel*. The equivalent resistance of a parallel combination is less than any of its components.

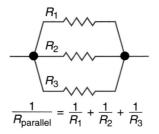

$$R_{series} = R_1 + R_2 + R_3$$

$$\frac{1}{R_{parallel}} = \frac{1}{R_1} + \frac{1}{R_2} + \frac{1}{R_3}$$

4 Electric energy and power.

Since voltage is potential energy per unit charge, multiplying a voltage difference by charge yields energy. Since current is the rate of flow of charge, multiplying a voltage difference by current yields power, the rate of energy use. The power supplied by a source must equal the power dissipated in the resistances.

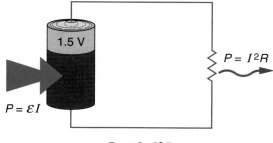

$$P = \mathcal{E}I = I^2R$$

5 Alternating current and household circuits.

Household circuits use alternating current, which is continually reversing its direction, unlike direct current, which flows in one direction only. When we plug in appliances, we connect them in parallel with other appliances on the same circuit. The fuse or circuit breaker used to limit the current is in series with one leg of the circuit.

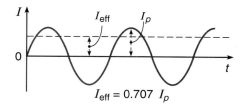

$$I_{eff} = 0.707 \ I_p$$

key terms

Circuit, 254
Electric current, 255
Ohm's law, 258
Resistance, 258

Conductivity, 258
Electromotive force, 258
Series circuit, 259
Parallel circuit, 261

Direct current, 266
Alternating current, 266
Sinusoidal (sine) curve, 266

questions

*Questions identified with an asterisk are more open-ended than the others. They call for lengthier responses and are more suitable for group discussion.

Q1. Two arrangements of a battery, bulb, and wire are shown below. Which of the two arrangements, if either, will light the bulb? Explain.

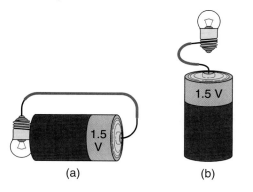

Q2. Suppose that you have two wires, a battery, and a bulb. One of the wires is already in place in each of the arrangements shown in the next column. Indicate with a drawing

where you would place the second wire to get the bulb to light. Explain your decision in each case.

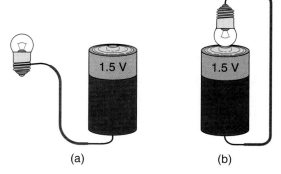

Q3. In a simple battery-and-bulb circuit, is the electric current that enters the bulb on the side nearer to the positive terminal of the battery larger than the current that leaves the bulb on the opposite side? Explain.

Q4. Are electric current and electric charge the same thing? Explain.

Q5. Consider the circuit shown, where the wires are connected to either side of a wooden block as well as to the light bulb. Will the bulb light in this arrangement? Explain.

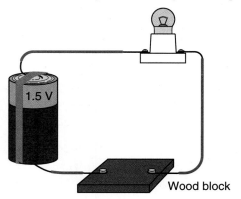

Wood block

*Q6. Consider the circuit shown. Could we increase the brightness of the bulb by connecting a wire between points A and B? Explain.

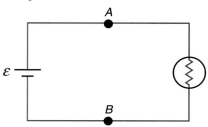

*Q7. Two circuit diagrams are shown. Which one, if either, will cause the light bulb to light? Explain your analysis of each case.

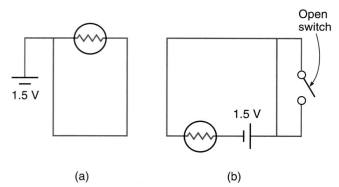

(a)   (b)

Q8. Suppose that we use an uncoated metal clamp to hold the wires in place in the battery-and-bulb circuit shown. Will this be effective in keeping the bulb burning brightly? Explain.

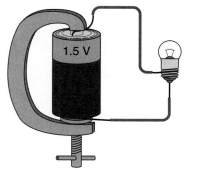

Q9. Consider the two signs shown, located in different physics labs. Which of the two would be reason for greater caution? Explain.

Q10. If we decrease the potential difference across a resistance in a circuit, will the current flowing through that resistance increase, remain the same, or decrease? Explain.

Q11. A dead battery will still indicate a voltage when a good voltmeter is connected across the terminals. Can the battery still be used to light a light bulb? Explain.

Q12. When a battery is being used in a circuit, will the voltage across its terminals be less than that measured when there is no current being drawn from the battery? Explain.

Q13. Two resistors are connected in series with a battery as shown in the diagram. R_1 is less than R_2.
   a. Which of the two resistors, if either, has the greater current flowing through it? Explain.
   b. Which of the two resistors, if either, has the greatest voltage difference across it? Explain.

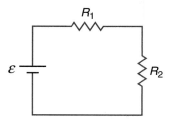

Q14. In the circuit shown below, R_1, R_2, and R_3 are three resistors of different values. R_3 is greater than R_2, and R_2 is greater than R_1. $\mathcal{E}$ is the electromotive force of the battery whose internal resistance is negligible. Which of the three resistors has the greatest current flowing through it? Explain.

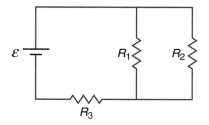

*Q15. In the circuit shown in question 14, which of the three resistors, if any, has the largest voltage difference across it? Explain.

*Q16. If we disconnect R_2 from the rest of the circuit shown in the diagram for question 14, will the current through R_3 increase, decrease, or remain the same? Explain.

Q17. When current passes through a series combination of resistors, does the current get smaller as it goes through each successive resistor in the combination? Explain.

Q18. In the circuit shown, the circle with a *V* in it represents a voltmeter. Which of the following statements is correct? Explain.
   a. The voltmeter is in the correct position for measuring the voltage difference across *R*.
   b. No current will flow through the meter, so it will have no effect.
   c. The meter will draw a large current.

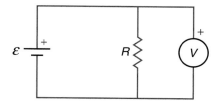

Q19. In the circuit shown, the circle with an *A* in it represents an ammeter. Which of these statements is correct? Comment on each.
   a. The meter is in the correct position for measuring the current through *R*.
   b. No current will flow through the meter, so it will have no effect.
   c. The meter will draw a significant current from the battery.

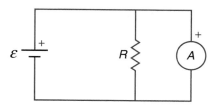

Q20. Which will normally have the larger resistance, a voltmeter or an ammeter? Explain.

Q21. Is electric energy the same as electric power? Explain.

Q22. If the current through a certain resistance is doubled, does the power dissipated in that resistor also double? Explain.

Q23. Does the power being delivered by a battery depend on the resistance of the circuit connected to the battery? Explain.

Q24. What energy source increases the potential energy of the water behind the dam of a hydroelectric power plant? Explain.

*Q25. Does a battery connected to an electric motor represent a perpetual-motion machine (see chapter 11)? Explain.

Q26. In using a dc voltmeter, it is important to connect the positive terminal of the meter in the correct direction in the circuit relative to the positive terminal of the battery. Is this likely to be true for the use of an ac voltmeter? Explain.

Q27. Which of these appliances is most likely to cause an overload problem when connected to a circuit that already has other appliances drawing current from it: an electric shaver, a coffeemaker, or a television set? Explain.

Q28. Would it make sense to connect a fuse or circuit breaker in parallel with other elements in a circuit? Explain.

Q29. Suppose that the appliances connected to a household circuit were connected in series rather than in parallel. What disadvantages would there be to this arrangement? Explain.

Q30. How does a bimetallic strip break a circuit when things heat up? Explain.

exercises

E1. A charge of 30 C passes at a steady rate through a resistor in a time of 5 s. What is the current through the resistor?

E2. A current of 2.5 A flows through a battery for 1 min. How much charge passes through the battery in that time?

E3. A 24-Ω resistor in a circuit has a voltage difference of 6 V across its leads. What is the current through this resistor?

E4. A current of 2.5 A is flowing through a resistance of 18 Ω. What is the voltage difference across this resistance?

E5. A current of 0.6 A flows through a resistor with a voltage difference of 120 V across it. What is the resistance of this resistor?

E6. Three 20-Ω resistors are connected in series to a 6-V battery of negligible internal resistance.
   a. What is the current flowing through each resistor?
   b. What is the voltage difference across each resistor?

E7. A 40-Ω resistor and a 60-Ω resistor are connected in series to a 12-V battery.
   a. What is the current flowing through each resistor?
   b. What is the voltage difference across each resistor?

E8. In the circuit shown, the 1-Ω resistance is the internal resistance of the battery and can be considered to be in series, as shown, with the battery and the 9-Ω load.
   a. What is the current flowing through the 9-Ω resistor?
   b. What is the voltage difference across the 9-Ω resistor?

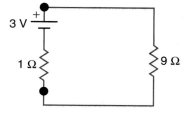

E9. Three resistors are connected to a 6-V battery as shown. The internal resistance of the battery is negligible.
a. What is the current through the 15-Ω resistance?
b. Does this same current flow through the 25-Ω resistance?
c. What is the voltage difference across the 20-Ω resistance?

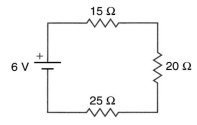

E10. Two resistors, each having a resistance of 6 Ω, are connected in parallel. What is the equivalent resistance of this combination?

E11. Three resistors of 3 Ω, 6 Ω, and 2 Ω are connected in parallel with one another. What is the equivalent resistance of this combination?

E12. Three identical resistances, each 24 Ω, are connected in parallel with one another as shown. The combination is connected to a 12-V battery whose internal resistance is negligible.
a. What is the equivalent resistance of this parallel combination?
b. What is the total current through the combination?
c. How much current flows through each resistor in the combination?

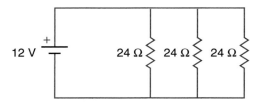

E13. A 9-V battery in a simple circuit produces a current of 1.5 A through the circuit. How much power is being delivered by the battery?

E14. A 24-Ω resistor has a voltage difference of 3 V across its leads.
a. What is the current through the resistor?
b. What is the power being dissipated in this resistor?

E15. A 60-W light bulb operates on an effective ac voltage of 110 V.
a. What is the effective current through the light bulb?
b. From Ohm's law, what is the resistance of the light bulb?

E16. A toaster draws a current of 7 A when it is connected to a 110-V ac line.
a. What is the power consumption of this toaster?
b. What is the resistance of the heating element in the toaster?

E17. A clothes dryer uses 5500 W of power when connected to a 220-V ac line. How much current does the dryer draw from the line?

challenge problems

CP1. In the circuit shown, the internal resistance of the battery can be considered negligible.
a. What is the equivalent resistance of the two-resistor parallel combination?
b. What is the total current flowing through the battery?
c. What is the current flowing through the 6-Ω resistor?
d. What is the power dissipated in the 8-Ω resistor?
e. Is the current flowing through the 8-Ω resistor greater or less than that flowing through the 6-Ω resistor? Explain.

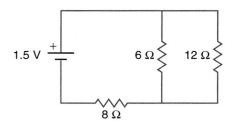

CP2. Three 30-Ω light bulbs are connected in parallel to a 1.5-V battery with negligible internal resistance.
a. What is the current flowing through the battery?
b. What is the current flowing through each bulb?
c. If one bulb burns out, does the brightness of the other two bulbs change? Explain.

CP3. In the circuit shown on the next page, the 6-V battery is opposing the 9-V battery as they are positioned. The total voltage of the two batteries will be found by subtracting.
a. What is the current flowing around the circuit?
b. What is the voltage difference across the 20-Ω resistor?

c. What is the power delivered by the 9-V battery?

d. Is the 6-V battery discharging or charging in this arrangement?

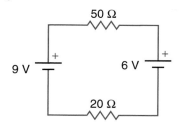

CP4. In the combination of 3-Ω resistors shown in the diagram, there are two different parallel combinations that, in turn, are in series with the middle resistor.

a. What is the resistance of each of the two parallel combinations?

b. What is the total equivalent resistance between points A and B?

c. If there is a voltage difference of 6 V between points A and B, what is the current flowing through the entire combination?

d. What is the current flowing through each of the resistors in the three-resistor parallel combination?

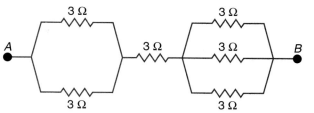

CP5. A 600-W toaster, a 1200-W iron, and a 500-W food processor are all connected to the same 115-V household circuit, fused at 15 A.

a. What is the current drawn by each of these appliances?

b. If these appliances are all turned on at the same time, will there be a problem? Explain.

c. What is the resistance of the heating element in the iron?

home experiments and observations

HE1. If you have a flashlight handy, you have a working example of a simple battery-and-bulb circuit.

a. Open the flashlight and remove the batteries and bulb. Can you see how the switch operates? What makes contact when the switch is closed? Sketch the switch to make this clear.

b. How are the connections made to the bulb in the flashlight?

c. Leaving the bulb in its socket, can you get it to light by using external wires to make connections to the batteries? Sketch the arrangement that works.

HE2. If you have two flashlight bulbs, a flashlight battery, and some wires available, you can try your hand at building your own series and parallel circuits. The wires can be looped around the side of the bulb to hold the bulbs, and the contact with the base may be made to any metal object to which you can touch another wire. The bulb sockets from a flashlight may be useful here.

a. Construct a circuit with the two bulbs in series with one another and with the battery. Note the brightness of the bulbs when they are lit. (They may barely be glowing if you are using just one battery.)

b. Now construct a circuit with the two bulbs in parallel with one another. Note the brightness of the bulbs in this case.

c. Construct a circuit with just one bulb and compare the brightness of the bulb in this situation to the brightness noted in parts a and b.

HE3. Some light bulbs are designed to operate at three different brightness levels. Locate such a bulb either in your home or residence hall, or in a local store.

a. Unscrew the bulb from the socket and examine the base of the bulb. Sketch carefully the points at which electrical contact can be made with the base of the bulb.

b. *Unplug the lamp* and examine the socket. How does it make electrical contact with the bulb? (If you do not have such a lamp handy, your local hardware store will have three-way sockets for sale that you can examine.)

c. If you can see inside the bulb, how many filaments can you identify?

d. Develop an explanation for how the bulb can produce three different brightness levels depending upon the switch position.

HE4. Locate all of the electrical appliances that you use in your residence-hall room or in a single room in your house.

a. Can you find current or power ratings on these appliances? What are they?

b. Even if power ratings cannot be found for each device, can you guess (with the help of table 13.2) approximately how much current each appliance is likely to draw? Rank the appliances in order from the one likely to draw the most current to the one that would draw the least.

c. Is there likely to be a problem if all of the appliances are turned on at the same time on a single circuit fused at 20 A?

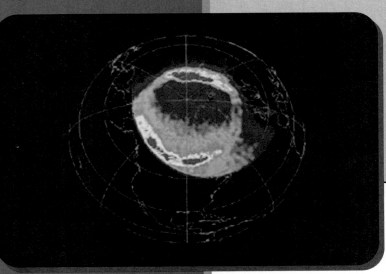

Magnets and Electromagnetism

chapter
14

unit three

chapter overview

After a description of the behavior of magnets that builds on similarities to the electrostatic force, we will explore the relationship between electric currents and magnetic forces. We then introduce Faraday's law, which describes how electric currents can be generated by interaction with magnetic fields (electromagnetic induction). Our main objectives will be to get a grasp of the nature of magnetic forces and fields and of Faraday's law of electromagnetic induction.

chapter outline

1 **Magnets and the magnetic force.** What are the poles of a magnet? How is the magnetic force similar to the electrostatic force? Why does a compass work, and what is the nature of the earth's magnetic field?

2 **Magnetic effects of electric currents.** What is the relationship between electric currents and magnetic forces and fields? How can we describe the magnetic force in terms of moving charges?

3 **Magnetic effects of current loops.** What are the magnetic characteristics of current loops? In what way is a loop of electric current like a bar magnet?

4 **Faraday's law: electromagnetic induction.** How can electric currents be produced using magnetic fields? What is Faraday's law of electromagnetic induction?

5 **Generators and transformers.** How can Faraday's law be used to explain how electric generators and transformers work? What role do these devices play in the production and transmission of electric power?

Do you remember playing with refrigerator magnets as a child? Plastic letters and figures, backed with small magnets, can be arranged at your pleasure on the refrigerator door. They stick to the steel door because of the magnets, a fact that you have accepted since you were four years old or so. They are a wonderful teaching tool as well as a plaything, since the letters can be arranged to spell simple words (fig. 14.1).

Because of such toys, we are generally more familiar with magnetic forces than with electrostatic forces. We have played with small horseshoe magnets and compasses—maybe even with simple *electromagnets* made from a steel nail, some wire, and a battery. We know that there is a force that attracts some types of metal but not others to magnets. Like the gravitational force and the electrostatic force, this force acts even when the objects are not in direct contact with one another.

Although the magnetic force is familiar, it is also mysterious. Beyond these simple facts just listed, your understanding of the magnetic force is likely to be limited. In what ways is it like the electrostatic force, and how does it differ? On a more fundamental level, is there a relationship between electrical effects and magnetism? Our mention of electromagnets hints at such a connection.

There is indeed a connection between electric currents and magnetic forces. These relationships were discovered in the early nineteenth century and culminated in a theory of electromagnetism developed during the 1860s by the Scottish physicist James Clerk Maxwell. In the early part of the twentieth century, it became clear

figure **14.1**   A child playing with magnet-backed letters that stick to the refrigerator door. What is the force that holds the letters to the door?

that the electrostatic force and the magnetic force are really just different aspects of one fundamental electromagnetic force.

Our understanding of the relationships between electricity and magnetism has led to numerous inventions that play enormous roles in modern technology. These include electric motors, electric generators, transformers, and many other devices.

14.1 Magnets and the Magnetic Force

If you gather up a few magnets from around the home or office (fig. 14.2), you can establish basic facts about their behavior with some simple experiments. You probably already know that magnets attract paper clips, nails, or any metallic item made of iron or steel. They do not attract items made of silver, copper, or aluminum or *most* nonmetallic materials. Magnets also attract each other but only if the ends are properly aligned. If the ends are not properly aligned, the force between them is repulsive rather than attractive.

The three common magnetic elements are the metals iron, cobalt, and nickel. The small magnets for sticking things to refrigerator doors or metal cabinets are made of combinations of these three metals, with other elements added to obtain desired properties. The first known magnets were a form of iron ore called *magnetite,* which occurs naturally and is often weakly magnetized. The existence of magnetite was known in antiquity, and its properties had long been a source of curiosity and amusement.

What are magnetic poles?

If you look closely at some of the magnets in your collection, you may find their ends labeled with the letters *N* and

figure **14.2**   A collection of magnets. Where are the poles likely to be found on each magnet?

S for *north* and *south.* If we explored the origin of these labels, we would discover that they originally stood for *north-seeking* and *south-seeking.* If you suspend a simple bar magnet by a thread tied around its middle and allow it to turn freely about the point of suspension, it eventually comes to rest with one end pointing approximately north-ward. This is the end or **magnetic pole** that we label *N;* the other is then labeled *S.*

If your magnets are labeled in this manner, you will quickly discover that the opposite poles of two magnets at-tract one another. The north pole of one attracts the south pole of the other. If you hold the magnets firmly in each hand and bring two north poles near one another, you can feel a repulsive force pushing them apart. The same is true of two south poles. In fact, if one of the magnets is lying on the table and you try to bring two like poles together, the magnet on the table will suddenly flip around so that the op-posite pole comes into contact with the pole of the approach-ing magnet.

These simple observations can be summarized in a rule that you probably first learned as part of a science unit in grade school (fig. 14.3):

> Like poles repel one another, and unlike poles attract one another.

There is an obvious similarity here to the rules for the elec-trostatic forces between like and unlike charges. The rule for magnets, however, was known well before the electro-static rule.

It is fun to play with magnets, chasing them around with each other. Small disk or ring magnets usually have their poles on opposite sides of the disk. If you bring a small bar magnet nearby with like poles facing each other, the disk may hop away or do a quick flip and attach itself to the bar magnet. Sometimes the flip is so quick that it is difficult to see. If the magnets are shaped like small rings, you can get them to levitate easily on a thin wooden post, as shown in figure 14.4.

Some magnets seem to have more than two poles: there may be a south and north pole somewhere in the middle of the magnet. Iron filings sprinkled on a piece of paper lying on top of the magnet will show where its poles are. The fil-ings will group most densely near the poles.

The magnetic force and Coulomb's law

The similarity between the behavior of magnetic poles and electrostatic charges goes well beyond the rules for attrac-tion and repulsion. If we attempt to measure how the force

figure 14.4   Ring magnets levitating on a vertical wooden rod. The rod keeps the magnets from flying off to the side.

that two poles exert on one another varies with distance or pole strength, we find a behavior similar to Coulomb's law for the electrostatic force. In fact, Coulomb himself made the initial measurements and first stated the force law for magnets.

Coulomb made his measurements with magnets by sub-stituting long, thin bar magnets for the metal balls in his torsion balance (see section 12.3). The magnets have to be long so that the opposite poles are far enough away from the point of measurement that their effect on the force can be ignored.

Coulomb's experiments showed that the magnetic force between two poles decreases with the square of the distance between the two poles, just as the electrostatic force does. The force is also proportional to a quantity called the *pole strength* of each pole. Some magnets are stronger than oth-ers, and the magnitude of the force exerted by one magnet on the other depends on the pole strength of both magnets.

Can we associate field lines with magnets?

A magnet is always at least a **magnetic dipole:** we cannot completely isolate a *single* magnetic pole. A dipole con-sists of two opposite poles separated by some distance. Al-though there can be more than two poles, apparently there can never be fewer than two. Physicists have invested con-siderable effort looking for **magnetic monopoles** (particles consisting of a single isolated magnetic pole), but there is

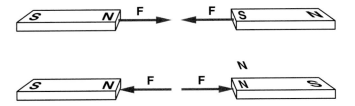

figure 14.3   Unlike poles attract one another, and like poles repel one another.

no conclusive evidence that they exist. Breaking a magnetic dipole in half always produces two dipoles.

In this one respect, then, magnetic poles are *not* similar to electric charges. Positive and negative charges can be isolated. Electric dipoles do exist, consisting of equal but opposite charges separated by a small distance, as in figure 14.5. The electric field lines produced by an electric dipole originate on the positive charge and end on the negative charge.

Does a magnetic dipole have similar field lines? We can indeed define a *magnetic field* whose field lines for a magnetic dipole are similar to the electric field lines for an electric dipole. (The definition of magnetic field is considered in more detail in section 14.2.) The magnetic field lines emerge from the north pole of the dipole and go into the south pole (fig. 14.5). Unlike electric field lines, however, the magnetic field lines do not end—they form continuous loops. You can make this pattern visible by sprinkling iron filings on a piece of paper covering a bar magnet. The filings line up in the direction of the field.

If we place an electric dipole in an electric field produced by other charges, the dipole will line up with this field. The forces acting on each charge in the dipole combine to produce a torque on the dipole that turns it toward the direction of the field (fig. 14.6). Similarly, a magnetic dipole lines up with an externally produced magnetic field. This is why iron filings line up with the field lines around a magnet. The filings become magnetized in the presence of the field—each one becomes a small magnetic dipole.

Is the earth a magnet?

A compass needle is a magnetic dipole. The first compasses were made by balancing a thin crystal of magnetite on a support so that it could turn freely. As you are aware, the north (or north-seeking) pole of the magnet will point north, although not exactly due north. The invention of the compass became a tremendous aid to ocean navigation in the early Renaissance. Before then, sailors were unable to navigate on cloudy days or nights when they could not see the sun or stars.

Is the earth a magnet? What kind of magnetic field is the compass needle responding to? The compass was invented by the Chinese, but an English physician named William Gilbert (1540–1603) was the first to study these phenomena thoroughly. Gilbert suggested that the earth behaves like a large magnet. You can picture the magnetic field produced by the earth by imagining that inside the earth there is a large bar magnet oriented as shown in figure 14.7. (Do not take this picture literally. It is merely a device for describing the field.)

Since unlike poles attract, the south pole of the earth's magnet must point in a northerly direction. The north-seeking pole of the compass aligns itself northward along the field lines produced by the earth. The axis of the earth's magnetic field, though, is not aligned exactly with the earth's axis of rotation. Since the rotational axis defines geographic or true north, the compass needle does not point exactly north at most locations. On the east coast of the United States, magnetic north is a few degrees to the west of true north, while on the west coast it is several degrees to the east of true north. Somewhere in the middle of the United States, magnetic north and true north are identical. The precise location of this line varies slowly with time.

We do not know exactly how the earth's magnetic field is produced, although models have been developed that account for many of its features. Most of these models assume that electric currents associated with the motion of fluids in the core of the earth are responsible. But what do electric currents have to do with magnetic fields?

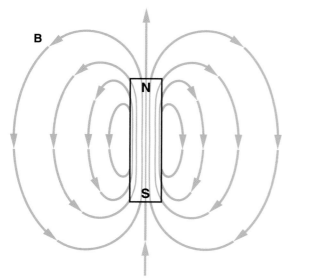
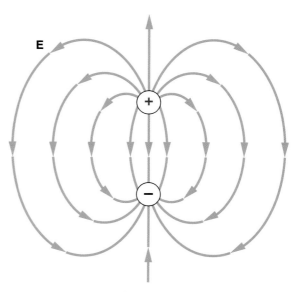

figure **14.5** Magnetic field lines produced by a magnetic dipole form a pattern similar to the electric field lines produced by an electric dipole. However, the magnetic field lines form continuous loops.

Magnetic field **B**   Electric field **E**

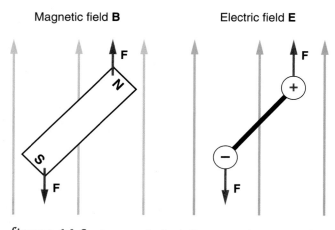

figure 14.6   A magnetic dipole lines up with an externally produced magnetic field just as an electric dipole lines up with an electric field.

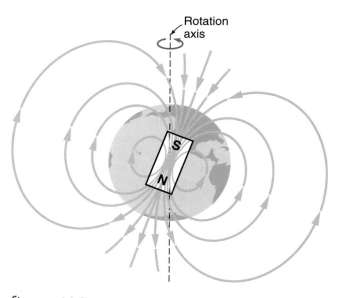

figure 14.7   The magnetic field of the earth can be pictured by imagining a bar magnet inside the earth (there is not, of course), oriented as shown here.

Simple magnets have two poles, usually labeled north and south, which obey force rules like those for electric charges. Like poles repel and unlike poles attract, and the force exerted by one on another is inversely proportional to the square of the distance between them. As far as we know, isolated magnetic monopoles do not exist: the simplest magnet is a dipole. Magnetic field lines for a magnetic dipole form a pattern similar to the electric field lines for an electric dipole. The earth itself resembles a large magnetic dipole with its south magnetic pole pointing more or less northward.

14.2 Magnetic Effects of Electric Currents

The invention of the battery by Volta in 1800 made it possible to produce steady electric currents for the first time. Previously, currents could be produced only as rapid discharges of charge accumulated in electrostatic experiments. Connecting a long, thin wire across the terminals of Volta's battery permitted a much steadier flow of charge. Scientists were then able to study electric currents in ways not previously possible.

The similarities between magnetic and electrostatic effects discussed in section 14.1 led scientists to suspect that a direct connection between electricity and magnetism might exist. The same people were often involved in the study of both areas. William Gilbert explored electrostatic effects as well as magnetism, and Charles Coulomb measured the force law for both magnetic poles and electric charges. Twenty years after the invention of the battery, a striking discovery was made by the Danish scientist Hans Christian Oersted (1777–1851).

An unexpected effect

Oersted's initial discovery of the magnetic effect of an electric current was made during a lecture demonstration in 1820. Demonstrations often fail to go exactly as planned, but this is one case when failure was fortuitous. Oersted was showing the effects of electric currents and happened to have a compass handy. He noticed that the compass needle deflected when he completed a circuit consisting of a long wire and a battery.

Oersted had used a compass near a current-carrying wire before but had not noticed any effect. Other scientists had also looked for such effects without success, so the deflection of the compass needle during Oersted's lecture was unexpected. Not wishing to make a fool of himself in front of his students, he decided to explore the situation more carefully after the lecture was over. He found a reproducible effect on the compass needle as long as the current was sufficiently strong and the compass and wire were situated in certain ways.

The strange directional aspects of this newly discovered effect may explain why it had not been observed earlier. To get the maximum effect with a horizontal wire, the wire must be oriented along a north-south line (along which the compass needle would point in the absence of current). When the current is turned on, the needle deflects away from north (fig. 14.8). Apparently, the magnetic field produced by the current in the wire is *perpendicular* to the direction of the current.

Further study by Oersted and others showed that the magnetic field lines produced by a straight, current-carrying wire form circles centered on the wire. Oersted did not talk about field lines but about the direction that the compass needle pointed at various locations. When he placed the

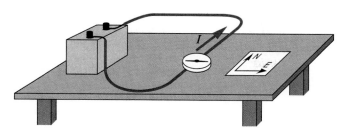

figure **14.8** With the wire oriented along a north-south line, the compass needle deflects away from this line when there is current flowing in the wire.

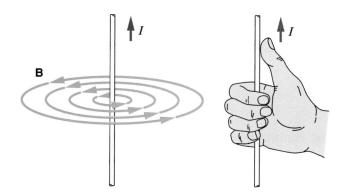

figure **14.9** The right-hand rule gives the direction of the magnetic field lines that encircle a current-carrying wire. The thumb points in the direction of the current and the fingers curl in the direction of the field lines.

compass below the wire, the needle deflected in the opposite direction from what it did when it was above the wire. A simple right-hand rule describes the direction of the field. If you imagine holding the wire in your right hand with your thumb pointing in the direction of the electric current, your fingers curl around the wire in the direction of the magnetic field lines (fig. 14.9).

Not surprisingly, the effect gets weaker as the compass is moved away from the wire. A current of a few amperes is needed to get a large deflection of the compass needle even when the compass is just a few centimeters from the wire. Early batteries were not capable of sustaining large currents. This limitation, together with the unexpected direction of the effect, probably delayed its discovery.

The magnetic force on a current-carrying wire

If we can produce magnetic fields with electric currents, do currents behave like magnets in other respects? Does an electric current experience a magnetic force in the presence of a magnet or another current-carrying wire? This question was explored by many scientists who were excited by Oersted's discovery, including André Marie Ampère in France.

Ampère discovered that there is indeed a force exerted on one current-carrying wire by another. He carefully demonstrated that this force is related to the magnetic effect discovered by Oersted and cannot be explained as an electrostatic effect. A current-carrying wire is usually electrically neutral, with no net positive or negative charge. Ampère measured the strength of the magnetic force between two parallel current-carrying wires and studied how it varies with the distance between the wires and the amount of current flowing in each (fig. 14.10).

Ampère's experiments showed that the magnetic force between two parallel wires is proportional to the two currents (I_1 and I_2) and inversely proportional to the distance r between the two wires. We usually state this relationship as

$$\frac{F}{l} = \frac{2k' I_1 I_2}{r},$$

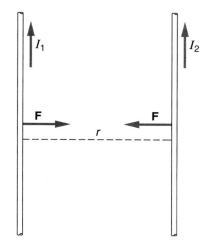

figure **14.10** Two parallel current-carrying wires exert an attractive force on each other when the two currents are in the same direction.

where the constant k' is equal to 1×10^{-7} N/A^2, and F/l represents the force per unit length of the wire. The longer the wires, the greater the force. The force exerted by one wire on the other is attractive when the currents are flowing in the same direction and repulsive when the currents are flowing in opposite directions.

The strangely simple value of the constant k' is no accident. We define the unit of current, the ampere, by using this relationship and measuring the force between two current-carrying wires:

One ampere (A) is the amount of current flowing in each of two parallel wires separated by a distance of 1 meter that produces a force per unit length on each wire of 2×10^{-7} N/m (newtons per meter).

The ampere is the basic unit of electromagnetism. It is standardized by making measurements of the magnetic force between two wires. The coulomb is defined from the ampere. Since current is defined as charge per unit time ($I = q/t$), charge is the product of current multiplied by time ($q = It$). Therefore, 1 coulomb equals 1 ampere-second (1 C = 1 A·s).

The magnetic force on a moving charge

What, then, is the basic nature of the **magnetic force**? Magnetic forces are exerted by magnets on other magnets (section 14.1), by magnets on current-carrying wires, and by current-carrying wires on each other. Since electric current is the flow of electric charge, we apparently obtain magnetic forces when charges are moving. Is the motion of charge somehow a fundamental criterion for magnetic forces to exist? Ampère also addressed this question during the 1820s.

Think about the direction of the force exerted by one current-carrying wire on another: this force is perpendicular to the direction of the current. Since the direction of electric current is the direction of flow of positive charge, we might suppose that a magnetic force is exerted on the moving charges in the wire. This force must be perpendicular to the velocity of the charges to be consistent with the observed force on a given wire (fig. 14.11).

The magnetic force on a current-carrying wire has its maximum value when the wire is perpendicular to the direction of the magnetic field. All these facts suggest that a magnetic force is exerted on a moving charge. This force is proportional to the quantity of the charge and the velocity of the moving charge (which are related to the current for the charges flowing in the wire) and to the strength of the magnetic field. In symbols, we write this as $F = qvB$, where q is the charge, v is the velocity of the charge, and B is the strength of the magnetic field. For this relationship to be valid, however, the velocity must be perpendicular to the field. The charge need not be confined in a wire, however.

This expression for the magnetic force actually defines the magnitude of the magnetic field. Dividing by q and v, it can be put in the form

$$B = \frac{F}{qv_\perp} ,$$

where $v_\perp$ stands for the component of the velocity that is perpendicular to the field. From this definition, the unit of magnetic field is equal to 1 newton per ampere-meter (N/A·m), which is now called a *tesla* (T).

Just as electric field is the electrostatic force per unit of charge, the **magnetic field** is the magnetic force per unit of charge and unit of velocity. If the velocity of the charge is zero, there is no magnetic force.

What is the direction of the magnetic force on a moving charge?

Another right-hand rule, illustrated in figure 14.12, is often used to describe the direction of the magnetic force on a moving charge. If you point the index finger of your right hand in the direction of the velocity of a positive charge, and the middle finger in the direction of the magnetic field, the thumb points in the direction of the magnetic force on the moving charge. This force is always perpendicular to both the velocity and the field direction. The force on a negative charge is in the opposite direction of a positive charge moving in the same direction.

Because the magnetic force on a moving charge is perpendicular to the velocity of the charge, this force does no work on the charge and cannot increase its kinetic energy. The resulting acceleration of the charge is a centripetal acceleration that changes the direction of the charge's velocity. If the charge is moving in a direction perpendicular to a uniform magnetic field, the magnetic force will bend the path of the charged particle into a circle. The radius of the circle is determined by the mass and speed of the

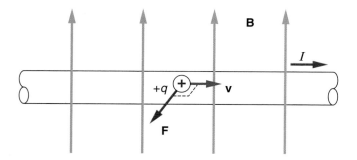

figure 14.11   The magnetic force exerted on the moving charges of an electric current is perpendicular to both the velocity of the charges and to the magnetic field.

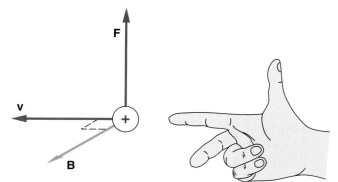

figure 14.12   If the index finger of the right hand points in the direction of the velocity of the charge, and the middle finger in the direction of the magnetic field, the thumb indicates the direction of the magnetic force acting on a positive charge.

particle and can be found by applying Newton's second law. (See challenge problem 2.)

Since the direction of electric current is the same as the flow of positive charge, the right-hand rule used to find the direction of the force on a moving charge can also be used to describe the direction of the force on a current-carrying wire. In this case, the index finger points in the direction of the current and the middle finger and the thumb have the same meanings as before.

The magnetic force on a segment of wire can be expressed in terms of the magnetic field as $F = IlB$, where I is the current, l is the length of the segment of wire, and B is the strength of the magnetic field (see Try This Box 14.1). The direction of the current must be perpendicular to the field for this expression to be valid. This expression is merely another way of writing the relationship $F = qvB$. The product of the charge times the velocity (qv) is replaced by the product of the current times the length of the segment of wire (Il).

The magnetic force is exerted by *moving* charges on one another. Since electric currents are moving charges, we can also say that the magnetic force is a force exerted by electric currents on one another, and we can always replace the product qv with the product Il in expressions involving magnetic fields or forces.

Oersted discovered that, when properly oriented, an electric current can cause a deflection of a compass needle. Following up on this discovery, Ampère showed that a magnetic force is exerted on one current-carrying wire by another. Since an electric current consists of moving charges, the size of a magnetic field can be defined as the force per unit charge per unit velocity of the charge, provided that the velocity is perpendicular to the magnetic field.

14.3 Magnetic Effects of Current Loops

Up to this point, the current-carrying wires that we have considered have been straight, although they must bend somewhere to complete an electric circuit. Many applications of electromagnetism involve loops of wire. Wire coils are used in electromagnets, electric motors, electric generators, transformers, and a host of other applications.

What happens when we bend a current-carrying wire into a coil? What does the magnetic field look like, and how is the coil affected by other magnetic fields? The intense experimental activity of the 1820s following Oersted's discovery explored these questions. Ampère, as well as many other scientists, was active in this work.

The magnetic field of a current loop

As discussed in section 14.2, the magnetic field lines around a straight wire form circles centered on the wire. Imagine the process of bending such a wire into a circular loop: what happens to the field lines? Very near the wire, the field lines presumably are still circles. Toward the center of the loop, however, the field contributions produced by different segments of the wire are all approximately in the same direction. They add to give a strong field at the center.

The resulting field is depicted in figure 14.13. The field lines are close together near the center of the loop, indicating a strong field. The field is also strong near the wire itself. Farther from the loop, the field weakens, as should be expected. Like the field lines for a straight wire, the magnetic field lines for the current loop curve around until they meet themselves, forming closed loops.

Note that the field shown in figure 14.13 resembles the field of a bar magnet (fig. 14.5). In fact, if the bar magnet is short enough, as in figure 14.14, the field patterns are

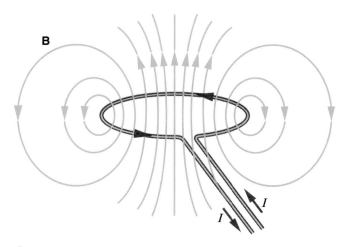

figure **14.13**   When a current-carrying wire is bent into a circular loop, the magnetic fields produced by different segments of the wire add to produce a strong field near the center of the loop.

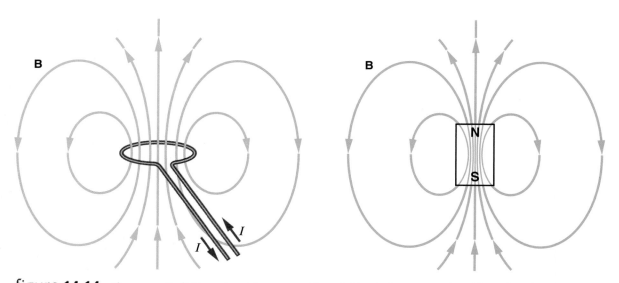

figure 14.14   The magnetic field produced by a current loop is identical to one produced by a short bar magnet (a magnetic dipole).

identical. We might conclude that a current loop is a magnetic dipole, because its field is the same as the more familiar dipole, the bar magnet.

Is there a magnetic torque on a current loop?

The similarity between a current loop and a bar magnet goes beyond the fields produced. If you place a current loop in an external field, it experiences a torque, just as a bar magnet would if it were not initially aligned with the field. It is possible to use a loop or several loops of current-carrying wire as a compass needle, because its axis (perpendicular to the plane of the loop) would line up with an external field, just as a normal compass needle does.

We can see the origin of the torque on a current loop most readily by considering a loop in the form of a rectangle, as in figure 14.15. Each segment of the rectangular loop is a straight wire, and the force on each segment is given by the expression $F = IlB$, introduced in section 14.2. The directions of the forces are given by the right-hand rule in figure 14.12. Limber up your right hand, and verify the force directions in the diagram.

Each of the four sides of the loop in figure 14.15 experiences a magnetic force. The forces on the two ends of the loop (F_1 and F_2) produce no torque about the center of the loop, because their lines of action pass through the center of the loop. Their lever arms about any axis passing through the center of the loop are zero, and they produce no torque about the center. (See chapter 8 for a discussion of torque and rotational motion.)

The forces on the other two sides (F_3 and F_4) have lines of action that do not pass through the center of the loop, provided that the plane of the loop is not perpendicular to

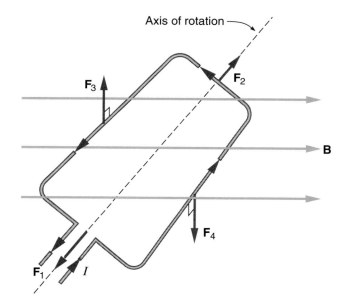

figure 14.15   The forces on each segment of a current-carrying rectangular loop of wire combine to produce a torque that tends to rotate the coil until its plane is perpendicular to the external magnetic field.

the external magnetic field. From the diagram, we see that these two forces combine to produce a torque that tends to rotate the loop about the axis shown, so that its plane ends up perpendicular to the magnetic field.

Since the magnetic forces on the loop segments are proportional to the electric current flowing around the loop, the magnitude of the torque on the loop is also proportional to the current. This fact makes the torque on a current-carrying coil useful for measuring electric current. Most

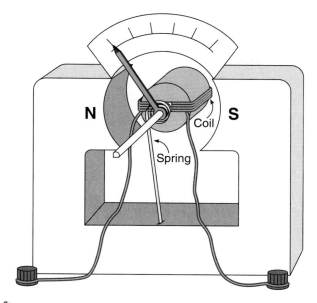

figure **14.16** An electric meter consists of a coil of wire, a permanent magnet, and a restoring spring to return the needle to zero when there is no current flowing through the coil.

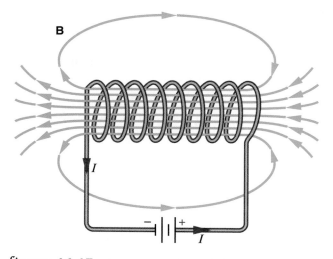

figure **14.17** A current-carrying coil of wire produces a magnetic field greater than a single loop and is proportional in strength to the number of loops in the coil.

simple electric meters have a coil of wire with a needle attached to it at their center. A permanent magnet provides the magnetic field, and a spring returns the needle to zero when there is no current (fig. 14.16).

This torque is also the basis of operation for electric motors. To keep the coil turning, however, the current in the coil must reverse directions every half turn. Otherwise, the coil would come to rest with the plane of its loop perpendicular to the external magnetic field. An alternating current is well suited to operating electric motors, but we can also design motors that operate on direct current (see Everyday Phenomenon Box 14.1). Electric motors are found everywhere, from the starting motor in your car to the motors in kitchen appliances, vacuum cleaners, washers and dryers, and electric shavers.

How do we make an electromagnet?

We have been talking about single loops of current and coils, which are several loops, all wound in the same direction. When we add several loops to form a coil, is the magnetic field stronger than for a single loop? What effect do we get by winding a coil of wire on an iron or steel core?

We can readily conclude that the magnetic field produced by a coil of wire will be stronger than one produced by a single loop carrying the same current. The magnetic fields produced by each loop are all in the same direction near the axis of the coil, and thus they add. The resulting field strength is proportional to the number of turns N that are wound on the coil (fig. 14.17). The torque on the coil, when placed in an external field, is also proportional to both the current and the number of turns in the coil.

The effectiveness of a coil of wire as a magnet was discovered by the French scientist Dominique-François Arago (1786–1853) soon after Oersted's discovery. Arago first noticed that iron filings are attracted to a current-carrying copper wire much as they are to a magnet. Following a suggestion by Ampère, he found that the effect is enhanced if the wire is wound into the shape of a helix or coil. An even larger effect is obtained, however, if the coil is wound around a steel needle or nail. The ability to attract iron filings in this case is better than most natural magnets.

Arago showed that his simple **electromagnet** had a north pole and a south pole like those of a bar magnet. In fact, the magnetic field of a bar magnet is identical to one produced by an electromagnet of the same length and strength. This similarity led Ampère to suggest that the source of the magnetism in a naturally magnetic material like magnetite is current loops in the atoms that make up the material. If, for some reason, these atomic current loops all line up with one another and lock into those positions, a permanent magnet results.

In Ampère's day, nothing was known about the structure of atoms. Atomic structure was not understood until the beginning of the twentieth century, almost a hundred years later. Even so, Ampère's theory is remarkably close to our modern view of what happens in ferromagnetic metals like iron, nickel, and cobalt and their alloys. We now know that the current loops are associated with the spins of the electrons in the atoms. We have only recently understood why these spins tend to line up in ferromagnetic materials but not in other materials.

Within ten years of Oersted's initial discovery of the magnetic effect of an electric current in 1820, many of the phenomena of electromagnetism had been thoroughly explored and described. Ampère was the leader in much of this work,

Everyday Phenomenon

box 14.1

Direct-Current Motors

The Situation. As a child, you may have built a simple direct-current motor that ran on flashlight batteries. Inexpensive kits are available, and building motors is a common science activity in the middle grades. How does a direct-current motor work? How can we keep the rotor moving in the same direction without using an alternating-current power source? How can we vary the speed of the motor?

The Analysis. There are many ways to build a dc motor, but the simplest ones usually take the form illustrated in the drawing. A coil of wire is wound on a rotor mounted so that it can rotate between the pole faces of a permanent horseshoe magnet. The coil is connected to a battery by sliding contacts that come up on either side of a split ring mounted on the axle of the rotor.

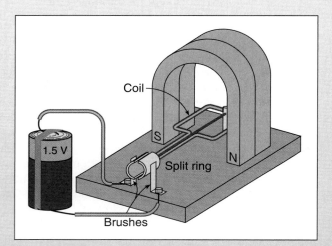

A simple dc motor consists of a wire-wound rotor mounted on an axle between the pole faces of a permanent magnet. The split ring causes the current to reverse directions every half turn.

Viewing the rotor as an electromagnet is a straightforward way to understand how the motor works. The south pole of the rotor is attracted to the north pole of the horseshoe magnet, causing the rotor to turn. When the rotor reaches the position where these two poles are closest, the current through the rotor is reversed. The contacts from the battery cross the gap between the two half-rings of the split ring attached to the coil, so that opposite ends of the coil are now connected to the positive and negative terminals of the battery.

This reversal in current makes the polarity of the electromagnet reverse. What had been the south pole becomes the north pole and vice versa. As its momentum carries the rotor

past the vertical position, its newly created north pole will now be attracted to the south pole of the horseshoe magnet, causing the rotor to continue to turn in the same direction that it had been moving. The torque on the rotor comes from the attraction of opposite poles of the two magnets, the rotor electromagnet and the permanent horseshoe magnet. Every half-turn, the poles of the electromagnet flip due to the effect of the split-ring contacts.

This split-ring arrangement is a crucial feature of direct-current motors and is called a *split-ring commutator.* It consists of two half-circle metal bands, each connected to one end of the coil, mounted on an insulating cylinder. The metal half-rings make sliding contact with fixed-position metal strips or *brushes* on either side of the cylinder. An alternating-current motor does not need a split-ring commutator because the current reverses direction every half-cycle.

Although we have viewed the rotor as an electromagnet, we could also view the torque on the rotor as a result of the magnetic forces on current elements on either side of the coil (see figure 14.15 in section 14.3). When the current in the coil reverses direction, these forces also reverse direction, so that the torque can be maintained in the appropriate direction.

The speed of a dc motor is related to the applied voltage, a convenience in situations where a variable-speed motor is needed. At first glance, this voltage dependence might not seem surprising, but its explanation actually follows from Faraday's law (section 14.4) rather than Ohm's law. If you compare our description of a motor to the description of the generator in section 14.5, you may be struck by the similarity of these two devices. In fact, we could use a motor as a generator if we supplied mechanical energy from an external source to turn the rotor. This happens when a hybrid car is braking as described in Everyday Phenomenon Box 11.1.

Because a motor behaves like a generator, there is a "back" voltage induced in the coil of the rotor due to Faraday's law and the changing magnetic flux through the coil. The magnitude of this induced voltage increases as the rotational velocity of the coil increases, just as in a generator. A larger voltage from the power source is required to overcome the larger back voltage induced as the rotational velocity of the rotor increases. To reach higher speeds, therefore, the applied voltage must be increased. An alternating-current motor usually runs at a fixed speed, but the speed of a direct-current motor can be continuously varied if we vary the voltage of the power supplied to the motor.

and he was also responsible for developing the mathematical theory relating the magnetic forces on current-carrying wires and coils. His theory relating the magnetism in natural magnets to atomic current loops completed the link between electricity and magnetism. Magnetic effects can all be regarded as the action of electric currents—or moving charges.

The magnetic field of a single loop of wire carrying an electric current is identical to one of a short bar magnet or magnetic dipole. Like a bar magnet, a current loop also experiences a torque when placed in an externally produced magnetic field. This torque is the basis of operation of simple electric meters and electric motors. The effect is enhanced if we wind the wire into a coil and is greater still if we place an iron core in the center of the coil. A coil wound around an iron core becomes an electromagnet if we run a current through the coil. Natural magnets can be thought of as consisting of self-aligned current loops associated with the spins of electrons in the atoms of the magnetic material.

14.4 Faraday's Law: Electromagnetic Induction

The work of Oersted, Ampère, and others firmly established that magnetic forces are associated with electric currents. Using the field concept introduced by Maxwell, we now say that an electric current produces a magnetic field. What about the reverse? Can magnetic fields produce an electric current?

Michael Faraday got his start in science as an assistant to the English chemist Sir Humphry Davy. Davy's primary interest, and much of Faraday's early work, was on the chemical action of electric currents, or electrolysis. Later, Faraday began a series of experiments to explore the possibility of producing an electric current from magnetic effects.

What did Faraday's experiments show?

Faraday was aware that the magnetic effects of electric current are enhanced by coils, so he started his experiments by winding two unconnected coils of wire on the same wooden cylinder. One coil was connected to a battery and the other coil to a galvanometer. (A galvanometer measures both the direction and the size of an electric current.) Faraday wanted to see whether he could detect a current in the coil connected to the galvanometer when a current was flowing in the other coil connected to the battery.

The results of these first experiments were negative: no current was detected in the second coil. Undeterred, Faraday persisted by winding longer and longer coils. Finally, with coils of about 200 feet of copper wire, he noticed an effect. There was still no evidence of a steady current in the second coil, but there was a very brief and feeble deflection of the galvanometer needle when he connected the first coil to the battery. There was another momentary deflection in the opposite direction when he broke contact with the battery.

Faraday had not expected this effect, but experimentalists cannot be choosy about results. Through further experiments, Faraday found that he could get a considerably stronger deflection (although still momentary) if he wound both coils on an iron core (fig. 14.18) rather than on a wooden cylinder. Faraday wound two coils on either side of a welded iron ring. One coil, the *primary,* was connected to a battery, the other, the *secondary,* to his galvanometer. Again, the galvanometer needle deflected in one direction when the primary circuit was closed by making contact with the battery and in the opposite direction when the primary circuit was broken.

No current was detected in the secondary coil when there was a steady current in the primary coil. Only if this primary current was *changing* as the circuit was completed or broken was a deflection noted. The strength of the deflection was proportional to the number of turns of wire wound on the secondary coil and to the strength of the battery used with the primary coil. Faraday began to formulate the idea that an electric current could be induced by magnetic effects that change with time, rather than by the steady-state presence of a magnet or electric current.

Faraday pursued this idea with many other experiments. One of these involved moving a permanent bar magnet in and out of a hollow helical coil of wire attached to a galvanometer (fig. 14.19). When the magnet moved in, the galvanometer needle deflected in one direction. When the magnet moved out, the needle deflected in the opposite direction. When the magnet was not moving, no deflection resulted. This is an easy effect to demonstrate with equipment available in most physics labs.

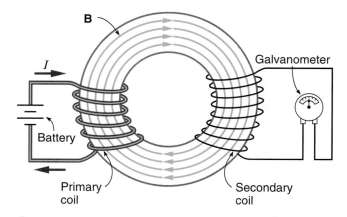

figure 14.18 The magnetic field produced by the primary coil was channeled through the secondary coil by the iron ring used in one of Faraday's experiments.

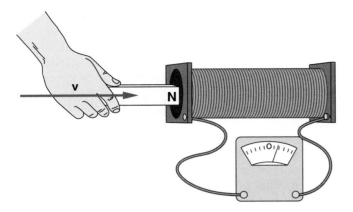

figure **14.19**   A magnet moved in or out of a helical coil of wire produces an electric current in the coil.

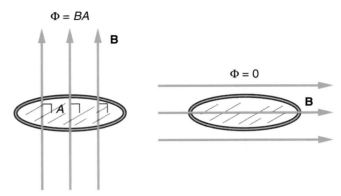

figure **14.20**   The magnetic flux through the loop of wire has its maximum value when the field lines are perpendicular to the plane of the loop. It is zero when the field lines are parallel to the plane of the loop and do not cross the plane.

What is Faraday's law?

The results of all of Faraday's experiments indicate that an electric current is induced in a coil or circuit when the magnetic field passing through that circuit is changing. Because the amount of current flowing in the secondary circuit depends on the resistance of that circuit, we generally express these results in terms of the induced voltage rather than the current. To construct a quantitative statement to describe this effect, however, we need somehow to define how much of the magnetic field passes through the circuit. This latter problem was solved by introducing the concept of **magnetic flux.** The magnetic flux is related to the number of magnetic field lines passing through the area bounded by the loop of wire. For a simple loop of wire lying in a plane perpendicular to the magnetic field, the magnetic flux is the product of the magnetic field B and the area A bounded by the loop (fig. 14.20). In symbols, the flux definition takes the form

$$\Phi = BA.$$

The Greek letter Φ (phi) is the standard symbol used for flux.

We must place an important qualification on this definition. Maximum flux is obtained when the field lines pass through the circuit in a direction perpendicular to the plane of the circuit. If the field lines are parallel to this plane, there is no flux because the field lines do not pass through the circuit (fig. 14.20). To take this fact into account, we use only the component of B that is perpendicular to the plane of the loop to calculate the flux.

We can now make a quantitative statement of **Faraday's law** that summarizes the results of his experiments:

> A voltage (electromotive force) is induced in a circuit when there is a changing magnetic flux passing through the circuit. The induced voltage is equal to the rate of change of the magnetic flux. In symbols,
>
> $$\varepsilon = \frac{\Delta\Phi}{t}.$$

The rate of change of flux is found by dividing the change in flux $\Delta\Phi$ by the time t required to produce this change. The process of inducing a voltage as described in Faraday's law is called **electromagnetic induction.**

The more rapidly the magnetic flux through the circuit changes, the larger the induced voltage, which can be readily observed in the experiment involving the moving magnet. If we move the magnet in and out of the coil quickly, we get larger deflections of the galvanometer needle than if we move it more slowly. The magnetic flux passing through a coil of wire runs through each loop in the coil, so the total flux through a coil is equal to the number of turns of wire in the coil times the flux through each turn, or $\Phi = NBA$. (Think about professional basketball to remember this expression.) The more turns of wire in the coil, the larger the induced voltage.

A coil of wire can be used to assess the strength of a magnetic field. The magnetic flux through the coil can be quickly reduced to zero by either removing the coil from the field or by giving it a quarter turn so the field lines lie parallel to the plane of the coil. If we know the time required to make this change, we can find the strength of the magnetic field by measuring the induced voltage. The application of Faraday's law in Try This Box 14.2 reverses this procedure by predicting the induced voltage from a known magnetic field.

What is Lenz's law?

Can we predict the direction of the induced current in the coil? The rule for doing so, **Lenz's law,** goes hand in hand with Faraday's law and is credited to Heinrich Lenz (1804–1865):

> The direction of the induced current generated by a changing magnetic flux produces a magnetic field that opposes the *change* in the original magnetic flux.

try this box 14.2

Sample Exercise: How Much Voltage Is Induced?

A coil of wire with 50 turns has a uniform magnetic field of 0.4 T passing through the coil perpendicular to its plane. The coil encloses an area of 0.03 m². If the flux through the coil is reduced to zero by removing it from the field in a time of 0.25 second, what is the induced voltage in the coil?

N = 50 turns   The initial flux through the coil is:

B = 0.4 T $\Phi = NBA$

A = 0.03 m² $= (50)(0.40\ \text{T})(0.03\ \text{m}^2)$

t = 0.25 s $= 0.60\ \text{T·m}^2$

The induced voltage is equal to the rate of change of flux:

$$\varepsilon = \frac{\Delta\Phi}{t}$$

$$= \frac{(0.60\ \text{T·m}^2 - 0)}{(0.25\ \text{s})}$$

$$= 2.4\ \text{V}$$

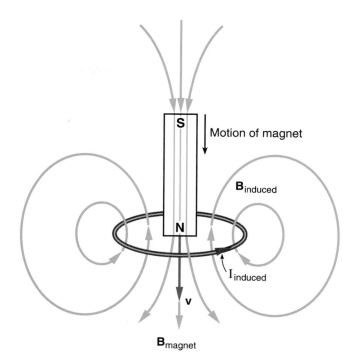

figure 14.21   The induced current in the loop of wire produces an upward magnetic field inside the loop that opposes the increase in the downward field associated with the moving magnet.

If the magnetic flux decreases with time, the magnetic field produced by the induced current will be in the same direction as the original external field, thus opposing the decrease. On the other hand, if the magnetic flux increases with time, the magnetic field produced by the induced current will be in the opposite direction to the original external field, thus opposing the increase.

Lenz's law is illustrated in figure 14.21. When the magnet moves into the coil and the flux increases with time, the induced current is in a counterclockwise direction in the coil. The magnetic field produced by this induced current is upward through the area bounded by the coil, thus opposing the increase in the downward field associated with the moving magnet. Lenz's law also explains why the galvanometer needle deflected in opposite directions when Faraday closed or opened the primary circuit in his experiments.

What is self-induction?

Faraday first reported the results of his discoveries on electromagnetic induction in 1831. One year later, the discovery of a related effect was reported by Joseph Henry (1797–1878), who was working at Princeton University in the United States. Henry was experimenting with electromagnets when he noticed that the spark or shock obtained when an electromagnet was connected to a battery was larger than one obtained by touching the terminals of the battery with an uncoiled wire. The biggest sparks were obtained when the circuit was broken.

Henry explored this phenomenon in more detail. He found that a long wire yields a larger spark than a short wire and that coiling the wire into a helix yields an even larger spark. If the wire is wound on a steel nail to make an electromagnet, the effect is larger still. He roughly calibrated the strength of the shock by noting how far up his arms it could be felt. If it could be felt just at the fingers, it was not very strong—one that could be felt at the elbows was obviously stronger. You can get a much stronger shock when you connect the wires of an electromagnet to a battery than when you touch the terminals of the battery directly with your fingers.

Henry had discovered **self-induction.** The changing magnetic flux through a coil of wire produced when the coil is connected or disconnected from the battery produces an induced voltage in the same coil. The induced current in the coil and its associated magnetic field oppose the changing magnetic flux. The magnitude of the induced voltage is described by Faraday's law and is proportional to the number of turns in the coil and to the rate of change of current in the coil. The voltage induced when the circuit is broken can be several times larger than the voltage of the battery itself.

Self-induction has many applications. Coils of wire are used in circuits to smooth out changes in current. The voltage induced when the current is increasing causes the change to take place more slowly than it would without the coil.

Everyday Phenomenon

box 14.2

Vehicle Sensors at Traffic Lights

The Situation. When you approach a traffic light in your car, you probably often notice a circular or diamond-shaped line in the pavement as shown in the photograph. You may be aware that this pattern in the pavement has something to do with causing the traffic light to change. If your car is not positioned over this pattern, the light may not change.

How does this detector work? What lies underneath the pattern in the pavement? Why are these patterns becoming more and more common in traffic control applications?

The circular pattern in the pavement indicates the presence of a vehicle sensor. How does this sensor work?

The Analysis. The pattern in the pavement conceals a large loop of wire with multiple turns. This loop or coil is an inductor. It is usually placed in the pavement after the paving is completed. A saw is used to cut a circular or diamond-shaped groove in the pavement and the coil is laid in this groove. A rubberlike sealing compound is then used to cover the wire loop.

The wire-loop inductor is part of a circuit. The rest of this circuit is usually located somewhere on the side of the road. You may see a line in the pavement that contains wires leading to this circuit or they may have been put in place before the paving was completed. The circuit containing the inductor is connected to the timing controls for the traffic light as shown in the diagram.

How is this wire loop involved in detecting the presence of your car? When your car is located over the loop, the steel in the frame of your car increases the magnetic field being produced by the current in the coil. The effect is similar to that of placing a piece of iron inside the coil of an electromagnet. The presence of the iron strengthens the magnetic field, thus increasing the inductance of the coil. Iron is a component of steel, which is used extensively in the frames of automobiles or trucks. Any metal will have some effect on the inductance, though, due to its ability to conduct an induced current.

(continued)

This induced *back voltage* opposes the voltage difference that is producing the increase in current. An *inductor* or coil effectively adds electrical inertia to the system, which reduces rapid changes in current. (See Everyday Phenomenon Box 14.2 for an application of this effect.) Induction coils are also used to generate large voltage pulses such as those needed to power the spark plugs in an automobile.

Michael Faraday discovered that an electric current can be induced in a coil of wire when the magnetic field passing through that coil is changing. Faraday's law states that the induced voltage in the coil is equal to the rate of change of the magnetic flux through the coil. Magnetic flux is defined as the magnetic field times the area enclosed by the coil. Lenz's law says that the direction of the induced current generates a magnetic field that opposes the change in the original magnetic flux. Joseph Henry discovered the related effect of self-induction, in

which a voltage is induced in the same coil that is producing the changing magnetic flux.

14.5 Generators and Transformers

Our everyday use of electric power is so commonplace that we do not usually stop to think about where the energy comes from or how it is generated. We know that it comes to us through overhead or underground wires and that it powers our appliances, lights our homes and offices, and sometimes heats or cools our homes. We may also be aware of the transformers visible on utility poles and in electric substations.

Electric generators and transformers both play crucial roles in the production and use of electric power. Both are based on Faraday's law of electromagnetic induction. How do they work, and what roles do they play in the power-distribution system?

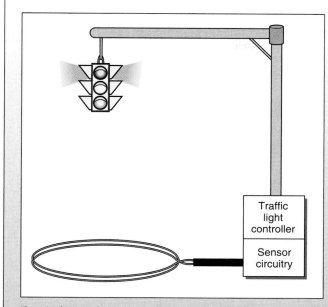

The wire loop in the pavement is part of a sensor circuit that sends a signal to the traffic-light timing controls.

Since the wire loop is part of an electric circuit, the change in inductance will affect the behavior of the circuit. There are many possible circuit designs that could be used to convert this change in inductance to a signal that controls the traffic light. For example, the inductor could be part of an oscillating circuit whose frequency depends upon the value of the inductance. The effect of increasing the inductance in the circuit is similar to that of increasing the mass of a mechanical oscillator. A larger mass (or inductance) leads to a smaller frequency of oscillation. This change in frequency can be used to signal other circuitry that a vehicle is present.

Depending upon how the timing controls for the traffic light are programmed, the signal indicating the presence of a vehicle will cause the light to change, perhaps after a short delay. Additional detectors farther back from the intersection can indicate how many vehicles are present. This information can be used to determine how long the light should remain green for a turn lane or side road.

These inductive detectors have become very common for managing the flow of traffic. If traffic is heavy in all directions at all times, then a simple timer system is used to control the traffic lights. If the traffic patterns vary with time of day and other factors, then the detectors can be used to optimize the timing of the lights to the current traffic conditions. If there is no traffic on a side road, for example, there is no point in stopping the traffic on a more major street or highway. If we can increase the efficiency of traffic flow using such techniques, we can reduce driver frustration and decrease the need for more expensive highway-building solutions.

How does an electric generator work?

We have already mentioned the use of electric generators in our discussion of energy resources in chapter 11. The basic function of a **generator** is to convert mechanical energy, which might be obtained from a water turbine at a dam or a steam turbine in a power plant, to electrical energy. How is this done? How is Faraday's law involved in describing how a generator works?

A simple electric generator is shown in figure 14.22. Mechanical energy supplied by turning the crank rotates the coil of wire located between the pole faces of permanent magnets. The magnetic flux through the plane of the coil has its maximum value when this plane is perpendicular to the magnetic field lines passing between the poles of the magnets. As the coil is turned to a position where the plane of the coil is parallel to the field lines, the flux becomes zero. If we continue to turn the coil past this point, the field lines pass through the coil in the direction opposite to the initial direction (relative to the coil).

The coil's rotation causes the magnetic flux passing through the coil to change continuously from a maximum in one direction, to zero, to a maximum in the opposite direction, and so on, as shown in the upper graph of figure

figure **14.22**   A simple generator consists of a coil of wire that generates an electric current when turned between the pole faces of permanent magnets.

14.23. By Faraday's law, a voltage is induced in the coil because of this changing magnetic flux. The magnitude of the induced voltage depends on the rate of change of the magnetic flux and on factors associated with the magnet and coil that determine the size of the maximum magnetic flux. The faster the coil is turned, the larger the maximum value of the induced voltage, since increased speed causes the magnetic flux to change more rapidly.

Figure 14.23 shows graphs of the continuously varying magnetic flux and the induced voltage plotted against time. If the axle of the coil is turned at a steady rate, we get a smooth variation of the flux, as in the top graph. Below the flux graph, the graph of the induced voltage is plotted on the same time scale. By Faraday's law, induced voltage is equal to the rate of change of magnetic flux, so its maximum values occur where the slope of the flux curve is the greatest, and its zero values occur where the flux is momentarily not changing (zero slope). The resulting curve for the induced voltage has the same shape as for the flux but is shifted relative to the flux curve.

As the graphs show, the voltage that is normally produced by a generator is an *alternating* voltage: this is one reason that we use alternating current in our power-distribution system. For power production, the rotational velocity of the generator coils must be maintained at a specific value to produce the 60-Hz (60-cycles/s) alternating current that is the standard frequency in the United States.

The electric generators used in power plants resemble the simple one we have described here. Usually, they have more than one coil, and the magnets are electromagnets rather than permanent magnets, but the principle of operation is the same. There is also an electric generator in your car—the electric power generated keeps the battery charged, operates the lights, and powers other electrical systems.

What does a transformer do?

Another advantage of alternating current in power distribution is that **transformers** can change the voltage. Transformers are familiar sights. You see them on utility poles, at electrical substations, and as voltage adapters for low-voltage devices like model electric trains (fig. 14.24). Their function is to adjust the voltage up or down to suit the needs of a particular application.

Some of Faraday's early experiments involved prototypes of the modern transformer. His apparatus with primary and secondary coils wound on either side an iron ring (fig. 14.18) is an example of a simple transformer. By Faraday's law, a voltage will be induced in the secondary coil if the current (and the associated magnetic field) of the primary coil is changing: that condition is present if we power the primary coil with an alternating-current source (fig. 14.25).

What determines the size of the change in voltage that a transformer can produce? A simple relationship follows from Faraday's law. The voltage induced on the secondary coil is proportional to the number of turns on the secondary coil, since the number of turns determines the total magnetic flux passing through this coil. The induced voltage is also proportional to the voltage on the primary coil, since this determines the size of the primary current and its associated magnetic field. The induced voltage is inversely proportional, however, to the number of turns on the primary coil.

Stated in symbols, this relationship takes the form

$$\Delta V_2 = \Delta V_1 \left(\frac{N_2}{N_1} \right),$$

where N_1 and N_2 are the number of turns of wire on the primary and secondary coils, and ΔV_1 and ΔV_2 are the

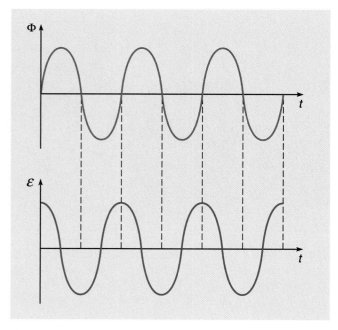

figure 14.23 Graphs of the magnetic flux Φ and induced voltage ε for an electric generator, plotted on the same time scale.

figure 14.24 Transformers on utility poles and at electrical substations are familiar sights. Smaller ones are common components of electrical devices.

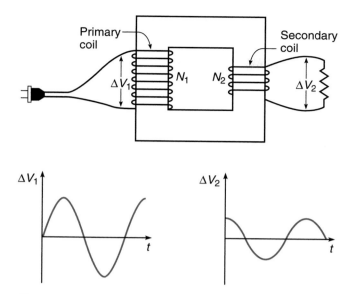

figure 14.25 An alternating voltage applied to the primary coil of a transformer produces a changing magnetic flux, which induces a voltage in the secondary coil.

associated voltages. This relationship is often written as the proportion

$$\frac{\Delta V_2}{\Delta V_1} = \frac{N_2}{N_1}.$$

The ratio of the number of turns on the two coils determines the ratio of the voltages.

Self-induction is the reason that the voltage induced on the secondary coil is inversely proportional to the number of turns on the primary coil. The more turns on the primary coil, the harder it is to produce a rapid change in current in the primary coil because of the back voltage produced by self-induction. This effect limits the current and the size of the magnetic field produced by the primary coil, which limits the magnetic flux passing through the secondary coil.

Suppose that you want to run an electric train at 12 volts, but the electric power provided at the wall socket is 120 volts. You need a transformer with ten times as many turns on the primary coil as on the secondary coil, which yields a voltage in the secondary coil $\frac{1}{10}$ of the primary coil, or 12 volts. A television set needs much higher voltages than 120 volts to power the picture tube, so we use a transformer with many more turns on the secondary than on the primary.

If the output voltage is higher than the input voltage, are we somehow getting more power out of the transformer than what we put in? The answer, of course, is no. The power delivered to the secondary circuit is always less than, or at best equal to, the power provided to the primary coil. Since electrical power can be expressed as the product of the voltage times the current, conservation of energy

provides a second relationship useful for analyzing transformers:

$$\Delta V_2 I_2 \le \Delta V_1 I_1.$$

A high-output voltage is associated with a low-output current. The output power does not exceed the input power. On the other hand, if we are stepping down to a smaller voltage in the secondary circuit, the secondary current can be larger than what was provided to the primary coil.

Transformers and power distribution

High voltages are desirable for long-distance transmission of electric power. The higher the voltage, the lower the current for a given amount of power transmitted. Since the heat losses caused when current flows through a resistance are related most directly to the current ($P = I^2R$), smaller currents mean less energy lost to resistive heating in the transmission wires. Transmission voltages as high as 230 kV (230 kilovolts or 230 000 volts) are not unusual.

Such high voltages are not safe or convenient for power distribution in a city or for use in homes or buildings. Transformers at electrical substations reduce the voltage to 7200 volts for distribution in town. Transformers on utility poles or partially underground lower this voltage further to 220 to 240 volts for entry into buildings. This ac voltage can be split inside the building to yield the 110 volts commonly used for household circuits, while keeping 220 volts available for stoves, dryers, and electric heating.

The original electrical distribution system in the United States was a 110-volt direct-current system designed in 1882 by Thomas Edison for a portion of New York City. There was controversy for several years afterward about whether direct-current or alternating-current systems were the most appropriate for the delivery of electric power. In the end, the proponents of the ac system prevailed. The abilities to transmit electrical power at high voltages and to use transformers to step these voltages down at the point of use were major reasons for this choice.

Direct current is occasionally used to transmit power over large distances because it does not lose energy by radiation of electromagnetic waves. Radiation is a drawback of alternating current. The power line acts as an antenna that radiates electromagnetic waves when the current is oscillating (see section 16.1). The static that you hear on your car radio when you pass near an ac power line is caused by this radiation.

A transformer works best with alternating current. Faraday's law requires that magnetic flux *change* for there to be an induced voltage. If we tried to use a transformer with a battery or other dc source, we would have to continually make and break the primary circuit somehow to induce a voltage in the secondary circuit. There are means of accomplishing this, but they add to the complexity and cost of a system and are less efficient than using alternating current.

Electric generators convert mechanical energy to electrical energy by rotating a coil through a magnetic field, thus inducing a voltage due to the changing magnetic flux. Transformers adjust voltages up or down by passing a changing magnetic field produced by an alternating

current in a primary coil through a secondary coil. By Faraday's law, this induces a voltage in the secondary coil. The operation of generators and transformers is based on Faraday's law, and both play major roles in power-distribution systems.

summary

The discovery that magnetic forces are associated with electric currents unified the study of electricity and magnetism in the early part of the nineteenth century. The fundamental nature of magnetism was shown to be due to the motion of electric charges. Electric currents can be produced by changing magnetic fields, as described by Faraday's law.

1　Magnets and the magnetic force. Like electrostatic forces, the magnetic force between the poles of two magnets obeys Coulomb's law and the rule that like poles repel and unlike poles attract. The earth itself behaves like a large magnet with its south pole pointing northward.

2　Magnetic effects of electric currents. Oersted's discovery that magnetic effects are associated with electric currents led to the description of the magnetic field of a current-carrying wire. A magnetic force is exerted on either a current-carrying wire or a moving charge when they are located in a magnetic field.

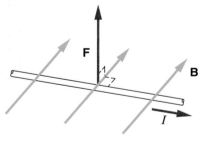

$$F = IlB$$

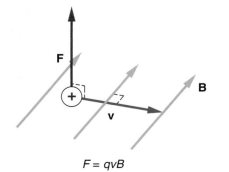

$$F = qvB$$

3　Magnetic effects of current loops. A current-carrying loop of wire produces a magnetic field identical to the field of a short bar magnet. Current loops and coils are magnetic dipoles, and the magnetism of natural magnets can be attributed to atomic current loops. A magnetic torque exerted on a current loop in an external field is similar to that exerted on a bar magnet.

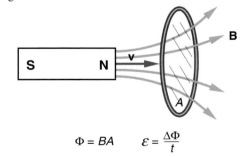

4　Faraday's law: electromagnetic induction. Faraday discovered that a voltage is induced in a circuit equal to the rate of change of magnetic flux passing through the circuit (Faraday's law). Magnetic flux is equal to the product of the magnetic field and the area bounded by the circuit when the field lines are perpendicular to the plane of the loop. Lenz's law describes the direction of the induced current: the current opposes the change producing it.

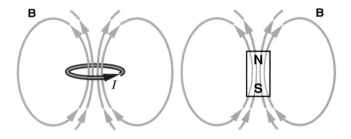

$$\Phi = BA \qquad \varepsilon = \frac{\Delta\Phi}{t}$$

5　Generators and transformers. A generator converts mechanical energy to electrical energy by electromagnetic induction and naturally produces an alternating current. Transformers are used to adjust alternating voltages up and down. Faraday's law is the basic principle of operation for both of these devices, and both are commonly used in our power-distribution system.

key terms

Magnetic pole, 278
Magnetic dipole, 278
Magnetic monopole, 278
Magnetic force, 282
Magnetic field, 282

Electromagnet, 285
Magnetic flux, 288
Faraday's law, 288
Electromagnetic induction, 288

Lenz's law, 288
Self-induction, 289
Generator, 291
Transformer, 292

questions

*Questions identified with an asterisk are more open-ended than the others. They call for lengthier responses and are more suitable for group discussion.

Q1. The north pole of a hand-held bar magnet is brought near the north pole of a second bar magnet lying on a table. How will the second magnet tend to move? Explain.

Q2. If the distance between the south poles of two long bar magnets is reduced to half its original value, will the force between these poles be doubled? Explain.

Q3. In what respects is the force between two magnetic poles similar to the force between two charged particles? Explain.

Q4. Is it possible for a bar magnet to have just one pole? Explain.

Q5. Does a compass needle always point directly northward in the absence of other nearby magnets or currents? Explain.

*Q6. If we regard the earth as magnet, does its magnetic north pole coincide with its geographical north pole? What defines the position of the geographical north pole? Explain.

Q7. We visualized the magnetic field of the earth by imagining that there is a bar magnet inside the earth (fig. 14.7). Why did we draw this magnet with its south pole pointing north? Explain.

Q8. A horizontal wire is oriented along a north-south line, and a compass is placed above it. Will the needle of the compass deflect when a current flows through the wire, and if so, in what direction? Explain.

Q9. A horizontal wire is oriented along an east-west line, and a compass is placed above it. Will the needle of the compass deflect when a current flows through the wire from east to west, and if so, in what direction? Explain.

Q10. Is the force exerted by one current-carrying wire on another due to electrostatic effects or to magnetic effects? Explain.

Q11. A uniform magnetic field is directed horizontally toward the north, and a positive charge is moving west through this field. Is there a magnetic force on this charge, and if so, in what direction? Explain.

Q12. A positively charged particle is momentarily at rest in a uniform magnetic field. Is there a magnetic force acting on this particle? Explain.

Q13. If a uniform magnetic field is directed horizontally toward the east, and a negative charge is moving east through this field, is there a magnetic force on this charge, and if so, in what direction? Explain.

Q14. Why does the magnetic force on a current-carrying segment of wire behave like one on a positive charge traveling in the same direction as the current? Explain.

Q15. If we look down at the top of a circular loop of wire whose plane is horizontal and that carries a current in the clockwise direction, what is the direction of the magnetic field at the center of the circle? Explain.

Q16. If we were to represent the current loop of question 15 as a bar magnet or magnetic dipole, in what direction would the north pole be pointing? Explain.

*Q17. A current-carrying rectangular loop of wire is placed in an external magnetic field with the directions of the current and field as shown in the diagram. In what direction will this loop tend to rotate as a result of the magnetic torque exerted on it? Explain.

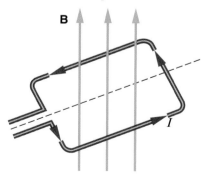

*Q18. If the rectangular loop of wire shown in question 17 were oriented so that the plane of the loop is parallel to the magnetic field lines, would there be a net torque acting on the loop? Explain.

Q19. Since the magnetic fields of a coil of wire and a bar magnet are identical, are there loops of current inside natural magnetic materials like iron or cobalt? Explain.

Q20. In what respect is a simple ammeter designed to measure electric current like an electric motor? Explain.

Q21. Does an ac motor require a split-ring commutator to work? Explain.

Q22. If Faraday wound enough turns of wire on the secondary coil of his iron ring, would he have found that a large steady-state current in the primary coil induced a current in the secondary coil? Explain.

Q23. Is a magnetic flux the same as a magnetic field? Explain.

*Q24. A horizontal loop of wire has a magnetic field passing upward through the plane of the loop. If this magnetic field increases with time, is the direction of the induced current clockwise or counterclockwise (viewed from above) as predicted by Lenz's law? Explain.

Q25. Two coils of wire are identical except that coil A has twice as many turns of wire as coil B. If a magnetic field increases with time at the same rate through both coils, which coil (if either) has the larger induced voltage? Explain.

*Q26. Suppose that the magnetic flux through a coil of wire varies with time, as shown in the graph. Using the same time scale, sketch a graph showing how the induced voltage varies with time. Where does the induced voltage have its largest magnitude? Explain.

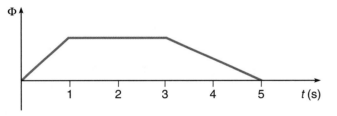

Q27. If the magnetic field produced by the magnets in a generator is constant, does the magnetic flux through the generator coil change when it is turning? Explain.

Q28. Does a simple generator produce a steady direct current? Explain.

Q29. A simple generator and a simple electric motor have very similar designs. Do they have the same function? Explain.

Q30. Can a transformer be used, as shown in the diagram below, to step up the voltage of a battery? Explain.

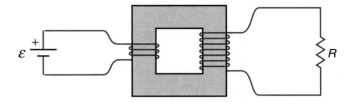

Q31. By stepping up the voltage of an alternating current source using a transformer, can we increase the amount of electrical energy drawn from the source? Explain.

exercises

E1. Two long bar magnets lying on a table with their south poles facing one another exert a force of 10 N on each other. If the distance between these two poles is doubled, what is the new value of this force?

E2. Two long parallel wires, each carrying a current of 5 A, lie a distance of 10 cm from each other. What is the magnetic force per unit length exerted by one wire on the other?

E3. If the distance between the two wires in exercise 2 is tripled, how does the force per unit length change?

E4. Two parallel wires, each carrying a current of 2 A, exert a force per unit length on each other of 1.6×10^{-5} N/m. What is the distance between the wires?

E5. A wire carries a current of 4 A. How much charge flows past a point on the wire in a time of 5 s?

E6. A particle with a charge of 0.06 C is moving at right angles to a uniform magnetic field with a strength of 0.5 T. The velocity of the charge is 600 m/s. What is the magnitude of the magnetic force exerted on the particle?

E7. A straight segment of wire has a length of 10 cm and carries a current of 5 A. It is oriented at right angles to a magnetic field of 0.6 T. What is the magnitude of the magnetic force on this segment of wire?

E8. The magnetic force on a 40-cm straight segment of wire carrying a current of 5 A is 2.5 N. What is the magnitude of the component of magnetic field perpendicular to the wire?

E9. A coil of wire with 80 turns has a cross-sectional area of 0.04 m². A magnetic field of 0.6 T passes through the coil. What is the total magnetic flux passing through the coil?

E10. A loop of wire enclosing an area of 0.03 m² has a magnetic field passing through its plane at an angle to the plane. The component of the field perpendicular to the plane is 0.4 T, and the component parallel to the plane is 0.6 T. What is the magnetic flux through this coil?

E11. The magnetic flux through a coil of wire changes from 6 T·m² to zero in a time of 0.25 s. What is the magnitude of the average voltage induced in the coil during this change?

E12. A coil of wire with 60 turns and a cross-sectional area of 0.02 m² lies with its plane perpendicular to a magnetic field of magnitude 1.5 T. The coil is rapidly removed from the magnetic field in a time of 0.2 s.
   a. What is the initial magnetic flux through the coil?
   b. What is the average value of the voltage induced in the coil?

E13. A transformer has 15 turns of wire in its primary coil and 60 turns on its secondary coil.
   a. Is this a step-up or step-down transformer?
   b. If an alternating voltage with an effective voltage of 110 V is applied to the primary, what is the effective voltage induced in the secondary?

E14. If 8 A of current are supplied to the primary coil of the transformer in exercise 13, what is the effective current in the secondary coil?

E15. A step-down transformer is to be used to convert an ac voltage of 120 V to 6 V to power an electric train. If there are 300 turns in the primary coil, how many turns should there be in the secondary coil?

E16. A step-up transformer is designed to produce 1380 V from a 115-V ac source. If there are 400 turns on the secondary coil, how many turns should be wound on the primary coil?

challenge problems

CP1. Two long parallel wires carry currents of 5 A and 10 A in opposite directions as shown in the diagram. The distance between the wires is 5 cm.
   a. What is the magnitude of the force per unit length exerted by one wire on the other?
   b. What are the directions of the forces on each wire?
   c. What is the total force exerted on a 30-cm length of the 10-A wire?
   d. From this force, compute the strength of the magnetic field produced by the 5-A wire at the position of the 10-A wire ($F = IlB$).
   e. What is the direction of the magnetic field produced by the 5-A wire at the position of the 10-A wire?

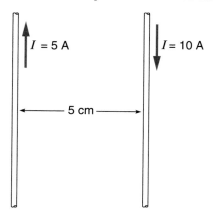

CP2. A small metal ball with a charge of +0.05 C and a mass of 25 g (0.025 kg) enters a region where there is a magnetic field of 0.5 T. The ball is traveling with a velocity of 200 m/s in a direction perpendicular to the magnetic field, as shown in the diagram.
   a. What is the magnitude of the magnetic force acting on the ball?
   b. What is the direction of the magnetic force exerted on the ball when it is at the position shown?
   c. Will this force change the magnitude of the velocity of the ball? Explain.

   d. From Newton's second law, what is the magnitude of the acceleration of the charged ball?
   e. Since centripetal acceleration is equal to v^2/r, what is the radius of the curve the ball will move through under the influence of the magnetic force?

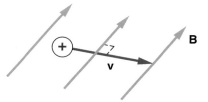

CP3. A rectangular coil of wire has dimensions of 3 cm by 6 cm and is wound with 60 turns of wire. It is turned between the pole faces of a horseshoe magnet that produces an approximately uniform field of 0.4 T, so that sometimes the plane of the coil is perpendicular to the field and sometimes it is parallel to the field.
   a. What is the area bounded by the rectangular coil?
   b. What is the maximum value of the total magnetic flux that passes through the coil as it is turned?
   c. What is the minimum value of the total flux through the coil as it turns?
   d. If the coil makes one complete turn each second and turns at a uniform rate, what is the time involved in changing the flux from the maximum value to the minimum value?
   e. What is the average value of the voltage generated in the coil as it passes from the maximum to the minimum value of flux?

CP4. A transformer is designed to step down line voltage of 110 V to 22 V. The primary coil has 400 turns of wire.
   a. How many turns of wire should there be on the secondary coil?
   b. If the current in the primary coil is 5 A, what is maximum possible current in the secondary coil?
   c. If the transformer gets warm during operation, will the current in the secondary coil equal the value computed in part b? Explain.

home experiments and observations

HE1. Look around your home or residence hall to see how many magnets you can find. Small packages of magnets can often be found in variety stores or hardware stores for hanging tools or other uses.

a. If you have a bar magnet, find its north-seeking pole by suspending it from a string and seeing which end points north.

b. Use this magnet, or another one whose north and south poles can be identified, to find the poles of your other magnets and label them north or south with a crayon.

c. Verify that two like poles repel and two unlike poles attract for any combination of your available magnets.

d. Try to determine which of your various magnets is the strongest by seeing over what distance it will attract a paper clip or other small steel item. (The largest magnet may not be the strongest!)

HE2. If you have a small compass, you can use it to explore the magnetic field of magnets and currents. The compass needle will always point in the direction of the net magnetic field, wherever it is located.

a. Using the compass near a bar or horseshoe magnet, find out how far you must move the compass from the magnet before the field of the magnet is no longer noticeable and the compass needle simply points north.

b. Bringing the compass close to the magnet, determine the direction of the magnetic field at various points around the magnet. Make a sketch showing the direction of the field at selected points around the magnet.

HE3. Using a long insulated wire, wrap several turns (as many as 50 to 100) around a large steel nail. If the free ends of the wire are connected to a flashlight battery, you have an electromagnet. (To avoid depleting the battery, do not leave the wire connected to the battery very long.)

a. Compare the strength of your electromagnet to some of the permanent magnets you may have available. (At what distance can each pick up a paper clip?)

b. Note whether you see a spark when you break the connection between the electromagnet and the battery. Do you get the same effect with an uncoiled length of wire?

unit Four

Wave Motion and Optics

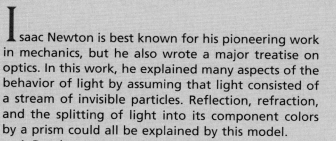

Isaac Newton is best known for his pioneering work in mechanics, but he also wrote a major treatise on optics. In this work, he explained many aspects of the behavior of light by assuming that light consisted of a stream of invisible particles. Reflection, refraction, and the splitting of light into its component colors by a prism could all be explained by this model.

A Dutch contemporary of Newton, Christian Huygens (1629–1695), held an opposing view. He assumed that light was a wave and he successfully explained the same phenomena treated by Newton. For the next one hundred years, these two views competed for acceptance, but because of Newton's prestige, the particle view was more widely held.

In 1800, a British physician, Thomas Young (1773–1829), performed his famous double-slit experiment, which demonstrated interference effects involving light. Interference is a wave phenomenon, so Young's experiment tipped the scales in favor of a wave model of light. During the next fifty years, physicists and mathematicians developed a detailed mathematical description of the behavior of waves and successfully explained many new features of light interference. In 1865, James Clerk Maxwell predicted the existence of electromagnetic waves having the same velocity as

light. This suggested that light was an electromagnetic wave and reinforced the wave model of light by describing the nature of these waves.

Wave motion is now recognized as a universal feature of a wide range of phenomena. Sound waves, light waves, waves on springs and ropes, water waves, seismic waves (involved in earthquakes), and gravitational waves are actively studied. The theory of quantum mechanics, developed in the twentieth century, even applies wave concepts to particles such as electrons. Reflection, refraction, and interference, which we explore in chapters 15, 16, and 17, are all characteristic of wave motion.

Ironically, developments in the twentieth century have shown that light sometimes demonstrates particle-like behavior. We now view a light wave as consisting of a stream of *photons,* particles of light having some characteristics of waves. Although this concept differs from Newton's particle model, Newton may be partially vindicated by these more recent developments. Light (or any wave motion) can exhibit both wave and particle-like behavior, and conversely, any particle has wavelike features. Waves are everywhere—they have become a central theme in physics.

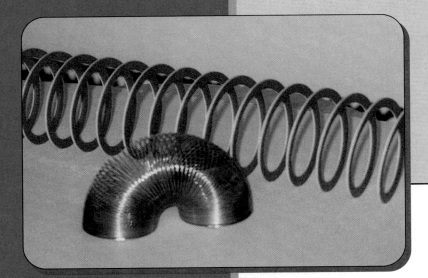

Making Waves

chapter

15

chapter overview

Our main objective in this chapter is to investigate the basic nature of waves and their properties. The properties of waves include speed, wavelength, period, and frequency and the phenomena of reflection, interference, and energy transmission, too. We will examine a few types of wave motion in more detail, including waves on a string and sound waves. We will also explore features of musical sound waves.

If you have ever visited a beach at an ocean or large lake, you have probably delighted in the idle pastime of watching the waves come in. A cliff or some other high spot is a good vantage point for this amusement. From such a height, you can track a single wave crest as it moves in and finally breaks on the shore (fig. 15.1). Because of their regularity, watching waves can be relaxing, even hypnotic.

As you watched the waves, you may have been struck by a curious aspect of their behavior. Although the water seems to move toward the shore, no water accumulates on the beach. What is happening then—what is actually moving? Is this apparent motion of the water somehow deceiving us?

If you stand in the surf and let the waves break over you, you certainly get the sense that the waves are carrying energy. A big wave can knock you over or carry you along. A person on a surfboard can achieve a large velocity by riding the crest of a wave. The kinetic energy gained in this manner comes from the wave.

Although the behavior of water waves near the shore is complex, water waves exhibit many of the general features associated with wave motions of all types. Light,

figure **15.1**   The waves move in and break on the shore. Why does the water not accumulate there?

sound, radio waves, and waves on guitar strings are all examples of wave motion that have much in common with the waves we observe at the beach. Wave motion has implications for all areas of physics, including atomic and nuclear physics. An enormous range of phenomena can be explained in terms of waves.

15.1 Wave Pulses and Periodic Waves

Although water waves have characteristics that are common to any wave motion, the details of their motion are often quite complex, particularly near a beach. That complexity adds to their beauty and allure, but for a beginning discussion of the nature of waves, we need a simpler example. A Slinky, that toy spring that walks down stairs, is an ideal medium for studying simple waves.

The original Slinkies were made of metal, but plastic ones are also available now. They are standard equipment in most physics storerooms, but you may have one left over from your childhood or that you can borrow from a younger acquaintance. Having your own Slinky handy can help you develop a feeling for wave phenomena. What kinds of waves can you generate with the Slinky?

How do wave pulses travel down a Slinky?

If a Slinky is laid out on a smooth table with one end held motionless, you can easily produce a single traveling pulse on the Slinky. With the Slinky slightly stretched, move the free end back and forth once along the axis of the Slinky. As you do so, you will see a disturbance move from the free end of the Slinky to the fixed end (fig. 15.2).

The motion of the **wave pulse** created in this manner is easy to see. The pulse travels down the Slinky and may be reflected at the fixed end and come back toward its starting point before dying out. But what is actually moving? The pulse moves through the Slinky, and portions of the Slinky move as the pulse passes through them. After the pulse dies

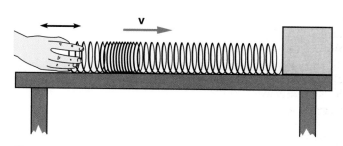

figure **15.2**   With one end of the Slinky fixed, a simple forward and back motion of the other end produces a traveling pulse.

out, though, the Slinky is exactly where it was before the pulse began. The Slinky itself has gone nowhere.

A closer look at what is happening within the Slinky can clarify the basic features of the pulse. Moving one end of the Slinky back and forth as in figure 15.2 creates a local compression where the rings of the spring are closer together than in the rest of the Slinky. This region of compression moves along the Slinky and constitutes the pulse that you see. Individual loops of the spring move back and forth as this region of compression goes by.

Some general features of wave motion

A pulse in a Slinky shows some general features of wave motion. The wave or pulse moves through the medium (in this case, the Slinky) but the medium itself goes nowhere.

Water waves move toward the beach, but water does not accumulate. What moves is a disturbance within the medium, which may be a local compression, a sideways displacement, or some other kind of local change in the state of the medium. The disturbance moves with a definite velocity determined by properties of the medium.

In a Slinky, the speed of the pulse is determined partly by the tension in the Slinky. You can easily confirm this if you stretch the Slinky by different amounts and note the changes in the pulse velocity. The pulse travels faster when the Slinky is highly stretched than when it is only slightly stretched. The other factor that determines the pulse speed for the Slinky is the mass of the spring: for the same tension, a pulse travels more slowly on a steel Slinky than on a plastic one because the mass for a given length of the spring is greater for steel than for plastic.

Transmission of energy through the medium is another general feature of wave motion. The work done in moving one end of the Slinky increases both the potential energy of the spring and the kinetic energy of individual loops. This region of higher energy then moves along the Slinky and reaches the opposite end. There, the energy could be used to ring a bell or to perform other types of work. Energy carried by water waves does substantial work over time in eroding and shaping a shoreline. Energy transmission is an extremely important aspect of any wave motion.

How do longitudinal and transverse waves differ?

The wave pulse that we have described in the Slinky is called a *longitudinal* wave. In a longitudinal wave, the displacement or disturbance in the medium is *parallel* to the direction of travel of the wave or pulse. In the Slinky, the loops of the spring move back and forth along the axis of the Slinky, and the pulse also travels along this axis.

You can also produce a *transverse* wave on a Slinky or spring. In a transverse wave, the disturbance or displacement is *perpendicular* to the direction the wave is traveling (fig. 15.3). If you move your hand back and forth perpendicular to the axis of the spring, you create a transverse pulse. This works best with a Slinky when it is highly stretched. A long thin spring is actually more effective than a Slinky for producing transverse pulses or waves. Like the longitudinal pulse, the transverse pulse moves down the spring and also transmits energy along the spring.

Waves on a rope, discussed in section 15.2, are generally transverse, as are electromagnetic waves, which will be discussed in the first section of chapter 16. Polarization effects (section 16.5) are associated with transverse waves but not longitudinal waves. Sound waves (section 15.4) are longitudinal and are similar in many ways to longitudinal waves on a Slinky. Water waves have both longitudinal and transverse properties.

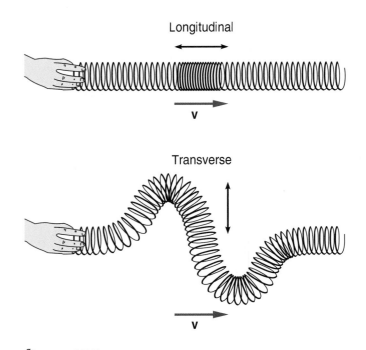

figure 15.3 In a longitudinal pulse, the disturbance is parallel to the direction of travel. In a transverse pulse, the disturbance is perpendicular to the direction of travel.

What is a periodic wave?

Up to this point, we have discussed single wave pulses on a Slinky. If instead of moving your hand back and forth just once, you continue to produce pulses, you will send a series of longitudinal pulses down the Slinky. If equal time intervals separate the pulses, you produce a **periodic wave** on the Slinky (fig. 15.4).

The time between pulses is called the *period* of the wave and is often represented by the symbol T. The *frequency* is the number of pulses or cycles per unit of time and is equal to the reciprocal of the period:

$$f = \frac{1}{T},$$

where the symbol f represents the frequency. (The same symbols and meanings were given to these quantities in chapter 6 when we discussed simple harmonic motion.)

As the pulses move down the Slinky, they are spaced at regular distance intervals if they have been created at regular time intervals. The distance between the same points on successive pulses is called the **wavelength.** This distance is shown in figure 15.4 and is labeled with the Greek letter λ, lambda, the commonly used symbol for wavelength.

A pulse in a periodic wave travels a distance of one wavelength in a time equal to one period before the next pulse is created. The speed of the wave can be expressed in terms of these quantities. The speed is equal to one wavelength

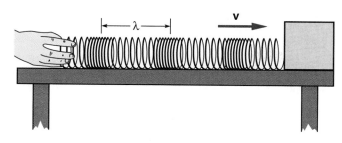

figure 15.4 A series of pulses produced at regular time intervals generates a periodic wave on the Slinky. The wavelength λ is the distance between the pulse centers.

(the distance traveled between pulses) divided by one period (the time between pulses), or

$$v = \frac{\lambda}{T} = f\lambda.$$

The last portion of this equality is true because the frequency f is equal to $1/T$, the reciprocal of the period.

This relationship is valid for any periodic wave and is useful for finding the frequency or wavelength when one of these quantities is known. The wave speed depends only on the properties of the medium. Often, the speed is known from other considerations: the speed of electromagnetic waves (section 16.1) in free space, for example, has a fixed value (the speed of light) that is independent of wavelength or frequency. The relationship between speed, wavelength, and frequency will be illustrated many times in the following sections.

A wave pulse is a disturbance that moves through some medium while the medium itself goes nowhere. If we move the end of a Slinky back and forth along its axis, we create a longitudinal pulse whose disturbance is an area of compression on the spring. If we wave the end of the Slinky up and down in a direction perpendicular to its axis, we create a transverse pulse. A periodic wave consists of several pulses spaced at regular intervals in time (the period) and space (the wavelength). The frequency is the reciprocal of the period, and the pulse speed is equal to the frequency times the wavelength.

15.2 Waves on a Rope

Imagine a heavy rope tied at one end to a wall or post. If you moved the free end of the rope up and down, you could create either a transverse wave pulse or a periodic transverse wave that would travel down the rope toward its fixed end. The situation is the same as producing a transverse wave on a Slinky or spring: think of the rope as a very stiff spring.

What does the wave look like as it travels down the rope? In this case, the disturbance is a vertical displacement of the rope from its straight-line position. This disturbance is easy to visualize and graph, which is why the study of this type of wave is so useful for understanding wave phenomena.

What does the graph of the wave look like?

Suppose that we give the end of the rope an up-and-down motion, producing a pulse like the one depicted in figure 15.5. The right edge of this pulse corresponds to the beginning of the motion. The left edge is the end of that motion. Like the pulse on the Slinky, this disturbance travels down the rope. In this case, however, the picture of the rope can be thought of as a graph, with the vertical axis representing the vertical displacement y of the rope, and the horizontal axis the horizontal position x of a point on the rope.

A single picture of the pulse on the rope does not tell the whole story—it is like a snapshot showing the displacement of the rope at only one instant in time. The pulse is moving, so at some later time the pulse will be farther down the rope at a different horizontal position. We would have to draw a series of graphs at different times to represent this. The pulse may gradually decrease in size because of frictional effects, but the shape remains basically the same as the pulse moves along the rope.

If, instead of giving the rope just one up-and-down motion, you repeat a series of identical pulses at regular time intervals, you might produce a periodic wave like the one shown in figure 15.6. The wavelength λ (shown on the graph) is the distance covered by one complete cycle of the wave. This wave pattern moves to the right along the rope, retaining its shape, just as the single pulse did. When the leading edge of the wave reaches the fixed end of the rope, it will be reflected and start to move back toward your hand. As this happens, the reflected wave interferes with the wave still traveling toward the right, and the picture becomes more complex.

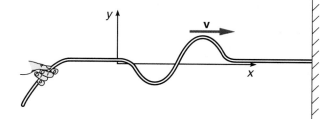

figure 15.5 At any instant a transverse pulse moves along a stretched rope, the shape of the rope can be thought of as a graph of the vertical displacement of the rope plotted against horizontal position.

The shape of the wave in figure 15.6 depends on the exact motion of the hand or other oscillator generating the wave. The wave could be much more complex than the shape shown. One particularly simple shape plays an important role in the analysis of wave motion. If you move your hand up and down smoothly in simple harmonic motion, the displacement of this end of the rope will vary sinusoidally with time, as discussed in chapter 6. The resulting periodic wave also has a sinusoidal form, and we call it a **harmonic wave** (fig. 15.7).

Like other waves, the sinusoidal wave shown in figure 15.7 travels along the rope until it is reflected at the fixed end. If you move the end of the rope carefully, you can produce a wave that looks like the one shown in figure 15.7. The individual segments of a rope or spring tend to move with simple harmonic motion, because the restoring force pulling the rope back toward the center line is roughly proportional to its distance from the center line. This was the condition for simple harmonic motion discussed in chapter 6.

Harmonic waves play an important role in the discussion of wave motion for another reason—it turns out that any periodic wave can be represented as a sum of harmonic waves with different wavelengths and frequencies. We call this *Fourier,* or *harmonic analysis,* which is the process of breaking a complex wave down into its simple harmonic components (see section 15.5). Harmonic waves can be thought of as building blocks of more complex waves.

What determines the speed of a wave on a rope?

Like the waves on a Slinky, the waves on a string or rope move along the rope with a speed independent of the shape or frequency of the pulses. What determines this speed? To answer this question, we need to think about what causes the disturbance to propagate along the rope. Why do the pulses move?

If we picture just a single pulse moving along the rope, we note that segments of the rope lying in front of this pulse are at rest before the pulse gets there. Something must cause these segments to accelerate as the pulse approaches. The reason the pulse moves is that lifting the rope causes the rope's tension (which is a force acting along the line of the rope) to acquire an upward component. This upward component acts on the segment of the rope to the right of the raised portion, as in figure 15.8. The resulting upward force causes this next segment to accelerate upward, and so on down the rope.

The speed of the pulse depends on the rate of acceleration of succeeding segments of the rope—the faster they can be started moving, the more rapidly the pulse moves down the rope. By Newton's second law, this acceleration is proportional to the magnitude of the force, and inversely proportional to the mass of the segment ($a = F/m$). The tension in the rope provides the accelerating force, so a larger tension produces a larger acceleration. The acceleration is also related to the mass of the segment—a greater

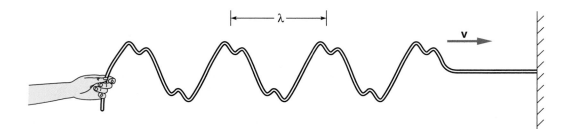

figure **15.6**   A periodic wave moving along a stretched rope. The distance between pulses is the wavelength λ.

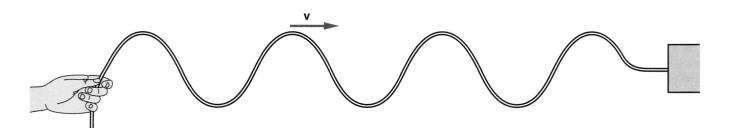

figure **15.7**   A harmonic wave results when the end of the rope is moved up and down in simple harmonic motion.

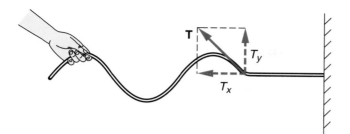

figure **15.8**   As the raised portion of a pulse approaches a given point on the rope, the tension in the rope acquires an upward component. This causes the next segment to accelerate upward.

mass produces a smaller acceleration. These ideas suggest that the speed of a pulse on the rope will increase with the tension in the rope and decrease with the mass per unit of length of the rope.

Actually the square of the speed is directly related to the ratio of the tension to the mass per unit of length, $v^2 = F/\mu$, so the expression for the wave speed involves a square root:

$$v = \sqrt{\frac{F}{\mu}}\ ,$$

where F is the magnitude of the tension in the rope (a force), and μ is the Greek letter mu often used to represent mass per unit of length. This quantity is found by dividing the total mass of the rope by its length, or $\mu = m/L$.

If you increase the tension by pulling harder on the rope, you can expect the wave speed to get larger, just as it did for the Slinky. On the other hand, a thick rope, with a large mass per unit of length, will produce a slower wave speed than a lighter rope. For this reason, a heavy rope is more effective for demonstrating wave motion than a light string. The waves move too swiftly on a light string for us to follow them visually.

What determines the frequency and wavelength of the wave?

The expression relating the wave speed to the frequency and wavelength ($v = f\lambda$, from section 15.1) is useful for predicting what wavelength will result for a wave on a rope. As we have just seen, the speed depends on the tension and the mass per unit of length. Once these quantities have been fixed, the wave speed is constant but the frequency and wavelength may vary. A given frequency will determine the wavelength, and vice versa.

These ideas are demonstrated in Try This Box 15.1. The numerical values in this example are realistic for producing waves that can be followed visually. The assumed tension of 50 N should be large enough to keep the heavy rope

---

try this box 15.1

Sample Exercise: Making Waves

A rope has an overall length of 10 m and a total mass of 2 kg. The rope is stretched with a tension of 50 N. One end of the rope is fixed, and the other is moved up and down with a frequency of 4 Hz.
   a. What is the speed of waves on this rope?
   b. What is the wavelength for the frequency of 4 Hz?

a. $L = 10$ m $\mu = \dfrac{m}{L} = \dfrac{2\ \text{kg}}{10\ \text{m}}$
    $m = 2$ kg
    $F = 50$ N $= 0.2$ kg/m (mass per unit of length)
    $v = ?$ $v = \sqrt{\dfrac{F}{\mu}}$

$$= \sqrt{\frac{50\ \text{N}}{0.2\ \text{kg/m}}}$$

$$= \sqrt{250\ \text{m}^2/\text{s}^2}$$

$$= \mathbf{15.8\ m/s}$$

b. $f = 4$ Hz $v = f\lambda$
    $\lambda = ?$ $\lambda = \dfrac{v}{f}$

$$= \frac{15.8\ \text{m/s}}{4\ \text{Hz}}$$

$$= \mathbf{3.95\ m}$$

---

from sagging too much while also producing a relatively slow wave speed. The wave frequency is determined by the frequency of motion of your hand. This frequency is 4 cycles per second (Hz) in Try This Box 15.1, which gives a wavelength of 3.95 meters. Since this is almost 4 meters, two-and-a-half complete cycles of the wave will fit along the 10-meter length of the rope. Lower frequencies would result in longer wavelengths, higher frequencies in shorter wavelengths.

With a wave speed of almost 16 m/s, it takes less than a second for a pulse to travel the 10-meter length of the rope. To observe these waves, you would need to look quickly, because within a second the wave will reach the fixed end of the rope and be reflected. This reflected wave will interfere with the wave still traveling in the original direction. You would need either a longer rope or some means of damping out the reflected wave if you wanted a more leisurely view.

The real advantage of waves on a rope is the ease with which we can picture them graphically and get a physical sense of how they are produced. In lecture or laboratory demonstrations, a long but not very stiff spring is often used instead of the rope to provide a larger mass per unit of length and a slower wave speed.

A snapshot of a wave moving on a rope is like a graph showing the vertical displacement of the rope plotted against position. This picture shows the wave at only one instant in time, however, because the wave pattern is moving. The speed of the wave increases with increasing tension and decreases with increasing mass per unit of length of the rope. The frequency is determined by how rapidly you move your hand. Along with the speed, frequency determines the wavelength.

15.3 Interference and Standing Waves

When a wave on a rope reaches the fixed end of the rope, it is reflected and travels in the opposite direction back toward your hand. If only a single pulse is involved, you can see the returning pulse quite clearly. If the wave is a longer periodic wave, though, the reflected wave *interferes* with the incoming wave. The resulting pattern becomes more complex and confusing.

When waves of water approach a beach, waves reflected at the beach interfere with those coming in and create a more complex pattern than the waves at some distance from the beach. This process, in which two or more waves combine, is called **interference.** What happens when waves interfere? Can we predict what the resulting wave pattern will be?

How do two waves on a rope combine?

Waves on a rope give examples of interference that are easy to visualize and useful in highlighting the basic concepts. Imagine a rope that consists of two identical segments smoothly spliced to form a single rope *of the same mass per unit length* as the original two segments (fig. 15.9). If you hold one segment in your left hand and the

other in the right, you can generate waves on each segment that will combine when they reach the junction.

If you move both hands up and down in the same way, the waves generated on each segment of the rope should be identical. What happens when they reach the junction? Since each wave by itself would generate a disturbance equal to its own height, we might assume that the combined effect of the two waves will produce a wave with the same frequency and wavelength but *twice* the height of the initial two waves. In fact, this is what happens: the double-height wave proceeds down the single rope to the right of the splice.

The idea that we can find the total effect when two or more waves interfere by simply adding their individual displacements is called the **principle of superposition:**

When two or more waves combine, the resulting disturbance or displacement is equal to the sum of the individual disturbances.

This principle is valid for most types of wave motion. In some situations, the resulting disturbance might be so large that the medium in which the wave is traveling cannot fully respond. The net disturbance in this case is less than what the principle of superposition predicts. For most situations, however, the principle of superposition holds. It is the basis for analyzing all interference phenomena.

When two waves are moving the same way at the same time, as in figure 15.9, they are said to be *in phase*. When they reach the junction, both waves are going up or down at the same time. We can, however, produce waves that are not in phase with one another. If you move the two segments so that one is going up while the other is going down, for example, the two resulting waves are said to be *completely out of phase* with one another (fig. 15.10). What happens in this situation? If the two waves have the same height, the principle of superposition predicts that the net disturbance will be zero. When the displacements of the two waves are added at the junction, the displacement of

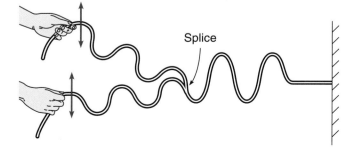

figure 15.9 Identical waves, traveling on two identical ropes that are spliced together, combine to produce a larger wave.

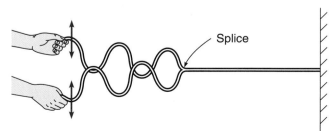

figure 15.10 Two waves, exactly out of phase in their up-and-down motions, combine to produce no net disturbance on the rope beyond the splice.

one is positive (up) while the other is negative (down), and they cancel one another. The sum is always zero, and no wave is propagated beyond the junction.

The result of adding two or more waves depends on their phases as well as on their *amplitude* or height. The two situations that we have depicted with the spliced rope are the extremes—in one case, the waves are completely in phase, and in the other, they are completely out of phase. In the first case, we get complete addition, or *constructive interference*. In the other, we get complete cancellation, or *destructive interference*. It is also possible for the two waves to be neither completely in phase nor out of phase but somewhere in between. In these situations, the resulting wave has a height somewhere between zero and the sum of the two initial heights.

What is a standing wave?

Although it is difficult to get two waves traveling in the same direction on separate ropes to combine, interference of two or more waves traveling in the same direction is quite common for water waves, sound waves, or light waves. The difference in phase between the waves determines whether the interference will be constructive, destructive, or somewhere in between, just like the waves on the rope. What does often occur with waves on ropes or strings is the interference of two waves traveling in *opposite* directions, which happens when the waves are reflected at a fixed support. How do the waves combine in this situation?

Figure 15.11 shows two waves of the same height and wavelength traveling in opposite directions on a string. We can apply the principle of superposition by selecting different points on the string and considering how these two waves add at different times. At point A, for example, the

two waves will cancel each other at all times. One wave is positive, while the other is negative by an equal amount at all times as the two waves approach this point from opposite sides. At this point, the string will not oscillate at all.

If we move one-quarter of a wavelength in either direction from point A, we see a very different result. At point B, for example, both waves will be in phase at all times as they approach from opposite sides. When one is positive, so is the other, and so on. At this point, the two waves always add, producing a displacement twice that of each wave by itself.

What is notable about these two points is that they remain fixed in space along the string. Point A, at which there is no motion, is called a **node.** There are nodes at regular intervals along the string, separated by half the wavelength of the two traveling waves. You can confirm this by moving half a wavelength in either direction from point A and seeing that the two waves cancel at these points also. These nodes do not move.

The same is true for points such as B, where the waves add to yield a large height or amplitude. These points are called **antinodes,** and they are also found at fixed locations separated by half a wavelength along the string. The resulting pattern is shown in figure 15.12, which shows the string position at several different times. The two waves traveling in opposite directions produce a fixed pattern with regularly spaced nodes and antinodes. At the antinodes, the string is oscillating with a large amplitude. At the nodes, it is not moving at all. At points between the nodes and antinodes, the amplitude has intermediate values.

This pattern of oscillation of the string is called a **standing wave** because the pattern does not move. The two waves that produce this pattern are *moving* in opposite directions. They interfere, however, in a way that produces a standing or fixed pattern. Standing waves can be observed for all types of wave motion and always involve the interference of waves traveling in opposite directions. The interference of a reflected wave with an incoming wave is usually how this happens.

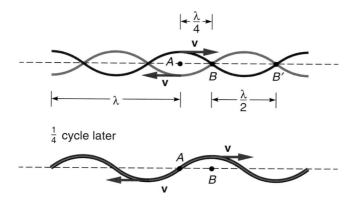

figure 15.11 Two waves of the same amplitude and wavelength are shown traveling in opposite directions on a string. The lower drawing shows the two strings at a time a quarter period after the upper drawing. A node results at point A and an antinode at point B when the waves combine.

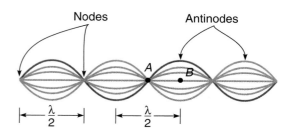

figure 15.12 The pattern produced by two waves traveling in opposite directions in figure 15.11 is called a *standing wave.* (The position of the string is shown at several different times.) The distance between adjacent nodes or antinodes is half the wavelength of the original waves.

What determines the frequency of a wave on a guitar string?

The author plays the guitar and has found much enjoyment over the years in generating standing waves on strings. A guitar, a piano, and any other stringed instrument is made of strings or wires of different weights that are fixed at both ends, with tuning pegs to adjust the tension. A wave generated by plucking the string is reflected back and forth on the string, producing a standing wave.

The frequency of the sound wave produced by the string equals the string's frequency of oscillation, and this frequency is related to the musical **pitch** that we hear. A higher frequency represents a higher-pitched note. What conditions determine the frequency of the guitar string? How are standing waves involved?

The standing wave on a plucked guitar string has nodes at both ends. The string is fixed at both ends and cannot oscillate at these points. The simplest standing wave is one with nodes at either end and an antinode in the middle (fig. 15.13a). This standing wave usually results when the string is plucked. Since the distance between nodes is half the

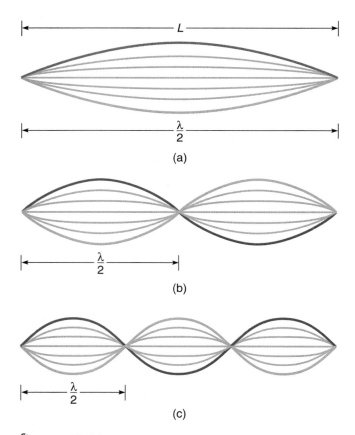

figure 15.13   The first three harmonics are the three simplest standing-wave patterns that can be generated on a guitar string fixed at both ends. String positions are shown at several different times.

wavelength of the waves interfering to produce the standing wave, the wavelength of these interfering waves must be twice the length of the string (2L) in this case.

This simplest standing wave is called the *fundamental* wave, or the *first harmonic* (fig. 15.13a). The wavelength of the interfering waves is determined by the length of the string. The frequency can be found from the relationship between speed, frequency, and wavelength, $v = f\lambda$. The speed is determined by the tension in the string F, and the mass per unit of length of the string μ as discussed in section 15.2.

The frequency of the fundamental wave is given by:

$$f = \frac{v}{\lambda} = \frac{v}{2L}.$$

A string with a longer length L will result in a lower frequency, which is why the bass strings on a piano are much longer than the treble strings. On a guitar, you can change the effective length of the strings by *fretting* them; that is, by placing your finger firmly on the string along the neck of the guitar. Shortening the effective length of the string produces a higher frequency and a higher-pitched tone.

Other factors that affect the frequency are the tension in the string and the mass per unit length of the string, which together determine the wave speed. A higher tension leads to a higher wave speed and a higher frequency. You can easily confirm this by tightening a tuning peg on a guitar. A heavier string, on the other hand, produces a lower wave speed and a lower frequency. The bass strings on a steel-string guitar or piano are made by wrapping wire around a core wire to produce a larger mass per unit of length.

Although the fundamental frequency for a guitar string dominates if you just pluck the string near the middle, you can also produce the other two patterns shown in figure 15.13, as well as patterns with even more nodes. To produce the second harmonic, for example, you touch the string lightly at the midpoint at the same time that you pluck the string. This creates the pattern shown in figure 15.13b, with nodes at the center and at either end. The wavelength of the interfering waves for this pattern is equal to the length L of the string. Since this wavelength is half the fundamental, the resulting frequency is twice the fundamental, as shown in Try This Box 15.2. Musically, the pitch produced by doubling the frequency is an octave above the fundamental (see section 15.5).

If you touch the string lightly at a position one-third the length of the string from one end, you get the pattern in figure 15.13c, which has four nodes (counting those at either end) and three antinodes. The resulting frequency is three times the fundamental and 3/2 that of the second harmonic. Musically, this is not a complete octave above the second harmonic but rather an interval called a *fifth* above that pitch.

Guitars are common fixtures in many residence-hall rooms. If you have one handy, try generating some of these harmonics and noting the pitch that results. It requires no

try this box 15.2

Sample Exercise: Waves and Harmonics

A guitar string has a mass of 4 g, a length of 74 cm, and a tension of 400 N. These values produce a wave speed of 274 m/s.
  a. What is its fundamental frequency?
  b. What is the frequency of the second harmonic?

a. $L = 74$ cm $= 0.74$ m $f_1 = \dfrac{v}{\lambda_1} = \dfrac{v}{2L}$
  $v = 274$ m/s
  $\lambda = 2L$ $= \dfrac{274 \text{ m/s}}{1.48 \text{ m}}$
  $f_1 = ?$
                                        $= \mathbf{185\ Hz}$

b. $\lambda = L$ $f_2 = \dfrac{v}{\lambda_2} = \dfrac{v}{L}$
  $f_2 = ?$
                                        $= \dfrac{274 \text{ m/s}}{0.74 \text{ m}}$

                                        $= \mathbf{370\ Hz}$

(Notice that $f_2 = 2f_1$.)

figure **15.14** The strings on a steel-string guitar have different weights. The bass string has been plucked, producing a blur near the middle where the amplitude is greatest.

job. Small children are experts at finding ways of generating sound waves—the louder the better.

Since sound waves reach our ears, they must be able to travel through air. How do they travel through air? How fast do they travel? Can they interfere like waves on a string to form standing waves?

What is the nature of a sound wave?

If you look closely at one of the speakers used in your stereo system or car radio, you will see a mechanism like that shown in figure 15.15. A flexible, cardboardlike material (the diaphragm) is mounted in front of a permanent magnet fixed to the housing of the speaker. A coil of wire

particular skill as a musician and will help to clarify these ideas. If you look closely at the string, you can see the standing wave patterns (fig. 15.14). The patterns are often easier to see when illuminated by a fluorescent light.

When two or more waves combine, their disturbances add according to the principle of superposition. If they are in phase, the resulting interference is constructive. If they are completely out of phase, the waves cancel one another and the interference is destructive. If the waves are traveling in opposite directions, this interference produces a standing-wave pattern with fixed positions for the nodes and antinodes. This is what happens on a guitar string. The length of the string and the form of the standing-wave pattern determine the wavelength. The wavelength determines the frequency, since the wave speed is set by the tension and mass of the string.

15.4 Sound Waves

Sound waves can be generated by an oscillating string on a guitar or piano, but you certainly can think of many other ways of producing such waves. Firing a pistol, using your voice, or banging on a metal pot with a stick will all do the

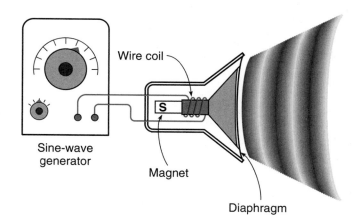

figure **15.15** An oscillating current applied to the coil of wire attached to the diaphragm of a speaker makes the diaphragm oscillate as it is attracted to or repulsed by the magnet, generating a sound wave.

is attached to the base of the diaphragm so that it is centered on the end of the permanent magnet. An oscillating current applied to the wire coil causes it to behave as an electromagnet, alternately attracted to and repelled by the permanent magnet. The diaphragm then oscillates with the same frequency as the applied electrical current.

What effect does the oscillating diaphragm have on nearby air? As the diaphragm moves forward, it compresses the air in front of it. As it moves backward, it produces a region of lower pressure. The compressed region, in turn, pushes against air in front of it, increasing the pressure there. This region of increased pressure propagates through the air, as do the regions of reduced pressure. The disturbance could take the form of a single pulse, but if the diaphragm is moving back and forth repeatedly, it produces a continuous periodic wave consisting of pressure variations.

This wave of pressure variations is a **sound wave.** In the regions of elevated pressure, the molecules in the air are closer together, on the average, than they are in the regions of reduced pressure. Figure 15.16 shows a sound wave with the variations in air density exaggerated. Below the picture of the wave is a graph of pressure versus position. For a simple harmonic wave, this pressure graph has a simple sinusoidal form. To complete the illustration, imagine that the whole pattern is moving away from the source, where new regions of elevated and reduced pressure are being constantly generated.

The molecules making up air are in constant motion in all directions, as is true in any gas. Besides random motion,

however, there must be a back-and-forth motion of the molecules along the line of the wave's path to create the regions of higher and lower density. A sound wave is therefore a *longitudinal* wave. The displacement of the molecules is parallel to the direction of propagation of the wave, much like on a Slinky when one end is moved back and forth. The coils of the Slinky move back and forth along the direction of propagation just as the molecules in the air do, and there are moving regions of increased density (compression) in the Slinky just as in the air.

What determines the speed of sound?

How fast do sound waves travel, and what factors determine the speed of sound? The first half of this question turns out to be easier to answer than the second. In room-temperature air, sound waves travel with a speed of approximately 340 m/s (1100 ft/s) or roughly 750 MPH. If you have ever watched from a distance as someone pounds a nail, you probably have noticed that the sound reaches your ear a split second after you see the collision of hammer and nail. If you stand at the finish line of a 100-meter dash, you see the flash of the starter's pistol before you hear the shot. Sound travels quickly but not nearly as fast as light (see section 16.1).

Similarly, you hear a clap of thunder a few seconds after you see the flash of lightning. Since light travels extremely fast, the light flash reaches you almost instantaneously. The sound wave, on the other hand, takes about 3 seconds to cover 1 kilometer (or 5 seconds to cover 1 mile), given the value for the speed of sound just stated. Counting seconds between the flash and the thunder tells you how far away the lightning strike happened (fig. 15.17). If the flash and the thunder clap occur almost simultaneously, you may be in trouble!

The factors that determine the speed of sound are related to how rapidly one molecule transmits changes in velocity to nearby molecules to propagate the wave. In air, temperature is a major factor since air molecules have higher average velocities at higher temperatures and collide more frequently. An increase in temperature of 10°C increases the speed of sound by about 6 m/s.

For gases other than air, the masses of the molecules or atoms make a difference in the speed of propagation. Hydrogen or helium molecules, with their small mass, are easier to accelerate than the nitrogen or oxygen molecules that are the main ingredients of air. The speed of sound in hydrogen is almost four times larger than in air for similar pressures and temperatures.

Sound waves can also travel through liquids and solids, often with considerably higher speeds than for gases. The speed of sound in water, for example, is four to five times faster than in air. Molecules of water are much closer together than molecules in a gas, so the propagation of a wave does not depend on random collisions. Sound travels even more rapidly through a steel bar or other metals in

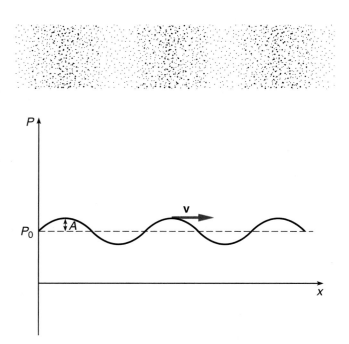

figure 15.16   Variations in air pressure (and density) move through the air in a sound wave. The graph shows pressure plotted against position.

figure **15.17**   Timing the interval between a lightning flash and the associated clap of thunder provides an estimate of your distance from the lightning strike.

figure **15.18**   Placing your lower lip at the rim of a soft-drink bottle and blowing softly across the opening produces a standing wave in the bottle.

which the atoms are rigidly bound within a crystal lattice. Sound waves in rock or metal generally have speeds four to five times faster than in water and fifteen to twenty times faster than in air.

If someone strikes a long steel rail with a hammer at one end and you listen for the sound at the other end, you may hear two bangs. The first one comes to you through the steel rail itself and reaches your ear a moment before the second one, which comes through the air. The actual difference in time will depend on how far you are from the hammer blow. Motion of the source or observer can also affect what we hear, as is described in Everyday Phenomenon Box 15.1.

Making music with soft-drink bottles

Can we observe interference phenomena such as standing waves in sound waves? We can indeed—playing many musical instruments depends on creating standing waves in a tube or pipe. An organ pipe, a clarinet barrel, and the several meters of metal tubing in a sousaphone all serve that purpose. A soft-drink bottle is the handiest example for observing this phenomenon.

If you place your lips near the edge of a soft-drink bottle, as in figure 15.18, and blow softly across the opening, the sound wave reflected from the bottom of the bottle interferes with the incoming wave to produce a standing wave in the bottle. Since the bottle is closed at one end, there should be a displacement node at the bottom of the bottle.

(A displacement node is one where there is no longitudinal motion of the air.) On the other hand, we expect a displacement antinode somewhere near the opening of the bottle, since that is where we are exciting the oscillation.

The simplest standing wave that can exist in a pipe open at one end varies in displacement amplitude like the one plotted in figure 15.19a. The curved line is a graph of the displacement amplitude plotted against position. It is a measure of how far the molecules move back and forth on the average. There is a node at the closed end and an antinode near the open end. Since there is just a quarter-wavelength distance between a node and an antinode, the wavelength of the sound waves interfering to form this standing wave must be approximately four times the length of the tube.

We can determine the frequency of this standing wave from the speed of sound in air (about 340 m/s) and the wavelength. For a tube 25 cm in length (more or less the length of a 16-oz soft-drink bottle), the wavelength of the interfering sound waves is about 1 meter ($4\,L$). Since $f = v/\lambda$, this should produce a frequency of

$$f = \frac{340 \text{ m/s}}{1 \text{ m}} = 340 \text{ Hz.}$$

Since a soft-drink bottle is not a simple pipe with straight sides, the frequency produced by a bottle may differ from this estimate.

Everyday Phenomenon

box **15.1**

A Moving Car Horn and the Doppler Effect

The Situation. We have all had the experience of standing near a busy street and hearing someone lean on their car horn as they are going by. If you remember how that sounds and try to mimic the sound, you will hum at one pitch to represent the car horn as it is nearing you and then at a lower pitch after the car has passed. In other words, you hear a lower-frequency sound wave after the car has passed than when it is approaching, as in the drawing.

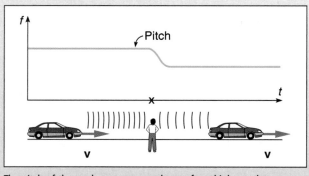

The pitch of the car horn seems to change from higher to lower as the car passes.

Does the frequency of the sound wave produced by the car horn actually change? This hardly seems likely. Something about the motion of the car and the horn must affect the frequency of the pitch that we hear. How can we explain this change in frequency?

The Analysis. The top half of the second diagram shows the wave crests or wavefronts of the car horn when the car is not moving. Each curve represents a surface along which air pressure is at its maximum in the pressure variations associated with the sound wave. The distance between these curves is the wavelength of the sound wave. The wavelength is determined by the frequency of the horn and the speed of sound in air. The wave speed dictates how far a crest will move in a given time, while the frequency of the horn determines when the next crest will appear.

The frequency that we hear is equal to the rate at which the wave crests reach our ear, which is determined by the distance between the wave crests (the wavelength) and by the wave speed. You can think of these wave crests as impinging on your ear in much the same way that water waves wash up on the shore. The greater the speed, the greater the rate at which the wave crests reach your ear. The longer the wavelength, however, the smaller the rate (frequency) at which they reach the ear ($v = f\lambda$, or $f = v/\lambda$).

What happens when the horn is moving? The lower portion of the diagram shows the case in which the horn is moving

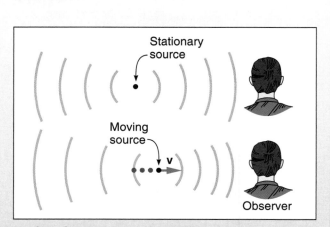

Wavefronts for a stationary car horn (top) and for one that is moving toward the observer (bottom). Motion of the source changes the wavelength on both sides of the source.

toward the observer. Between the time that one crest and the next are emitted by the horn, the horn has moved a short distance. As the diagram shows, this movement shortens the distance between successive wave crests. Even though the horn is still emitting the same frequency as before, the wavelength of the sound wave traveling toward the observer is now shorter.

Because the wavelength of the waves traveling toward the observer is shorter than what is produced by the stationary horn, the wave crests now reach your ear at a higher rate. There is less distance between wave crests, but the waves are still moving at the same speed as before. The higher rate at which the wave crests reach your ear is detected by the ear as a higher frequency. The frequency of the horn that you hear when the horn is moving toward you is higher than what you hear when the horn is stationary. This change in the detected frequency of a wave resulting from the motion of either the source or the observer is called the **Doppler effect.**

Using similar reasoning, you can see that the wavelength in air will get longer if the horn is moving away from you. A longer wavelength will produce a lower frequency as detected by the observer. For the moving car, the frequency that you hear as the car is approaching is higher than the natural frequency of the horn and what you hear as the car is receding from you is lower than the natural frequency.

There is also a Doppler effect if the observer is moving relative to the air in which the wave is traveling. If the observer is moving toward the wave source, he or she will intersect wave crests more rapidly than if stationary and detect a higher frequency than the natural frequency. A receding observer detects a lower frequency. The Doppler effect occurs for light and other types of wave motion as well, but it is most familiar in the common experience of listening to moving vehicles.

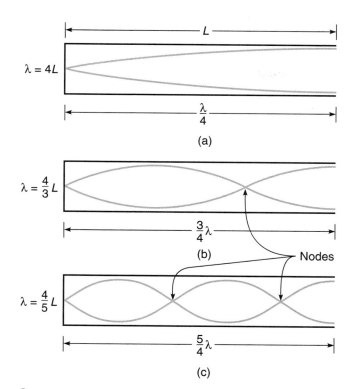

figure **15.19** The standing-wave patterns for the first three harmonics are shown for a tube open at one end and closed at the other. The curves represent the amplitude of back-and-forth molecular motion at each point in the tube.

With some practice, you can generate higher harmonics than the simplest, or fundamental, frequency. Figure 15.19b shows the standing-wave pattern for the next highest harmonic. This one has an antinode near the opening and another antinode and two nodes within the tube. Since three-fourths of the wavelength is contained within the tube, the wavelengths of the interfering waves should be approximately four-thirds the length of the tube, or about 33 cm in a 25-cm tube. The frequency of the sound waves generated for this harmonic is then about 1020 Hz (340 m/s divided by 0.33 m). This frequency is three times the fundamental frequency and corresponds to a tone in the next octave.

Similar reasoning can be used to predict the frequencies of higher harmonics. We can also analyze the standing-wave patterns produced in a tube open at both ends or in one closed at both ends. The actual tone produced by a bottle or a musical instrument is usually a mixture of the various possible harmonics. The mix determines the quality or richness of the resulting sound waves.

Standing waves are one type of interference that is readily observable with sound waves. Sound waves traveling in the same direction can also interfere, producing either constructive or destructive interference, depending on the phase relationship between the interfering waves. "Dead spots" in auditoriums are sometimes produced by destructive in-

terference. The acoustic design of concert halls is a complex art that must take interference into account.

Waves of sound produced by musicians' tubes and strings may wash upon our ears. These sound waves are longitudinal waves involving regions of compression and decompression in the air. The air molecules must move back and forth along the axis of the wave to produce these changes. Sound waves can interfere like other waves. The tones produced by wind instruments or soft-drink bottles involve standing waves of sound.

15.5 The Physics of Music

In sections 15.3 and 15.4, we discussed standing waves formed on a guitar string and in a pipe closed at one end. The length of the string or pipe determined the frequencies that are produced, and these frequencies are related to our perception of musical pitch. These ideas relating physics and music were known in antiquity.

There is a lot more to the physics of musical sounds than this, however. For example, why does the same note played on a clarinet sound very different from that played on a trumpet or almost any other musical instrument? Why do you get different-sounding notes when we pluck a guitar string at different points? Why do certain combinations of notes (*chords*) sound better than others?

The concept of harmonic or frequency analysis plays a big role in understanding many of these issues. Determining the mix of frequencies present in a note played by a musical instrument can explain the quality of the tone produced. The relationships between the different frequencies or harmonics explains why different notes played together sound harmonious. Cultural factors also play a big role in our appreciation of music. Heavy metal may sound good to some, but others will prefer Bach.

What is harmonic analysis?

Recall the different standing waves that we could generate on a guitar string discussed in section 15.3. The first harmonic or *fundamental* had nodes at either end and an antinode at the middle of the string. The second harmonic had two lobes with an additional node at the center of the string. The third harmonic had three lobes with two nodes along the string in addition to the two at either end, and so on. Each of these harmonics has a different frequency, which are simple ratios of the fundamental.

If we simply pluck the string with a pick or finger somewhere along the string, it turns out that we do not get just a single harmonic. Instead there will be a mix of different standing wave modes or harmonics. If we can somehow determine the amplitude of each of the harmonics that make up the complex standing wave on the string, we have then

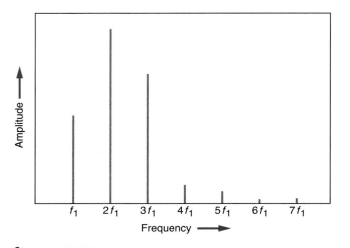

figure **15.20**   When a guitar string is plucked in the usual position, the second and third harmonics often dominate the harmonic spectrum. f_1 is the frequency of the first harmonic.

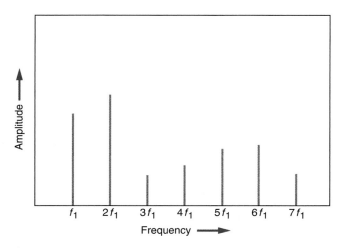

figure **15.21**   When a guitar string is plucked near the bridge, many higher harmonics are present in the harmonic spectrum. This results in a twangy sound.

performed a *frequency* or **harmonic analysis.** The two terms mean the same thing.*

In the case of a guitar string, the results of performing a harmonic analysis can be somewhat surprising. We usually pluck the string about a quarter of the way up from the *bridge,* where the string effectively terminates. The second harmonic ($f_2 = 2f_1$) has an antinode at this position, so the second harmonic is strongly stimulated. The fundamental is not so strongly stimulated since its antinode is at the middle of the string. For an acoustic guitar, the body of the guitar also determines which harmonics will be reinforced when the sound wave is generated. The resulting mix of harmonics might look like that pictured in figure 15.20.

Each peak in the graph in figure 15.20 represents a different harmonic. The height of the peak represents the amplitude of that harmonic. Such a graph can be easily generated with equipment and computer software available in most physics departments. Notice that the second and third harmonics are stronger than the fundamental. Interestingly, when we pluck the string in this manner we still identify the pitch associated with this mix of harmonics as being that of the fundamental, not the second or third harmonics. Our ears and brains perform their own analysis and interpretation.

If we pluck the string much nearer to the bridge, we get a different mix of harmonics. The tone quality is also very different—the sound can be described as twangy. A frequency analysis would show a lot of higher harmonics (figure 15.21) because these have antinodes nearer to the bridge. The amplitude of the second and third harmonics may be much reduced. We still interpret the pitch of the

note as being that of the fundamental, but the note sounds different. This twangy sound is often used in country and western music.

Different musical instruments generate tones that have very different harmonic mixes or graphs. A trumpet usually produces a lot of higher harmonics in its frequency spectrum. It is these higher harmonics that give it a "bright" or "brassy" sound. A flute, on the other hand, can be played such that the fundamental dominates the frequency spectrum with almost no higher harmonics being present. This yields the very "pure" tone that we associate with the flute. How the instrument is played can also have a large effect on the harmonics produced. If you have access to the equipment and software needed to generate a harmonic spectrum, it can be fun to test these ideas with different musical instruments (including your voice).

How are musical intervals defined?

We noted in section 15.3 that doubling the frequency of a note (going from the first to the second harmonic) produces a change in pitch that we call an octave jump. The word *octave* has as its root the number eight. In Western music, we have traditionally used an eight-tone musical *scale.* When learning to sing, we identify the eight tones in the scale by the syllables *do, re, mi, fa, sol, la, ti, do* (fig. 15.22). The two *do*s at either end are an octave apart.

If you play or sing two notes an octave apart, they sound very similar. In fact, we often have difficulty telling the difference between the notes. This is partly due to the fact that, except for the fundamental of the lower note, two notes an octave apart have most of the same higher harmonics present when played on the same instrument. In identifying the note, our ears and brains use these higher harmonics.

When we play the third harmonic on a guitar, the frequency is three times the fundamental frequency, but only

---

* A frequency analysis of this sort is also called a *Fourier analysis,* after the French mathematical physicist Jean-Baptiste Fourier (1768–1830), who developed the mathematical techniques involved.

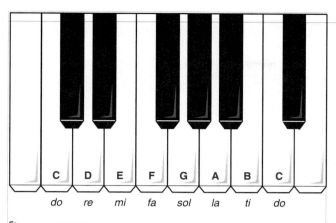

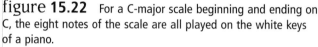

figure **15.22** For a C-major scale beginning and ending on C, the eight notes of the scale are all played on the white keys of a piano.

³⁄₂ times the second harmonic. We call the *musical interval* between the second and third harmonics a *fifth*. In singing, this is the difference between *do* and *sol*, the first and fifth notes in the eight-tone scale. It is the interval between the first pair of notes and the second pair of notes in the tune "Twinkle, Twinkle, Little Star."

The fourth harmonic has a frequency four times the fundamental frequency and thus twice that of the second harmonic. It is an octave above the second harmonic and two octaves above the fundamental. The third and fourth harmonics have the frequency ratio ⁴⁄₃ and this is the interval between *sol* and the *do* at the top of the scale. This interval is called a *fourth*, which is also the interval between *do* and *fa*, the first and fourth notes on the scale. (See Try This Box 15.3.)

The fifth harmonic has a frequency five times the fundamental, but only ⁵⁄₄ times the fourth harmonic. The musical interval between the fourth and fifth harmonics is called a *major third*. In singing, it is the interval between *do* and *mi*, the first and third notes in the scale. Other intervals can be defined in a similar manner.

The other three notes in the scale can also be defined as simple frequency ratios of other notes in the scale. (See challenge problem 5.) *Ti* is a musical third above *sol* (⁵⁄₄), *re* is a musical fourth below *sol* (³⁄₄), and *la* is a musical third above *fa* (⁵⁄₄). All of the notes in the scale can therefore be related to length ratios of a stretched string on an instrument like a guitar. The role of these length ratios in music was recognized by the Greek mathematician Pythagoras as early as 530 B.C., although Pythagoras limited himself to the octave and the fifth in building his scale.

There is a problem, however, with tuning a guitar or particularly a piano in this way. Although the instrument may sound great in one key—that for which it was tuned—it will sound terrible in others. Thus if we tune the piano so that the ratios are all correct (called **just tuning**) when we start our scale on middle C ($f = 264$ Hz for just tuning), the ratios will not be correct if we start our scale on the

---

next note up (called D). So unless you want to play your piano in just one key, you need to compromise on the ratios between the notes. The most common compromise is called **equally-tempered** tuning. In this method of tuning, the ratios are all approximately correct, but not perfect, for any key. The ratios between adjacent half steps on the scale are all identical, however, so the scales sound correct regardless of where you start. (See challenge problem 4.)

Table 15.1 shows the frequencies for both equally-tempered tuning and just tuning for a C scale. Both are based upon the standard 440 Hz frequency for A above middle C. The equally-tempered scale shows the sharps (♯) and flats (♭), which together with the notes of the major scale make up a scale consisting of 12 equal-ratio half steps. The frequency ratio of each note to that of the preceding half step is 1.05946, which is the twelfth root of two ($\sqrt[12]{2}$). Multiplied by itself 12 times it equals 2, which is the appropriate ratio for a full octave. We have not shown the frequencies for the sharps and flats for just tuning. In some versions of just tuning, the flat can have a different frequency than the sharp of the note just below—for example, A-flat (A♭) might have a different frequency than G-sharp (G♯).

Historically, many physicists and mathematicians have been involved in debates over the ideal tuning method. Pythagoras, Ptolemy, Kepler, and Galileo all contributed ideas. Claudius Ptolemy, who is primarily known for his geocentric model of the solar system, was instrumental in the introduction of just tuning. Galileo's father, Vincenzo Galilei, was a music theorist. Galileo himself was actively involved in these issues at the time that equal temperament

table 15.1

Frequencies and Ratios for Different Tuning Methods

Equal Temperament			Just Tuning		
Note	f (Hz)	Ratios	Note	f (Hz)	Frequency Ratios
C	261.6		C (do)	264.0	
C# (D♭)	277.2	1.05946			9/8
D	293.7	1.05946	D (re)	297.0	4/3 · 5/4
D# (E♭)	311.1	1.05946			3/2
E	329.6	1.05946	E (mi)	330.0	4/3
F	349.2	1.05946	F (fa)	352.0	6/5
F# (G♭)	370.0	1.05946			
G	392.0	1.05946	G (sol)	396.0	5/4
G# (A♭)	415.3	1.05946			4/3
A	440.0	1.05946	A (la)	440.0	5/4 · 6/5
A# (B♭)	466.2	1.05946			
B	493.9	1.05946	B (ti)	495.0	
C	523.3	1.05946	C (do)	528.0	

was being proposed and debated. Equal temperament now dominates for tuning pianos and we have become used to the compromises involved.

Why do some combinations of notes sound harmonious?

Why do some notes sound pleasing to us when played together, while others do not? The issue is partly cultural, but there is also a physical basis for what we call *harmony*. Not surprisingly, it can be *partly* explained by the harmonics present in the notes.

For example, if we play the notes *do* and *sol* together, the sound is pleasing to most people. These two notes sound like they belong together, even for people from different cultures. Remember that when you play or sing any note, the resulting sound will usually contain higher harmonics of the fundamental frequency. Because of the simple frequency ratio between these two notes a fifth apart, many of these higher harmonics will be the same. The third harmonic for *do,* for example, is identical to the second harmonic for *sol.* They reinforce one another in the higher harmonics.

If we add two more notes to yield the *major chord do, mi, sol, do,* the sound is even more harmonious. Once again, there is strong overlap of the higher harmonics in these notes. The same is true for many other simple chords. On a piano or a guitar, we make strong use of chords such as these to build our musical structures. In a band or chorus, different instruments or voices provide the notes in the chords.

Some notes do not sound good together, at least to people accustomed to classical music. If we play a *do* and a *re* together, for example, the sound is *dissonant.* Although a lack of strong overlap in higher harmonics can partly ex-

plain the harsh sound that results, another physical phenomenon called *beats* is at work.

When two waves of different frequencies are combined, they interfere. Because their frequencies differ, they come in and out of phase with one another as time progresses. When they are in phase, the combined wave has a large amplitude. When they are out of phase, the amplitude is much smaller. This fluctuation in amplitude of the combined wave is called **beats** (fig. 15.23).

The frequency of the variation in amplitude (called the *beat frequency*) is equal to the difference in frequency of the two waves. If the two notes are very close in frequency, the beat frequency is slow enough to be heard as a variation in amplitude. In other words, you can hear the sound getting louder and softer in a repetitive wah-wah pattern. This effect can be useful in tuning one instrument to another. When the beats become very slow or disappear altogether, the two instruments are in tune. (Beats are also very useful in tuning the double strings on a mandolin or twelve-string guitar.)

When the two notes differ by a full step on the scale, as with *do* and *re,* the beat frequency is rapid enough that we hear a harsh buzz. This buzz produces, what for many of us, is an unpleasant sound. In modern music, however, this dissonant sound is sometimes used to produce desired effects.

As the difference in frequency becomes even larger, the beat frequency can sometimes be heard as a separate tone. The pitch of this tone corresponds to the beat frequency. For example, if we play *do* and *sol* together, the beat frequency is

$$\frac{3}{2}f - f = \frac{1}{2}f$$

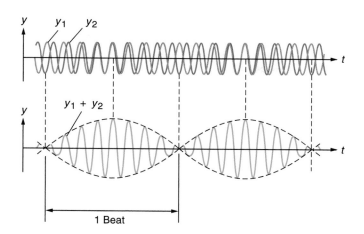

figure 15.23 The two waves with slightly different frequency in the top drawing interfere to produce the beats shown in the lower drawing. The two waves come in and out of phase as time progresses.

where f is the frequency of the lower note, *do*. A note with a frequency one-half that of *do* is an octave below *do*, so it fits nicely within a chord containing *do* and *sol*. Thus beats can help to explain the richness of the harmony that we get when we play major chords.

There is much more to the physics of music than these basic ideas. The acoustics of a good concert hall, for example, involves wave interference, reflection, sound absorption, and many other effects. A good musician develops a discriminating "ear," and can detect and identify musical intervals and chords much more readily than an untrained person, but most of us have had enough exposure to music to appreciate the effects produced. Playing around with a guitar or piano can demonstrate many of the ideas we have just discussed.

The frequency of a sound wave is associated with its musical pitch. When we play a note on a musical instrument the sound will contain the fundamental frequency as well as higher harmonics with frequencies that are integer multiples of the fundamental. A harmonic analysis of the sound determines the amplitude of each of the frequencies present. Harmonics play an important role in defining the musical intervals in the scale used in Western music. Different combinations of these notes sound harmonious when the higher harmonics in the notes reinforce one another as they do in a major chord. Two notes of different frequency interfere to produce a beat frequency equal to the difference in frequency between the two notes. The beat frequency also plays a role in whether the combination of two notes will be pleasing or dissonant.

summary

A wave is a moving disturbance that propagates energy through a medium. Water waves, waves on a Slinky, waves on a string or rope, and sound waves all share features like reflection and interference, which are common to any wave motion. These waves differ in the type of medium involved, the nature of the disturbance that is propagating, and in the wave speed. Standing waves and harmonic analysis play an important role in the physics of music.

1   Wave pulses and periodic waves. Basic features of wave motion can be demonstrated on a Slinky, including single pulses as well as continuous waves. For longitudinal waves, the disturbance is along the line of travel. For transverse waves, it is perpendicular to the direction of travel. The wave speed is equal to the frequency times the wavelength of the wave.

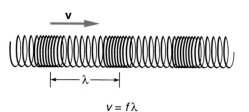

$$v = f\lambda$$

2   Waves on a rope. Transverse waves can be generated on a rope or string. If the end of the rope is moved in simple har-monic motion, the wave has a sinusoidal shape. The wave speed depends on the tension in the rope and the mass per unit of length ($\mu = m/L$) of the rope.

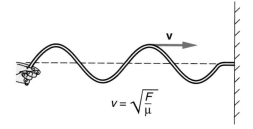

$$v = \sqrt{\frac{F}{\mu}}$$

3   Interference and standing waves. When two or more waves combine, the disturbances add to form a new wave. The interference can be constructive, producing a larger amplitude if the waves are in phase, or it can be destructive, producing a smaller or zero amplitude if they are out of phase. Two waves traveling in opposite directions produce a standing wave.

4   **Sound waves.** Sound waves are longitudinal waves involving the propagation of pressure variations through air or other media. The speed of sound is about 340 m/s in room-temperature air. Standing sound waves can be formed in pipes or soft-drink bottles. The length of the pipe determines the wavelength and the frequency of the various harmonics.

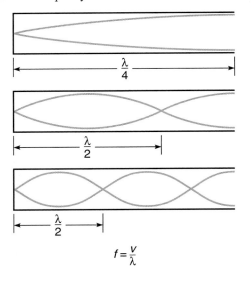

$$\frac{\lambda}{4}$$

$$\frac{\lambda}{2}$$

$$\frac{\lambda}{2}$$

$$f = \frac{v}{\lambda}$$

5   **The physics of music.** Harmonic analysis shows that musical notes played by most instruments contain a mix of higher harmonics along with the fundamental frequency. Musical scales and intervals are based upon the ratios between these higher harmonics. Combinations of notes sound harmonious when the higher harmonics overlap. When two notes are too close in pitch, beats can produce a dissonant buzz.

key terms

Wave pulse, 301
Periodic wave, 302
Wavelength, 302
Harmonic wave, 304
Interference, 306
Principle of superposition, 306

Node, 307
Antinode, 307
Standing wave, 307
Pitch, 308
Sound wave, 310

Doppler effect, 312
Harmonic analysis, 314
Just tuning, 315
Equally-tempered, 315
Beats, 316

questions

*Questions identified with an asterisk are more open-ended than the others. They call for lengthier responses and are more suitable for group discussion.

Q1. A wave pulse is transmitted down a Slinky, but the Slinky itself does not change position. Does a transfer of energy take place in this process? Explain.

Q2. Waves are traveling in an eastward direction on a lake. Is the water in the lake necessarily moving in that direction? Explain.

Q3. A slowly moving engine bumps into a string of coupled railroad cars standing on a siding. A wave pulse is transmitted down the string of cars as each one bumps into the next one. Is this wave transverse or longitudinal? Explain.

Q4. A wave can be propagated on a blanket by holding adjacent corners in your hands and moving the end of the blanket up and down. Is this wave transverse or longitudinal? Explain.

Q5. If you increase the frequency with which you are moving the end of a Slinky back and forth, does the wavelength of the wave on the Slinky increase or decrease? Explain.

Q6. If you increase the speed of a wave on a Slinky by increasing the tension but keep the same frequency of back-and-forth-motion, does the wavelength increase or decrease? Explain.

Q7. Is it possible to produce a transverse wave on a Slinky? Explain.

*Q8. At sporting events the crowd sometimes generates a "wave" that propagates around the stadium. Is this wave transverse or longitudinal? What causes the wave to travel through the crowd? Explain.

Q9. Is it possible to produce a longitudinal wave on a rope? Explain.

Q10. Suppose that we double the mass per unit of length of a rope by twining two ropes together. What effect does this have on the speed of a wave on this rope? Explain.

Q11. What force causes individual segments of a rope to accelerate when a transverse pulse travels down the rope? Explain.

Q12. Why is it easier to observe transverse waves on a heavy rope than on a light string? Explain.

Q13. Suppose that we increase the tension in a rope, keeping the frequency of oscillation of the end of the rope the same. What effect does this have on the wavelength of the wave produced? Explain.

Q14. Is it possible for two waves traveling in the same direction to produce a wave (when they interfere) that has a smaller height (amplitude) than either of the individual waves? Explain.

Q15. Two ropes are joined smoothly to form a single rope, which is attached to a wall. If the ends of the two ropes are moved up and down in phase, but one rope is half a wavelength longer than the other to the point where they join, will the interference of the two waves when they join be constructive or destructive? Explain.

*Q16. When two waves on separate ropes reach the spliced junction out of phase with one another, they interfere destructively producing no wave beyond the splice. What happens then to the energy carried by the waves? Will there by reflected waves? Explain.

Q17. We can form standing waves on a rope attached to a wall by moving the opposite end of the rope up and down at an appropriate frequency. Where does the second wave come from that interferes with the initial wave to form the standing wave? Explain.

Q18. A standing wave is produced on a string fixed at both ends so that there is a node in the middle as well as at either end. Will the frequency of this wave be greater than, equal to, or less than the frequency of the fundamental frequency, which has nodes only at the ends? Explain.

Q19. Is the distance between the antinodes of a standing wave equal to the wavelength of the two waves that interfere to form the standing wave? Explain.

Q20. If we increase the tension of a guitar string, what effect does this have on the frequency and wavelength of the fundamental standing wave formed on that string? Explain.

Q21. If we wrap a second wire around a guitar string to increase its mass, what effect does this have on the frequency and wavelength of the fundamental standing wave formed on that string? Explain.

Q22. Why is it much easier to produce longitudinal waves traveling in air than it is to produce transverse waves? Explain.

Q23. Is it possible for sound to travel through a steel bar? Explain.

Q24. Suppose that we increase the temperature of the air through which a sound wave is traveling.
a. What effect does this have on the speed of the sound wave? Explain.
b. For a given frequency, what effect does increasing the temperature have on the wavelength of the sound wave? Explain.

Q25. If the temperature in an organ pipe increases above room temperature, thereby increasing the speed of sound waves in the pipe but not affecting the length of pipe significantly, what effect does this have on the frequency of the standing waves produced by this pipe? Explain.

Q26. Is the wavelength of the fundamental standing wave in a tube open at both ends greater than, equal to, or less than the wavelength for the fundamental wave in a tube open at just one end? Explain.

Q27. A band playing on a flat-bed truck is approaching you rapidly near the end of a parade. Will you hear the same pitch for the various instruments as someone down the street who has already been passed by the truck? Explain.

Q28. Is it possible for sound waves to travel through a vacuum? Explain.

Q29. When you pluck a guitar string, are you likely to get a sound wave containing just a single frequency? Explain.

*Q30. Why is the second harmonic of a plucked guitar string likely to be stronger than the first harmonic or fundamental when the string is plucked in the usual position? Explain.

Q31. What are we measuring when we perform a harmonic analysis of a sound wave? Explain.

Q32. How is the musical interval that we call a fifth related to the third harmonic of a plucked string? Explain.

Q33. Why do two notes an octave apart sound so much alike? Explain.

*Q34. Frequency and pitch are related, but are they the same thing? When we identify a note as having a certain pitch, is it likely to contain a single frequency? How does perceived pitch differ for someone who cannot carry a tune from that for a trained musician? Discuss.

Q35. Two notes close together on the scale such as *do* and *re* produce a buzz when played together. What is the source of this buzz? Explain.

exercises

E1. Suppose that water waves coming into a dock have a velocity of 1.2 m/s and a wavelength of 2.4 m. With what frequency do these waves meet the dock?

E2. Suppose that water waves have a wavelength of 1.4 m and a period of 0.8 s. What is the velocity of these waves?

E3. A longitudinal wave on a Slinky has a frequency of 5 Hz and a speed of 2.0 m/s. What is the wavelength of this wave?

E4. A wave on a rope is shown in the diagram.
   a. What is the wavelength of this wave?
   b. If the frequency of the wave is 2 Hz, what is the wave speed?

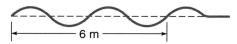

E5. A wave on a string has a speed of 12 m/s and a period of 0.4 s.
   a. What is the frequency of the wave?
   b. What is the wavelength of the wave?

E6. Suppose that a guitar string has a length of 0.8 m, a mass of 0.12 kg, and a tension of 135 N.
   a. What is the mass per unit of length of this string?
   b. What is the speed of a wave on this string?

E7. A string with a length of 0.8 m is fixed at both ends.
   a. What is the longest possible wavelength for the traveling waves that can interfere to form a standing wave on this string?
   b. If waves travel with a speed of 120 m/s on this string, what is the frequency associated with this longest wavelength?

E8. Suppose that the string in exercise 7 is plucked so that there are two nodes along the string in addition to those at either end. What is the wavelength of the interfering waves for this mode?

E9. Sound waves have a speed of 340 m/s in room-temperature air. What is the wavelength of the sound waves for the musical tone concert A, which has a frequency of 440 Hz?

E10. What is the frequency of a sound wave with a wavelength of 0.68 m traveling in room-temperature air ($v = 340$ m/s)?

E11. An organ pipe closed at one end and open at the other has a length of 0.5 m.
   a. What is the longest possible wavelength for the interfering sound waves that can form a standing wave in this pipe?
   b. What is the frequency associated with this standing wave if the speed of sound is 340 m/s?

E12. Suppose that we start a major scale on concert A, which is defined to have a frequency of 440 Hz. If we call this frequency *do,* what is ideal-ratio frequency of
   a. *mi?*
   b. *sol?*

E13. If *sol* on a given scale has a frequency of 396 Hz, what is the ideal-ratio frequency of *do* at the bottom of this scale?

E14. In just tuning, the ratio for a major third is ⁵⁄₄. In equally-tempered tuning the ratio is 1.260. If we start a scale on a frequency of 440 Hz for *do,* what is the difference in frequency for *mi* (a major third above *do*) on an equally-tempered piano and a justly-tuned piano?

E15. If *do* has a frequency of 263 Hz and *re* a frequency of 323 Hz, what is the beat frequency produced when these two notes are played together?

E16. If one guitar string is tuned to a frequency of 440 Hz and a string on another guitar produces 8 beats per second when played together with the first string, what are the possible frequencies of the second string?

E17. What is the beat frequency that results when an 880-Hz note is played with a 660-Hz note? If this beat frequency is heard as a musical tone, how is this tone related to the original two notes? What are the intervals?

challenge problems

CP1. A certain rope has a length of 8 m and a mass of 2.4 kg. It is fixed at one end and held taut at the other with a tension of 30 N. The end of the rope is moved up and down with a frequency of 2.5 Hz.
   a. What is the mass per unit of length of the rope?
   b. What is the speed of waves on this rope?
   c. What is the wavelength of waves on this rope having a frequency of 2.5 Hz?
   d. How many complete cycles of these waves will fit on the rope?
   e. How long does it take for the leading edge of the waves to reach the other end of the rope and start coming back?

CP2. A guitar string has an overall length of 1.25 m and a total mass of 40 g (0.04 kg) before it is strung on the guitar. Once on the guitar, however, there is a distance of 64 cm between its fixed end points. It is tightened to a tension of 720 N.
   a. What is the mass per unit of length of this string?
   b. What is the wave speed for waves on the tightened string?

   c. What is the wavelength of the traveling waves that interfere to form the fundamental standing wave (nodes just at either end) for this string?
   d. What is the frequency of the fundamental wave?
   e. What are the wavelength and frequency of the next harmonic (with a node in the middle of the string)?

CP3. A pipe that is open at both ends will form standing waves, if properly excited, with antinodes near both ends of the pipe. Suppose we have an open pipe 40 cm in length.
   a. Sketch the standing-wave pattern for the fundamental standing wave for this pipe. (There will be a node in the middle and antinodes at either end.)
   b. What is the wavelength of the sound waves that interfere to form the fundamental wave?
   c. If the speed of sound in air is 340 m/s, what is the frequency of this sound wave?
   d. If the air temperature increases so that the speed of sound is now 350 m/s, by how much does the frequency change?
   e. Sketch the standing-wave pattern and find the wavelength and frequency for the next harmonic in this pipe.

CP4. For standard tuning, concert A is defined to have a frequency of 440 Hz. On a piano, A is five white keys above C, but 9 half steps above C counting both the white and black keys. (See fig. 15.22.) A full octave consists of 12 half steps (semitones). In equally-tempered tuning, each half step has the ratio of 1.0595 above the preceding step. (This ratio is the 12th root of 2.0.)
   a. What is the frequency of A-flat, one half step below A for equal temperament?
   b. Working down, find the frequency of each succeeding half step until you get down to C. (Carry your computations to four figures, to avoid rounding errors. For each half step, divide by 1.0595.)
   c. In just tuning, middle C has a frequency of 264 Hz. How does your result in part b compare to this value?

   d. Working up, find the frequency of C above concert A in equal temperament. Is this frequency twice that obtained in part b for middle C?

CP5. Using the procedure outlined in section 15.5 where the ideal ratios for a justly-tuned scale are described, find the frequencies for all of the white keys between middle C (264 Hz) and the C above middle C (a C-major scale). If you have worked challenge problem 4, compare the frequencies for just tuning to those for equal temperament.
   a. G (*sol*) is a fifth above C ($3/2$).
   b. F (*fa*) is a fourth above C ($4/3$).
   c. E (*mi*) is a major third above C ($5/4$).
   d. B (*ti*) is a major third above G (*sol*).
   e. D (*re*) is a fourth below G (*sol*).
   f. A (*la*) is a major third above F (*fa*).

home experiments and observations

HE1. If you have access to a Slinky, either through a younger brother or sister or through your local physics lab, try producing some of the effects described in section 15.1.
   a. Can you estimate the speed of single longitudinal pulses produced on the Slinky?
   b. How does this speed change as the Slinky is stretched?
   c. Does a transverse pulse travel with the same speed as a longitudinal pulse?
   d. Can you produce a continuous longitudinal wave on the Slinky?

HE2. Water waves can be created easily by moving your hand in a bathtub or other small pool of water. The wave crests can be directly observed.
   a. Try moving your hand at different frequencies. How does the wavelength vary with frequency?
   b. Can you estimate the speed of a wave pulse?
   c. By using both hands, two waves can be created that come together and interfere. If you move your hands in unison, you should observe constructive interference along the center line between the two waves. What effect does this have on the wave that you observe along the center line?
   d. Can you produce destructive interference by moving your hands so that one is going up while the other is going down? Describe what happens when you attempt this.

HE3. Empty a soft-drink bottle and practice blowing over the opening as described in section 15.4 until you are able to produce a consistent tone.
   a. How does the pitch of this tone vary if you put water in the bottle? What is the relationship of the pitch when the bottle is half-filled to when it is empty?
   b. Try producing higher harmonics by blowing harder and reducing the opening in your lips. (This is easy for a flute player but takes some practice for other mortals.) How is the pitch of a higher harmonic related to the pitch of the fundamental?
   c. By filling eight bottles to different levels, you can produce all the notes of a one-octave scale. Gather a few friends and try playing "Three Blind Mice" or some other simple tune.

HE4. If you have access to a guitar, try generating some of the higher harmonics using the technique discussed in section 15.3.
   a. How many higher harmonics can you produce? Do the locations of dots on the neck of the guitar provide guidance on where to place your finger?
   b. Try plucking the guitar in the usual fashion and also much nearer to the bridge. Can you hear higher harmonics in the tone produced? How does the tone differ in these two situations?
   c. If you have access to harmonic analysis equipment in your physics lab, compare the harmonic spectrum produced by plucking a string in the two positions described in part b.

HE5. If you have never played a piano, a lot can be learned by sitting down at a keyboard and trying a few scales and intervals.
   a. Play a C-major scale as illustrated in figure 15.22.
   b. Play a D-major scale, beginning and ending on D. Which black keys did you have to use to make the scale sound right?
   c. Play a G-major scale beginning and ending on G. Which black keys did you use?
   d. Play a major chord (*do, mi, sol, do*) starting on the different notes listed in parts a, b, and c as well as others. Which keys on the piano did you use?
   e. Play C in several different octaves. Do these notes all sound similar? Can you hear different harmonics?

Light Waves
and Color

chapter 16

chapter overview

What is light, what are its properties, and how can we explain the rich array of color phenomena that are a part of our everyday experience? Starting from the recognition that light is a form of electromagnetic wave, we explore its behavior, including the properties of absorption, selective reflection, interference, diffraction, and polarization. Our perception of color is affected by all of these processes.

unit four

Have you ever wondered why our world appears so colorful to us? Light is certainly involved in what we see, but what is light? Why do soap films appear so multicolored and why is the sky blue? All of these phenomena are related to the wave nature of light.

All of us have played with soap bubbles at some point in our lives. You may have also been fascinated with soap films in the metal loop that is often used to create bubbles. If you hold the loop still, the film will settle into a pattern of colored bands (fig. 16.1). As you watch these bands, the colors change until finally the film breaks.

How do we explain this? The colorful behavior of soap films and bubbles, as well as many other phenomena involving color, involves interference of light waves. Many ideas regarding waves, including interference, were introduced in chapter 15, but light waves provide some surprising and interesting examples of these ideas.

Light is an electromagnetic wave, so we will have to explore what that means. Light waves can be reflected, refracted, polarized, scattered, and absorbed. They also can interfere with one another to produce some striking effects. Reflection and refraction can be described in terms of ray optics as we will see in chapter 17. Ray optics describes the behavior of lenses and mirrors, which are used to make many optical instruments such as microscopes and telescopes.

This chapter will focus on aspects of light that are directly dependent on the wave properties of light. Wave

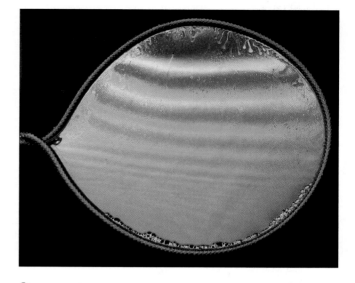

figure **16.1**   A soap film viewed in reflected light displays striking interference colors.

optics describes interference and diffraction phenomena as well as properties such as absorption, scattering, and polarization. All of these phenomena are involved in producing the colors that we see.

16.1 Electromagnetic Waves

What do light, radio waves, microwaves, and X rays have in common? They are all forms of electromagnetic waves. Together they represent an enormous range of phenomena that have become extremely important in our modern technological world.

The prediction of the existence of electromagnetic waves and a description of their nature was first published by James Clerk Maxwell in 1865. Maxwell was an enormously talented theoretical physicist who made important contributions in many areas of physics including electromagnetism, thermodynamics, the kinetic theory of gases, color vision, and astronomy. He is best known, however, for his treatise on electric and magnetic fields, which we have already mentioned. His description of electromagnetic waves, with a prediction of their speed, was just one feature of this work.

What is an electromagnetic wave?

To understand **electromagnetic waves,** we need to review the concepts of electric field and magnetic field. Both fields can be produced by charged particles. *Motion* of the charge is necessary to generate a magnetic field, but an electric field is present regardless of whether the charge is moving. These fields are a property of the space around the charges

and are useful for predicting the forces on other charges, as discussed in chapters 12 and 14.

Suppose that charge is flowing up and down in two lengths of wire connected to an alternating-current source, as in figure 16.2. If the current reverses direction rapidly enough, an alternating current will flow in this arrangement even though it appears to be an open circuit. Charge of one sign will begin to accumulate in the wires, but before the accumulated charge gets too large, the current reverses, the charge flows back, and the opposite-sign charge begins to build. We thus have both a changing amount of charge and a changing electric current in the wires.

The magnetic fields generated by this arrangement can be depicted by circular field lines centered on the wires, as shown. This field, however, is constantly changing in both magnitude and direction as the current changes. From Faraday's law, Maxwell knew that a changing magnetic field would generate a voltage in a circuit whose plane is perpendicular to the magnetic field lines. A voltage implies an electric field, and even in the absence of a circuit, a changing magnetic field will generate an electric field at any point in space where the magnetic field is changing.

Thus, we expect a changing electric field to be generated by the changing magnetic field, according to Faraday's law. Maxwell saw a symmetry in the behavior of electric and magnetic fields: a changing electric field also generates a magnetic field. Maxwell predicted this phenomenon

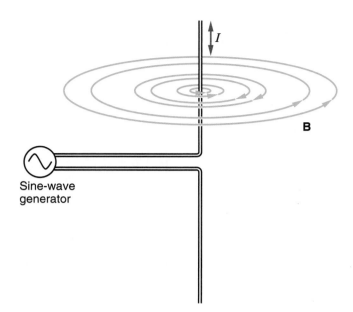

figure **16.2**   A rapidly alternating electric current in the wires generates magnetic fields whose direction and magnitude change with time.

Although Maxwell predicted the existence of such waves in 1865, the first experiment to produce and detect them with electric circuits was performed by Heinrich Hertz (1857–1894) in 1888. Hertz's original antennas were circular loops of wire instead of straight wires, but he also used straight wires in later work. He could detect a wave produced by the source circuit with another circuit at a considerable distance from the source. Hertz discovered radio waves by doing these experiments.

Figure 16.3 presents a closer look at the nature of simple electromagnetic waves. If the magnetic field is in the horizontal plane, as in figure 16.2, the electric field generated by the changing magnetic field is in the vertical direction. These two fields are perpendicular to each other, and they are also perpendicular to the direction of travel away from the source antenna. Electromagnetic waves are therefore transverse waves. The magnitudes of the electric and magnetic fields are pictured here as varying sinusoidally—and in phase with one another—but more complex patterns are also possible.

Like the other types of waves that we have studied, the sinusoidal wave pattern moves. Figure 16.3 shows the field magnitudes and directions at a single instant in time and along only one line in space. The same kind of variation occurs in all directions perpendicular to the antenna. As the sinusoidal pattern moves, the field values at any point in space alternately increase and decrease. As the fields go through zero, they change direction and begin to increase in the opposite direction. These coordinated changes of the electric and magnetic fields make up the electromagnetic wave.

in his equations describing the behavior of electric and magnetic fields. Experimental measurements confirmed its existence.

Maxwell realized that a wave involving these fields could propagate through space. A changing magnetic field produces a changing electric field, which, in turn, produces a changing magnetic field, and so on. In a vacuum, the process can go on indefinitely and affect charged particles at much greater distances from the source than would be possible with static fields generated by nonchanging currents or charges. This is how an electromagnetic wave is produced. The wires in figure 16.2 serve as a transmitting antenna for the waves. A second antenna can be used to detect the waves.

What is the speed of electromagnetic waves?

In predicting the existence of electromagnetic waves from his theory of electric and magnetic fields, Maxwell could also predict their speed. The speed of these waves in a vacuum can be computed from just two constants, the Coulomb constant k in Coulomb's law and the magnetic force constant k' in Ampère's expression for the force between

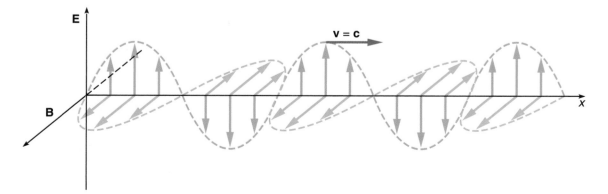

figure **16.3**   The time-varying electric and magnetic fields in an electromagnetic wave are in directions perpendicular to each other as well as to the wave speed.

two current-carrying wires. Maxwell's theory predicted that the wave speed should be equal to the square root of the ratio of these two numbers ($v = \sqrt{k/k'}$), which yields a value of 3×10^8 m/s, or 300 million meters per second.

The striking fact about this value, other than its incredible size, is that it corresponded with the known value of the speed of light. The speed of light had been accurately measured by different scientists not too many years before Maxwell's work. This coincidence led Maxwell to suggest that light itself was a form of electromagnetic wave—the first direct connection between the fields of optics and electromagnetism.

Measuring the speed of light was no easy task in Maxwell's day. Galileo was one of the first to attempt a measurement 250 years earlier. He sent an assistant with a shuttered lantern to a distant hill with instructions to open his lantern when he first saw the light from a similar lantern operated by Galileo. Galileo planned to measure the time required for the light to travel to his assistant and back. This attempt was doomed to failure. The reaction times involved in opening the lanterns were much greater than the actual time of flight for the beam of light.

Although astronomers had made estimates of the speed of light, the first successful land-based measurement was made by Armand-Hippolyte Fizeau (1819–1896) in 1849. He used a toothed-wheel apparatus like the one pictured in figure 16.4. A light beam passes through the gap between the teeth of the rotating wheel and is reflected from a distant mirror. The beam will be blocked on its return if the wheel has rotated just far enough so that a tooth is in the place where the gap had been. By measuring the rotational speed of the wheel and knowing the distance that the light beam traveled to get back to the wheel, Fizeau could calculate the speed of light.

The speed of Fizeau's wheel allowed a tooth to move into the former position of a gap in less than $1/10\,000$ of a second. Even at that rate, he had to place his reflecting mirror at a distance more than 8 kilometers (about 5 miles) from the wheel. Knowing that the light beam traveled more than 10 miles in less than $1/10\,000$ of a second in Fizeau's experiment may give you some appreciation of the enormous magnitude of the speed of light.

The speed of light is an important constant of nature, so we give it its own symbol, c, its value in a vacuum. This value is now defined as $c = 2.99792458 \times 10^8$ m/s, very close to the 3×10^8 m/s value that we usually quote and remember. Light (and other forms of electromagnetic wave) travel somewhat more slowly in other media like glass or water, but the speed of electromagnetic waves in air is very close to their speed in a vacuum.

Are there different kinds of electromagnetic waves?

We have already noted that both radio waves and light waves are electromagnetic waves. Are they the same, or do they differ in some significant respect? The main difference between radio waves and light waves lies in their wavelengths and frequencies. Radio waves have long wavelengths, several meters or more, but light waves have very short wavelengths, less than a micron (one-millionth of a meter).

Since different types of electromagnetic waves all travel with the same speed in a vacuum (and also in air, approximately), their frequencies are related to their wavelengths by the relationship $v = f\lambda$, where the speed v is equal to c. The frequencies for typical wavelengths of radio waves and light waves are computed in Try This Box 16.1. With their shorter wavelengths, light waves have much higher frequencies than radio waves.

If the frequency is known, we can also use the relationship between wavelength and frequency to compute the wavelength. An AM radio station broadcasting at a frequency of 600 kilohertz produces radio waves with a wavelength of 500 m, which we find by dividing the speed of light by the frequency ($\lambda = c/f$). Radio waves in the AM band have very long wavelengths.

Figure 16.5 shows the wavelength and frequency bands for various parts of the **electromagnetic spectrum.** The waves in different parts of this spectrum differ not only in their wavelength and frequency but also in how they are generated and what materials they will travel through. X rays, for example, will pass through materials that are opaque to visible light. Radio waves will also pass through walls that light cannot penetrate.

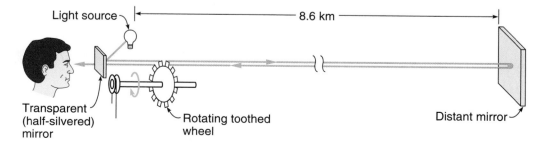

figure 16.4 A diagram of Fizeau's toothed-wheel apparatus for measuring the speed of light. As the rapidly spinning wheel turns a small fraction of a revolution, the returning light beam is blocked by a tooth on the wheel.

Light wavelengths are associated with color and range from roughly 3.8×10^{-7} m at the violet end of the visible spectrum to 7.5×10^{-7} m at the red end. The colors change progressively from violet through blue, green, yellow, orange, and red as the wavelengths lengthen. Electromagnetic waves with wavelengths somewhat longer than the red end of the visible spectrum are called **infrared light,** and waves with shorter wavelengths than the violet end are called **ultraviolet light.** Although X rays and gamma rays have even shorter wavelengths than ultraviolet light, they too are electromagnetic waves.

Electromagnetic waves undergo interference phenomena like other kinds of waves. Interference of light can produce striking effects, some of which are discussed in section 16.3. Radio waves reflected from belts of charged particles in the atmosphere can interfere with those coming directly from the transmitter, causing the station to fade in and out.

How different types of electromagnetic waves are produced varies enormously for the different parts of the spectrum, but they all involve accelerated charged particles. The accelerated charges can be in an oscillating electrical circuit as in radio waves, or within atoms as in light, X rays, and gamma rays. Like any warm body, your body is radiating electromagnetic waves in the infrared part of the spectrum. In this case, oscillating atoms within the molecules of your skin serve as the antennas.

---

try this box 16.1

Sample Exercise: Frequencies of Two Kinds of Electromagnetic Waves

What are the frequencies of
   a. radio waves with a wavelength of 10 m?
   b. light waves with a wavelength of 6×10^{-7} m?

a. $\lambda = 10$ m $v = f\lambda = c$
   $v = c = 3 \times 10^8$ m/s $f = \dfrac{c}{\lambda}$
   $f = ?$
   $\qquad\qquad\qquad = \dfrac{3 \times 10^8 \text{ m/s}}{10 \text{ m}}$
   $\qquad\qquad\qquad = \mathbf{3 \times 10^7 \text{ Hz}}$

b. $\lambda = 6 \times 10^{-7}$ m $f = \dfrac{c}{\lambda}$
   $f = ?$
   $\qquad\qquad\qquad = \dfrac{3 \times 10^8 \text{ m/s}}{6 \times 10^{-7} \text{ m}}$
   $\qquad\qquad\qquad = \mathbf{5 \times 10^{14} \text{ Hz}}$

The frequency of light waves is over 10 million times larger than the frequency of the 10-m radio waves.

Maxwell's theory of electric and magnetic fields predicted that a wave involving these fields could be propagated through a vacuum. We call these waves electromagnetic waves. Their speed in a vacuum is approximately 300 million meters per second, as predicted by Maxwell. Since this value was known to be the speed of light, light was identified as an electromagnetic wave, along with radio waves, microwaves, X rays, and gamma rays, which were discovered after Maxwell's work. Various kinds of electromagnetic waves differ from one another in their wavelengths and frequencies, and in how they are generated. They all are waves, however, and they exhibit the general properties of wave motion.

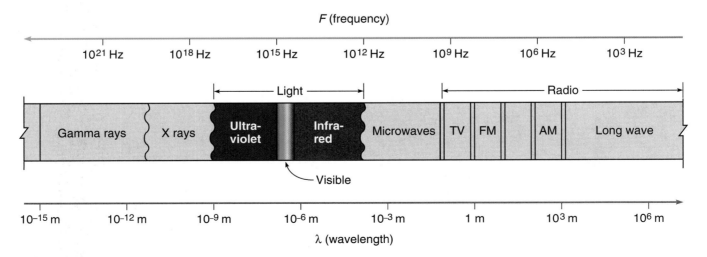

figure **16.5**   The electromagnetic wave spectrum. Both wavelengths and frequencies are shown for different parts of the spectrum.

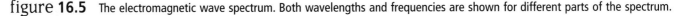

16.2 Wavelength and Color

We live in a colorful world, and we learn to distinguish colors when we are in preschool. How do we do this? What causes different objects to have different colors? Why is the sky blue? These phenomena are related to the wavelength of light, to the properties of different materials, and to how we see. In this section, we will explore a variety of aspects of color and color vision.

Does light consist of different colors?

If you have ever played with a prism, you are aware that a prism can produce a rainbow of color. If we allow a beam of white light from a small light bulb or from the sun to strike a prism, the prism will bend the beam. The emerging beam does not appear white, however. Instead it is multi-colored with violet light appearing on one side of the emerging beam and red light on the other end. In between, we find blue, green, yellow, and orange.

One of the first people to study this phenomenon systematically was Isaac Newton. Newton is best known for his work in mechanics, but he also did extensive work in optics. In one experiment, he produced a narrow beam of sunlight by passing it through a hole in his window shade. He passed this beam through a glass prism and displayed the colorful *spectrum* of light that we have just described. This seemed to demonstrate that white light had within it light of different colors.

Newton did not stop with that observation, however. He passed the light emerging from the prism through a second identical prism that was inverted relative to the first prism (fig. 16.6). The light emerging from the second prism was white like the original sunbeam. The different colors when recombined produced white light. These studies of Newton demonstrated that white light was a mixture of different colored components.

We now know that the different colors of light that Newton observed are associated with the wavelength of light. As mentioned in section 16.1, violet light has shorter wavelengths than red light with the other colors of the spectrum having intermediate values. These wavelengths can be measured by interference experiments that we will describe in sections 16.3 and 16.4. The wavelengths are extremely short, however, and it is not surprising that Newton was not convinced that they existed. The wavelengths of visible light are roughly one-hundredth the diameter of a human hair!

We usually express the wavelengths of visible light in nanometers (nm). One nanometer is equal to 10^{-9} meters or one-billionth of a meter. The wavelengths of visible light range from 380 nm on the violet end of the visible spectrum to 750 nm on the red end. Table 16.1 shows approximate wavelength ranges for the different colors of the visible spectrum.

How do our eyes distinguish different colors?

Even though the wavelengths of visible light are extremely short, our visual systems can readily distinguish different colors. Our eyes are the front end of our visual system, but our brains are also heavily involved in how we see. Figure 16.7 shows a simplified diagram of some of the structures of the eye.

table 16.1	
Colors Associated with Different Wavelengths of Light	
Color	Wavelength (nm)
violet	380–440
blue	440–490
green	490–560
yellow	560–590
orange	590–620
red	620–750

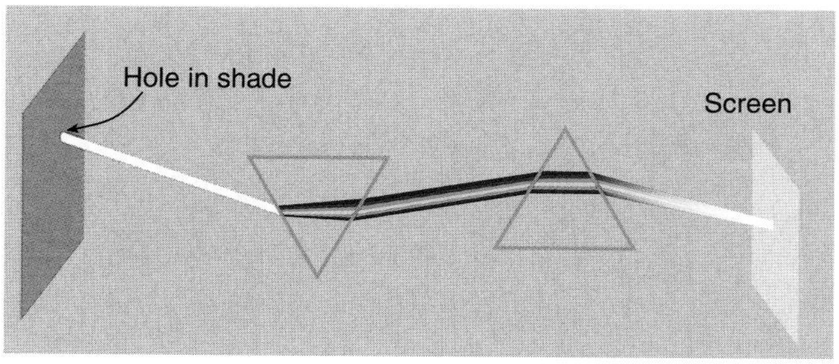

figure **16.6**   Newton showed that white light from the sun, after being split into different colors by one prism, could be recombined by a second prism to form white light again.

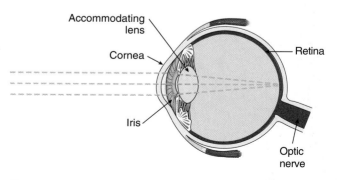

figure **16.7**   Light entering the eye is focused by the cornea and accommodating lens to form an image on the retina. The retina contains light-sensitive cells that send signals to the brain via the optic nerve.

Light is focused by the cornea and crystalline lens onto the retina. The retina is made up of light-sensitive cells of two types: *rods* and *cones*. The cones are concentrated near the center of the retina in an area called the fovea and are responsible for our daylight and color vision. The rods are distributed throughout the retina and are responsible for our nighttime and peripheral vision. The rods do not provide any information on color—we are color blind at night or in other low-light conditions.

The cones dominate our vision in well-lit conditions and provide us with our ability to see fine detail as well as color. There are actually three types of cones: S cones, M cones, and L cones that are sensitive to light in different parts of the spectrum. The S cones are most sensitive to the shorter (hence S) wavelengths, the M cones to medium wavelengths, and the L cones to longer wavelengths (fig. 6.8). The sensitivity ranges overlap, however, so that light near the middle of the visible spectrum will stimulate all three cone types.

How do we identify different colors, then? Suppose light of 650 nm wavelength enters our eyes. From the cone sensitivity curves, we can see that this light will stimulate the L cones more strongly than the M cones, which in turn are stimulated much more than the S cones. From childhood experience in identifying colors, we have learned to identify this mix of signals as the color red. In a similar fashion, light of 450 nm will stimulate the S cones most strongly, and we identify that color as blue.

Light of 580 nm wavelength stimulates both the M and L cones strongly, and we identify this color as yellow. However, a mixture of red light and green light will produce a similar response, and we will also perceive this mixture as yellow. This is essentially the process underlying **additive color mixing.** Combining the three primary colors blue, green, and red in different amounts can produce responses in our brains corresponding to all of the colors we are used

figure **16.9** Additive color mixing is demonstrated by projecting blue, green, and red light from separate projectors onto overlapping circles on a screen.

to identifying. Red and green produce yellow, blue and green produce cyan (blue-green), and blue and red produce magenta as shown in figure 16.9.

Combining all three primary colors in appropriate amounts produces a response that we identify as white. This is true despite the fact that all of the wavelengths that make up the white light of the sun may not be present. We perceive white, though, because the cones are being stimulated in a similar proportion to the response that sunlight produces. When light levels are very low compared to the background, we see black, which is essentially the absence of light.

Why do objects have different colors?

Why does a blue dress appear blue or a green shirt green? Most objects are not producers of light; they merely reflect or scatter light coming from other sources. The color that we perceive depends on the wavelengths present in the light source as well as on the manner in which the object reflects or scatters the light.

The color mixing done by artists on their palette is another way of producing different colors. The pigments used in paints or dyes used on fabrics work by **selective absorption.** By this we mean that they absorb some wavelengths of light more than others. When light is absorbed, the energy contained in the light wave is converted to other forms of energy, usually thermal energy. The object doing the absorbing becomes warmer.

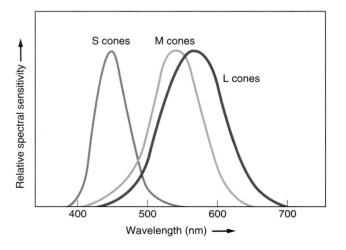

figure **16.8** The three different types of cones are sensitive to different wavelengths of light, but their ranges overlap.

The selective absorption of light is a subtractive process that can also be used to produce a complete range of colors. Suppose that we illuminate a blue-green (also called cyan) book with an incandescent light bulb. An incandescent lamp produces light by heating a tungsten filament to a very high temperature causing it to glow. A continuous range of wavelengths results, similar to the white light produced by the sun but usually containing a higher intensity of light in the red end of the spectrum.

When this light strikes the book, a number of things can happen. If the book cover is glossy, some of the light undergoes *specular* reflection (fig. 16.10). Specular reflection is mirrorlike; the light is reflected in a specific direction defined by the law of reflection. (The law of reflection and the behavior of mirrors are discussed in chapter 17.) Light that is reflected in this manner will usually appear as a white glare, the color of the light bulb.

The rest of the light will be reflected *diffusely,* meaning that it is reflected in all directions. This may be due to the surface being somewhat rough, but it also results from penetration of the light a small distance into the surface of the book as shown in figure 16.10. When this happens, some of the light may be selectively absorbed by particles of pigment in the surface coating. If the book appears blue-green, the pigments are selectively absorbing red light, leaving an excess of blue and green wavelengths in the diffusely reflected light. The pigments are subtracting some wavelengths coming from the white light source producing an altered mix of wavelengths in the reflected light.

Subtractive color mixing has its own set of rules. In color printing, three primary pigments are used: cyan, yellow, and magenta. (Black ink is also used to darken some colors.) Cyan (blue-green) absorbs strongly in the red portion of the spectrum, but transmits and reflects blue and green wavelengths. The yellow pigments absorb in the blue, but transmit and reflect green and red. Magenta absorbs at intermediate wavelengths, but transmits and reflects blue and red.

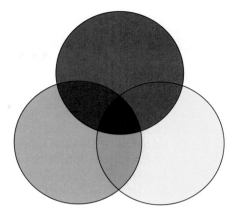

figure **16.11** Subtractive color mixing is demonstrated by overlaying yellow, magenta, and cyan pigments used in color printing.

When light is reflected from a surface coated with just one of these pigments, we see the colors appropriate to that pigment: cyan, yellow, or magenta. If we mix these pigments, however, we can get a complete range of colors. For example, if we mix cyan and yellow pigments, both blue and red light are absorbed (fig. 16.11), allowing only intermediate wavelengths to be transmitted. This results in green light being strongly reflected. Likewise, cyan and magenta produce blue, and yellow and magenta produce red. These resulting colors (green, red, and blue) are the three primary colors for *additive* color mixing, which generate the responses in our eyes described earlier.

There is much more to color perception than this basic discussion can cover. Adding white reduces the *saturation* of the color producing pink from red, for example. The wavelengths present in the light source illuminating an object will affect the perceived color. Although the details may be complex, the same basic phenomena are at work in all of these effects. Different mixes of wavelengths stimulate the different cones in the retina by different amounts, and our brains identify that response as a certain color. (See Everyday Phenomenon Box 16.1 for some related color effects.)

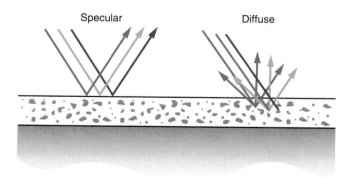

figure **16.10** Specular reflection obeys the law of reflection with all colors reflected equally. In diffuse reflection, light rays penetrate a short distance and some wavelengths are absorbed.

The white light coming from the sun is actually a mixture of light of different wavelengths. This was first demonstrated by Isaac Newton using two prisms to first separate sunlight into a spectrum of different colors and then to recombine these into white light with the second prism. Three different types of light-sensitive cells (called cones) in the retina of our eye allow us to distinguish different colors. The light reaching our eyes consists of mixtures of wavelengths that determine the color we perceive. The rules of color addition and subtraction are based on how these different mixtures stimulate the three types of cones.

Everyday Phenomenon

box 16.1

Why Is the Sky Blue?

The Situation. "Mommy, why is the sky blue?" This is a question almost any parent is likely to hear from their child at some point. Related questions such as "Why is there a sky?" or "Why is the sun red at sunset?" can also stymie parents.

We know that light coming directly from the sun appears white at midday. It contains a continuous mix of visible wavelengths peaking in the green-yellow portion of the spectrum. Why do we see a blue sky, then? What causes the spectacular oranges and reds that we see at sunset? These phenomena are all related to a process called *scattering*.

The Analysis. Before we consider the color of the sky, we should ask why we see a sky at all. Where is the light coming from? Skylight comes from the sun, but it is *scattered* out of the direct beam of sunlight. **Scattering** can be described as a process in which light is absorbed by small particles in the atmosphere and quickly re-emitted at the same wavelength. The scattered light travels in a different direction from that of the incident light, though.

If there were no atmosphere, then there would be no scattering. In this case the sky would appear black, the absence of light. The earth does have an atmosphere that extends a few miles beyond the earth's surface. This atmosphere consists primarily of molecules of nitrogen and oxygen as well as smaller quantities of other gases. In addition there are often particles of smoke, volcanic ash, or other particulate matter.

These particles are very small, but still considerably larger than individual gas molecules.

It is scattering by the gas molecules that is primarily responsible for the blue sky. When the particles doing the scattering are smaller than the wavelength of light, the process is called *Rayleigh scattering*, named after Lord Rayleigh (William Thompson). Rayleigh scattering depends upon the wavelength—shorter wavelengths are scattered much more effectively than longer wavelengths.

We can think of the molecules as tiny antennas. Since gas molecules consist of charged particles (chapter 18), when an electromagnetic wave strikes the molecules, these charges will oscillate at the frequency of the wave. Just as radio waves are produced by oscillating electric currents, the scattered light wave is produced by oscillating currents in the gas molecules. This process is most efficient when the wavelength of the wave is approximately the same size as the antenna. Since gas molecules are just a few nanometers in size, and the wavelengths of visible light are a few hundred nanometers, the scattering process is not very efficient. It is more efficient, however for the shortest wavelengths—those in the blue region of the visible spectrum.

This is why the sky is blue. Blue light is scattered out of the direct beam from the sun more effectively than red or intermediate wavelengths. To reach our eyes, it must be scattered again, perhaps several times, as shown in the first drawing.

(continued)

16.3 Interference of Light Waves

During the seventeenth and eighteenth centuries, scientists debated whether light was a wave phenomenon or a stream of particles. Most of the known effects could be explained by either model. Interference, however, is inherently a wave effect. The question would be settled if interference effects involving light could be produced.

In 1800, a British physician, Thomas Young, performed his famous double-slit experiment demonstrating interference of light. Why did it take this long for light interference to be recognized? The very short wavelengths of visible light make the effects subtle and difficult to observe. Once this difficulty was recognized, the door was opened for a series of predictions and experiments that firmly established that light is a wave.

Young's double-slit experiment

In any interference experiment, we need at least two waves that have a consistent phase relationship with one another,

like the two waves on the ropes discussed in chapter 15. The phase of ordinary light waves is continually changing, however, so we need to start with an isolated light wave and split it into two or more parts to meet this condition. Young accomplished this by passing a light beam through two narrow, closely spaced slits.

A diagram of Young's arrangement is shown in figure 16.12. Light from the source first passes through a single slit to isolate a small portion of the beam. This light then strikes the double slit, and light passing through the two slits is viewed on a screen to the right of the slits. Young made his double slits by depositing carbon black on a microscope slide and inscribing two fine lines in this black layer with a thin knife or razor. To be effective for visible light, the spacing of the two slits should be less than a millimeter.

What determines whether the light waves coming from the two slits will interfere constructively or destructively? At the point where the center line between the two slits meets the screen, the two light waves have traveled equal distances from each slit. The two waves were in phase

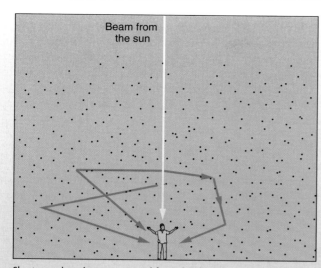

Short wavelengths are scattered from the direct beam from the sun at midday more effectively than longer wavelengths. These scattered waves produce the blue sky.

When we see the sky, we are not looking directly at the sun (which would be painful). We are seeing light that has been scattered multiple times, which concentrates the shorter blue and violet wavelengths in the light that reaches our eyes. Since the spectrum of sunlight contains more blue than violet, and our eyes respond more strongly to blue wavelengths than to violet, the color we identify is blue.

Why does sunlight appear orange or red near sunset or sunrise? The light reaching our eyes from the direct beam of

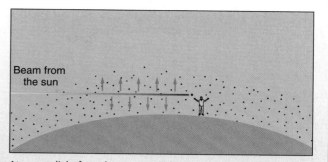

At sunset, light from the sun travels farther through the atmosphere than at midday. The shorter wavelengths are scattered out of the beam leaving the longer red wavelengths.

the sun at sunset travels a much longer distance through the atmosphere than it does at midday, as shown in the second drawing. Since the blue light and intermediate wavelengths are scattered out of the beam more effectively than red light, the direct beam is left with predominately red wavelengths. The closer the sun gets to the horizon, the redder the sun appears.

Scattering can also occur from larger particles such as water droplets in clouds. Water droplets are usually larger than the wavelengths of visible light. In this case, the amount of scattering does not depend strongly on the wavelength. Light scattered from clouds therefore appears white or gray. All wavelengths are equally scattered and the resulting color is the same as the incoming sunlight, but less intense.

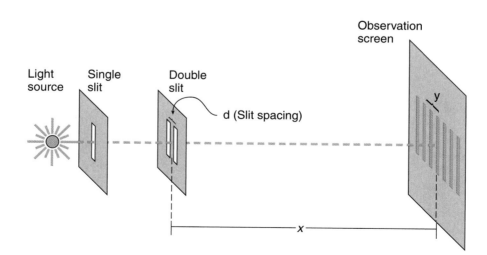

figure 16.12   Light from the source passes through a single slit before striking the double slit in Young's experiment. Interference fringes are observed on the screen. (The spacings of the double slit and the fringes are exaggerated in this drawing.)

initially, since they came from a single wave. Having traveled equal distances to reach the center point, they are still in phase when they strike the screen. They interfere constructively and produce a bright spot or line at the center of the screen.

What about other points on the screen? The two waves reaching points on either side of the center line have traveled *unequal* distances, which means that they may no longer be in phase with one another. Because it takes more time to travel a longer distance, when they reach the screen, the wave traveling the longer distance is at a different point in its cycle than the one traveling the shorter distance. If the difference in distance is just half a wavelength, the two waves will be half a cycle out of phase, and destructive interference results (fig. 16.13). This produces a dark spot or line on the screen.

Suppose that the distances traveled by the two waves differ by a full wavelength. The waves are then back in phase, since a full-cycle difference brings the wave oscillation back to the initial stage in its cycle. Differences in the distance traveled of one wavelength, two wavelengths, three wavelengths, and so on, produce constructive interference and bright lines on the screen. At positions halfway between the bright lines, the waves are half a cycle out of phase, so destructive interference produces dark lines at those points.

The resulting interference pattern of alternating bright and dark lines on the screen is called a *fringe* pattern. If *monochromatic* (single-wavelength) light is used, this pattern can extend to several fringes. If white light (which consists of a mixture of wavelengths) is used, the pattern is visible for only one or two fringes on either side of the center, because the different wavelengths produce constructive interference at different points on the screen. (Why is this so?) Although limited in extent, the fringes are colored and display a visually striking result.

What determines the spacing of the fringes?

We can predict the positions of the bright and dark fringes on the screen by knowing the wavelength of the light and the geometry of figure 16.13. The path difference between the two waves is the critical quantity. If the screen is not too close to the double slit, the path difference can be found from the similar triangles in the drawing and is given approximately by

$$\text{path difference} = d\,\frac{y}{x},$$

where d is the distance between the centers of the slits, y is the distance from the center point on the screen, and x is the distance between the screen and the slits. As indicated earlier, if this path difference is equal to an integer times the wavelength, constructive interference results (see Try This Box 16.2).

As the result in Try This Box 16.2 indicates, the second bright fringe from the center is only 2.5 mm from the center of the screen, for a slit separation of half a millimeter. The first bright fringe from the center would be found at half this distance, or only 1.25 mm from the center. These fringes are very closely spaced, which can make them difficult to see. A smaller slit spacing d yields a larger fringe separation, as you can see from the equation for y in Try This Box 16.2. In Young's day, considerable experimental ingenuity was needed to produce slit spacings as small as half a millimeter. Today, we can do this readily by using photographic reduction and other techniques.

What is thin-film interference?

If you have ever observed bands of different colors on a soap film or on the oil slick on a parking lot puddle, you have observed **thin-film interference.** How do these

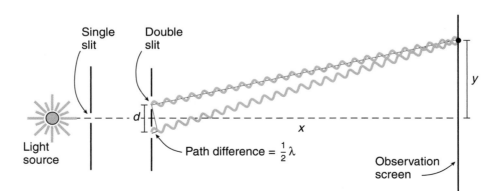

figure **16.13**   When the path difference between the two waves is equal to half a wavelength, the two waves arrive at the screen half a cycle out of phase, producing destructive interference.

try this box 16.2

Sample Exercise: Working with the Double-Slit Experiment

Red light with a wavelength of 630 nm strikes a double slit with a spacing of 0.5 mm. If the interference pattern is observed on a screen located 1 m from the double slit, how far from the center of the screen is the second bright line from the central (zeroth) bright line?

$\lambda = 630 \text{ nm} = 6.3 \times 10^{-7} \text{ m}$
$d = 0.5 \text{ mm} = 5 \times 10^{-4} \text{ m}$
$x = 1 \text{ m}$
$y = ?$

Since we are interested in the second bright fringe from the center, the path difference is:

$$d\frac{y}{x} = 2\lambda$$

rearranging this expression yields:

$$y = \frac{2\lambda x}{d}$$

$$= \frac{2(6.3 \times 10^{-7} \text{ m})(1 \text{ m})}{(5 \times 10^{-4} \text{ m})}$$

$$= \textbf{0.0025 m} = \textbf{2.5 mm}$$

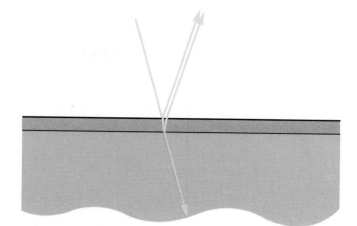

figure **16.14** Two waves interfere after reflection from a thin film of oil lying on water. One wave is reflected from the top surface of the film and the other from the bottom surface.

colors arise? Light waves reflected from the top and bottom surfaces of the film interfere to produce these effects.

Figure 16.14 shows a very thin film of oil lying on top of a water puddle. In order to be effective, the film must be no thicker than a few wavelengths of light. The underlying water is assumed to be much thicker (often several millimeters) than the oil film, which is a fraction of a micrometer. Two waves are shown being reflected from the film, one from the top surface and the other from the bottom surface. When these two waves reach our eyes, the one reflected from the bottom of the film has traveled slightly farther than that reflected from the top. Just as with the double-slit experiment, this path difference will cause a difference in phase between the two waves.

Since the two reflected waves are almost perpendicular to the surface of the film, the difference in distance traveled will be twice the thickness of the film. The wave reflected from the bottom surface passes through the film twice. The effect of this path difference is related to the wavelength of light in the film, which is shorter in the film than it is in air. (The wavelength in the film is given by λ/n, where n is the index of refraction of the film, discussed in chapter 17, and λ is the wavelength in air.)

The phase difference between the two reflected waves depends upon how many wavelengths farther the wave reflected from the bottom surface has traveled. If, for exam-

ple, the wave reflected from the bottom surface travels a half wavelength farther than the wave reflected from the top surface, then the two waves would be out of phase, and we would expect to observe destructive interference.*

Why do we see different colors? The interference condition depends on the thickness of the film and on the wavelength of the light. Some wavelengths may interfere destructively while others interfere constructively (or somewhere in between). If the thickness of the oil film is such that we have destructive interference for wavelengths near the middle of the spectrum (green-yellow), then those wavelengths will be missing and we will see a mixture of blue and red (magenta). If, on the other hand, the thickness is such that we get destructive interference for red light, the film will appear blue-green. The usual rules of color addition apply.

The colors that are reflected by the oil film will thus depend upon the thickness of the film. As the thickness changes, the color will change (fig. 16.15). The path difference between the two reflected waves also depends upon the angle of the incoming and reflected light. If light is incident on the film from multiple angles, the colors will change with viewing angle as the path difference changes.

There are many other ways of producing thin-film interference. Soap bubbles and soap films can produce colorful displays. The thin film in this case is a soap-water solution with air on either side. Figure 16.1 shows a soap film viewed in reflected light where the different colors form horizontal bands. The bands are produced by gravity—the film is thicker near the bottom of the loop than near the top.

---

* This is true if there are no other factors affecting the phase of the wave. The reflection process itself sometimes changes the phase, so this effect must also be taken into account in a complete analysis.

figure **16.15**   The colorful fringes from oil films are a common sight in parking lots and streets. The different colors result from changes in thickness of the film or from differing angles of view.

figure **16.16**   Colored interference fringes are produced by light reflected from an air film between two glass plates. Each colored band represents a different thickness of the film.

When the soap film gets extremely thin (just before it breaks), the path difference between the two surfaces should approach zero. In this case, we would expect the light waves reflected from the two surfaces to be in phase producing a bright reflection. Instead, however, the film appears quite dark for this condition, indicating destructive interference. The two waves are out of phase because of a phase difference introduced by the reflection process itself—the wave reflected from the top surface is shifted in phase by a half cycle, while that reflected from the bottom surface is not. The path difference between the two waves produces no phase change, but the reflection process does. This occurs whenever the medium on either side of the film is the same (air in this case), as well as in some other circumstances.

The thin film in some cases can be air. This happens when we place one glass plate on top of another. If the plates are flat and clean, this leaves a very thin film of air between the two plates. As the thickness of the air film varies we see alternating light and dark fringes (figure 16.16). The color of these fringes depends on the nature of the light source as well as on the thickness of the film and the angle of view. Thin-film interference is also involved in the anti-reflection coating on eyeglasses as is described in Everyday Phenomenon Box 16.2

Interference is observed for many different types of waves and light is no exception. The wave nature of light was first established by the double-slit interference experiment performed by Thomas Young. If the two waves coming from the two slits travel different distances, they may arrive at the screen either in or out of phase. If the path difference is an integer number of wavelengths, the waves are in phase and we get a bright fringe. If the path differ-

ence is a half-integer number of wavelengths, we get dark fringes. Light reflected from the two surfaces of a thin film will also travel different distances. This produces the colorful interference effects observed on soap films, oil films, or the air film between two glass plates.

16.4 Diffraction and Gratings

If you looked closely at a double-slit interference pattern like that described in section 16.3, you might notice that the bright fringes are not all equally bright. Going out from the center of the pattern, the fringes become less bright and seem to fade in and out. This effect is due to another aspect of interference that we usually label **diffraction.** Diffraction involves interference of light coming from different parts of the same slit or opening.

How does a single slit diffract light?

Diffraction from a single slit is the easiest to describe. If we replace the double slit shown in figure 16.12 with a sufficiently narrow single slit, a pattern like that shown in figure 16.17 results. A series of dark and light fringes lie on either side of a bright central fringe.

The bright central fringe is not surprising. Light waves reaching the center of the screen from different parts of the slit will travel roughly the same distance. They will therefore reach the screen in phase with one another and interfere constructively. They add together to yield the bright fringe that we see.

Explaining the other fringes calls for a more elaborate analysis. We can think of the two dark fringes on either side

Everyday Phenomenon

box 16.2

Antireflection Coatings on Eyeglasses

The Situation. When you select new eyeglasses, your optometrist or optician will show you a variety of lens and frame styles. They are also likely to recommend that you order antireflection (AR) coatings on your lenses. They tell you that this will help you see better in low-light or night driving conditions. You will also look better because people will be able to see your eyes without the distracting interference of reflections from the lenses.

Can you tell which of the two lenses in these eyeglasses has the antireflection coating?

How do these coatings work? How is it possible to reduce the amount of light that is reflected by glass or plastic lenses? Thin-film interference provides the answers.

The Analysis. When light strikes the surface of a glass or plastic lens, some light is inevitably reflected. For the types of glass and plastic used in spectacle lens, approximately 4% of the incident light is reflected at each surface (front and back) when the light comes in perpendicular to the surface. When the light comes in at an angle, the amount of light reflected is even larger.

We therefore lose at least 8% of the incoming light with uncoated lenses. These reflections can be distracting to people you are talking to, since your eyes are an important part of interpersonal communication. The reflections also reduce the amount of light that reaches your retina in low-light conditions.

An antireflection coating is designed to reduce these reflections. A thin film of transparent material is deposited on both surfaces of the lens. The film must be hard and durable, which limits our selection of the materials that will work. Magnesium fluoride is often used for glass lenses.

The thickness of the film must be carefully controlled. The film is designed to produce destructive interference for reflected light at wavelengths near the middle of the visible spectrum. For a single-layer coating, we use a film that is just a quarter of a wavelength thick. (The wavelength in question is that inside the thin film, which is somewhat shorter than the wavelength in air.) If the design wavelength is 550 nm (in air), the appropriate thickness of the film would be about 100 nm, which is very thin indeed. This is equal to one ten-thousandth of a millimeter!

Because the wave reflected from the bottom surface of the film travels through the film twice, the quarter-wavelength thickness results in a half-wavelength path difference between the waves reflected from the top and bottom surfaces. This produces destructive interference for the design wavelength. If the two reflected waves were equal in amplitude, the destructive interference would be total and there would be no light reflected at that wavelength. In practice, this condition cannot be achieved and there will still be a small amount of reflected light.

The coating is most effective for wavelengths near the middle of the visible spectrum. At the red or blue ends of the spectrum, the condition for destructive interference is not met, so we get stronger reflections at either end of the spectrum. The eye is not as sensitive there, however, so these reflections are less noticeable. It does result in the lens having a purplish appearance when viewed in reflected light.

Better results can be achieved by using multiple layers of thin films made from different materials rather than a single layer. The more layers used, the more expensive the process, however. The coatings now available for spectacle lenses are usually multiple-layer coatings. Besides the additional cost, the disadvantage of antireflection coatings is that they scratch and show dirt more readily than noncoated lenses. The buyer must decide whether the advantages of reduced reflections outweigh the disadvantages.

of the central fringe as arising from light coming from the two halves of the slit. If the light coming from the bottom half of the slit travels half a wavelength farther than that coming from the top half, these waves will interfere destructively (fig. 16.18). This produces a dark fringe at the point on the screen for which this condition holds.

A more complete argument involves dividing the slit into several segments. For fringes farther from the center of the pattern, this procedure is necessary to understand the results. Moving out from the center of the pattern, light coming from different segments of the slit comes in and out of phase producing the fringes that we see. The different

figure **16.17** The diffraction pattern for a narrow single slit has a series of light and dark fringes on either side of a bright central fringe.

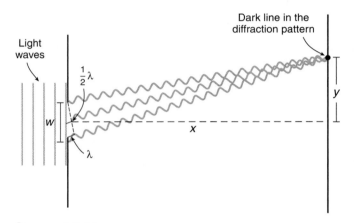

figure **16.18** When the path difference between light coming from the top half of the slit and that coming from the bottom half is $\frac{1}{2}\lambda$, a dark line appears in the single-slit diffraction pattern.

waves are never all in phase at the same point, though, as they were at the center of the pattern, so the central fringe is much brighter than the secondary bright fringes on either side of the center.

For the first dark fringe on either side of the center, the path difference between light coming from the two halves of the slit is half a wavelength. This requires that the path difference for light coming from the top and bottom *edges* of the slit be a full wavelength as indicated in figure 16.18. From the same geometry used to describe the double-slit pattern, the position y of the first dark fringe is then given by

$$y = \frac{\lambda x}{w},$$

where w is the width of the slit and, as before, λ is the wavelength of the light and x is the distance to the screen. (See Try This Box 16.3.)

try this box 16.3

Sample Exercise: How Broad Is the Central Fringe of a Single Slit?

Light with a wavelength of 550 nm strikes a single slit that is 0.4 mm wide. The diffraction pattern produced by the slit is observed on a wall a distance of 3.0 m from the slit.
  a. What is the distance from the center of the pattern to the first dark fringe?
  b. How wide is the central bright fringe of this diffraction pattern?

a. $\lambda = 550$ nm $y = \dfrac{\lambda x}{w}$

  $w = 0.4$ mm

  $x = 3.0$ m $y = \dfrac{(5.50 \times 10^{-7} \text{ m})(3.0 \text{ m})}{0.4 \times 10^{-3} \text{ m}}$

  $y = \textbf{0.0041 m} = \textbf{4.1 mm}$

b. $2y = ?$ $2y = 2(4.1 \text{ mm}) = \textbf{8.2 mm}$

The central bright fringe extends out to the first dark fringe on either side, so its width is just twice y.

This result is important because it describes the width of the central bright fringe. This width is $2y$ where y is the distance from the center of the screen to the first dark fringe on either side. As the width w of the slit decreases, the breadth $2y$ of the central maximum increases. Thus as the slit is narrowed, the diffraction pattern spreads out. A beam of light cannot be narrowed indefinitely by passing it through increasingly narrow slits. For a very narrow slit, the central bright fringe is much broader than the slit itself. Light is bent by diffraction into regions that we would expect to be shadowed.

How is light diffracted by other shapes?

You have probably observed diffraction effects without being aware of it. If you look at a star or a distant streetlight through a window screen, you will observe a diffraction pattern similar to that shown in figure 16.19. A window screen has multiple square openings, but figure 16.19 shows the diffraction pattern produced by just a single square opening in an otherwise opaque object. If we replaced the single slit described earlier with a square aperture, this is what we would see on the screen.

The explanation of the square-opening pattern is similar to the reasoning we used for the single slit. The pattern varies in two dimensions, however, rather than just one dimension, so the explanation is more involved. A rectangular opening would produce a similar pattern, but the spacing of the bright spots would be different in the horizontal direction than in the vertical direction. The narrower dimension of the rectangular opening would produce broader spacing of the diffraction spots.

The openings or *apertures* in most optical instruments such as telescopes, microscopes, or our eyes are circular rather than square or rectangular. The diffraction pattern produced by a circular aperture is shown in figure 16.20. This bull's-eye pattern can be produced by poking a small hole in a piece of aluminum foil with a pin. If we then illuminate the hole with a laser pointer or other bright light source, we can see the ringed diffraction pattern on the wall.

Just as with the single slit, the breadth of the central bright spot increases as we decrease the diameter of the

figure **16.20**   The diffraction pattern produced by a circular opening has a bright central spot surrounded by a series of dimmer rings.

pinhole. This implies that if we make the aperture of our instrument too small in diameter, light from pointlike objects such as stars will be spread out due to diffraction effects. This will produce fuzzy images and will cause nearby stars to blur into one another. We then say that these stars are not *resolved* by our telescope. For this reason, we generally use large mirrors for the opening element of high-quality telescopes. Telescopes will be described more fully in chapter 17.

The pupils of your eyes can vary in size depending on light levels. At high light levels, when our pupils are small in diameter, diffraction effects can limit our visual acuity. A small pupil spreads the light from point objects into larger blur circles due to diffraction. Our ability to see fine detail may actually improve at somewhat lower light levels that allow a larger pupil size.

We also see diffraction effects produced by our pupils when we look at stars. Instead of looking like points, the stars often appear to have little spikes of light radiating out from the center. These spikes are produced by diffraction from straight-edge segments of our otherwise circular pupils. These straight edges produce effects that are more similar to those of a single slit than those of a circular aperture.

What is a diffraction grating?

A **diffraction grating** is a multiple-slit interference device useful for viewing the color spectrum of a light source. Although the term *diffraction* is used in the name, it would

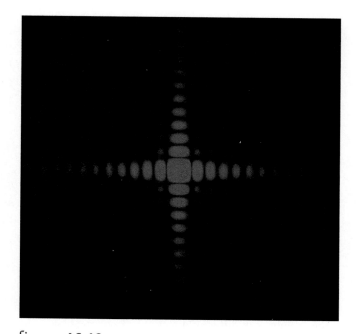

figure **16.19**   The diffraction pattern produced by a square opening has an array of bright spots. A longer exposure would show more off-axis spots.

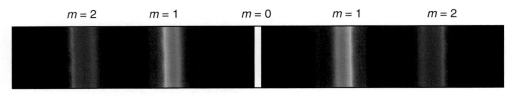

$m = 2$ $m = 1$ $m = 0$ $m = 1$ $m = 2$

figure 16.21   A diffraction grating produces a colorful spectrum from a white-light source. The second-order spectrum ($m = 2$) spreads out more than the first-order spectrum ($m = 1$).

be more appropriate to call these devices *interference* gratings. The primary effect is that of interference of light coming from different slits. Although the terms interference and diffraction are sometimes interchangeable, we usually use interference when referring to effects of separate slits or openings and diffraction for effects from a single opening.

When you increase the number of slits in an interference experiment, an interesting pattern emerges. If the spacing of the slits remains the same, the bright fringes become brighter and more narrow as the number of slits increases. At the same time, a few much dimmer fringes are seen in between the primary bright fringes. For example, if we have four slits, there are two *secondary* dimmer fringes lying between each pair of bright fringes. In general, the number of dimmer fringes is given by $N - 2$, where N is the number of slits. Ten slits would have eight secondary dimmer fringes between each pair of bright fringes.

As we continue to increase the number of slits, these dimmer secondary fringes become very dim, while the bright fringes become very narrow. A diffraction grating has a very large number of slits very closely spaced. A good grating may have several hundred slits in the space of just 1 mm. Precision machines have been designed to produce these closely spaced slits or *lines* needed for producing high-quality gratings. Nowadays, good gratings can be made much more simply using lasers and holographic techniques. (See Everyday Phenomenon Box 21.1.)

The condition for locating the bright fringes produced by a diffraction grating is essentially the same as that introduced for the double slit. If the distance between adjacent fringes is d, then just as for the double slit, the path difference between waves coming from adjacent slits is $d \cdot y/x$, where y is the distance from the center of the viewing screen and x is the distance from the grating to the viewing screen. Whenever this path difference is equal to an integer multiple of the light wavelength, we get a strong bright fringe for that wavelength. This condition can be expressed* as

$$d \, \frac{y}{x} = m\lambda,$$

where m is an integer having possible values 0, ± 1, ± 2, ± 3, etc.

Since the condition locating the fringes depends on the wavelength of the light, different wavelengths will appear at different points on the screen for a given order, m. Thus passing light through a diffraction grating will spread the light into its spectrum of colors (fig. 16.21). A good grating produces a wider separation of colors than a prism and also allows direct computation of the wavelength from the condition for interference.

Diffraction gratings are used to separate and measure the wavelengths of light in instruments we call *spectrometers*. They are a common piece of apparatus in chemistry and physics laboratories. Holographically produced gratings are now also seen in many novelty products including "space glasses" and reflective gift wrappings. The colorful effects that we see when viewing a compact disc (CD) are also a grating phenomena. The disc contains a continuous spiral track that circles the disc from the inside to the outside. Adjacent turns of this spiral track are very closely spaced and act as a reflecting diffraction grating.

Diffraction involves the interference of light waves coming from different parts of the same opening. Diffraction from a single slit produces a bright central fringe with a series of weaker dark and light fringes on either side of the broader central fringe. A circular aperture produces a bull's-eye pattern. Making the aperture smaller causes the diffraction pattern to spread out, frustrating efforts to narrow a beam of light. A diffraction grating is a multiple-slit interference device that allows us to separate and measure wavelengths of light.

16.5 Polarized Light

Many of us have used polarizing sunglasses or camera filters to reduce glare and darken the sky. How do these work? You may have also worn special polarizing glasses to view a 3-D video or movie. In this case, your two eyes are receiving light with different directions of polarization.

What do we mean when we talk about polarization of light? How does polarized light differ from ordinary light? To answer these questions, we need to revisit the nature of electromagnetic waves.

---

* This condition is only valid for relatively small angles from the center of the screen. For larger angles we must use a more precise condition involving the sine of an angle θ to a point on the screen. The exact condition is $d \sin \theta = m\lambda$. For small angles, $\sin \theta \approx y/x$.

What is polarized light?

Take a look at the electromagnetic wave pictured in figure 16.3 in section 16.1. Light is an electromagnetic wave consisting of oscillating electric and magnetic fields. The wave pictured in figure 16.3 is actually polarized. The oscillating electric field vector is in the vertical plane in this diagram, and the magnetic field is horizontal.

This is not the only possibility, however. We could have pictured the electric field oscillating in the horizontal direction with the magnetic field in the vertical plane. Or, we could have pictured the electric field oscillating at some angle to the horizontal. These choices would all represent different directions of polarization. To define the direction of polarization of a light wave, we specify the direction of oscillation of the electric field vector. Figure 16.22 pictures different states of polarization. In these diagrams, the light wave is coming straight out of the page toward you. The electric field vector is represented by a two-headed green arrow because it is oscillating back and forth as the wave progresses.

How does unpolarized light differ from polarized light? The last diagram in figure 16.22 represents unpolarized light. The electric field vector in this case is shown pointing in several different directions. (In reality, there are an infinite number of possible directions.) **Unpolarized light** is a mixture of many waves having different orientations for the electric field oscillations. Light coming from an ordinary light bulb is unpolarized. In order to produce **polarized light,** something must occur to select just one direction of field oscillation from these multiple possibilities.

Light or other electromagnetic waves are not the only type of wave that can be polarized. Any transverse wave, including waves on a rope or guitar string, can be polarized.

If you wave a rope up and down to produce a traveling wave on the rope, you have produced a vertically polarized wave. If you wave it back and forth in a horizontal plane, you produce a horizontally polarized wave. If you move your hand in random directions, you produce an unpolarized wave.

How do we produce polarized light?

The most common way of producing polarized light is with a polarizing filter (sometimes called a *Polaroid* after the original manufacturer). The early polarizing filters employed *dichroic* crystals. These were special materials that absorbed light with one direction of polarization, while transmitting light with the perpendicular direction of polarization. The trick in manufacturing the filter was to get a large number of these small crystals all lined up in the same direction.

Later processes used special plastics or *polymers* to make polarizing filters. By stretching these polymers in the manufacturing process, they become dichroic. The stretching process also assures that the molecules are all aligned in the same direction. Modern polarizing filters are made in this manner. Polarizing filters are routinely available at camera stores, but also as polarizing sunglasses.

What happens when unpolarized light passes through a polarizing filter? Does it only let through that tiny fraction of light that happens to have the correct orientation of the electric field? Remember that electric fields are vectors. The polarizing filter selects that portion or *component* of each electric field vector that has the appropriate orientation. Figure 16.23 illustrates this process. The component of the electric field vector that gets through is the one in

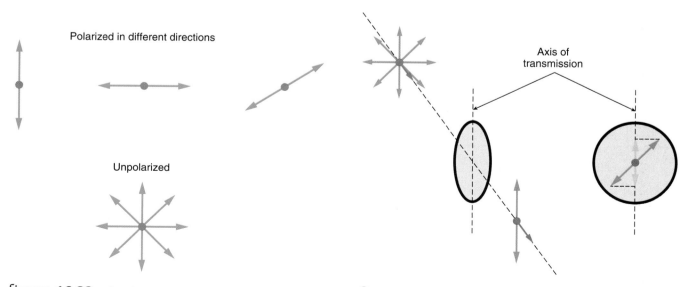

figure **16.22** The electric field vector oscillates in a single direction for polarized light. Unpolarized light has random directions of oscillation.

figure **16.23** A polarizing filter selects the components of the electric field vector that are in the direction of the transmission axis of the filter.

the direction of the transmission axis of the filter. The component perpendicular to this axis is absorbed.

An ideal polarizing filter therefore transmits 50% of an unpolarized light beam. The light that emerges is polarized with the electric field in the direction of the transmission axis of the filter. In practice, somewhat less than 50% of the light makes it through the filter because some absorption takes place even in the direction of the transmission axis.

How can we tell whether a beam of light is polarized or not? We place a polarizing filter in the path of the beam and then rotate the filter. If the intensity of the transmitted light varies as we rotate the filter, then we know that the beam is either partially or completely polarized. If the intensity becomes effectively zero for one orientation of the filter, then the light beam being analyzed is completely polarized.

Why do we use polarizing sunglasses?

There are various ways in which light can become polarized without passing it through a polarizing filter. Reflection from a smooth surface of a transparent material such as glass or water can produce polarization, for example. When sunlight is reflected from a lake surface at just the right angle, the reflected wave can be completely polarized.

Figure 16.24 pictures light striking a lake surface at the polarizing angle. The light coming in is unpolarized as is represented by the arrows both perpendicular to the page (dots) and those lying in the plane of the drawing. When the angle between the reflected wave and the transmitted wave is a right angle (90°), the reflected wave is completely polarized, with the direction of polarization perpendicular to the page (dots) and parallel to the surface of the lake. For water, the polarizing angle is approximately 37° above the horizontal.

How can polarizing sunglasses help? Light reflected from a lake surface will produce a strong glare that can be quite annoying for boaters or water skiers. Since this light is

strongly polarized when the sun is at or near the polarizing angle, polarizing sunglasses can virtually eliminate this glare. Because the reflected light is polarized horizontally, we want the transmission axis of our sunglasses to be vertical. Even when the sun is not at the polarizing angle, the reflected light is partially polarized so the sunglasses will still help.

Polarizing sunglasses can help reduce glare from sunlight reflected by a wet road surface or a polished car hood. They can also help skiers experiencing glare from sunlight reflected from snow. Even skylight is polarized, although here the process involves scattering rather than reflection (see Everyday Phenomenon Box 16.1). It turns out that skylight is partially polarized with the electric field direction horizontal, so polarizing filters make the sky appear darker. This is the primary reason they are used in photography.

What is birefringence?

Many interesting and colorful effects of polarized light are related to the phenomenon of **birefringence.** Birefringence is also called *double refraction.* Calcite crystals are usually used to demonstrate the effect.

If you draw a small dot on a piece of paper and view this dot through a calcite crystal, you will see two dots rather than one. This is the origin of the term double refraction. One of the two dots that we see obeys the ordinary rules of refraction (light bending), which are discussed in chapter 17. The light wave associated with this dot (called the *ordinary* wave) passes straight through the crystal.

The wave associated with the second dot is called the *extraordinary* wave, and it does not obey the usual rules of refraction. Instead, the amount of light bending for this wave depends upon the direction of the light beam relative to the crystal lattice. The extraordinary wave does not pass straight through the crystal, and thus this dot appears to be offset from the other dot. The photograph in figure 16.25 shows a series of straight lines viewed through a calcite crystal. Each line appears to be doubled due to birefringence.

A simple test using a polarizing filter demonstrates that the two waves associated with the two dots (or lines) are polarized at right angles to each other. If we view the two lines through the filter, and rotate the filter slowly about our line of sight, first one line disappears. As we continue to rotate the filter an additional 90°, the second line disappears while the first one reappears. The direction of polarization must have something to do with the double-refraction effect.

How can we explain these effects? A calcite crystal is an example of an anisotropic crystal. Quartz and many other crystals are also anisotropic. In an isotropic crystal, where the atoms are arranged is a simple array (often cubic), light travels through the crystal in the same manner in all directions. Anisotropic crystals have more complicated

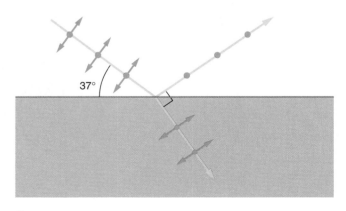

figure 16.24 Unpolarized light incident on a water surface at 37° above the horizontal. The reflected wave is completely polarized with the electric field parallel to the water surface.

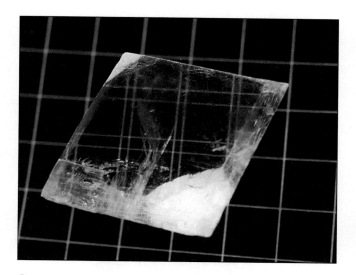

figure **16.25**   Lines are doubled when viewed through a calcite crystal. This effect is called double refraction or birefringence.

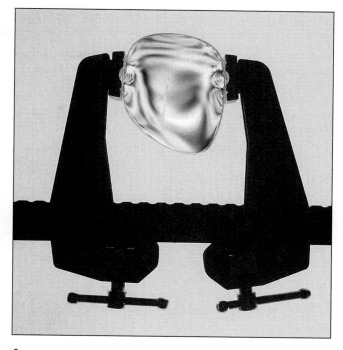

figure **16.26**   A plastic lens being compressed between two clamps shows stress birefringence when viewed between crossed polarizers.

arrangements of the atoms in the crystal lattice. This causes light to travel with different velocities in different directions. The direction of polarization also affects how light travels through an anisotropic crystal.

If two polarizing filters are placed with their transmission axes perpendicular to one another, no light gets through. The light passing through one filter emerges polarized, and that direction of polarization is completely blocked by the second filter with its axis at 90° to the first one. However, if we put a piece of birefringent material between the two filters, some light usually gets through. The birefringent material modifies the state of polarization of the light.

The degree to which the polarization of the light is changed depends upon the thickness of the material and also on the wavelength of the light. For this reason, placing thin pieces of birefringent material between two *crossed* (axes at 90°) polarizing filters will often produce a colorful display. A full analysis of the resulting colors is complex, but the colored bands give a clear indication that the inserted material is indeed birefringent.

Plastic materials provide interesting applications of these phenomena. Although plastics (polymers) are not crystalline, stressing a plastic by compressing it or pulling on it will often produce birefringence. Engineers sometimes analyze a structure they are designing by first building a small

plastic model of the structure. When forces are applied to the structure, they can then see where the stress is greatest by placing the plastic model between crossed polarizers and viewing the resulting patterns (figure 16.26).

Light is polarized if the electric vector is oscillating in only one direction rather than in random directions. We can produce polarized light by passing unpolarized light through a polarizing filter. The filter transmits light with one direction of polarization and absorbs light polarized at 90° to the transmission axis. Reflection of light from a smooth surface of water, glass, or plastic will also produce complete or partial polarization of the reflected beam. Polarized sunglasses are designed to block the reflected beam, thus reducing glare. Light with different polarizations travels with different velocities when passing through a birefringent material. This causes colorful displays when the birefringent material is viewed through crossed polarizers.

summary

Light is an electromagnetic wave having very short wavelengths. The wave properties of light are responsible for many phenomena that produce colorful effects. Absorption, scattering, interference, diffraction, and polarization effects are all explained in terms of waves. Our color vision provides us with the ability to appreciate these phenomena.

1   Electromagnetic waves.   James Clerk Maxwell predicted the existence of electromagnetic waves from his theory of electromagnetism. Oscillating electric charges produce variations in electric and magnetic fields that propagate through space with a speed of 3×10^8 m/s. Radio waves, microwaves, light, and X rays are all forms of electromagnetic waves.

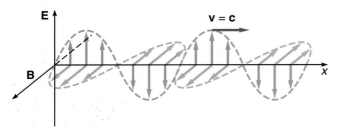

2   Wavelength and color.   Different colors are associated with different wavelengths of light. Three types of cones in the retina of our eye are sensitive to different ranges of wavelength. The combined responses of these cones define the color that we see and explain the rules of additive and subtractive color mixing.

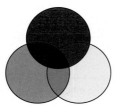

3   Interference of light waves.   Just as with other types of wave, two or more light waves can interfere constructively or destructively depending upon their relative phases. Thomas Young's double-slit experiment was the first to conclusively demonstrate interference of light. Interference of light waves reflected from thin films such as soap film produce colorful effects.

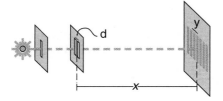

4   Diffraction and gratings.   Interference of light coming from different parts of the same opening is called *diffraction*. Light diffracted by a single slit produces a bright central spot of light with dimmer spots of light on either side of the center. More complex diffraction patterns are produced by square or circular apertures. Diffraction gratings use interference from multiple slits to separate and measure wavelengths of light.

5   Polarized light.   A polarized light wave has the electric field oscillating in just a single direction, but the electric field in an unpolarized wave oscillates in random directions. Light can be polarized by passing it through a polarizing filter or by reflecting it from a smooth surface of water, glass, or plastic. Birefringence effects are the result of light with different polarizations traveling with different speeds in an anisotropic crystal or a piece of stressed plastic.

Polarized in different directions

Unpolarized

key terms

questions

*Questions identified with an asterisk are more open-ended than the others. They call for lengthier responses and are more suitable for group discussion.

Q1. What characteristic of the electromagnetic waves predicted by Maxwell's theory led him to suggest that light might be an electromagnetic wave? Explain.

Q2. Is the electric field associated with an electromagnetic wave constant in time? Explain.

Q3. Is it possible for an electromagnetic wave to travel through a vacuum? Explain.

Q4. For which of the following characteristics—speed, wavelength, and frequency—is light similar to microwaves, and for which does it differ? Explain.

Q5. A starter's pistol is fired at the beginning of a race. If you are at the other end of the track, which will you perceive first, the sound of the pistol or the flash associated with its firing? Explain.

Q6. What is the color of light with a wavelength of 470 nm? Explain.

Q7. Do the L cones in the retina of the eye respond to just a single wavelength? Explain.

Q8. If we combine red light and green light in equal proportions, what color do we see? Explain.

Q9. A certain pigment absorbs green light, while reflecting blue and red wavelengths. If a surface coated with this pigment is illuminated by white light, what is the color of the light reflected from this surface? Explain.

*Q10. A color TV uses red, green, and blue phosphors to produce the colors that we see. In printing, however, we use magenta, yellow, and cyan as the primary colors. How do these two situations differ? Explain.

Q11. Skylight is produced by scattering of light from the direct beam coming from the sun. Why is the color of the sky different from the color of the light of the sun itself? Explain.

Q12. Two waves interfere to form fringes in Young's double-slit experiment. Do these two waves come from the same light source? Explain.

Q13. If two waves start out in phase with one another, but one wave travels half a wavelength farther than the other before they come together, will the waves be in phase or out of phase when they combine? Explain.

Q14. If two waves start out in phase with one another, but one wave travels two wavelengths farther than the other before they come together, will the waves be in phase or out of phase when they combine? Explain.

Q15. When light is reflected from a thin film of oil on a water puddle, the colors we see are produced by interference. What two waves are interfering in this situation? Explain.

Q16. Thin-film interference occurs when one clean glass plate is placed on top of another glass plate. What does the thin film consist of in this case? Explain.

*Q17. Suppose that white light is reflected from a thin soap film. If the thickness is such that destructive interference is occurring for red light, what color will the film appear to be when viewed in reflected light? Explain.

Q18. An antireflection coating on eyeglasses employs a thin-film coating on the lenses. If the coating is designed properly, does light reflected from the film undergo constructive or destructive interference? Explain.

Q19. Is diffraction the same as interference? Explain.

*Q20. A light beam passes through a slit and forms a spot of light on a screen at some distance from the slit. Can we make this spot of light as small as we wish by making the slit very narrow? What happens in this process? Explain.

Q21. What is a diffraction grating? Explain.

Q22. Suppose that light consisting of just two wavelengths, one blue and the other green, is passed through a diffraction grating. Which of these two colors will lie farther from the center of the screen in the first-order ($m = 1$) spectrum produced by the grating? Explain.

Q23. How does polarized light differ from unpolarized light? Explain.

Q24. Can a wave on a guitar string be polarized? Explain.

Q25. If you pass an unpolarized light beam through a polarizing filter, will the light beam emerging from the filter be weaker or stronger than the incoming beam? Explain.

Q26. Besides passing light through a polarizing filter, is there any other way that unpolarized light can become polarized? Explain.

Q27. If you use polarized sunglasses to eliminate glare from water surfaces, should the axis of transmission of the polarizer in the sunglasses be oriented in the vertical direction or in the horizontal direction? Explain.

Q28. A double image of a dot is seen when viewing the dot through a calcite crystal. Is there any difference in the polarization of the light associated with these two images? Explain.

Q29. Birefringence is associated with anisotropic crystals but plastics are not usually crystalline. Is there any way in which a plastic material can exhibit birefringence? Explain.

exercises

E1. Microwaves used in laboratory experiments often have a wavelength of about 1 cm. What is the frequency of these waves?

E2. What is the wavelength of the radio waves from a station broadcasting at 900 kilohertz?

E3. What is the frequency of green light waves with a wavelength of 520 nm?

E4. X rays often have a wavelength of about 10^{-10} m. What is the frequency of such waves?

E5. Light with a wavelength of 500 nm (5×10^{-7} m) is incident upon a double slit with a separation of 0.4 mm (4×10^{-4} m). A screen is located 2.0 m from the double slit. At what distance from the center of the screen will the first bright fringe beyond the center fringe appear?

E6. For the same conditions described in exercise 5, at what distance from the center of the screen will the second dark fringe appear?

E7. A green fringe produced by double-slit interference lies 2.2 cm from the center of a screen placed 1.2 m from the double slit. If the screen is moved back so that it is now a distance of 3.6 m from the double slit, how far from the center of the screen will this green fringe lie?

E8. Light of 500 nm is reflected from a thin film of air between two glass plates. The thickness of the film is 1 μm (1000 nm).
   a. How much farther does the light reflected from the bottom surface of the film travel than that reflected from the top surface?
   b. How many wavelengths of light does this represent?

E9. An antireflection coating is designed with a thickness of a quarter of the wavelength of the light traveling in the film.
   a. How many wavelengths farther does the light reflected from the bottom surface of the coating travel than that reflected from the top surface?
   b. Does this produce constructive or destructive interference?

E10. Light with a wavelength of 600 nm (6×10^{-7} m) strikes a single slit that is 0.5 mm (5×10^{-4} m) wide. The diffraction pattern produced by the slit is observed on a wall a distance of 2.0 m from the slit.
   a. What is the distance from the center of the pattern to the first dark fringe?
   b. How wide is the central bright fringe of this diffraction pattern?

E11. When illuminated with light of 500 nm (5×10^{-7}), the first dark fringe produced by a single slit lies a distance of 1.2 cm from the center of the screen placed 4.0 m from the slit. How wide is the slit?

E12. A diffraction grating has 1000 slits or lines ruled in a space of 1.4 cm. What is the distance d between adjacent slits?

E13. Light of 546 nm (5.46×10^{-7} m) wavelength from a mercury lamp passes through a diffraction grating with a spacing between adjacent slits of 0.005 mm (5×10^{-6} m). A screen is located a distance of 2.5 m from the grating.
   a. How far from the center of the screen will the first-order bright fringe lie?
   b. How far from the center of the screen will the second-order bright fringe lie?

E14. When passed through a diffraction grating with a slit spacing of 0.004 mm (4×10^{-6} m), the first-order fringe for light of a single wavelength lies a distance of 29 cm from the center of a screen located 2.0 m from the grating. What is the wavelength of the light?

challenge problems

CP1. The visible spectrum of colors ranges from approximately 380 nm in wavelength at the violet end to 750 nm at the far red end.
  a. What are the frequencies associated with these two wavelengths?
  b. When light passes into glass, the speed of light is reduced to $v = c/n$ where n is called the *index of refraction*. For many types of glass or plastic, n is approximately 1.5. What is the approximate speed of light in glass?
  c. If the frequency of light does not change when light enters glass, the wavelength must change to account for the reduction in speed. What are the wavelengths for the two ends of the visible spectrum in glass?

CP2. Light with a wavelength of 600 nm (6×10^{-7} m) passes through two slits separated by just 0.03 mm (3×10^{-5} m). A fringe pattern is observed on a screen placed 1.2 m from the double slit.
  a. How far from the center of the screen will the first bright fringe appear on either side of the central fringe?
  b. Where will the second bright fringe appear on either side?
  c. At what distance from the center of the screen will the first dark fringe appear on either side?
  d. Sketch a diagram showing the positions of the seven central bright fringes (the center fringe and three fringes on either side of the center fringe). Clearly indicate the distance of each fringe from the center of the screen.

CP3. Standing waves (see section 15.3) can be formed by reflecting light from a mirror and allowing the reflected wave to interfere with the incoming wave. Suppose that we do so using light with a wavelength of 500 nm (5×10^{-7} m). Assume that there is a node at the mirror surface.
  a. How far from the mirror will the first antinode lie?
  b. What is the separation distance between adjacent antinodes (right fringes) as we move away from the mirror?
  c. Will it be easy to observe the dark and bright fringes associated with the nodes and antinodes of the standing wave? Explain.

CP4. A soap film has an index of refraction (see challenge problem 1) of $n = 1.333$. This implies that the wavelength of light in the film is shorter than in air by a factor of $1/n$. The index of refraction of air is approximately 1.0, so there is little difference between the wavelength in air and that in a vacuum.
  a. If light with an air or vacuum wavelength of 600 nm enters a soap film, what is the wavelength of this light in the film?
  b. If the film is 900 nm thick, how many wavelengths farther does the wave reflected from the bottom surface of the film travel than that reflected from the top?
  c. Would you be surprised to find that this thickness produces destructive interference for reflected light?*

---

* The two waves are out of phase because of a phase difference introduced by the reflection process itself—the wave reflected from the top surface is shifted in phase, while that reflected from the bottom surface is not. The path difference between the two waves produces no phase change, but the reflection process does.

home experiments and observations

HE1. If you have access to a color ink jet printer, print a color test pattern with the printer. View the test pattern through a strong magnifying glass.
  a. Can you see the individual colored dots that form the colors on the page? Which of the colors in the test pattern appear to be made up of just single colored dots?
  b. With the help of the magnifying glass, describe how the other colors in the test pattern are produced.
  c. Use the magnifying glass to observe the lines on the screen of a color television set (*with the set turned off!*). What colors do you observe? How do these colors differ from those used in color printing?

HE2. It is a simple matter to produce soap films that demonstrate thin-film interference phenomena. Place a few drops of liquid dish-washing detergent in a shallow bowl and add enough water to form a reasonably fluid solution. Bend a piece of wire into a closed loop, leaving a small length to serve as a handle. (Alternatively, you can purchase bubble-making solution with the wire loop included for a very small cost.)
  a. Dip the loop in the soap solution to form a soap film in the plane of the loop. Reflect light from a desk lamp from the film. Describe the colored patterns that you see.
  b. If you hold the loop in a vertical plane, the colors should settle into a pattern of horizontal bands. Sketch the bands that you see indicating both their color and width. Explain why we get horizontal bands.
  c. Try illuminating your soap film first with a frosted incandescent lamp and then with a fluorescent lamp. Describe the differences that you see.

HE3. Find two flat glass plates (microscope slides will work nicely). Wipe the glass plates clean and place one on top of the other. View the plates in reflected light from the overhead lighting in your room.
   a. Describe the pattern of light and dark fringes that you see. Are there any colors associated with these fringes?
   b. Place a very thin object such as a human hair between the two plates near one end. Does this change the fringe pattern? Can you explain the pattern that you see?

HE4. View a distant point light source (such as a distant streetlight or a bright star) at night through a window screen.
   a. Sketch the appearance of the light when viewed through the screen. How does this differ from the appearance of the light when viewed directly (not looking through the screen)?
   b. Contrast your observations with the square-aperture pattern pictured in figure 16.19. What similarities and differences can be identified?

HE5. Locate two polarizing filters. (Your physics instructor is probably your best source. Most sunglasses are not polarizers.)
   a. View the blue sky through one polarizer while rotating the polarizer around your line of sight. What changes in intensity do you observe?
   b. Place one polarizer on top of the other and rotate one of them. When the polarizers are crossed (axes at 90°), no light should get through. How does the transmitted light vary as you rotate the second polarizer?
   c. Place various plastic items (rulers, protractors, plastic spectacle lenses, etc.) between crossed polarizers. Describe the patterns that you observe.

Light and Image Formation

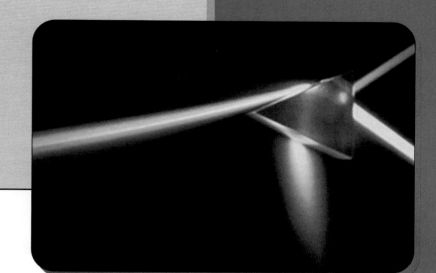

chapter overview

Our main goal in this chapter is to provide an understanding of the laws of reflection and refraction and how they are used to explain image formation. To do this, we will need to discuss the relationship between waves and rays and to show how rays can be traced to define and locate images. In the process of exploring image formation, we will examine the behavior of mirrors and lenses and the workings of simple optical instruments such as cameras, magnifiers, microscopes, and telescopes.

chapter outline

1 **Reflection and image formation.** What is the relationship between rays and waves? What is the law of reflection, and how are images formed by a plane mirror?

2 **Refraction of light.** How does the law of refraction describe the bending of light rays when they pass from one medium to another? How can we explain the separation of colors by a prism?

3 **Lenses and image formation.** How can the law of refraction explain image formation by simple lenses? How do positive lenses differ from negative lenses?

4 **Focusing light with curved mirrors.** How can curved mirrors be used to focus light? How do concave mirrors differ from convex mirrors in their image-forming properties?

5 **Eyeglasses, microscopes, and telescopes.** How do eyeglasses help us to see better? How can lenses and mirrors be combined to produce microscopes and telescopes?

Have you ever looked in a mirror and wondered how you are able to stare yourself in the face? How does that early-morning, mussed-up face pretending to be you appear behind the glass plate (fig. 17.1)? You know that you are seeing an image that, depending on the quality of the mirror, may or may not be an accurate reflection of reality. You are free to believe what you wish.

You may also wear eyeglasses, and you probably have used optical instruments that incorporate lenses like binoculars, overhead projectors, and microscopes. Images formed by mirrors involve the reflection of light. Images formed by lenses involve the refraction or bending of light. Both reflection and refraction are at work in producing rainbows.

In chapter 16, we introduced the idea that light is an electromagnetic wave and described the general features of these waves. How are such waves involved in the formation of images in a mirror, by a slide projector, or in your eye itself? What are the basic principles of image formation? Can we predict where the images will be formed and how they will appear?

Such questions lie in the realm of **geometric optics,** in which we describe the behavior of light waves by using rays that are perpendicular to the wavefronts. The laws of reflection and refraction are the basic principles

figure **17.1**   An early-morning look in the mirror. How does that discouraging image get there?

of geometric optics. They allow us to trace the paths of light rays, and to predict how and where images will be formed. **Physical optics** (chapter 16) treats phenomena such as interference and diffraction, which involve the wave aspects of light more directly.

17.1 Reflection and Image Formation

How is your image in the bathroom mirror produced? You know that light is involved somehow, as you can easily verify by turning off the bathroom light. If the room is completely dark, the image disappears, only to reappear instantly when the light is turned back on. Light waves from the bathroom light must bounce off your face, travel from there to the mirror, and then reflect back to your eyes. How does this process create the image that we see?

How are light rays related to wavefronts?

If we consider just one point on your face and trace what happens to the waves that are reflected from that point, we get a clearer idea of what is happening. Since the skin on your face is somewhat rough (at least on a microscopic scale), light that reaches your face is reflected or *scattered* in all directions from any given point. The tip of your nose, for example, behaves as though it were a source of light waves that spread out uniformly from that point (fig. 17.2). These waves are like the ripples that spread on a pond when a rock is dropped into the water.

The light waves scattered from your face are electromagnetic waves, not waves of water, but they have crests (where the electric and magnetic fields are the strongest) that move outward from the source point just as water waves do. If we connect the points on the wave that are all at the same point in their cycle, we define a **wavefront.** We

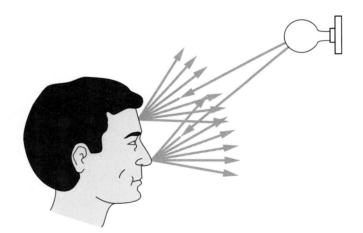

figure **17.2**   Any point on your face acts as a secondary source of light rays that are reflected in all directions from that point.

often choose the crest of the wave (the point of maximum positive amplitude) for this purpose, since it is clearly visible in water waves. The next wavefront behind the leading one is the next point at which the waves are at their crest, and it is separated from the previous wavefront by a distance of one wavelength, as in figure 17.3. For light waves, these wavefronts move away from the source point at the speed of light.

We could describe almost everything that happens to these waves by tracing what happens to the wavefronts. It is easier, however, to examine their behavior using rays perpendicular to the wavefronts. If the waves are traveling in the same medium (air, for example), the wavefronts move forward uniformly and the resulting rays are straight lines (fig. 17.3). These rays are easier to draw than the curved wavefronts and can be traced more readily.

Since each point on your face scatters light in all directions, these points act as sources of diverging light rays, as in figure 17.2. The light rays travel from your face to the mirror, where they are reflected in a predictable manner because of the smoothness of the mirror. Your eyes are receiving light from the light bulb, but that light has been reflected by both your face and the mirror before getting back to your eyes.

What is the law of reflection?

What happens when light rays and wavefronts strike a smooth reflecting surface like a flat mirror? The waves are reflected, and after reflection, they travel away from the mirror with the same speed that they had before reflection. Figure 17.4 depicts this process for plane wavefronts (no curvature) that are approaching the plane mirror at an angle rather than head-on.

Since the wavefronts approach the mirror at an angle, some parts of the wavefront are reflected sooner than others

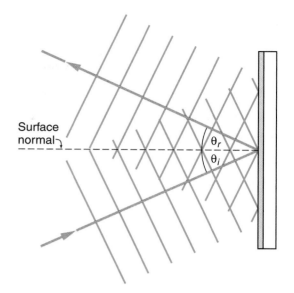

figure **17.4**   Plane light waves approaching a mirror at an angle travel with the same speed both before and after striking the mirror. The angle of reflection equals the angle of incidence.

(fig. 17.4). These wavefronts now travel away from the mirror with the same spacing and speed but in a new direction. The angle between the wavefront and the mirror is the same, however, for the emerging wave. Because the outgoing wavefronts travel at the same speed and cover the same distance in a given time as the incoming waves, this produces equal angles between the wavefronts and the surface of the mirror.

This result is usually stated using rays. The angle that a ray makes to a line drawn perpendicular to the surface is the same angle that the wavefront makes to the surface of the mirror. Using the word *normal* to mean *perpendicular* (as in chapter 4 when we discussed normal forces), we call the line drawn perpendicular to the surface of the mirror the **surface normal.** The equal angles the wavefronts make to the mirror dictate that the angle the reflected ray makes to the surface normal is equal to the angle that the incoming, or *incident,* ray makes to the surface normal (fig 17.4).

What we have just described is the **law of reflection,** which can be stated concisely as

> When light is reflected from a smooth reflecting surface, the angle the reflected ray makes with the surface normal is equal to the angle the incident ray makes with the surface normal.

In other words, the angle of reflection (θ_r in figure 17.4) is equal to the angle of incidence (θ_i). The reflected ray also lies within the plane defined by the incident ray and the surface normal. It does not deviate in or out of the plane of the page in our diagram.

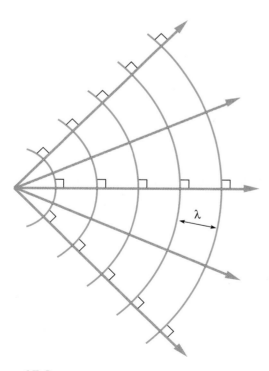

figure **17.3**   Light rays are drawn so that they are everywhere perpendicular to the wavefronts. If the waves travel with uniform speed, the rays are straight lines.

How are images formed by a plane mirror?

How can the law of reflection help us to explain how an image is formed in a plane mirror? What happens to the light rays that are scattered from your nose? Tracing these rays from their origin at the nose and following them through reflection from the mirror using the law of reflection shows what happens to individual rays (fig. 17.5).

If we extend the reflected rays backward from the mirror, they intersect at a point behind the mirror, as shown. As your eye collects a small bundle of reflected rays, it perceives an image that appears to lie at this point of intersection. In other words, as far as your eye can tell, these light rays are coming from that point. You see the tip of your nose as lying behind the mirror. The same argument holds for any other point on your face—they all seem to lie behind the mirror.

Using simple geometry, you can see that the distance of this image behind the mirror (measured from the mirror surface) is equal to the distance of the original object from the front of the mirror. These equal distances follow from the law of reflection, as figure 17.6 illustrates. Here we have taken just two rays coming from the top of a candle and traced them as before. The ray that comes into the mirror in the horizontal direction and perpendicular to the mirror is reflected back along the same line. The angle of incidence for this ray is zero—so is the angle of reflection.

The other ray is shown as being reflected from a point on the mirror even with the base of the candle. This ray is reflected at an angle equal to its angle of incidence. When these two rays are extended backward, their intersection locates the image position. Any other rays traced from the top of the candle would also appear to come from this point. The two rays shown form identical triangles on either side of the mirror (fig. 17.6). Since the angles are equal, and the short side of both triangles is equal to the height of the candle, the long sides of these identical trian-

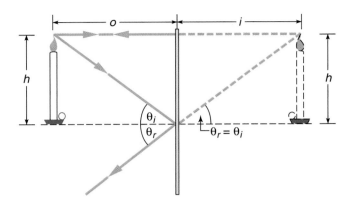

figure **17.6** Two light rays coming from the top of the candle appear to diverge after reflection from a point as far behind the mirror as the candle is in front of the mirror.

gles must also be equal. The image is therefore located behind the mirror at an *image distance, i,* equal to the *object distance, o,* of the candle from the front of the mirror.

Since we all have access to plane mirrors, you can verify some of these ideas by observing your own image and images of other objects as you move before the mirror. The mirror does not have to be as tall as you, for example, for you to see your entire height. Do you see more of you as you move toward the mirror or away from the mirror? What about other objects? Where must you be positioned to see various other objects in the room? Can you explain these observations by the law of reflection and by which rays reach your eyes?

The image formed by a plane mirror is called a *virtual image* because the light never actually passes through the point where the image is located. In fact, the light never gets behind the mirror at all—it just appears to come from points behind the mirror as it is reflected. The image can also be characterized as being upright (right side up) and as having the same size as the object (not magnified). There is a reversal, however, of right and left: what appears to be the right hand of your mirror image is actually the image of your left hand, and vice versa. Take another look. You see such images every day, but you probably have not given them much thought.

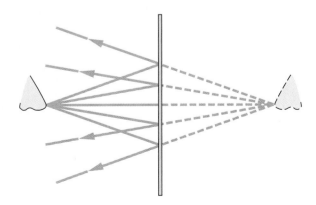

figure **17.5** A few rays are traced from the tip of the nose to show their reflection from the mirror. They diverge after reflection as though they were coming from a point behind the mirror.

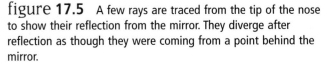

Light rays are drawn perpendicular to the wavefronts. Tracing these rays often provides an easier way of seeing what happens to light than following the wavefronts themselves. The law of reflection states that when light is reflected from a smooth reflecting surface, the angle of reflection equals the angle of incidence (both measured from the surface normal). The image formed by a plane mirror lies at the point from which the light rays appear to be diverging after reflection, which is as far behind the mirror as the object is in front of the mirror.

17.2 Refraction of Light

The most familiar images are those formed by plane mirrors. Images can be formed in other ways, though, with prisms, lenses, or maybe just a tank of water. These examples all involve substances that are transparent to visible light. What happens to light rays when they encounter the surface of a transparent object? Why do we get misleading impressions about the location of underwater objects? The *law of refraction* helps us answer these questions.

What is the law of refraction?

Suppose that light waves encounter a plane surface of a piece of glass after traveling initially through air. What happens to these waves as they pass into the glass and continue traveling through the glass? Experimental measurements have shown that the speed of light in glass or water is less than the speed of light in a vacuum or air. (The speed of light in air is very close to its speed in a vacuum.) The distance between wavefronts (the wavelength) will be shorter in glass or water than in air (fig. 17.7), since the waves travel a smaller distance in one cycle given their smaller speed.

The difference in the speed of light in different substances is usually described by a quantity called the **index of refraction,** represented by the symbol *n*. The index of refraction is defined as the ratio of the speed of light *c* in a vacuum to the speed of light *v* in some substance, $n = c/v$. The speed of light *v* in the substance is then related to the speed of light *c* in a vacuum by

$$v = \frac{c}{n}.$$

In other words, to find the speed of light in some transparent material, we divide the speed of light in a vacuum ($c = 3 \times 10^8$ m/s) by the index of refraction of that material. Typical values for the index of refraction of glass are between 1.5 and 1.6, so the speed of light in glass is approximately two-thirds the speed of light in air.

What effect does this reduction in speed and wavelength have on the direction of light rays as they pass into glass? If we consider wavefronts and their corresponding rays approaching the surface at an angle, as in figure 17.8, we can see that the rays will bend as the waves pass from air to glass. The bending occurs because the wavefronts do not travel as far in one cycle in the glass as they do in air. As the diagram shows, the wavefront halfway into the glass travels a smaller distance in glass than it does in air, causing it to bend in the middle. Thus, the ray, which is perpendicular to the wavefront, is also bent. The situation is like a marching band marching onto a muddy field at an angle to the edge of the field. The rows bend as their speed is reduced by the mud.

The amount of bending depends on the angle of incidence and on the indices of refraction of the materials involved, which determine the change in speed. A larger difference in speed will produce a greater difference in how far the wavefronts travel in the two substances. A larger difference in indices of refraction of the two substances therefore produces a larger bend in the wavefront and ray.

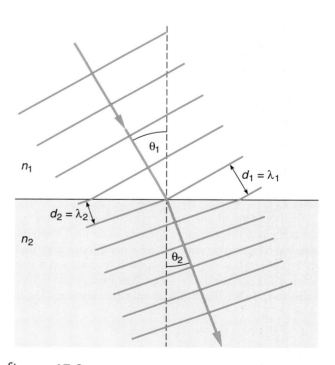

figure **17.8** Wavefronts approaching a glass surface at an angle to the surface are bent as they pass into the glass. The angle of refraction, θ_2, is less than the angle of incidence, θ_1. (n_2 is larger than n_1.)

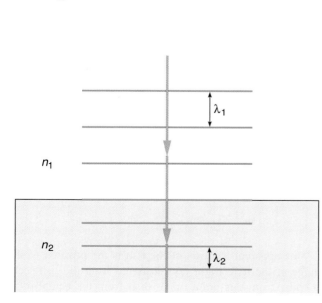

figure **17.7** The wavelength of light waves traveling from air into glass is shorter in the glass than in air because of the smaller speed of light waves in glass.

The bending described by the **law of refraction** can be stated in qualitative* terms as:

> When light passes from one transparent medium to another, the rays are bent toward the surface normal (the axis drawn perpendicular to the surface) if the speed of light is smaller in the second medium than in the first. The rays are bent away from this axis if the speed of light in the second medium is greater than in the first.

If the angles are small, the quantitative statement of the law of refraction takes a simple form. For small angles, the sine function is proportional to the angle itself, so

$$n_1\theta_1 = n_2\theta_2 \, .$$

The product of the index of refraction of the first medium times the angle of incidence is *approximately* equal to the product of the index of refraction of the second medium times the angle of refraction. As the index of refraction of the second medium increases, the angle of refraction must decrease, which means that the ray is bent closer to the axis (the surface normal) for larger indices of refraction.

For light waves traveling from glass to air, the bending is in the opposite direction: the rays are bent away from the surface normal, according to the law of refraction. Simply reversing the directions of the rays and wavefronts in figure 17.8 will make this clear. The increase in speed as the wave travels from glass to air causes the ray to bend away from the axis.

Why do underwater objects appear to be closer than they are?

The bending of light rays at the interface of two transparent substances is responsible for some deceptive appearances. Suppose, for example, that you are standing on a bridge over a stream looking down at a fish. Water has an index of refraction of about 1.33, and air has an index of refraction of approximately 1. Light traveling from the fish to your eyes is bent away from the surface normal, as in figure 17.9.

This bending of the light rays coming from the fish causes them to diverge more strongly in air than when they were traveling in the water. If we extend these rays backwards, we see that they now appear to come from a point closer to the surface of the water than their actual point of origin (fig. 17.9). Since this is true for any point on the

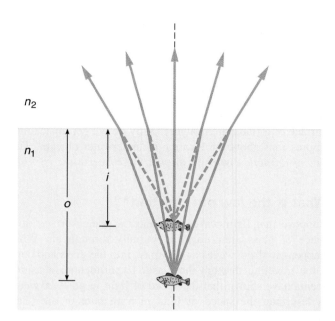

figure **17.9** Light rays coming from the fish are bent as they pass from the water to air, so that the rays appear to diverge from a point closer to the surface.

fish, the fish appears to be closer to the surface than it actually is. If you were attempting to shoot the fish (illegal in most places), you would likely miss unless you were shooting straight down.

The apparent distance of the fish beneath the surface can be predicted from the law of refraction if we know the actual distance. The argument involves some geometry and the assumption that the angles of incidence and refraction are small. We find that the apparent distance as seen from the air (the image distance i) is related to the actual distance under water (the object distance o) and the indices of refraction by

$$i = o\left(\frac{n_a}{n_w}\right),$$

where the index of refraction of air ($n_a = 1$) is less than that of water ($n_w = 1.33$). The image distance is therefore less than the actual distance, as figure 17.9 clearly illustrates. If the fish is actually 1 m below the surface, its apparent distance below the surface will be

$$i = 1\text{ m}\left(\frac{1}{1.33}\right) = 0.75\text{ m.}$$

This apparent location of the fish is the position of the *image* of the fish. Light rays scattered from the fish seem to come from this point rather than from the actual position of the fish. We see a virtual image, like the image seen in a mirror, since the light rays do not actually pass through the image position. They only appear to come from that point.

---

* The law of refraction is stated quantitatively using the trigonometric function, the sine, which will only be familiar if you have some background in trigonometry. It is usually written in symbolic form as $n_1\sin\theta_1 = n_2\sin\theta_2$, where n_1 and n_2 are the indices of refraction of the two media, and θ_1 and θ_2 are the angle of incidence and the angle of refraction (fig. 17.8).

The misleading position of objects viewed underwater is something we observe daily but often fail to notice simply because the experience is so common. A straight stick or rod seems to bend or break if part of it is above water and the rest below. When viewed from the top, each point of the underwater object appears to lie closer to the surface than its actual distance. A straw or spoon in a glass of water or other beverage likewise appears to bend, as in figure 17.10. We are used to the deception and seldom give it a second thought.

Total internal reflection

Another interesting phenomenon occurs when light rays travel from either water or glass to air. As we have already indicated, the light rays bend away from the axis as they pass into the medium with the lower index of refraction (air in these examples). What happens, though, if the rays are bent so much that the angle of refraction is 90°? The angle of refraction cannot be any larger and still result in rays passing into the second medium.

This situation is depicted in figure 17.11. As the angle of incidence for rays traveling in the glass gets larger, so does the angle of refraction (ray 1). Ultimately, we reach the point at which the angle of refraction is 90° (ray 2). At this point, the refracted ray would just skim the surface as it emerged from the glass. For any angle of incidence inside the glass larger than this angle (ray 3), the ray does not escape the glass at all—it is reflected instead. The angle of incidence for which the angle of refraction is 90° is called the *critical angle* θ_c. Rays incident at angles greater than the critical angle are reflected back inside the glass and obey the law of reflection rather than the law of refraction.

This phenomenon is called *total internal reflection.* For angles equal to or greater than the critical angle, 100% of the light is reflected inside the material with the larger index

figure **17.10**   When viewed from above, a straw appears to bend when part of it is above water and the rest below. Viewed from the side, the straw is magnified.

of refraction. Under these conditions, the glass-air interface makes an excellent mirror. For glass with an index of refraction of 1.5, the critical angle is approximately 42°. Glasses with larger indices of refraction have even smaller critical angles. A prism cut with two 45° angles, as in figure 17.12, can be used as a reflector. Light incident perpendicular to the first surface strikes the long surface at an angle of incidence of 45°, which is greater than the critical angle. It is totally reflected at this surface, so the surface acts as a plane mirror.

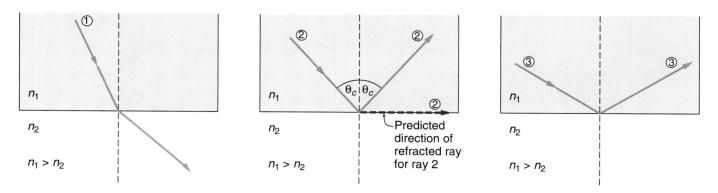

figure **17.11**   When light travels from glass to air, the angle of incidence that would produce an angle of refraction of 90° (ray 2) is called the critical angle θ_c. Rays incident at equal or greater angles than θ_c are totally reflected.

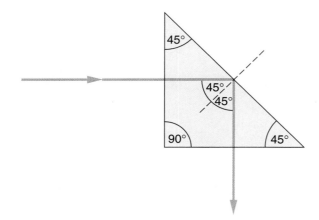

figure **17.12**   A prism cut with two 45° angles can be used as a mirror, since the light is totally reflected.

How do prisms bend light, and what is dispersion?

You know that white light can be separated into different colors by a prism, producing an effect like a rainbow. How do we get the colors of the rainbow when we start with ordinary white light?

The index of refraction of a material varies with the wavelength of light: different wavelengths are bent by different amounts. The wavelength, in turn, is associated with the color that we perceive as discussed in chapter 16. Red light, at one end of the visible spectrum, has longer wavelengths and lower frequencies than violet light, at the opposite end of the spectrum. The index of refraction for violet or blue light is greater than for red light for most types of glass and other transparent substances like water. Blue light is bent more than red light, and the intermediate wavelengths associated with the colors green, yellow, and orange are bent by intermediate amounts.

Everyday Phenomenon

box **17.1**

Rainbows

The Situation. Figure 1.1 in chapter 1 is a photograph of a rainbow. We have all seen such sights and been awed by their beauty. We know that rainbows occur when the sun is shining and it is raining nearby. If conditions are right, we can see an entire semicircular arc of color with red on the outside and violet on the inside. Sometimes we can also see a fainter bow of color forming a secondary arc outside of the primary one. The colors in the secondary rainbow are in reverse order of those in the primary rainbow, as the photo shows.

How is a rainbow formed? What conditions are necessary for observing a rainbow? Where should we look? Can the laws of reflection and refraction be used to explain this

phenomenon? We can now address these questions, some of which were first raised in chapter 1.

The Analysis. The secret to understanding the rainbow lies in considering what happens to light rays when they enter a raindrop, as in the first drawing. When a light ray strikes the first surface of the raindrop, some of it is refracted into the drop. Since the amount of bending depends on the index of refraction, which depends on the wavelength of the light, blue light is bent more than red light at this first surface. This effect is the same as the dispersion that occurs when light passes through a prism.

The primary rainbow has red on the outside and violet on the inside. The secondary rainbow, sometimes visible, has the colors reversed.

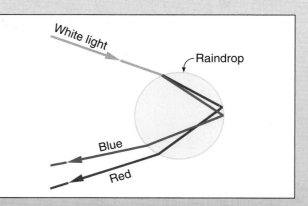

Light rays entering a raindrop are refracted by different amounts at the first surface, reflected at the back surface, and refracted again as they leave.

(continued)

When light passes through a prism at an angle, the light is bent as it enters the prism and again as it leaves. The minimum deflection of the light occurs when it passes through the prism symmetrically, as in figure 17.13. The rays bend toward the surface normal as they enter the prism and away from the normal at the second surface, consistent with the law of refraction. Since the index of refraction varies for different wavelengths, the violet and blue rays are bent the most and lie at the bottom of the resulting spectrum of colors, as figure 17.14 shows.

This variation of index of refraction with wavelength is called **dispersion,** and it exists for all transparent materials including water, glass, and clear plastics. Dispersion is responsible for the colors that you see when light passes through a fish tank or around the edges of a lens, as in an overhead projector. It is also responsible for the beautiful displays of color seen in rainbows, as explained in Everyday Phenomenon Box 17.1.

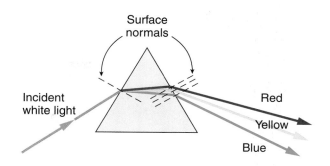

figure **17.13** Light rays passing through a prism are bent at both surfaces, with blue light being bent more strongly than red.

After being refracted at the first surface, the light rays travel through the drop and strike its back surface, where they are partially reflected. Some of the light passes out of the drop, and some is reflected back toward the front surface as shown. At the front surface, the rays are refracted again, causing more dispersion as the rays leave the drop.

It may seem paradoxical that red light, which is bent the least in the refractions, is actually diverted through a larger angle than blue light in its overall path through and back out of the raindrop. The reason can be understood from the first drawing. The smaller bending at the first surface causes the red rays to strike the back surface of the drop at a greater angle of incidence than for the violet rays. The red rays are reflected through a greater angle, according to the law of reflection. This larger angle of reflection dominates in determining the overall deflection of the ray.

When we view a rainbow, the sun must be behind us, since we are observing reflected rays. We see different colors at different points in the sky because, for a given color, the raindrops must be at the appropriate height for the light rays to reach our eyes. Since red rays are deviated the most, we see red light reflected from raindrops at the top of the rainbow or at the greatest angle from the center of the arc. Violet light comes to us from raindrops lying at smaller angles. The other colors lie in between, producing the colorful arc that we see.

The secondary rainbow, which is usually much fainter than the primary rainbow, is produced by a double reflection inside the raindrops, as the second drawing shows. Light rays entering near the bottom of the first surface of the drop may strike the back surface at a large enough angle of incidence to be

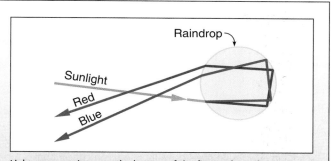

Light rays entering near the bottom of the first surface of a raindrop may be reflected twice before reemerging. These rays produce the secondary rainbow.

reflected twice before getting out of the drop. The light rays cross over in this case and are deviated through larger overall angles than any of the colors in the primary bow. For this reason, the secondary bow is seen outside the arc of the primary bow. Blue light is bent through a larger angle than red light so blue light lies at the top of the secondary rainbow.

The sun must be reasonably low in the sky for either the primary or secondary rainbow to be observed. During the summer, a late afternoon shower usually provides the best opportunity for viewing a rainbow. Rainbows can be seen at almost any time during the day from a high vantage point such as an airplane, however, and sometimes make a complete circle rather than just an arc. If you can understand why this is so, you have mastered the explanations provided here.

figure **17.14**   White light incident on one side of the prism emerges as a spectrum of different colors because of dispersion.

Light rays are bent when they pass from one transparent substance to another because the speed of light changes at the boundary between the two substances. The law of refraction describes how much bending occurs and whether the bending is toward or away from the surface normal. Because of this bending, the image of an underwater object appears to lie closer to the surface of the water than the actual position of the object. For light traveling initially inside glass or water, there is a critical angle of incidence beyond which the light is totally reflected rather than being refracted. The index of refraction varies with wavelength, producing the dispersion of colors that we see when light passes through a prism.

17.3 Lenses and Image Formation

We encounter the images formed by mirrors daily and seldom give them a second thought. We may be even less aware of the images formed by lenses, although many of us have lenses hanging on our noses or sitting on our eyeballs in the form of corrective eyeglasses or contact lenses. We also encounter lenses in cameras, overhead projectors, opera glasses, and simple magnifying glasses.

How do lenses form images? Lenses are usually made of glass or plastic, so the law of refraction governs their behavior. The bending of light rays as they pass through a lens is responsible for the size and nature of the images formed. We can understand the basics of this process by tracing what happens to just a few of these rays.

Tracing rays through a positive lens

A lens shaped so that both sides are spherical surfaces with the convex sides facing out is pictured in figure 17.15. According to the law of refraction, the light rays are bent toward the surface normal at the first surface (going from air to glass) and away from the surface normal at the second surface (going from glass to air). If both surfaces are convex, as shown, each of these refractions bends the ray toward the axis (a line passing through the center of the lens and perpendicular to the lens). Such a lens causes light rays to converge. A *converging* lens is called a **positive lens.**

The easiest way to see that the light will bend as pictured in figure 17.15 is to imagine that each section of the lens behaves like a prism. The prism angle (the angle between the two sides) becomes larger toward the top of the lens so that light coming through the lens near the top is bent more than light passing through near the middle. Because the prism effect gets stronger farther from the axis, light rays coming in parallel to the axis are bent by different amounts as they pass through the lens. This causes them to all pass approximately through a single point F on the opposite side of the lens, which we call the **focal point.** The focal point is the point where rays traveling parallel to the axis when they enter the lens are focused after leaving the lens.

The distance from the center of the lens to the focal point is called the **focal length** f. Focal length is a property of the lens that depends on how strongly the surfaces are curved and on the index of refraction of the lens material. There is also a focal point a distance f on the other side of the lens associated with parallel light rays coming from the opposite direction. Since the paths of rays are reversible, rays that diverge from either focal point and pass through the lens will emerge parallel to the axis.

How can we use ray-tracing techniques to show how images are formed by such a lens? The process is illustrated

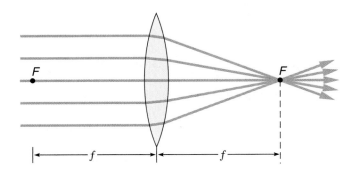

figure **17.15**   Parallel light rays passing through a simple convex lens are bent toward the axis so that they all pass approximately through the focal point F. A lens has two focal points, one on either side.

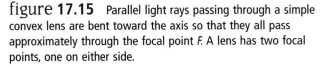

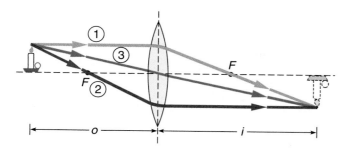

figure 17.16 Three rays can be traced from the top of an object through a positive lens to locate the image. If the object is beyond the focal point, an inverted real image is formed on the opposite side of the lens.

in figure 17.16 for an object lying beyond the focal point of the lens. Three rays (labeled on the diagram) are traced, taking advantage of the properties of the focal points:

1. A ray coming from the top of the object traveling parallel to the axis is bent so that it passes through the focal point on the far side of the lens.
2. A ray coming through the focal point on the near side emerges parallel to the axis.
3. A ray coming in through the center of the lens is undeviated and passes through the lens without being bent.

The sides of the lens near the center are parallel (like a window pane), which is why ray 3 is not bent. When tracing rays, we usually show the bending taking place at a vertical line through the center of the lens.

The image lies on the opposite side of the lens from the object in figure 17.16 and is a *real image,* since the light rays pass through the image point. Light rays diverging from the object are converged by the lens to this image point. If you placed a screen at that point, you would see the upside-down (*inverted*) image on the screen. When a slide projector is used, the slides must be inserted upside down to produce upright images on the screen.

How is the image distance related to the object distance?

Can we predict where an image will be found for a given location of an object? One way to do so is to carefully trace the rays, as we have already illustrated. Using triangle relationships and the law of refraction, we can also develop a quantitative relationship between the object distance o, the image distance i, and the focal length f of a lens. (These distances are all measured from the center of the lens.) The relationship involves the reciprocals of these distances—the reciprocal of the object distance plus the re-

ciprocal of the image distance is equal to the reciprocal of the focal length. Stated in symbols,

$$\frac{1}{o} + \frac{1}{i} = \frac{1}{f} .$$

In the case in figure 17.16, these distances are all positive quantities. In the most frequently used sign convention, object and image distances for *real* objects or images are positive, but the image distance for a *virtual* image is negative. The focal length is positive for a *converging* (positive) lens, one that bends the rays toward the axis, and is negative for a *diverging* (negative) lens, one that bends the rays away from the axis.

The geometry of figure 17.16 can also be used to find a relationship between the magnification of the image m and the object and image distances. **Magnification** is defined as the ratio of the image height h_i to the object height h_o, or

$$m = \frac{h_i}{h_o} = -\frac{i}{o} .$$

The sign in this equation indicates whether the image is upright or inverted. A negative magnification represents an inverted image, as is the case when the image and object distances are both positive (fig. 17.16). Depending on the object and image distances, the image can be either magnified or reduced in size.

If the object lies "inside" the focal point (closer to the lens) of a positive lens, we can get a virtual image with a positive magnification, as in figure 17.17. In this case, the image distance is a negative quantity, since it is associated with a virtual image. Light rays appear to diverge from the image point but do not actually pass through this point. A virtual image lies on the side of the lens from which the

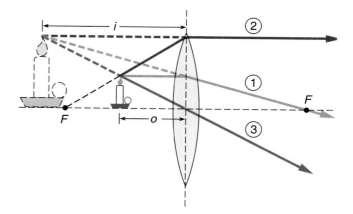

figure 17.17 A magnified virtual image is formed when the object lies inside the focal point of a positive lens. The emerging light rays appear to diverge from a point behind the object.

try this box 17.1

Sample Exercise: A Virtual Magnified Image

An object 2 cm in height lies 10 cm to the left of a positive lens with a focal length of 20 cm.
 a. Where is the image located?
 b. What is its magnification?

a. $f = +20$ cm
$o = +10$ cm
$i = ?$

$$\frac{1}{o} + \frac{1}{i} = \frac{1}{f}$$

$$\frac{1}{i} = \frac{1}{f} - \frac{1}{o}$$

$$= \frac{1}{20 \text{ cm}} - \frac{1}{10 \text{ cm}}$$

$$= \frac{1}{20 \text{ cm}} - \frac{2}{20 \text{ cm}}$$

$$= -\frac{1}{20 \text{ cm}}$$

$$i = -20 \text{ cm}$$

The image lies 20 cm to the left of the lens, on the same side as the object (fig. 7.17).

b. $m = ?$

$$m = -\frac{i}{o}$$

$$= -\frac{-20 \text{ cm}}{10 \text{ cm}}$$

$$= +2$$

The image is magnified to twice the height of the object as shown in figure 17.17. It is upright, since the magnification is positive.

If we imagine that each section of the lens behaves like a prism (as discussed on page 356), these prism sections are upside down compared to the convex lens. As you can see in figure 17.18, these prism sections bend light rays away from the axis rather than toward it. The lens is therefore a *diverging* or **negative lens.**

Light rays coming in parallel to the axis are bent away from the axis by the negative lens so that they all appear to be diverging from a common point *F*, the focal point (fig. 17.18). This point is one of two focal points of the negative lens. The other lies on the opposite side at the same distance from the center of the lens as the first one. Light coming in toward the focal point on the far side of the lens is bent so that it comes out parallel to the axis. As mentioned earlier, the focal length *f* of a negative lens is defined as a negative quantity.

We can trace the same three rays that we traced for the positive lens to locate an image for the negative lens, as in figure 17.19. The ray coming from the top of the object parallel to the axis (ray 1) is bent away from the axis in this case, so that it appears to come from the focal point on the near side of the lens. A ray coming in toward the focal point on the far side (ray 2) is bent to come out parallel to the axis. Ray 3 passes through the center of the lens undeviated, as before.

The resulting image lies on the same side of the lens as the object and is upright and reduced in size (fig. 17.19). This can be verified by using the object-image distance formula, treating the focal length as a negative quantity. The image is virtual because the rays appear to come from the image point but do not (except for the undeviated ray) actually pass through that point. This is true regardless of the object distance—a negative lens used by itself always forms a virtual image smaller than the object.

light is coming, the left side in this case. This situation is dealt with in Try This Box 17.1.

When we view the image of an object placed inside the focal point of a positive lens, we are using the lens as a magnifying glass. The image is magnified, but it also lies behind the object, farther from our eyes. This greater distance makes it easier for us to focus on the image than on the object itself. This is an advantage, particularly for older people who have lost the ability to accommodate their focus to view objects that are close to their eyes.

Tracing rays through negative lenses

As we have seen, a simple convex lens is a positive, *converging* lens—it bends light rays toward the axis. What happens if we change the direction of the curvature of the lens surfaces so that they are concave rather than convex?

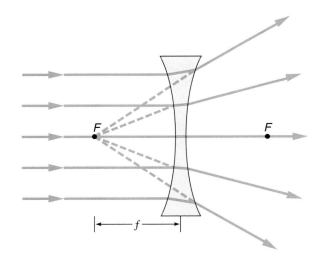

figure **17.18** Light rays traveling parallel to the axis are bent away from the axis by a negative lens so that they appear to diverge from a common focal point *F*.

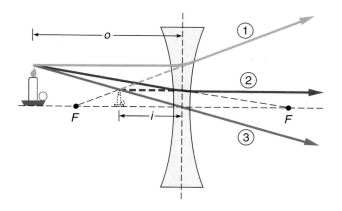

figure 17.19 Three rays are traced from the top of an object to locate the image formed by a negative lens. The virtual, upright image lies on the same side of the lens as the object and is reduced in size.

If you hold a negative lens near a printed page and view the page through the lens, the letters will appear smaller than their actual size. Distinguishing a negative lens from a positive lens is easy: one makes the print smaller and the other (the positive lens) magnifies the print when held close to the page. If you move the positive lens away from the page, the image disappears when you reach a distance equal to the focal length and then reappears upside down.

A positive lens converges light toward the axis. If the incoming rays are parallel to the axis, they converge approximately to a single point, the focal point. The properties of the focal points can be used to trace rays to locate and characterize the images formed. The distance of the focal points from a thin lens is called the focal length, which can be used to find the image position for any given object position. A negative lens diverges light rays and always forms a reduced-in-size virtual image of a real object. The object-image distance formula and the associated ray-tracing techniques can be used to find and describe the images formed by both positive and negative lenses.

17.4 Focusing Light with Curved Mirrors

Most of us have had the experience of using a shaving or makeup mirror that magnifies features on your face. This experience can be even more disconcerting early in the morning than that provided by an ordinary mirror. What is going on here? How is the magnification accomplished?

Magnifying mirrors involve curved surfaces rather than plane surfaces. The curvature is usually spherical in nature—the surface of the mirror is a portion of a sphere. A spherical reflecting surface has the ability to focus light rays in a manner similar to a lens. Simple ray-tracing techniques can provide an understanding of the resulting images.

Ray tracing with a concave mirror

Mirrors that produce magnification are **concave mirrors,** which means that light is being reflected from the inside of a spherical surface. Their focusing properties can be understood by following rays that approach the mirror parallel to the axis, as shown in figure 17.20. The center of curvature of the spherical surface lies on the axis, as shown, and the law of reflection dictates where each ray will go.

Each ray that we have traced in figure 17.20 obeys the law of reflection—the angle of reflection equals the angle of incidence. The surface normal for each ray is found by drawing a line from the center of curvature of the sphere to the reflecting surface. A radius of a sphere is always perpendicular to its surface.

As you can see from the diagram, each ray is reflected so that it crosses the axis at approximately the same point as all the other rays. This point of intersection is the focal point, labeled with the letter *F*. Like the focal point of a lens, it is the point where rays coming in parallel to the axis are focused. Since any of these rays could be reversed in direction and still obey the law of reflection, rays that pass through the focal point on their way to the mirror will emerge parallel to the axis of the mirror.

We can also see from the diagram in figure 17.20 that the distance of the focal point *F* from the mirror is approximately half the distance of the center of curvature *C* from the mirror. These two points, *F* and *C*, can be used in tracing

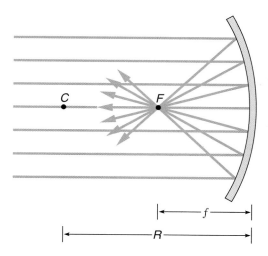

figure 17.20 Light rays approaching a spherical concave mirror traveling parallel to the axis are reflected so that they all pass approximately through a common focal point *F*. The focal length *f* is half the radius of curvature *R*.

rays to locate and describe images formed by the mirror. When you use such a mirror to examine your face, you generally place your face "inside" the focal point (closer to the mirror). This produces the magnified image that you normally observe.

Figure 17.21 shows how rays can be traced to find the image of an object placed inside the focal point of the mirror. There are three rays that we can easily trace, but any two of these rays would be sufficient to locate the image. The three rays are:

1. A ray coming from the top of the object parallel to the axis and reflected through the focal point.
2. A ray coming in through the focal point and reflected parallel to the axis.
3. A ray coming in along a line passing through the center of curvature and reflected back along itself.

The third ray comes in and out along the same line because it strikes the mirror perpendicular to its surface. The angles of incidence and reflection are both zero in this case.

When these three rays are extended backward, we see that they all appear to come from a point behind the mirror. This point defines the position of the top of the image. The bottom of the image lies on the axis. We see the image as lying behind the mirror and, as is obvious from the diagram, magnified.

The resulting image is upright, magnified, and virtual. It is a virtual image because, just as in the case of the plane mirror, the light rays do not actually pass through the image. The rays never get behind the mirror, but the image appears to lie behind the mirror.

It is also possible to form real images with a concave mirror. This occurs when the object is located beyond the focal point of the mirror, as is illustrated in figure 17.22. In this case, when the three rays are traced, we see that they

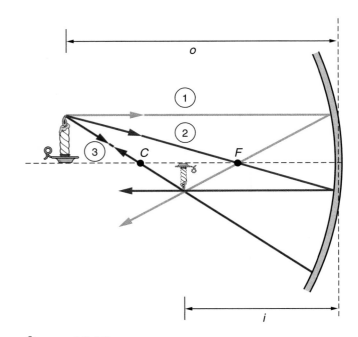

figure **17.22** Light rays coming from an object located beyond the focal point of a concave mirror converge to intersect in front of the mirror forming an inverted real image.

converge rather than diverge as they leave the mirror. The rays coming from the top of the object intersect at a point on the same side of the mirror as the object and then diverge again from this point. If our eyes collect these rays, we see an image that lies in front of the mirror.

Since we are used to looking at images that lie behind a mirror, it is a little harder for us to focus on one lying in front. This image can be observed, however, using a curved makeup or shaving mirror. As you move the mirror away from your face, the original magnified image becomes larger and finally disappears. It is replaced by an upside-down image that grows smaller as you continue to move away from the mirror. This image is a real image because the light rays do pass through the image and then diverge again from these points. As is also clear from the diagram, the image is inverted (upside down).

Object and image distances

Can we predict where an image will be found for a given location of an object? One way of doing this is simply to carefully trace the rays as already illustrated. Using triangle relationships and the law of reflection, it is also possible to develop a quantitative relationship between the object and image distances. The relationship turns out to be the same as that stated earlier for a thin lens.

$$\frac{1}{o} + \frac{1}{i} = \frac{1}{f} \ .$$

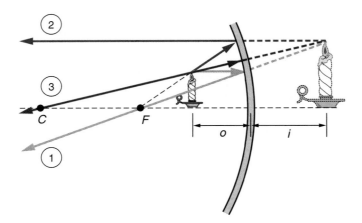

figure **17.21** Three rays are traced from the top of the candle placed in front of the mirror. Extending the reflected rays backward locates the top of the image behind the mirror.

The object distance *o*, the image distance *i*, and the focal length *f* are all measured from the vertex, the point where the axis meets the mirror. As shown in figure 17.20, the focal length is half the radius of curvature.

In the case shown in figure 17.22 these distances are all positive quantities. In general, distances are positive if they lie on the same side of the mirror as the light rays themselves. If they lie behind the mirror, the distances are negative. The example exercise in Try This Box 17.2 demonstrates the use of the relationship between object and image distances.

The triangles in figures 17.21 and 17.22 can also be used to find a relationship between the image height and object height—or, in other words, the magnification, *m*. This relationship is also the same as that obtained earlier for a thin lens,

$$m = -\frac{i}{o}.$$

In the example in Try This Box 17.2, where the image distance is −10 cm and the object distance +5 cm, the magnification is +2.0. In other words, the image height is twice that of the object. The fact that this magnification is positive indicates that the image in this case is upright.

Convex mirrors

Up to this point, we have been considering concave mirrors, which curve inward toward the viewer. What happens if the mirror curvature is in the opposite direction? **Convex mirrors,** for which light is reflected from the outside of the spherical surface, are used as wide-angle mirrors in store aisles or as the side-view mirror on the passenger side of a car. These mirrors produce a reduced-in-size image, but a large field of view. It is this wide-angle view that makes them useful.

Figure 17.23 pictures rays approaching a convex mirror traveling parallel to the axis of the mirror. The center of curvature *C* lies behind the mirror in this case. Lines drawn from the center of curvature are perpendicular to the surface of the mirror, as before. The law of reflection dictates that the parallel rays will be reflected away from the axis as shown. When the reflected rays are extended backward, they all appear to have come from the same point *F*, the focal point.

A convex mirror is therefore a diverging or *negative* mirror. Parallel light rays diverge as they leave the mirror rather than converging as they do with a concave mirror. We can use the same ray-tracing techniques, however, to locate an image. Figure 17.24 illustrates this process. The ray coming from the top of the object traveling parallel to the axis (1) is reflected as though it came from the focal point. A ray coming in toward the focal point (2) is reflected

---

try this box 17.2

Sample Exercise: Finding an Image for a Concave Mirror

An object lies 5 cm to the left of a concave mirror with a focal length of +10 cm. Where is the image? Is it real or virtual?

$o = 5$ cm

$f = 10$ cm

$i = ?$

$$\frac{1}{o} + \frac{1}{i} = \frac{1}{f}$$

$$\frac{1}{i} = \frac{1}{f} - \frac{1}{o}$$

$$\frac{1}{i} = \frac{1}{10 \text{ cm}} - \frac{1}{5 \text{ cm}}$$

$$\frac{1}{i} = \frac{1}{10 \text{ cm}} - \frac{2}{10 \text{ cm}}$$

$$\frac{1}{i} = -\frac{1}{10 \text{ cm}}$$

$$i = -10 \text{ cm}$$

Since the image distance is negative, the image lies 10 cm behind the mirror and is virtual. The situation is much like that pictured in figure 17.21.

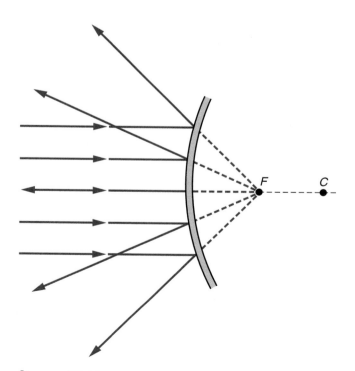

figure **17.23** Light rays traveling parallel to the axis of a convex mirror are reflected so that they appear to come from a focal point, *F*, located behind the mirror.

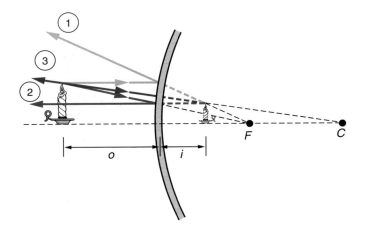

figure **17.24**   Three rays are traced to locate the virtual image of an object placed in front of a convex mirror. The reflected rays diverge as though coming from the image point behind the mirror.

parallel to the axis, and the ray coming in toward the center of curvature (3) is reflected back along itself since it is perpendicular to the surface.

When extended backward, these three rays all appear to come from a point behind the mirror, thus locating the top of the image. The image is virtual since it lies behind the mirror, and it is obviously upright and reduced in size. Check it out the next time you are in a store that has one of these mirrors mounted above an aisle. The images are small but cover a broad area.

The object-image distance formula introduced for concave mirrors and lenses can also be used to locate images for a convex mirror. The one difference is that the focal length of convex mirror must be treated as a negative quantity, like that for a negative lens. The image formed by a convex mirror always lies behind the mirror and therefore the image distance will always be negative for any object distance that you choose.

The side mirror on the passenger side of a car is usually a convex mirror. It produces a wide-angle view of the traffic lane to the right. Since it is a negative mirror, the image is reduced in size and lies just behind the mirror as in figure 17.24. However, there is usually a warning written on the mirror saying: "OBJECTS IN MIRROR ARE CLOSER THAN THEY APPEAR." How can this be if the image being viewed is very close to the mirror itself?

The answer lies in the fact that our brains use many different cues to determine distance. In this case, since the size of the vehicles viewed in the mirror is small, our brains interpret this as meaning that the vehicles must be farther away than they actually are. We know the actual size of that truck or car, and our brains use this size information to determine distance. If you viewed some object of unknown size, your binocular vision might place the image of this object at the actual image location behind the mirror.

Mirrors with curved surfaces can be used to focus light rays and form images. We can locate and describe these images by tracing rays associated with the focal point of the mirror and its center of curvature. The same formula used for lenses can be used to relate object and image distances for mirrors. A concave mirror produces a magnified virtual image when the object is inside the focal point and real images when the object is beyond the focal point. A convex mirror produces a reduced-in-size image for any object position, but provides a wide angle of view.

17.5 Eyeglasses, Microscopes, and Telescopes

Lensmaking was an art that developed during the Renaissance. Before then, it was not possible to correct visual problems such as nearsightedness or farsightedness or to magnify objects with a magnifying glass. Once lenses became common, though, it did not take long for people to discover that they could be combined to make optical instruments like microscopes and telescopes. Both were invented in Holland in the early 1600s.

Correction of visual problems is still the most familiar use of lenses. Most of us will wear eyeglasses at some time in our lives, and many of us have worn them since adolescence or even earlier. What goes wrong with our vision that requires corrective lenses? To answer that question, we need to explore the optics of the eye itself.

How do our eyes work?

Our eyes contain positive lenses that focus light rays on the back surface of the eyeball when working properly. As shown in figure 17.25, the eye actually contains two positive lenses—the *cornea,* which is the curved membrane forming the front surface of the eye, and the *accommodating lens* attached to muscles inside the eye. Most of the bending of light occurs at the cornea. The accommodating lens is more for fine-tuning.

There is a good analogy between the eye and a camera. A camera uses a compound positive lens system to focus light rays coming from objects being photographed onto the film at the back of the camera. The lens system in a camera can be moved back and forth to focus on objects at different distances from the camera. In the eye, the distance between the lens system and the back surface of the eye is fixed so that we need a variable focal-length lens, the accommodating lens, to focus on objects at different distances.

The positive lenses form an inverted real image on the *retina,* the layer of receptor cells on the back inside surface of the eye. The retina plays the role of the film in a camera and is the sensor that detects the image. The light reaching the cells in the retina initiates nerve impulses that are

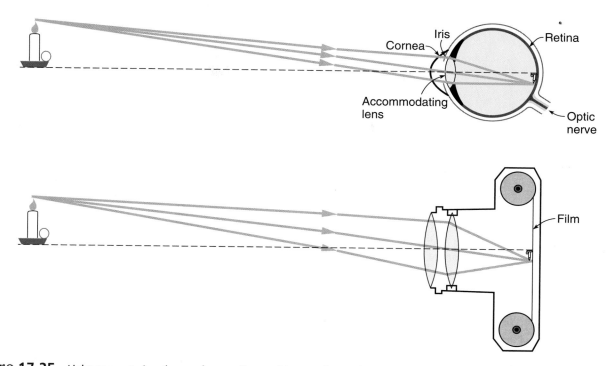

figure 17.25 Light rays entering the eye from a distant object are focused on the back surface of the eye (the retina) by the cornea and the accommodating lens, much as light rays are focused on the film by a camera lens.

carried to our brain. The brain processes the nerve impulses received from both eyes and interprets the image according to its experience. Most of the time this interpretation is straightforward, and what we see is what we get, but at times, the brain's interpretation can produce misleading impressions.

Even though the image on the retina is upside down, our brains interpret the scene as right side up. Interestingly, if we fit people with inverting lenses that turn the image on the retina right side up, they initially see things as being upside down. After some time, the brain makes an adjustment, and they begin to see things right side up again. Everything is fine until they take the inverting lenses off. Then everything appears to be upside down again until the brain readjusts! A great deal of processing takes place between the signal received by your eyes and what you actually perceive.

What problems are corrected with eyeglasses?

The most common visual problem of people who do a lot of reading is nearsightedness or **myopia.** The eyes of a nearsighted person bend the light rays from a distant object too strongly, causing them to focus in front of the retina, as in figure 17.26a. By the time the rays reach the retina, they are diverging again and no longer form a sharp focus. Things appear fuzzy, although we sometimes fail to notice, because we grow accustomed to this indistinct view. A my-

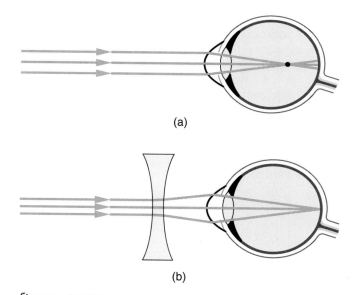

figure 17.26 For a nearsighted person, parallel light rays from a distant object are focused in front of the retina. A negative lens placed in front of the eye can correct this problem.

opic person can see near objects distinctly, because the incident light rays are diverging more strongly from a near object than from a distant object.

Negative eyeglass lenses correct for the tendency of the eye itself to converge the light rays too strongly (fig. 17.26b).

Since a negative lens diverges light rays, it compensates for the excessive convergence by the lenses in the eye and forms distinct images of distant objects on the retina. For a nearsighted person who has not worn glasses before, the difference can be striking.

A farsighted person has the opposite problem. The eye does not converge light rays strongly enough, and images of near objects are formed behind the retina. Positive lenses correct this problem. Laser refractive surgery (see Everyday Phenomenon Box 17.2) can correct both nearsightedness and farsightedness by reshaping the cornea, eliminating the need for eyeglasses.

As we age, the accommodating lenses lose their flexibility. We gradually lose the ability to change the converging power of our eyes and cannot focus on near objects, since light rays diverge more strongly from near objects than from distant objects. At this point, we need bifocals, in which the top half of the lens has one focal length and the bottom half another. We look through the bottom half to do close work and through the top half to view distant objects.

How does a microscope work?

How are lenses combined to form a microscope? A **microscope** consists of two positive lenses spaced as shown in figure 17.27. They are usually held together by a connecting tube, which is not shown in the diagram. If you have ever used a microscope, you know that the object being viewed is placed near the first lens, called the *objective lens.*

The objective lens forms a real, inverted image of the object, provided that the object lies beyond the focal point of the objective lens. If the object lies just beyond this focal point, the real image has a large image distance and the image is magnified. This can be verified by tracing rays or by using the object-image distance formula.

Since light rays actually pass through a real image and diverge again from that point, this real image becomes the object for the second lens in the microscope. The eyepiece lens, or *ocular,* is used like a magnifying glass to observe the real image formed by the objective lens. This real image is focused just inside the focal point of the eyepiece, which then produces the magnified virtual image that we see. The virtual image is also located farther from your eye, so that it can be focused on more readily (fig. 17.27).

Both lenses in a microscope cooperate to produce the desired magnification. The objective lens forms a magnified real image, and this image is magnified again by the eyepiece. The overall magnification of the microscope is found by multiplying these two magnifications together, sometimes achieving magnifications of several hundred times the original object size.

Since eyepiece powers have a limited range, the magnification power of a microscope is determined primarily by the power of the objective lens. A high-power objective lens has a very short focal length, and the object must be placed very close to the objective lens. Microscopes often have two or three different objective lenses of different powers mounted on a turret (fig. 17.28).

The invention of the microscope opened up a whole new world for biologists and other scientists. Microorganisms too small to be seen with the naked eye or with a simple magnifying glass became visible when viewed through a microscope. Seemingly clean pond water was revealed to be teeming with life. The structure of a fly's wing and various kinds of human tissue suddenly became apparent. The microscope is a striking example of how developments in one area of science have a dramatic impact on other areas.

How does a telescope work?

The development of the microscope opened up the world of the very small. The earlier invention of the telescope had an equally dramatic impact in opening up the world of distant objects. Astronomy was the primary beneficiary. A simple astronomical **telescope,** like a microscope, can be

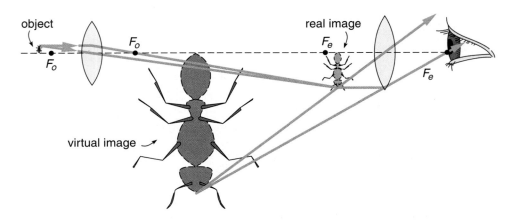

figure **17.27**   A microscope consists of two positive lenses separated by a connecting tube (not shown). The real image formed by the first lens is viewed through the second lens. Both lenses produce magnification.

Everyday Phenomenon

box 17.2

Laser Refractive Surgery

The Situation. Megan Evans has been nearsighted since her early teens. She has worn contact lenses for several years after first using spectacle lenses. Now in her twenties, she has heard friends talk about a new procedure called laser refractive surgery that can allow people to see well without corrective lenses. She is intrigued and wants to know more about it.

How can bombarding her eye with a laser beam improve her vision? She knows that lasers can be dangerous in other situations. Is this procedure safe? How does it work, and can it help her situation?

The Analysis. In our culture, myopia or nearsightedness is the most common visual problem. It may develop from doing a lot of near work such as reading during childhood, although there are also hereditary factors. As is described in figure 17.26, the lens system of the myopic eye is too strong, which causes light from distant objects to focus in front of the retina rather than on the retina.

Most of the optical power of the eye is produced by the front surface of the cornea. Optical power is measured in diopters, which is the reciprocal of the focal length measured in meters ($P = 1/f$) when the lens is surrounded by air. The shorter the focal length, the stronger the optical power because a short focal length implies that the light rays are being strongly bent by the lens. The overall power of the lens system of the eye is about 60 diopters, but the front surface of the cornea produces 40 to 50 diopters by itself.

The optical power of the cornea (or of any lens) is determined by two things—how strongly the surface is curved and the difference in index of refraction on either side of the surface. For a nearsighted person, the surface of the cornea is too strongly curved for the length of the eyeball. It is not unusual for a person like Megan to have an optical power of the cornea that is too strong by 4 to 5 diopters. She then requires a corrective lens of −4 to −5 diopters to allow her to see distant objects clearly.

The purpose of laser refractive surgery is to reshape the cornea by vaporizing different portions of the cornea by different amounts. The most commonly used procedure is called LASIK, which is an acronym for *laser assisted in situ keratomileusis*. In this procedure, the surgeon cuts a circular flap of the outer layer of the cornea with a surgical scalpel and pulls this flap to the side as shown in the drawing. She then uses a pulsed *excimer laser* to vaporize small amounts of corneal tissue to produce a predetermined new shape for the central portion of the cornea. When finished, the flap of the outer layer is replaced.

The excimer laser used has a wavelength of 192 nm, which lies in the ultraviolet portion of the spectrum. This wavelength

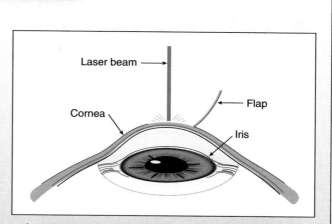

In the LASIK procedure a circular flap of the outer layer of the cornea is pulled aside. Controlled pulses from the laser reshape the central region of the cornea.

is strongly absorbed by corneal tissue, so it vaporizes or *ablates* this tissue without heating the surrounding tissue. The laser operates in a pulsed mode, with each pulse delivering a definite amount of energy. The surgeon can then control how much tissue is ablated by the number of pulses that are delivered to each section of the cornea. This is all controlled by a computer program to achieve the desired new shape.

The LASIK procedure is done on an outpatient basis, and the cornea heals in just a few days. When successful, the reshaped cornea generally allows a person to discard their glasses or contact lenses. Sometimes a weak correction is still needed because the cornea does not heal to quite the desired power. Older people who have lost the ability to accommodate will generally still need reading glasses unless one eye is shaped to have a stronger power than the other. The LASIK procedure is most commonly used to cure myopia where the goal is to flatten the shape of the cornea. It can also be used, though, for farsightedness (hyperopia) or astigmatism. In the case of astigmatism, the cornea is not spherical and this can also be addressed by reshaping with the laser.

Is the procedure safe? The jury is still out on possible long-term effects, but most patients experience only minor problems, if any. There is always a small risk of infection or poor healing, as with any surgical procedure. People sometimes experience problems with night vision after undergoing LASIK. This is because only the central portion of the cornea is reshaped so there is then a circular boundary between the reshaped and untreated portions of the cornea. At night when light levels are low, the pupil of the eye opens more widely and some light may get through this boundary region producing blurring of the image.

constructed from two positive lenses. How does a tele-scope differ from a microscope in its design and function?

Distant objects, such as stars, are very large but so far away that they appear to be tiny. One obvious difference between the uses of a microscope and a telescope is that objects viewed with a telescope are much farther from the objective lens. As shown in figure 17.29, the objective lens of a telescope, like a microscope, forms a real image of the object, which is then viewed through the eyepiece. Unlike the microscope, however, the real image formed by a telescope is reduced in size rather than magnified.

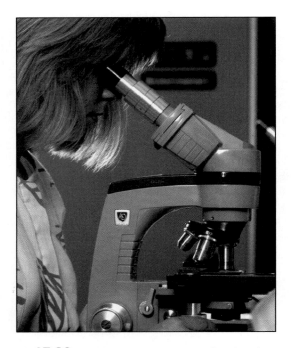

figure 17.28 A laboratory microscope often has three or four objective lenses mounted on a rotatable turret. Light passes through the object slide from a source below the slide.

If the real image formed by the objective lens is smaller than the object, how can there be an advantage to using a telescope? The answer is that this image is much closer to the eye than the original object. Even though this image is smaller than the object itself, it forms a larger image on the retina of the eye when viewed through the eyepiece. Figure 17.30 shows two objects of equal height at different distances from the eye. By assuming that the images of both objects are focused on the retina and tracing just the central undeviated ray from the top of each object, we see that the nearer object forms a larger image on the retina.

When you want to see fine detail on an object, you bring the object closer to your eye to take advantage of the larger image formed on the retina. Since the size of the image on the retina is proportional to the angle that the object forms at the eye, we say that we have achieved an *angular magnification* by bringing the object nearer. We are limited in how close we can bring the object by the focusing power of the eye. The eyepiece of either a telescope or microscope solves this problem by forming a virtual image farther from the eye but at the same angle as the original real image.

The magnifying effect of a telescope is basically an angular magnification. Because it is closer to the eye, the image seen through the telescope forms a larger angle at the eye than the original object. This larger angle produces a larger image on the retina and allows us to see more detail on the object, even though the real image is much smaller than the actual object.

The overall angular magnification produced by a telescope is equal to the ratio of the focal lengths of the two lenses,

$$M = (-)\frac{f_o}{f_e},$$

where f_o is the focal length of the objective lens, f_e is the focal length of the eyepiece lens, and M is the angular magnification. A minus sign is sometimes included in this

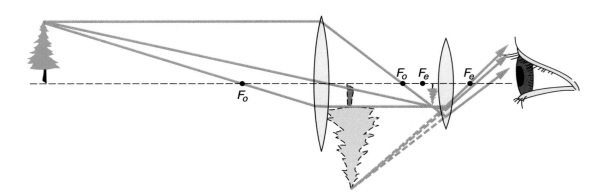

figure 17.29 The objective lens of a telescope forms a real, reduced image of the object, which is then viewed through the eyepiece. The real image is much closer to the eye than the original object. (Not drawn to scale.)

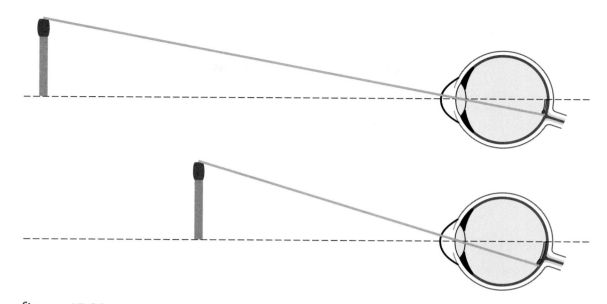

figure **17.30** Two objects of the same size but at different distances from the eye form different-sized images on the retina. We can see more detail on the nearer object.

figure **17.31** Opera glasses and prism binoculars employ different means of reinverting the image so that it is upright.

ing, for the incoming light. Since it is easier to make and physically support large mirrors than large lenses, concave mirrors are used in the telescopes at most observatories.

Binoculars and opera glasses

The image formed by an astronomical telescope is inverted, like the image formed by a microscope. The inverted image is not a big problem for viewing stars or planets, but it can be confusing for viewing objects on land. The most familiar form of land or *terrestrial* telescope is a pair of prism binoculars, which use multiple reflections in the prisms to reinvert the image (fig. 17.31).

Opera glasses are a simpler form of terrestrial telescope. The two tubes are straight, and the image is re-inverted by using negative instead of positive lenses for the eyepieces. Using negative lenses has the additional advantage of making the tubes shorter because the negative lenses must be placed in front of where the real image would be formed. The disadvantage of opera glasses is their narrow field of view and weak magnification. They fit into a purse or pocket more readily than prism binoculars, though.

The two tubes in binoculars and opera glasses allow us to use both eyes when viewing distant objects. Using both eyes preserves some of the three-dimensional aspects of what we see. In normal vision, your two eyes form slightly different images of what you are viewing, because each eye sees objects from a slightly different angle. Your brain interprets these differences as being produced by three-dimensional features of the scene. Try closing one eye when you are viewing near objects, and then reopen that eye.

relationship to indicate that the image is inverted. From this relationship, we see that it is desirable to have a large focal length for the objective lens of a telescope to produce a large angular magnification. A microscope, on the other hand, uses an objective lens with a very short focal length. This is the fundamental difference in the design of telescopes and microscopes.

The large telescopes used in astronomy have concave mirrors instead of lenses for the objective lens. The objects astronomers study are often very dim, and the telescope must collect as much of their light as possible. This requires an objective lens or mirror with a large *aperture,* or open-

Can you see the difference? A person with just one functional eye sees a flatter world at first, although the brain can make use of head movements and other cues for judging distances.

Our eyes are similar to cameras. They use positive lenses to focus an inverted image on the retina or film. If the point of focus does not lie on the retina, we need corrective lenses. Negative lenses are used to correct nearsight-

edness and positive lenses to correct farsightedness. A microscope uses a combination of positive lenses to produce a magnified virtual image. The overall magnification is the product of the magnifications produced by each lens. A telescope produces an angular magnification of distant objects by bringing the image that we view closer to our eyes. Binoculars and opera glasses are terrestrial telescopes that reinvert the image and allow us to use both eyes.

---

summary

For many purposes, the propagation of light can be studied using rays drawn perpendicular to the wavefronts. The laws of reflection and refraction are the basic principles governing these rays. Using these ideas, we can explain how images are formed by mirrors and lenses and how these elements can be combined to make optical instruments.

1   Reflection and image formation. The law of reflection states that the angle that the reflected ray makes to an axis drawn perpendicular to the surface equals the angle made by the incident ray. The image formed by a plane mirror is the same distance behind the mirror as the object is from the front of the mirror. Light rays appear to diverge from this image.

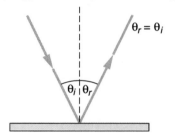

$$\theta_r = \theta_i$$

2   Refraction of light. A light ray passing into glass or water from air is bent toward the axis by an amount that depends on the index of refraction n. Because of this bending, the image of an underwater object seems to lie closer to the surface than it actually does. The index of refraction depends on the wavelength of the light causing *dispersion* or different amounts of bending for different colors.

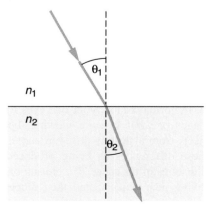

3   Lenses and image formation. Lenses can focus light rays to form either real or virtual images. A convex or *positive* lens converges light rays and can be used as a magnifying glass. A concave or *negative* lens diverges light rays and forms reduced images. Image positions can be predicted by ray tracing or by using the object-image equation.

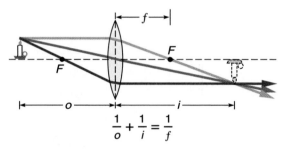

$$\frac{1}{o} + \frac{1}{i} = \frac{1}{f}$$

4   Focusing light with curved mirrors. A mirror with a spherical curved surface can focus light so that incoming parallel rays pass through or appear to come from a single focal point. A concave mirror can form real images or magnified virtual images depending upon the object position. A convex mirror forms reduced virtual images with a wide angle of view.

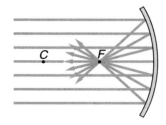

5   Eyeglasses, microscopes, and telescopes. Lenses can be used to correct vision problems and can also be combined to make optical instruments. Negative lenses are prescribed for nearsightedness and positive lenses for farsightedness. A microscope forms a magnified real image of the object with the objective lens. This real image is then magnified again when viewed through the eyepiece. A telescope produces an angular magnification by forming an image of a distant object that is much nearer to the eye than the original object.

key terms

questions

*Questions identified with an asterisk are more open-ended than the others. They call for lengthier responses and are more suitable for group discussion.

Q1. Does either the velocity or the speed of light change when a beam of light is reflected from a mirror? Explain.

Q2. Does light actually pass through the position of the image formed by a plane mirror? Explain.

Q3. How can an image lie behind a mirror hanging on a wall when no light can reach that point? Explain.

*Q4. When you view your image in a plane mirror, your right hand appears to be your left hand and vice versa. Explain how this reversal occurs.

Q5. Can a plane mirror focus light rays to a point like a positive lens does? Explain.

Q6. If you want to view your full height in a plane mirror, must the mirror be as tall as you are? Explain using a ray diagram.

Q7. Objects A, B, and C lie in the next room hidden from direct view of the person shown in the diagram. A plane mirror is placed on the wall of the passageway between the two rooms as shown. Which of the objects will the person be able to see in the mirror? Explain using a ray diagram.

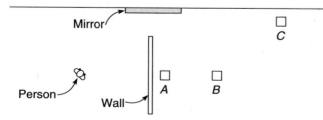

*Q8. When two plane mirrors are joined at right angles to one another, three images of an object can be seen. The image of the object formed by each mirror can serve as object for the other mirror. Where is the third image located? Explain using a ray diagram.

Q9. A light ray traveling in water ($n = 1.33$) passes from the water into a rectangular piece of glass ($n = 1.5$). Is the light ray bent toward the surface normal (the axis drawn perpendicular to the surface) of the glass or away from that axis? Explain.

Q10. Does either the speed or the velocity of light change when light passes from air into a glass block? Explain.

Q11. When we view an underwater object, is the image we see a real image or a virtual image? Explain.

Q12. A fish swimming in a pond looks up at an object lying a couple of feet above the surface of the water. Does this object appear to the fish to lie nearer to the surface or farther from the surface than its actual distance? Explain.

Q13. A light ray traveling in glass for which the critical angle is 42° strikes a surface between the glass and air at an angle of 45° to the surface normal. Is this ray refracted into the air at this surface? Explain.

Q14. Do light waves of different colors all travel at the same speed in glass? Explain.

Q15. Is reflection or refraction responsible for the separation of colors in a rainbow? Explain.

Q16. Can we see a rainbow looking eastward if it is raining in the early morning? Explain.

Q17. Is it possible to form a real image with a positive (converging) lens? Explain.

Q18. An object is located at a distance twice the focal length from a positive lens. Trace three rays from the top of the object to locate the image. Is the image real or virtual, erect or inverted?

Q19. An object is located at the left-side focal point of a negative lens. Trace three rays from the top of the object to locate the image. Is the image real or virtual, erect or inverted?

*Q20. Is there any position that an object could be placed in front of a negative (diverging) lens that will result in the formation of a real image? Explain.

Q21. Suppose that light rays approach a negative lens so that they are converging toward the focal point on the far side of the lens. Will these rays be diverging when they leave the lens? Explain.

Q22. Do rays traveling parallel to the axis of a concave mirror pass through the center of curvature of the mirror after they are reflected? Explain.

Q23. An object is located at the center of curvature of a concave mirror. Trace two rays from the top of the object to locate the image formed by the mirror. Is the image real or virtual, upright or inverted? Explain.

Q24. An object is located inside the focal point of a concave mirror. Will the image of the object be nearer or farther from the observer than the object itself? Explain.

Q25. Is there any distance at which an object can be located in front of a convex mirror that will produce a real image? Explain.

Q26. Why would you use a convex mirror rather than a concave or plane mirror for viewing activities in a store aisle? Explain.

*Q27. When a convex mirror is used as a side-view mirror on an automobile, where is the image located? Why does printing on the mirror warn you that vehicles may be closer than they appear to be when viewed in the mirror? Explain.

Q28. Does a nearsighted person have trouble seeing near objects? Explain.

Q29. Would you use a positive lens or a negative lens to correct the vision of a farsighted person? Explain.

Q30. Does each of the two lenses used in a microscope produce a magnification of the object being viewed? Explain.

Q31. Does each of the two lenses used in a telescope produce a magnification of the object being viewed? Explain.

Q32. Is it possible to produce an angular magnification of an object by simply bringing the object closer to your eye? Explain.

Q33. Is the objective lens of a microscope likely to have a longer focal length than that of the objective lens of a telescope? Explain.

Q34. What advantages might there be to using binoculars rather than an astronomical telescope for viewing distant objects on land? Explain.

exercises

E1. A man with a height of 1.8 m stands 3.0 m in front of a plane mirror viewing his image. How tall is the image, and how far from the man is the image located?

E2. A fish lies 60 cm below the surface of a clear pond. If the index of refraction of water is assumed to be 1.33 and that of air is approximately 1, how far below the surface does the fish appear to a person looking down from above?

E3. A rock appears to lie just 24 cm below the surface of a smooth stream when viewed from above the surface of the stream. Using the indices of refraction given in exercise 2, what is the actual distance of the rock below the surface?

E4. An insect is embedded inside a glass block ($n = 1.5$) so that it is located 2.4 cm below a plane surface of the block. How far from this surface does this insect appear to a person looking at the block?

E5. A positive lens has a focal length of 6 cm. An object is located 24 cm from the lens.
a. How far from the lens is the image?
b. Is the image real or virtual, erect or inverted?
c. Trace three rays from the top of the object to confirm your results.

E6. A positive lens has a focal length of 12 cm. An object is located at a distance of 4 cm from the lens.
a. How far from the lens is the image?
b. Is the image real or virtual, erect or inverted?
c. Trace three rays from the top of the object to confirm your results.

E7. A positive lens forms a real image of an object placed 8 cm to the left of the lens. The real image is found 16 cm to the right of the lens. What is the focal length of the lens?

E8. A negative lens has a focal length of -10 cm. An object is located 20 cm from the lens.
a. How far from the lens is the image?
b. Is the image real or virtual, erect or inverted?

E9. A magnifying glass with a focal length of $+4$ cm is placed 2 cm above a page of print.
a. At what distance from the lens is the image of the page?
b. What is the magnification of this image?

E10. A concave mirror has a focal length of 12 cm. An object is located 6 cm from the surface of the mirror.
a. How far from the mirror is the image of this object?
b. Is the image real or virtual, upright or inverted?

E11. A concave mirror has a focal length of 10 cm. An object is located 30 cm from the surface of the mirror.
a. How far from the mirror is the image of this object?
b. Is the image real or virtual, upright or inverted?
c. Trace three rays from the top of the object to confirm your numerical results.

E12. A convex mirror has a focal length of -10 cm. An object is located 20 cm from the surface of the mirror.
a. How far from the mirror is the image of this object?
b. Is the image real or virtual, upright or inverted?
c. Trace three rays from the top of the object to confirm your numerical results.

E13. A convex mirror used in a store aisle has a focal length of -60 cm. A person in the aisle is 3.0 m from the mirror.
a. How far from the mirror is the image of this object?
b. If the person is 1.8 m tall, how tall is the image viewed in the mirror?

E14. The objective lens of a microscope has a focal length of 0.4 cm. An object on the microscope slide is placed at a distance of 0.5 cm from the lens.
   a. At what distance from the lens is the image formed by the objective lens?
   b. What is the magnification of this image?

E15. The objective lens of a telescope has a focal length of 1.0 m. An object is located at a distance of 10 m from the lens.
   a. At what distance from the objective lens is the image formed by this lens?
   b. What is the magnification of this image?

E16. A telescope has an objective lens with a focal length of +40 cm and an eyepiece with a focal length of +2.5 cm. What is the angular magnification produced by this telescope?

E17. A telescope that produces an overall angular magnification of 20× uses an eyepiece lens with a focal length of 2.5 cm. What is the focal length of the objective lens?

challenge problems

CP1. A fish is viewed through the glass wall of a fish tank. The index of refraction of the glass is 1.5 and that of the water in the tank is 1.33. The fish lies a distance of 6 cm behind the glass. Light rays coming from the fish are bent as they pass from the water to the glass and then again as they pass from the glass to air ($n = 1$). The glass is 0.4 cm thick.
   a. Considering just the first interface between the water and the glass, how far behind the glass does the image of the fish lie? (This is an intermediate image formed by bending of light at just the first surface.)
   b. Using this image as the object for the second interface between the glass and air, how far behind the front surface of the glass does this "object" lie?
   c. Considering the bending of light at this second interface between the glass and air, how far behind the front surface of the glass does the fish appear to lie?

CP2. An object is located at the focal point of a positive lens with a focal length of 12 cm.
   a. What is the image distance predicted by the object-image distance formula?
   b. Trace two rays to confirm the conclusion of part a.
   c. Will the image be in focus in this situation? Explain.

CP3. An object with a height of 2.5 cm lies 10 cm in front of a lens with a focal length of 6 cm.
   a. Using the object-image distance formula, calculate the image distance for this object.
   b. What is the magnification of this image?
   c. Trace three rays to confirm your conclusions of parts a and b.
   d. Suppose that this image serves as the object for a second lens that has a focal length of +4 cm. The second lens is placed 6 cm beyond the image serving as its object. Where is the image formed by this second lens, and what is its magnification?
   e. What is the overall magnification produced by this two-lens system?

CP4. An object 2 cm tall is located 30 cm from a concave mirror with a focal length of 15 cm. Since the focal length is half the radius of the curvature, the object is located at the center of curvature of the mirror.
   a. Using the object-image distance formula, find the location of the image.
   b. Calculate the magnification of this image.
   c. Is the image real or virtual, upright or inverted?
   d. Trace two rays from the top of the object to confirm your results.

CP5. Suppose that a microscope has an objective lens with a focal length of 0.8 cm and an eyepiece lens with a focal length of 2.5 cm. The object is located 1.0 cm in front of the objective lens.
   a. Calculate the position of the image formed by the objective lens.
   b. What is the magnification of this image?
   c. If the eyepiece lens is located 2 cm beyond the position of the image formed by the objective lens, where is the image formed by the eyepiece lens? (The image formed by the objective serves as the object for the eyepiece.)
   d. What is the magnification of this image?
   e. What is the overall magnification produced by this two-lens system? (This is found by multiplying the magnifications produced by each lens.*)

---

*This is not the usual way of calculating the magnifying power of a microscope. The standard method compares the size of the angle subtended at the eye with and without the use of the microscope (called angular magnification).

home experiments and observations

HE1. Fill a clear glass almost to the top with water and insert various objects into the water.
   a. Do the objects appear to be shorter than their actual length when viewed from above the glass?
   b. Do the objects appear to be shorter than their actual length when viewed through the sides of the glass? What distortions do you notice when viewing the objects through the sides?

HE2. Locate two small plane mirrors like the ones often carried in a purse. Place the two mirrors next to each other so that they touch along one edge, making an angle of 90° between the two mirrors. Place a small object like a paper clip in front of the two mirrors.
   a. How many images do you see in the two mirrors when the angle between the mirrors is a right angle (90°)?
   b. As you decrease the angle between the two mirrors, describe what happens to the number of images that you can see.
   c. Using the idea that each of the images formed can serve as an object for the other mirror, can you explain your observations? (Ray diagrams may be useful.)

HE3. If you have a magnifying (concave) mirror available, such as a shaving or makeup mirror, try moving the mirror slowly away from your face.
   a. Describe the changes in the image of your face as the mirror is moved away from your face.
   b. The image should become blurred and indistinct when your face is at the focal point of the mirror. Can you estimate the focal length of the mirror by finding the distance from your face where the image disappears?

HE4. The passenger-side side-view mirror on most cars is a convex mirror. (The warning that objects may be closer than they appear indicates that the mirror is convex.)
   a. View some object of known height in the mirror. A friend will serve nicely. Estimate the height of the image viewed in the mirror. What is the approximate magnification produced by the mirror?
   b. Using your binocular depth perception, estimate the distance behind the mirror that the image is located. (You first have to convince your brain that the image is behind the mirror.) Estimate also the distance of the object from the mirror. Using these values, calculate the focal length of the mirror. (It should be negative.)

HE5. If you have access to an overhead projector, examine the device carefully so that you can describe the optical system involved.
   a. What optical elements (lenses or mirrors) are present?
   b. What is the function of each of these elements? (Holding a white card or stiff paper at various places between the elements when the projector is in use may help you analyze their function.)
   c. Can you produce a ray diagram showing how the rays coming from the object (the transparency being viewed) converge or diverge on their way to the screen?

unit Five

The Atom and Its Nucleus

The idea that matter is made up of tiny particles called atoms has a long history dating at least to the early Greeks, a few hundred years before the birth of Christ. We knew virtually nothing about the structure of atoms, though, until the early part of the twentieth century. In fact, just before the turn of that century, physicists debated whether atoms existed at all—or were merely a convenient fiction used mainly by chemists. At that time, the evidence for the existence of atoms was not overwhelming.

From about 1895 to 1930, a series of discoveries and theoretical developments revolutionized our view of the nature of the atom. We went from knowing almost nothing about the structure of atoms (and even questioning their existence) to a firmly based theory of their structure capable of explaining an enormous range of physical and chemical phenomena. This revolution is surely one of the greatest achievements of the human intellect, with wide-ranging implications for our economy and technology. Its story deserves to be understood by more than a small fraction of our population.

The discovery of the electron in 1897, followed by the discovery in 1911 that an atom has a nucleus, were critical breakthroughs that provided building blocks for atomic models. Niels Bohr's model of the atom put some of these pieces together and stimulated research that led to the more complete and highly successful theory that we now call *quantum mechanics.* Quantum mechanics serves as the basis for most work in theoretical physics and chemistry; its detailed predictions about the nature of the atom have spawned many advances in science and technology.

The nucleus of the atom, that tiny center containing all of the positive charge and most of the atom's mass, has also been found to have an underlying structure. Advances in nuclear physics, discussed in chapter 19, led to the invention of nuclear reactors and nuclear weapons, which thrust physics into world politics. The story of the development of the atomic bomb during World War II is a fascinating mixture of human ingenuity and conflict. Science and world politics have both been irrevocably changed in the process.

The twentieth century has seen a revolution in our understanding of the atom and in the role of science in modern life. This revolution may be a mixed blessing, but it cannot be ignored: it involves chemistry and molecular biology as well as physics. Chapters 18 and 19 look at how this revolution began—where it will lead is an open question.

The Structure
of the Atom

chapter

18

chapter overview

Our principal goals in this chapter are to investigate some of the evidence for the existence of atoms and to describe several discoveries that led to an understanding of the structure of atoms. We will begin with evidence from chemistry and proceed to the discoveries of the electron, X rays, natural radioactivity, the nucleus of the atom, and atomic spectra. We will then discuss the Bohr model of the atom and its relationship to the modern view given by the theory of quantum mechanics.

chapter outline

1 **The existence of atoms: evidence from chemistry.** What information does the study of chemical reactions offer about the existence and nature of atoms? How was the periodic table of the elements developed?

2 **Cathode rays, electrons, and X rays.** How are cathode rays produced, and what are they? How did the study of cathode rays lead to the discovery of the electron and of X rays?

3 **Radioactivity and the discovery of the nucleus.** How was natural radioactivity discovered, and what is it? What role did natural radioactivity play in the discovery of the nucleus of the atom?

4 **Atomic spectra and the Bohr model of the atom.** What are atomic spectra? What role did they play in understanding atomic structure? What are the basic features of Bohr's model of the atom?

5 **Particle waves and quantum mechanics.** What were the limitations of the Bohr model, and how does the theory of quantum mechanics address these problems? What do we mean when we say that particles have wave properties?

Have you ever seen an atom? You have certainly heard people talk about atoms and have probably seen pictures of atomic models like the one in figure 18.1, but do you know why we think atoms exist? Perhaps the question that should be posed is, Do you believe in atoms? And, if so, why?

Most of us have accepted the existence of atoms based on the pronouncements of textbooks or teachers dating back to our elementary-school days. You may be shocked to learn that many of those teachers never seriously questioned why they believed in atoms or understood where our evidence for the existence of atoms originated. Why, then, should you believe in atoms or in descriptions of their structure?

Although we cannot see atoms directly and we may not recognize them as part of our everyday experience, atomic phenomena are evident in our everyday world. The operation of a television set, chemical changes that occur in our bodies, the use of diagnostic X rays, and many other common phenomena can all be understood by relying on our modern knowledge of atomic behavior.

Most important, we will consider the question of why we believe in the existence of atoms and in our models of atomic structure. How have these ideas developed? Understanding how our knowledge of the atom originated can make atoms themselves seem more real.

figure **18.1**   A stylized atom. Does an atom really look like this?

18.1 The Existence of Atoms: Evidence from Chemistry

Why should we believe in the existence of things that we have never personally seen? Why did many nineteenth-century scientists talk with confidence of the existence and properties of the atoms of different substances when they knew nothing about their actual structure? What in our everyday experience may cause us to believe in atoms?

Much of modern science involves things that we cannot see directly. We infer their existence from observations that, taken together, provide convincing evidence of their behavior and characteristics. In the case of atoms, much early evidence came from the study of chemistry. Chemical processes are very common in our daily life, although we may not give them much thought. If the concept of atoms were not already available, you might have to invent the idea just to explain these phenomena.

What did early studies in chemistry reveal about atoms?

Chemistry is the study of the differences in substances and how they can be combined to form still other substances. During early Greek civilization, philosophers tried to identify the elementary substances from which all things are made. Fire, earth, water, and air were the early candidates. Clearly, those choices required some refinement. Earth, in particular, was capable of taking on many forms.

figure **18.2**   A tablet of food coloring is dropped into a glass of water. How might we explain what happens?

One of the most striking demonstrations in elementary chemistry involves taking a small tablet of dye or food coloring and dropping it into a glass of water (fig. 18.2). Rather quickly the color diffuses, until the originally clear water becomes a uniformly colored fluid. Evidently, a change has taken place, but how might we explain that change?

Even without instruction in the language of chemistry, we might find ourselves thinking that tiny particles of the dye migrate in the water and move between particles of the water through spaces that are not apparent. Similar explanations might account for the disappearance of sugar or salt when placed in water or other fluids.

We also know that almost any solid substance can be crushed into a fine powder. By subjecting that powder to heat (or fire), we often can form a solid again, although perhaps modified compared to the original substance. If the powder is combined with some other powdered substance and heated, the resulting product can be quite different from either of the original substances. Baking is a familiar example of that process, but experiments in refining and modifying metals may have been the most important examples in early chemistry. Alchemists were enticed by the elusive prospect of making gold from more common metals.

Is it possible to reduce something into a powder of ever-smaller particles? Early scientists were tempted to think that this was not the case. Since certain elementary substances always seemed to be retrievable from their experiments, they assumed that irreducible particles of these substances retained their form. The notion that each of the elementary substances or **elements** was made up of tiny particles or **atoms** was an attractive model for explaining chemical phenomena. Such atoms might then combine with atoms of other elements to form different substances but could always be retrieved from these substances by sufficient heating or other processes.

Systematic study of how elements combined with each other revealed regularities and rules that formed the basis of early chemical knowledge. Clearly, certain elements were more alike in their properties and reactions than others. Elements could thus be grouped or classified, which further suggested that the atoms of these elements must have similarities in structure. The details of that structure, and even the size of the atoms, were completely unknown and seemingly inaccessible.

Is mass conserved in chemical reactions?

The birth of modern chemistry is often dated to the work of the French scientist, Antoine Lavoisier (1743–1794). Lavoisier discovered that the total mass of chemical reactants and products is conserved in chemical reactions. This discovery established the importance of weighing the reactants and products, which has since become a routine procedure in most chemical experiments.

Although the idea that mass is conserved in chemical changes might seem self-evident now, conservation of mass was not at all clear in Lavoisier's time. The reason was simple: most chemical experiments were performed in open air, and oxygen and other gases in the air were involved in the reactions. Since the quantities of these reacting gases were not recognized and measured, the masses

figure 18.3   Burning wood is a chemical reaction. What substances are reacting? What is produced?

of the solid or liquid substances in the reactions did not seem to be conserved. Air itself is not a simple substance, a fact only beginning to be understood at that time.

One of the most common of all chemical reactions is the burning or **combustion** of carbon compounds such as wood or coal (fig. 18.3). That reaction combines oxygen from the air with the carbon in the coal or wood to form carbon dioxide (a gas) and water vapor (also a gas). If we are not aware of the role of the gases, we can be easily misled into thinking that we have lost some mass during combustion.

Lavoisier performed a series of experiments in which he carefully controlled and weighed the quantities of gas that participated either as reactants or products. The results of these experiments showed clearly that the total mass of the products was equal to the total initial mass of the reactants—no mass was lost or gained. In the process, he was able to distinguish oxygen (or "highly respirable air") from carbon dioxide and water vapor and to provide the first accurate description of combustion reactions. The results of this work were published in 1789 as the *Traité Elémentaire de Chimie*.

How did the concept of atomic weight emerge?

Although Lavoisier's brilliant career was tragically cut short (literally) when he lost his head to the guillotine in the French Revolution, his discoveries were soon followed by another important insight from the English chemist, John Dalton (1766–1844). Dalton tried to make sense of the regularities observed in the ratios of weights of chemical reactants and products. Dalton's ideas depended heavily on the experimental work of other chemists, who were engaged in the new practice established by Lavoisier of carefully weighing all reactants and products.

Dalton was intrigued by the fact that when a chemical reaction took place, the reactants always seemed to combine in the same proportions by mass. For example, when carbon combined with oxygen to form the gas now called carbon dioxide, the ratio of the mass of oxygen to the mass of carbon that would react was always the same: 8 to 3. In other words, if 3 grams of carbon were involved, 8 grams of oxygen were required to complete the reaction, no more and no less. If more than 8 grams of oxygen were present, some would be left over; if less than 8 grams were present, some carbon would be left over.

This idea of specific mass ratios held for other reactions, although with different ratios for each reaction. When hydrogen combines with oxygen to form water, the mass ratio is 8 to 1. That is, 8 grams of oxygen are required to react completely with 1 gram of hydrogen. Each reaction required a specific proportion to react completely. This observation is often referred to as **Dalton's law of definite proportions.**

Dalton recognized that a model using atoms explained these observations. Dalton thought that each element was made up of tiny atoms that were all identical in mass and form, but that different elements had different atomic masses. A chemical compound might then be the combination of a few atoms of one element with a few atoms of another to form **molecules**—several atoms of different elements bound together somehow. The characteristic masses of the atoms were responsible for the regular ratios of mass observed in the reactions.

This idea is illustrated in figure 18.4. Suppose (as we now know) that the chemical compound of water is formed by two atoms of hydrogen combining with one atom of oxygen to form a water molecule (H_2O in modern notation). If the mass of an oxygen atom is 16 times the mass of a hydrogen atom, this would account for the observed proportion of oxygen to hydrogen of 8 to 1 (16 to 2), since each water molecule has two atoms of hydrogen for every one of oxygen.

The 8-to-3 proportion for the carbon dioxide reaction can likewise be explained if the carbon atom has a mass 12 times that of hydrogen, and two atoms of oxygen (each one 16 times the mass of hydrogen) combine with one atom of carbon to form a carbon dioxide molecule (CO_2). The mass ratio would then be 32 parts of oxygen (2×16) to 12 parts of carbon, or 8 to 3 (dividing each number by 4). One reaction by itself is not enough to establish the relative atomic masses, but the study of several reactions provides a consistent picture, which Dalton showed in his treatise, *A New System of Chemical Philosophy,* published in 1808.

Dalton's atomic hypothesis did not establish the actual mass of individual atoms—he still did not know how large

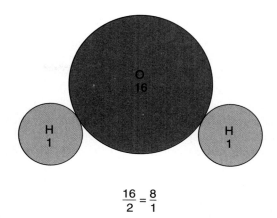

$$\frac{16}{2} = \frac{8}{1}$$

figure 18.4   Two atoms of hydrogen combine with one of oxygen to form water (H_2O). The mass ratio is 8 to 1 because the atomic mass of oxygen is 16 times the mass of hydrogen.

one atom was. Dalton's hypothesis is a means of determining the masses of the atoms of one element relative to those of another, a task that occupied chemists through much of the remainder of the nineteenth century. Like any good theory, Dalton's model was a productive guide for further chemical research. It also presented the concept of atoms in a new and more detailed light: some atoms were heavier than others, and **atomic mass** was a property of an element.

Table 18.1 compares the masses of the atoms of several common elements. Traditionally referred to as *atomic weight,* atomic mass is a more appropriate term, given our distinction between mass and weight. A complete list is found in the periodic table in the inside back cover of this text. Note that many of the relative masses are approximately whole numbers, but others are not. Therein lay another intriguing mystery and a clue to the structure of the atom.

table 18.1		
Atomic Masses of Some Common Elements Compared		
Element	Chemical symbol	Atomic mass
hydrogen	H	1.01
helium	He	4.00
carbon	C	12.01
nitrogen	N	14.01
oxygen	O	16.00
sodium	Na	22.99
chlorine	Cl	35.45
iron	Fe	55.85
lead	Pb	207.2

How was the periodic table developed?

As chemists gathered more and more information on the atomic masses and chemical properties of the various elements, some other interesting regularities began to emerge. It had been known for some time that families of elements displayed similar chemical properties. Chlorine (Cl), fluorine (F), and bromine (Br) (called *halogens*), for example, formed similar compounds when combined with highly reactive metals such as sodium (Na), potassium (K), or lithium (Li) (*alkali metals*). The atomic weights of the elements within any given family, however, were very different.

When all of the elements were listed in order of increasing atomic weight, the members of a family seemed to pop up at more-or-less regular intervals in the list, particularly the lighter elements. Although others had tried to make sense of these regularities, the person who succeeded in producing the most useful organization was the Russian chemist, Dmitri Mendeleev (1834–1907). Mendeleev's scheme, first published in 1869, is now called the **periodic table of the elements.**

To understand Mendeleev's table, imagine listing all of the known elements by increasing atomic weight on a long strip of paper. Then, to make the table, we cut the strip at various points and lay the strips out in rows. We begin by cutting at every place that we encounter an alkali metal in the list. We line these strips up in our table so that the alkali metals all lie in a column on the left side of the table (fig. 18.5).

To get the halogens to line up, we need to cut the remaining strips again, somewhere near their midpoints, because there are more elements in some rows than in others. The halogens are also arranged above one another in a column near the right side of the table. In Mendeleev's original table, the halogens formed the column on the far right, because the noble gas elements (helium (He), neon (Ne), Argon (Ar), krypton (Kr), xenon (Xe), and radon (Rn)) had not yet been discovered. In the finished table, elements with common chemical properties line up in columns above one another, but the order of atomic weights is preserved in the rows and throughout the table. (See the complete periodic table on the inside back cover.)

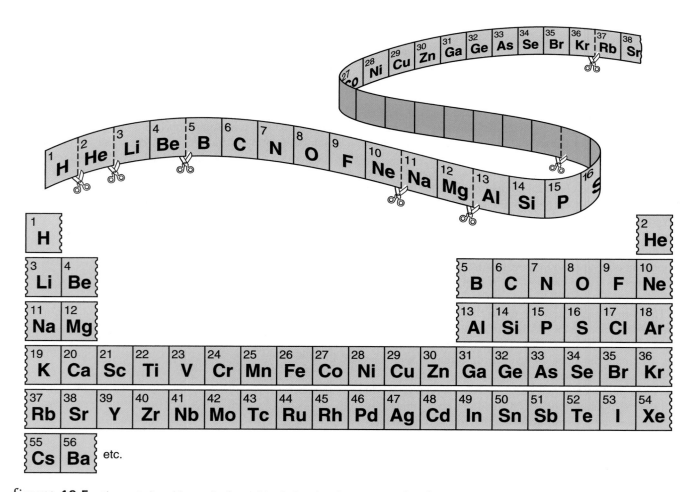

figure **18.5**   The periodic table can be formed by listing the elements in order of increasing atomic weight and then cutting the list at certain points. Elements with similar chemical properties are then aligned in columns.

Although the periodic table was an intriguing way of organizing knowledge of the chemical elements, it raised more questions than it answered. Atoms of the elements in a given column must somehow have similar properties, but chemists still knew virtually nothing about the structure of atoms. They were driven to drawing little hooks and rings on their atoms as they attempted to explain how atoms combined, but they knew that these pictures were unlikely to be accurate. A body of knowledge was building that called for explanation, but the explanation did not come until the early part of the twentieth century.

From many observations on how substances combine to form other substances, scientists speculated that each different substance consisted of tiny particles, or atoms, with properties characteristic of that substance. Lavoisier's discovery of the conservation of mass in chemical reactions established the importance of weighing the reactants and products. Dalton's law of definite proportions introduced the property of atomic weight for atoms of different substances. The periodic table of the elements was developed by Mendeleev in the 1860s. He listed the elements in rows of increasing atomic weight and then divided the list into rows so that elements with similar properties sat above one another in the columns. These regularities suggested recurring similarities in atomic structure.

18.2 Cathode Rays, Electrons, and X Rays

By the end of the nineteenth century, chemists were quite comfortable with the concept of atoms and knew a good deal about their relative masses and properties, if not their actual structure. Physicists, on the other hand, were less convinced. Many physicists were not aware of the details of the chemical evidence, and some even denied that atoms existed.

Near the end of the nineteenth century, several discoveries were made in physics that would prove crucial to understanding atomic structure. This part of the story begins with the study of *cathode rays,* the focus of much curiosity and research in the latter half of the century.

How are cathode rays produced?

You use **cathode rays** almost every day, although you may not be aware of it. The heart of most television sets, its picture tube, is a cathode-ray tube (or CRT, as it is known in the electronics industry). The discovery of cathode rays resulted from merging two different technologies: the ability to produce good vacuums with improved vacuum pumps and the growing understanding of electrical phenomena. (The cathode-ray tube used in a television set is described in Everyday Phenomenon Box 18.1.)

Johann Hittorf (1824–1914) was one of the first to observe cathode rays. In a paper published in 1869, he described in detail what happens when a high voltage is placed across two electrodes sealed in a glass tube connected to a vacuum pump (fig. 18.6). As the air is pumped out of the tube, a colorful glow first appears in the gas near the *cathode,* the negative electrode. As the gas pressure in the tube is reduced, the glow spreads through the entire volume between the two electrodes. The colors of this glow discharge depend on the kind of gas originally in the tube.

As the tube is evacuated to still lower pressures, the glow discharge disappears. A dark region starts to form near the cathode and then moves across the tube toward the *anode,* the positive electrode, as the pressure is further reduced. When the dark region has moved completely across the tube, a new phenomenon appears. Instead of the gas glowing, there is now a faint glow on the glass wall of the tube opposite the cathode.

Since the darkening began near the cathode and spread across the tube, scientists surmised that something emitted from the cathode was responsible for the glow on the opposite wall of the tube. For this reason, the invisible radiation was called *cathode rays.*

One of the simpler experiments that can be performed with cathode rays is deflecting the beam with a magnet. If cathode rays are focused into a narrow beam by appropriate shaping and positioning of the cathode and anode, the beam can be moved around with a magnet. If the north pole of a magnet is brought down from the top, as in figure 18.7, the spot of light created by the beam is deflected to the left on the face of the tube. This result is consistent with the assumption that the cathode rays are negatively charged particles, which you can confirm using the right-hand rule for magnetic forces introduced in chapter 14.

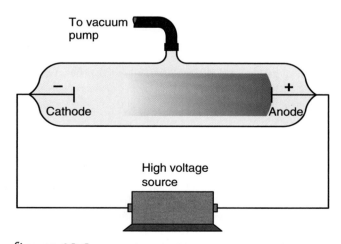

figure **18.6** A simple cathode-ray tube consists of two electrodes sealed in a glass tube. A glow discharge appears in the gas between the electrodes as the tube is evacuated. With further evacuation, the discharge disappears, and a glow appears on the end of the tube opposite the cathode.

Everyday Phenomenon

box 18.1

Electrons and Television

The Situation. Television plays an enormous role in modern life. It is the primary source of entertainment and news for much of our population. For many of us, spending some time watching the dancing light of the TV is an everyday pursuit.

A family performing the common ritual of communing with a television set.

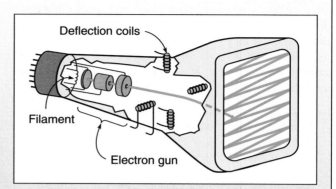

A cut-away view of a modern picture tube shows the electron gun and other electrodes used to deflect the beam to different points on the screen.

How does a television set work? Could you explain its basic principles of operation to a younger brother or sister? How are cathode rays involved in producing the picture that we see?

The Analysis. The heart of most television sets is the cathode-ray tube or CRT. As described earlier, a CRT is an evacuated tube containing electrodes across which we place a high voltage. This produces a beam of electrons in the tube, which create flashes of light when they strike the glass wall of the tube.

The type of cathode-ray tube used in television sets is shown in the drawing. The electrodes that produce and focus the electron beam are all located near the tube socket on the left side of the diagram and are called the *electron gun*. The gun contains a cathode (negatively charged), behind which lies a *filament*. An electric current passes through the filament to heat the cathode, which increases the rate of emission of electrons.

Beyond the heated cathode lies the anode, which is positively charged and has a hole in its center. Electrons are accelerated from the cathode to the anode by the high voltage placed across these two electrodes. Electrons passing through the hole in the anode make up the electron beam. These electrons are focused into a narrow beam by further electrodes located beyond the anode. The filament, cathode, anode, and focusing electrodes together constitute the electron gun.

After leaving the electron gun, the beam of electrons travels across the tube, producing a bright spot of light when it strikes the glass face of the tube. This effect is enhanced by coating the inside of the front surface of the tube with a special *phosphor*, a material that emits light when struck by fast-moving particles. Magnetic coils, usually arranged in a yoke that fits around the tube, are used to deflect the electron beam so that it strikes different points on the face of the tube at different times.

The electron beam can be moved quickly from one point to another on the face of the tube, and the intensity of the beam varied to produce degrees of brightness at different points. The pattern of varying brightness of the different spots makes up the picture that we watch. Usually, the beam moves in a zigzag scan pattern back and forth across the face of the tube in a fraction of a second. In the system used in the United States, 525 horizontal scans are required to make one picture, and this process is repeated 30 times a second.

The process just described produces a black-and-white picture. To produce a colored picture, three phosphors are used for three different colors. Each spot on the face of the tube is, in fact, three closely spaced spots or lines, and three different electron guns are used, one for each of the colors. Different combinations of these three colors produce the range of color that we see. If you look closely at the face of a color-television picture tube (with the set turned off!), you can detect the pattern of vertical lines containing the three different phosphors.

The information used to produce both the pictures and the sound is carried to the set by electromagnetic waves lying in the shorter-wavelength portion of the radio-wave spectrum. The signals can also be transmitted by cables or reflected from satellites (as microwaves) and picked up by dish antennas at remote locations. The availability of multiple stations and programming and the range of technologies used to record and transmit the signals would have seemed like an absurd dream to people just a hundred years ago when radio waves were first studied.

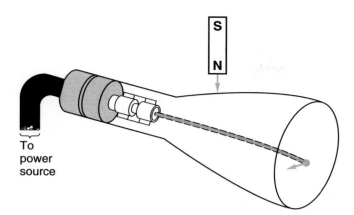

figure **18.7**   If the north pole of a magnet is brought down toward the top of a cathode-ray tube, the spot of light is deflected to the left across the face of the tube.

How was the electron discovered?

The questions about the nature of cathode rays were largely resolved by J.J. Thomson (1856–1940). Thomson performed a series of experiments to measure the masses of the negatively charged particles thought to be present in the cathode-ray beam. In one experiment, Thomson passed the beam through crossed electric and magnetic fields of known strength (fig. 18.8). The combined effect of the electric and magnetic fields on the beam allowed him to estimate the velocity of the particles, since the magnetic force depends on velocity, but the electric force does not.

Knowing the velocity and the extent of deflection produced by the magnetic field alone, he could estimate the mass of the particles. By Newton's second law, the acceleration of the particles by the magnetic force that causes their deflection is inversely proportional to their mass.

Since the magnetic force (**F** = qvB) also depends on the charge of the particles, which was not known, he actually ended up measuring the ratio of the charge to the mass, q/m. He published the results of this work in 1897.

The striking features of Thomson's results were the apparently small mass of these particles and the fact that all of them seemed to have the same ratio of charge to mass, which suggested that the particles were identical. The lightest element in the periodic table is hydrogen. If hydrogen ions and cathode-ray particles had the same charge, the mass of the hydrogen atom was nearly 2000 times larger than the mass of a cathode-ray particle.

Not only were these particles identical for a given cathode, they had the same charge-to-mass ratio even if the cathode was made of a different metal. Thomson checked this result by repeating the experiment with cathodes made of various metals. The same particles seemed to be present in all those that he tested. This fact, together with their small mass, suggested that these particles must be common constituents of different types of atoms.

We now call the negatively charged particles of the cathode-ray beam **electrons,** and Thomson is credited with discovering the electron in these experiments. A cathode-ray beam is a beam of electrons. Each electron is now known to have a mass of 9.1×10^{-31} kg and a charge of -1.6×10^{-19} C. Thomson's discovery provided the first known *subatomic* particle, a particle smaller than the smallest known atom. The electron became the first possible candidate for a building block of atoms.

How were X rays discovered?

The study of cathode rays produced other dividends besides the discovery of the electron. A German physicist, Wilhelm Roentgen (1845–1923), discovered another type of radiation associated with the cathode-ray tube. His discovery

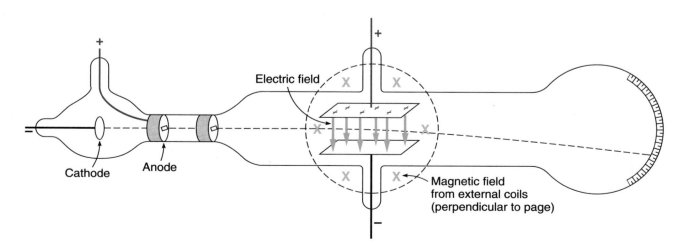

figure **18.8**   Thomson used both electric and magnetic fields to deflect the cathode-ray beam in a tube specially designed to measure the mass of the cathode-ray particles.

created a sensation in the popular press and in the scientific community.

As is often the case, Roentgen made his discovery partly by accident. For reasons that he never made completely clear, he was experimenting with a cathode-ray tube that he had covered with black paper. Nearby on his workbench was a piece of paper coated with a fluorescent material, barium platinocyanide. Roentgen noticed that the paper glowed in the dark when the cathode-ray tube was turned on, even though no light was escaping from the tube (fig. 18.9). The glow stopped when the tube was turned off.

It was already well known that cathode rays could not travel far in air—nor could they travel through the glass walls of the tube. The fluorescence, though, appeared even when the paper was located as far as 2 meters from the tube. The new radiation causing the fluorescence could not be the cathode rays. Not knowing exactly what they were, Roentgen called them **X rays,** because the letter *X* is often used to represent an unknown quantity.

The most striking feature of these X rays was their penetrating power. They apparently passed readily through the glass walls of the tube, and also through other obstacles in their path. In some of his earliest experiments, Roentgen showed that he could produce a shadow of the bones in his hand by placing his hand between the end of the cathode-ray tube and the fluorescent screen (fig. 18.10). He also showed that the X rays were capable of exposing a covered photographic plate. He took pictures of the outlines of brass weights inside a wooden box. Roentgen published the results of his initial experiments with X rays in 1895.

This ability to "see" through objects opaque to visible light excited the imagination of the popular press. Everyone, scientists included, wanted to see an X-ray tube at work. Within a year, doctors were using X rays to take pictures of broken bones and other dense tissue. Unfortunately, they knew little about the hazards of repeated exposure to X rays, so many doctors and dentists suffered from severe radiation effects in the early years of their use.

Roentgen performed an extensive series of experiments with his newly discovered radiation as he tried to ascertain

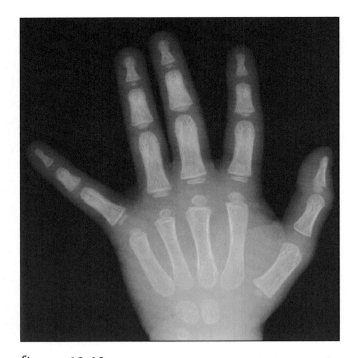

figure **18.10**   Roentgen discovered that a shadowgram of the bones in a hand could be produced by passing X rays through the hand.

what it was. As a result of the work done by Roentgen and other scientists, it was eventually determined that X rays are a form of electromagnetic wave with very short wavelengths and very high frequencies. X rays are produced by collisions of the cathode rays (electrons) with the walls of the cathode-ray tube or with the anode of the tube. The strongest X-ray beams are produced by placing the metal anode at a 45° angle to the electron beam and using high voltages to excite the tube (fig. 18.11).

Although the discovery of X rays was important to medicine, it was also important to physicists, partly because it led directly to the discovery of yet another kind of radiation and thus furthered the exploration of the atom's structure. This new type of radiation, called *natural radioactivity,* actually is three distinct forms of radiation. In section 18.3, we will describe how this discovery provided a powerful probe for getting inside an atom.

Cathode rays are produced by placing a high voltage across two electrodes sealed in an evacuated tube. Experiments with these rays showed that they were comprised of negatively charged particles, all having an identical charge-to-mass ratio. The mass of these particles apparently was smaller than the mass of the smallest atom, and they seemed to be present in different types of metal, which suggested that they could be constituents of all atoms. Study of cathode rays also resulted in the

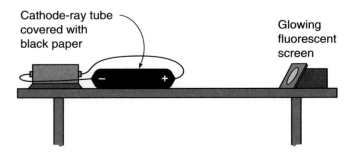

figure **18.9**   Roentgen noticed that a fluorescent material would glow when placed near his covered cathode-ray tube. The glow appeared only when the cathode-ray tube was turned on.

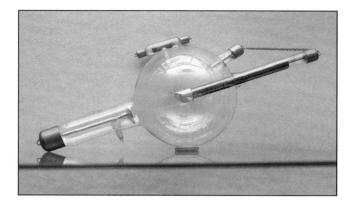

figure **18.11** An X-ray tube in a diagnostic X-ray machine uses an angled anode to project X rays through the side of the tube.

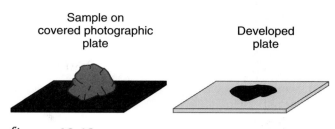

figure **18.12** When Becquerel placed a piece of phosphorescent material on a covered photographic plate, the developed plate showed a silhouette of the sample, indicating that the plate had been exposed by radiation passing through the black paper cover.

discovery of X rays, which are highly penetrating electromagnetic waves with a very short wavelength. This discovery, in turn, led to the discovery of natural radioactivity and the nucleus of the atom.

18.3 Radioactivity and the Discovery of the Nucleus

Most of us have heard of radioactivity and perhaps have come to fear it because of publicity about nuclear power, nuclear weapons, and radon in our homes and other buildings. For most of the time that humans lived on earth, however, we were blissfully unaware of its presence. How was radioactivity discovered, and how did that lead to the discovery of the nucleus of the atom? The fields of atomic and nuclear physics, which did not exist before the beginning of the twentieth century, arose from these events.

How was radioactivity discovered?

A French scientist, Antoine-Henri Becquerel (1852–1908), discovered natural radioactivity in 1896. His experiments were directly motivated by Roentgen's discovery of X rays the previous year. For many years, Becquerel had studied phosphorescent materials, which glow in the dark after being exposed to visible or ultraviolet light. Many of the phosphorescent materials that Becquerel was studying were compounds containing uranium, the heaviest element known at that time.

Becquerel wondered whether penetrating radiation like Roentgen's X rays was emitted by his phosphorescent compounds. He tried a simple experiment in which he exposed some of these compounds to sunlight for a while and then placed them on top of photographic plates wrapped in black paper so that no light could reach them. Sure enough, the photographic plates were exposed near

the pieces of phosphorescent material (fig. 18.12). Radiation apparently was passing from these materials through the black paper to expose the film.

Although an interesting discovery in itself, there was more to come. Further experiments by Becquerel showed that not all phosphorescent materials could expose a photographic plate, only those that contained uranium or thorium. Furthermore, somewhat by accident, Becquerel discovered that it was not necessary to expose these materials to light to produce the effect. Becquerel had prepared samples one day intending to expose them to sunlight. The sun was not shining that day, however, so he put them away in a drawer for a few days, together with the covered photographic plate. When he returned to the project several days later, he decided to develop the plate before proceeding, just to be safe, fully expecting that it would not be exposed. To his great surprise, he discovered that the plate was very heavily exposed near the uranium samples. Apparently, earlier exposure to sunlight (necessary for phosphorescence) was not needed to produce the radiation that was exposing the plates.

Becquerel was even more surprised to discover that the uranium samples retained the ability to expose film indefinitely, even if kept in a dark box or drawer for weeks. The phosphorescent effect, on the other hand, disappeared swiftly (in just a few minutes) after the samples were removed from the source of light. The penetrating radiation coming from his uranium samples did not seem to be connected with the phosphorescence at all.

Becquerel named this new radiation **natural radioactivity**, because it seemed to be produced continuously by compounds containing uranium or thorium without a need for special preparation. Natural radioactivity puzzled physicists of that time, because there was no apparent source of energy to produce the radiation. Where did these rays come from? How could rays continue to be emitted when no energy was being added to the samples in any obvious manner? Was this radiation somehow a property of the atoms themselves?

Is more than one type of radiation involved in radioactivity?

Along with the discovery of X rays, the discovery of natural radioactivity generated much new experimental activity and theoretical speculation. Many scientists were involved, most notably Marie (1867–1935) and Pierre (1859–1906) Curie and a young Australian-born physicist, Ernest Rutherford (1871–1937). The Curies, by painstaking chemical techniques, were able to isolate two more radioactive elements, radium and polonium. Both were contained in samples of uranium and thorium but were much more radioactive than uranium or thorium themselves.

Rutherford became interested in the nature of the radiation. One of his earliest experiments with this new phenomenon showed that at least three kinds of radiation came from the uranium samples. By placing a uranium sample at the base of a hole drilled in a piece of lead, he could produce a beam of radiation. When this beam was passed through a magnetic field produced by a strong magnet, it split into three components, as in figure 18.13.

Rutherford used the first three letters of the Greek alphabet—α (alpha), β (beta), and γ (gamma)—to name these three components. One of the components, alpha, deviated slightly to the left (fig. 18.13). This is the direction we obtain, using the right-hand rule for the magnetic force, for radiation consisting of positively charged particles. The location of the beam could be detected with photographic

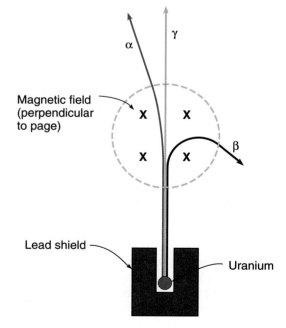

figure **18.13** When the beam of radiation coming from a uranium sample passes through a magnetic field, it splits into three components, which are named—α (alpha), β (beta), and γ (gamma).

film or, more conveniently, with a zinc-sulfide screen, which produces flashes of light when struck by the beam.

The second component of the beam, beta, was bent in the opposite direction (as would be expected for negatively charged particles) and much more strongly than the alpha rays. Further study indicated that these beta rays were electrons, recently discovered by J. J. Thomson. The gamma rays, the third component, were undeviated by the magnetic field. These rays turned out to be a variety of electromagnetic wave similar to X rays but with even shorter wavelengths.

Exactly what the alpha rays were remained a mystery, however, until clarified by an experiment performed by Rutherford and a student assistant, T. D. Royds. That these rays or particles were deviated only slightly by the magnetic field suggested that they were much more massive than the electrons in the beta portion of the beam. They were also the primary component emitted by the radium isolated by Marie and Pierre Curie.

In 1908, Rutherford and Royds established that the alpha rays were helium atoms stripped of their electrons. They determined this by placing a small sample of radium in a very thin-walled tube sealed inside a somewhat larger tube. The alpha particles could escape from the thin-walled tube but not from the larger tube. The larger tube contained electrodes across which a high voltage was introduced to produce a glow discharge in the alpha-particle gas that accumulated. The colors of this discharge were characteristic of helium, which had not been present initially in the tube. (See section 18.4 for a discussion of atomic spectra.)

How was the nucleus of the atom discovered?

Rutherford quickly realized that alpha particles would make effective probes for studying the structure of the atom. Because they were much more massive than electrons, and also highly energetic, it seemed possible to get alpha particles inside the atom. By firing a beam of alpha particles at a thin metal foil and noting what happens to the beam, Rutherford thought he might deduce features of atomic structure. Such an experiment is called a *scattering* experiment.

The basic scheme of Rutherford's scattering experiments is illustrated in figure 18.14. An alpha-emitting substance such as radium or polonium is placed at the bottom of a hole in a lead shield to produce a beam of alpha particles. This beam is directed at a very thin foil of gold or some other metal. The scattering of the alpha particles is then detected by a small hand-held scope with a zinc-sulfide screen at one end and a magnifying eyepiece at the other. The experimenter counts the flashes of light (scintillations) produced by alpha particles striking the screen at various angles from the initial direction of the beam.

The initial results of these experiments were not surprising or informative. Most of the alpha particles went straight

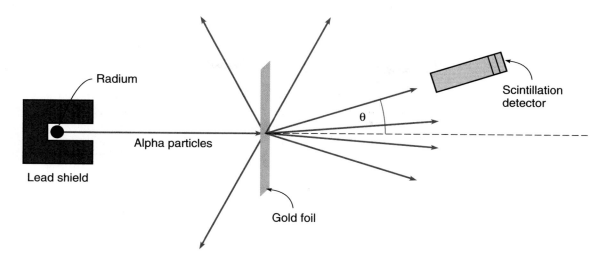

figure **18.14**   A beam of alpha particles was scattered from a thin gold foil in the experiments performed by Rutherford's assistants.

through the gold foil without deviating much. A few were scattered through larger angles, but the number fell off rapidly as the angle from the initial direction of the beam was increased. These results seemed consistent with the prevailing view of the atom at that time: the mass and positive charge of the atom were seen as being distributed uniformly throughout the volume of the atom. Electrons, known to be present in atoms, were thought to be strewn here and there within this volume, much like the raisins in a plum pudding (a British dish). Such an arrangement would not be dense enough to affect a beam of energetic alpha particles.

Just to be sure, however, Rutherford suggested to one of his students, an undergraduate named Ernest Marsden, that he look for scattered alpha particles reflected back from the foil. After a few days in the dark lab squinting at occasional flashes of light through the detecting scope, Marsden reported to Rutherford that a few alpha particles did indeed scatter at these much larger angles. Rutherford was tempted not to believe him, but further checking by Marsden and a more senior research associate, Hans Geiger (1882–1947), verified their presence.

Much later, Rutherford said that the backward scattering was as if someone had fired bullets into a piece of tissue paper and the bullets bounced back. The result was totally unexpected. An analogy often used to explain this scattering experiment is illustrated in figure 18.15. We are trying to learn what is inside a Christmas present without opening the box. We may lift and shake the package to get some sense of its weight and nature. Another (somewhat more destructive) test could be made by firing a rifle at the package and noting what happens to the bullets as they emerge (fig. 18.15). This is a scattering experiment similar to what Rutherford and his assistants performed.

If we had already determined that the package is not heavy, we would be surprised to find some of the bullets,

even just a few, coming back toward us. Somewhere in the package, small but dense objects with enough mass to reverse the momentum of a rapidly moving bullet must be present. Since the package is not heavy and many of the bullets go right through, the objects responsible for the large-angle scattering must be small. Small steel balls held in a light but rigid packing material might do the job.

Similar reasoning applies to the atom. If most of the alpha particles go through, but a few are scattered through large angles, there must be very dense but small centers somewhere within the atoms massive enough to reverse the momentum of rapidly moving alpha particles. To explain the quantitative results of the scattering experiment, Rutherford had to assume that these massive centers were very small indeed. By this time, atoms were known to have a diameter of approximately 10^{-10} m. The diameter of the tiny, but massive centers had be just a ten-thousandth of the diameter of the atom in order to explain the data!

The discovery of the **nucleus** of the atom followed from the analysis of these scattering experiments. The nucleus was presumed to be a very dense center of the atom that contained most of the mass of the atom and all of its positive charge. The rest of the atom consisted of the negatively charged electrons arranged somehow around this center. The electrons were responsible for most of the size of the atom, but for very little of its mass. To get a sense of the scale, imagine that the atom is enlarged to the size of a football field (about 100 m, counting the end zones). The nucleus would be roughly the size of a pea on the 50-yard line.

Rutherford's analysis of the alpha-particle scattering experiments performed by Geiger and Marsden was published in 1911. The idea that the atom has a tiny nucleus containing most of its mass and all of its positive charge presented a radical new view of the atom.

figure **18.15** The contents of a Christmas present could be probed by firing a rifle into it and noting how the bullets are scattered by the contents.

Becquerel discovered that a penetrating radiation, which he called natural radioactivity, was emitted by phosphorescent materials containing uranium or thorium. Rutherford showed that this radiation had three components: alpha (helium ions), beta (electrons), and gamma (short-wavelength X rays). Using the alpha particles as probes in scattering experiments, Rutherford and his assistants learned that the atom must have a tiny, massive center, which we now call the nucleus. This set the stage for the first successful model of the atom.

18.4 Atomic Spectra and the Bohr Model of the Atom

If the atom has a positively charged nucleus, and electrons (with their negative charges) are arranged somehow around this nucleus, it is natural to compare the atom to the solar system. In the solar system, the planets are held in orbit about the sun by the gravitational force, which is proportional to the inverse square of the distance ($1/r^2$) between the planets and the sun (see chapter 5). In an atom, the electrons are attracted to the nucleus by the electrostatic force, which by Coulomb's law is also proportional to the inverse square of the distance. Maybe an atom is like a miniature solar system.

Although the comparison was intriguing, there were some problems. Since an orbiting electron should act like a transmitting antenna and radiate electromagnetic waves, the atom would lose energy, and the electron would spiral into the nucleus, causing the atom to collapse. Physicists were aware, though, that atoms did sometimes emit electromagnetic waves in the form of light. The patterns of the

light emitted by the smallest atom, hydrogen, were particularly interesting because of their simplicity.

Niels Bohr (1885–1962) was working with Rutherford when the nucleus was discovered. Bohr's model of the atom first suggested answers to these problems and provided an explanation of the wavelengths of light (the *spectrum*) emitted by hydrogen. The publication of Bohr's model of the atom in 1913 opened a tremendously exciting period of research that resulted in our current understanding of atomic structure.

What is the nature of the hydrogen spectrum?

The study of the light emitted by different substances began more than fifty years before Bohr's work. If a substance is heated in the flame of a Bunsen burner and the emitted light observed through a prism, each substance produces characteristic colors or wavelengths. These characteristic wavelengths are the **atomic spectrum** of that substance.

For gases, the most convenient way of producing this spectrum was in a gas-discharge tube. (We encountered this phenomenon in section 18.2 when we discussed cathode rays.) When a high voltage is placed across electrodes sealed inside a tube containing a gas at low pressure, a colorful discharge is observed (fig. 18.16). This is what happens in a fluorescent light, too, although the light tube has a fluorescent coating to produce a more uniform distribution of wavelengths.

If you observe the light emitted by a gas discharge through a prism or diffraction grating, you see that the spectrum consists of a series of discrete bright lines at specific wavelengths. (As discussed in section 16.4, diffraction gratings use interference effects to separate wavelengths.)

If the source itself is long and thin (fig. 18.16), or if the light passes through an entrance slit, the separate wavelengths show up as colored lines. Each kind of gas has its own spectrum, which can be used as a reliable means of identifying the substance.

The spectrum of hydrogen is quite simple. The visible portion has just four wavelengths—a red line, a blue line, and two violet lines, one of which is quite difficult to see (fig. 18.17). In 1884, a Swiss teacher, J.J. Balmer (1825–

1898), discovered that the wavelengths of these four lines could all be computed from a simple formula. Balmer's formula is not based on underlying theory but is just a numerical way to compute the wavelengths of the observed lines. When other lines were discovered in the near-ultraviolet portion of the spectrum, they also were correctly described by Balmer's formula.

Somewhat later, other series of spectral lines were discovered for hydrogen in the infrared and ultraviolet regions. All of these lines could be predicted from a generalized form of Balmer's formula published in 1908 by Rydberg and Ritz. This formula is usually written as

$$\frac{1}{\lambda} = R\left(\frac{1}{n^2} - \frac{1}{m^2}\right),$$

where n and m are both integers and R is called the **Rydberg constant**, $R = 1.097 \times 10^7 \, \text{m}^{-1}$. Setting $n = 2$, we get the Balmer series of lines lying in the visible and near ultraviolet. For $n = 1$, we get a series in the ultraviolet portion of the spectrum, and for $n = 3$ or 4, we get lines in the infrared. The integer m is always greater than n for a given series. For the Balmer series, $n = 2$, and m can be 3, 4, 5, and so on. Each value of the integer m generates a different line in the series.

The formula developed by Rydberg and Ritz pointed out a simple regularity in the spectrum of hydrogen that cried out for explanation. Thomson, in creating his so-called plum-pudding model of the atom, had attempted to explain this regularity without success. Bohr, working with the new view of the atom provided by Rutherford, took a fresh approach to the problem.

Quantization of light energy

Although the discovery of the nucleus and the regularities in the spectrum of hydrogen were crucial to Bohr's model, another new idea was at least as important. Introduced tentatively by Max Planck (1858–1947) in 1900

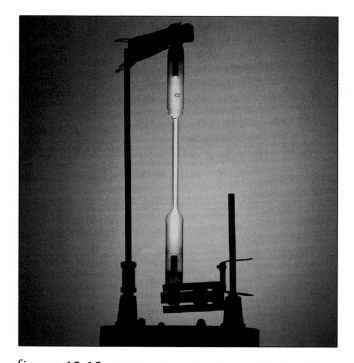

figure **18.16**   A high voltage placed across the electrodes of a gas-discharge tube produces a colorful glow. The colors are characteristic of the type of gas in the tube.

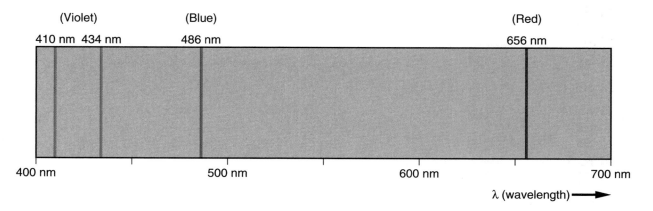

figure **18.17**   The hydrogen gas-discharge spectrum has four lines in its visible portion: a red line, a blue line, and two violet lines.

and later strengthened by Albert Einstein (1879–1955), it also had its origin in the study of spectra, in this case the spectrum produced by a heated **blackbody.**

A blackbody is best represented by a hole or cavity carved into metal or ceramic material that can be heated to high temperatures (fig. 18.18). Such a hole appears black at room temperature. The spectrum that it emits when heated depends only on the temperature and not on the material in which the cavity is carved. The spectrum is continuous (no discrete lines), but the average wavelength emitted becomes shorter as the temperature is increased. At high temperatures, the wavelengths become short enough to be visible: the cavity first appears "red hot," and at even higher temperatures "white hot," meaning that the average wavelength is near the middle of the visible spectrum.

Planck and other theorists were trying to explain the distribution of wavelengths that emerged from a heated blackbody. Planck arrived at a formula that succeeded in predicting the proper distribution and its dependence on temperature. In giving a rationale for his formula, however, he was forced to a radical conclusion: apparently, light could not be absorbed or emitted from the surface of the blackbody in continuously varying energies but only in discrete chunks, or *quanta,* whose energy depended on the frequency or wavelength.

To be more precise, at a given frequency, the only energies allowed are integer multiples of the energy:

$$E = hf$$

where *f* is the frequency and *h* is a constant called **Planck's constant.** The value of this constant is extremely small. In metric units, it is

$$h = 6.626 \times 10^{-34} \text{ J·s.}$$

According to Planck's theory, for a particular frequency *f*, light could be emitted with energies of *hf*, 2*hf*, 3*hf*, and so on, but not at any energy between these values.

This idea disturbed Planck himself as well as other physicists at that time. There had previously been no reason to suspect that light waves could not be emitted in continuously varying energies, depending only on how much energy was available. The idea that this process was **quantized,** meaning that it could only happen in discrete energy chunks, was indeed radical. In 1905, Einstein showed that the quantization of light energy could be used to explain a number of other phenomena. The idea of light quanta (or particles of light that we now call **photons**) having energies $E = hf$, was thus available, if not fully accepted, when Bohr began to develop a new model of the atom.

What were the features of Bohr's model?

Bohr's accomplishment was to combine all of these ideas— the discovery of the nucleus, knowledge of the electron, the regularities in the hydrogen spectrum, and the new quantum ideas of Planck and Einstein—into a new model of the atom. He started with the miniature-solar-system model mentioned on p. 386, in which the electron in the hydrogen atom orbits about the nucleus. The electrostatic force provides the necessary centripetal acceleration.

Bohr's first bold step departed from classical physics: he assumed certain stable orbits that do not continuously radiate electromagnetic waves as expected from classical physics. Instead, he imagined that light is emitted from the atom when the electron jumps from one stable orbit to another (fig. 18.19). Since the energy of a quantum of light or photon, as given by Planck and Einstein, is $E = hf$, the energy of the emitted photon is equal to the difference in

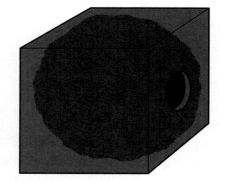

figure **18.18**   A blackbody radiator consists of a hole carved in material that can be heated to high temperatures. When heated, it emits a continuous spectrum of electromagnetic radiation.

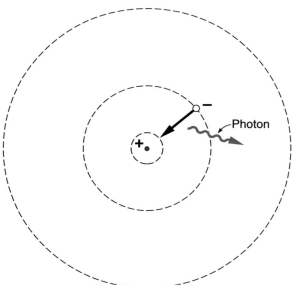

figure **18.19**   Bohr pictured the electron as orbiting the nucleus in certain quasi-stable orbits. Light is emitted when the electron jumps from one orbit to another.

energies of the two stable (or almost-stable) orbits. Expressed in symbols,

$$E = hf = E_{initial} - E_{final},$$

where $E_{initial}$ is the energy of the electron in the initial orbit, and E_{final} is the energy of the electron in the final orbit. These energies could be calculated for a specific orbital radius with ordinary Newtonian mechanics.

This energy difference could then be used to compute the frequency or wavelength of the emitted photon. Comparison of the resulting formula to the Rydberg-Ritz formula for the lines in the hydrogen spectrum showed that the energies of the stable orbits must all be given by a constant divided by an integer squared, $E = E_0/n^2$. This relationship placed a condition on the orbits: the only orbits allowed were those whose angular momentum, L, was equal to

$$L = n\left(\frac{h}{2\pi}\right),$$

where n is an integer and h is Planck's constant.

These are the essential features of Bohr's model:

1. Electrons are pictured as orbiting the nucleus in certain quasi-stable orbits, given by the condition $L = n(h/2\pi)$.
2. Light is emitted when an electron jumps from a higher-energy orbit to a lower-energy orbit.
3. The frequencies and wavelengths of the emitted light are computed from the energy differences between the two orbits, yielding the wavelengths in the hydrogen spectrum.

Figure 18.20 shows an energy-level diagram computed from Bohr's model for hydrogen. Try This Box 18.1 uses

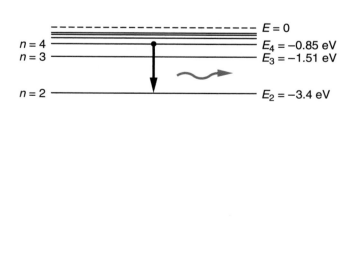

figure **18.20** The energies for the different orbits are shown in an energy-level diagram for hydrogen. The blue Balmer line is produced by the indicated jump.

try this box 18.1

Sample Exercise: Energy Levels in a Hydrogen Atom

Using the energy values shown in figure 18.20, calculate the wavelength of the photon emitted in the transition from the $n = 4$ energy level to the $n = 2$ energy level in the Bohr model of the hydrogen atom.

$E_2 = -3.4$ eV   The energy difference is
$E_4 = -0.85$ eV $\Delta E = E_4 - E_2$
$\lambda = ?$ $= -0.85$ eV $- (-3.4$ eV$)$
                                  $= 2.55$ eV

Using $h = 6.626 \times 10^{-34}$ J·s $= 4.14 \times 10^{-15}$ eV·s, the frequency of the emitted photon is given by

$$E = hf$$
$$f = \frac{E}{h}$$
$$= \frac{2.55 \text{ eV}}{4.14 \times 10^{-15} \text{ eV·s}}$$
$$= 6.16 \times 10^{14} \text{ Hz}$$

From $v = c = f\lambda$, the wavelength of the emitted photon is then

$$\lambda = \frac{c}{f}$$
$$= \frac{3 \times 10^8 \text{ m/s}}{6.16 \times 10^{14} \text{ Hz}}$$
$$= 4.87 \times 10^{-7} \text{ m} = \textbf{487 nm}$$

This is the blue line in the Balmer series of the hydrogen spectrum.

these values to find the wavelength of one line in the Balmer series of the hydrogen spectrum. The energy values in the diagram and exercise are expressed in *electron volts* rather than joules. An electron volt (eV) is the amount of kinetic energy gained when an electron is accelerated through a potential difference of 1 volt. It has the value 1 eV $= 1.6 \times 10^{-19}$ J. The energy levels in figure 18.20 can all be found by dividing the bottom level (-13.6 eV) by n^2, as predicted by the Bohr model. The energy values are all negative because the potential energy associated with the opposite-sign charges is negative.

One of the most striking successes of Bohr's model was that it could predict the correct value of the Rydberg constant from quantities like the mass of the electron, the charge of the electron, Planck's constant, and the speed of light. Bohr's theory was an instant and controversial sensation in the physics community. Its introduction spurred intense activity in both experimental and theoretical physics. Much of the experimental work focused on making more accurate measurements of the atomic spectra of different elements. The theoretical work sought to extend Bohr's model to atoms other than hydrogen and to try to understand the periodic properties observed in the periodic table of the elements.

Despite its impressive successes, the Bohr model left many unanswered questions. The most bothersome question was why just those few orbits described by the Bohr condition should be stable and not others. Attempts to extend the Bohr model to elements other than hydrogen met with limited success. Physicists now recognize that the Bohr model is inaccurate in many of its details. Its historic significance was that it opened the door to research that ultimately led to our modern theory of the atom.

The atomic spectrum of hydrogen has a particularly simple and regular form; the measured wavelengths are accurately described by the Rydberg formula. Bohr used these results, together with Rutherford's discovery of the nucleus and the Planck–Einstein condition for quantization of light energy, to develop a model of the hydrogen atom. Bohr assumed that there were just a few stable orbits for the electron about the nucleus. Light was emitted when the electron jumped from a higher-energy orbit to a lower-energy orbit. His model accurately described the wavelengths in the hydrogen spectrum and predicted the value of the Rydberg constant from fundamental quantities.

18.5 Particle Waves and Quantum Mechanics

The unanswered questions and the intense activity generated by Bohr's model of the atom attracted many young physicists to the field of atomic physics. A more comprehensive model of the atom that could explain why only certain orbits were stable obviously was needed. That need was filled when quantum mechanics was developed in 1925. Quantum mechanics was actually developed from two independent approaches that were quickly shown to be fundamentally the same in their structure and predictions.

The approach that is usually described followed from the work of Louis de Broglie (1892–1987) and Erwin Schrödinger (1887–1961).

De Broglie lit the spark by asking a simple but radical question: if light waves sometimes behave like particles (as shown by Planck and Einstein), could particles sometimes behave like waves? That question produced a revolution in our thinking about basic physical principles.

What are de Broglie waves?

The question posed by de Broglie was inspired by the concept of the photon introduced by Planck and Einstein. In 1865, Maxwell had shown that light could be described as an electromagnetic wave. On the other hand, light sometimes behaved as though it were made up of discrete and localized particlelike bundles of energy, now called photons. Certain experiments involving the interaction of light with electrons were most simply explained by thinking of light as a particle.

Einstein was the leader in pointing out this aspect of light. His 1905 paper discussed a number of phenomena that could be treated this way, the simplest being the photoelectric effect, a phenomenon in which light shining on an electrode in an evacuated tube causes an electric current to flow across the tube. This effect is frequently used in electric-eye devices that open doors when a person interrupts the light beam.

Einstein showed that the photoelectric effect could be explained by assuming that one photon of light, with energy $E = hf$ as suggested by Planck's work, ejected one electron on hitting the electrode. This simple model predicted the observed frequency dependence of the photoelectric effect as well as its other features. Other effects could also be treated in this manner by attributing an energy, $E = hf$, and a momentum, $p = h/\lambda$, to the photon.

Although this idea was simple, physicists were slow to accept it because particles and waves were thought to be very different phenomena. It was hard to understand how light could behave as a particle in some respects and a wave in others. An ideal wave extends indefinitely in space, but an ideal particle is completely localized, a simple point in space (fig. 18.21). *Real* waves have a finite length, of course, and *real* particles have some extension in space, but the concepts are still quite different.

De Broglie suggested that certain things traditionally thought of as particles, such as the electron, might sometimes behave like waves. In particular, he proposed that we reverse the relationships that described the energy and momentum of a photon to find the frequency and wavelength associated with a particle. Inverting the energy relationship yields a frequency for the particle of $f = E/h$. Photons have

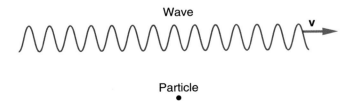

figure **18.21**   An ideal wave goes on indefinitely, but an ideal particle is just a point with no volume or extension in space.

a momentum $p = h/\lambda$, and inverting this relationship yields the *de Broglie wavelength*

$$\lambda = \frac{h}{p},$$

where p is the momentum and h is Planck's constant. If we knew the energy and momentum of an electron, for example, we could compute its frequency and wavelength from these relationships.

De Broglie's suggestion might have passed unnoticed if not for a striking result that he obtained. If he treated the electron as a wavelike entity orbiting the nucleus of the hydrogen atom, he could explain the condition for the quasi-stable orbits in Bohr's atomic model. He pictured the electron wave forming a standing wave wound around the circular orbit, as in figure 18.22.

To form a circular standing wave, the wavelength would have to be restricted to values such that an integer number of wavelengths would fit onto the circumference of the circle. By using the de Broglie wavelength for the electron, he could derive the Bohr condition on the allowed values of angular momentum, $L = n(h/2\pi)$. In other words, by assuming that particles had wavelike properties and by visualizing a standing particle-wave winding around a circular orbit, de Broglie could explain why only certain orbits would be stable. He had answered one of the fundamental questions in Bohr's theory.

De Broglie's picture of a standing wave on a circular orbit should not be taken literally. In fact, both the Bohr model and the standing-wave explanation predict the wrong

value for the angular momentum of the various stable states in the hydrogen atom. The basic difficulty is that a circular orbit is two-dimensional and the atom itself is three-dimensional. We need a more sophisticated analysis to depict the standing waves properly.

The suggestion that particles had wavelike properties was quickly borne out by experiment. It was known that X rays, which are electromagnetic waves, could be diffracted by a crystal lattice to form interference patterns characteristic of the crystal structure. Various workers soon showed that electron beams could also be diffracted. The interference patterns that resulted looked just like those obtained with X rays for the same crystal, and the wavelengths needed to explain these patterns were exactly the ones predicted by de Broglie's relationship, $\lambda = h/p$.

How does quantum mechanics differ from the Bohr model?

Erwin Schrödinger had spent much of his professional life studying the mathematics of standing waves in two and three dimensions. He was well prepared to explore the implications of standing electron waves in the atom. In the year following de Broglie's suggestion, Schrödinger developed a theory of the atom that used three-dimensional standing waves to describe the orbits of the electron about the nucleus.

Within the next five years, Schrödinger and other scientists pursuing the same problem from different approaches worked out the details of the theory that we now call **quantum mechanics.** This new theory gave a much more complete and satisfactory view of the hydrogen atom than the Bohr model. Quantum mechanics predicted the same primary energy levels for the different orbits as the Bohr model, however, and retained the basic features of Bohr's idea of how the hydrogen spectrum is produced.

In quantum mechanics, the orbits are not simple curves, as pictured in the Bohr model. Instead, they are three-dimensional probability distributions centered on the nucleus. These distributions rely on treating the electron as a standing wave. These standing waves describe the probability of finding electrons at certain distances and orientations about the nucleus. The probability distributions for a few quasi-stable orbits of the hydrogen atom appear in figure 18.23. The dark or denser areas are the places where the electron is most likely to be found. The average distances of the electron from the nucleus for different quasi-stable orbits are consistent with the orbital radii given by the Bohr model.

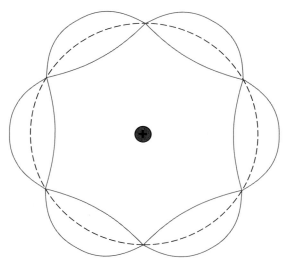

$$3\lambda = 2\pi r$$

figure **18.22** If the electron were pictured as a standing wave wrapped around a circular orbit, de Broglie showed that its wavelength could take on only certain values. These values yield the quasi-stable orbits predicted by Bohr.

What is the Heisenberg uncertainty principle?

Dealing with probability distributions rather than well-defined orbital paths is a fundamental, necessary feature of quantum mechanics. The waves associated with electrons

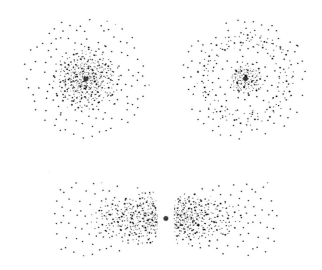

figure **18.23** The probability of finding the electron at different distances from the nucleus is given by these particle-density diagrams of a few of the quasi-stable hydrogen orbits predicted by quantum mechanics.

and other particles predict a probability for finding the electron at various positions, but they cannot tell us exactly where the particle is located. Likewise, electromagnetic waves give us a probability of finding photons at various positions. In situations where wave properties are dominant, we lose information about precise particle locations.

This limitation on what we can know about a particle's location is summarized in the famous **Heisenberg uncertainty principle,** introduced by Werner Heisenberg (1901–1976). This principle states that the position and momentum of a particle cannot both be known simultaneously with high precision. There will be an uncertainty in one that depends on how precisely we have determined the other. In symbols, this limitation takes the form:

$$\Delta p \Delta x \geq h,$$

where h is Planck's constant, Δp is the uncertainty in the momentum of the particle, and Δx is the uncertainty in its position. If the uncertainty in position is small, the uncertainty in momentum must be large, and vice versa.

Since the momentum p is related to the wavelength associated with the particle by the de Broglie relationship $\lambda = h/p,$ the uncertainty principle says that if we know the wavelength accurately, we cannot know the position of the particle accurately. The converse is also true: if we know the position accurately, we cannot know the wavelength accurately. Some experiments tend to bring out the particle-like aspects of photons or electrons (knowledge of position), and others tend to bring out the wave features (knowledge of wavelength).

Heisenberg's uncertainty principle is a fundamental limitation on what we can observe rather than a lack of experimental capability. The limitation is an inevitable fea-

ture of wave pulses. If we attempt to localize a wave by creating a brief pulse, the wavelength cannot be accurately defined. On the other hand, an extended wave, which permits accurate definition of the wavelength, gives us no precise information regarding position.

How does quantum mechanics explain the periodic table?

Quantum mechanics provides a means of answering most of the questions raised by the Bohr model about atomic structure and spectra. In particular, quantum mechanics is successful in predicting the structure and spectra of atoms with many electrons, although the computations are difficult. It also clarifies other features of the spectra that cannot be understood using the Bohr model.

The picture of atomic structure that emerges from quantum mechanics had already been partially assembled from attempts to explain the regularities in the periodic table of the elements. The theory provides us with **quantum numbers** that describe the various possible stable orbits. One of these is the *principal quantum number, n,* needed to compute the energies in the Bohr model. Quantum mechanics provides three others, however, associated with the magnitude and orientation of the angular momentum and with the spin of the electron.

No two electrons in an atom can have the same set of quantum numbers. Once an orbit is filled, other electrons must take on new, and generally higher, values for at least one of the quantum numbers. The number of possible combinations increases rapidly as the principal quantum number n increases. For $n = 1$, there are only two possible combinations corresponding to two different orientations of the electron-spin axis, but for $n = 2$, there are eight, for $n = 3$, eighteen, and so on. Once the two possible states for $n = 1$ are filled, the next electron added must go into an $n = 2$ level, or *shell.*

From this conception, we can explain certain regularities of the periodic table. The first two elements, hydrogen (H) and helium (He), have one and two electrons, respectively. Two electrons fill the $n = 1$ shell. The next element, lithium (Li), which has three electrons, must have its third electron in the $n = 2$ shell. Since lithium has one electron beyond the filled $n = 1$ shell, its chemical properties are quite similar to those of hydrogen, which has just one electron. Likewise, the next element in that column of the periodic table, sodium (Na), has one electron beyond the filled $n = 2$ shell. Its other ten electrons fill the $n = 1$ and $n = 2$ levels, two in the first shell and eight in the second shell. Figure 18.24 shows a schematic representation of the shell structure of hydrogen, lithium, and sodium.

The element immediately preceding sodium in the periodic table is neon (Ne), which has ten electrons, two in the $n = 1$ shell and eight in the $n = 2$ shell. Like helium, therefore, it has a closed-shell arrangement and does not react readily with other elements. Helium and neon are both **noble gases,** which are chemically nonreactive.

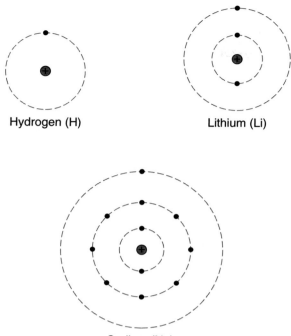

figure **18.24**   The chemical properties of sodium, with one electron in the $n = 3$ shell, are similar to those of hydrogen and lithium, which also have a single electron in their outermost shells.

Fluorine (F), however, has nine electrons, one short of a filled shell, and is very reactive. It forms compounds with elements such as hydrogen or sodium, which can contribute an electron to close a shell.

The principles used in explaining the entire periodic table are the same as those we have just outlined, although the details become more complicated with higher numbers of electrons. The theory explains the regularities of the periodic table and is highly successful in predicting the ways different elements combine to form chemical compounds. Quantum mechanics has become the fundamental theory of chemistry as well as atomic, nuclear, and condensed-matter physics.

Louis de Broglie suggested that particles such as electrons might have wavelike properties. Using this idea, he was able to explain the Bohr condition for stable orbits in the hydrogen atom. Thinking of particles as having wavelike properties leads directly to the Heisenberg uncertainty principle, which tells us that we cannot precisely determine both the position and the momentum of a particle at the same time. Quantum mechanics treats the standing waves associated with electrons in the atom as three-dimensional probability distributions. This theory has been successful in predicting the spectra and chemical properties of atoms with many electrons. Shells explain why we get the regularities described by the periodic table. Quantum mechanics is now the fundamental theory underlying most areas of physics and chemistry.

summary

In a period of less than fifty years, we progressed from knowing virtually nothing about the structure of the atom to having detailed knowledge of that structure. Some of the critical discoveries that led to this knowledge were discussed, starting from chemical evidence for the existence of atoms and culminating in the theory called quantum mechanics that explains atomic structure.

1   The existence of atoms: evidence from chemistry.   Recognition of the importance of weighing chemical reactants and products led to the statement of the law of definite proportions and the concept of atomic mass. If each element consists of atoms all having the same mass, we could explain the mass ratios observed in chemical reactions. The periodic table shows regularities in the properties of different elements when they are organized in order of increasing atomic weight.

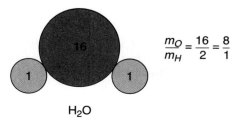

2   Cathode rays, electrons, and X rays.   The study of cathode rays, produced by placing a high voltage across two electrodes in an evacuated tube, led to the discovery of the electron and X rays. The electron is a negatively charged particle with a mass much smaller than the smallest atom, so it was the first known subatomic particle available for building atomic models.

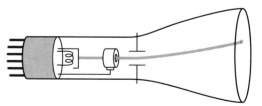

3   Radioactivity and the discovery of the nucleus.

Natural radioactivity, discovered shortly after the discovery of X rays, has three components: alpha (helium ions), beta (electrons), and gamma (short-wavelength X rays). The alpha rays were used to probe the structure of the atom in scattering experiments, which led to the discovery of the nucleus of the atom.

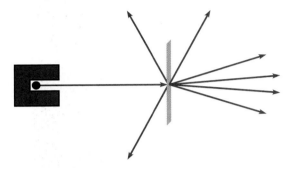

4   Atomic spectra and the Bohr model of the atom.

Bohr explained the regularities in the observed spectrum of hydrogen (the colors of light emitted by excited hydrogen atoms) with a model that incorporated the new quantum ideas introduced by Planck and Einstein. Bohr pictured light as being emitted when an electron jumped from one stable orbit to a lower-energy orbit. The energy difference explained the frequency and wavelength of the emitted photons.

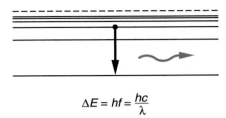

$$\Delta E = hf = \frac{hc}{\lambda}$$

5   Particle waves and quantum mechanics.

De Broglie's suggestion that particles such as electrons could have wavelike properties characterized by a wavelength related to the momentum of the particle, $\lambda = h/p$, was one path that led to the development of quantum mechanics. The stable orbits of the electrons in atoms can be described in terms of three-dimensional standing waves in this theory. The resulting probability distributions can explain the atomic spectra and chemical properties of multi-electron atoms.

key terms

Element, 376
Atom, 376
Combustion, 376
Dalton's law of definite proportions, 377
Molecule, 377
Atomic mass, 377
Periodic table of the elements, 378
Cathode rays, 379

Electron, 381
X rays, 382
Natural radioactivity, 383
Nucleus, 385
Atomic spectrum, 386
Rydberg constant, 387
Blackbody, 388

Planck's constant, 388
Quantized, 388
Photon, 388
Quantum mechanics, 391
Heisenberg uncertainty principle, 392
Quantum number, 392
Noble gas, 392

questions

*Questions identified with an asterisk are more open-ended than the others. They call for lengthier responses and are more suitable for group discussion.

Q1. Is a chemical element the same as a chemical compound? Explain.

Q2. Can the element iron (Fe) be changed to gold (Au) by heating it to a high enough temperature? Explain.

Q3. When a substance is burned, are all of the products of that reaction solid substances that can be easily weighed? Explain.

Q4. Is mass conserved in a chemical reaction? Explain.

Q5. In a chemical reaction, do the elements involved change into different elements as the reaction proceeds? Explain.

Q6. Does an atom of carbon (C) have the same mass as one atom of oxygen (O)? Explain.

Q7. Is it possible for any number of hydrogen atoms to combine with just one atom of oxygen? Explain.

*Q8. Can the law of definite proportions be explained by a model in which different atoms of the same element have widely varying masses? Explain.

Q9. Do cathode rays consist of electromagnetic waves? Explain.

Q10. Do X rays consist of electromagnetic waves? Explain.

Q11. Assuming that cathode rays are a beam of charged particles, how could you demonstrate that these particles are negatively charged? Explain.

Q12. What characteristics of the negatively charged particles that make up cathode rays suggested to Thomson that they might be atomic building blocks? Explain.

Q13. Would you expect X rays to be produced by a television picture tube? Explain.

Q14. If the electron beam in a television tube is striking just one point on the screen at a time, how can we get a full picture? Explain.

Q15. Following Roentgen's discovery of X rays, Becquerel discovered a seemingly similar type of radiation given off by phosphorescent materials containing uranium or thorium. Was this new radiation the same as X rays? Explain.

Q16. Was it necessary for Becquerel's phosphorescent materials to be exposed to sunlight for them to exhibit natural radioactivity? Explain.

Q17. What are two important differences that distinguish alpha particles from beta particles when they are passed through a magnetic field? Explain.

*Q18. When alpha particles are scattered from a thin piece of gold foil, why do most of them go through with very little deflection? Explain.

Q19. Does most of the mass of the atom reside inside or outside of the nucleus? Explain.

*Q20. What role did Rutherford's scattering experiment play in our developing understanding of atomic structure? Explain.

Q21. Would you expect electrons to be effective in deflecting an alpha-particle beam? Explain.

*Q22. How are the atomic spectra of hydrogen or other gaseous elements generated experimentally? How are they measured? Explain.

Q23. Does the spectrum of hydrogen consist of randomly spaced wavelengths or is there a pattern to the spacing? Explain.

Q24. According to Planck's theory, can light be emitted from a blackbody radiator in continuously varying amounts of energy for a given wavelength or frequency? Explain.

Q25. According to Bohr's theory of the hydrogen atom, is it possible for the electron to orbit the nucleus with any possible energy? Explain.

Q26. What happens to the excess energy when the electron jumps from a higher-energy orbit to a lower-energy orbit in the hydrogen atom? Explain.

Q27. Does an electron have a wavelength? Explain.

Q28. According to the theory of quantum mechanics, is it possible to pinpoint exactly where an electron is located in an atom? Explain.

Q29. The Bohr model of the hydrogen atom predicts a circular orbit for the electron about the nucleus; the theory of quantum mechanics predicts a three-dimensional probability distribution for locating the electron. Which of these views provides the more realistic picture of the hydrogen atom? Explain.

Q30. The chemical properties of sodium (Na), with eleven electrons, are similar to those of hydrogen (H), which has just one electron. How do we explain this fact?

Q31. Does helium (He), with two electrons (one more than hydrogen), react chemically with other substances more readily or less readily than hydrogen? Explain.

Q32. Why does the second row in the periodic table have more elements than the first row containing hydrogen and helium? Explain.

exercises

E1. If sodium (Na), with an atomic weight of 23, combines with oxygen (O), with an atomic weight of 16, to form the compound Na_2O, what is the ratio of the mass of sodium to oxygen that you would expect to react completely in this transformation?

E2. If carbon (C), with an atomic weight of 12, combines with oxygen (O), with an atomic weight of 16, to form carbon dioxide (CO_2), how many grams of carbon would react with 96 g of oxygen?

E3. If 38 g of fluorine (F) react completely with 2 g of hydrogen to form the compound hydrogen fluoride (HF), what is the atomic weight of fluorine?

E4. If aluminum (Al), with an atomic weight of 27, combines with oxygen (O), with an atomic weight of 16, to form the compound aluminum oxide (Al_2O_3), how much oxygen would be required to react completely with 54 g of aluminum?

E5. If the mass of a hydrogen atom is 1.67×10^{-27} kg and the mass of an electron is 9.1×10^{-31} kg, how many electrons would be required to have a mass equivalent to one hydrogen atom?

E6. How many electrons would be required to produce 1 microcoulomb (10^{-6} C) of negative charge? ($e = -1.6 \times 10^{-19}$ C)

E7. Suppose that an X-ray beam has a wavelength of 1.5×10^{-10} m. What is the frequency of these X rays? ($v = c = f\lambda$)

E8. Using the Rydberg formula, find the wavelength of the line in the Balmer series of the hydrogen spectrum for $m = 3$. ($n = 2$ for the Balmer series.)

E9. Using the Rydberg formula, find the wavelength of the spectral line for which $m = 3$ and $n = 1$. Would this line be visible to the unaided eye? Explain.

E10. Suppose that a photon has a wavelength of 520 nm (green).
   a. What is the frequency of this photon?
   b. What is the energy of this photon in joules?

E11. Suppose that a photon has an energy of 3.6×10^{-19} J.
a. What is the frequency of this photon?
($h = 6.626 \times 10^{-34}$ J·s)
b. What is the wavelength of this photon?

E12. An electron in the hydrogen atom jumps from an orbit in which the energy is 1.89 eV higher than the energy of the final lower-energy orbit.
a. What is the frequency of the photon emitted in this transition? ($h = 4.14 \times 10^{-15}$ eV·s. See Try This Box 18.1.)
b. What is the wavelength of the emitted photon?

challenge problems

CP1. An electron beam in a cathode-ray tube passes between two parallel plates that have a voltage difference of 300 V across them and are separated by a distance of 2 cm, as shown in the diagram.
a. In what direction will the electron beam deflect as it passes between these plates? Explain.
b. Using the expression for a uniform field, $\Delta V = Ed$, find the value of the electric field in the region between the plates.
c. What is the magnitude of the force exerted on individual electrons by this field? ($F = qE$, $q = 1.6 \times 10^{-16}$ C)
d. What are the magnitude and direction of the acceleration of an electron? ($m = 9.1 \times 10^{-31}$ kg)
e. What type of path will the electron follow as it passes through the region between the plates? Explain.

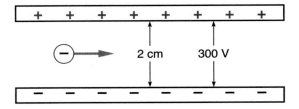

CP2. Study the energy-level diagram shown in figure 18.20. The Balmer series of spectral lines all involve transitions to the $n = 2$ energy level, and the Lyman series in the ultraviolet involves transitions to the $n = 1$ level. The energies are all

negative as a result of the negative potential energy for two charges of opposite sign.
a. Which transition in the Balmer series produces the smallest frequency photon (and the largest wavelength)?
b. What is the energy difference in joules for the two levels involved in the transition of part a?
c. What are the frequency and wavelength of the photon emitted in this transition?
d. Similarly, find the frequency and wavelength of the photon with the longest wavelength in the Lyman series.

CP3. When an electron is removed completely from an atom, we say that the atom is ionized. An ionized atom has a net positive charge since an electron has been removed.
a. From the energy-level diagram in figure 18.20, how much energy would be required to ionize a hydrogen atom when it is in its lowest energy level?
b. How much energy would be required to ionize the atom when it is in the first excited state above the lowest level?
c. If an electron with zero kinetic energy was "captured" by an ionized hydrogen atom and went immediately to the lowest energy level, what wavelength would you expect to observe for the photon emitted in this transition?

CP4. Suppose that an electron ($m = 9.1 \times 10^{-31}$ kg) is moving with a velocity of 1500 m/s.
a. What is the momentum of this electron?
b. What is the de Broglie wavelength of this electron?
c. How does this wavelength compare to those of visible light? (See figure 16.5.)

home experiments and observations

HE1. Obtain some red or blue liquid food coloring from your kitchen or local grocery. Prepare two glasses or transparent plastic cups containing equal amounts of water, one drawn from the cold tap and the other drawn from the hot tap. Drop one drop of food coloring into each cup and observe what happens. (Do not stir once the food coloring has been added.)
a. Describe the changes that take place over several minutes in time until the food coloring is well dispersed in both cups. What differences do you note between the cups with the cold and hot water?
b. Develop an explanation for your observations. Do they suggest the presence of tiny particles such as molecules or atoms?

HE2. With the television set turned off, take a close look at the screen of a color-television set. If you have a magnifying glass available, it will help to get a good view of the detail.
a. Describe the pattern of lines that you observe. Produce a careful sketch showing the arrangement of the lines. How many lines are there, approximately?
b. If there is a black-and-white television set handy, compare the pattern on its screen to the color set. What differences can you describe?

The Nucleus and Nuclear Energy

PROFESSIONAL RADON TEST KIT

chapter overview

The story of how physicists explored the nucleus and its structure is told in this chapter, including the discovery of nuclear fission just before World War II and the wartime effort to invent the atomic bomb. The postwar desire to find peaceful uses for the atom led to the development of commercial power plants, although this same era also saw the invention of the hydrogen (fusion) bomb and a rapid buildup of the nuclear arsenals of the major world powers. Our goal is to understand the underlying science of these issues.

chapter outline

I n 1986, the news media were full of reports and comments on a serious nuclear accident at the Chernobyl nuclear power plant in Ukraine, then part of the Soviet Union. Radioactivity was dispersed across parts of Europe, several firefighters and reactor employees were killed, and the fears of the public about nuclear power were dramatically reawakened.

Closer to home, in the United States we have about seventy nuclear reactors producing electric power, submarines that use nuclear reactors as their power source, and many smaller reactors used for research and other purposes. Most of these reactors have operated with only minor problems and have had minimal impact on their surroundings. Nevertheless, the environmental consequences and economics of nuclear power have been extremely controversial issues over the past three decades.

What goes on inside a nuclear reactor? How do we derive power from uranium, and what nuclear wastes are generated? Need we fear those benign-looking clouds of steam that billow from the cooling towers (fig. 19.1)? Can a reactor explode like a nuclear bomb? What is the difference between nuclear fission and nuclear fusion? If you know the answers to these questions, you will be less at the mercy of the extremists on either side of the issue who assert simple but misleading views.

The development of our knowledge of the nucleus of the atom is one of the most fascinating tales of twentieth-century science. The political consequences of nuclear

figure 19.1   The large cooling tower is often the most prominent feature of a modern nuclear power plant. What is the source of energy in such a plant?

weapons and nuclear power have been critical components of that story. These issues, more than any other, have thrust science and physics into the cauldron of national and international policy. Nuclear issues have become part of our common concern as citizens.

19.1 The Structure of the Nucleus

We began to understand the structure of the nucleus only in the twentieth century. Even the existence of an atomic nucleus was not suspected until Rutherford's famous alpha-particle scattering experiments performed between 1909 and 1911 (see section 18.3). The idea that this tiny center of the atom also has a structure that we can decipher may seem amazing.

What are the building blocks from which the nucleus is constructed? Ernest Rutherford, who is credited with the discovery of the nucleus, also played a major role in answering this question. The evidence came from more scattering experiments. Scattering experiments of one sort or another are the major tool for probing the nucleus and other subatomic particles.

How was the proton discovered?

Rutherford performed the experiment that uncovered the first nuclear building block in 1919. Once again, he used alpha particles as his probe. Figure 19.2 shows a conceptual diagram of this scattering experiment. A beam of alpha particles was used to bombard a cell containing nitrogen gas. As expected, some of the alpha particles went through the sample without hitting anything, and others were de-

flected (scattered) by the nuclei of the nitrogen atoms. The deflected particles could be observed with scintillation detectors (see section 18.3).

The unexpected result of this experiment was the emergence of a different particle from the cell containing the nitrogen. These new particles were positively charged like the alpha particles but could be distinguished by how far the particles traveled in air and by other features. In fact, the new particles behaved like the nuclei of hydrogen atoms that Rutherford had observed in earlier experiments when he bombarded hydrogen gas with alpha particles. The mass of the hydrogen atom is approximately one-fourth that of an alpha particle, which is the nucleus of a helium atom, as noted in chapter 18.

Finding hydrogen nuclei emitted from a cell that contained no hydrogen hinted at an exciting possibility: perhaps the hydrogen nucleus was a basic constituent of the nucleus of other elements. It was already known that the atomic masses of many elements were close to being integer multiples of the atomic mass of hydrogen. The atomic mass of nitrogen, for example, is approximately 14 times the atomic mass of hydrogen, while carbon is approximately 12 times and oxygen 16 times the atomic mass of hydrogen. These masses could be explained if the nuclei of these elements were made up of 12, 14, and 16 hydrogen nuclei for carbon, nitrogen, and oxygen, respectively.

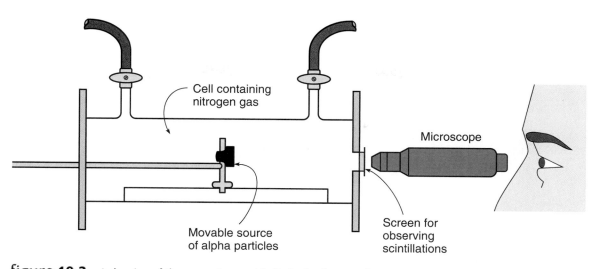

figure 19.2 A drawing of the apparatus used in Rutherford's scattering experiment, which resulted in the discovery of the proton. Nitrogen gas was the target for the alpha particles.

Further experiments by Rutherford and others showed that hydrogen nuclei could also be ejected from sodium and other elements by bombarding them with alpha particles. We now call this particle a **proton**—it is the nucleus of the hydrogen atom as well as a component of other nuclei. A proton has a charge $+e = 1.6 \times 10^{-19}$ C, opposite in sign but identical in magnitude to the charge of the electron $-e$. Its mass is much larger than the electron's, however, and is approximately equal to the mass of the hydrogen atom, 1835 times the mass of the electron.

How was the neutron discovered?

The hypothesis that the nuclei of different elements could be made simply of protons had some serious problems, the most obvious of which was the charge of the nucleus. From nitrogen's place in the periodic table and other evidence, the charge of the nitrogen nucleus should be $+7e$ rather than $+14e$. If there were 14 protons in the nucleus of the nitrogen atom, the nuclear charge would be too large. Likewise, carbon and oxygen have nuclear charges of $+6e$ and $+8e$, respectively, rather than 12 or 16 times e.

For a while, physicists considered the possibility that electrons were also present in the nucleus, partially neutralizing the extra charge of the protons. This view had some serious drawbacks, however, in light of the new insights of quantum mechanics. The energies of electrons confined to the very small region of the nucleus would have to be much larger than the measured energies of the electrons emerging as beta rays in radioactive decay. Therefore, it did not seem likely that electrons existed as separate particles in the nucleus.

It took several years to solve this riddle. Yet another scattering experiment, performed by Walther Bothe and Wil-

helm Becker in Germany around 1930, provided the breakthrough. Bothe and Becker bombarded thin beryllium samples with alpha particles and found that a very penetrating radiation was emitted. Since gamma rays were the only radiation known to be so penetrating, Bothe and Becker originally assumed that gamma rays were involved. Other experiments, however, showed that this new emission had an even greater ability to pass through lead than gamma rays and possessed other properties quite unlike gamma rays.

In 1932, a British physicist, James Chadwick (1891–1974), showed that this new emission from beryllium behaved like a neutrally charged particle with a mass roughly equal to the proton. Chadwick's experiment used the penetrating emission from the alpha bombardment of beryllium to bombard a piece of paraffin (fig. 19.3). Paraffin is a compound of carbon and hydrogen, and hydrogen nuclei (protons) emerged from the paraffin when placed in the path of the penetrating radiation coming from the beryllium. If a new neutral particle with a mass equal to the proton was colliding with protons in the paraffin, it neatly explained the energies of the protons emerging from the paraffin. This new particle was called a **neutron**—it has no charge, and its mass is very close to the proton's mass.

Chadwick's discovery of the neutron settled the question of the basic building blocks of the nucleus (fig. 19.4): If the nucleus is made of neutrons and protons, we can explain both the charge and the mass of the nucleus. Nitrogen, for example, can have a nucleus made up of 7 protons and 7 neutrons for a total mass 14 times the mass of hydrogen and a nuclear charge 7 times that of hydrogen. You can easily deduce the required numbers of protons and neutrons for carbon and oxygen. The numbers for several elements are shown in figure 19.5.

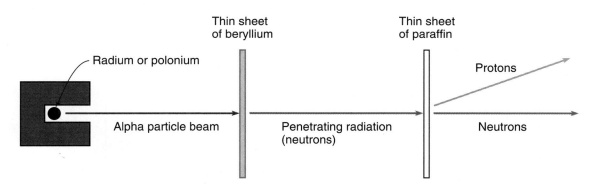

figure 19.3   A diagram of Chadwick's experiment. Radiation coming from the beryllium target was used to bombard a paraffin target.

figure 19.4   The basic building blocks of the nucleus are the proton and the neutron.

What are isotopes?

Another puzzle was also solved with the discovery of the neutron. It had been known for some time that atoms of the same element could have different values of nuclear mass. Nuclear masses were measured with high accuracy by passing nuclei of known velocity through a magnetic field and observing how much their paths were bent by the magnetic force on the positively charged nucleus. For example, chlorine was known from chemistry to have an average atomic mass 35.5 times that of hydrogen. When chlorine ions were passed through a magnetic field, however, two different masses were present, one 35 times hydrogen and the other 37 times hydrogen. Their chemical properties were identical—both behaved like chlorine.

Today we call different-mass versions of the same element **isotopes.** Different isotopes have the same number of protons in the nucleus, but different numbers of neutrons. The two common isotopes of chlorine, for example, both have 17 protons in the nucleus, but one has 18 neutrons for a total mass number of 35 and the other 20 neutrons for a total mass number of 37. (The mass number is the sum of the proton and neutron numbers.) Table 19.1 gives other examples.

The chemical properties of an element are determined by the proton number, also called the **atomic number.** Since the net charge of an atom is normally zero, the atomic

table **19.1**			
Neutron and Proton Numbers for Different Isotopes of the Same Elements			
Name	Symbol*	Protons	Neutrons
Hydrogen-1	$_1H^1$	1	0
Hydrogen-2 (deuterium)	$_1H^2$	1	1
Hydrogen-3 (tritium)	$_1H^3$	1	2
Carbon-12	$_6C^{12}$	6	6
Carbon-14	$_6C^{14}$	6	8
Chlorine-35	$_{17}Cl^{35}$	17	18
Chlorine-37	$_{17}Cl^{37}$	17	20
Uranium-235	$_{92}U^{235}$	92	143
Uranium-238	$_{92}U^{238}$	92	146

* See section 19.2 for an explanation of the notation used here.

number also gives the number of electrons orbiting the nucleus. Nitrogen, for example, has an atomic number of 7: there are 7 protons in the nucleus and 7 electrons orbiting the nucleus. There also happen to be 7 neutrons in the nucleus, but, in general, the neutron number is *not* equal to the atomic number. Except for the lightest elements, the neutron number is generally larger than the atomic number.

With the discovery of the neutron in 1932, many pieces of the puzzle fell into place. Atomic masses as well as the chemical properties of atoms could now be explained. Physicists began inventing models of the nucleus and designing new experiments to test these models. Perhaps most important, the neutron provided a powerful new probe for exploring the structure of the nucleus. Since it has no charge,

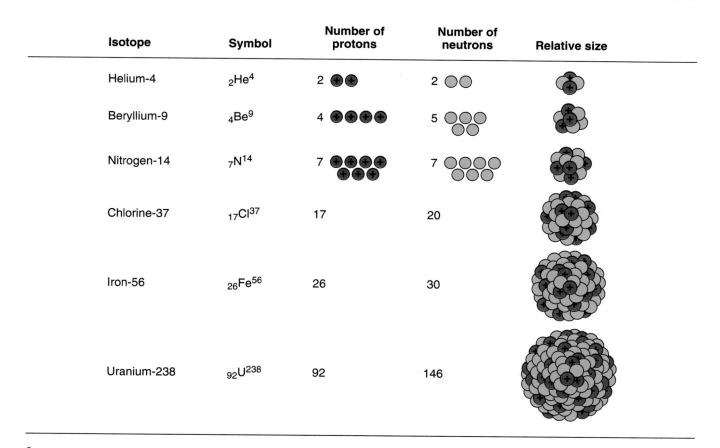

Isotope	Symbol	Number of protons	Number of neutrons	Relative size
Helium-4	$_2\text{He}^4$	2	2	
Beryllium-9	$_4\text{Be}^9$	4	5	
Nitrogen-14	$_7\text{N}^{14}$	7	7	
Chlorine-37	$_{17}\text{Cl}^{37}$	17	20	
Iron-56	$_{26}\text{Fe}^{56}$	26	30	
Uranium-238	$_{92}\text{U}^{238}$	92	146	

figure 19.5 The proton and neutron numbers for the most common isotopes of several elements. The nucleus gets larger as the number of protons and neutrons increases.

a neutron can penetrate the nucleus and begin to rearrange it. The proton and the alpha particle, on the other hand, are both positively charged and, therefore, repelled by the positive charge of the nucleus. A host of new experiments were begun, and some produced even more spectacular surprises than those we have been describing here.

A series of scattering experiments performed by Rutherford and other physicists gave clues to the building blocks of the nucleus. The proton, which is the nucleus of the hydrogen atom, was found in other nuclei as well. The neutron, which has a mass almost equal to the proton but zero charge, could also be generated in scattering experiments. Together, the proton and neutron account for both the mass and charge of different nuclei, as well as the existence of different isotopes of the same element.

19.2 Radioactive Decay

Becquerel discovered natural radioactivity in 1896, as described in section 18.3. By 1910, Rutherford and others had demonstrated that one element was actually being changed into another during **radioactive decay.** The nucleus of the atom itself is modified when a decay occurs. How can our new insights into nuclear structure help to illuminate this phenomenon?

What happens in alpha decay?

Radium, which was isolated and identified at the turn of the century by Marie and Pierre Curie, was one of the first radioactive elements to be studied extensively. Radium was found in the uranium ore, pitchblende, but was soon shown to be much more radioactive than uranium itself. Alpha particles, which Rutherford identified as the nuclei of helium atoms, were the primary radiation emitted in the decay of radium.

The dominant isotope of radium found in pitchblende contains a total of 226 **nucleons** (neutrons and protons) in its nucleus. We call this isotope *radium-226* and often write it as $_{88}\text{Ra}^{226}$, where Ra is the chemical symbol for radium, the subscript 88 is the atomic number, and the superscript 226 is the **mass number,** the total number of neutrons and protons. Since the atomic number is 88, there are 88 protons and 138 neutrons $(226 - 88)$ in the nucleus of this isotope.

If we know that radium-226 emits alpha particles, we can figure out what element results from its decay. The process is straightforward: we know how many protons and neutrons are contained in both the radium and the alpha particle (a helium nucleus), so we also know how many of each are left in the decay product or *daughter* element. We often write a reaction equation to help keep track of the numbers and to serve as a shorthand for describing the reaction:

$$_{88}Ra^{226} \Rightarrow \; _{86}X^{222} \; + \; _2He^4.$$

Here we have found the atomic number (86) of the unknown element X by subtracting the atomic number of helium, 2, from radium, 88. The mass number is found similarly: subtracting 4 from 226 yields the mass number of 222. Both the atomic numbers and the mass numbers must add to the same total on either side of the reaction equation. The unknown element can then be identified by looking in a periodic table (see the inside back cover) to find which element has an atomic number of 86. It turns out to be the noble gas radon (Rn), so the daughter nucleus that we have temporarily labeled X is radon-222 ($_{86}Rn^{222}$).

The alpha decay of radium-226 is illustrated in figure 19.6, where the alpha particle is shown emerging with a much larger velocity than the recoil velocity of the radon nucleus as required by conservation of momentum (see chapter 7). If the initial momentum of the system was zero, the alpha particle and the radon nucleus must have equal but oppositely directed momentums after the decay. Since the alpha particle has a much smaller mass than the radon nucleus, its velocity must be much larger than that of the helium nucleus for the momenta to be equal in magnitude ($p = mv$).

Even though we did not obtain gold, we find that the alchemist's dream of turning one element into another does happen in radioactive decay and other nuclear reactions.

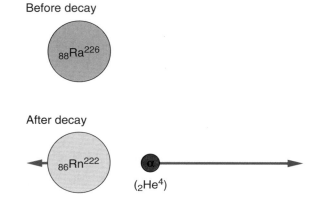

figure 19.6   Alpha decay of radium-226. The daughter isotope is radon-222.

The daughter isotope, radon-222, is itself radioactive and undergoes alpha decay to produce polonium-218, which, in turn, undergoes alpha decay to yield lead-214. Although lead-214 is not a stable isotope, lead is often the end product of the radioactive decay of heavier elements.

What happens in beta and gamma decay?

Lead-214 undergoes beta decay—the particle emitted in beta decay is either an electron or a *positron* (a positively charged version of the electron). In the case of lead-214, an ordinary (negatively charged) electron is emitted. The mass of an electron is so small that it can be ignored on the scale of nuclear masses—its mass number is effectively zero. Since an electron is negatively charged, its charge or atomic number is -1, so the reaction equation takes the form

$$_{82}Pb^{214} \Rightarrow \; _{83}X^{214} \; + \; _{-1}e^0 \; + \; _0\bar{\nu}^0.$$

The third particle appearing on the right-hand side of the beta-decay reaction equation is called an *antineutrino* and is represented by the Greek letter nu (ν). The bar over the symbol indicates an **antiparticle.** All elementary particles also have antiparticles. Antiparticles have identical masses but opposite-sign charges if they are charged. For example, the positron is the antiparticle of an electron. Antiparticles will annihilate one another, releasing energy in other forms, if they interact.

The antineutrino had not been directly observed in beta decay but was included to conserve energy. Since the electrons in beta decay emerge with a range of energies, physicists reasoned that something else must be involved to account for the remaining energy. Neutrinos were not actually detected until 1957, but physicists believed in their existence for many years before, because of their faith in the principle of conservation of energy. Neutrinos (and antineutrinos) have an extremely small mass and no charge, so they do not affect the charge and mass numbers in the reaction equation.

Here again we see that the mass numbers and charge numbers add up on either side of the reaction equation. The atomic number of the resulting element is 83, since $83 - 1 = 82$ is the original atomic number of the lead (Pb) isotope. Looking up the atomic number 83 in the periodic table shows that bismuth-214 is the daughter element in this decay (fig. 19.7). One of the neutrons inside the nucleus of lead-214 has been changed to a proton, yielding a nucleus with a higher atomic number. Thus, we could substitute $_{83}Bi^{214}$ for the unknown X in the reaction equation.

Through a series of further beta and alpha decays, bismuth-214 decays finally to a stable isotope of lead, lead-206. Some of the isotopes involved in this decay chain also emit gamma rays, which are high-energy X rays. Since the emitted particle in this case is a photon, which has no charge or mass, neither the mass number nor the charge number changes in a gamma decay. We are merely left with a more stable version of the original isotope (fig. 19.8).

Before decay

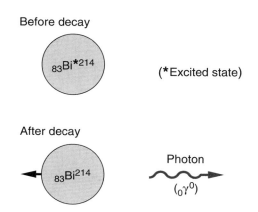

figure **19.7**   Beta decay of lead-214. The daughter isotope, bismuth-214, has a higher atomic number than lead.

figure **19.8**   Gamma decay of bismuth-214. The daughter isotope is a more stable (lower-energy) version of the original bismuth-214.

How do we describe the rate of decay?

How long does it take for these different decays to happen? The different kinds of radioactive decay are all spontaneous events that happen randomly. There is no way of predicting exactly when a specific unstable nucleus will throw out a particle and change to a different isotope.

Different radioactive isotopes have different average or characteristic times that elapse before they decay. The concept of **half-life** describes this characteristic time: the half-life is the time required for half of the original number of atoms to decay. For example, the half-life of radon-222 is about 3.8 days. If we started with 20 000 atoms of radon-222, 3.8 days later we would have 10 000 remaining. The other half would have decayed to polonium-218 ($_{84}Po^{218}$). In two half-lives, or 7.6 days, half of the remaining atoms would have decayed, leaving only 5000 atoms of radon-222. In three half-lives, we would be down to 2500, and in four half-lives this number would be halved again, yielding 1250.

Because the number of radon-222 atoms is reduced by half each time 3.8 days passes, it does not take many half-lives to reduce the remaining number of atoms of an isotope to a tiny fraction of the original amount. After 10 half-lives, or 38 days, the number of atoms remaining from the original 20 000 would be just 20 atoms or a thousandth of the original number.

This decay process is graphed in figure 19.9. The curve that results is called an *exponential decay curve*, because mathematically it can be represented by an exponential function. **Exponential decay** or growth occurs in nature whenever the number of decays or additional events is proportional to the total number of candidates for decay or growth. This is true for many random processes.

The half-lives of different radioactive isotopes vary enormously. The half-life of radium-226 is 1620 years, for example, which is quite short compared to the 4.5-billion-year half-life of the common isotope of uranium, uranium-238. At the other extreme, polonium-214 has a half-life of just 0.000164 second. It does not stick around long! The longer the half-life, the more stable the isotope.

On the other hand, the shorter the half-life, the greater the rate of radioactivity. From an environmental standpoint, isotopes with intermediate half-lives pose the greatest problem. An isotope with a very short half-life is highly radioactive while it lasts but decays quickly and does not remain dangerous. An isotope with an extremely long half-life like uranium-238 is not highly radioactive, although it can be a hazard if enough is present. An isotope such as strontium-90 with a half-life of 28.8 years is much more radioactive than uranium-238 and remains in the environment long enough to pose serious problems. Strontium-90 is sometimes present in fallout from bomb tests or nuclear accidents.

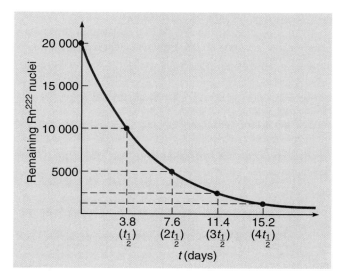

figure **19.9**   Decay curve for radon-222. The amount remaining decreases by one-half every 3.8 days, the half-life.

Why is radioactivity hazardous to our health?

Radioactivity is dangerous because the emitted particles—alpha particles, beta rays (electrons), and gamma rays—can penetrate the body and alter chemical compounds that make up our cells. These alterations can cause cancer and other damage, including mutations in offspring. In high-enough doses, these alterations produce radiation sickness and death. The effects of very low doses are still being debated. Some scientists believe that any dose is potentially damaging, but others hold that very low levels may have beneficial effects that counter the negative effects.

Because their half-lives are similar to the estimated age of the earth, uranium-238 and thorium-232 have not completely decayed and are still present naturally in our environment. They appear in trace amounts in all rocks and soils and in more concentrated amounts in uranium ores. We are constantly exposed to very low levels of radioactive emissions from trace amounts of these isotopes and their decay products as well as from cosmic rays that impinge on earth from space. (See Everyday Phenomenon Box 19.1.) Only when levels of radioactivity become larger than natural background radiation is there serious cause for concern.

In radioactive decay, the nucleus of the radioactive atom is modified as it emits different kinds of particles. In alpha decay, helium nuclei are emitted, leaving the daughter isotope with lower atomic and mass numbers. In beta decay, electrons or positrons are emitted, which changes the atomic number but not the mass number. In gamma decay, a high-energy X ray is emitted, which does not change either the atomic or mass number of the original isotope. Unstable isotopes are like time bombs waiting to go off. They have a characteristic decay time that we often describe in terms of the half-life.

Everyday Phenomenon   box 19.1

Radiation Exposure

The Situation. People are exposed to radiation from various sources in their everyday lives. How do we quantify this exposure? How much exposure is likely to be dangerous?

The Analysis. Although a variety of units are used for measuring amounts of ionizing radiation and their impact on human tissue, the unit most commonly used for comparison of doses of differing kinds of ionizing radiation (X rays, natural radioactivity, and so on) is the *rem*. Rem is an acronym standing for *roentgen equivalent in man*—the *roentgen* is a unit that describes the amount of ionization produced by radiation. The rem takes into account differing effects of different types of radiation on human tissue.

A whole-body dose of 600 rems usually is lethal. Much smaller doses can also produce damage, however, and these are generally quoted in *millirems* (mrems). One millirem is one-thousandth of a rem. On the average, people in the United States receive about 295 mrems yearly from natural sources and another 64 mrems yearly from human-produced sources. As the table shows,* the largest human-produced source is diagnostic X rays and radioactive isotopes used in medicine.

*For further discussion, see "Health Effects of Low-Level Radiation" (*Physics Today*, August 1991, pp. 34–39), from which this table was adapted.

Natural sources	mrems/yr
inhaled radon	200
cosmic rays	27
terrestrial radioactivity	28
internal radioactivity	40
	295
Human-produced sources	
medical	53
consumer products	10
other	1
	64

Both the natural background and the human-produced sources vary widely depending on where you live and what medical procedures you undergo. The average dose received by people in the United States from nuclear-power sources is not significant on this scale, but individuals working in the industry may be exposed to larger amounts. Current standards place a limit of 5 rems a year (5000 mrems/yr) on nuclear workers, X-ray technicians, or other people exposed to radiation in their occupations.

19.3 Nuclear Reactions and Nuclear Fission

We have seen that the nucleus can change spontaneously in radioactive decay. One element changes to another when this happens. Is it possible for us to cause such changes to occur experimentally rather than waiting for spontaneous decays?

The discovery of the neutron in 1932 provided a new probe for rearranging the nucleus. As a result of this discovery, nuclear physics became a field of intense activity during the 1930s. This work produced some unanticipated results with enormous implications for both science and human affairs.

What are nuclear reactions?

We have already encountered experimentally produced changes in the nucleus: Rutherford's discovery that protons are emitted from nitrogen nuclei when they are bombarded by alpha particles is an example of such a change. We could write the reaction equation for Rutherford's experiment as

$$_2He^4 + \,_7N^{14} \Rightarrow \,_8O^{17} + \,_1H^1.$$

The alpha particle is a helium nucleus, and the emitted proton is a hydrogen nucleus, as indicated in the equation. The other product of the reaction is an element with an atomic number of 8, which turns out to be oxygen (fig. 19.10). Oxygen-17 is not the most common isotope of oxygen (which is oxygen-16), but it is found in nature as a stable isotope.

This is an example of a **nuclear reaction.** Note that the charge and mass numbers add up to the same total on either side of the equation: the total charge or atomic number is 9 and the total mass number is 18. (This was also true of the radioactive-decay equations in section 19.2.)

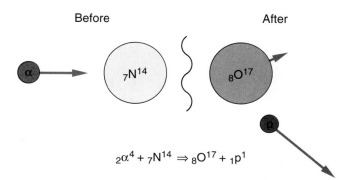

Before   After

$$_2\alpha^4 + \,_7N^{14} \Rightarrow \,_8O^{17} + \,_1p^1$$

figure **19.10**   The collision of an alpha particle and a nitrogen nucleus results in a proton being emitted and an oxygen-17 nucleus remaining in place of nitrogen.

The mass and atomic numbers are used to identify the other reaction product, oxygen-17, but we could also confirm this result by analyzing the gas found in the cell after the experiment. Oxygen would be found, even though none was present at the beginning of the experiment. Unlike a chemical reaction, which involves rearrangement of the electrons orbiting the nucleus, the elements themselves can change in a nuclear reaction.

The emission of neutrons from beryllium samples bombarded by alpha particles is another example of a nuclear reaction. The reaction equation in this case would be

$$_2He^4 + \,_4Be^9 \Rightarrow \,_6C^{12} + \,_0n^1.$$

The neutron has no charge, so its atomic or charge number is 0, and its mass number, like the proton, is 1. The other reaction product, carbon-12, is identified by looking in the periodic table to find which element has an atomic number of 6. The resulting isotope of carbon turns out to be the most common one. After the experiment has been performed, we will find small amounts of carbon present in the originally pure beryllium target.

How are energy and mass involved in nuclear reactions?

One of the original mysteries of radioactivity was the question of where the energy comes from. The alpha particles, beta rays, or gamma rays emerged with large kinetic energies even though Becquerel's uranium samples were stored in a dark drawer for weeks at time. What was the source of this energy?

The answer was found in Einstein's famous $E = mc^2$ relationship, developed as part of his theory of relativity about ten years after the discovery of radioactivity. The meaning of this equation is that mass and energy are equivalent: mass is energy and energy is mass. (This equation will be explained in more detail in chapter 20.) The constant c^2, the speed of light squared, is a unit-conversion factor that allows us to convert mass units to energy units, and vice versa. If the mass of the products is less than the reactants, the energy represented by this mass difference shows up in other forms, usually as the kinetic energy of the emerging particles.

This idea is illustrated for the beryllium reaction in the computations shown in Try This Box 19.1. The masses for the isotopes and particles involved are given in atomic mass units (or unified mass units, u), which are based on the mass of the carbon-12 atom. The mass of carbon-12 is exactly 12.000 000 u, by definition of the atomic mass unit. (One atomic mass unit is equal to 1.661×10^{-27} kg.) The mass difference is converted to kilograms and then multiplied by c^2 to find the energy released in joules. This energy will appear as kinetic energy of the emerging neutron and the recoiling carbon-12 nucleus.

Although the amount of energy released in a single reaction may seem small, it is roughly a million times larger

try this box 19.1

Sample Exercise: Transforming Mass Energy into Kinetic Energy

The nuclear masses for the reactants and products of the reaction

$$_2\text{He}^4 + {}_4\text{Be}^9 \Rightarrow {}_6\text{C}^{12} + {}_0\text{n}^1$$

are provided here. Using these values and Einstein's $E = mc^2$ relationship, calculate the energy released in this reaction.

Reactants		*Products*	
Be^9	9.012 186 u	neutron	1.008 665 u
He^4	+4.002 603 u	C^{12}	+12.000 000 u
	13.014 789 u		13.008 665 u

$E = ?$ The mass difference is

$$\begin{aligned} & 13.014\ 789\ \text{u} \\ & \underline{-13.008\ 665\ \text{u}} \\ \Delta m = {} & 0.006\ 124\ \text{u} \end{aligned}$$

$$1\ \text{u} = 1.661 \times 10^{-27}\ \text{kg}$$

$$\begin{aligned} \Delta m &= (0.006\ 124\ \text{u})(1.661 \times 10^{-27}\ \text{kg/u}) \\ &= 1.017 \times 10^{-29}\ \text{kg} \end{aligned}$$

$$\begin{aligned} E &= \Delta m c^2 \\ &= (1.017 \times 10^{-29}\ \text{kg})(3.0 \times 10^8\ \text{m/s})^2 \\ &= \mathbf{9.15 \times 10^{-13}\ J} \end{aligned}$$

than the typical energy released (per atom) in a chemical reaction. Such changes in mass are the source of the particle energies involved in radioactive decay and other nuclear reactions. From these ideas formulated in the early 1900s, physicists became aware of the possibility of releasing large quantities of energy in nuclear reactions.

How was nuclear fission discovered?

Before 1932, alpha particles or protons were the primary probes available for attempting to rearrange the nucleus. The discovery of the neutron immediately offered a powerful new probe, since its zero charge means that it is not repelled by the positive charge of the nucleus. Because alpha particles and protons are positively charged and are repelled by the charge on the nucleus, they require high initial energies and head-on collisions to produce a reaction. The neutron, on the other hand, can slip right into the nucleus at low energies.

The Italian physicist Enrico Fermi (1901–1954) was one of the first to explore the potential of the neutron for producing nuclear reactions. In a series of experiments from 1932 to 1934, he attempted to produce new heavy elements. The element with the largest mass and atomic number then known was uranium, so Fermi decided to bombard uranium samples with neutrons produced from the beryllium reaction (fig. 19.11). He then analyzed the samples to see whether he could detect elements with atomic numbers higher than 92.

At first, the results were both confusing and disappointing. Fermi and his chemist colleagues were able to predict the likely chemical properties of the elements that they were seeking by knowing what should come next in the periodic table, but attempts to isolate such elements were

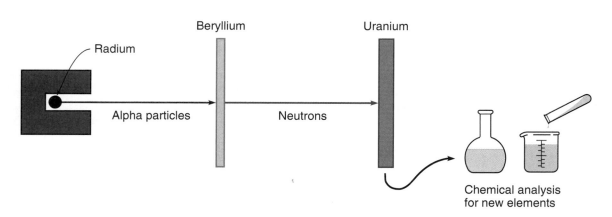

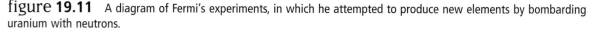

figure **19.11**   A diagram of Fermi's experiments, in which he attempted to produce new elements by bombarding uranium with neutrons.

not successful in these early experiments. Since the expected quantities would be small and the exact chemical properties were not known, the lack of clear-cut results may not have been surprising.

Others took up the effort, and the first real breakthrough in this line of research occurred in 1938, when two German scientists, Otto Hahn and Fritz Strassmann, isolated the element barium from uranium samples that had been bombarded with low-energy neutrons. This was an astonishing result that Hahn and Strassmann carefully rechecked before announcing their findings. The result was unexpected because barium is nowhere near uranium in the periodic table—barium has an atomic number of 56, just a little more than half that of uranium.

What kind of reaction could produce barium from uranium? Two other German scientists, then working in Denmark and Sweden because of the growing persecution of Jews in Germany, provided a possible answer.

Lise Meitner and her nephew, O. R. Frisch, speculated that the uranium nucleus might be splitting into two smaller nuclei, a process that we now call **nuclear fission.** If one of these nuclei was barium with an atomic number of 56, the other should have an atomic number of 36 in order to add up to 92, the atomic number of uranium. This element happens to be krypton, a noble gas (fig. 19.12). Thus, one possible reaction equation might be

$$_0n^1 + {}_{92}U^{235} \Rightarrow {}_{56}Ba^{142} + {}_{36}Kr^{91} + 3{}_0n^1.$$

This equation has been written with some extra neutrons being emitted. We will indeed end up with an excess of neutrons if we split a large nucleus into two smaller ones because the ratio of neutrons to protons gets larger and larger with increasing atomic number for the heavier elements in the periodic table (see fig. 19.5). The isotopes of barium and krypton that we have proposed in the reaction equation also contain an excess of neutrons. Therefore, they are unstable and undergo beta decay, which means that the reaction products are radioactive.

We have jumped ahead of the story in writing uranium-235 as the isotope of uranium involved. Naturally occurring uranium is mostly uranium-238, and only 0.7% is uranium-235. Uranium-235 most readily undergoes nuclear fission, however. Low-energy neutrons are absorbed more readily by uranium-235 than by uranium-238, and fission is more likely for uranium-235 than for uranium-238 when a neutron is absorbed.

The two elements that emerge from a fission reaction are called *fission fragments;* in this case, barium and krypton. Many other elements can result, however, all having atomic numbers between 30 and 60, near the middle of the list of known elements. These fission fragments are generally radioactive because of the excess of neutrons and make up the bulk of nuclear wastes from applications of nuclear fission.

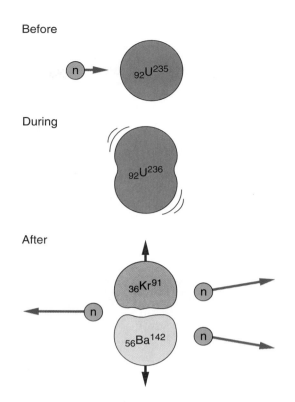

figure **19.12**   Barium-142 and krypton-91 are two possible fission fragments produced when a neutron is absorbed by uranium-235, causing a fission reaction.

The excitement generated among scientists by these discoveries and ideas was far-reaching. In 1939, Niels Bohr, who discussed these speculations with Meitner and Frisch in Denmark, suggested that uranium-235 was the isotope involved. Bohr then traveled to the United States, where he spread the word to a growing community of nuclear scientists. Many of these scientists were refugees who had fled the precarious situation in Europe created by the beginning of World War II and the persecution of Jews in areas under Nazi control.

The excitement and concern were engendered, in part, by the immediate recognition that a **chain reaction** involving nuclear fission could be produced. Since the fission reaction is initiated by neutrons and several more neutrons are emitted in the reaction itself, creating a rapidly expanding chain reaction seemed a real possibility (fig. 19.13). A chain reaction would release enormous quantities of energy, as predicted by Einstein's mass-energy equation.

Among the European-born scientists working in the United States by 1939 were Enrico Fermi, Edward Teller, and Albert Einstein. Although Einstein was not primarily interested in nuclear physics, he was recognized throughout the world as a brilliant theoretician. The general public was familiar with his name, and for this reason, some of

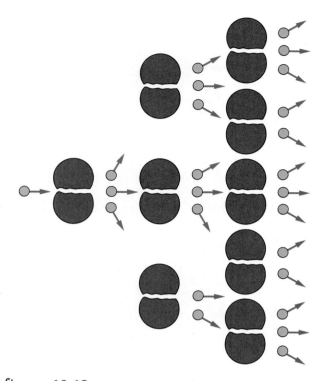

figure **19.13** A chain reaction involving nuclear fission. Neutrons are produced in each fission of a uranium-235 nucleus, which, in turn, can initiate more fission reactions.

his colleagues prevailed on Einstein to write a letter to President Franklin D. Roosevelt suggesting the need for a crash research program to explore the military implications of nuclear fission. The Manhattan Project, begun shortly afterward, led to the development of both nuclear reactors and nuclear weapons.

Nuclear reactions involve changes in the nucleus of atoms, which often result in one element changing into another. Radioactive decay and the reactions involved in the discoveries of the proton and neutron are examples. The amount of energy released or absorbed can be predicted from the difference in the masses of the reactants and products using Einstein's famous mass-energy equation. The discovery of the neutron provided a new probe for initiating nuclear reactions. Experiments using this probe led to the discovery of nuclear fission in the late 1930s. The possibility of a chain reaction spurred research that produced both nuclear reactors and nuclear weapons.

19.4 Nuclear Reactors

By 1940, the possibility of producing a chain reaction by nuclear fission was apparent to physicists in both Europe and the United States. If the fission reaction is initiated by neutrons, and each reaction produces several additional neutrons, why are chain reactions not occurring all the time in uranium samples? What conditions are necessary to produce a chain reaction? These questions demanded answers because war was raging in Asia and in Europe, and the United States and its allies feared that Germany might be developing a nuclear bomb.

How can we achieve a chain reaction?

Understanding the conditions necessary for a chain reaction is a crucial step to knowing how both nuclear reactors and nuclear bombs work. The key is to trace what happens to the neutrons produced in the fission reaction. If enough neutrons are captured by other uranium-235 nuclei, new fissions will occur and the reaction will be sustained. If too many of the neutrons produced are absorbed by other elements or escape from the reactor or bomb without colliding with other uranium-235 nuclei, the reaction dies.

Keep in mind that natural uranium consists mainly of uranium-238 (99.3%)—only 0.7% is uranium-235. Uranium-238 also absorbs neutrons, but this does not usually result in fission. The main reason that a chain reaction does not occur in natural uranium is that the neutrons produced in one fission reaction are more likely to be absorbed by uranium-238 than by uranium-235 nuclei. One way of increasing the likelihood of a chain reaction is to increase the proportion of uranium-235 in the sample.

Unfortunately (or, fortunately, depending on your point of view), it is extremely difficult to separate uranium-235 from uranium-238. Different isotopes of the same element have identical chemical properties, so chemical separation techniques are useless. Only the very small difference in mass between the two isotopes can be used as the basis for separation. Various techniques were tested during the war years—the most promising was a gas-diffusion technique tried in a plant at Oak Ridge, Tennessee. After enormous effort and expense, scientists succeeded in separating only enough uranium-235 for one bomb (just a few kilograms) by the end of the war.

A different strategy for achieving a chain reaction is used in nuclear reactors designed to run on natural uranium or on uranium only slightly enriched in uranium-235. The trick is to slow the neutrons down between fission reactions. Slow neutrons have a much greater probability of being absorbed in collisions with uranium-235 nuclei than with uranium-238 nuclei. The faster neutrons emitted in the fission reactions have almost equal probabilities of absorption when they encounter the two isotopes. Thus, if the neutrons can be slowed down before encountering additional uranium nuclei, we have a better chance of producing additional fission reactions.

Using this idea, Enrico Fermi and his co-workers at the University of Chicago achieved the first controlled chain reaction in 1942. Fermi constructed a *nuclear pile* of blocks of pure graphite interspersed with small pieces of natural

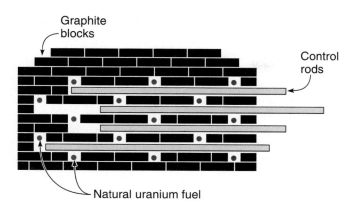

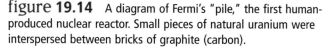

figure **19.14** A diagram of Fermi's "pile," the first human-produced nuclear reactor. Small pieces of natural uranium were interspersed between bricks of graphite (carbon).

uranium (fig. 19.14). Graphite is a solid form of the element carbon, which has a relatively small mass number, 12. Since carbon does not absorb neutrons readily, neutrons that collide with carbon-12 nuclei just bounce off. In each collision, the neutrons lose energy, while the carbon nuclei gain kinetic energy. The graphite slows down the neutrons without absorbing them—such a material is called a **moderator.**

Besides the moderator and the uranium fuel, one other feature is required in any reactor. Because a chain reaction can grow very rapidly under the proper conditions, we need some means of controlling the rate of reaction. **Control rods** containing a neutron-absorbing material can be inserted or removed from the pile to maintain the desired level of reaction. In Fermi's pile, the control rods were made of cadmium, but boron is the material most commonly used now.

On December 2, 1942, Fermi and his colleagues slowly removed some of the control rods from their carefully constructed pile. By monitoring the rate of neutron flow at different points in the pile, they established that a self-sustaining chain reaction had occurred. The reactor had gone *critical;* that is, for each fission reaction, one of the new neutrons generated went on to be absorbed by another uranium-235 nucleus, producing another fission reaction. If more than one new fission reaction is produced for each initial reaction, we say that the reactor is *supercritical.* If less than one is produced, the reactor is *subcritical.*

In starting up a reactor, we let the reactor go just slightly supercritical at first, until the desired reaction level is reached. Then the control rods are reinserted slightly to maintain the reactor at a critical or steady-state level. Inserting the control rods farther decreases the reaction level—the control rods are like the gas pedal in an automobile. Frequent adjustment of some of the control rod positions fine-tunes the level of reaction. Other control rods are designed to be rapidly inserted to shut down the reactor.

Why is plutonium produced in nuclear reactors?

Up to this point, we have not considered what happens when uranium-238 absorbs a neutron other than to indicate that this does not usually result in fission. Fermi's original objective of producing new elements heavier than uranium does indeed occur: a series of nuclear reactions produces plutonium, now the primary material in fission bombs.

The reactions that generate plutonium-239 from uranium-238 are summarized in table 19.2 and figure 19.15. The first step is the absorption of a neutron by a uranium-238 nucleus to produce uranium-239. Uranium-239, in turn,

table **19.2**
Reactions Involved in the Production of Plutonium
1. *Neutron absorption by uranium-238* $_0n^1 + _{92}U^{238} \Rightarrow _{92}U^{239}$ 2. *Beta decay of uranium-239* $_{92}U^{239} \Rightarrow _{93}Np^{239} + _{-1}e^0 + _0\bar{\nu}^0$ 3. *Beta decay of neptunium-239* $_{93}Np^{239} \Rightarrow _{94}Pu^{239} + _{-1}e^0 + _0\bar{\nu}^0$

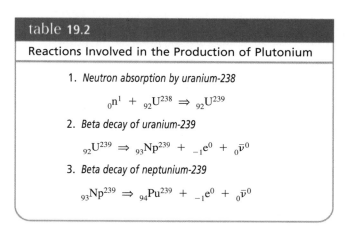

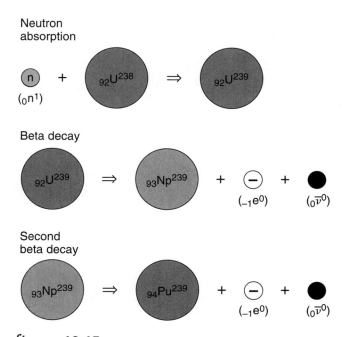

figure **19.15** Absorption of a neutron by uranium-238 followed by two beta-decay reactions produces plutonium-239, which can also be used as a fission fuel.

undergoes beta decay (half-life, 23.5 minutes) to produce neptunium-239, a new element with an atomic number of 93, one higher than uranium. Neptunium-239 also undergoes beta decay (half-life, 2.35 days), producing yet another new element with an atomic number of 94, plutonium-239. Plutonium-239 is relatively stable—its half-life is roughly 24 000 years. Like uranium-235, it readily undergoes nuclear fission when it absorbs a neutron.

Since nuclear reactors use either natural uranium or uranium only slightly enriched in uranium-235, the production of plutonium-239 is a natural by-product of the operation of a reactor. Also, because it is a different element, its chemical properties are not the same as uranium, and it can be separated from uranium using chemical techniques. Nuclear reactors can be used to produce plutonium, a fissionable material, for nuclear weapons. A crash program was begun at Hanford, Washington, during the closing years of World War II to build nuclear reactors for precisely this purpose.

What are the design features of modern power reactors?

Most modern reactors designed for power production do not use graphite as the moderator—they use ordinary (*light*) water instead. Water (H_2O) contains nuclei of both hydrogen (H) and oxygen (O), and the hydrogen nuclei are effective in slowing the neutrons. Unfortunately, hydrogen also absorbs neutrons to form its heavier isotopes, deuterium (H^2) and tritium (H^3). *Heavy water,* made using deuterium in place of the ordinary isotope of hydrogen, absorbs neutrons less readily than light water, so it is used sometimes as a moderator.

Because the neutrons absorbed by ordinary hydrogen are removed from circulation, ordinary water is not effective as a moderator when natural uranium is the fuel. **Enrichment** of the uranium-235 concentration from the 0.7% concentration in natural uranium to approximately 3% compensates for the neutrons lost to absorption by hydrogen, however, and a chain reaction can then be achieved. Light-water reactors must use slightly enriched uranium as the fuel, but reactors using heavy water as the moderator can use natural uranium.

The advantage of using water as a moderator is that it can also serve as a **coolant,** removing heat from the reactor core. The kinetic energy of the neutrons and fission fragments released in the fission reactions shows up as heat when this energy is randomized by collisions with other atoms. Any large reactor must have some means of cooling the core, or the temperature will increase to the point where some of the reactor components melt. The coolant circulates through the reactor carrying the energy generated in the fission reactions to the steam turbines, which turn the generators to produce electric power. The turbines themselves must be cooled to operate efficiently. Heat from water used to cool the turbines is released into the atmosphere by the cooling towers. This water never passes through the reactor itself.

Figure 19.16 is a diagram of a modern nuclear power reactor. The reactor core is contained within a thick-walled

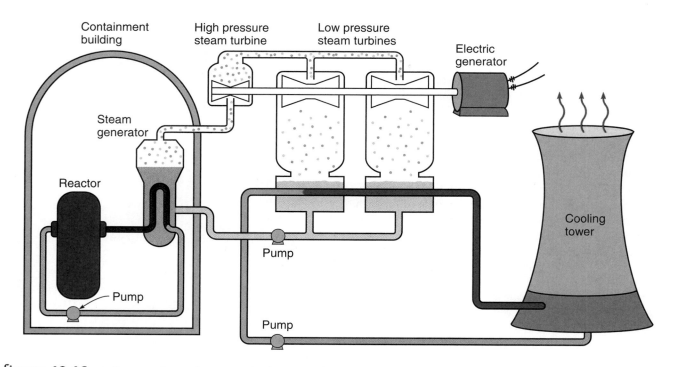

figure 19.16   A diagram of a modern pressurized-water nuclear reactor. Hot water coming from the reactor is converted to steam when the pressure is reduced in the steam generators. The steam turns the turbines, which power the electric generator.

steel reactor vessel through which the coolant circulates. The reactor vessel is housed in a heavily reinforced concrete containment building designed to withstand strong pressure variations and to shield the reactor from external influences as well as to contain radioactivity resulting from accidents. The coolant passes through steam generators, and the steam passes through the steam turbines that turn electrical generators. There is a lot of plumbing in a nuclear power plant.

Within the plant, but outside the containment building, is a control room to oversee the pumps, control rods, and other equipment for running the reactor. Here we also find temperature and radiation gauges and other monitoring equipment that tell the operators what is happening inside. Most reactors are designed with a lot of redundancy in safety equipment, and reactor operators are trained to deal with many contingencies. Because of the complexity of reactor operation, however, operator error is often a factor when accidents do occur, either in the initial event or in the response to a problem. Everyday Phenomenon Box 19.2 describes the Chernobyl reactor accident.

Environmental issues surrounding nuclear power

Any material like hydrogen, which absorbs neutrons in a reactor, is called a *poison*. Impurities in the moderator or fuel elements, or even the fission products themselves, can act as poisons that slow the reaction. The longer a reactor is operated, the more the poisons build up in the fuel elements and, of course, the more the uranium-235 fuel is depleted. The fuel rods must be replaced from time to time (fig. 19.17). The spent fuel rods containing uranium, plutonium, and the radioactive fission fragments must be stored or disposed of somehow. Radioactive elements in the spent fuel rods make up most of the nuclear wastes produced by nuclear reactors.

Our current national policy on waste disposal set by the federal government proposes burying these radioactive materials in a solid rock formation without separating the plutonium and remaining uranium from the fission fragments. This policy avoids the need for the expensive and environmentally hazardous processing involved in chemical separation. The disadvantage, however, is that it wastes fissionable material in the form of plutonium and uranium. The disposal site also must remain stable for thousands of years, because plutonium and uranium have much longer half-lives than most of the fission fragments. Plutonium-239 has a half-life of 24 000 years, but most of the fission fragments have half-lives of several years or less. They decay more rapidly and do not have to be isolated nearly as long.

When nuclear power was first introduced in the late 1950s, it was seen as a clean and inexpensive means of generating electric power. Concerns about reactor safety and waste disposal have caused many people to modify their views, but nuclear power still causes far less atmospheric

figure **19.17** An assembly containing fuel elements and control rods for use in a modern power reactor.

pollution than burning fossil fuels such as coal or oil to generate power. Disposal of nuclear wastes has become a political issue, however, and reactor safety and economic issues have also been a source of debate.

The development of nuclear power in the United States is at a virtual standstill, because of economic issues. The high costs and long time required for plant construction, and concerns about public acceptance, have made utilities back away from ordering new reactors. In Japan, Europe, and other parts of the world where fossil fuels are less available, the use of nuclear power continues to expand. As we enter the twenty-first century, the future of nuclear power in the United States is uncertain. The design of new, smaller reactors that are inherently stable may someday bring new life to this industry.

Nuclear reactors are used for many other purposes besides power production. One important application is the production of radioactive isotopes for use in nuclear medicine. These isotopes are involved in diagnostic procedures as well as for treatment of various types of cancer. Radioactive isotopes produced in nuclear reactors are also used as tracers in industrial processes and environmental studies, and for many other applications in industry and research.

To sustain a chain reaction, on the average at least one neutron released in each fission must be absorbed by a uranium-235 nucleus to initiate a new fission reaction. If more than one neutron is absorbed, the reaction is

Everyday Phenomenon

box 19.2

What Happened at Chernobyl?

The Situation. In 1986, a serious accident happened at a nuclear power plant at Chernobyl in Ukraine, then part of the Soviet Union. In terms of loss of life and radiation releases into the environment, the Chernobyl incident is the worst reactor accident that has occurred anywhere in the world to date. Worldwide publicity about the Chernobyl accident raised many questions and inflamed public debate about the safety of nuclear power. What type of reactor was involved, and how did the accident occur? Could a similar accident happen in the United States?

The damaged reactor at Chernobyl. Rapid buildup of heat caused explosions that ignited the graphite moderator.

The Analysis. The nuclear reactor at Chernobyl, like many others in the former U.S.S.R., was a dual-purpose reactor designed both to generate power and to produce weapons-grade plutonium. It used graphite as the moderator and also circulated water through the core as a coolant. The presence of ordinary water in the core necessitates some enrichment of uranium-235 (to about 2%) but not as much as if water were also the moderator, as in the commercial reactors in the United States.

Because of the use of water as the coolant but not as the moderator, the reactor had an unusual design characteristic. In the event of loss of the coolant or increase in core temperature, the rate of reaction actually increases, because the water used as the coolant absorbs neutrons. However, if the water is lost (or changes to steam, becoming less dense), it absorbs fewer neutrons, and the chain reaction accelerates. This result is not possible in a reactor in which water also serves as the moderator, because loss of the moderator reduces the rate of reaction.

In a reactor used to produce weapons-grade plutonium, the fuel rods must be removed from the reactor about 30 days after being inserted to avoid consuming the plutonium-239 through fission and to avoid the buildup of plutonium-240 (another, less fissionable isotope of plutonium). Reactors of the Chernobyl type were designed for easy access to the top of the reactor for replacement of fuel rods, as in the second photograph. The building housing the reactor is not designed to withstand strong pressure variations, as are the containment buildings housing most reactors in the United States.

When the accident occurred, an experiment was being conducted to see whether the electric generators could be used to provide power to the reactor pumps in the event of loss of external power. The experiment required that the reactor be run at a low power level to simulate conditions in which the reactor was being shut down. The person in charge of the experiment was an electrical engineer who was primarily interested in the generator response and was not an expert on the operation of the reactor.

(continued)

supercritical and will grow rapidly. If less than one is absorbed, the reaction is subcritical and will die. In a nuclear reactor, the neutrons are slowed down by the moderator, increasing their likelihood of being absorbed by other uranium-235 nuclei. Control rods absorb neutrons and permit the rate of reaction to be adjusted. Fermi's original reactor used graphite as the moderator, but modern reactors use water, which also serves as a coolant. Fission fragments and plutonium build up in the fuel rods, which must eventually be removed from the reactor, becoming nuclear wastes.

19.5 Nuclear Weapons and Nuclear Fusion

In a nuclear reactor, the objective is to release energy from fission reactions in a controlled manner, never letting the chain reaction get out of hand. In a bomb, on the other hand, the objective is to release energy very quickly—a supercritical chain reaction is what is sought. How can this state be achieved? What conditions are necessary for a nuclear explosion? What is the difference between fission weapons and weapons that use *nuclear fusion*?

The top of a Chernobyl-type reactor, showing a technician working on the square tops of the fuel-rod assemblies.

The fire in the graphite had to be put out, so firefighters from nearby towns were called in. Of the 31 deaths directly caused by the accident, many were firefighters who were exposed to high levels of radiation. Radiation, in the form of fission fragments carried by emissions from the fire, spread over the surrounding countryside and, at decreasing levels, over much of Europe. An increase in cancer deaths among those exposed in the surrounding communities is expected as a result.

Certain features of commercial reactors in this country make an accident like Chernobyl impossible. Most importantly, a loss of coolant slows the chain reaction because water is used as the moderator in our commercial reactors. An explosive increase in the chain reaction is not possible. The partial meltdown at Three Mile Island in Pennsylvania in 1979 was caused by heat generated from residual radioactivity of the fission fragments following shutdown of the chain reaction. Residual heat can still be a serious problem but does not result in an explosive buildup of the chain reaction.

Secondly, most reactors throughout the world are built with heavily reinforced concrete containment buildings, so the fission fragments are highly unlikely to escape from the containment building (or even from the reactor vessel itself) in the event of a partial meltdown or other accident. Although serious, the accident at Three Mile Island released little radioactivity into the environment. The economic impact of the accident was considerable, however, because of the loss of the reactor and the cleanup costs.

Operator error was a major factor in the accident at Chernobyl. We would like to think that the training given to reactor operators in the United States precludes the serious errors in judgment that happened at Chernobyl. Operator misjudgments have also been a problem, however, in the accidents in our own nuclear industry. The complex details of reactor behavior make it hard to plan and train for all possible contingencies.

Here is a summary of the complex series of events and errors. First, the reactor was partially shut down to perform the experiment. To bring it back to a desired power level, most of the control rods were removed, and several other safety features disengaged to achieve the conditions called for by the experiment. In the initial stages of the experiment, water flow to the reactor core was reduced, and this caused a rapid increase in fission reactions. The heat generated caused explosions that blew open the top of the reactor building and ignited the graphite moderator (which can burn much like coal).

What do we mean by critical mass?

The initial approach to producing nuclear weapons involved separating uranium-235 from uranium-238 to produce a highly enriched sample of uranium-235. If most of the uranium-238 is removed, neutrons emitted in the initial fission reactions would be much more likely to encounter other uranium-235 nuclei, and a rapidly increasing chain reaction should result. The size of the uranium mass is also important: if it is too small, neutrons will escape through its surface before they encounter other uranium-235 nuclei.

A **critical mass** of uranium-235 is a mass just large enough for a self-sustaining chain reaction. For a mass smaller than the critical mass, too many of the neutrons generated in the initial fission reactions escape through the surface of the uranium without encountering other nuclei.

For a mass larger than the critical mass, more than one of the neutrons produced in each fission reaction will be absorbed by other uranium-235 nuclei and produce additional fission reactions. The chain reaction will grow very rapidly, because the time between reactions is very short. This is the *supercritical* state necessary for an explosion.

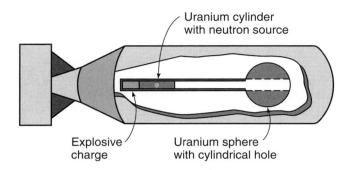

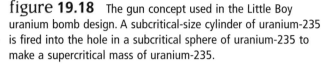

figure 19.18   The gun concept used in the Little Boy uranium bomb design. A subcritical-size cylinder of uranium-235 is fired into the hole in a subcritical sphere of uranium-235 to make a supercritical mass of uranium-235.

How can we create a supercritical mass without having it blow apart prematurely? The energy released in a fission chain reaction causes rapid heating and expansion. If a supercritical mass were built up slowly from subcritical pieces, the mass would begin to come apart as soon as it became supercritical. The bomb would fizzle. One approach to solving this problem is to bring two subcritical pieces of almost-pure uranium-235 together rapidly to produce a strongly supercritical mass. A gun design, as in figure 19.18, was developed for the first uranium bomb. A subcritical cylinder of uranium-235 is fired into a subcritical sphere of the same material containing a cylindrical hole, thus quickly assembling a supercritical mass. A neutron source must also be present to initiate the reaction.

The major problem in producing a bomb like this is the extreme difficulty of separating uranium-235 from the much more abundant isotope, uranium-238. The gas-diffusion plants at Oak Ridge, Tennessee, were able to produce only enough pure uranium-235 for one bomb during the war years. Building a nuclear arsenal at that rate would have been slow and expensive. It became apparent that plutonium-239 might be a better nuclear fuel than uranium-235.

How are plutonium bombs designed?

Plutonium-239 is a natural by-product of nuclear reactors that use uranium for fuel. As noted in section 19.4, the reactors built at Hanford, Washington, during World War II were designed to produce plutonium for weapons. The fission reaction of plutonium-239 is different from uranium-235, however, so that the gun design used for the uranium bomb will not work with plutonium. Plutonium-239 absorbs fast neutrons much more readily than uranium-235, which causes the chain reaction to grow even more swiftly than for uranium, and two subcritical pieces cannot be brought together quickly enough to avoid the fizzle produced by premature disintegration.

The design used for plutonium bombs relies on *implosion:* explosives arranged around a subcritical mass of plutonium and fired together create a tremendous inward pressure on the plutonium. This pressure increases the density of the plutonium sample enough to make the mass supercritical (fig. 19.19). The same number of atoms are now confined in a smaller volume, increasing the probability of absorption of neutrons by other plutonium nuclei.

By the end of World War II, enough weapons-grade material had been assembled to produce just three nuclear bombs. Two of them were plutonium bombs, dubbed *Fat Men* because of their shape. The third bomb was a uranium bomb, called *Little Boy* because of its slimness. In a historic test, one of the plutonium bombs was exploded at White Sands, New Mexico, in the summer of 1945, producing the first of the awesome mushroom clouds of nuclear explosions. Shortly thereafter, the other two bombs were dropped on the cities of Hiroshima and Nagasaki in Japan.

During the war years, the effort to build and test nuclear bombs was seen as a race with Nazi Germany, where the fission reaction had originally been discovered. Many of the scientists working on the Manhattan Project were European refugees. The thought that Germany might acquire the bomb first was a horrifying possibility. As the war was winding down, and the success of the bomb-building effort was approaching, it became evident that Germany no longer had the resources to produce a nuclear bomb. A debate began among the scientists on the project about how

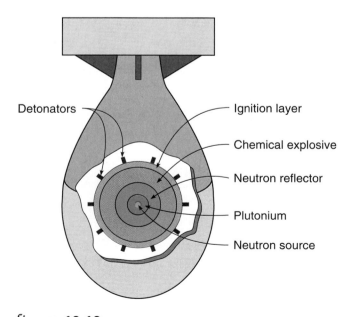

figure 19.19   The Fat Man plutonium bomb used chemical explosives arranged around a subcritical mass of plutonium-239. When imploded by the explosives, the increased density makes this mass supercritical.

to use the bomb. Many scientists favored a demonstration of the bomb's effects without actually using one on military targets.

A demonstration involved serious problems, however, especially since there were only two bombs available after the test of the plutonium bomb. One of these, the uranium bomb, could not be tested beforehand because only one had been built. The decision to drop the bombs on Japan was made by the top military authorities, including the President, Harry S Truman. Although still controversial, that decision probably hastened the end of the war, avoiding the need for a costly invasion of Japan.

After the war, weapons production went on, as the Hanford reactors continued to generate more plutonium. Soon Russia had the ability to manufacture fission bombs, and the pressure increased to proceed to the next step in the race, the development of the hydrogen bomb, which was first successfully tested in 1952. A hydrogen bomb involves nuclear fusion rather than fission.

What is the fusion reaction?

What happens in a hydrogen bomb? **Nuclear fusion** is another kind of nuclear reaction that also releases large quantities of energy. Fusion is the energy source of the sun and other stars as well as of *thermonuclear* bombs. How does nuclear fusion differ from fission, and how can we generate a chain reaction involving fusion?

Nuclear fusion combines very small nuclei to form somewhat larger nuclei. In a sense, it is the opposite of nuclear fission, the splitting of large nuclei into smaller fission fragments. The fuel for fusion consists of very light elements, usually isotopes of hydrogen, helium, and lithium. (Lithium has an atomic number of 3.) The end product is often the particularly stable nucleus, helium-4, which we have already encountered as the alpha particle.

As long as the mass of the helium-4 nucleus and other reaction products is slightly less than the sum of the masses of the isotopes combining to produce the reaction, this mass difference will show up as kinetic energy, as predicted by Einstein's formula $E = mc^2$. One possible reaction is the combination of two isotopes of hydrogen, deuterium (H^2) and tritium (H^3), to form helium-4 plus a neutron:

$$_1H^2 + {}_1H^3 \Rightarrow {}_2He^4 + {}_0n^1.$$

As shown in figure 19.20, the sum of the masses of the two particles on the right side of this equation is less than the sum of the masses of the two particles on the left side. The total kinetic energy of the alpha particle and the neutron will be greater than the kinetic energy of the initial two particles.

What makes this reaction so difficult to produce is that all the nuclei are positively charged and repel one another. Large initial kinetic energies are needed to overcome the

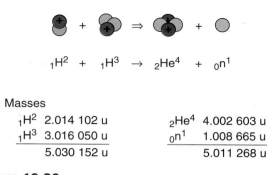

$$_1H^2 + {}_1H^3 \rightarrow {}_2He^4 + {}_0n^1$$

Masses

$_1H^2$	2.014 102 u	$_2He^4$	4.002 603 u
$_1H^3$	3.016 050 u	$_0n^1$	1.008 665 u
	5.030 152 u		5.011 268 u

figure 19.20   A deuterium nucleus and a tritium nucleus combine to form a helium-4 nucleus and a neutron. The difference in the masses is converted to the kinetic energy of the emerging particles.

repulsive electrostatic force so that the two nuclei can combine. One way of assuring large kinetic energies is to heat the reactants to a very high temperature. High densities of the reacting isotopes are also necessary to increase the probability of the reactions occurring. The reactants must be confined in a very small space at a very high temperature, two requirements hard to achieve simultaneously.

The chain reaction that results under these conditions is called a *thermal chain reaction*. Very high temperatures are needed to initiate the reaction, and the energy released in the reaction raises the temperature even more. The temperatures required for nuclear thermal chain reactions are a million degrees celsius or more. Chemical explosions are also thermal chain reactions, but the temperatures required for chemical reactions are much lower and more easily attainable.

The easiest way to get both the high temperatures and the high densities required for a fusion chain reaction is to explode a fission bomb to initiate the fusion reaction. The high temperature produced by the fission bomb creates the high kinetic energies necessary for fusion. The fission explosion also compresses the fusion fuel momentarily—the time is very short but still long enough for considerable additional energy to be released from fusion reactions (fig. 19.21). Basically, this is how a hydrogen bomb (also called a thermonuclear bomb) works.

Hydrogen bombs can be made in various sizes, unlike fission bombs, which are restricted to a size dependent on the critical mass of the fissionable material. Much larger energy yields are possible with fusion bombs than with pure fission bombs. Even though the energy released per reaction is smaller for fusion reactions than for the typical fission reaction, pound for pound the fusion reactions pack more wallop because the fusion fuels consist of very light elements. Hydrogen bombs can be made with an explosive power equivalent to 20 million tons of TNT or more. (Tons of TNT is the standard basis for comparison in quoting bomb yields.) Both fission and fusion bombs now make up the arsenals of the nuclear powers.

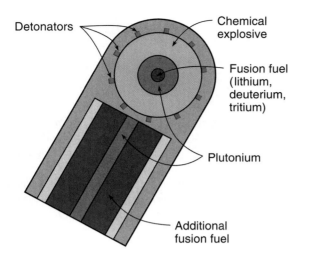

figure **19.21**   A fission bomb is exploded around the fusion fuel to produce the high temperatures and density required for a fusion chain reaction in a thermonuclear bomb.

figure **19.22**   The Tokamak Fusion Test Reactor at Princeton, New Jersey, designed to confine and heat fusion fuels using magnetic fields.

Can we generate power from controlled fusion?

Producing fusion reactions for a commercial power source has not yet been accomplished. Confining the fuel at very high temperatures in a very small space for a long enough time to release a significant amount of energy presents extreme difficulties. Since any solid will melt at temperatures well below those required for fusion reactions, magnetic fields or some other scheme must be used to confine the fuel (fig. 19.22). Another approach involves bombarding a small pellet of fusion fuel with laser beams (or particle beams) from several directions to both heat and compress the pellet.

Projections made in the 1970s that we would have working fusion reactors by the 1990s were highly optimistic. It now seems that this goal may not be achieved until well into the twenty-first century. We cannot be completely confident that an economically feasible reactor will ever be built, but we have already invested heavily in the effort, and someday, the goal should be reached. Experimental reactors, such as the Tokamak (fig. 19.22), have generated energy from fusion, but they have not reached the break-even point, where as much energy is released as is required to initiate the reaction. We need to improve yields to make the process commercially viable.

Considerable excitement was generated a number of years ago (1988–1989) when scientists working in Utah claimed to have achieved *cold fusion* in a cell that did not require extraordinary temperatures or densities. Their cell involved a palladium electrode immersed in a beaker containing heavy water (in which deuterium replaces the ordinary isotope of hydrogen). By passing a current through the cell, deuterium atoms are drawn into the spaces between the atoms in the palladium electrode. The Utah group claimed to have observed excess heat (presumed to be produced by fusion) that could not be explained by other chemical or physical processes occurring in the cell.

Although occasional fusion reactions occur under these circumstances, not enough fusion results to produce usable quantities of energy. If it were possible, though, the commercial potential would be enormous. The claims of the Utah group piqued a great deal of publicity and public interest. Many workers have attempted to reproduce their experiments, but so far the results have been disappointing. Most physicists do not believe that usable fusion energy is likely to come from cold fusion.

We do not yet know the best approach to achieving energy from fusion. Work continues on magnetic containment and on the particle-beam and laser-beam techniques. A few scientists are still exploring cold fusion. If a breakthrough occurs in the near future, some of today's students of science and engineering will be involved in a new expansion of applications of nuclear power.

To produce a nuclear explosion using fission, a supercritical mass of either uranium-235 or plutonium-239 must be assembled quickly from subcritical components. In a uranium bomb, firing a subcritical cylinder of uranium into a subcritical sphere with a cylindrical hole achieves critical mass. In a plutonium bomb, this is achieved by firing chemical explosives arranged around a subcritical sphere of plutonium, compressing it to a supercritical condition. Nuclear fusion releases energy by combining small nuclei to form larger nuclei. Fusion is the energy source of the sun and of hydrogen or thermonuclear bombs. Attempts to produce power from controlled fusion are not yet commercially viable, but research on this problem may someday reach that goal.

summary

We have traced how identifying the nuclear constituents, the proton and the neutron, led to an understanding of isotopes and of what happens in radioactive decay. The discovery of the neutron also gave us a new tool for probing and modifying nuclei, which led to the discovery of nuclear fission. Nuclear fission and fusion are nuclear reactions capable of releasing large amounts of energy.

1 **The structure of the nucleus.** Experiments in which alpha particles were scattered from various nuclei led to the discovery of both the proton and neutron as constituents of the nucleus. The number of protons in a nucleus is the atomic number of an element. The total number of protons and neutrons (nucleons) is the mass number of a given isotope. The same element can have isotopes of different masses.

Lithium-7 ($_3Li^7$)

3 protons
4 neutrons
7 nucleons

2 **Radioactive decay.** Radioactive decay is a spontaneous nuclear reaction in which a particle or gamma ray is emitted and the structure of the nucleus changes. One element is transformed into another in alpha or beta decay. The half-life is the time required for half of the original number of radioactive nuclei to undergo decay.

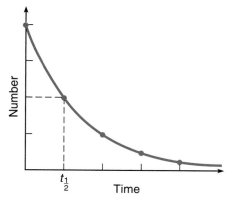

3 **Nuclear reactions and nuclear fission.** In any nuclear reaction, one element may change into another, but the total charge number (atomic number) and mass number (nucleon number) are conserved. Fission was discovered by bombarding uranium samples with neutrons, which leads to the splitting of the uranium-235 nucleus into fission fragments and the emission of additional neutrons.

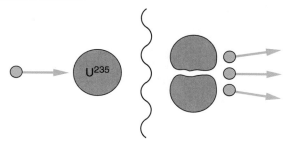

4 **Nuclear reactors.** Nuclear reactors generate energy by allowing a controlled nuclear-fission chain reaction to take place. The moderator slows down the neutrons to increase the probability of absorption by uranium-235, the fissionable isotope of uranium. Control rods absorb neutrons to control the reaction level, and the coolant carries off the energy generated. Spent fuel rods contain radioactive fission fragments as well as uranium and plutonium.

5 **Nuclear weapons and nuclear fusion.** Bombs can be built from either uranium or plutonium. Plutonium is produced in reactors when neutrons are absorbed by uranium-238. In nuclear fusion, small nuclei combine to form larger nuclei, which also can release energy. Hydrogen bombs use a fission bomb to trigger the high temperatures and densities needed for fusion.

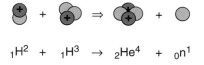

$$_1H^2 + {_1}H^3 \rightarrow {_2}He^4 + {_0}n^1$$

key terms

questions

*Questions identified with an asterisk are more open-ended than the others. They call for lengthier responses and are more suitable for group discussion.

Q1. In 1919, Rutherford bombarded a sample of nitrogen gas with a beam of alpha particles.
   a. In addition to alpha particles, what other particle emerged from the nitrogen gas in this experiment?
   b. What conclusion did Rutherford draw from this experiment? Explain.

Q2. When beryllium is bombarded with alpha particles, a very penetrating radiation is emitted from the beryllium sample. Does this radiation consist of X rays? Explain.

Q3. Is it possible for two atoms of the same chemical element to have different masses? Explain.

Q4. Is it possible for atoms of the same chemical element to have different chemical properties? Explain.

Q5. Which number, the mass number or the atomic number, determines the chemical properties of an element? Explain.

Q6. Why is the atomic weight of chlorine, as determined from chemical experiments, not a whole-number multiple of the atomic weight of hydrogen? Explain.

*Q7. In a nuclear reaction, can the total mass of the products of the reaction be less than the total mass of the reactants? Explain.

Q8. In alpha decay, do we expect the atomic number of the daughter nucleus to be equal to, greater than, or less than the atomic number of the isotope undergoing decay? Explain.

Q9. In beta decay, do we expect the atomic number of the daughter nucleus to be equal to, greater than, or less than the atomic number of the isotope undergoing decay? Explain.

Q10. What is a neutrino and why did scientists believe in its existence long before it was ever detected? Explain.

Q11. In gamma decay, do we expect the atomic number of the daughter nucleus to be equal to, greater than, or less than the atomic number of the decaying isotope? Explain.

Q12. Do all radioactive substances decay at the same rate? Explain.

Q13. In a time equal to two half-lives of a radioactive isotope, would you expect all of that isotope to have decayed? Explain.

Q14. In chemical reactions, the individual elements present in the reactants are the same as in the products of the reaction. Is this also true in a nuclear reaction? Explain.

Q15. Chemical reactions and nuclear reactions can both release energy. On the average, would you expect the energy released per unit of mass in a chemical reaction to be greater than, equal to, or less than what is released in a nuclear reaction? Explain.

*Q16. Why do we expect fission fragments to have higher neutron numbers than stable isotopes of the same element and, therefore, to be radioactive? Explain.

Q17. Suppose that you light a match to a mixture of oxygen and hydrogen, which then reacts explosively to form water. Is this a chemical reaction or a nuclear reaction? Explain.

Q18. The most common isotope of uranium is uranium-238. Is this the isotope that is most likely to undergo fission? Explain.

Q19. What property of the fission reaction leads to the possibility of a chain reaction? Explain.

*Q20. What is the function of the moderator in a nuclear reactor? Explain why the moderator is needed to obtain a chain reaction using natural uranium.

Q21. Do the control rods in a nuclear reactor absorb or emit neutrons? Explain.

Q22. If you wanted to slow down the chain reaction in a nuclear reactor, would you remove or insert the control rods? Explain.

*Q23. Will a reactor that uses ordinary water as the moderator be able to operate using unenriched uranium as a fuel? Explain.

Q24. If a reactor goes subcritical, will the chain reaction speed up? Explain.

*Q25. If plutonium and uranium are removed from the spent fuel of a nuclear reactor, will the remaining nuclear wastes need to be stored for thousands of years before they become nonradioactive? Explain.

Q26. In Fermi's original experiments in which he bombarded uranium samples with neutrons, he was trying to produce new elements heavier than uranium. Is it possible to do so? Explain.

Q27. What was the purpose of the nuclear reactors built at Hanford, Washington, during World War II? Explain.

Q28. How does nuclear fusion differ from nuclear fission? Explain.

Q29. Is nuclear fission the main process involved in the energy generated in the sun? Explain.

Q30. Do we currently have commercial nuclear reactors that use nuclear fusion as their energy source? Explain.

*Q31. Which can produce larger yields of energy, a fission weapon or a fusion weapon? Explain.

exercises

E1. Sodium has an atomic number of 11 and an atomic weight of approximately 23. How many neutrons would you expect to find in the nucleus of the most common isotope of sodium?

E2. $_{94}Pu^{239}$ is an isotope of plutonium produced in nuclear reactors.
a. How many protons are in the nucleus of this isotope?
b. How many neutrons are in the nucleus of this isotope?

E3. A certain isotope has 13 protons and 14 neutrons in its nucleus. Identify the element involved and write its symbol in the standard notation including the atomic and mass numbers.

E4. Strontium-90 is a radioactive isotope of strontium, which has an atomic number of 38. How many protons and how many neutrons are present in the nucleus of this isotope?

E5. Thorium-232 undergoes alpha decay. Complete the reaction equation for this decay and identify the daughter nucleus.

$$_{90}Th^{232} \Rightarrow ? + \alpha$$

E6. The fission fragment iodine-131 undergoes negative beta decay. Complete the reaction equation and identify the daughter nucleus.

$$_{53}I^{131} \Rightarrow ? + _{-1}e^{0} + _{0}\bar{\nu}^{0}$$

E7. Nitrogen-13 is a radioactive isotope of nitrogen that undergoes positive beta decay in which a positive electron (or positron) is emitted. Complete the reaction equation and identify the daughter nucleus.

$$_{7}N^{13} \Rightarrow ? + _{+1}e^{0} + _{0}\bar{\nu}^{0}$$

E8. Suppose that we have 10 000 atoms of a radioactive substance with a half-life of 2 hours.
a. How many atoms of that element remain after 4 hours?
b. How many atoms remain after 8 hours?

E9. When we measure the rate of radioactivity of a given isotope 18 days after making an initial measurement, we discover that the rate has dropped to one-eighth of its initial value. What is the half-life of this isotope?

E10. How many half-lives must go by for the radioactivity of a given isotope to drop to
a. One-sixteenth ($^{1}/_{16}$) of its original value?
b. One-sixty-fourth ($^{1}/_{64}$) of its original value?

E11. Suppose that we discover that one of the fission fragments for a given fission reaction of uranium-235 is tin-130 and that four neutrons are emitted in this reaction. Complete the reaction equation and identify the other fission fragment.

$$_{0}n^{1} + _{92}U^{235} \Rightarrow ? + _{50}Sn^{130} + 4\,_{0}n^{1}$$

E12. Suppose that two deuterium nuclei ($_{1}H^{2}$) combine in a fusion reaction in which a neutron is emitted. Complete the reaction equation and identify the resulting nucleus.

$$_{1}H^{2} + _{1}H^{2} \Rightarrow ? + _{0}n^{1}$$

challenge problems

CP1. Using the periodic table found in the inside back cover, we can get some idea of how the number of neutrons increases compared to the number of protons as the atomic number increases. By rounding the atomic weight to the nearest whole number, we can estimate the total number of nucleons (neutrons and protons).
a. What are the neutron and proton numbers for carbon (C), nitrogen (N), and oxygen (O)?
b. What is the ratio of neutrons to protons for the stable isotopes of these three elements? (Ratio = N_n/N_p)
c. Taking three elements near the middle of the table, silver (Ag), cadmium (Cd), and indium (In), find the number of neutrons and protons for each the same way.
d. Compute the ratio of neutrons to protons for the elements in part c and find the average ratio.
e. Repeat the process of parts c and d for thorium (Th), protactinium (Pa), and uranium (U).
f. Compare the ratios of parts b, d, and e. Can you see why there are extra neutrons when uranium or thorium undergo fission?

CP2. Uranium and thorium are the radioactive elements found in some abundance in the earth's crust. As each isotope of these elements decays, new radioactive elements are created that have much shorter half-lives than uranium or thorium. A series of alpha and beta decays occurs that leads to a stable isotope of lead (Pb). One such series begins with the isotope thorium-232 and proceeds through these elements:

Th $\Rightarrow$ Ra $\Rightarrow$ Ac $\Rightarrow$ Th $\Rightarrow$ Ra $\Rightarrow$ Rn $\Rightarrow$ Po $\Rightarrow$ Pb $\Rightarrow$ Bi $\Rightarrow$ Po $\Rightarrow$ Pb

a. Using the periodic table, find the atomic numbers for all of these elements.
b. Identify which of the reactions in this series involve alpha decay and which involve beta decay. (The change in atomic number provides all the information you need.)
c. Write the reaction equations for the first three decays in this series.
d. Fill in the mass numbers for all of the isotopes in this series.

CP3. Consider the fusion reaction: $_1H^2 + {}_1H^2 \Rightarrow {}_2He^3 + {}_0n^1$. From tables of nuclear masses, we can find the masses for the reactants and products in this reaction:

H^2 2.014 102 u

He^3 3.016 029 u

n   1.008 665 u

a. Find the mass difference Δm between the reactants and the products for this reaction.

b. Following the procedure used in Try This Box 19.1, convert this mass difference to energy units.

c. Is energy released in this reaction, and if so, where does it go? Explain.

CP4. Nuclear power has been a constant source of controversy over the last few decades. Although the use of nuclear power has grown during this time, we still get over half of our electric power by burning fossil fuels. The environmental and economic impacts differ for these energy sources.

a. Burning fossil fuels produces carbon dioxide as a natural by-product. Carbon dioxide is one of the gases that contributes to the greenhouse effect and global warming, as discussed in chapter 10. Is this a problem with nuclear power also? Explain.

b. What environmental problems associated with nuclear power are not present in the burning of fossil fuels? Explain.

c. What environmental problems associated with fossil fuels are not present in the use of nuclear power? Explain.

d. On balance, which of these power sources would you choose to develop further if other alternatives were not available? Explain. (Reasonable people may differ here!)

home experiments and observations

HE1. The concept of half-life, and the associated exponential decay curve, can be made more vivid by using piles of pennies (or other stackable objects) to represent atoms.

a. Collect as many pennies as you can find on dresser tops and from coin purses. Fifty to one hundred should suffice.

b. Divide your pile into two equal stacks, placed side by side. The left pile represents the original number of atoms.

c. Divide the right pile in half. Place one of the resulting stacks next to the original left stack. This represents the number of atoms remaining after one half-life has passed.

d. Continue this process, always dividing the remaining right stack in half and placing the stack obtained from division next to those stacks already accumulated. The resulting row of stacks, each one smaller than the preceding one, forms an exponential decay curve. How many half-lives do you obtain before you are down to one penny a stack?

unit Six

Relativity and Beyond

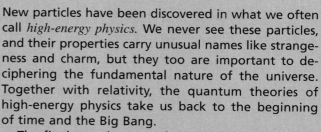

Throughout this book, we have tried to build your understanding of physics concepts by stressing their origin in, and application to, everyday phenomena—things that happen around us all the time. Certainly, we have strayed at times from everyday events, particularly in discussing the structure of the atom and its nucleus. These ideas have their origins in simple experiments, however, and they have many applications to familiar technology.

Some of the most fascinating ideas of modern physics are a bit more difficult to relate to everyday experience, for example, the theories of special and general relativity developed by Albert Einstein in the early years of the twentieth century. Relativity is fun to explore, though, because it challenges us to rethink basic concepts like space and time. Einstein's ideas can stretch your mind.

The development of quantum mechanics and its application to nuclear physics have also led to areas of research far removed from everyday experience.

New particles have been discovered in what we often call *high-energy physics*. We never see these particles, and their properties carry unusual names like strangeness and charm, but they too are important to deciphering the fundamental nature of the universe. Together with relativity, the quantum theories of high-energy physics take us back to the beginning of time and the Big Bang.

The final two chapters of this book briefly explore the theories of relativity (chapter 20) and recent developments in modern physics (chapter 21). Chapter 21 includes discussions of the particle zoo and cosmology, as well as developments in condensed-matter physics that brought about the revolution in microelectronics and computers.

The discoveries of fission and fusion occurred over 50 years ago, during the 1930s and 1940s. What has happened since these breakthroughs? What can we expect to hear from physics in the future? Chapters 20 and 21 may serve as a preview.

Relativity

chapter 20

chapter overview

After reexamining how relative motion is handled in classical physics, we introduce Einstein's postulates of special relativity and explore their consequences for our views of space and time. We then consider how Newton's laws of motion must be modified to be valid for very high velocities and explore the idea of mass-energy equivalence. Finally, we briefly discuss the general theory of relativity.

unit six

Have you ever been in a stationary bus peering out the window at another bus sitting alongside when suddenly the other bus moved forward and you had the distinct sensation that your own bus was moving backward (fig. 20.1)? The sensation lasts until the other bus moves out of your view, and you realize that you are not moving.

Your senses have deceived you because of the motion of your **frame of reference.** We normally measure our own motion with respect to objects that we expect to remain at rest. If these objects are fixed to the surface of the earth, our frame of reference is the earth: position, velocity, and acceleration are measured relative to that frame. If objects that we have identified with that fixed frame of reference suddenly move, though, we may perceive ourselves as moving.

All motion must be measured with respect to some frame of reference, and that frame of reference may also be moving. The earth rotates on its axis and also orbits the sun. The sun, in turn, is moving with respect to other stars, and so on. Something at rest in one frame of reference may be moving with respect to some other frame of reference—we must define our frame of reference to provide a complete description of motion.

The problem in defining a frame of reference, and in describing how a certain motion might look as it is seen from different frames of reference, was discussed by both Galileo and Newton. This is a simple problem if the relevant velocities are not large. Relative motion in this sense is part of our everyday experience. The motion of a boat relative to a flowing stream, for example, is familiar to many of us.

figure **20.1** The forward motion of an adjacent bus can give you the impression that your own bus is moving backward.

If we imagine ourselves moving along with a light beam, however, as Einstein may have as a boy, some very interesting questions arise. Addressing some of these questions led to the **special theory of relativity,** introduced by Einstein in 1905. This theory is mainly concerned with cases in which different frames of reference move at constant velocity with respect to one another. The **general theory of relativity,** which Einstein published roughly ten years later, deals with the relationship of gravity to accelerated frames of reference. Together these theories have revolutionized the way we view the universe.

20.1 Relative Motion in Classical Physics

Imagine that you have dropped a twig in the water of a moving stream and are watching its motion as it is carried along by the water (fig. 20.2). What is the velocity of the twig with respect to the bank of the stream? What is its velocity with respect to the water, and how are these two velocities related? How does this picture change if we consider a motorboat moving on a flowing stream? These questions can be addressed within the framework of classical mechanics developed by Galileo and Newton.

How do velocities add?

When you drop a twig in the water, the twig swiftly reaches the velocity of the stream. Once that happens, its velocity with respect to the bank is the same as the water's velocity with respect to the bank, but the twig's velocity with respect to the water is zero. If you watched the twig from a boat that is also floating with the current, the twig would not seem to move.

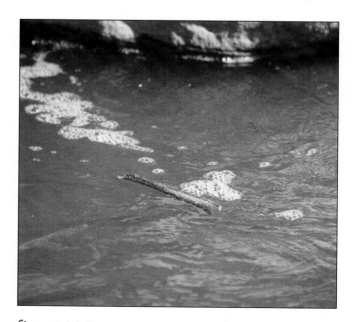

figure **20.2** A floating twig moves with the current. What is its velocity with respect to the stream bank?

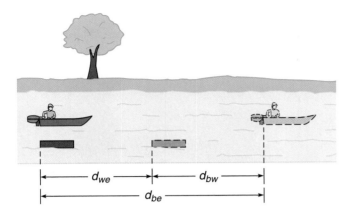

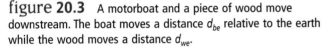

figure **20.3**   A motorboat and a piece of wood move downstream. The boat moves a distance d_{be} relative to the earth while the wood moves a distance d_{we}.

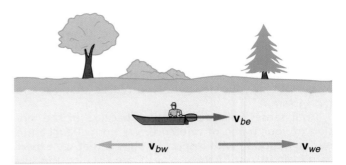

figure **20.4**   A motorboat moving upstream loses ground to the stream's current if it cannot travel fast enough relative to the water ($\mathbf{v}_{bw}$).

A more interesting situation arises if your boat is moving with respect to the water with the aid of oars, a motor, or a sail. In this case, the velocity of the boat relative to the water is not zero. The boat moves relative to the water, and the water moves relative to the stream bank (fig. 20.3). If the boat and the stream are moving in the same direction, we might assume that the velocity of the boat relative to the water should add to the velocity of the water relative to the earth to yield the overall velocity of the boat relative to the earth. It is like walking on a moving escalator or one of those traveling walkways found in airports.

If we express this idea in symbols, it takes the form

$$\mathbf{v}_{be} = \mathbf{v}_{bw} + \mathbf{v}_{we},$$

where $\mathbf{v}_{be}$ is the velocity of the boat relative to the earth, $\mathbf{v}_{bw}$ the velocity of the boat relative to the water, and $\mathbf{v}_{we}$ the velocity of the water relative to the earth. The velocity of the boat relative to the earth is the vector sum of the velocity of the boat relative to the water and the velocity of the water relative to the earth.

As shown in figure 20.3, our expectation that these velocities should add can be justified by considering the distances that the water and the boat travel in some fixed time. In this time, the piece of wood floating in the stream moves a distance $d_{we,}$ which is the distance that the water has moved relative to the earth. In that same time, however, the boat has moved a distance d_{bw} relative to the water, so that it is now that much ahead of the piece of wood. The total distance that the boat has moved relative to the earth in this time, $d_{be,}$ is the sum of the other two distances. Since the magnitude of the velocity (the speed) is distance divided by time, the velocities also add.

Although we have illustrated this idea for the simple case in which all three velocities were in the same direction, the velocity-addition result is valid more generally. For example, if we point the boat upstream, the different direction of the velocities of the boat and the current can be indicated by a difference in sign. If the boat travels at a speed of 4 m/s upstream relative to the water, and the water is moving at a speed of 6 m/s (downstream) relative to the earth, the velocity of the boat relative to the earth is −2 m/s (downstream). Since the result is negative, and we chose upstream to be positive, the boat is moving downstream (fig. 20.4).

How do velocities add in two dimensions?

Addition of relative velocities can also be extended to two or three dimensions. Suppose, for example, that you are traveling across stream, as in figure 20.5. If you point the motorboat directly toward a point straight across the stream, will the boat end up at that point? Not if the stream is moving.

At the same time that the boat is moving across the stream relative to the water, the water is moving downstream relative to the earth. The boat is carried downstream as it moves across the stream. As before, the two velocities add. (The vector addition process can be handled by the graphical method discussed in appendix C and illustrated in figure 20.5.) Notice that the size of the velocity of the boat relative to the earth (its speed) will not be equal to the simple numerical sum of the other two speeds in this case. Since this velocity vector is the hypotenuse of a right triangle in the vector diagram, it will be equal to the square root of the sum of the squares of the other two sides.

If you wanted to hit a point on the bank directly across the stream from your starting point, you would have to point the boat somewhat upstream, at an angle to the line drawn perpendicular to the bank (fig. 20.6). In this case, when the two velocities add, they produce a velocity of the boat relative to the earth that is straight across the stream. The magnitude of this velocity will be smaller than that of the boat relative to the water, however, as can be seen from the vector diagram.

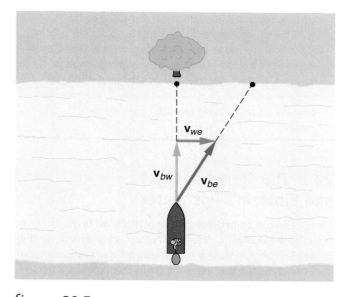

figure 20.5 A motorboat pointed straight across the stream ends up at a point on the opposite bank that is somewhat downstream.

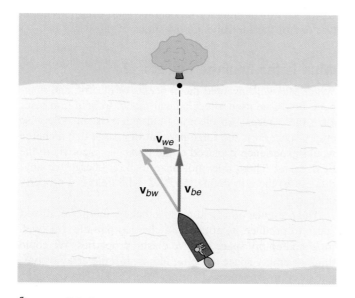

figure 20.6 The boat must be pointed somewhat upstream to travel straight across the stream.

We can apply the same analysis to an airplane. The velocity of the plane relative to the air adds to the velocity of the air relative to the earth (the wind velocity) to yield the velocity of the plane relative to the earth, $\mathbf{v}_{pe} = \mathbf{v}_{pa} + \mathbf{v}_{ae}$. A tailwind has a different effect than a head wind or a crosswind. The situation is directly analogous to the boat on the stream.

The principle of relativity

Velocity addition can also be applied to events happening in moving vehicles. Imagine, for example, that you are walking up the aisle of a large airliner traveling with a constant velocity relative to the earth. Your velocity relative to the plane must be added to that of the plane relative to the earth to find your own velocity relative to the earth. In practice, however, you are usually much more aware of your velocity relative to the plane than your velocity relative to the earth—the plane is your frame of reference.

As long as the plane is moving with constant velocity, you can move about a large airliner quite easily without much awareness of the motion of the plane. In fact, you can throw a ball back and forth or perform physical experiments in the airplane and obtain the same results as if these experiments were done in a stationary building. (Since the earth itself is rotating on its axis and orbiting about the sun, it is not truly stationary either.)

Even when a plane is moving with approximately constant velocity, we often have some sense of motion because of air turbulence causing it to bounce around. We also can look out the window and watch the clouds or earth go by. In smooth air and with the window shades closed, though, we lose the impression that we are moving. This loss is even more striking in an elevator moving up or down with constant velocity. Since elevators usually have no windows and can move very smoothly, it becomes hard to tell whether or not they are actually moving.

These ideas were discussed by both Galileo and Newton and are often summarized in the **principle of relativity:**

> The laws of physics are the same in any inertial frame of reference.

This principle means that we cannot tell whether our frame of reference is in motion or not by performing physical experiments. As long as our frame of reference is moving with constant velocity with respect to other inertial frames of reference, the results of our experiments are the same.

Inertial frames of reference

But what is an *inertial* frame of reference? At this point, logical difficulties arise, as Newton was well aware. Newton's second law of motion is valid only in an inertial frame of reference, that is, in any frame of reference for which Newton's first law is obeyed. If an object at rest remains at rest when the net force acting on the object is zero, that object is in an inertial frame of reference. Any frame of reference moving with constant velocity with respect to some valid inertial frame will also be an inertial frame of reference.

If some other frame of reference is accelerated with respect to a valid inertial frame, the accelerated frame is *not* an inertial frame of reference. If, for example, your airplane lurches up or down because of accelerations produced by air turbulence, your experimental results (and your ability to walk a straight line) are modified. Likewise, if an elevator is accelerating up or down, your apparent weight changes. If you were standing on a bathroom scale, as discussed in chapter 4, this change would register on the dial. Other experiments would also have modified results due to this acceleration.

If we want to apply Newton's second law in these situations, we have to modify the law by adding imaginary or *inertial* forces that arise as a result of the acceleration of the frame of reference. The *centrifugal force* that we sometimes talk about feeling in a rotating frame of reference is such an imaginary force. A rotating frame has a centripetal acceleration and is not a valid inertial frame. We feel as though we are being pulled outward by a centrifugal force—but viewed from a valid inertial frame, we see that this apparent force is really just our own inertia at work, our tendency to continue moving in a straight line while our reference frame is turning.

The centrifugal force that seems to be present in a rotating frame of reference is not a valid force in the Newtonian sense, because it does not derive from the interaction of the affected body with any other body. In other words, it does not obey Newton's third law, which is part of Newton's definition of force. It arises solely because of the acceleration of the frame of reference. It is like the increase or decrease in apparent weight observed in an accelerating elevator.

Defining an inertial frame of reference seems easy enough: it is one that is not accelerated. But with respect to what? For many purposes, we can treat the surface of the earth as an inertial frame of reference, since its acceleration is small. But the earth is rotating and also orbiting (in a curved path) about the sun, so it is accelerated with respect to the sun. The sun itself is accelerated with respect to other stars, so it is not a completely valid inertial frame of reference either.

Our problem stems from the apparent impossibility of establishing a frame of reference that is absolutely at rest, or at least not accelerated in any sense. Maxwell's prediction and description of electromagnetic waves brought new attention to this problem in the latter half of the nineteenth century. The possibility that measuring the velocity of light could help to establish an absolute inertial frame of reference was an exciting idea. Questions about the appropriate reference frame for measuring the velocity of light led to Einstein's special theory of relativity.

The velocity of any object must always be measured with respect to some frame of reference. For ordinary motions, the surface of the earth often provides that frame. If our frame of reference is moving with respect to some other frame, as with a boat on a flowing stream, the velocity of the boat relative to the stream can be added to the velocity of the stream relative to the earth to obtain the velocity of the boat relative to the earth. This process works in two or three dimensions as well as for straight-line motion. Galileo and Newton recognized that the laws of physics take the same form in any inertial frame of reference. The difficulty comes in trying to establish an absolute inertial frame.

20.2 The Speed of Light and Einstein's Postulates

Light is an electromagnetic wave, which was originally predicted by Maxwell's theory of electromagnetism, as discussed in chapter 16. An electromagnetic wave consists of oscillating electric and magnetic fields that propagate through empty space, as well as through air, glass, and other transparent materials. Light waves can travel through a vacuum.

Is there a medium that light waves travel through even when they are passing through a vacuum? Most waves travel through some medium or material, sound waves through air (and other materials), water waves in water, waves on a rope on the rope itself, and so on. Does light also have a medium? At the end of the nineteenth century (and even now), this was one of the fundamental questions of physics.

What is the luminiferous ether?

When Maxwell invented the concepts of electric and magnetic fields, he used a mechanical model to help him visualize these ideas. An electric field can exist in otherwise empty space: it is the force per unit of charge that a charge would experience if placed at the point in space where the field exists. There does not have to be a charge (or anything else) there to define the field—the field is a property of space.

The presence of a field must somehow modify or distort space to produce its effect on a charged particle. Maxwell imagined empty space to have elastic properties. We could think of empty space as being like an infinite array of tiny, massless, interconnected springs (fig. 20.7). The changing electric and magnetic fields of an electromagnetic wave could be viewed as a distortion of this array of springs. Although Maxwell did not believe this was an accurate depiction of empty space, such a model may have helped him to think about fields and the process of wave propagation.

Even if this model was not to be taken literally, attributing elastic properties to empty space seemed necessary to explain the propagation of a wave through space. Otherwise, nothing oscillates as the wave passes through. This invisible, elastic, and apparently massless medium that could exist in a vacuum was called the *luminiferous ether*—it was the medium light waves and other electromagnetic waves supposedly traveled through. Whether or not it was needed to

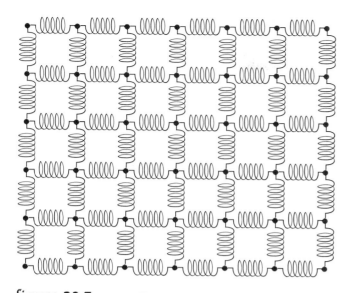

figure 20.7 Maxwell imagined empty space to have elastic properties. An array of massless, interconnected springs can serve as a crude model of this idea.

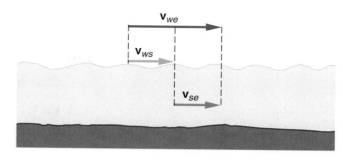

$$\mathbf{v}_{we} = \mathbf{v}_{ws} + \mathbf{v}_{se}$$

figure 20.8 The velocity of wave relative to the stream adds to the velocity of the stream to yield the velocity of the wave relative to the earth.

explain the propagation of electromagnetic waves was a matter of debate. Maxwell himself was not totally convinced.

Could the ether serve as a universal reference frame?

The supposed existence of the ether opened an exciting possibility for solving the problem of inertial frames of reference mentioned in the preceding section. Perhaps, the ether could serve as an absolute or universal reference frame for measuring any motion. Any other valid inertial frame of reference would then be traveling with constant velocity relative to the ether. The ether itself could be pictured as being embedded and fixed somehow in empty space.

How could we measure motion relative to the ether? Simply by measuring the velocity of light. If the earth is moving relative to the ether, the velocity of light should be affected by this motion. The velocity-addition formula introduced in section 20.1 should apply. This idea is easier to visualize if we consider a water wave on a flowing stream, as in figure 20.8.

If the wave is moving with a velocity $\mathbf{v}_{ws}$ relative to the stream, and the stream is flowing in the same direction with a velocity $\mathbf{v}_{se}$ relative to the earth, the velocity of the wave relative to the earth, $\mathbf{v}_{we}$, should be $\mathbf{v}_{we} = \mathbf{v}_{ws} + \mathbf{v}_{se,}$ just like a boat going downstream. The velocity of the wave in the medium (the stream, in this case) adds to the velocity of the medium relative to the earth to yield the overall velocity of the wave relative to the earth.

Extending these ideas to light waves traveling in the ether is not difficult. If the earth is moving through the ether, the ether is also flowing past the earth. The velocity of light

that we measure should then be the vector sum of the velocity of light relative to the ether and the velocity of the ether relative to the earth. Accurate measurement of the velocity of light relative to the earth at different times of the year would then let us determine whether or not the earth is moving in a certain direction relative to the ether.

The Michelson-Morley experiment

The most famous experiment designed to detect the possible motion of the earth relative to the ether was performed by Albert Michelson (1852–1931) and Edward Morley (1838–1923) during the 1880s at what is now Case Western Reserve University in Cleveland. To detect small differences in the velocity of light, they used a special instrument designed by Michelson, now called the *Michelson interferometer.* As the name suggests, this instrument uses interference phenomena to detect small differences in the velocity of light or in the distance that the light travels (see section 16.3).

A Michelson interferometer is shown in figure 20.9. Light introduced from the source on the left is split into two beams by a partially silvered mirror, or beam splitter. Roughly half of the light striking this mirror passes through the mirror and half is reflected, producing two beams of equal intensity. These two beams travel along perpendicular paths, as shown, and are reflected by fully silvered mirrors, returning through the beam splitter. Here, the beams are partially transmitted and partially reflected again.

The observer looks at images of the source along a path on the fourth side of the interferometer at the bottom of the diagram. Because the light the observer sees has come from the same source but has traveled along different paths, it may be either in or out of phase for different points on the image of the source. If one of the end mirrors is slightly tilted so that it is not exactly perpendicular to the light beam, these phase differences will create a pattern of dark and light fringes, as in figure 20.9. The dark fringes are

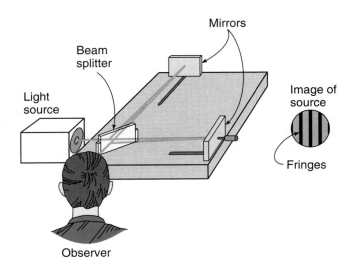

figure **20.9**   A Michelson interferometer. Light waves traveling along the two perpendicular arms interfere to form a pattern of light and dark fringes.

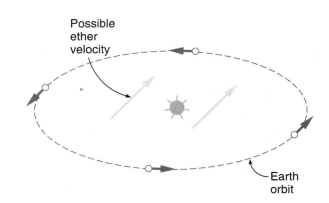

figure **20.10**   Regardless of the direction of the ether's motion relative to the sun, at some time during the year the earth should be moving relative to the ether.

produced by destructive interference, the light fringes by constructive interference.

If anything happens to change the time required for either one of the light beams to make its trip to the end mirror and back, the phase difference will change and the fringe pattern will shift. Michelson and Morley reasoned that if the ether was moving in a direction parallel to one of the arms, the time interval for the beam moving parallel to the ether stream would be slightly different from the time interval of the beam moving perpendicular to the ether stream. These time differences could be computed by finding the effective wave velocities for light traveling along each arm. The computation is similar to the one for a boat moving parallel or perpendicular to the current of a stream.

To see the expected shift of fringes, the interferometer must be rotated through 90° (along with the observer), so that the arm that had been parallel to the ether stream is now perpendicular to the ether stream, and vice versa. Michelson and Morley mounted the interferometer on a rock slab and floated the slab in a vat of mercury to let the interferometer rotate smoothly. (Mercury was the only obtainable fluid dense enough to float a heavy rock slab.)

Michelson and Morley based their computations on the assumption that the velocity of the earth relative to the ether would be due, in part, to the orbital motion of the earth around the sun. Since they could not assume that the ether was fixed with respect to the sun, they had to do the experiment at different times of the year. At some time during the year, the motion of the earth should be parallel to a component of the ether's motion, and six months later it should be antiparallel (fig. 20.10). They assumed that the minimum velocity of the earth relative to the ether would

be equal to the orbital velocity of the earth about the sun. It would have this value if the ether was fixed with respect to the sun. If the ether was moving relative to the sun, the relative velocity of the ether to the earth would be even larger at some time during the year.

The results of the Michelson-Morley experiment were disappointing: no fringe shift was observed when the interferometer was rotated at any time during the year. Although the shift was expected to be small (of the order of half a fringe width), it should have been observable according to the assumptions the experiment was based on. The experiment failed to detect any motion of the earth relative to the ether. Often in science, though, failing to find what is expected can be an important result.

Einstein's postulates of special relativity

The failure of the Michelson-Morley experiment to detect any motion of the earth relative to the ether raised new questions about the ether. Why could we not detect its motion? Maybe the ether was being dragged along with the earth, much like the atmosphere, so that an experiment performed on the surface of the earth would not detect motion. This assumption seemed to be precluded, however, by other observations involving shifts in the apparent positions of stars viewed at different times of the year.

Although Einstein was just a child at the time of the Michelson-Morley experiment, he later became familiar with the debate that followed. His solution to the dilemma was both simple and radical. He merely took as a basic postulate what seemed to be the case experimentally, namely, that the velocity of light is not affected by motion of the source or of the frame of reference.

Einstein actually stated two postulates in his introductory paper on special relativity published in 1905. The first was a reaffirmation of the principle of relativity stated similarly

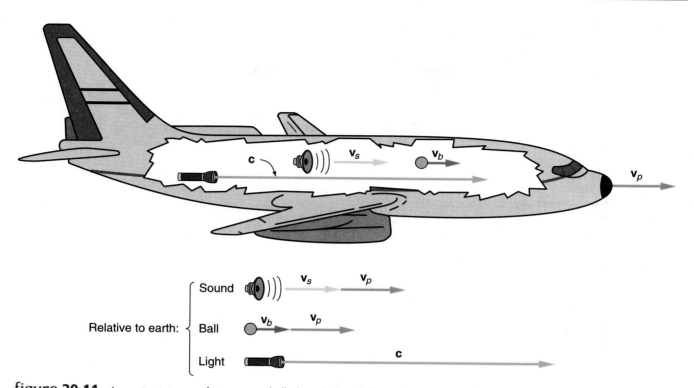

figure 20.11   In contrast to sound waves or a ball, the velocity of the airplane does not add to the velocity of a flashlight beam to yield the speed of light relative to the earth. The speed of light is the same for all observers. (Velocity vectors are not drawn to scale.)

more than 200 years earlier by Galileo and Newton and discussed in section 20.1:

> **Postulate 1:** The laws of physics are the same in any inertial frame of reference.

The second postulate involved the speed of light:

> **Postulate 2:** The speed of light in a vacuum is the same in any inertial frame of reference, regardless of the relative motion of the source and observer.

Although both postulates are important to Einstein's theory, the second one calls for a radical change in our thinking. What he was saying, in essence, is that light (or any electromagnetic wave) does not behave like most waves or moving objects. If we throw a ball on a moving airplane, the velocity of the ball relative to the earth is the vector sum of the velocity of the ball relative to the airplane plus the airplane's velocity relative to the earth. If the pilot speaks on the sound system, the sound wave travels with a velocity relative to the earth that is the vector sum of the velocity of sound in the air of the airplane plus the airplane's velocity relative to the earth (fig. 20.11).

If we shine a flashlight on the airplane, however, the velocity of light measured on the plane must be the same, according to Einstein's second postulate, as what is measured for the same flashlight beam by an observer at rest on the earth. The classical velocity-addition formula does *not* hold for light—not an easy idea for physicists to accept in 1905. In fact, if we examine this second postulate more closely, we find that it requires us to rethink space and time themselves. This aspect of relativity really challenges our minds. We will begin to explore some of these consequences in section 20.3.

The ether was assumed to be the medium for electromagnetic waves. Since these waves could travel in a vacuum, the ether was thought to exist even in a vacuum. If the earth moves through the ether, we might be able to establish an absolute frame of reference associated with the ether. The Michelson-Morley experiment was designed for this purpose but failed to detect any motion of the earth relative to the ether. In response to this and other experiments, Einstein postulated that the speed of light is the same in any inertial frame of reference, denying the existence of the ether. This assumption holds radical implications for our concepts of space and time.

20.3 Time Dilation and Length Contraction

The units for velocity are always a ratio of distance (a measure of space) divided by time—for example, meters per second. The law of velocity addition depends on the assumption that space and time can be measured by different observers in the same manner and with the same results, regardless of whether these observers are moving relative to one another. This assumption is consistent with our daily experience.

If the velocity of light does not add like ordinary velocities, there must be some problem with how different observers measure space or time. If we accept Einstein's second postulate that the speed of light is the same for all observers, we must give up the ideas that space and time are the same for all observers. This goes against our intuition or common sense and requires us to throw out ideas that seem to be inherently true.

To approach these questions, Einstein devised *thought experiments,* experiments that are impractical to perform because of the tremendous velocities involved but that can be readily imagined and their consequences explored. Thought experiments allow us to see how the concepts of space and time must be altered to accept Einstein's second postulate. Anyone can do thought experiments; no physical equipment is required.

Measurements of time by different observers

Suppose that you wish to measure time using the velocity of light as your standard of measurement. Imagine, too, that you are riding in a spaceship moving with a large velocity with respect to the earth. If your spaceship has a large glass window on one side, another observer standing on earth could also watch your experiments and make measurements.

How would you go about using the speed of light as a standard for a time measurement? One way would be to send a light beam at a mirror directly overhead and use the time required for the beam to travel to the mirror and back as a basic unit of time. This arrangement would be a **light clock**—it uses the speed of light to establish a time standard. If the distance from the light source to the mirror is d (fig. 20.12), the time interval required for the light to travel to the mirror and back (a distance of $2d$) is

$$t_0 = \frac{2d}{c},$$

where c is the speed of light. This quantity becomes your basic measure or unit of time as measured in the spaceship.

The observer standing on the earth as you flash by in your glass-walled spaceship can also see the time taken for the light beam to make its trip to the mirror and back. She views the events somewhat differently, though. If the spaceship is moving with a velocity **v** relative to the earth, the

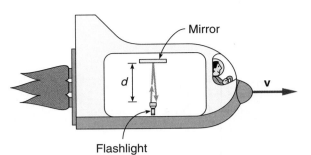

figure 20.12 In a light clock, the time taken for light to travel the distance $2d$ to the overhead mirror and back becomes the basic measure of time t_0.

mirror also moves at that velocity. For the light beam to be reflected from the mirror and return to the source (which, meanwhile, has moved), it must travel along the diagonal path shown in figure 20.13.

If the observer on earth uses the same light clock to establish a measure of time, her basic measure t will be larger than t_0. She sees the light beam traveling a longer distance at the *same* speed c than the distance measured by the observer on the spaceship (yourself). We assume that she measures the vertical distance d in the same manner as you do, because this distance is perpendicular to the direction of relative motion and should be unaffected by the motion. The longer path for the light beam yields a greater time.

The difference in the time intervals measured by the two observers can be found by considering the geometry and distances in figure 20.13. (See challenge problem 5.) The time t measured by the earth observer can be expressed in terms of the time t_0 measured on the spaceship as

$$t = \frac{t_0}{\sqrt{1 - \dfrac{v^2}{c^2}}}.$$

This is the *time-dilation* formula. In it, t will always be larger than t_0, since the number in the denominator is always less than one. The observer on earth measures a longer, *dilated* time with the light clock.

The time t_0 is often called the **proper time.** In this case, it is the time interval measured in the spaceship where the light starts and finishes its trip at the same point in space.

A proper time interval is the elapsed time between two events measured in a frame of reference in which the two events occur at the same place within that frame of reference.

This is true for the time interval measured by the person on the spaceship but not for the observer standing on earth. She sees the light leaving the source at one point in space

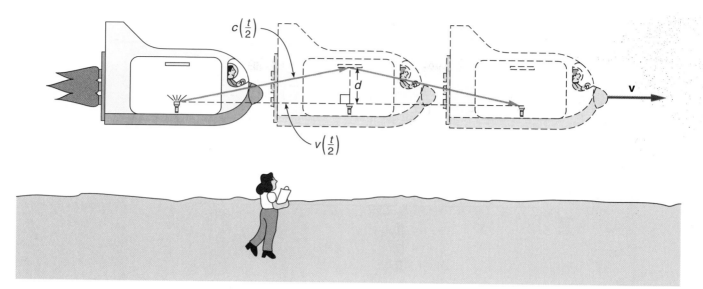

figure 20.13 To an observer standing on earth, the light takes a diagonal path to the mirror and back to the source. This yields a longer time interval as measured by the light clock.

and returning to the source at a different point in space and measures a dilated time interval for the elapsed time between the start and finish of the beam's flight.

Step back and try to see what we have actually done in this thought experiment. We have used the speed of light as a standard for measuring time in a light clock. By insisting that the two observers moving relative to one another observe the same value for the speed of light c, we find that they arrive at different measures of time using the same clock. If they agree on the value of c, they cannot agree on the travel time of the light beam. Time does not pass at the same rate in the different frames of reference.

For ordinary speeds of relative motion, the difference in these two time intervals would be extremely small. For a speed of one-hundredth the speed of light ($0.01c$), for example (still the enormous speed of 3 million meters per second), the quantity v/c in the time-dilation formula is 0.01. The dilated time t in this case is just 1.00005 multiplied by t_0, so the difference between t and t_0 is very small. The relative speed v of the two observers must be almost as large as the speed of light for the difference in these time intervals to be noticeable.

The quantity containing the square root that appears in the time-dilation formula is involved in many relativistic expressions, and the Greek letter γ (gamma) is used as its symbol, where

$$\gamma = \frac{1}{\sqrt{1 - \dfrac{v^2}{c^2}}} .$$

Table 20.1 shows the value of γ for a few values of the relative speed v. These values are very close to 1 for small

table **20.1**	
Values of γ for Different Values of Relative Speed v	
$v = 0.01c$	$\gamma = 1.00005$
$v = 0.1c$	$\gamma = 1.005$
$v = 0.5c$	$\gamma = 1.155$
$v = 0.6c$	$\gamma = 1.250$
$v = 0.8c$	$\gamma = 1.667$
$v = 0.9c$	$\gamma = 2.294$
$v = 0.99c$	$\gamma = 7.088$

values of v but increase rapidly as the magnitude of v approaches c. (The relative speeds are all expressed here as fractions of the speed of light c.) Written in terms of gamma, the time-dilation relationship takes the form $t = \gamma t_0$, that is, the proper time t_0 must be multiplied by the factor γ to obtain the dilated time t.

How do length measurements vary for different observers?

Our two observers disagree on the elapsed time for the flight of the light beam, and they will also disagree on the distance the spaceship and the mirror traveled during this time. We can see this by extending our thought experiment to measure the distance that the spaceship travels in the time required for one round-trip of the light beam.

This distance can be measured most readily by the observer standing still on the surface of the earth. With the

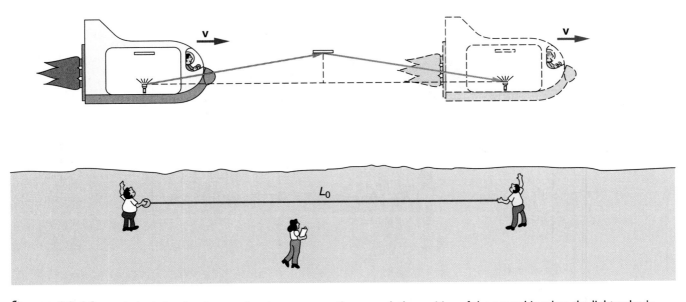

figure 20.14 With the help of assistants, the observer on earth can mark the position of the spaceship when the light pulse is emitted and when it returns. The distance between these positions L_0 can be easily measured.

help of assistants spaced along the path of the spaceship, she can mark both where the spaceship is when the light pulse is emitted and when it returns to the source. She can then measure the distance between these two points at her leisure, since this distance is fixed on the earth's surface (fig. 20.14).

The observer in the spaceship has a somewhat more difficult task in measuring this distance. He sees the earth as moving past him and must somehow locate the end points of the distance simultaneously. If he could measure the speed of the spaceship independently of this distance, he could then compute the distance by multiplying the speed v by the time of flight of the light beam. This quantity is the time t_0, the proper time, since that is the time measured by the astronaut. Using this method, he would measure a length $L = vt_0$ for the distance covered by the spaceship during the flight of the light beam.

Using the same reasoning, the observer on earth could also compute this distance. She finds $L_0 = vt$, where t is the dilated time that she measures for the flight of the light beam. We have used the symbol L_0 because this is the **rest length**, the length measured by the observer who is at rest relative to the distance being measured. Since we have already discovered that t is greater than t_0, we see that the rest length L_0 must be larger than the length L measured by the observer in the spaceship. He measures a *contracted,* or shorter, length than the rest length.

Since $t = \gamma t_0$, the contracted length can be expressed as

$$L = \left(\frac{1}{\gamma}\right) L_0.$$

This is the *length-contraction* formula. Since γ is always greater than 1, L is always less than L_0, the rest length.

Again, for the effect to be noticeable, v must be very large, as illustrated in Try This Box 20.1. In this example, the spaceship is traveling with a velocity of $0.6c$. The spaceship's pilot measures a contracted length of 720 km for the trip distance while the earth observer measures the trip distance as 900 km. The pilot also measures a shorter time of 4 milliseconds (the proper time) for the trip than the 5-millisecond dilated time measured by the earth observer.

Although these effects seem strange, they have been observed in a wide variety of circumstances. While ordinary-sized objects seldom move at speeds large enough to show noticeable effects, subatomic particles routinely travel at such speeds. A particle that has a certain lifetime at rest in the lab seems to have a longer (dilated) lifetime when it is moving at velocities near the velocity of light. It travels farther before decaying, as seen from the perspective of an observer at rest in the laboratory.

From the perspective of an observer traveling with the particle, however, the particle has its proper lifetime t_0 and travels a contracted distance L shorter than the rest length measured by the laboratory observer. The situation is basically the same as for the spaceship in Try This Box 20.1. All of these observations are consistent if we treat them according to Einstein's theory.

The problem of measuring distances from a moving spaceship, if examined closely, boils down to a problem of simultaneously locating the end points of the length being measured. Not only will two observers disagree on the elapsed time, they will also disagree about whether events are simultaneous or not. Two events separated in space may be seen as being simultaneous by one observer but as occurring at different times by an observer moving relative to the first observer.

try this box 20.1

Sample Exercise: A Contraction in Length

A spaceship traveling at the velocity of 1.8×10^8 m/s ($0.6c$) covers a distance of 900 km as measured by an observer on the earth.

    a. What is the distance traveled in this time as measured by the pilot of the spaceship?

    b. How much time does it take to cover this distance as measured by the observer on earth and as measured by the pilot?

a. $v = 0.6c$ From table 20.1,

$c = 3 \times 10^8$ m/s $\gamma = 1.25$

$L_0 = 900$ km $\dfrac{1}{\gamma} = \dfrac{1}{1.25}$

$L = ?$

$$\dfrac{1}{\gamma} = 0.8$$

$$L = \left(\dfrac{1}{\gamma}\right)L_0$$

$$= (0.8)(900 \text{ km})$$

$$= \mathbf{720 \text{ km}}$$

b. $t = ?$ As seen by the observer on earth: $L_0 = vt$

$t_0 = ?$

$$t = \dfrac{L_0}{v}$$

$$= \dfrac{9 \times 10^5 \text{ m}}{1.8 \times 10^8 \text{ m/s}}$$

$$= 5 \times 10^{-3} \text{ s} = \mathbf{5 \text{ ms}}$$

As seen by the spaceship pilot: $L = vt_0$

$$t_0 = \dfrac{L}{v}$$

$$= \dfrac{7.2 \times 10^5 \text{ m/s}}{1.8 \times 10^8 \text{ m/s}}$$

$$= 4 \times 10^{-3} \text{ s} = \mathbf{4 \text{ ms}}$$

These space and time effects are explored further in Everyday Phenomenon Box 20.1 on the famous *twin paradox*. The paradox involves the difference in aging rates of two twins, one who makes a space trip to a distant star and back and the other who remains on earth. The fact that the traveling twin ages less than the one remaining on earth can be understood using the concept of time dilation. It is not just science fiction!

If we accept Einstein's second postulate that the speed of light has the same value for different observers, regardless of their relative motion, we must give up some cherished ideas about space and time. Using the speed of light as a standard for time measurements, we find that a dilated or longer time is measured between two events for an observer who does not see these events as occurring at the same place in space. Also, an observer who is moving relative to a distance being measured finds a contracted or shorter length. Different observers cannot even agree on whether two events are simultaneous or not.

20.4 Newton's Laws and Mass-Energy Equivalence

Accepting Einstein's postulates requires some major changes in how we think about space and time. Since space and time measurements are involved in velocity and acceleration, and acceleration plays a key role in Newton's laws of motion, we might suspect that Newton's laws must also be modified to be consistent with Einstein's postulates. Does Newton's second law of motion still apply when objects are moving at very large velocities?

In addressing these questions, Einstein discovered that it was necessary to modify Newton's second law by redefining the concept of momentum. As he explored the consequences of this new approach to dynamics in his early papers on relativity, he was also led to a striking conclusion about the relationship between mass and energy summarized in the often-quoted equation $E = mc^2$.

How must Newton's second law be modified?

When Einstein examined Newton's second law in the light of his postulates, he discovered problems: an acceleration measured in one frame of reference is not the same as the acceleration of the same object measured in some other frame of reference. If a spaceship pilot fires a projectile with a certain acceleration, the observer on earth would measure a different acceleration.

At ordinary velocities, there is no difference in the accelerations measured by different observers. We can apply Newton's second law, $\mathbf{F} = m\mathbf{a}$, in any inertial frame of reference using the same forces to explain the acceleration of the object. Newton's second law apparently will not work when the velocities are very large, because the accelerations are no longer equal in different frames of reference. We then have a violation of Einstein's first postulate: Newton's second law does not seem to take the same form in different inertial frames of reference.

As discussed in chapter 7, the most general form of Newton's second law is stated in terms of momentum rather

Everyday Phenomenon

box 20.1

The Twin Paradox

The Situation. One of the most discussed phenomena of relativity is the so-called *twin paradox*. One of a pair of identical twins, Adele, journeys at very large velocities to a distant star and then returns to earth. The second twin, Bertha, remains on earth the entire time that her twin is traveling.

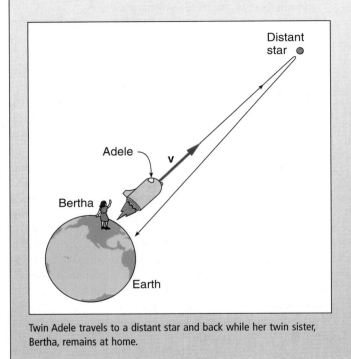

Twin Adele travels to a distant star and back while her twin sister, Bertha, remains at home.

Since one of the twins is traveling at a speed approaching the speed of light, the twins should measure time as passing at different rates because of time dilation. When the traveling twin returns, will she find that she is younger than the twin who stayed at home? Since each twin can regard the other as moving while she herself is standing still (as long as the velocity is constant), shouldn't the other twin also see herself as younger? This question lies at the heart of the apparent paradox.

The Analysis. Suppose that Adele's entire trip is made at the speed $v = 0.6c$. From table 20.1, the factor γ that appears in the time-dilation formula is equal to 1.25. If Adele perceives the trip as taking 12 years, this is a proper time in her frame of reference, the spaceship. She has lived 12 years and experienced an appropriate number of heartbeats (or other suitable biological measures of time) during her trip. In other words, as far as she is concerned she is 12 years older than when she left.

Her twin, Bertha, on the other hand, experiences a dilated time for this same time interval. She lives a time $t = \gamma t_0$, which is 15 years (1.25×12 years) for the trip. Bertha has aged 15 years waiting for her sister to return. Since Adele has aged only 12 years during the trip, Adele is 3 years younger than her identical twin at the end of the trip!

(continued)

---

than acceleration, $\mathbf{F} = \Delta\mathbf{p}/\Delta t$—the net force equals the rate of change of momentum. Momentum is defined as the product of the mass multiplied by the velocity, $\mathbf{p} = m\mathbf{v}$. We also find problems at large velocities when we attempt to use Newton's second law in this form. Even the law of conservation of momentum does not seem to work when viewed by observers in different frames of reference.

To salvage the law of conservation of momentum at relativistic (large) velocities, Einstein found that he had to redefine momentum as

$$\mathbf{p} = \gamma m\mathbf{v}.$$

Here $\mathbf{v}$ is the velocity of an object with respect to a given frame of reference and γ is the relativistic factor defined in section 20.3, which also depends on the magnitude of the velocity v. Using this new definition of momentum, Einstein was able to show that different observers could agree

that momentum was conserved in a collision, even though these different observers would measure different values for the velocities and momenta. At low velocities, this revised definition of momentum reduces to the ordinary definition of momentum, $\mathbf{p} = m\mathbf{v}$, because the factor γ is then approximately equal to 1.

Since the law of conservation of momentum follows directly from Newton's second law, this revised definition of momentum must be used there also. In other words, Einstein found that he could make Newton's second law conform to his postulates by using the new definition of momentum in the general form of Newton's second law, $\mathbf{F} = \Delta\mathbf{p}/\Delta t$. At ordinary velocities, Newton's second law works the usual way because the relativistic momentum reduces to the classical definition. At very large velocities, we are forced to use the relativistic definition of momentum. Einstein's special theory of relativity is a significant revision of Newton's theory of mechanics.

Suppose, however, that we did the same analysis with the spaceship as fixed and the earth as moving. Would we arrive at the reverse conclusion to what we just computed? Would such an analysis suggest that Bertha should be 3 years younger than Adele? Surely we cannot hold both of those results to be true. Therein lies the paradox.

The resolution of the paradox lies in the fact that we have ignored the accelerations and the resulting changes in reference frame. To make a space trip such as this, Adele's spaceship must accelerate away from earth until it reaches the enormous speed of 0.6c that we have assumed. When it reaches the distant star, it must turn around, which involves a deceleration and acceleration in the opposite direction (and into a different frame of reference). It must decelerate again when it reaches the earth. Our situation is not really completely symmetric: the spaceship exists in two different frames of reference, each having speed v relative to the earth but with velocities in opposite directions.

Although accelerations can be handled using general relativity, this is not really necessary for our purposes. If we assume that the accelerations take place in time intervals that are small compared to the overall time of flight, the computation done above using special relativity produces the correct result—Adele does age less than Bertha. This can be confirmed by doing a thought experiment using the basic assumptions of special relativity and carefully treating the behavior of clocks in either frame of reference. We must assume, however, that the spaceship changes frames of reference, not the earth.

At a speed of 0.995c for v, the differences in aging of Adele and Bertha could be quite striking.

This difference in the passage of time and the resulting difference in aging of the twins is a real effect that has been confirmed experimentally using highly accurate clocks and the much slower speeds of a jet plane. If we would reach velocities as high as 0.995c, the difference in aging of the twins would be quite striking. At a velocity of 0.995c, the time-dilation factor is approximately 10 rather than 1.25. A trip that took 10 years for Adele would take 100 years for Bertha. Adele would return to earth 100 earth-years later having aged only 10 years herself.

How did the idea of mass-energy equivalence emerge?

As he revised Newton's second law, Einstein discovered that mechanical energy also took on a new meaning. In classical physics, the kinetic energy of an object is found by computing the work done to accelerate the object to a given speed, resulting in the familiar expression $KE = \frac{1}{2}mv^2$. (See chapter 6 and figure 20.15.) Using the same procedure, we can compute the kinetic energy for an object accelerated to a very large velocity. In this case, however, we must use the modified version of Newton's second law to describe the process of acceleration.

When Einstein computed the kinetic energy using the relativistic modification of Newton's second law, he obtained the result

$$KE = \gamma mc^2 - mc^2.$$

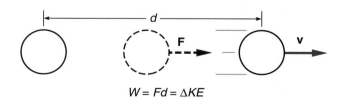

$$W = Fd = \Delta KE$$

figure 20.15 As before, the work done by the net force used to accelerate an object equals the increase in the kinetic energy of the object.

Note that only the first term in this expression depends on the speed of the object, since the factor γ contains the speed. The second term is independent of the speed of the object.

Arriving at this result was a simple process for Einstein (the computation is not difficult for someone experienced in using calculus). Interpreting the result, however, offered a significant challenge. The new expression for kinetic energy is the difference between two terms, one that depends on the speed and another that does not. Apparently, accelerating an object increases the energy of the object above an energy that it already possesses by virtue of its mass, mc^2.

The quantity mc^2 is often called the **rest energy** and is given its own symbol, $E_0 = mc^2$. The expression for the kinetic energy obtained by Einstein can be rearranged, putting the rest energy on the other side of the equation with the kinetic energy, $KE + E_0 = \gamma mc^2$. The expression γmc^2 is the total energy, the sum of the kinetic energy and the rest energy. When an object is accelerated, the total energy and the kinetic energy increase because the factor γ increases as the speed increases.

How do we interpret the rest energy?

The rest energy term was the most interesting feature of Einstein's computation of the kinetic energy. Since c is a constant of nature, in multiplying the mass of an object by c^2 to obtain an energy value (mc^2), we are just multiplying the mass by a constant. What this seems to indicate is that mass is *equivalent to* energy. If we increase the mass of an object or system, we increase its energy—if we increase the energy of a system, we increase its mass. This is the essence of the $E_0 = mc^2$ relationship.

Mass-energy equivalence is illustrated in figure 20.16, which shows a Bunsen burner heating a beaker of water.

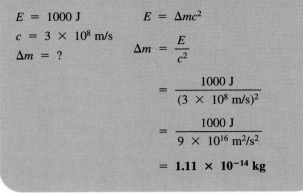

try this box 20.2

Sample Exercise: Adding Energy Adds Mass

A Bunsen burner adds 1000 J of heat energy to a beaker of water. What is the increase in the mass of the water?

$E = 1000$ J

$c = 3 \times 10^8$ m/s

$\Delta m = ?$

$E = \Delta mc^2$

$\Delta m = \dfrac{E}{c^2}$

$\qquad = \dfrac{1000 \text{ J}}{(3 \times 10^8 \text{ m/s})^2}$

$\qquad = \dfrac{1000 \text{ J}}{9 \times 10^{16} \text{ m}^2/\text{s}^2}$

$\qquad = \mathbf{1.11 \times 10^{-14} \text{ kg}}$

Since heat flow is a flow of energy, we increase the internal energy of the water by heating it. We are also increasing the mass of the water, since energy is mass. The amount the mass increases in this example would be very small and extremely difficult (if not impossible) to measure. If we add 1000 joules of heat energy, the increase in mass is just 1.1×10^{-14} kg, as shown in Try This Box 20.2. Since a beaker of water would normally contain a few tenths of a kilogram of water, a change of 10^{-14} kg would be utterly negligible.

Because we are so used to thinking of mass and energy as different, the idea that mass is equivalent to energy can be hard to accept. The principle itself has been thoroughly confirmed, because it correctly predicts the amount of energy released in nuclear reactions such as fusion or fission, as described in chapter 19. It is sometimes stated that mass is converted to energy in such reactions, but it would be better to say that rest-mass energy has been transformed to kinetic energy. In other words, mass cannot be converted to energy because it already is energy—we are merely transforming one type of energy into another.

Mass-energy equivalence, like the other ideas that we have been describing, is just another result of applying Einstein's postulates carefully and consistently to mechanics. The surprising results have led to fundamental revisions in our understanding of the concepts of energy and mass, as well as of space and time.

figure 20.16   A Bunsen burner adds mass to a flask of water by increasing the internal energy of the water. Energy and mass are equivalent.

Further exploration of Einstein's postulates showed that different observers could not agree on acceleration values, or even that momentum is conserved. A modification of Newton's second law of motion was required, which was accomplished by changing the definition of momentum in the general form of the second law. Using the modified form of the second law also changes the expression for kinetic energy. One of the terms in the new expression for kinetic energy does not depend on the speed of the object, suggesting the concept of a rest energy associated with the mass of the object. Mass-energy equivalence has since been demonstrated dramatically in nuclear reactions.

20.5 General Relativity

Our discussion so far has been restricted to cases involving inertial frames of reference, that is, reference frames moving at constant velocity relative to one another. What happens if our frame of reference is accelerating? Can we extend the type of thinking used in special relativity to this situation?

Einstein addressed these questions shortly after his introduction of the theory of special relativity, but it took him some time to refine the ideas. He did not publish the resulting general theory of relativity until 1915, about ten years after his first paper on special relativity. Once again, Einstein's ideas led to radical adjustments in our view of the universe.

What is the principle of equivalence?

We discussed accelerating reference frames earlier, when we considered how things appear to someone inside an accelerating elevator, at the beginning of this chapter and in chapter 4. If the elevator is moving with constant velocity, Einstein's first postulate (the principle of relativity) tells us that the laws of physics will behave exactly as they would if the elevator were at rest. In other words, no experiment that we can do inside the elevator could establish whether or not we are moving with respect to the earth.

If the elevator is accelerating, however, we expect differences from what we would see if the elevator were at rest or moving with constant velocity. In particular, as discussed in chapter 4, a person standing on a bathroom scale while the elevator is accelerating upward will register a greater weight than if the elevator were not accelerating (fig. 20.17). From Newton's second law, this greater apparent weight results from the scale's exerting a larger upward force on your feet (the normal force) than your actual weight. The net upward force makes you accelerate upward along with the elevator.

This change in the scale reading could be used as an indication that the elevator is accelerating. If the elevator is accelerating upward, the reading will be higher than normal. If the elevator is accelerating downward, the reading will be lower than normal. If the cable of the elevator is cut and the elevator accelerates downward with an acceleration g (free fall), the scale will read zero—apparent weightlessness. Until things come to a crashing halt at the bottom of the shaft, you can float around inside the elevator much like an astronaut in an orbiting space shuttle.

figure 20.17 If the elevator is accelerating upward, the scale reads a value N, which is larger than the person's usual weight W.

Other experiments will also lead to results that you would not expect if the elevator were not accelerated. For example, a dropped ball will approach the floor of the elevator with an apparent acceleration that is different from $g = 9.8$ m/s^2. If the elevator is accelerating upward, the apparent acceleration of the ball will be larger than g: it is the sum of the magnitude of the acceleration of the elevator a and the gravitational acceleration g. If the elevator is accelerating downward, the apparent acceleration of the ball will be less than g, as figure 20.18 shows. The period of a swinging pendulum will also differ from the one you would observe if the elevator were not accelerating.

These experiments all have one thing in common: they can be interpreted in terms of an apparent acceleration of gravity that differs from $g = 9.8$ m/s^2. Since weight is equal to mass times the acceleration of gravity (mg), the changes in apparent weight can be attributed to a change in the apparent value of the gravitational acceleration. We explain the changes in the acceleration of the ball or the period of the pendulum in the same way. Although we can detect the acceleration of the elevator, we cannot distinguish these effects from what would happen if the acceleration due to gravity were being increased or decreased somehow.

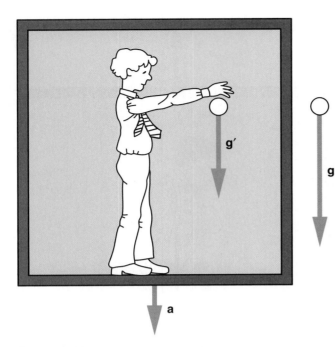

figure **20.18**   A ball dropped in an elevator accelerating downward approaches the floor with an apparent acceleration **g′** that is less than **g**.

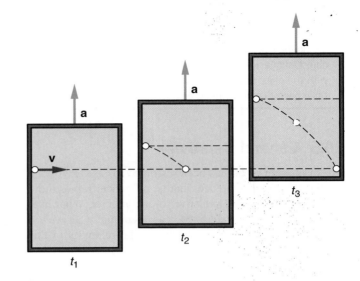

figure **20.19**   A ball thrown horizontally in an accelerating elevator in outer space (where the earth's gravitational pull is negligible) falls toward the floor in the same way as a projectile near the earth's surface.

Our inability to distinguish these accelerations underlies Einstein's basic postulate of general relativity, the **principle of equivalence:**

> It is impossible to distinguish an acceleration of a frame of reference from the effects of gravity.

From inside the elevator, you cannot tell whether the elevator is accelerating or whether the gravitational acceleration g is increasing or decreasing. Since we do not expect gravity to change, we would usually interpret the effects as being due to an acceleration of our frame of reference.

Let's move our elevator to outer space, where gravitational effects will be much smaller than near the surface of the earth. If the elevator is not accelerating, we would be weightless. If the elevator is accelerating upward, however, any experiments performed in the elevator will behave as though a gravitational acceleration is acting in the direction opposite the acceleration of the elevator.

If a ball is thrown horizontally in the elevator (fig. 20.19), its trajectory will be the same as a ball thrown on the surface of the earth. From the perspective of someone inside the elevator, the upward acceleration of their frame of reference is equivalent to the presence of a downward gravitational acceleration of the same magnitude as the acceleration of the elevator. (This is the principle of equivalence at work.) The ball "falls" toward the floor of the elevator, and we can predict its motion by the same methods used in chapter 3 to describe projectile motion.

If the acceleration of the elevator were equal to 9.8 m/s², mechanical experiments done in the elevator and on the surface of the earth would have identical results. It has often been proposed that a space station have a constant acceleration to mimic the effects of gravity. A straight-line acceleration would send the space station out of orbit, so we usually imagine the space station with a centripetal acceleration associated with a constant rotational velocity. Since the direction of a centripetal acceleration is toward the center of rotation, that direction would be up (fig. 20.20).

Does a light beam bend in a strong gravitational field?

The principle of equivalence also has implications for the propagation of light. Imagine an experiment similar to the one in figure 20.19 but using a beam of light instead of a ball. If the elevator were not accelerating, the beam would trace a horizontal line across the elevator. From the theory of special relativity, we know that this is true whether or not the elevator is moving with constant velocity relative to any other inertial frame of reference.

If the elevator is accelerating upward, however, we get a different result. If the acceleration is large enough, the path of the light beam will be curved, as viewed from within the elevator, like the ball's path in figure 20.19. This bending can be visualized by superimposing the positions of the accelerated elevator on the straight-line light beam observed from outside of the elevator, just as we did for the ball. As shown in figure 20.21, the path traced by the beam relative to the elevator is curved.

By the principle of equivalence, however, we cannot distinguish the acceleration of our frame of reference from the

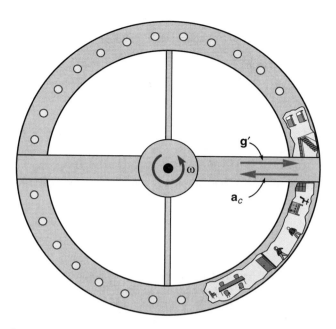

figure **20.20** The centripetal acceleration **a**$_c$ of a rotating, wheel-like space station can produce an artificial gravitational acceleration **g′** for the astronauts.

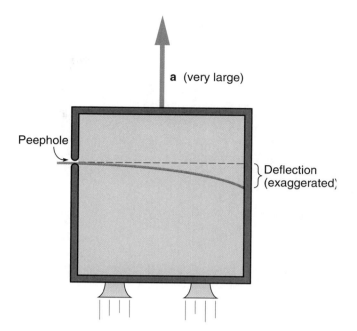

figure **20.21** The path traced by a beam of light relative to a rapidly accelerating elevator is curved because of the motion of the elevator.

presence of a gravitational acceleration. We should therefore expect the path of a light ray to be bent in passing through a strong gravitational field. It takes a very large acceleration of the frame of reference or a very large gravitational acceleration to produce a noticeable bending because of the extremely large velocity of light. The relatively puny field of the earth is not enough to produce much of an effect.

When light from a distant star passes near our sun, however, the gravitational field of the sun should be large enough to have a measurable effect. Einstein was able to predict how much bending is produced by the gravitational field of the sun—and how the true position of stars is distorted when light from these stars passes near the sun. The effect is small but measurable.

Unfortunately, it is difficult to observe stars from the surface of the earth during the daytime. Light from the sun is scattered in the earth's atmosphere and completely washes out the much more feeble light of the stars. Such observations are only feasible during a total eclipse of the sun, when the light from the sun is blocked by the moon. Einstein suggested that such measurements be attempted during a total eclipse, and this has since been done almost every time the opportunity has arisen. These measurements have confirmed Einstein's predictions.

What are the space and time effects of general relativity?

Special relativity tells us that different observers moving with respect to one another will disagree on measurements of time. In the time-dilation effect, the time interval mea-

sured by observers who see the start and finish of some process occurring at the same place in their frame (the proper time) is shorter than the time interval measured by observers moving with respect to that frame. The astronaut in the twin paradox measures the proper time for events occurring in her frame, which is shorter than the dilated time measured by her stay-at-home twin. The astronaut's clock runs slower than her twin's clock and the astronaut measures a shorter time.

In general relativity we also find that an accelerated clock runs more slowly than a nonaccelerated clock. By the principle of equivalence, we also expect a clock in a strong gravitational field to run more slowly than one in a weaker gravitational field. This time effect predicted by general relativity is often referred to as the **gravitational red shift.** If the period (the time for one cycle) of a light wave is increased, the frequency is decreased. A lowered frequency shifts the light toward the red end of the visible spectrum.

The general theory of relativity is largely about the nature of gravity. Gravity affects a straight-line path as well as time—it has an impact on how we measure both space and time. To develop a self-consistent mathematical framework for handling these effects, Einstein resorted to a non-Euclidean or curved space-time geometry.

Briefly, in Euclidean, or ordinary geometry, two parallel lines never meet, but in non-Euclidean geometry two parallel lines *can* meet. An example is parallel lines drawn on the surface of a sphere, such as the lines of longitude on maps. The parallel lines drawn perpendicular to the equator meet at the poles because the surface they are drawn on is a sphere

figure 20.22   Parallel lines of longitude drawn on the globe meet at the poles of the sphere.

(fig. 20.22). It is all a matter of how we define the rules of geometry.

Einstein's theory of special relativity showed that the measurement of time depends on spatial measures, while the measurement of length depends on time measurements. Therefore, we can no longer regard space and time as independent of one another. To fully represent these ideas using geometry, we must use four dimensions or coordinates: three perpendicular spatial coordinates and a fourth one representing time. To describe a motion or event, we need to locate its path in this **space-time continuum.**

Although the space-time continuum is four-dimensional and somewhat difficult to visualize, diagrams like figure 20.23 illustrate how space might be curved near a very strong gravitational field. The diagram shows only two dimensions on a curved surface, but it does suggest how things might be pulled drainlike into the center of the field. Since light rays are bent by strong gravitational fields, they, as well as particles having some mass, can be pulled into the center of the field.

What is a black hole?

Figure 20.23 is a two-dimensional representation of a **black hole.** Black holes are thought to be very massive collapsed stars, which generate an extremely strong gravitational field and, therefore, a strong curvature of space in their vicinity. This field is so strong that light rays coming in at certain angles are bent into the center and do not reemerge.

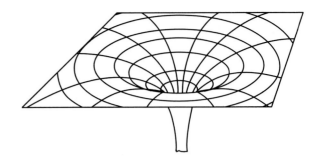

figure 20.23   The gravitational effects of a black hole can be represented by a strong curvature of space near the black hole.

Light gets in but cannot get out. A black hole is a perfect absorber of light and appears black.

Although black holes cannot be observed directly because they neither emit nor reflect light, their presence can be inferred from the effects of their gravitational fields on nearby stars and other matter. For example, if a binary star consists of two stars, one visible and the other a black hole, the motion of the visible star indicates the presence of its partner. Astronomers have found several good candidates for this type of black hole. Many other observations suggest the presence of black holes, but convincing proof of their existence is difficult to establish.

Einstein's theories of special and general relativity have had an enormous impact on modern physics. The predicted effects are well confirmed, from the energy released in nuclear reactions to astronomical effects such as the bending of starlight. Our fundamental concepts of space and time have been modified and intermixed by these ideas. Although removed from everyday experience, these ideas certainly excite the imagination.

While Einstein's special theory of relativity is primarily concerned with inertial frames of reference, his general theory of relativity treats accelerated frames of reference. The additional basic postulate of general relativity is the principle of equivalence, which states that we cannot distinguish the acceleration of a frame of reference from the effects of a gravitational field. General relativity predicts the bending of light by a strong gravitational field, the slowing of clocks in accelerated reference frames or gravitational fields, and the curvature of space-time produced by gravitational effects. The concept of black holes emerged from these ideas. The study of general relativity and the nature of gravity remains an active area of research.

---

summary

The velocity of light apparently does not add to the velocity of the source or frame of reference in the way that ordinary velocities do in classical mechanics. This idea led Einstein to a radical new way of looking at the nature of space and time. We have described the basic postulates of Einstein's special and general theories of relativity in this chapter and explored some of the consequences of accepting these postulates.

1   Relative motion in classical physics. If an object is moving relative to a frame of reference (such as a stream) that is itself moving, classical mechanics predicts that the velocities of these motions will add as vectors. Newton's laws are valid in any inertial frame of reference, which are frames that are not accelerated relative to other inertial frames.

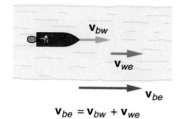

$$\mathbf{v}_{be} = \mathbf{v}_{bw} + \mathbf{v}_{we}$$

2   The speed of light and Einstein's postulates. The failure of experiments to detect any motion of the ether (the supposed medium for light waves) relative to the earth led Einstein to his two basic postulates of special relativity: First, the laws of physics have the same form in any inertial frame of reference (the principle of relativity), and second, the speed of light is the same in any inertial frame of reference, regardless of the motion of the source.

3   Time dilation and length contraction. Application of Einstein's postulates to the measurement of time and length shows that observers in different frames of reference will not agree on these measurements. A person observing a moving clock will see a longer (or dilated) time than the time measured by an observer for whom the clock is at rest. An observer measuring a moving length will observe a shorter (contracted) length than the rest length.

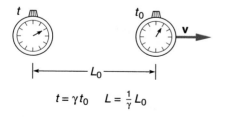

$$t = \gamma t_0 \qquad L = \frac{1}{\gamma} L_0$$

4   Newton's laws and mass-energy equivalence. Extending these ideas to dynamics, Einstein found that Newton's second law of motion could be preserved only if we redefined momentum and used the general form of Newton's second law written in terms of momentum. A computation of kinetic energy then led to the recognition that mass is equivalent to energy.

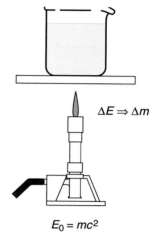

$$\Delta E \Rightarrow \Delta m$$

$$E_0 = mc^2$$

5   General relativity. Special relativity is primarily concerned with inertial frames of reference. Accelerated frames of reference and gravity are treated in general relativity. The basic new postulate of general relativity is the principle of equivalence: an acceleration of a frame of reference cannot be distinguished from the presence of a gravitational field. This additional postulate leads to new effects involving the bending of light and the modification of time by gravitational fields. It also requires the use of non-Euclidean, curved-space geometries to describe the space-time continuum.

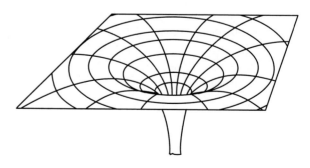

key terms

questions

*Questions identified with an asterisk are more open-ended than the others. They call for lengthier responses and are more suitable for group discussion.

Q1. If a boat is moving downstream, will the velocity of the boat relative to the water be greater than the velocity of the boat relative to the stream bank? Explain.

Q2. If a boat is moving upstream, will the velocity of the boat relative to the bank of the stream be greater than the velocity of the boat relative to the water? Explain.

Q3. If an airplane is flying in the same direction as the wind, will the velocity of the plane relative to the ground be greater than, less than, or equal to the velocity of the plane relative to the air? Explain.

Q4. Is it possible that a person in a rowboat will not be able to move upstream? Explain.

Q5. If a boat is moving across a stream, will the speed of the boat relative to the bank equal the numerical sum of speed of the boat relative to the water and the speed of the water relative to the bank? Explain.

Q6. If an airplane is flying in a crosswind at an angle of 90° to the direction the airplane is pointed, will the speed of the plane relative to the ground be less than the speed of the plane relative to the air?

Q7. Does the addition of the velocities of things like airplanes and wind speed require use of the special theory of relativity? Explain.

*Q8. Would it be appropriate, from the perspective of special relativity, to add the velocity of light relative to the earth to the velocity of the earth relative to the sun to obtain the velocity of light relative to the sun? Explain.

Q9. Was the ether (the assumed medium for light waves) presumed to exist in a vacuum? Explain.

*Q10. What was the Michelson-Morley experiment designed to detect? Why did they expect that the speed of light might vary at different times of the year? Explain.

Q11. Did the Michelson-Morley experiment succeed in measuring the velocity of the ether relative to the earth? Explain.

Q12. Do either of Einstein's postulates contradict the classical assumptions of how relative velocities add? Explain.

Q13. Which of Einstein's postulates deals most directly with the failure to detect motion of the earth relative to the ether? Explain.

Q14. Is it possible that two different observers will not agree on the time taken for a light beam to bounce off a mirror and return to its source? Explain.

Q15. A chess game taking place on earth is seen by observer A, who is passing by in a spaceship. Observer B is standing on earth looking over the shoulder of the players. Which of these two observers measures the longer time for the interval between moves in the game? Explain.

Q16. A radioactive isotope with a certain half-life is moving with a high speed in a particle accelerator. Does an observer at rest in the lab measure the proper time for the half-life of this isotope? Explain.

Q17. A spaceship is moving with a large velocity past observer A, who is standing on earth. Observer B is aboard the spaceship. Which of these observers measures the longer length for the length of the spaceship? Explain.

*Q18. Is it theoretically possible for a father to be younger (to have aged less) than his son or daughter? Explain.

*Q19. Is it possible for an astronaut to leave on a space trip and to return a year before her twin sister was born? Explain.

Q20. Is Newton's second law, written in the form $\mathbf{F} = m\mathbf{a}$, valid for objects traveling at velocities near the velocity of light? Explain.

Q21. Could we use the relativistic momentum expression $\mathbf{p} = \gamma m\mathbf{v}$, for objects moving at small velocities? Explain.

Q22. If we compress a spring and lock it into its newly compressed configuration, have we changed the mass of the spring? Explain.

Q23. Is the increase in kinetic energy of an object equal to the work done to accelerate the object for an object moving at a very high speed? Explain.

*Q24. Is it completely correct to say that mass is converted into energy in a nuclear reaction such as a fission reaction? Explain.

Q25. If the velocity of an object is reduced to zero, does all of its energy disappear? Explain.

Q26. If an elevator is accelerating downward, will your apparent weight (as measured by a bathroom scale) be greater than your weight measured when the elevator is not accelerating? Explain.

Q27. When you are inside a closed space vehicle, is it possible for you to tell whether the vehicle is accelerating or whether you are simply near some massive body such as the sun or the earth? Explain.

Q28. Would your experiences inside a freely falling elevator be similar in any way to those inside a spaceship moving with constant velocity when it is a long distance away from any planet or star? Explain.

Q29. Does light traveling in empty space always travel in a straight line? Explain.

Q30. Would a clock located on the surface of the sun measure time at the same rate as a clock located a long distance away from any planet or star? Explain.

Q31. Is a black hole just a hole in space that contains no mass? Explain.

exercises

E1. A boat that can travel with a velocity of 12 m/s in still water is moving at maximum speed *against* the current of a stream that flows with a velocity of 5 m/s relative to the earth. What is the velocity of the boat relative to the bank of the stream?

E2. A plane that can travel at 460 MPH in still air is flying with a tailwind of 40 MPH. How long does it take for the plane to travel a distance of 750 miles (relative to the earth)?

E3. A swimmer swims upstream with a velocity of 4 m/s relative to the water. The velocity of the current is 3.5 m/s (downstream). What is the velocity of the swimmer relative to the bank?

E4. A ball is thrown with a velocity of 60 MPH down the aisle (toward the tail of the plane) of a jetliner traveling with a velocity of 260 MPH relative to the earth. What is the velocity of the ball relative to the earth?

E5. An astronaut aims a flashlight toward the tail of his spaceship, which is traveling with a velocity of 0.5c relative to the earth. What is the velocity of the light beam relative to the earth?

E6. The factor $\gamma = 1/\sqrt{1 - (v^2/c^2)}$ appears in many expressions derived from the theory of special relativity. Show that $\gamma = 1.25$ when $v = 0.6c$.

E7. An astronaut cooks a three-minute egg in his spaceship whizzing past earth at a speed of 0.6c. How long has the egg cooked as measured by an observer on earth? (See table 20.1.)

E8. An observer on earth notes that an astronaut on a spaceship puts in a 4-hour shift at the controls of the spaceship. How long is this shift as measured by the astronaut himself, if the spaceship is moving with a velocity of 0.8c relative to the earth? (See table 20.1. Be careful—which observer measures the proper time?)

E9. A spaceship that is 50 m long as measured by its occupants is traveling at a speed of 0.1c relative to the earth. How long is the spaceship as measured by mission control in Houston? (See table 20.1.)

E10. The crew of a spaceship traveling with a velocity of 0.5c relative to the earth measures the distance between two cities on earth (in a direction parallel to their motion) as 600 km. What is the distance between these two cities as measured by people on earth? (See table 20.1—which observer measures the rest length?)

E11. A spaceship is traveling with a velocity of 0.8c relative to the earth. What is the momentum of the spaceship if its mass is 5000 kg? ($p = \gamma mv$ and $c = 3 \times 10^8$ m/s. See table 20.1.)

E12. Suppose that an object has a mass-energy of 200 joules when it is at rest.
   a. What is its total energy when it is moving with a velocity of 0.9c? ($E = \gamma E_0$. See table 20.1.)
   b. What is the kinetic energy of the particle at this speed? ($KE = E - E_0$)

challenge problems

CP1. A boat capable of moving with a velocity of 6 m/s relative to the water is pointed straight across a stream flowing with a current velocity of 3 m/s. The width of the stream is 48 m.
   a. Draw a vector diagram to show how the velocity of the stream adds to the velocity of the boat relative to the water to obtain the velocity of the boat relative to the earth.
   b. Use the Pythagorean theorem to find the magnitude of the velocity of the boat relative to the earth.
   c. How long does it take for the boat to cross the stream? (Hint: We need to consider only the component of the velocity of the boat that is straight across the stream if we use the stream width for the distance.)

d. How far downstream from its starting point does it hit the opposite bank?

e. How far does the boat actually travel in reaching the opposite bank?

CP2. Suppose that a beam of π-mesons (or pions) is moving with a velocity of $0.9c$ with respect to the laboratory. When the pions are at rest, they decay with a half-life of 1.77×10^{-8} s.

a. Calculate the factor γ for the velocity of the pions relative to the laboratory.

b. What is the half-life of the moving pions as seen by an observer in the laboratory?

c. How far do the pions travel, as measured in the laboratory, before half have decayed?

d. As measured in a frame of reference that moves with the pions, how far do the pions travel before half have decayed?

CP3. Suppose that an astronaut travels to a distant star and returns to earth. Except for brief intervals of time when he is accelerating or decelerating, his spaceship travels at the incredible speed of $v = 0.995c$ relative to the earth. The star is 40 light-years away. (A light-year is the distance light travels in 1 year.)

a. Show that the factor γ for this velocity is approximately equal to 10.

b. How long does the trip to the star and back take as seen by an observer on earth?

c. How long does the trip take as measured by the astronaut?

d. What is the distance traveled as measured by the astronaut?

e. If the astronaut left a twin brother at home on earth while he made this trip, how much younger is the astronaut than his twin when he returns?

CP4. Suppose that a beaker of water contains 1kg (1000 g) of water. Heat is added to the water to raise its temperature from 0°C to 100°C.

a. How much heat energy in joules must be added to the water to raise its temperature? ($c_w = 1$ cal/g·C° and 1 cal = 4.186 J.)

b. By how much does the mass of the water increase in this process? ($E_0 = mc^2$)

c. Compare this mass increase to the original mass of the water. Would this increase in mass be measurable?

d. If it were somehow possible to convert the original mass of the water into kinetic energy, how many joules of kinetic energy could be produced?

CP5. Using the diagram shown in figure 20.13, derive the time-dilation formula. The steps are:

a. From the symmetry of the diagram, we assume that the total time measured by the earth observer is twice the time required to reach the mirror.

b. Using the right triangle shown in the diagram and the Pythagorean theorem, we can write

$$c^2 \left(\frac{t}{2}\right)^2 = d^2 + v^2 \left(\frac{t}{2}\right)^2 .$$

c. Grouping terms containing t on one side of the equation and taking the square root of both sides, we can solve this expression for the time t measured by the earth observer.

d. Since the quantity $2d/c$ is the time measured by the spaceship pilot t_0, this expression reduces to the time-dilation formula introduced in section 20.3.

home experiments and observations

HE1. If you have access to a small smooth-flowing stream, you can test the velocity-addition ideas in the first section of this chapter. A small battery-powered or wind-up boat is also necessary. These can be found in toy or variety stores.

a. Test your boat first in the bathtub or a pond to estimate how fast it can move in still water.

b. Find a place in the stream where the current is slow and smooth (no eddies). Drop a twig in the stream, and with the help of a watch, estimate the velocity of the current.

c. Place your boat, with its motor running, in the stream with the boat pointed downstream. Estimate its velocity relative to the bank. Does the result agree with what you would predict based on the addition of relative velocities?

d. Will the boat move upstream? (If the current is too strong, it may be difficult to keep it headed in this direction.)

e. Try pointing the boat across stream in a location where the current is as uniform as possible. What do you have to do to get the boat to cross the stream?

Beyond Everyday Phenomena

chapter overview

This chapter delves into a few broad areas of research in physics that have either excited the popular imagination or are likely to have a significant impact on developing technologies. The ideas discussed include elementary particles, the origins of the universe, semiconductor electronics, computers, and superconductors and other exotic materials. The descriptions are necessarily brief—the objective is to emphasize the fundamental ideas and issues.

chapter outline

chapter

21

unit six

One of the most intriguing aspects of science is that we can never be sure just where it will lead us. Like a good mystery, there are clues and indications, but the answers are elusive. Unlike a mystery novel, there is no final resolution. Successes in science increase our understanding of nature and often lead to advances in technology, but they always raise new questions.

Physics does make the news from time to time and will continue to do so. Questions are raised about spending several billion dollars to build a new particle accelerator or space station, and ordinary citizens are sometimes called on to make judgments about such issues. The proposed construction of the superconducting supercollider (a particle accelerator) was a political issue for several years until the project was cancelled by Congress in 1993 (fig. 21.1).

Our everyday lives are impacted in more ways than we realize by advances in physics. Most of us use personal computers and electronic devices like video recorders that have microcomputers built into them. Computers have brought about enormous changes in the way that we live, work, and play. The invention of the transistor, which made the modern computer feasible, came about through advances in solid-state physics and our understanding of semiconductors.

Modern physics is active on many fronts. Some research areas are driven by the need to make improvements in technology, while others are motivated simply by a desire to better understand the universe in which we live. Although we cannot touch on all of these areas,

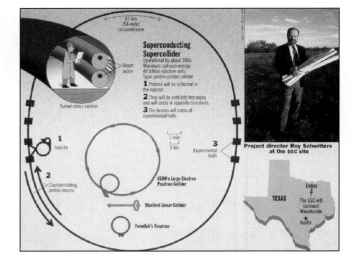

figure 21.1   Graphics for a *Time* magazine article about the superconducting supercollider (SSC) that was to have been built in Texas. Did its expected benefits to science justify the cost? *Copyright 1990 The Time Inc. Magazine. Reprinted by permission.*

we will describe a few that have received public attention and are likely to continue to do so. Ideas that seem well removed from common experience today may someday become part of everyday phenomena.

21.1 Quarks and Other Elementary Particles

One of the most enduring quests in science is the search for the building blocks of nature, the particles or entities from which everything else is constructed. Until the twentieth century, these building blocks were thought to be atoms, a view strengthened by the advances in chemistry during the nineteenth century (see chapter 18).

The discovery of the electron by J. J. Thomson in 1897 revealed the first subatomic particle apparently present in all atoms. This advance was followed in 1911 by the discovery of the nucleus of the atom and by the later recognition that nuclei are made up of protons and neutrons. We are now aware that protons and neutrons also have a substructure—they are composed of quarks (fig. 21.2).

Where will this all end? What are quarks and why do we believe that they exist? Will we someday discover that quarks also have a substructure? This final question cannot be answered with certainty, but recent advances in our theoretical understanding of high-energy physics have brought order to what seemed like a bewildering array of new particles. We will consider just a few features of this new theory, often called the **standard model.**

How are new particles discovered?

The electron, proton, and neutron, the basic constituents of the atom, were just the first in a long parade of subatomic particles discovered in the twentieth century. The positron, for example, was discovered in 1932 shortly after its existence was suggested on theoretical grounds by the British physicist Paul Dirac (1902–1984). This discovery was followed by the discovery of the muon and pion. The list grew rapidly during the 1950s and 1960s as work in high-energy physics intensified.

How are these discoveries made? Most of them involve scattering experiments, similar to those performed by Rutherford and his associates. Targets are bombarded with fast-moving particles, and particle detectors are used to study what emerges from these collisions. The emerging particles leave tracks in the photographic emulsions, cloud chambers, and bubble chambers that were used in early experiments and in other more sophisticated detectors in use today. In cloud chambers and bubble chambers, a rapidly moving charged particle nucleates water droplets or bubbles in a supersaturated vapor or superheated fluid (fig. 21.3).

Analyses of these tracks, along with other measurements, allow us to deduce the mass, kinetic energy, and charge of

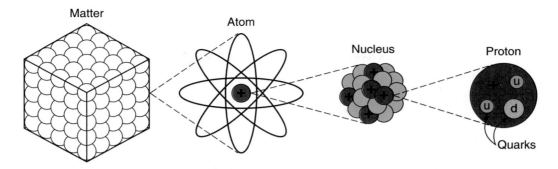

figure 21.2 Atoms, once thought to be the basic building blocks of all matter, are now known to consist of electrons, protons, and neutrons. Neutrons and protons also have a substructure made of quarks.

figure 21.3 Particle tracks in a bubble chamber provide information on the new particles produced in collisions or decays.

the particles created in the collisions. The path of a positively charged particle, for example, bends one way in a magnetic field, and the path of a negatively charged particle bends in the opposite direction. The degree of curvature of the path is related to the mass of the particle. The mass is particularly important—it is one of the major identifying characteristics of the particles.

The source of high-energy particles in these scattering experiments is usually a particle accelerator of some kind. Rutherford used alpha particles from radioactive substances, but the alpha particles have limited energies. Other early workers used the high-energy particles in *cosmic rays* that stream in from outer space. Particle accelerators, however, are capable of producing both high-energy and high-density beams of particles, which make interesting collisions more likely.

A modern particle accelerator uses electric and magnetic fields to accelerate and shape the beam. The beam it-

self is contained in a long evacuated tube that can either be straight (as in a linear accelerator) or bent into a large ring. Two beams of particles can be made to collide head-on at the point where the reactions are studied, providing a larger collision energy than from collisions with a stationary target. Beam energies are typically measured in electron volts (eV). Since mass is equivalent to energy, producing particles with masses much larger than the proton or neutron requires very high collision energies, which has stimulated building ever-bigger particle accelerators. Modern accelerators are capable of reaching collision energies up to 1000 GeV or more (1 GeV is 1 billion eV). The largest modern accelerators now operating include the Stanford Linear Accelerator Center (SLAC) in California, the CERN electron-positron collider in Switzerland, and the Fermilab proton-antiproton collider near Chicago.

Denizens of the particle zoo

As more and more particles were discovered, scientists tried to organize and classify them, guided by theoretical considerations. Although the models were often incomplete, they were sometimes successful in predicting the existence of new particles later discovered experimentally. The original classification schemes were based primarily on the masses of the particles.

The particles were grouped into three primary groups: leptons, mesons, and baryons. The **leptons** are the lightest particles and include electrons, positrons, and the neutrinos that are involved in beta decay. **Mesons** are intermediate in mass and include the pion (originally called the π-meson) and the kaon. **Baryons** are the heaviest—they include the neutron and proton as well as many heavier particles. A partial list of these particles is found in table 21.1.

Each particle has an antiparticle, which has the same mass as the particle, but opposite values of other properties such as charge. The positron, for example, is the antiparticle of the electron and has a positive charge instead of a negative charge. When a particle runs into its antiparticle,

table 21.1

Basic Characteristics of Leptons, Mesons, and Baryons

	Mass (MeV)	Charge	Spin	Lifetime
leptons				
electron neutrino	0?	0	½	
muon neutrino	0?	0	½	
electron	0.511	−e	½	
muon	105.7	−e	½	2.2×10^{-6} s
mesons				
pion	139.6	+e	0	2.6×10^{-8} s
neutral pion	135.0	0	0	8.3×10^{-15} s
kaon	493.7	+e	0	1.2×10^{-8} s
neutral kaon	497.7	0	0	9×10^{-11} s
eta	548.8	0	0	7×10^{-19} s
baryons				
proton	938.3	+e	½	
neutron	939.6	0	½	920 s
lambda	1115.6	0	½	2.5×10^{-10} s
sigma	1189.4	+e	½	8.0×10^{-11} s
neutral sigma	1192.5	0	½	5.8×10^{-20} s
xi	1321.3	−e	½	1.7×10^{-10} s
neutral xi	1314.9	0	½	3.0×10^{-10} s
omega	1672	−e	½	1.3×10^{-10} s

the two can annihilate each other, producing high-energy photons or other particles. Antiparticles are not shown in table 21.1.

The spin is listed in table 21.1 because it distinguishes mesons from leptons and baryons. All of the mesons have zero spin, but all of the leptons and baryons have a spin of one-half. Spin is a quantum property related to the angular momentum of the particle. If the particle is charged, the spin also generates a magnetic dipole, which affects how the particle interacts with other particles.

What are quarks?

Quantum electrodynamics and *quantum chromodynamics*, based on both quantum mechanics and relativity, are the theories that describe the interactions between these particles. Advances in these theories in the early 1970s suggested a more fundamental organization scheme for all of these particles, and this is where **quarks** come into the picture. Mesons and baryons (which together are now called **hadrons**) are all made up of quarks, new particles suggested by the theory. Each meson consists of two quarks—a quark and an antiquark—and baryons are groups of three quarks.

As the theories developed, it became evident that six types of quark were necessary (not counting the antiparticles) to account for all of the baryons and mesons. These have been dubbed the *up, down, charmed, strange, top,* and *bottom* quarks. Different combinations of these six quarks (and their antiparticles) account for all of the observed particles in the meson and baryon groups. Besides the quarks, only the leptons are still elementary that is, not made up of any more fundamental particles as far as we know.

The proton consists of three quarks: two up quarks each of charge $+\frac{2}{3}e$, and one down quark of charge $-\frac{1}{3}e$. A neutron is made of two down quarks and one up quark, for a total charge of zero (fig. 21.4). Scattering experiments in which extremely high-energy electrons are collided with protons provide strong evidence for this substructure. These experiments indicate the presence of hard scattering centers within the proton, with the appropriate charges for two up quarks and one down quark. The analysis is similar to Rutherford's discovery of the nucleus (chapter 18).

There are also similarities among the groups of leptons and quarks. We now group these particles into three families with similar properties. Each family consists of two leptons and two quarks, and one of the leptons in each family is a neutrino. Table 21.2 shows the particles that belong to each family. In this scheme, there are just twelve elementary particles (three families of four particles each), twenty-four counting the antiparticles.

At the time of this writing, the existence of these twelve particles has been confirmed experimentally, except for the tau neutrino. Neutrinos are extremely difficult to detect, and tau neutrinos are expected to be much rarer than the electron or muon neutrinos. The top quark was discovered

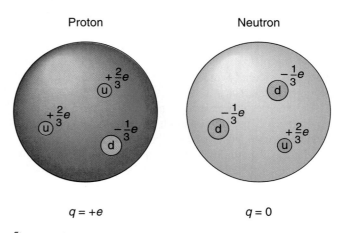

figure 21.4 A proton consists of two up quarks and one down quark. A neutron consists of two down quarks and one up quark.

most recently (1994) by physicists working at Fermilab near Chicago. Experiments are continuing to better define its mass. The successes of the overall model give us strong confidence in the existence of the tau neutrino as well.

Quarks are never present individually but always in combination with other quarks. For this reason, they cannot be directly observed as tracks in a particle detector. Their existence can be inferred, however, by scattering experiments and by observing reactions that are predicted by the quark model. High particle energies and beam densities are required to enhance the probability of observing such reactions. The cancelled superconducting supercollider project would have provided better conditions for observing these reactions, but refinements in other accelerators have continued to produce advances.

What are the fundamental forces?

What holds all of these particles together? The primary force responsible for binding the quarks in neutrons, protons, and other baryons (as well as mesons) is the **strong nuclear interaction.** This force also binds the neutrons and protons inside the nucleus of an atom and must be stronger than the electrostatic repulsion of the positively charged protons to keep the nucleus from flying apart. The

strong force has a very short range, however, and decreases rapidly at distances greater than nuclear dimensions.

In addition to the strong nuclear interaction, physicists had recognized three other fundamental forces—the electromagnetic force, the gravitational force, and the weak nuclear force. The **weak nuclear force** is involved in the interactions of leptons: the process of beta decay, which involves electrons and neutrinos, is an example. One of the goals of theoretical physics has been to unify all of these forces with a single theory. Since we usually describe these forces in terms of their fields, as we have done with electric and magnetic fields, such a theory is referred to as a **unified field theory.**

One of the major successes of the standard model of particle physics is that it has unified the weak nuclear force with the electromagnetic force. These two forces can now be viewed as different manifestations of the same fundamental force, the **electroweak force.** James Clerk Maxwell's theory of electromagnetism earlier had unified two seemingly independent forces, the electrostatic force and the magnetic force, into the electromagnetic force. Now, that force has been joined with the weak nuclear force.

Perhaps we should say that there are only three fundamental forces, the strong nuclear interaction, the electroweak force, and the gravitational force. This statement may also be misleading, since substantial progress has been made toward unifying the strong nuclear interaction with the electroweak interaction in extensions of the standard model (fig. 21.5). These theories are now referred to as **grand unified theories,** or *GUTs* for short.

One force, the gravitational force, has thus far resisted incorporation into a unified field theory. Gravity's theoretical basis is found in Einstein's theory of general relativity, but the mathematics of general relativity seem to be incompatible in some ways with quantum mechanics and the standard model. Fame and honor await those who succeed in unifying the gravitational force with the other fundamental forces in a *theory of everything.*

table 21.2		
The Three Families of Elementary Particles		
First family	Second family	Third family
electron	muon	tau particle
electron neutrino	muon neutrino	tau neutrino
up quark	charmed quark	top quark
down quark	strange quark	bottom quark

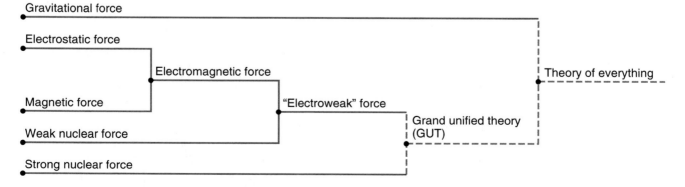

figure **21.5** Fundamental forces of nature once viewed as independent forces have been unified into fewer fundamental forces by advances in theoretical understanding. Complete unification of all fundamental forces may lie ahead.

Scattering experiments performed at ever-higher energies have uncovered an array of new subatomic particles. The standard model has succeeded in organizing these particles into three families, each with two leptons and two quarks (and their antiparticles). This model predicted the existence of new particles, including the top quark that was recently detected experimentally. Theoreticians continue to strive for a unified theory that will encompass all of the fundamental forces.

21.2 Cosmology and the Beginning of Time

Section 21.1 gave us a glimpse of advances in the physics of the very small—substructures of substructures of substructures. The quark is a building block of protons and neutrons, which form the nucleus. Atoms consist of the nucleus and the surrounding electrons. Atoms, in turn, make up molecules and the ordinary matter of our world.

What happens if we focus on the very large? Our world, the earth, is a part of the solar system (see chapter 5). The sun is one of a seemingly infinite number of stars grouped into galaxies, which themselves seem to come in clusters. What is the structure of the universe, and how is it changing? The answers to these questions may come from our knowledge of atoms, nuclei, and quarks.

Is the universe expanding?

Humans have long been fascinated by the night sky and by questions about the nature of the universe. The invention of the telescope around 1600 provided a new instrument for viewing the planets and stars. Using a crude telescope, Galileo discovered the moons of Jupiter and the phases of Venus, and helped to turn the tide in favor of the Copernican heliocentric model of the solar system (see chapter 5).

As telescopes improved, observers of the heavens became aware of many more objects out there than those visible to the unaided eye. Not all of these objects appeared to be pointlike stars. Some had a fuzzy appearance, and as the resolution of telescopes increased, it became obvious that they were not stars at all but collections of stars, what we now call a **galaxy.** Many galaxies have a spiral structure like the one shown in figure 21.6.

The galaxy that we see most readily with unaided eyes is our own, the Milky Way. On a clear night, the Milky Way is visible in what appears to be a continuous cloud of stars making a band across the sky (fig. 21.7). We are actually looking across the disk into our own spiral galaxy. The brighter stars lie on the same side of the spiral as our sun and are much closer to us. The sun is one of billions of stars that make up the Milky Way galaxy.

We have also come to think that the universe is expanding—the other galaxies are receding from us. This realiza-

figure **21.6**   A spiral galaxy viewed against a foreground of nearer stars. Our own Milky Way galaxy has a similar shape.

figure **21.7**   The Milky Way appears as a continuous cloud of stars that can be seen as a band across the sky on a clear night.

tion emerged from the spectrographic studies by the American astronomer Edwin Hubble (1889–1953) in the 1920s. Hubble was trying to estimate the distance to various stars and galaxies by measuring their relative brightness. To do so, he needed some assurance that he was looking at the same type of star. It was already known that different types of stars had characteristic colors or spectra—red giants are different from white dwarfs, and so on. Measuring the intensity of the distribution of wavelengths emitted from different stars gave Hubble a basis for comparing the size and temperature of the stars he was viewing.

When Hubble applied these techniques to galaxies, however, he noticed a startling feature. Specific *absorption lines* in the spectra of stars in these other galaxies were all shifted in wavelength and frequency towards the red portion of the spectrum. (Absorption lines make good reference points in the otherwise continuous spectrum of stars. They are produced by the absorption of light at specific wavelengths by gases in the outer portions of the star.)

The only reasonable explanation for this shift was that these stars must be moving away from us, producing a Doppler shift in the frequency of the light. The Doppler shift is the same phenomenon discussed in chapter 15 for sound waves. The frequency of a car horn is shifted to a lower value when it is moving away from us. For light, a lower frequency is a shift toward the red end of the visible spectrum. The cosmological *red shift* is now thought of as a relativistic effect of the expansion of space itself.

The farther away a galaxy is, the more swiftly it seems to recede from us, which is consistent with the hypothesis that the entire universe is expanding. From our knowledge of the curvature of space-time introduced in Einstein's theory of general relativity, we know that we do not need to be at the center of the universe to see things in this way. An often-used analogy is of spots on an expanding balloon. Viewed from any point on the surface of the balloon, all the other spots appear to recede as the balloon expands. Points farther away from the given point recede at a greater rate than closer points (fig. 21.8).

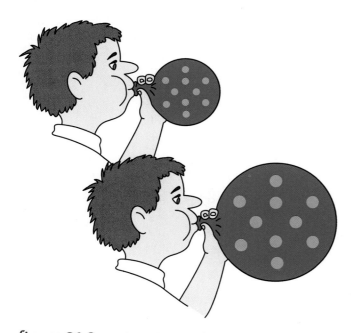

figure 21.8 As the balloon is inflated, spots on the surface recede from one another. Spots that are initially farther away from some chosen point recede more rapidly from that point than do nearer spots.

What was the Big Bang?

If the universe is expanding, at some point in time long ago the entire mass of the universe must have been much more compressed than it is now. If we could run a motion picture of the expanding universe in reverse, the universe should revert to a very small volume. The beginning of the expansion (and perhaps the beginning of time) was an explosion from which the universe has been spreading ever since. This initial rapid expansion or explosion is called the **Big Bang.**

As we run the film backward and confine a large quantity of matter in a very small space, matter no longer consists of individual atoms and molecules. The electrons get stripped from the atoms, and what is left is a dense plasma of electrons, protons, and neutrons. At even higher densities, the protons and electrons combine to form neutrons. This process may occur in the gravitational collapse of stars that have used up their fusion fuel, resulting in very small and dense *neutron stars.* If such a star has sufficient mass, it might collapse still further to form a *black hole* (see section 20.5). At even higher densities, the matter would exist as a sea of quarks, in which individual quarks would not belong to specific neutrons or protons.

In the earliest stages of the Big Bang (just a microsecond or so after the beginning), all of the matter of the universe was an extremely hot sea of quarks. As the expansion proceeded, the matter behaved like a gas, cooling off and condensing. The quarks condensed into mesons and baryons, including neutrons and protons. At approximately 3 minutes after the beginning, the protons and neutrons probably began to fuse into nuclei, primarily isotopes of hydrogen and helium.

At a much later point (roughly half a million years), the universe should have cooled down enough for electrons to begin to orbit about the nuclei to form atoms. Gravitational attraction produced clumps of matter that became galaxies, and matter within these galaxies condensed into individual stars. The synthesis of larger nuclei by fusion reactions began to take place within the stars.

The standard model of high-energy physics, discussed in section 21.1, has been able to predict how some of these steps could have occurred. The model has had success in explaining certain astronomical observations, including the ratio of helium to hydrogen observed in stars and galaxies. Another confirming observation has been the detection of the uniform background of microwave radiation, predicted as a residual effect of the Big Bang itself. Although irregularities in this background have been detected recently, many physicists consider the uniform background radiation one of the strongest pieces of evidence confirming the Big Bang hypothesis.

Our success in describing the world of the very small (nuclei and quarks) plays a large role in our understanding of the universe. Much of this success has been achieved in only the last twenty years or so, but more remains to be done. Advances in the theory of fundamental forces are quickly applied to models of the universe to test their implications.

There are still many unanswered questions. Since we still have not achieved a completely unified field theory, we cannot model the very earliest stages of the Big Bang. Therefore, we cannot describe with assurance the initial conditions of the universe. There may be many universes, some of which have evolved differently from our own. These questions hold a tremendous fascination for physicists, astronomers, philosophers, and the general public.

> Our own solar system is a small part of the Milky Way galaxy, just one of many observable galaxies. The discovery that distant galaxies seem to be receding from us led to the Big Bang theory of the expanding universe. Models of the early phases of this expansion are based on our knowledge of the elementary particles and forces discussed in section 21.1. These models have succeeded in explaining and predicting many astronomical observations.

21.3 Semiconductors and Microelectronics

Most of us have used a hand-held calculator. In our homes, we have microwave ovens, video recorders, and perhaps home computers. We probably also own other electronic devices, including television sets, radios, digital watches, stereo systems, and electronic ignition systems in automobiles. All of these devices use solid-state electronics, and many of them incorporate microcomputers.

What do we mean by *solid-state electronics*? What led to the current revolution in technology? Although these electronic devices are part of our everyday experience, how they work is invisible to us. Despite their enormous importance to our economy, most people have little understanding of how they function.

What are semiconductors?

In chapter 12, we discussed the distinction between electrical conductors and insulators. Good electrical conductors, mostly metals, permit a relatively free flow of electrons or other charge carriers through the material. Good insulators do not. There is an enormous difference in the values of electrical **conductivity** between these two types of materials. Conductivity is a property of the material that, together with its length and width, determines its electrical resistance.

Table 12.1 listed some conductors and insulators, as well as a few members of a third category called *semiconductors*. Semiconductors have a much higher conductivity than good insulators but a considerably lower conductivity than

good conductors. What causes these differences in electrical conduction? Can we predict which materials will be good conductors, insulators, or semiconductors?

If you examine the periodic table in the inside back cover of this book, you will see that the metals all lie on the left side of the table or in the transition regions. These elements have just one, two, or sometimes three electrons outside of a closed shell of electron states (see chapter 18). These outer electrons are responsible for the chemical properties of a particular element. Less tightly bound to the nucleus of the atom than the other electrons, they are relatively free to migrate within the material as conduction electrons.

On the other hand, elements that make good insulators lie on the right side of the periodic table. These elements are lacking one, two, or three electrons needed to complete a shell. They readily accept electrons from other elements when they combine to form chemical compounds. When they bond together in their pure state to form solids or liquids, there are no loosely bound electrons to contribute to electrical conduction.

The elements that we commonly list as semiconductors (carbon, germanium, and silicon) are found in column IVA of the periodic table. These elements have four outer electrons beyond a closed shell. When these elements bond together in a solid, electrons are shared with neighboring atoms, as in the two-dimensional depiction in figure 21.9. (The actual crystal structure is three-dimensional, of course, which is harder to show.) These shared electrons are more closely tied to their corresponding nuclei than in a metal, but they are freer to migrate through the substance than in a good insulator. The conducting properties of these materials are thus intermediate between metals and good insulators.

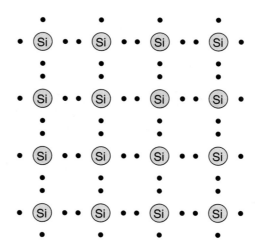

figure **21.9**   A two-dimensional representation of the sharing of the four outer electrons of silicon with neighboring atoms in solid silicon.

Although carbon, germanium, and silicon are semiconductors, they do not conduct well in their pure form. Carbon is often used to make resistors for circuit applications because a short piece of carbon has a much higher resistance than a much longer and thinner metal wire. Semiconductors' importance in electronics results from our ability to modify their conductivity by *doping* them with small amounts of impurity.

Suppose, for example, that we add a small amount of phosphorus or arsenic to silicon. These elements lie in column VA of the periodic table and have five outer electrons (fig. 21.10). Four of these five electrons will participate in the bonding of the impurity with neighboring silicon atoms. The fifth electron, however, is not needed in these bonds and will be free to migrate through the material. The doping therefore introduces conduction electrons into the material, making it a better conductor than pure silicon.

Doping with phosphorus, arsenic, or antimony produces an *n-type* semiconductor, because the charge carriers are negatively charged electrons. We can also produce *p-type* doping, for which the charge carriers are positive, by adding impurity atoms of elements from column IIIA in the periodic table, most commonly boron, gallium, or indium. Since atoms of these elements have just three outer electrons, they leave a *hole* in one of the bonds between the impurity atom and the neighboring silicon atoms (fig. 21.10).

A hole is the absence of an electron—but these holes can also migrate through the material. A moving hole behaves as a positive charge carrier because it leaves an excess positive charge (associated with the charge on the nucleus of the silicon atoms) wherever it goes. Electrons from neighboring silicon atoms move in to fill the hole, leaving an excess positive charge somewhere else in the material.

How does a semiconductor diode work?

Besides improving the conducting properties of semiconductors by amounts that can be carefully controlled, doping has other advantages. The boundaries, or *junctions*, between *p-* and *n*-type materials have properties that have proved extremely useful in electronics. These junctions are essential to the operation of diodes, transistors, and related devices.

A **diode** is a device that allows electric current to flow in one direction but not in another: it is a one-way valve for electric current. The diagrams in figure 21.11 illustrate why a diode behaves as it does. The essential feature of a semiconductor diode is the junction between the *n*-type and *p*-type materials. When the positive terminal of a battery is connected to the *p*-type material and the negative terminal to the *n*-type side of the diode (fig. 21.11a), electrons are introduced from the battery into the *n*-type side. These electrons will flow to the junction between the *n*-type and *p*-type materials. Here, the electrons attract holes in the *p*-type material to the junction, and these holes are eliminated when electrons move across the junction to fill them. The positively charged holes move through the *p*-type material from the positive side of the battery, and a continuous current will flow. This manner of connecting the battery is called *forward bias* of the diode.

A different situation exists if we reverse the connections of the battery to the diode (fig. 21.11b). Holes are now pulled away from the junction by the negative charges from the negative terminal of the battery (now connected to the *p*-type side of the diode). Likewise, electrons in the *n*-type material are attracted toward the positive terminal of the battery. Since the holes and electrons are both pulled away from the junction, no recombination of holes and

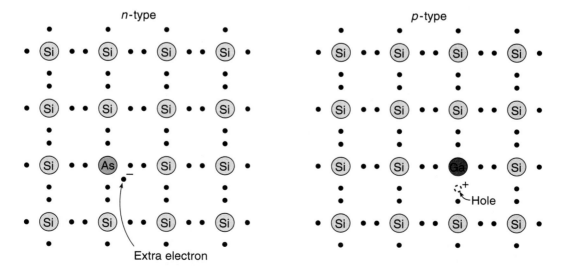

Extra electron

figure **21.10**   Doping silicon with phosphorus or arsenic provides an extra electron, making an *n*-type semiconductor. Doping with boron or gallium leave holes and produces *p*-type material.

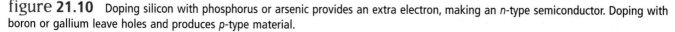

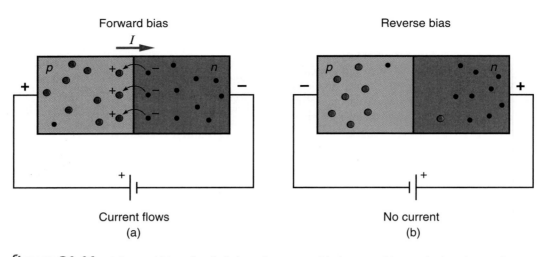

Forward bias

I

Reverse bias

Current flows
(a)

No current
(b)

figure 21.11   A forward bias of a diode lets electrons and holes recombine at the junction, and an electric current flows across the junction. Reverse bias produces no recombination and no current.

electrons occurs there. In other words, there is no flow of current across the junction in *reverse bias.*

Diodes have many applications in electric circuits. One of the easiest to understand is *rectification,* the process of converting an alternating current to a direct current. Since a diode lets current flow in only one direction, the simplest rectifier would be a single diode. Combinations of diodes, however, produce a steadier flow of current.

How does a transistor work?

Transistors are probably the most important semiconductor devices. For many years, the most commonly used type was a *bipolar* transistor made up of two pieces of semiconductor material heavily doped in the same manner and separated by a thin piece of oppositely doped material. Depending on which type of doping is used in the two outer pieces of the sandwich, either *p-n-p* transistors or *n-p-n* transistors can be made. The diagram in figure 21.12 shows how a *p-n-p* transistor works.

The transistor is a combination of two diodes that share the middle portion, the *base* of the transistor. When connected in the usual manner, holes are introduced to the *emitter* of the *p-n-p* transistor from the positive terminal of a battery. The junction between the emitter and the base behaves as a forward-biased diode, and the holes flow into the thin base layer. Because the base layer is very thin and only lightly doped compared to the emitter and *collector,* these holes can flow across the base and into the collector as long as not too many recombine with electrons in the base layer. The number of free electrons in the base layer is a critical property in determining how many holes get through. This number can be controlled by the current allowed to flow between the base and the emitter.

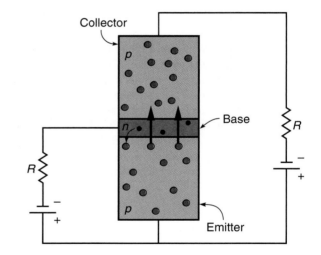

figure 21.12   The rate of flow of holes from the emitter to the collector of a *p-n-p* transistor depends on how much current is allowed to flow to the *n*-type base of the transistor.

A small change in the current from the base to the emitter can produce a large change in the current flowing between the collector and the emitter, which is why a transistor is so effective as an amplifier. Small variations in the signal applied to the base can produce large variations in the current that flows through the collector. A weak signal picked up by a radio antenna, for example, can be turned into a larger signal by using transistor amplifiers. This is done routinely in radios, television sets, and stereo amplifiers.

Another important application of transistors is as voltage-controlled switches. One value of the voltage applied across the base and emitter can cause a large flow of current through the collector, while another value produces a very small flow. The transistor is then either on or off depending on the value of voltage applied to the base. This feature of transistor operation is most useful in computers, which we will discuss in section 21.4.

A second type of transistor, called a *field-effect transistor* (*FET* for short), is often used in computer circuitry. In a field-effect transistor, the current flowing through a thin channel of *n*-type material is controlled by the voltage applied across two pieces of *p*-type material on either side of the channel. The strength of the resulting electric field determines how much current will flow through the channel.

The transistor was invented in 1947 and 1948 by scientists at Bell Laboratories in New Jersey, including William Shockley, who invented the bipolar junction transistor, and John Bardeen and Walter H. Brattain, who first demonstrated transistor action in a simpler but less effective *point-contact* transistor. These inventions resulted from the growth in our understanding of the physics of semiconductors in the previous half century. By 1960, transistors were used routinely in many electronic and switching applications.

Before then, electronic amplification and switching were accomplished with vacuum tubes, which required much higher voltages and generated much more heat than semiconductor diodes and transistors. Vacuum tubes were also considerably larger than transistors, so they took up more space, resulting in bulky instruments. During the 1950s and 1960s, vacuum tubes were replaced by solid-state electronics in all kinds of electronic devices.

What are integrated circuits?

Another major revolution in technology, the development and rapid growth in use of miniaturized *integrated circuits,* took place during the 1960s. An **integrated circuit** consists of several transistors, diodes, resistors, and electrical connections all built into a single tiny *chip* of semiconductor material, usually silicon. This advance allowed the production of circuits much smaller than circuits made from individual transistors or vacuum tubes. A computer that would fill a large room using vacuum tubes could now be reduced to the size of a hand-held calculator.

Producing integrated circuits begins with the growth of a large cylindrical crystal of doped silicon. This crystal is sliced into *wafers,* which are generally several centimeters in diameter but just a few millimeters thick (fig. 21.13). The wafers are polished and run through a long process in which insulating oxides are layered on the wafer, and circuit patterns are overlaid on the wafer by photographic methods. Some regions are masked, and the unmasked portions are doped opposite to the underlying silicon crystal to produce diodes and transistors. Metal strips are overlaid to provide conducting connections between elements.

Several identical circuits are usually imprinted on a single silicon wafer. Near the end of the process, the wafer is

figure 21.13 The starting point in producing integrated circuits is a polished wafer of single-crystal silicon. The wafer shown here has been processed to produce tiny circuits on its surface.

figure 21.14 A magnified view of the circuit on a single integrated-circuit chip. Millions of circuit elements may be contained on such a chip, and many chips can be produced from a single silicon wafer.

cut into individual chips, each containing a miniature circuit. A single wafer may yield a hundred or more chips (fig. 21.14). The final steps involve making electrical connections to the chip, packaging the chip in a sealed plastic enclosure (fig 21.15), and testing the resulting circuit. Producing integrated circuits (or ICs) has become a major industry.

Competition to produce ever smaller and faster circuitry for computers and other applications continues to push the technology forward. Physics and chemistry are central to the invention and improvement of new processing techniques. For some applications, silicon is being replaced by

figure 21.15   Rows of packaged microchips arranged on the circuit board of a computer.

semiconducting compounds like gallium arsenide. Research in the condensed-matter physics of semiconducting elements and compounds has become one of the largest areas of activity in modern physics. The revolution in electronics technology is still proceeding.

Semiconductors are materials with conducting properties intermediate between good conductors and good insulators. We can affect their conductivity by doping them with impurity elements that donate extra electrons or leave electron gaps (holes) in the bonds between atoms, permitting us to build diodes and transistors. A diode allows the flow of current in only one direction. Transistors can produce large variations in current from small changes in current or voltage. Integrated circuits combine many diodes, transistors, and other circuit elements on a tiny semiconductor chip.

21.4 Computers and Artificial Intelligence

The revolution in microelectronics has gone hand in hand with the development of computers and a revolution in the processing of information. Computers play an ever-

increasing role in our daily lives. Some of us use them directly for computation, word processing, or games, and others use them less directly as components of appliances such as microwave ovens and video recorders.

What is a digital computer? How has the development of computers been linked to the development of integrated circuits? Can computers think? Will they replace many forms of human employment? These and other questions accompany the computer revolution that is producing enormous changes in the way we work, live, and think.

What are digital computers?

The first true electronic computers were built during the 1940s, using vacuum tubes as the basic switching devices. Univac 1, considered to be the first fully operational digital computer, filled an entire room and required a large air-conditioning system to remove the heat generated by the tubes. A crew of technicians was needed to replace vacuum tubes as they burned out. It did use *digital logic,* however, and was capable of being programmed.

Computers much more powerful than Univac 1 and its immediate descendents now fit into small consoles that sit on a desk or lap (fig. 21.16). They use transistors contained in integrated circuits as their basic switching devices. Most computers still use digital logic, though, and must be programmed to carry out their functions.

What is **digital logic**? All information in a digital computer is stored as individual binary numbers. In other words, each discrete *bit* of information has only two possible values,

figure 21.16   A modern microcomputer is self-contained and easily sits on a desk or on the user's lap.

either 0 or 1. Logical operations that perform functions such as addition or multiplication are done by changing the values of each bit step by step to accomplish the task. This is what we mean by digital logic.

Inside a computer, numbers are in binary form, or *base two,* unlike ordinary numbers, which are decimal (base ten) numbers. In a decimal number, each digit stands for a different power of 10. The number 238, for example, represents

$$(2 \times 10^2) + (3 \times 10^1) + (8 \times 10^0) = 200 + 30 + 8.$$

Any number raised to the zero power equals 1. In a binary number, each digit in the number is a different power of 2 rather than a power of 10. For example, the binary number 1011 is equal to

$$(1 \times 2^3) + (0 \times 2^2) + (1 \times 2) + 1 = 8 + 0 + 2 + 1 = 11.$$

Can you show that the binary number 11010 is the decimal number 26? How would you write the number 5 in binary form?

Writing the number 11 as 1011 might seem cumbersome, but its advantage is that only 0 and 1 are needed to represent the number. Using only two values is ideally suited to simple switches like transistors that have just two states, off or on, 0 or 1. Transistors can also be combined to produce memory elements that can "remember" which state they were last in. These memory elements allow us to store numbers and program codes. In a digital computer, information is both stored and manipulated in binary form.

How are computers programmed?

Programming a digital computer is very tedious if you have to do it all in binary codes, as in the early days of computer development. Before too long, programming languages were invented to let a computer user code instructions to the computer less awkwardly. Special programs translate or *compile* these instructions from the programming language into binary code.

Higher-level programming languages have made the task much easier, but careful attention to detail is still needed to get the instructions exactly right to meet the desired objectives. Computers are very stupid in one sense: they generally do exactly what you tell them to do, even if those instructions make no sense. "Garbage in, garbage out" has become a common expression among computer users.

The author took his first computer-programming course as a first-year graduate student in physics in 1963, using an IBM 1620, a standard early computer, and the FORTRAN programming language. He used the computer extensively in his research to model the kinetics of phase transitions and to analyze experimental data. Computers are very good at doing repetitive computations quickly, without ever losing patience. Their use has been a tremendous boon to science—computations that would have been forbiddingly te-

dious if done by hand can be readily performed on the computer.

Later, the author learned BASIC, a high-level language commonly used on minicomputers in the late 1960s and 1970s. (Modern versions of FORTRAN and BASIC are still used extensively.) He bought his first home microcomputer in the early 1980s and now has three of them around the house, one of which was initially purchased for his children and was used almost exclusively for playing computer games. Another is devoted to word processing and was heavily used in writing this textbook. The third is connected to the Internet and is used by family members for e-mail and surfing the Web. For uses like word processing, we do not program at all, relying instead on sophisticated **application programs** or **software** designed for the personal computer. Application programs are written to be *user-friendly,* so that they are more forgiving of errors in input than traditional computer programs.

During the 1960s, the input of data and programs was usually done with punched cards or paper tape that had to be prepared on card-punch machines or teletypes, which were often quite slow (and noisy). Output was handled similarly and then fed into a separate printer, which was also generally slow. Today, most data input and output are handled by magnetic or optical laser disks that are much faster, less noisy, and easier to use. A printer must still be used to produce a hard copy, but modern laser printers give excellent letter-quality output in almost noiseless operation.

What is artificial Intelligence?

What can a computer do that we cannot? Perhaps a question of greater interest is, What can we do as thinkers that computers cannot? Does a computer actually think in some sense, or must it always just follow instructions? These questions are part of the new area of study called **artificial intelligence** and have broad implications for the future of computers and their applications.

Mostly, a computer does not think in the usual sense of the word. It computes and manipulates binary information following a set of instructions provided by the user or programmer. The output is only as good as the data that were entered and the particular programmed steps that were performed. People provide both the data and the programs. For this reason, you should always be wary of someone who says, "The computer did such and such." They are really telling you that they do not feel competent to question the program or the input data.

What computers can do very well is to execute a complicated set of computations or instructions at high speed and with very high accuracy. They can also store vast amounts of information in various forms of memory, so the information can be readily retrieved. In the speed and accuracy with which they do these tasks, modern computers far exceed human capabilities. All of these operations are done step by step, however, following instructions placed there by human programmers. An ordinary computer makes no intuitive leaps.

Although the speed and accuracy of computers are wonderful, thinking surely involves more. Thinking includes the ability to organize information, to recognize patterns in events, to learn from experience, and to create new ways of solving problems. The human brain can do step-by-step computations but also engages in these other aspects of thinking. (There are individual differences, of course, in how each of us handles different modes of thinking.)

In studying how the human brain functions, we have come to recognize that our brains are organized very differently from a normal digital computer. Instead of transistors, we have neurons as basic signal processors. Unlike transistors that can handle just one or two input signals, individual neurons receive and transmit information by interacting with many other neurons in a vast interconnected network called a *neural network* (fig. 21.17).

The neural network is an incredibly complex system. How individual neurons transmit signals is determined by what the neuron has experienced before, as well as by the strength of the signal and the interconnections of that particular neuron. As we interact with our environment, certain pathways in the neural network develop more than others: in a greatly oversimplified sense, this is how we learn concepts.

For example, input from nerve impulses transmitted from our eyes stimulates a set of neurons, which, when fired simultaneously, transmit signals along established pathways in the brain. These signals create a state in the brain that might correspond to the concept of *chair* (or whatever recognizable object we are looking at). These pathways and the response of the brain have developed through our previous experience with chairs. In some sense, the same process occurs when we recognize physical concepts such as acceleration or torque.

One of the exciting advances in computer science in recent years is the development of computers that mimic the neural network in processing information, either by special programs (software) or by electrical elements that actually connect to many other elements in the circuitry (hardware). Although the computers now in use are very crude approximations of a biological neural network, they do learn from prior experience similarly to learning in the brain. Such computers pose interesting new questions of how to go about programming, or training, the network.

Although we do not anticipate that neural-network computers will ever fully reflect the complexity of our brains, they have an added aspect of intelligence not present in an ordinary digital computer. They are particularly useful for pattern-recognition tasks not readily accomplished with ordinary computers. Of equal importance, perhaps, is that thinking about how to program neural-network computers gives us insights into teaching and learning that may carry over to humans. The entire field of artificial intelligence has caused us to reexamine the fundamentals of thinking and intelligence.

The field of artificial intelligence would not exist without computers. The continued refinement of computers depends on our ingenuity in inventing new ways of making and connecting semiconductor components. Physicists, along with electrical engineers and computer scientists, are heavily involved in this work. The revolution in information processing is still unfolding and may bring even greater changes into our lives.

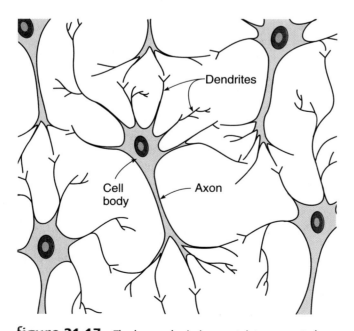

figure 21.17   The human brain is a vast, interconnected network of neurons that transmit and process electrical signals.

Digital computers, which process information in binary form, were first built in the 1940s. The development of integrated circuits in the 1960s began the process of reducing bulky computers to a size small enough to be put in hand-held or tabletop machines. A normal digital computer must be programmed with a detailed set of instructions to perform its functions. It can execute instructions with great speed and accuracy. Neural-network computers are patterned after the neurons in our brains, allowing the computer to do pattern-recognition tasks and other functions resembling how we think and learn. Although computers have been a tremendous boon to science and technology, whether they will ever rival human thought is an open question.

21.5 Superconductors and Other New Materials

A major scientific news item during the late 1980s was the discovery of so-called high-temperature superconductors. A series of news stories claimed that Japan was ahead in the superconductor race and speculated about exotic applications of superconductors. The news flap was almost as great as the one surrounding the supposed discovery of cold fusion a year or so later.

What was all the excitement about? What is superconductivity? What does temperature have to do with it? What other exotic materials are in the works? These questions stem from a field called *materials science,* a combination of metallurgy, chemistry, and condensed-matter physics. Materials research is another physics-related discipline that has already had a major impact on technology and the way we live.

What is superconductivity?

Physicists have been aware of **superconductivity** for some time: it was originally discovered in 1911 by a Dutch physicist, Heike Kamerlingh Onnes. Onnes found that if he cooled mercury to a temperature of about 4 K (4 degrees above absolute zero), the electrical resistance of his sample completely disappeared. An electric current, once started, would flow indefinitely with no continuing source of power.

The electrical resistance of most materials decreases with decreasing temperature, but in Onnes's mercury sample, the resistance completely disappeared at the temperature of 4.2 K, as illustrated in figure 21.18. The resistance drops abruptly to zero at the **critical temperature** T_c and is zero for any temperature below the critical temperature.

Further research showed that many metals became superconducting if cooled to a low-enough temperature. The metal niobium has one of the highest critical temperatures of a pure substance at 9.2 K. Some alloys have even higher critical temperatures. In 1973, an alloy of niobium and the semiconductor germanium was discovered to have a critical temperature of 23 K, still a very low temperature: 23 K is equal to $-250°C$ (or $-418°F$).

A theoretical explanation for the phenomenon of superconductivity, developed in 1957, applies quantum mechanics to the behavior of electrons in a low-temperature metal. This explanation is closely related to a theory that explains the behavior of **superfluids,** a phenomenon also observed at very low temperatures. Liquid helium becomes a superfluid below a critical temperature where it loses its viscosity (or resistance to the flow of the fluid), just as a superconductor loses its electrical resistance. Both superconductivity and superfluidity are macroscopic quantum phenomena: quantum mechanics explains their characteristics, but they are observable on the size scale of ordinary objects rather than at microscopic sizes.

What are high-temperature superconductors?

In 1986, a new type of superconducting compound—a ceramic material, a metal oxide containing various other elements—was discovered. The original ceramic superconductor had a critical temperature of 28 K, not much higher than the metal alloys. The discovery provoked a flurry of experimental activity, however, in which other combinations of elements were tried. By 1987, ceramic superconductors had been developed with critical temperatures first of 57 K and then of 90 K. Finally, in 1988 there were reports of materials with critical temperatures of over 100 K, and perhaps as high as 125 K.

These new ceramic superconductors are *high-temperature superconductors,* although these critical temperatures are still not what we would normally regard as high—100 K is $-173°C$, for example, still rather cold by most standards. The development of materials with critical temperatures around 90 K was a breakthrough, though, because these temperatures can be reached using liquid nitrogen. Liquid nitrogen is readily available for industrial and scientific uses. It boils at 77 K ($-196°C$), so a bath of liquid nitrogen can cool samples to that temperature.

Ceramic superconductors are made of various combinations of materials, and most have copper oxides as one of the components. The most commonly available superconducting ceramic material is a combination of yttrium (Y), barium (Ba), copper (Cu), and oxygen (O) in the proportions $Y_1Ba_2Cu_3O_7$. (The number of oxygen atoms in the structure varies depending on how the material is prepared.)

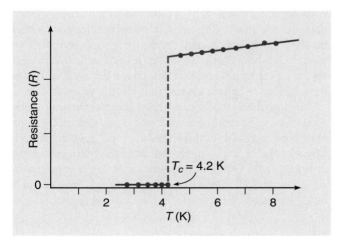

figure 21.18 The electric resistance of mercury decreases as the temperature is decreased. It drops abruptly to zero, however, at the critical temperature of 4.2 K.

This material can be prepared in undergraduate laboratories and has a critical temperature of approximately 90 K.

One striking property of a superconductor (called the *Meissner effect*) is that it will completely exclude magnetic field lines produced by an external magnet or electric current. A magnet brought near a superconducting material will be repelled. This property is commonly demonstrated by levitating a small magnet above a disk of superconducting material. (The materials for this demonstration have been widely distributed to science teachers.) A small amount of liquid nitrogen at the bottom of a Styrofoam cup is sufficient to cool the superconducting disk (fig. 21.19).

The theory that successfully explained superconductivity in pure materials was not adequate for explaining the superconductivity of these new ceramic materials. A common feature of the structure of many of these materials is that they contain layers of copper or copper oxide sandwiched between the atoms of the other elements (fig. 21.20). The superconduction is suspected to occur through these layers, and theoreticians have made good progress in understanding what is happening. Continued theoretical and experimental progress could point the way to designing materials with even higher critical temperatures.

High-temperature superconductors have many potential applications, especially in the use of electromagnets. A strong electromagnet requires large currents flowing in tightly wound coils of wire. With ordinary conductors, these large currents generate a great deal of heat and limit the amount of current that can be established (and the resulting strength of the electromagnet). Magnets already exist that use superconducting coils, but they must be maintained at

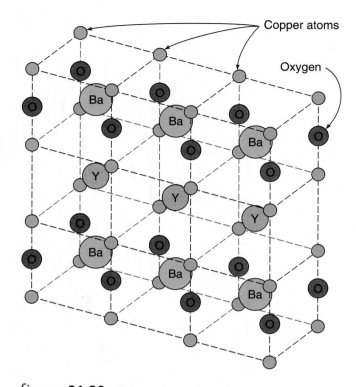

figure **21.20** The atomic structure of many superconducting ceramics has layers of copper atoms sandwiched between other elements, such as oxygen, yttrium, and barium.

temperatures below the critical temperature of the superconducting material. If we had materials that were superconducting near room temperature, such electromagnets would become much more feasible for general use.

The superconducting supercollider would have used superconducting magnets to control the particle beams. Such magnets are now being used in other accelerators. Other possible applications include the magnetic levitation of trains and other vehicles to reduce friction and attain higher speeds. Superconducting cables could be used in power transmission to reduce losses from electric resistance.

Most of these applications need superconducting materials that can be readily shaped into wires and cables, with even higher critical temperatures than those available now. Many of the ceramic superconductors are quite brittle and not suitable for cables or magnetic coils. Success in inventing more usable superconducting materials may await a new generation of scientists, engineers, and dreamers.

Other exotic materials

The discovery of the new superconductors was sparked by research on ceramic metal oxides already known to have interesting electrical properties. A compound of barium,

figure **21.19** A small magnet levitates above a superconducting disk cooled with liquid nitrogen. This simple test establishes the presence of superconductivity.

titanium, and oxygen, barium titanate ($BaTiO_3$), for example, had been used for many years to convert changes in pressure into electrical signals, or vice versa. This property allows a crystal of barium titanate to be used as a tiny microphone or speaker. Other metal oxides are important in integrated-circuit processing.

The search for new materials stems from our growing knowledge of how atoms interact in the solid or liquid states. We have become more able to design materials to meet specific needs. These needs may be special optical or electronic properties for use in electronics and communications or perhaps high-strength but lightweight materials for aircraft. Different elements can be combined in an infinite number of ways to make new materials, and the results cannot always be predicted.

Liquid crystals are one of the new materials that have found extensive application. Liquid crystals have a crystal-like organization in one direction but are disordered and free to flow along other directions in the material. They have some properties of both liquids and solids. Electric fields can affect how much light flows through the material. This property has led to their use in display screens of hand-held calculators and in very thin, flat television screens that do not require a bulky cathode-ray tube.

Liquid crystals are often made up of long organic (carbon-based) molecules that line up in layers, as in figure 21.21. These layers slide along one another, so that the material can flow in the directions parallel to the layers. The regular spacing perpendicular to these layers causes the crystal-like properties. The author has conducted research on another class of substances known as *plastic crystals* whose molecules are globular in shape and partially free to rotate in the solid crystal. Plastic crystals have many interesting properties, but so far, they have not led to the extensive applications (and financial rewards) that have grown out of research on liquid crystals.

It is hard to know just where this research will lead. We can be sure of two things: some new materials will have unexpected and exciting properties, and some of these ma-

figure 21.21 The long molecules in some liquid crystals line up in layers, allowing the liquid to flow along these layers but not in the perpendicular direction.

terials will lead to new products that you will encounter in your everyday activities.

Everyday Phenomenon Box 21.1 discusses another modern development resulting from research in physics, this one in the subfield of optics. We are now exploring the use of holograms for data storage in computers, which requires the development of special optical materials.

Superconductors are materials that lose all resistance to flow of an electric current below some critical temperature. For pure metals, these critical temperatures are only a few degrees above absolute zero, but more recently, superconducting compounds have been discovered with critical temperatures of 100 K or above. These high-temperature superconductors may someday find extensive applications in power transmission or in superconducting magnets. Materials science has invented many other exotic materials in recent years, including the liquid crystals that are used in display panels for calculators and laptop computers. Our increasing knowledge of how atoms interact in the solid and liquid states lets us design materials for specific applications, but their precise properties still hold surprises.

Everyday Phenomenon

box 21.1

Holograms

The Situation. We all have seen holograms on cereal boxes, toys, credit cards, and perhaps on simple jewelry, such as the pendants in the photograph. Although the credit cards and other surfaces are clearly two-dimensional, the images that we see in a hologram are three-dimensional. You can move your head from side to side and view these images from different perspectives just as you can with a real three-dimensional object. How are these three-dimensional images produced?

A hologram on a pendant viewed from two different angles. How is this three-dimensional image produced?

Holography seems like science fiction to many people—what can actually be achieved with holograms? What are holograms, and how do we go about making them? Could holograms be used to develop three-dimensional television or movies?

The Analysis. Although the idea was conceived earlier, the first good holograms were produced in the early 1960s following the invention of the laser. A hologram is an interference pattern produced by combining light waves reflected from some object with another wave coming directly from the laser. Lasers are highly *coherent* light sources—they produce much longer wave trains than ordinary light sources, which produce short, uncorrelated pulses of light. The high coherence of the laser is needed to produce interference patterns of light scattered from objects of ordinary size.

A common arrangement for making a hologram is shown in the drawing below. Light coming from the laser is split by a partially silvered mirror or beam splitter into two beams, one called the *object* beam and the other the *reference* beam. The object beam is scattered or reflected by the object back toward the photographic plate. The reference beam is directed to the photographic plate at some angle to the object beam, and these two beams of light then combine to produce an interference pattern on the photographic plate.

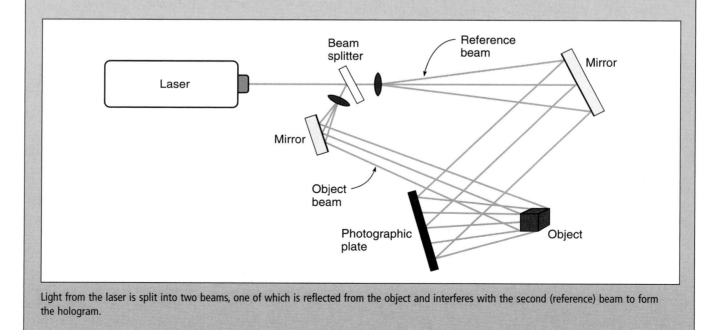

Light from the laser is split into two beams, one of which is reflected from the object and interferes with the second (reference) beam to form the hologram.

(continued)

When the photographic plate is developed, the recording of the interference pattern becomes the hologram. If light similar to the original laser light passes through the hologram, the interference pattern modifies the light so that one of the light waves transmitted is identical to the original light wave reflected from the object. This identical light wave is what we view when we look at a hologram. This light wave is diverging from an image of the original object—we are observing a three-dimensional virtual image. (See chapter 17 for a discussion of image formation.)

The most familiar holograms are reflection holograms designed to be viewed in light reflected from the hologram rather than light transmitted through the hologram. Reflection holograms have the additional advantage that we do not need a laser or other monochromatic light source to view them. The reflection process selects out only certain wavelengths of light. In the reflection holograms on cereal boxes or credit cards, the interference pattern representing the hologram is embossed onto a thin reflecting film. Holograms are used on credit cards because it is extremely difficult to produce counterfeit copies.

Originally, the process of making holograms required that the object be held completely still, with no vibrations, to produce an accurate interference pattern. More powerful lasers and better films now allow shorter exposure times, objects do not have to be kept quite so motionless. It is also possible to generate the interference patterns from mathematical computations on a computer, so that we can design computer-generated holograms of nonexistent objects. A single hologram contains an enormous amount of information, however, so moving holograms that can be transmitted by television signals are not yet feasible.

The invention of the laser in 1960 has spurred tremendous growth in the field of optics. Holography is just one of the applications that this amazing light source has made possible. Holography is now used in many technical applications as well as in art, special displays, and novelty items.

summary

In this chapter, we have gone beyond everyday phenomena to touch on some of the more exciting discoveries in the continually advancing areas of research in physics. The ideas explored included quarks and other elementary particles, cosmology and the Big Bang, integrated circuits, digital computers and neural-network computers, and superconductors and other "designer" materials.

1   Quarks and other elementary particles.   The standard model of high-energy physics can now describe all of the known particles as combinations of twenty-four elementary particles—six leptons, six quarks, and their antiparticles. Progress has been made in bringing the fundamental forces of nature into a single *unified field theory.*

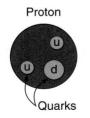

Proton

Quarks

2   Cosmology and the beginning of time.   Astronomical observations have shown that our sun is just one star in a large galaxy of stars and that there are many other galaxies, all receding from one another in an expanding universe. Knowledge of quarks is necessary to model the earliest moments of the universe following the Big Bang, which started the current expansion.

3   Semiconductors and microelectronics.   Our understanding of semiconductors and how their conductivity can be modified by doping with impurity atoms led to the invention of the transistor in the late 1940s. Since then, integrated circuits have been developed by combining hundreds of transistors and other elements on tiny silicon chips. A tremendous industry has grown from this ability to miniaturize electronic and computer circuitry.

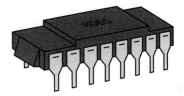

4   Computers and artificial intelligence. Invention of the transistor and integrated circuits led to the rapid growth in the use of digital computers. Ordinary computers perform programmed operations step by step, with high speed and accuracy. New neural-network computers are organized more like the connections between neurons in our brains. They "learn" from experience to recognize patterns in the data provided to the computer.

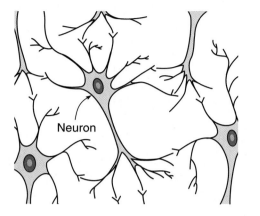

Neuron

5   Superconductors and other new materials. Superconductors are materials that lose all of their electrical resistance below a certain critical temperature. Recently, ceramic superconductors have been discovered with critical temperatures around 100 K, much higher than those previously known, but still well below room temperature. Research in the physics of solids and liquids has produced many other useful materials such as the liquid crystals that are used in calculator displays and similar applications.

key terms

Standard model, 446	Unified field theory, 449	Integrated circuit, 455
Lepton, 447	Electroweak force, 449	Digital logic, 456
Meson, 447	Grand unified theories, 449	Application programs, 457
Baryon, 447	Galaxy, 450	Software, 457
Quark, 448	Big Bang, 451	Artificial intelligence, 457
Hadron, 448	Conductivity, 452	Superconductivity, 459
Strong nuclear interaction, 449	Diode, 453	Critical temperature, 459
Weak nuclear force, 449	Transistor, 454	Superfluid, 459

questions

*Questions identified with an asterisk are more open-ended than the others. They call for lengthier responses and are more suitable for group discussion.

Q1. Are leptons generally heavier than protons or neutrons? Explain.

Q2. Do we now consider protons to be elementary particles that do not have any underlying structure? Explain.

Q3. Are quarks constituents of electrons? Explain.

Q4. Are baryons and mesons made up of the same number of quarks? Explain.

Q5. Why are high energies required to produce particles with larger masses than protons or neutrons? Explain.

Q6. Which fundamental force of nature is the most difficult to incorporate into a completely unified field theory? Explain.

*Q7. Why do physicists propose spending large amounts of money to build new particle accelerators? How can we justify these large expenditures? Explain.

*Q8. How do we know that the universe is expanding? Explain.

Q9. Is our own sun part of a galaxy? Explain.

Q10. Is the Milky Way a cloud of interstellar gases? Explain.

Q11. What force is responsible for the formation of atoms from individual nuclei and electrons? What force is responsible for the condensation of individual atoms into stars? Explain.

Q12. Does the term *Big Bang* refer to explosions of individual stars? Explain.

*Q13. Is it necessary to know anything about very small entities such as quarks to produce successful models of very large scale phenomena such as the beginning of the universe? Explain.

Q14. Is the electrical resistance of a semiconductor such as silicon increased when we dope it with impurity atoms of arsenic? Explain.

Q15. Does doping of silicon with gallium make the resulting semiconductor an *n*-type or *p*-type semiconductor? Explain.

Q16. Does the direction in which the battery is connected to a diode affect the amount of electric current that will flow through the diode? Explain.

Q17. Can a diode be made from material that is doped with just one type of impurity atom? Explain.

*Q18. What property of transistors makes them useful for amplifying an electrical signal? Explain.

Q19. In making ever-smaller electronic instruments, do integrated circuits have an advantage over the use of separate transistors and diodes? Explain.

Q20. Are decimal numbers better suited than binary numbers for processing information by circuits containing transistors? Explain.

Q21. Did the original electronic computers use transistors in their circuitry? Explain.

Q22. What is the purpose of a computer program designed for use with an ordinary digital computer? What does it do? Explain.

*Q23. Can an ordinary digital computer think? Explain.

Q24. Are the transistors in an ordinary digital computer connected like the neurons in our brains? Explain.

Q25. Is a neural-network computer programmed in the same way as an ordinary digital computer? Explain.

Q26. Does a superconductor have zero resistance only *above* a certain critical temperature? Explain.

Q27. With the high-temperature superconductors currently available, can we build superconducting magnets that will operate in a superconducting mode at room temperature? Explain.

Q28. Are superfluids the same as superconductors? Explain.

Q29. Are liquid crystals fluids or solids? Explain.

Q30. Does the production of a hologram involve the interference of light waves? Explain.

Q31. Can we make holograms using an ordinary light source rather than a laser? Explain.

exercises

E1. The average distance from the sun to the earth is approximately 1.5×10^8 km. How many seconds are required for light to travel from the sun to the earth? ($c = 3 \times 10^8$ m/s, 1 km = 1000 m)

E2. Convert these binary numbers to ordinary decimal numbers:
   a. 1010
   b. 11111

E3. Write these ordinary decimal numbers in binary form:
   a. 7
   b. 15

challenge problems

CP1. The nearest star to our sun is about 4 light-years away—a light-year is the distance that light travels in 1 year.
   a. How many seconds are there in a year?
   b. Since light travels at the rate of 3×10^8 m/s, how many meters are there in 1 light-year?
   c. How far is it to the nearest star in meters?
   d. How long would it take to travel to the nearest star if we were able to travel at a speed one-tenth the speed of light?

CP2. Suppose that we wished to develop a base-three number system to be used in place of either a decimal system or a base-two (binary) system.
   a. How many separate symbols would we need in this system? (In the binary system, we can get away with only two, 0 and 1. Will this work in a base-three system? How would you write the decimal number 2?)
   b. Write the decimal number 6 in base-three form.
   c. Write the decimal number 39 in base-three form.
   d. Can you think of any possible use of such a system?

home experiments and observations

HE1. On a clear night (preferably away from city lights), go outside and study the night sky.
   a. Can you see the Milky Way? (It will usually appear as a faint cloud of stars making an irregular band across the sky.) Make a sketch of its orientation.
   b. What are the brightest objects in the sky (other than the moon)? Are some of these objects planets? Some planets will produce a steadier-appearing light than stars and are quite bright.
   c. Can you pick out the Big Dipper (a part of the constellation Ursa Major) and other constellations? Make a sketch of the more prominent groupings of stars that you observe.

HE2. Find a hologram on a credit card or cereal box and examine it closely.
   a. As you move your head from side to side, can you see different features of the object? Is the image that you see clearly three-dimensional?
   b. Move your head up and down as you observe the hologram. Do the colors change? What sequence of colors do you observe? Is there any three-dimensional character to the hologram in the vertical (up-and-down) direction?

appendix a

Using Simple Algebra

In chapter 1, we described mathematics as part of the language of physics. It is a compact way of expressing relationships between physical quantities that makes manipulations of these relationships much easier than if they were expressed in words and sentences. People who are conversant in mathematics are very comfortable in interpreting and using such relationships.

Many people are not comfortable using simple mathematics, and therefore mathematics is used sparingly in this textbook. Most of what is used is basic algebra, to which most college students have been introduced in high school. Often that introduction fails to produce a firm understanding of the principles underlying algebraic operations, though, leaving students with little confidence in their use. This appendix presents the basic concepts underlying simple algebraic manipulations and provides illustrations of their application.

Basic Concepts

Three simple, but fundamental concepts form the basis for most algebraic manipulations. These concepts are the following:

■ **Concept 1: The letters used in algebra represent numbers.** Any operation performed with numbers (addition, subtraction, multiplication, division, and so on) can also be performed with these symbols.

In mathematics courses, the letters x and y are often used to represent unknown numbers, and other letters are used to represent constants or known numbers. In physics, specific letters are used to represent specific quantities: t for time, m for mass, d for distance, s for speed, and so on. They all represent numerical quantities, but some may be known and others may be initially unknown. The relationship $s = d/t$, for example, tells us that we can find the numerical value of speed by dividing a numerical value for distance by a numerical

value of time. This relationship holds for any possible values of the distance and time.

■ **Concept 2: If the same operation is performed on both sides of an equation, the equality expressed by that equation does not change.**

This principle is the basis for all algebraic manipulations performed to express a relationship in different forms. For example, if we multiply both sides of the equation $s = d/t$ by the quantity t, the equality still holds. This operation yields

$$st = d\,\frac{t}{t} = d$$

since t/t equals 1. Performing this operation expresses the original equation in a new form: $d = st$ tells us that the distance is equal to the speed multiplied by time. We can multiply both sides of an equation by the same quantity, divide both sides by the same quantity, add or subtract the same quantity from both sides, and the equation is still valid. We can also square both sides of the equation or perform various other operations, but the operations just listed are those most commonly used.

■ **Concept 3: When we solve an algebraic equation, we are merely rearranging the equation as just described so that the quantity we wish to know is expressed, by itself, on one side of the equation and everything else is on the other side of the equation.**

In the paragraph illustrating concept 2, we solved the equation $s = d/t$ for the quantity d, thus expressing the distance in terms of the other two quantities, speed and time. If we wanted to express the time of travel in terms of the speed and the distance, we could divide both sides of the equation $d = st$ by the quantity s:

$$\frac{d}{s} = \frac{st}{s} = t\left(\frac{s}{s}\right) = t$$

or, $t = d/s$, the distance divided by the speed. We see that the original equation $s = d/t$ can be expressed in two other forms, $d = st$ and $t = d/s$, that restate the original equality in forms suitable for computing a specific quantity when the other two quantities are known, an extremely useful thing to be able to do.

Since the letters represent numbers (concept 1), we can always check the validity of the operations we perform by inventing numbers for the quantities and checking to see that the equalities still hold in the new form. For example, if in the original equation, $d = 6$ cm and $t = 2$ s, then

$$s = \frac{d}{t} = \frac{6 \text{ cm}}{2 \text{ s}} = 3 \text{ cm/s}.$$

If we put the same numbers in the final equation, $t = d/s$, we have

$$t = \frac{d}{s} = \frac{6 \text{ cm}}{3 \text{ cm/s}} = 2 \text{ s}$$

or 2 s = 2 s, which is obviously an equality.

These concepts are straightforward, and their application is not difficult once the basic ideas are grasped. A little practice, obtained by following the additional examples given below and performing the exercises at the end of this appendix, should help to build confidence in their use. For most people who have trouble with mathematics, lack of confidence is the fundamental problem. Often, they have never fully accepted the idea that letters can represent numbers, and the manipulations and rules of algebra therefore seem arbitrary and mysterious.

Other Examples

1. Solve the equation $a = b + c$ for the quantity c.

Solution: We seek an expression in which c is by itself on one side of the equation and the other two quantities are on the other side. This can be accomplished by subtracting the quantity b from both sides of the equation, since doing so will leave c by itself on the right side:

$$a - b = b + c - b = c.$$

Thus we see that $c = a - b$. (It does not matter which side of an equality is stated first—the equality is the same in either case.) By subtracting b from the right side of the original equation, c now stands by itself, so we have achieved the desired result.

2. Solve the equation $v = v_0 + at$ for the quantity t.

Solution: This is best done in two steps. The first step is to subtract the quantity v_0 from both sides of the equation to isolate the product at:

$$v - v_0 = v_0 + at - v_0 = at$$

$$at = v - v_0.$$

Then we divide both sides of this equation by a to get t by itself:

$$\frac{at}{a} = \frac{v - v_0}{a}$$

$$t = \frac{v - v_0}{a}$$

If you can understand why each of these operations was performed (what was the motivation or objective?), you are well on your way to following the algebra used in this textbook.

3. Solve the equation $b = c + d/t$ for the quantity t.

Solution: Again, we first subtract the quantity c from both sides of the equation to isolate the term containing t:

$$b - c = c + \frac{d}{t} - c = \frac{d}{t}.$$

The quantity t is in the denominator, however, so we multiply both sides of the equation by t:

$$(b - c)t = \frac{d}{t} t = d.$$

Next, we divide both sides of the equation by $(b - c)$ to obtain t by itself on the left side of the equation:

$$\frac{(b - c)t}{b - c} = \frac{d}{b - c}$$

$$t = \frac{d}{b - c}$$

Although this is a more complex example, each of these steps has a specific objective. The first step isolates the quantity d/t, the second step removes t from the denominator so that we can more readily solve for t, and the final step leaves t by itself. These objectives must be recognized to gain confidence in performing such operations yourself. Even people who are familiar with algebra often forget just what they are trying to accomplish, or they get careless in making sure that they are doing the same thing to both sides of an equation.

Exercises

(Answers to odd-numbered exercises are found in appendix D.)

1. Solve the equation $F = ma$ for the quantity a.
2. Solve the equation $PV = nRT$ for the quantity P.
3. Solve the equation $b = c + d$ for the quantity d.
4. Solve the equation $h = g - f$ for the quantity g.
5. Solve the equation $a = bc + d$ for the quantity d.
6. Solve the equation in exercise 5 for the quantity b.

7. Solve the equation $a = b(c - d)$ for the quantity b.

8. Solve the equation in exercise 7 for the quantity c. (Hint: First rewrite the equation as $a = bc - bd$, multiplying both terms inside the parentheses by b. This does not change the equality.)

9. Solve the equation $a + b = c - d$ for the quantity b.

10. Solve the equation in exercise 9 for the quantity c.

11. Solve the equation $b(a + c) = dt$ for the quantity b.

12. Solve the equation in exercise 11 for the quantity c.

13. Solve the equation $x = v_0 t + \frac{1}{2} a t^2$ for the quantity v_0.

14. Solve the equation in exercise 13 for the quantity a.

appendix b

Decimal Fractions, Percentages, and Scientific Notation

In physics and many other fields in which numbers are important, we usually express fractions as decimal fractions and often use percentages as a means of expressing fractions or ratios. Because we need at times to deal with very large and very small numbers, we also use a means of expressing these numbers involving powers of ten or *scientific notation* to avoid writing out all of the zeros. Although scientific notation is used sparingly in this book, there are times when its use is highly desirable, if not essential, so it is important that you understand its meaning. It is part of the language of science.

Decimal Fractions

Although most college-level students are familiar with decimal fractions and percentages, they are not always completely sure of their meaning. Fractions involve ratios or proportions, which are not well understood by many people. One of the benefits of taking a course in physics is that it can strengthen your ability to think in terms of ratios or proportions and to understand how they are described.

A decimal fraction is just a fraction for which the number in the denominator is some multiple of the number 10 (10, 100, 1000, and so forth), with the appropriate multiple indicated by the location of the decimal point. For example, if we start with the fraction $\frac{1}{2}$ and divide 1 by 2 as the fraction indicates, a calculator will display the result as 0.5. The decimal point in front of the 5 is a shorthand notation for expressing the fraction $\frac{5}{10}$. The number 5 is half of 10, so the fraction $\frac{5}{10}$ is the same as the fraction $\frac{1}{2}$ (one-half). In other words, the *ratio* of 5 to 10 is the same as the ratio of 1 to 2.

If the fraction $\frac{3}{4}$ is evaluated on a calculator by dividing 3 by 4, the calculator will express the result as 0.75, which is equivalent to the fraction $\frac{75}{100}$ or 75 hun-

dredths. Thus, the first place or number after the decimal point represents tenths, the second place hundredths, the third place thousandths, and so on. The fraction 346 thousandths ($\frac{346}{1000}$) is expressed as 0.346, for example. We could read this as 3 tenths plus 4 hundredths plus 6 thousandths.

Decimal fractions are used very commonly, although we may not always stop to think about their meaning. In baseball, for example, we express a batter's hitting efficiency as a decimal fraction. A batter who has produced 35 hits in 100 official at-bats is said to be hitting 350. This is really $\frac{350}{1000}$ or 0.350, but the decimal point is often omitted. Most people understand that it should be there, however, and that we are merely expressing the fraction $\frac{35}{100}$ in decimal form and including three figures to the right of the decimal point.

What Are Percentages?

Another common way of expressing decimal fractions is to write them as percentages. The word *percent* means per one hundred, so a percentage is just a decimal fraction in which the denominator is 100. The fraction $\frac{1}{2}$, for example, is $\frac{5}{10}$ or $\frac{50}{100}$ and can be expressed as 50%—it is 50 hundredths. The fraction $\frac{3}{4}$ is 0.75 or 75% (75 hundredths), and the fraction $\frac{346}{1000}$ is 0.346 or 34.6% (34.6 hundredths). Thus, moving the decimal place two places to the right converts a decimal fraction to a percentage and is equivalent to multiplying the fraction by 100.

The use of percentages is even more common than the direct use of decimal fractions. Interest rates and tax rates are usually expressed as percentages, for example, so we should all have some understanding of their meaning. An interest rate of 7% means that you will receive or pay $7 each year for every $100 that you have invested or borrowed, $\frac{7}{100}$ of the total amount. (We are ignoring here the possible effects of compounded interest.) A tax rate of 28% means that we will owe to the government $28 of every $100 that we earn (after deductions are subtracted). A percentage is always per one hundred, by definition.

Although it is easy enough to understand how a percentage is calculated (compute the decimal fraction and multiply by 100), having a good feeling for the proportions represented by different percentages is another matter. Pie charts, like the one shown in figure B.1, are often used to provide a visual representation of these proportions. The slices of the pie should have sizes in proportion to the percentages or fractions being represented. If the graphic artist does not understand this (as sometimes happens), the resulting pie chart may be very misleading.

The pie chart in figure B.1 represents the average monthly expenditures of someone who takes home $2000 a month (after taxes and other deductions). If she spends $500 a month on rent, this is $500/2000$ or 0.25 (one-quarter) of her total income. Since 0.25 equals 25%, this is shown as 25% on the pie chart, and it takes up one-quarter of the total pie or circle. The size of the slice is in proportion to the percentage. Likewise, if she spends $800 a month on food, this is $800/2000$ or 0.40, which is 40% of her total take-home pay. The other slices represent smaller percentages and have correspondingly smaller sizes. If we have taken into account all of her normal expenses, the sum of the percentages in the chart should add to 100%.

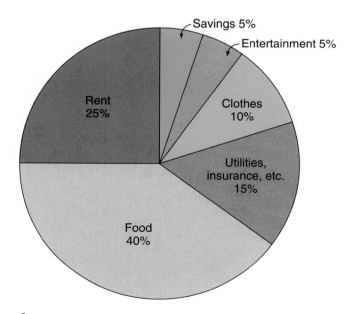

figure **B.1**   A pie chart showing the fractions of total take-home pay spent in different categories. The slices of the pie are in proportion to the percentages being represented.

Why Is Scientific Notation Used?

When we need to represent very large numbers or very tiny fractions, a lot of zeros are required to locate the decimal point properly. For example, 1.2 trillion dollars (corresponding roughly to the size of our accumulated national debt several years ago) can be written as

$1 200 000 000 000.

The zeros are there only to locate the decimal point; they do not imply that all of the other numbers to the right of the 1 and 2 are exactly zero. If we count the digits to the right of the 1, we see that there are 12 (11 zeros and the digit 2).

Another way of stating this number would be to say that it is 1.2 times 1 trillion, where 1 trillion is the number 1 followed by 12 zeros. One trillion is also the number that results when you multiply 1 by ten 12 times.

$$\begin{aligned} 1\,000\,000\,000\,000 = 1 &\times 10 \times 10 \times 10 \times 10 \\ &\times 10 \times 10 \times 10 \times 10 \\ &\times 10 \times 10 \times 10 \times 10 \end{aligned}$$

The shorthand notation for a number multiplied by itself 12 times is to say that has been raised to the *power* 12, which we write as 10^{12}. The superscript represents the power to which the number has been raised, which is the number of times that you have multiplied the number by itself. We read this number as "ten to the twelfth power" or often just "ten to the twelfth."

Thus we can write the number 1.2 trillion as

1.2×10^{12}.

This notation, in which we have written the number as some number times a power of ten, is called *scientific notation*. The number 12 (the power) simply tells us how many places to the right of the indicated decimal point we would move the decimal point if we wrote out all of the zeros. Scientific notation has several advantages: it saves space, it properly indicates the accuracy or precision of the number being represented by eliminating the zeros, and it makes the number easier to manipulate in calculations involving very large or small numbers.

Some examples involving smaller numbers may help to make the concept clear. The number 586, to choose a much smaller number than 1.2 trillion, can be expressed as 5.86 times 100, or 5.86×10^2 since $10 \times 10 = 100 = 10^2$ (10 squared). The number 6,180 can be expressed as 6.18×10^3, since 10^3 (10 cubed) is 1000. The number 5 400 000 (5.4 million) can be expressed as 5.4×10^6, since 10 to the sixth power is 1 million. Several other examples are provided in table B.1. The last number listed under the positive powers of ten is the approximate mass of the earth in kilograms.

Table B.1 also shows several decimal fractions written in scientific notation. A fraction will always have a negative exponent (negative power of ten) if the value of the fraction

table B.1		
Examples of Scientific Notation		
Positive powers of ten		
5 460	= 5.46 times 1 thousand	= 5.46×10^3
23 400	= 23.4 times 10 thousand	= 2.34×10^4
6 700 000	= 6.7 times 1 million	= 6.7×10^6
9 400 000 000	= 9.4 times 1 billion	= 9.4×10^9
5 980 000 000 000 000 000 000 000		= 5.98×10^{24}
Negative powers of ten (fractions)		
0.62	= 6.2 times one-tenth	= 6.2×10^{-1}
0.0523	= 5.23 times one-hundredth	= 5.23×10^{-2}
0.0082	= 8.2 times one-thousandth	= 8.2×10^{-3}
0.000 0024	= 2.4 times one-millionth	= 2.4×10^{-6}
0.000 000 0079	= 7.9 times one-billionth	= 7.9×10^{-9}
0.000 000 000 000 000 000 16		= 1.6×10^{-19}

is less than 1. For example, the fraction 0.000 000 000 001 2 is the very tiny fraction 1.2 trillionths. It can be expressed as 1.2×10^{-12}, which is equivalent to dividing the number 1.2 by 10^{12}, or by 1 trillion. The superscript -12 tells you that you have to move the decimal point 12 places to the *left* to express the number in normal decimal form.

Taking a simpler example, the decimal fraction 0.0346 is 3.46×10^{-2} or 3.46 hundredths. Moving the decimal points two places to the left, as indicated by the power of ten, yields the original decimal fraction. The fraction 0.0079 is 7.9 thousandths or 7.9×10^{-3}. Studying the other examples in table B.1 should make the pattern clear. The last number in table B.1 is the value of the charge on the electron in coulombs, a quantity that arises frequently in modern physics.

The prefixes used in the metric system of units (discussed in chapter 1) are another aid to expressing very large or very small numbers. Since the prefix *mega* stands for 1 million, the quantity 1.35 Mg (megagrams) is the same as 1.35×10^6 g (10^6 is one million). Likewise, 780 nm (nanometers) is the same as 780×10^{-9} m, since the prefix *nano* means one-billionth or 10^{-9}. The values of the commonly used metric prefixes are found in table 1.3 in chapter 1. These metric prefixes and the power-of-ten scientific notation are both types of scientific shorthand used to express numbers in briefer forms.

Multiplying and Dividing Using Powers of Ten

The process of multiplying or dividing numbers written in power-of-ten notation is straightforward if you understand what they mean. It is even easier if you have a calculator that handles scientific notation—you just punch the numbers in and push the appropriate function key. Some understanding of their meaning, though, can be useful for checking your results.

Suppose, for example, that we multiply the number 3.4×10^3 (3400) by 100 (10^2). Multiplying by 100 adds two zeros to the original number, yielding 340 000, as you can quickly check by doing this operation on a calculator or by direct multiplication. Thus

$$(3.4 \times 10^3) \times (10^2) = 3.4 \times 10^5.$$

In other words, the powers of ten add ($3 + 2 = 5$). If we divided by 100, we would remove two zeros:

$$\frac{3.4 \times 10^3}{10^2} = 3.4 \times 10^1 = 34.$$

In this case, the exponent of the denominator is subtracted from the exponent of the number being divided ($3 - 2 = 1$). The rules for these operations are thus

1. When numbers are multiplied, the powers of ten add.
2. When numbers are divided, the power of the denominator is subtracted from the power of the numerator.

These rules are valid regardless of whether the powers are positive or negative. Thus

$$(3 \times 10^6) \times (2 \times 10^{-4}) = 6 \times 10^2 = 600$$

since $6 + (-4) = 2$. This should make sense to you since multiplying by a fraction (a number with a negative power of ten) results in a smaller number than the number being multiplied.

Exercises

If any of these ideas are unfamiliar—or even if they are familiar but you are rusty in using them—working some or all of these exercises will help to build your confidence. The answers to the odd-numbered exercises are found in appendix D.

(Exercises 1 through 4) Express these numbers as decimal fractions:

1. **a.** $\dfrac{6}{10}$ **b.** $\dfrac{52}{100}$ **c.** $\dfrac{874}{1000}$ **d.** $\dfrac{5}{10\,000}$

2. **a.** $\dfrac{72}{100}$ **b.** $\dfrac{7}{10}$ **c.** $\dfrac{83}{10\,000}$ **d.** $\dfrac{45}{1000}$

3. **a.** $\dfrac{1}{4}$ **b.** $\dfrac{5}{8}$ **c.** $\dfrac{16}{52}$ **d.** $\dfrac{312}{914}$ (Use a calculator.)

4. **a.** $\dfrac{3}{7}$ **b.** $\dfrac{11}{15}$ **c.** $\dfrac{147}{654}$ **d.** $\dfrac{65}{150}$ (Use a calculator.)

5. Express the fractions in exercise 3 as percentages.

6. Express the fractions in exercise 4 as percentages.

7. Find: **a.** 50% of 105 **b.** 75% of 48
   c. 60% of 180 **d.** 85.2% of 100

8. Find: **a.** 40% of 120 **b.** 90% of 400
   c. 33.3% of 90 **d.** 70% of 540

(Exercises 9 through 12) Express these numbers in scientific notation (power-of-ten notation):

9. **a.** 5475 **b.** 200 000 **c.** 67 000 **d.** 35 000 000 000

10. **a.** 3560 **b.** 78 500 **c.** 622 000 **d.** 9 100 000

11. **a.** 0.0065 **b.** 0.000 333 **c.** 0.000 001 5
    d. 0.000 000 065

12. **a.** 0.075 **b.** 0.000 45 **c.** 0.000 003 2 **d.** 0.000 89

13. Express these numbers as decimal fractions:
    a. 6.7×10^{-3} **b.** 1.8×10^{-4}
    c. 5.77×10^{-6} **d.** 3.25×10^{-5}

14. Perform these operations:
    a. $(3.0 \times 10^2) \times (4.3 \times 10^5)$
    b. $(7.5 \times 10^3) \times (5.0 \times 10^6)$
    c. $(4.0 \times 10^8) \times (5.4 \times 10^{-5})$
    d. $(6.0 \times 10^8) \div (2.0 \times 10^3)$

15. Perform these operations:
    a. $(3.0 \times 10^5) \times (2.0 \times 10^4)$
    b. $(4.0 \times 10^7) \times (6.0 \times 10^{-3})$
    c. $\dfrac{3.6 \times 10^8}{2.0 \times 10^5}$
    d. $\dfrac{3.6 \times 10^{12}}{2.0 \times 10^{-6}}$

appendix c

Vectors and Vector Addition

Many of the quantities that we encounter in the study of physics are vector quantities, which is to say that their *direction* is important as well as their size or magnitude. Examples of vector quantities include velocity and acceleration (introduced in chapter 2), as well as force, momentum, electric field, and many others encountered in later chapters. Direction is an essential feature of these quantities: the result of traveling with a velocity of 20 m/s due north is very different from traveling with a velocity of 20 m/s due east.

Quantities for which direction is not an essential feature (or for which direction has no meaning at all) are called *scalar* quantities. Mass, volume, and temperature are examples of scalar quantities—it makes no sense to talk about the direction of a volume or a mass. Specifying the magnitude (the numerical value with appropriate units) of a scalar quantity is sufficient; no other information is needed. Vectors, on the other hand, require at least two pieces of information to describe both their size and direction.

How Do We Describe a Vector?

Suppose that we wished to describe the velocity of an airplane flying in a direction somewhat north of due east. The magnitude of the airplane's velocity can be specified by stating its speed as 400 km/h, for example. The direction of the airplane's velocity can be specified in a number of ways, but the simplest would be to specify an angle to some reference direction, 20° north of east, for example. These two numbers, 400 km/h and 20° north of east, are sufficient to describe the airplane's velocity, provided that its motion is two-dimensional (in a horizontal plane, not climbing or descending). If the plane is climbing or descending, a second angle, the angle of ascent or descent, must also be specified.

A diagram is often a more vivid way of describing this same vector. Figure C.1 shows the velocity of the airplane as an arrow, pointed in a direction 20° north of east. The magnitude of the velocity can be represented by the length of the arrow if we choose an appropriate scale factor when drawing the diagram. For example, if 2 cm is selected to represent 100 km/h, then we would draw the arrow with a length of 8 cm (4 × 2 cm) to represent the speed of 400 km/h. A smaller speed would be represented by a shorter arrow, and a larger speed would require a longer arrow.

An arrow is the universal symbol for representing vectors on diagrams. An arrow can clearly indicate direction, and can also be drawn to different lengths to indicate magnitude. We often use boldface type for the symbols that represent vector quantities: the symbol **v** tells us that we are dealing with the vector quantity, velocity. The symbol *v*, on the other hand, often represents the scalar quantity, speed.

How Do We Add Vectors?

We are often interested in the net result of combining two or more vectors. In chapter 4, for example, *net force* determines the acceleration of an object. This net force is the vector sum of whatever forces are acting on the object, which could include several forces. As a second example, an airplane's velocity relative to the ground is determined by the vector sum of its velocity relative to the air and the velocity of the air relative to the ground (the wind velocity), as discussed in chapter 20.

One of the most readily visualized examples of vector addition involves *displacements* of moving objects. Suppose, for example, that a student wishes to travel to an apartment complex located on North Main Street a few blocks north and west of campus. One way that she

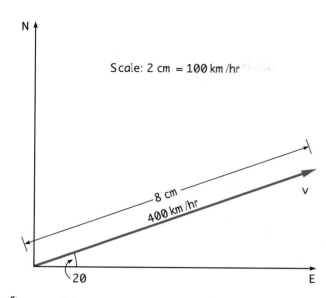

figure **C.1**   The velocity vector of 400 km/h in the direction 20° north of east is represented by drawing the arrow to scale (2 cm = 100 km/h) and at the appropriate angle (20°).

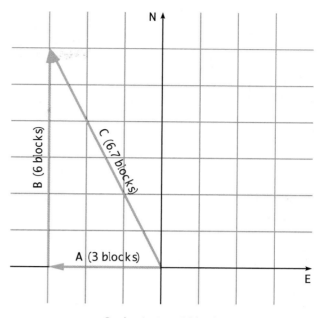

Scale: 1 cm = 1 block

figure **C.2**   The net result of adding displacement **A** (three blocks due west) and displacement **B** (six blocks due north) is the displacement **C** obtained by drawing a vector from the starting point to the final destination.

might get there from a starting point on the south side of campus is to walk three blocks due west along Pacific Avenue, and then six blocks due north along Main Street, as indicated in figure C.2. The result of these two motions can be represented by displacement vectors; the first motion displaces her three blocks due west, the second one six blocks due north.

The two displacements just described are drawn to scale and in the appropriate direction in figure C.2 (1 cm = 1 block on the drawing). The result of these combined motions is indicated by drawing displacement **C,** which is the vector drawn from the starting point to the final destination. Vector **C** is thus the vector sum of vectors **A** and **B:**

C = A + B.

The sum combines their individual effects into a single net displacement. The length of vector **C** is approximately 6.7 cm, which represents a distance of 6.7 blocks given the scale factor used in drawing the diagram. Measuring the angle with a protractor yields an angle of approximately 27° west of north for the displacement vector **C.**

This process of vector addition that we have just described can be used with any vectors. It is often referred to as the *graphical method* of adding vectors or, more descriptively, as the *tail-to-head technique.* Its steps are:

1. Draw the first vector to scale (1 cm equals so many units of the vector quantity) and in the appropriate direction using a ruler and a protractor.

2. Starting the second vector with its tail placed at the head of the first vector, draw the second vector to scale and in the appropriate direction.
3. If more than two vectors are involved, draw the succeeding vectors to scale and in the appropriate direction, starting each vector with its tail at the head of the previous vector.
4. To obtain the vector sum, draw a vector from the tail of the first vector to the head of the final vector. Measure the angle this vector makes to some reference direction with the protractor, and measure its length with a ruler. These two measurements represent the direction and magnitude of the vector sum. (The measured length must be multiplied by the scale factor used in drawing the original vectors to obtain the appropriate units.)

To illustrate this process in another example, we have added two velocity vectors in figure C.3. The first vector **A** is a velocity of 20 m/s at an angle of 15° north of east. The second vector **B** is a velocity of 40 m/s at an angle of 55° north of east. They have each been drawn to a scale of 1 cm = 10 m/s, so vector **A** is 2 cm long and vector **B** is 4 cm long. The tail of vector **B** is placed at the head or tip of vector **A** to add the vectors. The resulting sum, vector **C,**

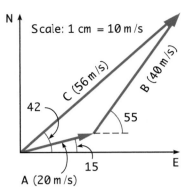

figure **C.3**   The velocity vectors **A** and **B** are added to obtain the vector sum **C**. A scale factor of 1 cm = 10 m/s is used, and the tail of vector **B** is placed at the head of vector **A**.

is approximately 5.6 cm long, determined by measuring with a ruler. Using the scale factor of 1 cm = 10 m/s, we have:

5.6 cm × (10 m/s per cm) = 56 m/s.

Measuring the angle that **C** makes to the horizontal axis (east), we find that **C** is approximately 42° north of east. Thus, the vector sum of vectors **A** and **B** is equal to 56 m/s at an angle of 42° north of east.

Note that in both this example involving velocities, and in the previous example involving displacements, the magnitude of the vector sum is not equal to the sum of the magnitudes of the two vectors being added. In the first case, the vector sum **C** had a magnitude of 6.7 blocks, which is less than the 9 blocks (3 + 6) that the student actually walked. In the velocity example, the vector sum has a magnitude of 56 m/s, less than the sum of 60 m/s obtained by adding the magnitudes of vectors **A** and **B.** This is a general feature of the process of vector addition. The *only* case in which the magnitude of the vector sum equals the sum of the magnitudes of the vectors being added (*A* + *B*) is when these vectors are in the same direction.

How Do We Subtract Vectors?

Once you have mastered the concept of vector addition, subtraction represents a straightforward extension of these ideas. Subtraction can always be represented as the process of adding to the original quantity the *minus* value (the negative of) the quantity being subtracted. Thus the process of subtracting 2 from 6 is the same as adding −2 to 6. If we want to subtract vector **A** from vector **B** to get the vector difference **B − A**, we add **−A** to **B**. To get the negative of a vector, we reverse its direction.

To illustrate this process, we have subtracted velocity **A** from velocity **B** in figure C.4, using the same two vectors that we added in figure C.3. We first draw vector **B** to scale and then add to it the negative of vector **A**. Note that we

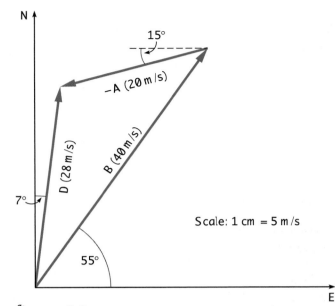

figure **C.4**   The velocity vector **A** is subtracted from the velocity vector **B** to obtain the difference vector **D**. The scale is 1 cm = 5 m/s.

have reversed the direction of **A** to get **−A**. The negative vector is 15° below the westward horizontal instead of 15° above the eastward horizontal. The difference vector **D** is then obtained by drawing the vector from the tail of the first vector (**B**) to the head of the second (**−A**). The vector **D** has a length of approximately 5.6 cm and makes an angle of approximately 7° to the vertical axis (north). The length of 5.6 cm represents a velocity of 28 m/s (5.6 cm × 5 m/s per 1 cm), since this is the scale factor used in figure C.4.

What Are Vector Components?

We often find it useful to describe vectors in terms of their horizontal and vertical *components,* rather than directly dealing with the entire vector. This is particularly true when we are discussing projectile motion, as in chapter 3, but it is also useful for computing work (chapter 6) and in many other applications.

> The components of a vector are any two (or more) vectors that when added together yield the vector of interest.

It is usually most productive to define these components as perpendicular to one another, often in the horizontal and vertical directions.

We can use graphical techniques to find the components of a vector, as well as to add or subtract vectors. The process is illustrated for a force vector in figure C.5. The force vector **A** has a magnitude of 8 N and a direction of 30°

above the horizontal. (The newton, N, is the metric unit of force.) Our first step in finding the horizontal and vertical components of this vector is to draw the vector to scale (1 cm = 1 N) and in the appropriate direction (30° above the horizontal) using a ruler and a protractor, as before.

The horizontal component of the vector **A** is then found by drawing a line from the tip of **A** to the horizontal (*x*) axis, such that the line makes a right angle (is perpendicular to) the *x*-axis. The distance along the *x*-axis measured from the tail of **A** (the origin) to the point where the perpendicular line meets the axis represents the magnitude of A_x, the horizontal component of **A**. The magnitude A_x is the portion of **A** that is in the horizontal direction.

A similar process yields A_y, the vertical component of **A,** but in this case, a dashed line is drawn from the tip of **A** to the vertical (*y*) axis, making a right angle to the vertical axis. Measuring the lengths of these components with a ruler yields magnitudes of 6.9 N (6.9 cm on the graph) and 4 N (4 cm on the graph) for A_x and A_y, respectively.

If we treat these two components of **A** as vectors and add them together in the usual tail-to-head manner, we obtain the original vector **A**, as shown in figure C.5. We can therefore use these two components to represent the vector, since, added together, they are identical to the original vector. Very often, however, we are really interested in only the horizontal effect or the vertical effect of the vector, and then we use just one of the components by itself. In the case of a force vector, for example, the effect of the force in moving an object in the horizontal direction will be de-

termined by the horizontal component of the force vector rather than by the total vector. In projectile motion, it is the horizontal component of the velocity that determines how far the object will travel horizontally in a given time, and so on.

The components of vectors can also be used in adding or subtracting vectors, as well as for many other purposes. In this book, however, our main use of the concept of vector components will be to break a vector down into its horizontal and vertical portions for the purpose of analyzing the horizontal and vertical motions separately. Knowing that this can be done will be a key to your understanding of projectile motion and many other physical processes.

Exercises

(Answers to the odd-numbered exercises are found in appendix D.)

Use the graphical tail-to-head technique to find the vector sums of the indicated vectors in exercises 1 through 4.

1. Vector **A** = a displacement of 20 m due east
   Vector **B** = a displacement of 30 m due north

2. Vector **A** = a velocity of 20 m/s at 30° north of east
   Vector **B** = a velocity of 50 m/s at 45° north of east

3. Vector **A** = an acceleration of 4 m/s² due east
   Vector **B** = an acceleration of 3 m/s² at 40° north of east

4. Vector **A** = a force of 20 N at 45° above the horizontal (to the right)
   Vector **B** = a force of 30 N at 20° to the left of vertical

5. Find the magnitude and direction of the difference vector **B** − **A** in exercise 1.

6. Find the magnitude and direction of the difference vector **A** − **B** in exercise 1.

7. Find the magnitude and direction of the difference vector **A** − **B** in exercise 2.

8. Find the magnitude and direction of the difference vector **A** − **B** in exercise 3.

9. Find the east and north components (*x* and `*l*) of vector **A** in exercise 2.

10. Find the horizontal and vertical components of vector **A** in exercise 4.

11. Find the horizontal and vertical components of vector **B** in exercise 4.

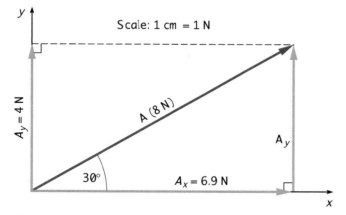

figure **C.5** The components of the force vector **A** are found by drawing the vector to scale and then drawing lines from the tip of the vector to the *x* and *y* axes, so that these lines make right angles (90°) with the axes.

appendix d

Answers to Odd-Numbered Exercises and Challenge Problems

Chapter 1

E1. 1000 mL **E3.** 10 **E5.** 135 in, 11.25 ft
E7. 8600 g, 8.6×10^6 mg **E9.** 1610 m
E11. 62 500 cm², 10 000 cm²

Chapter 2

E1. 57.5 MPH **E3.** 0.4 cm/day **E5.** 5 min
E7. 4.32 km **E9.** 93.3 km/h **E11.** 21 m/s
E13. a. 17 m/s **b.** 29 m **E15. a.** 21 m/s **b.** 76.5 m
E17. 9.09 s

CP1 a. 21 s **CP5. a.** car A: 10 m, 20 m, 30 m, 40 m;
car B: 2.25 m, 9 m, 20.25 m, 36 m **b.** 4.5 s

Chapter 3

E1. a. 8 m/s **b.** 16 m/s **E3.** 50 m/s, 112 MPH
E5. 22 m/s **E7. a.** +5 m/s (up) **b.** −5 m/s (down)
E9. 1.5 s **E11. a.** 0.167 s **b.** 13.9 cm **E13.** 1.25 m
E15. a. 1.0 s **b.** 6 m

CP1. a. 0 **b.** 1.6 s **c.** 12.8 m **d.** 12 m **e.** down
CP3. a. 0.4 s **b.** 1.2 m, 2.0 m **c.** no **CP5. a.** 40.2 m/s
b. 18.3 m **c.** 0.455 s **d.** 1.0 m

Chapter 4

E1. 8 m/s² **E3.** 5 kg **E5.** 1.5 m/s² **E7.** 5 kg
E9. 392 N **E11. a.** 490 N **b.** 50 kg **E13.** the 735-N
man **E15.** 14 N **E17. a.** 3.3 N **b.** 2.2 m/s²

CP1. a. 5 m/s² **b.** 15 m/s **c.** 22.5 m **CP3. a.** 0.375 m/s²
b. 0.075 m/s **c.** 0.75 cm **CP5. a.** 16 N **b.** 2.67 m/s²
c. 11.3 N **CP7. a.** 350 N **b.** 4.57 m/s² **c.** 75 m/s

Chapter 5

E1. 31.3 m/s² **E3.** 1.0 m **E5.** 1.0 N
E7. a. 18.2 m/s² **b.** 18.2 kN **E9.** 365/1 **E11.** 0.04 N
E13. 0.56 N **E15.** 300 lb

CP1. a. 26.7 m/s² **b.** 5.34 N **c.** 1.96 N
CP3. a. 10.4 m/s² **b.** 9.36 kN **c.** 8.82 kN
d. about 9.1 kN **e.** no **CP5. a.** 3.53×10^{22} N
b. 2.01×10^{20} N **c.** 175/1, no **d.** 4.34×10^{20} N, yes

Chapter 6

E1. 100 J **E3.** 5 m **E5. a.** 160 J **b.** 0 **c.** 160 J
E7. a. 80 J **b.** 80 J **E9.** 0.8 J **E11.** 78.4 J, 100 J,
accelerating the rock **E13.** 40 J **E15.** 520 kJ
E17. 4 Hz

CP1. a. 7.5 J **b.** 4.5 J **c.** 4.5 J **CP3. a.** 48 J **b.** 48 J
c. 43.8 m/s **CP5. a.** yes **b.** 34.9 m

Chapter 7

E1. a. 12 N·s **b.** 12 kg·m/s **E3.** the bowling ball
E5. 4.8 N·s **E7. a.** 1080 N·s **b.** 2700 N
E9. a. −7.5 kg·m/s **b.** −7.5 N·s **E11. a.** 0
b. −7.5 m/s **E13.** 20 000 kg·m/s **E15. a.** 2250 kJ
b. 450 kJ **c.** no

CP1. a. 12 kg·m/s **b.** yes **c.** 12 N·s **d.** 300 N
CP3. a. case A **b.** case A **CP5. a.** 30 000 kg·m/s, south
b. 5 m/s, south **c.** 975 kJ **d.** 75 kJ **e.** no

Chapter 8

E1. a. 0.167 rev/s **b.** 1.05 rad/s   **E3. a.** 18.8 rad
b. 4.7 rad/s   **E5. a.** 4.8 rev/s **b.** 9.6 rev   **E7. a.** 1.0 rev/s
b. 2.5 rev   **E9.** 15 cm   **E11. a.** 96 N·m **b.** −60 N·m
c. 36 N·m   **E13.** 13.5 N·m   **E15.** 0.1 kg·m^2
E17. a. 0.08 kg·m^2 **b.** 1.6 kg·m^2/s

CP1. a. 132 N·m **b.** 0.147 rad/s^2 **c.** 2.2 rad/s
d. −0.0133 rad/s^2, 165 s   **CP3. a.** 960 kg·m^2, 2460 kg·m^2
b. 1560 kg·m^2 **c.** 1.89 rad/s **d.** yes

Chapter 9

E1. 80 Pa   **E3.** 1.25 lb/in^2 **E5. a.** 400 kPa **b.** 80 kN
E7. 2250 Pa   **E9.** 26.7 kPa   **E11.** 500 kg/m^3
E13. 1960 N   **E15.** 0.5

CP1. a. 3.14 cm^2, 491 cm^2 **b.** 156/1 **c.** 13 720 N **d.** 88 N
CP3. a. 2.7 × 10^{-5} m^3 **b.** 0.21 kg **c.** 2.06 N **d.** 0.26 N
e. 1.8 N   **CP5. a.** 50.3 cm^2, 19.6 cm^2 **b.** 3.85 m/s
c. less than

Chapter 10

E1. 113°F   **E3.** 297.2 K   **E5. a.** 45°C **b.** 113°F
E7. 2.23 kcal   **E9.** 22°C   **E11.** 838 J   **E13.** 500 J
E15. −700 J (out of the system)   **E17. a.** +600 J
b. +1.43 C°

CP1. a. 90 F° **b.** 50 K **c.** no   **CP3. a.** 13.5 kcal
b. 3.75 kcal **c.** 17.25 kcal **d.** no   **CP5. a.** 2038 J
b. 486 cal **c.** 1.2 C° **d.** yes

Chapter 11

E1. 40%   **E3. a.** 300 J **b.** 33%   **E5. a.** 900 J **b.** 44%
E7. 59%   **E9.** 450 J   **E11.** 500 W   **E13.** no

CP1. a. 37.5 MJ **b.** 112.5 MJ   **CP3. a.** 6.7% **b.** 13.4 J
c. 187 J **d.** 14   **CP5. a.** 44.4% **b.** 35.5% **c.** 100 000 kW·h
d. 282 000 kW·h **e.** 166 barrels

Chapter 12

E1. 3 × 10^{13} electrons   **E3.** 4 μC per ball   **E5.** 3 N
E7. a. 1.8 N   **E9.** 100 N, down   **E11.** 2.67 × 10^6 N/C,
due east   **E13.** 12.5 J   **E15.** 20 kV

CP1. a. 4.5 × 10^6 N **b.** 7.2 × 10^6 N **c.** 2.7 × 10^6 N, to
the left **d.** 1.35 × 10^8 N/C, to the left **e.** 8.1 × 10^6 N, to
the right   **CP5. a.** −0.12 J **b.** upward **c.** upward
d. 3.33 × 10^4 N/C

Chapter 13

E1. 6 A   **E3.** 0.25 A   **E5.** 200 Ω   **E7. a.** 0.12 A
b. 4.8 V, 7.2 V   **E9. a.** 0.1 A **b.** yes **c.** 2 V   **E11.** 1 Ω
E13. 13.5 W   **E15. a.** 0.545 A **b.** 202 Ω   **E17.** 25 A

CP1. a. 4 Ω **b.** 125 mA **c.** 83.3 mA **d.** 125 mW
e. greater than   **CP3. a.** 42.9 mA **b.** 0.857 V **c.** 0.386 W
d. charging   **CP5. a.** 5.2 A, 10.4 A, 4.3 A **b.** yes **c.** 22 Ω

Chapter 14

E1. 2.5 N   **E3.** It is one-third the original value.
E5. 20 C   **E7.** 0.3 N   **E9.** 1.92 T·m^2 **E11.** 24 V
E13. a. step-up **b.** 440 V   **E15.** 15 turns

CP1. a. 2 × 10^{-4} N/m **b.** repulsive **c.** 6 × 10^{-5} N
d. 2 × 10^{-5} T **e.** into the plane of the page
CP3. a. 0.0018 m^2 **b.** 0.0432 T·m^2 **c.** 0 **d.** 0.25 s
e. 0.173 V

Chapter 15

E1. 0.5 Hz   **E3.** 0.4 m   **E5. a.** 2.5 Hz **b.** 4.8 m
E7. a. 1.6 m **b.** 75 Hz   **E9.** 77.3 cm   **E11. a.** 2 m
b. 170 Hz   **E13.** 264 Hz   **E15.** 60 Hz   **E17.** 220 Hz

CP1. a. 0.3 kg/m **b.** 10 m/s **c.** 4.0 m **d.** 2 cycles
e. 0.8 s **CP3. b.** 80 cm **c.** 425 Hz **d.** 12.5 Hz increase
CP5. a. 396 Hz **b.** 352 Hz **c.** 330 Hz **d.** 495 Hz
e. 297 Hz **f.** 440 Hz

Chapter 16

E1. 3 × 10^{10} Hz   **E3.** 5.77 × 10^{14} Hz   **E5.** 2.5 mm
E7. 6.6 cm   **E9. a.** 1/2 wavelength **b.** destructive
E11. 0.167 mm   **E13. a.** 27.3 cm **b.** 54.6 cm

CP1. a. 7.9 × 10^{14} Hz, 4.0 × 10^{14} Hz **b.** 2 × 10^8 m/s
c. 253 nm, 500 nm   **CP3. a.** 125 nm **b.** 250 nm **c.** no

Chapter 17

E1. 1.8 m, 6 m   **E3.** 32 cm   **E5. a.** 8 cm
b. real, inverted   **E7.** 3.43 cm   **E9. a.** −4 cm **b.** 2
E11. a. +15 cm **b.** real, inverted   **E13. a.** −50 cm
b. 30 cm   **E15. a.** 1.11 m **b.** −0.11   **E17.** 50 cm

CP1. a. 6.77 cm **b.** 7.17 cm **c.** 4.78 cm   **CP3. a.** 15 cm
b. −1.5 **d.** 12 cm, −2 **e.** 3   **CP5. a.** 4.0 cm **b.** −4
c. −10 cm **d.** 5 **e.** −20

Chapter 18

E1. 23/8 = 2.88/1   **E3.** 19   **E5.** 1835   **E7.** 2×10^{18} Hz
E9. 103 nm, no   **E11. a.** 5.43×10^{14} Hz **b.** 552 nm

CP1. a. upward **b.** 15 000 N/C **c.** 2.4×10^{-15} N
d. 2.64×10^{15} m/s^2 **e.** an upward-bending trajectory
CP3. a. 13.6 eV **b.** 3.4 eV **c.** 91.3 nm

Chapter 19

E1. 12   **E3.** aluminum, $_{13}$Al^{27} **E5.** $_{88}$Ra^{228} **E7.** $_{6}$C^{13}
E9. 6 days   **E11.** $_{42}$Mo^{102}

CP1. a. 6,6; 7,7; 8,8 **b.** 1.0/1 **c.** 61, 47; 64, 48; 66, 49
d. 1.30, 1.33, 1.35; ave = 1.33/1 **e.** 142, 90; 140, 91;
146, 92; 1.58, 1.54, 1.59; ave = 1.57/1
CP3. a. 0.003 51 u **b.** 5.25×10^{-13} J **c.** yes

Chapter 20

E1. 7 m/s   **E3.** 0.5 m/s   **E5.** c = 3×10^8 m/s
E7. 3.75 min   **E9.** 49.75 m   **E11.** 2×10^{12} kg·m/s

CP1. b. 6.71 m/s **c.** 8 s **d.** 24 m **e.** 53.7 m
CP3. b. 80.4 yr **c.** 8.04 yr **d.** 8 lt-yr **e.** 72.4 yr

Chapter 21

E1. 500 s   **E3. a.** 111 **b.** 1111

CP1. a. 3.15×10^7 s **b.** 9.45×10^{15} m **c.** 3.78×10^{16} m
d. 40 yr

Appendix A

1. $a = F/m$ **3.** $d = b - c$ **5.** $d = a - bc$
7. $b = a/(c - d)$ **9.** $b = c - d - a$ **11.** $b = dt/(a + c)$
13. $v_0 = \frac{x}{t} - \frac{1}{2}(at)$

Appendix B

1. a. 0.6 **b.** 0.52 **c.** 0.874 **d.** 0.0005   **3. a.** 0.25 **b.** 0.625
c. 0.308 **d.** 0.341   **5. a.** 25% **b.** 62.5% **c.** 30.8% **d.** 34.1%
7. a. 52.5 **b.** 36 **c.** 108 **d.** 85.2   **9. a.** 5.475×10^3
b. 2×10^5 **c.** 6.7×10^4 **d.** 3.5×10^{10} **11. a.** 6.5×10^{-3}
b. 3.33×10^{-4} **c.** 1.5×10^{-6} **d.** 6.5×10^{-8}
13. a. 0.0067 **b.** 0.00018 **c.** 0.000 005 77 **d.** 0.000 032 5
15. a. 6×10^9 **b.** 2.4×10^5 **c.** 1.8×10^3 **d.** 1.8×10^{18}

Appendix C

1. 36 m, 56° north of east   **3.** 6.6 m/s^2, 17° north of east
5. 36 m, 56° north of west   **7.** 31 m/s, 55° south of west
9. 17.3 m/s, 10 m/s   **11.** -10.3 N, $+28.2$ N

glossary

Absolute temperature   Temperature given in the Kelvin scale, which sets 0 K at absolute zero. Room temperature of 72°F or 22°C is about 295 K.

Absolute zero   The lowest possible temperature, 0 K (kelvin) or −273°C, the point where a gas would have no pressure or molecular motion.

Absorption lines   Dark bands or lines in a spectrum. In the spectra of stars, these bands or gaps indicate that light has been absorbed by the gases in the stars.

Acceleration   The rate of change of velocity.

Acceleration due to gravity   The uniform acceleration of an object in a gravitational field, $g = 9.8$ m/s^2 near the earth's surface (also called the *gravitational acceleration*).

Accelerator   One of several kinds of devices used to study subatomic particles by accelerating them to high velocities, causing them to collide and, if possible, break into previously undiscovered particles.

Action/reaction principle   For every action (force acting on one body) there is an equal but opposite reaction (force acting on another body), as described in Newton's third law of motion.

Additive color mixing   The production of different colors by combining light of the three primary colors blue, green, and red.

Adiabatic   An adjective that describes a thermodynamic change or process without gain or loss of heat.

Air foil   A surface designed to take advantage of air current to lift or steer an aircraft.

Air resistance   The frictional effects of the air or atmosphere on the motion of objects—especially noticeable on objects with large surface areas compared to their masses like leaves and feathers.

Alpha decay   See *radioactive decay.*

Alternating current (ac)   Electric current that continually reverses its direction. Current in use in North America is set at 60 Hz (60 back-and-forth cycles a second) and has an *effective* voltage of 115 V.

Ammeter   An instrument for measuring electric current.

Amplitude   The maximum swing from the point of equilibrium, for example, in the movement of a pendulum.

Angular magnification   Magnification of an object by moving the object (or its image) closer to the eye, thus forming a larger image on the retina of the eye.

Angular momentum   The rotational equivalent of linear momentum, found by multiplying the rotational inertia I by the rotational velocity ω, or $\mathbf{L} = I\omega$. Angular momentum helps to explain the orbits of planets, twirling ice skaters, and the spin of subatomic particles. Also called *rotational momentum.*

Anode   A positive electrode.

Antinode   The point, or points, in a standing wave with the greatest amplitude.

Antiparticle   An elementary particle with some properties, such as electric charge, opposite those of the corresponding particle. The antiparticle of an electron, for example, is a positron. When these two particles interact, they annihilate one another.

Aperture   An opening, especially the opening that allows light into a camera or other optical instrument.

Application program   Computer software designed to perform a specific task or set of tasks, for example, a word-processing program or graphics-design program.

Arc   A portion of the circumference of a circle or other curved line.

Archimedes' principle   The buoyant force acting on an object fully or partly submerged in a fluid equals the weight of the fluid displaced by the object.

Artificial intelligence   Computers and software programmed to replicate certain human thought processes and ways of perceiving.

Atmospheric pressure   The pressure of the layer of air that surrounds the earth. At sea level, the atmospheric pressure is 14.7 pounds per square inch but decreases with altitude.

Atom   From the Greek word for *undivided,* the smallest particle of an element, now known to be made up of a nucleus surrounded by one or more electrons.

Atomic mass   The average mass of an atom of an element, a distinguishing characteristic of an element.

Atomic number   The number of protons in an element's nucleus, which determines its place in the periodic table. Oxygen's atomic number is 8, krypton's is 36, and gold's is 79.

Atomic physics   The subfield of physics that studies the structure and behavior of atoms.

Atomic spectrum   The wavelengths of light emitted by a substance when heated. In the visible spectrum of light, each element produces a distinctive color display.

Atomic weight   The traditional term for atomic mass.

Average acceleration   Change in velocity divided by the time required to produce that change, $\mathbf{a} = \Delta\mathbf{v}/t$.

Average speed   Distance traveled divided by the time traveled (in symbols, $s = d/t$), the *rate* at which distance is covered.

Axis   An imaginary line running through an object, around which it may rotate.

Balmer series Characteristic absorption and emission lines that appear in the visible portion of the spectrum of hydrogen. There are several other series of lines in the other (invisible) portions of hydrogen's spectrum.

Barometer An instrument, originally a tube containing a column of mercury, for measuring atmospheric pressure.

Baryon The heaviest of subatomic particles, for example, neutrons and protons.

Battery Cells that produce electricity by chemical reaction between electrodes of different materials separated by a chemical solution.

Beats A regular variation in amplitude caused by the interference of two waves of different frequency.

Bernoulli's principle Pressure is lower where fluid speed is higher: the sum of the pressure plus the kinetic energy per unit of volume of a flowing fluid is constant.

Beta decay See *radioactive decay.*

Big Bang The theoretical beginning of the universe 10 to 20 billion years ago from an expansion, that still continues, of an extremely dense volume.

Birefringence A process (also called *double refraction*) in which light of different polarizations travels with different speeds in different directions within a material.

Bit A contraction of *binary digit,* a bit is the pair of choices, 0 or 1, in a single binary number. It is the basic unit for storing digitized information in a computer. Eight bits make a byte.

Blackbody Any body that absorbs all the radiation that falls on it, appearing perfectly black. Also, an instrument with a hole or cavity (dark at room temperature) that, when heated, emits a spectrum based on temperature and not on the composition of the material.

Black hole A very massive collapsed star with an extremely strong gravitational field that lets light in but not out, which makes it a perfect absorber of light.

Boltzmann's constant A universal constant k that relates the temperature, pressure, and volume of gas under the ideal gas law. Boltzmann's constant also applies to other systems.

Boyle's law The volume of a gas at constant temperature is inversely proportional to the pressure on it—doubling the pressure on a gas will halve its volume. In symbols, $PV =$ constant, where P is pressure on a gas and V is its volume.

Bubble chamber A device filled with a transparent liquid heated beyond the boiling point. A moving particle in a bubble chamber leaves a distinctive trail of bubbles in its wake, which is photographed for further study.

Buoyant force An upward force that lifts objects toward the surface of water or other fluid.

Caloric An invisible fluid once thought to flow from a hotter to a cooler object. This idea no longer is considered valid.

Capacitor A device for storing electric charge.

Carnot cycle An ideal reversible cycle devised by Carnot that is the model of a heat engine with maximum efficiency. The cycle consists of four steps, two isothermal and two adiabatic.

Carnot efficiency The efficiency of an ideal heat engine, or the maximum efficiency possible of a heat engine operating between two specified temperatures.

Carnot engine An ideal reversible heat engine using the Carnot cycle. Any heat engine with an irreversible step in its cycle will be less efficient.

Cathode A negative electrode.

Cathode rays Radiation emitted by a cathode in an evacuated tube, since discovered to be electrons.

Center of gravity The point on an object about which its weight exerts no net torque and the object will balance.

Centrifugal force An imaginary force that seems to be directed outward as a result of rotation.

Centripetal acceleration The rate of change in velocity of an object's direction on a circular or curved path. Centripetal acceleration is always perpendicular to the velocity vector and directed toward the center of the curve.

Centripetal force Any force or combination of forces that produces a centripetal acceleration.

Chain reaction A fission reaction that, by initiating several more reactions, becomes self-sustaining.

Change of phase The process of a substance going from one physical state—solid, liquid, or gas—to another. Also called *phase change.*

Chaos Unpredictability or instability. A system is chaotic if it is unstable or unpredictable: weather is often chaotic.

Circuit A closed or complete path for an electric current.

Classical physics The four branches of physics—mechanics, thermodynamics, electricity and magnetism, and optics—

that were already well developed by the beginning of the twentieth century.

Cloud chamber A device containing air or another gas saturated with water vapor. The chamber is cooled so that fog forms on a particle, and its track is then illuminated for study.

Collider A type of particle accelerator designed to cause circulating high-energy beams to collide, thus producing new particles for study.

Combustion Rapid combining of oxygen with a fuel; burning.

Concave mirror A mirror in which light is reflected from the inside of a curved surface.

Condensed-matter physics The subfield of physics that studies the properties of matter in the solid and liquid states.

Conduction (thermal) The ability of heat to flow through a material when objects at different temperatures are placed in contact.

Conductivity The ability of a material to carry electric current.

Conductor A material that readily allows charge to flow.

Conservation laws Principles of physics that show how, under certain conditions, a specified quantity in a system does not change regardless of actions that may take place. For example, in *conservation of angular momentum,* the total angular momentum of a system remains constant as long as the total torque on the system is zero. *Conservation of energy* means that the amount of mechanical energy of a system remains constant through physical changes or processes (if no work is done on the system). *Conservation of momentum* shows that the momentum of a system remains constant (changes of different parts of the system cancel) if no net external force acts on it.

Conservative force A force such as gravity or the elastic force that allows complete recovery of energy when work is done against it.

Control rods Rods that can be inserted or removed from the core of a nuclear reactor to maintain a desired rate of reaction. The rods are made of a neutron-absorbing substance, boron being the most common.

Convection The transfer of heat by the motion or circulation of a fluid (gas or liquid) that contains thermal energy.

Convex mirror A mirror in which light is reflected from the outside of a curved surface.

Coolant   The fluid (usually water) used to remove heat from a nuclear reactor or other heat engines.

Cosmology   Study of the structure and origins of the universe.

Coulomb's law   A description of how electrostatic force varies with quantity of charge and distance: electrostatic force is proportional to the size of each of the charges and inversely proportional to the square of the distance between two charges.

Critical   In describing nuclear reactors, the state when each fission reaction produces another fission reaction, which leads to a chain reaction. If each reaction produces more than one reaction, the reactor is *supercritical*. If less, it is *subcritical*.

Critical mass   A mass of fuel just large enough to produce a self-sustaining chain reaction.

Critical temperature   In electric resistance, the temperature below which a substance becomes a superconductor of electricity.

Dalton's law of definite proportions   Dalton's observation that the ratio of the masses of certain elements needed in chemical reactions did not vary, which helped to clarify the relationships of the masses of the atoms of elements.

de Broglie wave   The probability wave associated with particles of matter. De Broglie proposed that the electron, in particular, can be represented by a standing wave about the atomic nucleus.

de Broglie wavelength   A wavelength derived from the momentum of a particle and Planck's constant indicating that particles like electrons have wavelike features ($\lambda = h/p$).

Density   Mass per unit of volume.

Diffraction   Interference of light and other waves coming through different parts of the same slit or opening.

Diffraction grating   An instrument made of closely spaced parallel lines on a surface (glass or metal) used for diffracting light to produce spectra and other optical effects.

Digital logic   Operations performed on strings of discrete numbers, symbols, or characters, as in a computer.

Diode   An electronic device that allows electric current to flow in one direction only.

Dipole   An object with separated (polarized) positive and negative areas. In an *electric* dipole, electric charge is separated into positive and negative regions.

Magnetic dipoles have two magnetic poles, often labeled *N* and *S*. All magnets have at least two poles.

Direct current (dc)   Electric current that flows in a single direction.

Dispersion   The variation of the index of refraction of a transparent substance with wavelength. In a prism, this results in separation of light into its wavelengths, producing color displays.

Displacement   The linear or angular distance an object moves from its original position. Also, the amount of fluid displaced by a floating or submerged object.

Doping   Adding small amounts of a substance to a semiconductor to enhance, customize, or alter its properties.

Doppler effect   Change in the detected frequency of a wave because of movement by either the source or the observer.

Eclipse   A planet's passage into another planet's shadow. In a *lunar* eclipse, the moon passes into the earth's shadow. In a *solar* eclipse, the earth passes into the shadow of the moon.

Efficiency   The ratio of the work an engine produces to the input of heat energy, usually given as a percentage.

Elastic collision   A collision in which no energy is lost. The objects bounce off each other.

Elastic force   Force exerted by objects that can be deformed or stretched, such as a bowstring or a spring.

Elastic potential energy   Potential energy in a system that depends on the displacement from equilibrium of an elastic object like a spring.

Electric charge   The electromagnetic property of an object that produces the electrostatic force. A transfer of electrons to or from an object makes it electrically positive or negative.

Electric current   The rate of flow of electric charge.

Electric dipole   Consists of two equal-magnitude electric charges opposite in sign separated by a small distance.

Electric field   The electric force per unit of positive charge exerted on a charge if it were placed at that point. It is a property of space surrounding a distribution of electric charges.

Electric potential   The potential energy per amount of positive electric charge; voltage.

Electricity and magnetism   The subfield of physics that studies electric and magnetic forces and electric current.

Electromagnet   A current-carrying coil of wire with an iron core whose magnetic field is produced by electric current.

Electromagnetic induction   Production of electric current by a changing magnetic flux.

Electromagnetic spectrum   The array of electromagnetic waves of different wavelengths. At the longer wavelength end of the spectrum are radio waves and microwaves. Visible light begins with longer red waves and progresses to violet. Beyond violet (ultraviolet) are X rays and gamma rays.

Electromagnetic wave   A wave made up of changing electric and magnetic fields.

Electromagnetism   Study of the related phenomena of electricity, magnetism, electric fields, and magnetic fields, all of which are aspects of the electromagnetic force.

Electromotive force   Potential energy per unit of charge produced by a battery or other source of electric energy. The name is misleading, because it is a potential difference, or voltage—*not* a force.

Electron   Extremely small, negatively charged particles present in all atoms.

Electrostatic force   The force exerted by one stationary charge on another independently of their motion. The electrostatic force holds atoms together and binds one atom to another in liquids and solids.

Electroweak force   The electromagnetic force and the weak nuclear force unified into a single fundamental force.

Element   A basic chemical substance that consists of atoms of one kind.

Ellipse   An oval curve with two foci—the shape of the orbits of the planets. A circle is a special case of an ellipse with both foci at one point.

Emission lines   Bright lines in the spectrum of a substance indicating the emission of electromagnetic radiation.

Empirical law   A rule or generalization derived from experience, which in the sciences would come from, and be confirmed by, experiments and observations.

Enrichment   Increasing the proportion of the more reactive isotope of a nuclear fuel by separating the isotopes.

Entropy   The measure of the disorder or randomness of a system.

Epicycle   Imaginary circles made by the planets on their main orbits, used in the obsolete Ptolemaic model to explain planetary motions.

Equally tempered   Tuning system based upon equal ratios for all half-steps in a musical scale.

Equation of state   An equation that gives the thermodynamic relation between pressure, volume, and temperature for a specific kind of system.

Equilibrium   A state of balance when the net force on an object is zero or when a system stops undergoing change.

Ether   Before scientists understood that light may not require a medium, it was widely believed that light traveled through a massless medium pervading space called the ether. Also called the *luminiferous* (light-bearing) *ether*.

Experiment   Observations made under controlled conditions. An experiment must be able to be tested and repeated by other researchers.

Exponent   The superscript indicating what power a number or variable has been raised to, for example, x^3 or 10^9.

Exponential decay (or growth)   Decay (or growth) at a decreasing (or increasing) rate over time, which is graphed as an exponential curve.

Faraday's law   Induced voltage in a coil equals the rate of change of the magnetic flux through the coil, $\varepsilon = \Delta\phi/t$.

Field lines   Graphic illustrations of electric and magnetic fields.

Field theory   The study of phenomena that interact over space or distance, including phenomena that involve electric, gravitational, or magnetic fields.

First law of thermodynamics   The change in internal energy of a system equals the net amount of heat and work transferred into the system (conservation of energy).

Fission fragment   Either of the two lighter elements that emerge from the splitting of a large radioactive atom during nuclear fission. They tend to be highly unstable and radioactive.

Fluid   Something that flows—a gas or liquid.

Fluid pressure   Pressure exerted on or by a fluid.

Focal length   The distance from the center of a lens or mirror to the focal point or points.

Focal point   The point where a lens focuses or concentrates parallel beams of light.

Focus   In optics, the point where rays of light are concentrated by reflection or refraction and produce an image. In geometry, one of the two points that define the curve of an ellipse and give it its characteristic shape.

Force   The quantity that describes the mechanical interaction of two objects, causing the objects to accelerate as described in Newton's laws of motion.

Frame of reference   A standpoint or orientation from which we make measurements and observations of motion and to which we refer them. An *inertial* frame of reference does not accelerate in relation to other inertial frames.

Free-body diagram   A drawing commonly used in physics that identifies the interactions and directions of forces on objects.

Free fall   The motion of a falling object affected only by acceleration due to gravity.

Frequency   The number of pulses, repetitions, or cycles per unit of time.

Frictional force   A force that resists an object's movement.

Fulcrum   A pivot point or support for a lever.

Galaxy   A rotating assemblage of stars usually shaped like an ellipse or a disk with spiral arms. Our solar system is in the Orion arm of the Milky Way galaxy.

Galvanometer   An instrument that detects and measures the amount and direction of electric currents.

Gamma decay   See *radioactive decay*.

General theory of relativity   Einstein's generalization of his special theory of relativity to encompass accelerated frames of reference. Relying on the principle of equivalence, its basic postulate is that an acceleration of a frame of reference cannot be distinguished from the presence of gravity.

Generator   A device that converts the mechanical energy of a rotating coil to electric energy by electromagnetic induction.

Geocentric   An adjective used to describe a model of a solar system with the earth at the center and the planets and stars in orbit around it, later replaced by the *heliocentric* (sun-centered) model.

Geometric optics   A branch of optics that describes the behavior of light schematically using straight-line rays and the laws of reflection and refraction.

Geothermal   Relating to the earth's internal heat and to certain phenomena and processes in which the earth produces (usable) heat, such as geysers and hot springs.

Grand unified theories   Theories that seek to explain all of the elementary particles and the forces between them. One result of these theories is the uniting of the electroweak and the strong forces into a single fundamental force.

Gravitation   Mutual attraction of two objects proportional to the masses of the objects—also called gravity.

Gravitational acceleration   See *acceleration due to gravity*.

Gravitational potential energy   Stored energy linked with the position of an object in a gravitational field rather than with the object's motion.

Gravitational red shift   A lengthening in the wavelength of light toward the red end of the spectrum as photons move through a strong gravitational field.

Greenhouse effect   The trapping of radiation at long wavelengths (heat) in a system such as a greenhouse or parked car. Release of gases such as carbon dioxide into the atmosphere may lead to a similar outcome, increasing the earth's temperature and leading to changes in climate.

Hadron   A meson or baryon.

Half-life   The time needed for half of the original number of atoms in a radioactive isotope to decay. Each radioactive element has a different half-life.

Halogen   An element like iodine or chlorine that readily combines with a metal to form a salt.

Harmonic analysis   Breaking down a complex wave into its simple sine-wave components.

Harmonic motion, simple   The motion of a system whose energy changes smoothly from potential to kinetic energy and back again. The motion is symmetric about the point of equilibrium and is graphed as a sinusoidal (sine) curve.

Harmonic wave   A wave shaped like a sinusoidal curve.

Heat   Energy that flows from an object to another or to the surroundings when regions of different temperatures are involved.

Heat engine   A device or motor that takes in energy (heat), converts some of that heat to mechanical work, and releases leftover waste heat at a lower temperature into the surroundings.

Heat pump   The opposite of a heat engine, it moves heat from a colder reservoir to a warmer one by work supplied from an external source. The refrigerator is the best-known example.

Heisenberg uncertainty principle The position and momentum of a particle cannot both be known at the same time with high precision: if we are highly certain about the position of a particle, we will be almost completely uncertain about its momentum, and vice versa.

Heliocentric The revolutionary model of the solar system proposed by Copernicus and championed by Galileo that placed the sun rather than the earth at the center of the planets' orbits. Ptolemy's obsolete model, in which the sun and other planets orbit the earth, is a *geocentric* model.

High-grade heat Heat at temperatures of about 500°C or higher that can be used to run heat engines producing mechanical work or electrical energy.

Hole In a semiconductor, an absence of an electron that migrates through the material and carries a positive charge, a result of doping with certain elements.

Hologram A photograph of an interference pattern produced by a laser that forms a three-dimensional image when illuminated.

Hydraulics The science of the effects and applications of water and other fluids under pressure or in motion.

Hydrodynamics The study of the mechanics of fluids.

Hypothesis An educated guess or generalization grounded in experience and observation that can be tested to examine its consequences.

Ideal gas A gas in which the forces between atoms (and potential energy) are small enough to be ignored. The relation between properties like temperature, pressure, and volume is summarized in the *equation of state* of an ideal gas, $PV = NkT$. Most real gases behave like ideal gases at sufficiently low pressures and high temperatures.

Ideal heat engine A heat engine that conforms to the Carnot cycle, which would give it maximum efficiency.

Impulse The force acting on an object multiplied by the time interval over which the force acts.

Impulse/momentum principle The impulse acting on an object produces a change in the object's momentum equal in both size and direction to the impulse.

Incident Falling on or incoming, especially when applied to light.

Index of refraction A number that yields the speed of light in a transparent material ($v = c/n$). Different materials have different indexes of refraction.

Induction The ability of an object to produce electric charge or magnetism in another by the action of its field rather than by touching.

Inelastic collision A collision in which some kinetic energy is lost. In a *partially* inelastic collision, the objects do not stick together. A *perfectly* inelastic collision is a collision in which the objects stick together after colliding. The most energy is lost in this kind of collision.

Inertia An object's resistance to a change in its motion.

Inertial frame of reference A frame of reference that does not accelerate in relation to other inertial frames—Newton's laws of motion apply.

Infrared light Electromagnetic waves with wavelengths somewhat longer than red light of the visible spectrum.

Instantaneous acceleration The rate at which velocity is changing at a given instant in time.

Instantaneous speed How fast an object is moving at a particular instant. Instantaneous speed is related to *average speed* for very short time intervals.

Instantaneous velocity A vector quantity that is made up of an object's instantaneous speed and its direction at that instant.

Insulator A material that does not ordinarily permit a flow of charge through it.

Integrated circuit Many transistors, diodes, resistors, and electrical connections all built into a single tiny chip of semiconductor material.

Interference The combination or interaction of two or more waves. In *constructive* interference, the waves add together. In *destructive* interference, the waves cancel each other.

Interferometer An instrument that works by splitting a beam of light and bringing the rays back together to produce interference that can be used to make measurements of velocities, wavelengths, and distances.

Internal energy The sum of all kinetic and potential energies of the atoms and molecules inside a substance or system, uniquely determined by the state of the system.

Interstellar gas Gases that pervade the galaxy.

Inverted In optics, an adjective describing an image that is upside down.

Ion An atom or molecule that has gained or lost an electron, which changes its charge. *Ionization* is the formation of ions.

Ionize To add or remove electrons from an atom or molecule.

Isobaric An adjective describing any process in which pressure is held constant.

Isothermal An adjective describing any process in which the temperature remains the same.

Isotope A variety of an element that has a specific number of neutrons that may differ from other isotopes of the same element. Carbon-12, the most common isotope of carbon, has 6 neutrons, while carbon-14 has 8.

Just tuning Tuning system based upon the ideal frequency ratios between notes.

Kepler's laws of planetary motion Three descriptions of the movement of the planets that illuminate how the solar system works: The *first law* is that each planet moves in an orbit that is an ellipse with the sun at one focus. In the *second law*, each planet sweeps through equal areas in the ellipse of its orbit in equal intervals of time. The *third law* says that the square of the period of a planet's revolution around the sun is proportional to the cube of its average distance from the sun, so that T^2/r^3 is a constant.

Kinetic energy The energy of an object related to its motion—one-half the mass multiplied by the square of the speed, $KE = \frac{1}{2}mv^2$.

Kinetic force of friction The frictional force between two objects sliding at the point of contact of their surfaces.

Laminar Arranged in thin layers, as the parallel streamlines in smooth flow of a liquid.

Laser A source of a highly coherent beam of visible light. *Laser* stands for *l*ight *a*mplification by *s*timulated *e*mission of *r*adiation.

Latent heat The amount of heat needed to produce a phase change without a change in temperature. The *latent heat of fusion* melts a substance, and the *latent heat of vaporization* converts it to a gas.

Law of reflection When light is reflected from a smooth surface, the angle the reflected ray makes with the surface normal equals the angle the incident ray makes with the surface normal.

Law of refraction When light passes from one transparent medium to another, the rays are bent toward the surface normal if the speed of light is smaller in the second medium than in the first. The rays bend away if the speed of light in the second medium is greater than the first.

Length contraction   An effect of relativity— an object moving at a high speed is shortened compared with its length at rest.

Lens   A *positive* lens is a convex, converging lens and bends rays toward the axis. It makes things look larger if the object is inside the focal point. A *negative* lens is a concave, diverging lens that bends rays away from the axis. It makes things look smaller.

Lenz's law   A description of the direction of induced current: the current opposes the change in magnetic flux producing it.

Lepton   The lightest of subatomic particles, for example, electrons, positrons, and neutrinos.

Lever arm   The perpendicular distance from the fulcrum to the line of action of the force—a component of torque. Also called *moment arm*.

Light   In general, the visible portion of the electromagnetic spectrum, although *light* is sometimes used to encompass other electromagnetic waves as well.

Light clock   An "instrument" in a thought experiment about relativity that uses the speed of light over a set distance to define a basic unit of time.

Linear displacement   A vector representing how far, and in what direction, an object has moved.

Linear motion   An object's motion from one point to another in a straight line.

Longitudinal wave   A wave whose displacement or disturbance in the medium is parallel to the direction the wave travels. Sound waves are longitudinal waves.

Low-grade heat   Heat at temperatures of about 100°C or lower that is best used to heat homes or buildings.

Magnetic dipole   Consists of two equal magnetic poles, north and south, separated by a small distance, equivalent in its magnetic effects to a small electric-current loop.

Magnetic field   Magnetic force per unit of charge and unit of velocity. It is a property of space in the region surrounding moving charges or a magnet.

Magnetic flux   A measure of the number of magnetic field lines passing through an area bounded by a current loop or wire.

Magnetic force   A force exerted by moving electric charges or currents on one another.

Magnetic monopole   A particle with a single magnetic pole. Magnetic monopoles may have been present in the universe shortly after the Big Bang but probably no longer exist.

Magnetic north   North as indicated by a compass. Earth's magnetic north pole shifts and is not the same as the geographic North Pole.

Magnetic pole   A region of a magnet, usually labeled *N* or *S,* for *north-seeking* and *south-seeking,* that shows behavior similar to electric charge. Like poles repel, opposite poles attract.

Magnification   The ratio of height of an image produced by an instrument or lens to the object's actual height. The image can be either enlarged or reduced. In *negative* magnification, the image is inverted.

Magnitude   The size of something expressed as a quantity.

Mass   The measure of an object's inertia, the property that causes it to resist change in its motion. The kilogram is the basic metric unit for measurement of mass.

Mass-energy equivalence   Mass is energy: increasing the mass of an object increases its energy, and increasing the energy of an object increases its mass. This revolutionary idea was summarized by Einstein in his well-known mass-energy equation, $E = mc^2$ (in which E is energy, m is mass, and c is the velocity of light).

Mass number   The total number of neutrons and protons in an element or its isotopes. Carbon-14 has 6 protons and 8 neutrons yielding a mass number of 14.

Mechanical advantage   The ratio of the output force to the input force of a simple machine.

Mechanics   The branch of physics that studies forces and motion.

Meson   A subatomic particle of intermediate weight, for example, a pion or a kaon. All mesons lack spin.

Metal   Any of a number of elements that are opaque, shiny, conductive, and easy to shape. Their atoms have one to three electrons outside of a filled electron shell.

Metric system   The International System of Units (*Système international d'unités*), abbreviated SI, is a decimal system of measurement. The seven fundamental SI units ("base units") are the meter (length), kilogram (mass), second (time), ampere (electric current), kelvin (temperature), mole (amount of substance), and candela (intensity of a light source).

Microscope   An instrument that uses at least two positive lenses to produce a magnified image of a small object.

Microwave   An electromagnetic wave with wavelengths between the radio waves and infrared waves of the electromagnetic spectrum and commonly used in cooking and radar.

Moderator   Material used in a nuclear reactor to slow down the neutrons produced in fission reactions.

Modern physics   The subfields of physics—atomic, nuclear, particle, and condensed matter—that largely came into existence and made great advances in the 1900s.

Molecule   A combination of atoms of an element or of a chemical compound.

Moment of inertia   See *rotational inertia*.

Momentum   The product of the mass of an object and its velocity, **p** = *m***v,** also called *quantity of motion* by Newton.

Myopia   A condition in which the lens system of the eye is too strong, allowing people to see near objects clearly but not distant objects. Also called *nearsightedness.*

Natural radioactivity   Penetrating radiation produced without need of special preparation by ores or compounds containing radioactive elements.

Negative lens   A lens that causes light rays to diverge more than when the rays entered the lens.

Negative work   Work done by a force acting in a direction opposite to the object's motion.

Net force   The vector sum of the forces acting on an object.

Neutrino   An elusive particle that is one of the products of beta decay. Its antiparticle is the antineutrino.

Neutron   A particle with no charge, roughly the mass of a proton, found in atomic nuclei.

Newton   The metric unit of force: mass times acceleration, or kilograms times meters per second squared.

Newton's law of universal gravitation   The gravitational force between two objects is proportional to the mass of each object and inversely proportional to the square of the distance between the centers of the masses:

$$\mathbf{F} = \frac{Gm_1 m_2}{r^2}$$

It is attractive and acts along the line that joins the two objects.

Newton's laws of motion   *The first law:* Unless a force is applied, an object remains at rest or moves with constant velocity. *The second law:* The acceleration of an object is directly proportional to the magnitude of the imposed force and inversely proportional to the mass of the object, $a = F/m$, or $F = ma$. *The third law:* If object A exerts a force on object B, object B exerts a force on object A equal in magnitude and opposite in direction to the force exerted on B.

Noble gas   Gases that do not react readily with other elements because their outer shells are closed. (See *shell.*) Found in column VIIIA of the periodic table, the noble gases are helium, neon, argon, krypton, xenon, and radon.

Node   The point in a standing wave where there is no motion.

Normal force   The component of force on an object that acts perpendicular to the surface of contact, as opposed to the frictional force, which acts parallel.

Nuclear fission   A nucleus-splitting reaction accompanied by conversion of part of the mass of the nucleus to kinetic energy. The nucleus breaks into two roughly equal parts (the fission fragments) rather than just emitting a particle, as in most other nuclear reactions.

Nuclear fusion   A reaction that combines small nuclei to produce somewhat bigger nuclei and a large release of energy.

Nuclear physics   The subfield of physics devoted to the study of the nucleus of the atom.

Nuclear pile   Another name for a nuclear reactor, particularly the early reactors built using carbon bricks.

Nuclear reaction   Changes caused in the nucleus of atoms, which may result in one element turning into another.

Nucleon   A particle that inhabits the atomic nucleus: a proton or neutron.

Nucleus   The small, dense, positively charged center of an atom made up of protons and neutrons.

Ohm's law   Current flowing through a portion of a circuit equals the voltage difference across that portion divided by the resistance, $I = \Delta V/R$.

Optics   The subfield of physics devoted to the study of light and vision.

Organic   Referring to the chemistry of carbon compounds.

Oscillation   A repeated cycle, vibration, or movement about an equilibrium point, extremely common phenomena in nature.

Oscilloscope   An electronic device that plots changes in electric voltages on a screen.

Parallel circuit   A circuit with elements connected by more than one path so that the current divides and rejoins.

Partially inelastic collision   A collision in which some energy is lost, but the objects do not stick together after collision.

Particle   A single point of mass, often, a molecule, atom, or even smaller basic particle such as a quark.

Particle physics   The subfield of physics that studies subatomic particles (quarks, etc.).

Pascal's principle   Any change in fluid pressure is transmitted uniformly in all directions in the fluid.

Perfectly inelastic collision   A collision in which the greatest portion of energy is lost. Objects do not bounce at all but instead stick together after collision.

Period   In astronomy, the time it takes for an object to return to the point where it started, for example, a complete planetary orbit. In physics, a complete cycle, as of a wave.

Periodic table of the elements   A table that orders the elements by atomic weight and atomic number. Beginning with hydrogen, the lightest, the table also groups elements (in vertical columns in the typical format) with other elements with like properties.

Periodic wave   A wave made of pulses separated by equal time intervals.

Perpetual-motion machine   A *perpetual-motion machine of the first kind* would put out more energy as work or heat than it takes in (a violation of the first law of thermodynamics and the conservation of energy). A *perpetual-motion machine of the second kind* violates the second law of thermodynamics by claiming to convert heat completely to work or to surpass the Carnot efficiency.

Phase   From the word *phasis,* meaning manner, aspect, or stage of being. Also, a state of organization of matter, such as solid, liquid, or a gas.

Phases of the moon   The appearance or aspect of the moon and other planets at given times in their cycles.

Phosphor   A substance that emits light when struck by fast-moving particles.

Photon   A quantum of electromagnetic energy, especially a particle of light.

Physical optics   The branch of optics that treats light as an electromagnetic wave to examine its effects and properties.

Pitch   A musical term for height or depth of tone. The sound waves of notes with higher pitches have higher frequencies.

Planck's constant   A constant h, which determines the size of a quantum of light energy for a given frequency, as summarized in the equation $E = hf$.

Polarize   To separate positive and negative areas of charge on an object, for example, a magnetic or electrically charged body. In optics, to select a specific direction of oscillation of the electric field in a light wave.

Polarized light   Light for which the electric field vector oscillates in a specific direction.

Positive lens   A lens that causes light rays to converge more than when the rays entered the lens.

Positron   A positive electron; the antiparticle of an electron. Positrons are emitted in beta-plus decay.

Postulate   A fundamental statement (possibly one of several) that is the foundation of a theory.

Potential energy   Stored energy associated with the position of an object rather than the object's motion.

Power   The rate of doing work, found by dividing the amount of work done by the time, $P = W/t$.

Powers of 10   The basis of scientific notation, the powers of 10 give the multiples of 10 as superscripts or exponents: for example, 10^4 equals $10 \times 10 \times 10 \times 10$.

Pressure   The ratio of a force to the area over which it is applied, or force per unit of area.

Principle of equivalence   Part of the foundation of the general theory of relativity, the principle of equivalence states that it is impossible to distinguish an acceleration of a frame of reference from the effects of gravity. For example, the mass of an object measured by its inertia equals its mass measured by the action of a gravitational field on it.

Principle of relativity   The statement that the laws of physics are the same in any inertial frame of reference.

Principle of superposition   When two or more waves combine, the disturbance is equal to the sum of the individual disturbances.

Probability   The likelihood of an event happening.

Programming language   A set of systematized instructions to the computer that eases the programming of a computer.

A *high-level* programming language is adapted to handle certain kinds of applications by its designers. FORTRAN (Formula Translation), for example, is designed for mathematics.

Projectile motion   The trajectories and velocities of objects that have been launched, shot, or thrown.

Propagate   In physics, to transmit, especially a wave, over a distance.

Proper time   The time interval between two events measured in a frame of reference in which the two events occur at the same point in space.

Proportion   A comparison of the ratio of quantities to one another or to a whole.

Proton   A positively charged particle found in atomic nuclei. The number of protons in a nucleus determines an atom's chemical properties.

Quantization   Division into the smallest possible discrete amounts (quanta) of energy or mass.

Quantum   The smallest quantity of energy or mass of a particular kind. For example, a photon is a quantum of light.

Quantum mechanics   Mechanics (the study of forces and motion) applied to the atomic and nuclear level and dealing with photons and other quanta that show both wave and particle behavior.

Quantum number   One of a set of numbers that relate properties of electrons or other particles at the subatomic level. These quantum numbers determine an electron's orbit—no two electrons of an atom can have the same set of quantum numbers.

Quark   One of the particles that make up mesons and baryons (for example, neutrons and protons). There are six kinds of quarks.

Radian   A unit of angular measurement found by dividing the distance traveled along an arc length s by the radius of the circle r, or s/r.
One revolution $= 360° = 2\pi$ radians.

Radiation   The flow, emission, or propagation of energy by electromagnetic waves or particles.

Radioactive decay   The change of an element into another element or state as a result of changes in its unstable atomic nucleus. There are three kinds of decay, named after their radioactive emissions: alpha (helium ions), beta (electrons), and gamma (X rays). Any nucleus with more than 82 protons (the number in lead) is inherently unstable.

Radioactivity   Spontaneous emission of particles from the nucleus of the atoms of certain unstable elements.

Radius   A straight line from the center of a circle or sphere to its outer edge.

Rate   One quantity divided by another, especially a quantity divided by a unit of time. Rates of time like miles per hour and meters per second are important in measuring motion.

Reaction force   An oppositely directed force in response to the initial force in an interaction governed by Newton's third law of motion.

Real image   An image formed by light rays converging to the image point. Generally, real images formed by single lenses or mirrors are inverted.

Recoil   A brief interaction between two objects that causes them to move in opposite directions.

Reflection   Light waves bouncing off a surface (a mirror or the surface of a body of water, for example) in such a way that the angle of the incoming ray is the same as the angle of the reflected ray.

Refraction   Light waves passing through a transparent surface (such as a lens or a prism) and being bent by it.

Relativity   The discovery that matter, motion, space, and time are interdependent and that our frame of reference often governs how we view and explain their interactions.

Rem (roentgen equivalent in man)   The unit used to measure exposure to radiation. The average American receives about 360 millirems of radiation a year from various sources.

Resistance   The property in a component in a circuit of opposing the flow of electric current and generating heat.

Resistivity   The inherent property or tendency of a material to resist the flow of current as measured in a sample of the material.

Rest energy   Energy that an object has simply by virtue of its mass, $E_0 = mc^2$, an expression of mass-energy equivalence.

Rest length   Length measured by an observer at rest relative to the distance measured.

Restoring force   The force or system of forces that returns an object to equilibrium.

Retina   A layer of cells at the back of the eye that detects light. The cells are of two types, called rods and cones.

Retrograde motion   Backward movement of a planet in its orbit, an illusion caused by the position of the planet with respect to the earth.

Reversible   Describes a thermodynamic process that is always near equilibrium and can be turned around and run the other way at any point in the process.

Revolution   One complete cycle in the motion of an object about some point. The period of each of the earth's revolutions about its axis is one day.

Right-hand rule   To find the direction of a rotational velocity vector, curl your fingers in the direction of the rotation. Your thumb will then point in the direction of the vector. Other right-hand rules are used to describe the directions of magnetic forces and fields.

Rotational acceleration   The rate of change of rotational velocity.

Rotational displacement   The angular measure of how far an object has rotated, usually expressed in radians.

Rotational inertia   Resistance of an object to change in its rotational motion that depends on an object's mass and the distribution of mass around the axis. Also called *moment of inertia*.

Rotational motion   Motion of an object that turns on an axis.

Rotational velocity   The rate of rotation—how fast an object rotates. Revolutions per minute (rpm) is a common unit of measurement for it.

Rydberg constant   A constant used in computing the wavelengths and the lines in atomic spectra, particularly for hydrogen.

Satellite   An object that orbits a planet. The moon is the earth's only natural satellite, although many artificial satellites have recently been placed in orbit.

Scattering   A process in which light is absorbed and then re-emitted in different directions.

Scattering experiment   A kind of experiment on atomic nuclei in which particles are used as projectiles to produce changes in target nuclei, among them emission of other particles.

Scientific method   The systematic study of phenomena by people engaged in science. The way scientists go about investigating phenomena normally begins with observation and experiments to test generalizations and hypotheses that can be incorporated into tested, comprehensive theories. Theories often predict new areas to explore.

Scientific notation   A system for writing numbers succinctly and clearly that relies on the powers of ten, for example, 12 150 000 000 is written 1.215×10^{10}.

Second law of thermodynamics   No engine working in a continuous cycle can take heat from a reservoir at a single temperature and convert that heat completely to work (Kelvin's statement). Heat will not flow from a colder to a hotter body unless some other effect is also involved (Clausius's statement).

Selective absorption   A process in which some wavelengths of light are absorbed more than other wavelengths.

Self-induction   Voltage induced in the same coil that produces the changing magnetic flux.

Semiconductor   A substance with properties intermediate between electrical insulators and conductors. A fast-growing use of semiconductors is computer chips, where their properties can be customized.

Series circuit   A circuit connected in a single loop, so that current passes through each component in succession.

Shell   A grouping of electron orbits about the atomic nucleus defined by similar energy levels. Each shell accommodates a certain number of electrons.

Simple harmonic motion   See *harmonic motion, simple.*

Simple machine   Any elementary mechanical device that multiplies the effect of an applied force, for example, levers, wedges, and pulleys.

Sink   A place or device for disposing of energy in a system.

Sinusoidal curve   The graph of the trigonometric sine function that illustrates simple harmonic motion. The graph of alternating current is a sinusoidal curve. Also called a *sine curve.*

Slope   The change in the vertical coordinate of a graph divided by the change in the horizontal coordinate. In a graph of an object's position plotted against time, the slope at a point indicates how fast the object is moving.

Software   A computer program loaded into a computer to make it run or perform specific tasks, as opposed to *hardware,* the computer equipment (hard drive, monitor, computer chips, etc.).

Solar system   The sun and the planets, moons, and other objects that orbit it.

Solid-state electronics   Electronic circuits that contain semiconductor devices—for example, transistors that control current without relying on moving parts or vacuum tubes.

Sound wave   A longitudinal wave that propagates pressure variations through air or other media. Human beings generally hear sound waves from 16 Hz to 20 000 Hz in pitch.

Space-time continuum   A four-dimensional framework for describing an event that recognizes that the three dimensions of space and time are not independent of each other and that space is curved.

Special theory of relativity   Einstein's "limited" theory, based on two basic postulates: the laws of physics have the same form in any inertial frame of reference (the principle of relativity), and the speed of light is the same in any inertial frame of reference.

Specific heat capacity   A property of a material that is the quantity of heat needed to change a unit of mass by a unit of temperature (for example, to change 1 g of a substance by 1°C).

Spectrum, atomic and electromagnetic   See *atomic spectrum* and *electromagnetic spectrum.*

Speed   How fast something is moving without regard to direction of motion—not a synonym for *velocity.*

Speed of light   The speed of light is the maximum speed in the universe and a constant symbolized by *c.* The speed of light is roughly equal to 3×10^8 m/s or 186 282 miles per second.

Spring constant   A constant describing the relation between how far a spring is stretched or displaced and how much force it takes to do the stretching: a stiff spring has a large spring constant.

Standard model   The evolving framework currently in use for exploring the fundamental forces and particles at the subatomic level that includes theories of quarks, the electroweak force, and the strong nuclear interaction.

Standing wave   An unchanging pattern of waves or oscillations formed by interference of waves traveling in opposite directions.

Static electricity   The electric effects produced by charged objects that do not depend on motion of the charges.

Static force of friction   A frictional force that does not involve motion in the direction of the force.

Steady state   The relative equilibrium of a system if conditions are not changing with time.

Step down   To convert electric current from a higher to a lower voltage. To "step up" is the opposite process—from lower to higher voltage.

Streamline   Graphic illustrations of the flow of water, air, and other fluids.

Strong nuclear interaction   A fundamental force that binds together the quarks in neutrons and protons (and other baryons and mesons).

Subtractive color mixing   The production of different colors by absorbing different wavelengths from white light.

Superconductivity   The loss of all electrical resistance to the flow of current below a critical (usually very low) temperature.

Superfluid   A fluid that loses its viscosity at a very low temperature.

Surface normal   A line drawn perpendicular to a surface.

Symmetry   A common quality among objects, plants, and animals of having a similar shape or arrangement on either side of a line or lines drawn through them.

Synchronous orbit   An orbit timed to a planet's rotation to keep a satellite above the same point of the planet's surface.

System   A specific set of parts or objects that act together or on each other.

Telescope   An instrument that uses two or more lenses or mirrors to bring much closer to the eye an image of a distant object.

Temperature   A quantity that tells which direction heat will flow. If two objects are at different temperatures, heat flows from the higher to the lower temperature.

Terminal velocity   The point at which air resistance on a falling object equals the gravitational force, producing a net force of zero, which ends the object's acceleration. The object will continue down at constant velocity.

Terrestrial telescope   A type of telescope—spyglasses, binoculars, and opera glasses, especially—used mainly for making observations of earth-based phenomena rather than the heavens. Unlike astronomical telescopes, terrestrial telescopes are designed to turn the image right side up.

Theory   An organized body of knowledge used to explain and further investigate events, data, and facts.

Theory of everything   A theory that would succeed in unifying gravity with the strong and electroweak forces to encompass all of the forces of nature in a single force.

Thermal conductivity   The property of a substance that determines how swiftly it conducts heat. Metals are better conductors than wood or plastic.

Thermal equilibrium   The state at which the physical properties, like volume, of objects no longer vary with regard to each other, and the objects reach the same temperature.

Thermal power plant   Produces electricity by utilizing heat obtained from sources such as coal, oil, natural gas, nuclear fuels, or geothermal energy to run a heat engine.

Thermodynamics   The subfield of physics concerned with temperature, heat, and energy.

Thermonuclear   An adjective describing nuclear fusion chain reactions that go on at very high temperatures, as in stars or certain kinds of bombs.

Thin-film interference   Interference of light reflected from the top and bottom surfaces of a very thin film of a transparent material.

Thought experiment   A kind of experiment devised by Einstein to test relativistic concepts. Although the velocities involved are enormous, the consequences can be imagined and explored. As equipment has improved, a number of the results of thought experiments have been confirmed by observation.

Time dilation   An effect of relativity in which the observer of an extremely fast-moving object experiences a longer span of time than an observer moving with that object.

Tokamak   An experimental doughnut-shaped nuclear-fusion reactor.

Torque   The product of a force and its lever arm ($\tau = Fl$) that causes an object to rotate.

Total internal reflection   Reflection inside a transparent object that causes *all* light to be reflected. This occurs when the angle of incidence exceeds a certain value.

Trajectory   The path of a projectile or other object in motion.

Transformer   A device that steps (adjusts) alternating-current voltage up or down to suit the needs of a particular application.

Transistor   An electronic device made from semiconductors that carries out a number of functions, among them, producing variations in electric current, amplifying radio signals, and serving as a voltage-controlled switch.

Transverse wave   A wave whose disturbance or displacement is perpendicular to its direction of travel. Electromagnetic waves are transverse waves.

Triple point of water   The temperature at which ice, water, and water vapor are all in equilibrium, 0° in the Celsius scale.

Turbine   A device for generating power that is driven by a fluid (steam, for example) moving through a system of fixed and turning blades.

Turbulent flow   More complicated, less smooth flow of a fluid caused by random fluctuations usually resulting from an increase in speed or decrease in viscosity.

Ultraviolet light   Electromagnetic waves with wavelengths shorter than violet light of the visible spectrum.

Unified field theory   The theory seeking to unite the four fundamental forces—the strong nuclear interaction, electromagnetic force, gravitational force, and weak nuclear force—that relies on the behavior of their fields to explore and analyze them.

Uniform acceleration   Acceleration with a steady rate of change in velocity, the simplest kind of accelerated motion.

Universal gravitational constant   A constant G that equals 6.67×10^{-11} N·m²/kg². Because this constant is so small, gravitational forces between two objects of ordinary size are almost unnoticeable.

Unpolarized light   Light for which the electric field vector oscillates in random directions.

Vacuum   The absence of matter, especially air.

Vector   A quantity that has both magnitude and direction, often represented by an arrow. The length of the arrow is proportional to the size of the vector quantity and the angle shows its direction.

Vector quantity   Any quantity, such as velocity or a force, for which *both* the size and the direction are needed for a complete description.

Velocity   A vector quantity that describes how fast an object is moving *and* which direction it is moving.

Velocity of light   See *speed of light.*

Virtual image   An image formed by light rays that are diverging from where the image appears to be. Generally, virtual images formed by single lenses or mirrors are right side up.

Viscosity   A measure of the frictional forces between the layers of a fluid producing resistance to flow. Highly viscous liquids flow slowly.

Voltage   A change or difference in electric potential measured in volts.

Voltaic pile   A battery made of alternating disks of two different substances (Volta used zinc and silver) with a moistened soft substance (such as paper) layered between paired disks.

Voltmeter   An instrument for measuring voltage.

Waste heat   Heat released by a heat engine at a lower temperature into its surroundings.

Wave   A movement of energy through matter and space (and time).

Wavefront   A surface within a wave for which all points are at the same phase (stage of oscillation).

Wavelength   The distance between the same points of successive pulses in a wave. Its symbol is λ.

Wave pulse   A single brief wave traveling through a substance or system.

Weak nuclear force   A fundamental force involved in the interactions of leptons in beta decay.

Weight   The gravitational force on an object, in symbols, $\mathbf{W} = m\mathbf{g}$.

Work   Force applied to an object times the distance moved, $W = Fd$. The force acts along the object's line of motion. wThe joule is the metric unit of work.

X rays   Highly penetrating electromagnetic waves with a very short wavelength, now widely used in medicine to take pictures of internal organs and bones.

Zeroth law of thermodynamics   A basic assumption of thermodynamics: objects in thermal equilibrium have the same temperature.

credits

index

Note: Page references in *italics* indicate figures. Page references followed by a "t" indicate tables; page references followed by "b" indicate boxes.